New Trends in Quantum Structures

Mathematics and Its Applications

Volume 516

New Trends in Quantum Structures

by

Anatolij Dvurečenskij and Sylvia Pulmannová

Mathematical Institute,
Slovak Academy of Sciences,
Bratislava, Slovakia

SPRINGER-SCIENCE+BUSINESS MEDIA, B.V.

A C.I.P. Catalogue record for this book is available from the Library of Congress.

ISBN 978-90-481-5525-5 ISBN 978-94-017-2422-7 (eBook)
DOI 10.1007/978-94-017-2422-7

Published in co-edition with
Ister Science Ltd., Staromestská 6, 811 03 Bratislava,
Slovak Republic

Printed on acid-free paper

To the memory of my parents
Irena and Vladimír D.,
A.D.,

and dedicated to my children
Anna and Gejza J.,
and
Antónia and Richard P.,
S.P.

Preface

D. Hilbert, in his famous program, formulated many open mathematical problems which were stimulating for the development of mathematics and a fruitful source of very deep and fundamental ideas. During the whole 20th century, mathematicians and specialists in other fields have been solving problems which can be traced back to Hilbert's program, and today there are many basic results stimulated by this program. It is sure that even at the beginning of the third millennium, mathematicians will still have much to do.

One of his most interesting ideas, lying between mathematics and physics, is his sixth problem: *To find a few physical axioms which, similar to the axioms of geometry, can describe a theory for a class of physical events that is as large as possible.*

We try to present some ideas inspired by Hilbert's sixth problem and give some partial results which may contribute to its solution.

In the Thirties the situation in both physics and mathematics was very interesting. A.N. Kolmogorov published his fundamental work *Grundbegriffe der Wahrscheinlichkeitsrechnung* in which he, for the first time, axiomatized modern probability theory.

From the mathematical point of view, in Kolmogorov's model, the set L of experimentally verifiable events forms a Boolean σ-algebra and, by the Loomis-Sikorski theorem, roughly speaking can be represented by a σ-algebra $\mathcal{S}$ of subsets of some non-void set Ω.

The Kolmogorov model is very important, but it does not describe the situation that arises in quantum mechanics. A physical system is investigated through a measurement process. If a measurement reaches the microlevel, non-negligible indeterminacies appear that must be respected in the model. Therefore, in many cases, the use of classical probability theory is not sufficient, as follows, for example, from the Heisenberg uncertainty principle [Hei] which asserts that the position x and the momentum p of an elementary particle cannot be measured simultaneously with arbitrarily prescribed accuracy. If $\Delta_m p$ and $\Delta_m x$ denote the inaccuracies of the measurement of the momentum p and position x in a state m, then

$$(\Delta_m p)^2 \cdot (\Delta_m x)^2 \geq \frac{1}{4}\hbar^2 > 0,$$

where $\hbar = h/2\pi$ and h is Planck's constant.

J. von Neumann [vNeu] made the first fundamental step towards a rigorous mathematical apparatus, as well as an interpretational logic of the new mechanics on the basis of a complex separable Hilbert space.

In 1936, G. Birkhoff and J. von Neumann published their paper *The Logic of Quantum Mechanics.* In this historic paper the authors showed that the set of assertions of quantum mechanics has a different algebraic property from a Boolean algebra, and they suggested the structure of a projective geometry.

A turbulent development of quantum theories started in the Sixties, and there appeared many important results which were reflected in numerous monographs.

In the last decade, Slovak and Italian schools contributed with an important new notion, D-posets (difference posets), or equivalently, weak orthoalgebras. Later the American school presented effect algebras. All these equivalent algebraic partial structures combine algebraic as well as fuzzy set ideas. It was shown that the space of effect operators of a Hilbert space and also very important MV-algebras can be formalized within this frame.

The aim of the present book is to bring the reader some of the newest trends in this interesting interdisciplinary realm, and to indicate the many directions of their applications.

The book consists of seven chapters.

Chapter 1 is devoted to elements of D-posets and effect algebras. We start with basic definitions and properties and develop the theory further, up to recent results. The main notions are partial operations of subtraction and addition, respectively, which are endowed with some natural properties. They are introduced as an abstraction of the Hilbert space effects (i.e., Hermitian operators on a Hilbert space between O and I), which play an important role in the theory of quantum mechanics. We show that both structures are equivalent, and we describe their relations to partially ordered Abelian groups. Special attention is devoted to interval effect algebras and effect algebras with the Riesz decomposition property. We give a characterization of MV-algebras among lattice ordered effect algebras. We introduce the notions of compatibility and observables. We show that in lattice ordered effect algebras, maximal subsets of pairwise compatible elements form MV-algebras.

In the second Chapter, we introduce some basic properties of MV-algebras and their quantum counterpart, QMV-algebras. QMV-algebras are introduced as common abstraction of MV-algebras and orthomodular lattices, which also include the standard Hilbert space effect algebras. It turns out that MV-algebras and QMV-algebras have a common core of axioms, so called S-algebras, and they are obtained adding only one axiom, either the Łukasiewicz axiom in the case of MV-algebras or a weaker axiom in the case of QMV-algebras. MV-algebras and QMV-algebras give rise to effect algebras if we restrict the total operation $\oplus$ in them to orthogonal pairs. On the other hand, every effect algebra gives rise to a QMV-algebra by extension of the partial addition to a total operation by defining it to be 1 in the case where it is not defined. QMV-algebras arising in this way are called quasi-linear QMV-algebras. We describe ideals and congruences in MV-algebras and prove their representation theorem. In QMV-algebras, a hierarchy of ideals appears, which all coincide in the case of MV-algebras.

The third Chapter is devoted to congruences and ideals on partial Abelian monoids (PAM for short), which are generalizations of effect algebras. Special attention is devoted to R_1-ideals and Riesz ideals, which give rise to congruences. We show that in orthomodular lattices, Riesz ideals correspond to p-ideals. We give applications to dimension theory, and show how they are connected with the Grothendieck group K_0 of AF C*-algebras.

The tensor product of D-posets and effect algebras is studied in the fourth Chapter. We introduce D-test spaces and show their intimate relationship with difference posets. We present special types of tensor products (state tensor product, bounded Boolean power). In addition, we study partition logics and their connections with automata.

Chapter Five is devoted to a presentation of BCK-algebras which were introduced by Imai and Iséki in 1966. We show that these structures can also be found in the scheme of generalized D-posets and effect algebras, because compatible elements may form a commutative BCK-algebra. We present elements of BCK-algebras, and main attention is paid to commutative BCK-algebras with the relative cancellation property. We present a categorical equivalence of commutative BCK-algebras with a special category of Abelian ℓ-groups. As a special case, for bounded commutative BCK-algebras (= MV-algebras), we obtain the famous Mundici's categorical representation theorem for MV-algebras via unital Abelian ℓ-groups with strong (= order) unit. Our approach is different from that used by Mundici. We introduce products on MV-algebras and BCK-algebras and we show their ℓ-ring representation.

The sixth Chapter deals with applications of BCK-algebras. We study algebraic properties of BCK-algebra in more detail. We present semisimple BCK-algebras, simple BCK-algebras, and describe measures on them. Dedekind complete BCK-algebras sometimes can be presented as a direct product of basic linear commutative BCK-algebras. In addition, we present elements of pseudo MV-algebras, and it seems that they have a very close connection with (not necessarily Abelian) ℓ-groups.

The last Chapter is devoted to Loomis–Sikorski type representation theorems for σ-complete MV-algebras and BCK-algebras. We show that each σ-complete MV-algebra is a σ-homomorphic image of a tribe of fuzzy sets. In the case a of Boolean σ-algebra, we obtain the Loomis-Sikorski theorem. This result is generalized for σ-complete product MV-algebras and BCK-algebras. We characterize a weakly divisible MV-algebra by the set of all continuous fuzzy sets on a compact Hausdorff basically disconnected topological space. The Loomis–Sikorski representation theorem and weakly divisible MV-algebras enable us to characterize MV-observables and build up a calculus for MV-algebras.

Throughout the book, some sections conclude with exercises[1] that illustrate the previous facts and notions, and also present some problems.

Generally, we can say that this book gathers together facts on D-posets, effect algebras, MV-algebras, QMV-algebras, and BCK-algebras with connection to quantum structures and their applications, that have been provided by many specialists, as well as by the authors. Of course, the choice of material is subjective, and the book does not pretend to be a complete description of the status of quantum structures. For example, we did not include very important construction techniques of quantum structures, quantum structures on von Neumann algebras, quantum structures in the light of quantum computing, histories approach, etc.

The authors would like to express their gratitude to Profs. J.C. Carrega, G. Chevalier, F. Chovanec, M.L. Dalla Chiara, R. Frič, R. Giuntini, R. Hudson, A. Iorgulescu, J. Jakubík, G. Jenča, F. Kôpka, R. Mayet, R. Mesiar, D. Mundici, M. Navara, B. Riečan, T. Vetterlein, for reading and commenting on the manuscript, as well as to Dr. T. Žáčik for his LaTeX help. Last, but not least, the first author is grateful to his wife, Želmíra Dvurečenská and his children Dagmar and Andrej, and the second author thanks her husband Viliam Pulmann and children for their encouragement and patience.

It is also our pleasure to acknowledge the moral and financial support of the Mathematical Institute of the Slovak Academy of Sciences, Bratislava, grant VEGA No 2/7193/20; C.N.R.S. and Institut Girard Desargues, Université Lyon 1, Lyon; Dipartimento di Filosofia, Università di Firenze, Firenze; and University of Turku, Department of Physical Sciences, Turku.

Bratislava, March, 2000

[1]We note that e.g. Exercise 2.5.1.5 denotes Exercise 2 in Section 5.1.5.

Contents

CHAPTER 6 **BCK-algebras in Applications** 379

List of Figures and Tables

Introduction

One of the most important problems initiated by quantum theory is a mathematical description of the structure of random events of a quantum mechanical system. This problem was originally formulated in the famous paper by G. Birkhoff and J. von Neumann "The Logic of Quantum Mechanics" in 1936. The fundamental difficulty is the existence of quantum phenomena that cannot be described in terms of the event structure of the classical Kolmogorovian probability theory. It is noteworthy that similar situations seem to arise in other areas of sciences; e.g., psychology, design of computers, neural networks, and the biology of the human brain.

A number of different mathematical models have been proposed and studied to reflect various aspects of quantum mechanics. Historically the first approach was developed by von Neumann, Dirac based on a Hilbert space and its self-adjoint operators. This approach is still most frequently used in physical applications, but there still remains an open question of an adequate physical motivation, which was already raised by von Neumann. The quantum logic approach was motivated by the above mentioned paper of Birkhoff and von Neumann, and investigates the event structure of a physical system.

In the von Neumann approach, a quantum-mechanical system $\mathcal{S}$ is represented mathematically by a separable, complex Hilbert space H in such a way that the observables for $\mathcal{S}$ correspond to self-adjoint operators on H. The spectrum $sp(A)$ of a self-adjoint operator A consists of all those numerical values that can be obtained by measuring the corresponding observable. Thus, if the truth values "false" and "true" for two-valued propositions about $\mathcal{S}$ are encoded by 0 and 1, these propositions will correspond to self-adjoint operators P on H with $sp(P) \subseteq \{0, 1\}$. That is, two-valued propositions about $\mathcal{S}$ are represented by projection operators $P = P^* = P^2$ on the Hilbert space H. In this sense, the set L of all projections on H may be regarded as the "logic" of the quantum-mechanical system $\mathcal{S}$. [2]

In 1936, Birkhoff and von Neumann reconsidered the idea of a quantum logic, and proposed that the logic of a quantum-mechanical system should form an orthocomple-

---

[2]The use of the word "logic" here does not conform to the notion of a mathematical logic. It would be more accurate to think of L as the Lindenbaum-Tarski algebra affiliated with a quantum-propositional calculus.

mented modular lattice. The modularity was suggested because it seemed reasonable to suppose that L carries a normalized dimension function to serve as *a priory* probability measure. Also, complemented modular lattices were well understood because of their connections with projective geometry. But this approach suffered from some drawbacks. First, the lattice L of projections on a Hilbert space H is modular if and only if H is finite dimensional. Second, although von Neumann showed that there are "continuous" orthocomplemented modular lattices carrying an appropriate normalized dimension function, and these lattices can be realized as sublattices of the lattice of all projections on a Hilbert space, they are rather artificial objects, and relatively few mathematicians and physicists were moved to study his continuous geometries for the purposes of physics.

In 1937, it was observed by Husimi [Hus] that the lattice L of projection operators on a Hilbert space H always satisfies a weaker version of the modular law, namely the so-called orthomodular law: For all $P, Q \in L$,

$$P \leq Q \Rightarrow Q = P \vee (Q \wedge P'),$$

where $\vee$ and $\wedge$ denote the join and the meet in the lattice L and $P' = I - P$. The same condition was rediscovered by Loomis [Loo] and Maeda [Mae] in connection with their work on extending the Murray-von Neumann dimension theory of operator algebras to orthocomplemented lattices. The name "orthomodular" was suggested by Kaplansky, since it is equivalent to the condition that orthogonal pairs are modular pairs. Thus, an orthocomplemented lattice that satisfies the above weakened modular law is called an "orthomodular lattice".

In 1957, Mackey published an article on quantum mechanics and Hilbert space, and later a monograph [Mack] in which observables were introduced as functions assigning probabilities to pairs consisting of states and Borel subsets of the real line. Under a few rather weak and physically reasonable axioms, a special subclass of such observables, called "questions", with only two possible values 0 and 1, composes a system L reminiscent of a Birkhoff-von Neumann logic. Namely, this L is an orthomodular poset, that is, an orthocomplemented partially ordered set satisfying orthomodular law.

In the early Sixties, this approach inspired many authors: Foulis, Gudder, Greechie, Holland, Janowitz, MacLaren, Maeda, Pool, Ramsay, Suppes, Varadarajan, Piron, etc. Also, the famous Gleason theorem characterizing states on the projection lattice of a Hilbert space was inspired by Mackey's work. Because of the fundamental role played in quantum logic by orthomodular posets and orthomodular lattices, an extensive theory of orthomodular structures has been, and continues to be, developed.

The algebraic approach, also extensively developed (Emch, Haag and Kastler, Jordan, von Neumann and Wigner, Segal), studies sets of observables (physical quantities) as the basic concept, on the base of W*- and C*-algebras theory. In this approach, a quantum logic emerges as the set of all projections, which form an orthomodular poset, or even an orthomodular lattice.

The operational approach (Davies, Lewis, Edwards, Ludwig) is based on the convex structure of the state space, and methods of base-norm and order-unit spaces are used (Alfsen, Schultz). From an operational point of view, an observable should be defined in terms of an experimental procedure (or a class of them). The ensuing measurement statistics are to be described as probability measures which depend on the input states. As an illustration, let us consider a Hilbert-space framework. One needs to represent the observable, measured as a mathematical device that turns each state (represented by a density matrix) ρ into a probability measure $p_\rho : X \mapsto p_\rho(X)$ on the measurement outcome space. It is natural to assume that the map $\rho \mapsto p_\rho$ should respect the convex structure of the set of states. According to a basic result of Hilbert space operator theory, the numbers $p_\rho(X)$ can then be written as $tr[\rho M(X)]$, where $tr[.]$ denotes the trace and $M(X)$ is a positive linear operator for each (Borel) subset X of the outcome space. The normalization condition leads to $M(X) \leq I$. In summary, any observable is a (normalized) positive-operator-valued (POV) measure. The range of every such measure is a subset of the set of all self-adjoint operators A such that $O \leq A \leq I$, where O is the zero operator and I is the identity operator on H. These operators are called **effects**.

Ordinary observables, represented as self-adjoint operators with their spectral measures, arise as a special case of this definition, characterized by the additional property that the positive operators are projections. They are usually referred to as projection-valued (PV) measures.

The set of Hilbert space effects serves as a basis for more general algebraic structures.

In the Nineties, the Slovak school of quantum structures contributed to quantum structures with the important notion of a *D-poset*, which allows us to algebraize effect operators as well as POV measures within a nice structure. This was widely adopted by experts.

In 1992, in the study of axiomatic systems of fuzzy sets, Kôpka [Kop] defined a new structure, a so-called *D-poset of fuzzy sets*, which is closed under the formations of differences of fuzzy sets. The idea is simple: *If we have two comparable events, a and b ($a \leq b$), then our knowledge of a and b entails the complete knowledge of the rest of a in b, i.e., $b \ominus a$.* Firstly this idea was applied to fuzzy set models of quantum mechanics [Kop] and then was presented in a general algebraic form by Kôpka and Chovanec [KoCh]. These differences have the same properties as differences in "sharp" quantum logics, i.e., orthomodular posets and orthoalgebras.

A generalization of a D-poset of fuzzy sets to an abstract partially ordered set, where the basic operation is the difference, yields a very general and, at the same time, a very simple structure - a *D-poset* [KoCh]. The notion of a D-poset generalizes those of orthoalgebras (and hence orthomodular posets), MV-algebras (which were already introduced to mathematics by C. Chang in 1958), systems of effects, etc. The structure of a D-poset supports a noncommutative measure theory and allows the solution of some problems of noncommutative probability theory, including some problems of the theory of quantum measurement.

After that, it was recognized that *weak orthoalgebras*, introduced by R. Giuntini and H. Greuling [GiGr] in 1989 with a primary notion of a partial addition, are also equivalent to D-posets. Later D. Foulis and M.K. Bennett [FoBe 1] introduced so-called *effect algebras* with a partial addition as a primary operation which are also equivalent with D-posets. V.P. Belavkin also contributed to this discussion.

Recently Giuntini introduced QMV-algebras as another generalization of Hilbert space effect operators.

An effect algebra is a partial algebraic structure which has been introduced as a generalization of the set of the Hilbert-space effects, that is, self-adjoint operators between the zero and the identity operators on a Hilbert space.

Let us briefly recall the structure of effect algebras. An **effect algebra** is a partial algebraic structure $\mathcal{E} = (E; \oplus, 0, 1)$, where $\oplus$ is a partial binary operation, which is commutative and associative in a restricted sense (if one side exists then the other exists and equality holds). Moreover, to every element a there exists a uniquely defined element a' satisfying $a \oplus a' = 1$, and $a \oplus 1$ is defined if and only if $a = 0$. The element a' is called the orthosupplement of a. Such a structure can be endowed with a partial order by defining $a \leq b$ if and only if there exists c such that $b = a \oplus c$. The orthosupplementation $a \mapsto a'$ is order-reversing, and $a \oplus b$ is defined if and only if $a \leq b'$. We say that a and b are orthogonal ($a \perp b$) if $a \leq b'$ holds. Examples of effect algebras are the interval $[0, 1]$ of the real numbers with $\oplus$ defined as the usual addition if and only if the sum belongs to $[0, 1]$, and the set of Hilbert space effects, where the operation $\oplus$ is defined as the usual addition of operators if and only if the sum remains in the set of effects.

Hilbert-space effects play an important role in the theory of so-called unsharp quantum measurements, and in the frame of the orthodox approach to quantum theory, they are interpreted as the "unsharp" quantum events. Different from the "sharp" events (i.e., projection operators), the effects do not necessarily satisfy the non-contradiction principle (i.e., the conjuction of a and a' (= "non" a) may be different from zero). Motivation for passing from "sharp" to "unsharp" theory can be found even in classical probability theory. Namely, conditional expectations of "sharp" events, represented by characteristic functions of sets, are real functions with values in the unit interval $[0, 1]$, but not necessarily characteristic functions. Therefore, in order to take conditional expectations into account, we need to transfer from characteristic functions to functions with values between zero and one (in other words, from "crisp" sets to "fuzzy" sets). In quantum mechanics, the transition from sharp to unsharp theory is manifested by the transition from projector-valued measures (PV measures), representing the "sharp" observables and measurements, to positive-operator-valued measures (POV measures), representing the unsharp observables and measurements. Clearly, PV measures form a proper subset of POV measures.

Motivation for passing from PV measures to POV measures can also be given, for example, by considering the phase operator in quantum mechanics [BGL 1].

An electromagnetic field may be described by the photon annihilation and creation operators a, a^*, which yield the number operator $N = a^*a$ with the nondegenerate spectrum $\{0, 1, 2, \ldots\}$. Iterated application of the creation operator a^* to the ground state generates a complete orthogonal system of eigenvectors which span the photon Fock space. The operators a and a^* can be decomposed into real and imaginary parts such that $a = 2^{-\frac{1}{2}}(Q + iP)$, $a^* = 2^{-\frac{1}{2}}(Q - iP)$. The operators Q and P form a Schrödinger couple (in the sense of being connected via Fourier transform) and are therefore unitarily equivalent to position and momentum operators. In particular, the commutation relation $[a, a^*] = I$ is equivalent to the canonical commutation relation $[Q, P] = iI$.

The problem is whether there exists a self-adjoint operator (phase operator) Φ conjugate to the number operator N. This problem has been treated by many authors in different ways (see quotations in [BGL 1]). The difficulty is, in essence, caused by the fact that the operator N is nonnegative. This situation suggests considering an enlargemenet of the Hilbert space by extending N to an operator N' whose spectrum consists of all integers. This allows one to introduce a self-adjoint operator Φ' in the extended Hilbert space that is canonically conjugate to N' in the Weyl sense.

The Naimark projection of the spectral measure on Φ' to the original Hilbert space yields a POV measure $P_H E^{\Phi'}(X) P_H = M_0(X)$ for Borel sets X. This POV measure is canonically conjugate to the operator N in the sense of a weak Weyl relation expressing the shift covariance of the phase.

It turns out that there are infinitely many phase shift covariant POV measures which may serve as representing different possible measurement schemes in the framework of measurement theory.

It is an important question if the POV measures can be realized experimentally. In connection with measuring the phase of an electromagnetic field, POV measures have been realized by experiments in quantum optics.

Finally, we mention merits of the Slovak school of quantum structures for development of this field.

In Slovakia, there is a rather long history of development of quantum structures. In the middle of the Sixties in Bratislava, at the Institute of Measurement of the Slovak Academy of Sciences, a small group of physicists, I. Staríček, S. Pulmannová, and A. Pázman, became interested in theory of quantum measurements which lead to the study of quantum logics. The mathematical beauty of physical problems encouraged S. Pulmannová and A. Pázman to become professional mathematicians (A. Pázman declinated to mathematical statistics and design of experiments). In 1972 A. Dvurečenskij joined this group as a doctoral student of A. Pázman.

Another center in Bratislava, where the study of quantum logics began in the Sixties, is today's Faculty of Mathematics and Physics of the Comenius University. The first pioneers there were T. Neubrunn (1929–1990) and B. Riečan as well as T. Katriňák; among physicists, P. Bóna should also be mentioned. Neubrunn [Neu] solved a problem posed by S. Gudder, showing conditions which guarantee that a subsystem of subsets in a so-called quantum probability space is a Boolean algebra. These conditions, roughly speaking, describe situations when the physical model has a locally classical character. The problems of quantum logics and the famous book of Varadarajan [Var] were studied in the seminar of Professor Neubrunn, and attracted several young people to this area, among them J. Dravecký and J. Šipoš [DrSi], who studied the problems of quantum probability spaces, e.g. additivity of the so-called Gudder integral, and some measure theoretical problems.

In the autumn of 1979, we founded the Seminar on Quantum Logics at the Institute of Measurement, which in 1981 moved to the Mathematical Institute of the Slovak Academy of Sciences.

O. Nánášiová, B. Stehlíková, and P. Konôpka started their investigation of quantum logics under the leadership of Pulmannová at the Mathematical Institute. Nánášiová has been interested in the problem of conditional expectations on quantum logics, Stehlíková studied convergences of observables and laws of large numbers in noncommutative probability, and Konôpka investigated states and decomposition of quantum logics. A. Tirpáková from Nitra and Le Ba Long from Vietnam, as students of A. Dvurečenskij, also joined the group at the end of the eighties and studied fuzzy F-quantum spaces and applications of fuzzy set ideas to quantum mechanics. F. Kôpka and F. Chovanec (both from Liptovský Mikuláš) studied F-quantum spaces as well as quantum logics, under the leadership of A. Dvurečenskij, at the end of the Eighties.

We also appreciate very highly collaboration with the Prague school (P. Pták, M. Navara, J. Hamhalter, J. Tkadlec, V. Rogalewicz) which culminated in writing the joint monograph [PtPu].

In further developments, colleagues from the Slovak Technical University in Bratislava and colleagues from universities in other towns in Slovakia, in particular in Banská Bystrica and Liptovský Mikuláš, turned their interest to the area of quantum logics and noncommutative measure theory. Z. Riečanová and her students, V. Janiš, and V. Palko (all from the Slovak Technical University), enriched the scale of problems studied in Slovakia by topological aspects and relations between measures on quantum logics and topologies, and by a systematic study of Boolean powers of quantum logics (known also under the alternative name of Pták's sums). V. Olejček (from the Slovak Technical University) found a very interesting solution to a problem posed by Prof. Neubrunn about a quantum probability space generated by spheres in a plane. G. Jenča recently joined this team.

Later, in 1991–2, F. Chovanec and F. Kôpka, from the Military Academy in Liptovský Mikuláš, introduced the notion of a so-called difference poset (D-poset). Their colleagues J. Rybárik, E. Rybáriková, and M. Jurečková also contributed to this field.

At this time it is also necessary to mention the newest results of R. Frič and especially of J. Jakubík, both from the Mathematical Institute of the Slovak Academy of Sciences in Košice. J. Jakubík contributed in the last period to the theory of MV-algebras, and today it is clear that MV-algebras will play an analogous role to that of Boolean algebras in quantum logics. Moreover, today we can be fascinated with the fact that a fundamental connection of quantum structures with such an orthodox mathematical discipline as the theory of ordered groups was found.

Currently, these structures are widely studied by several groups in USA, Italy, Poland, Czech Republic, Germany, etc., mostly under the name effect algebras. In what follows, we will concentrate our attention mostly on these structures and their applications. Their importance follows from the fact, that in the theory of quantum measurements, the observables are, in general, represented by positive-operator-valued measures, in other words, by measures taking values in the set of effects. There are many areas of research where they find applications, e.g., the histories approach to quantum mechanics, K_0-theory of AF C*-algebras, etc.

We also recall an important area of applications of fuzzy set ideas in quantum structures in the work of R. Mesiar [Mes 3] (see also J. Pykacz [Pyk]).

For the reader's convenience, we give an extended list of references devoted to D-posets and effect algebras.

The year 1988 was very important, when, in January, in Liptovský Ján in the Tatra Mountains region of Slovakia, we organized the First Winter School on Measure Theory devoted to actual problems of quantum logics, measurement in quantum mechanics and fuzzy set ideas in quantum structures. This was the first regular biannual meeting in this field, not only in the former Czecho-Slovakia, but in the whole of Europe. The importance of this meeting is better understood, when we take into account that in those years it was in practice very complicated to travel abroad from Czecho-Slovakia and to make personal contact with colleagues all over the world. It was easier to invite them to our country. At our Winter School, there were many colleagues from nearly all "quantum structure relevant" countries. Further Winter Schools took place in 1990, 1993, and 1995. In addition, the first organizational meeting of the newly established International Quantum Structures Association took place in Slovakia, in Smolenice in Sept., 1991. In Aug. 30–Sept. 5, 1998, the Fourth Biannual Meeting of IQSA was held in Liptovský Ján.

The vigorous development in the last 30 years of quantum structures confirms that the theory of quantum structures is a typical example of an interdisciplinary theory having an influence into many branches, not only quantum physics. It was not possible to include all very important results and trends, like histories, new construction techniques of orthomodular posets, theory of quantum automata, quantum computing. In any event, the results show us that the efforts of mathematicians, physicists and logicians in building up the theory of quantum structures have been successful. Hence, there is a real hope that investigation of new trends in quantum structures can bring useful results for mathematicians, physicists, and logicians, as well as for experts in fuzzy sets, and quantum computing.

Chapter 1

D-posets and Effect Algebras

In this Chapter, we first introduce and investigate basic properties of D-posets and effect algebras as partial algebraic structures. We introduce the notions of their morphisms and prove a categorical equivalence of D-posets and effect algebras. We show relations of these structures with partially ordered Abelian groups and prove the existence of a universal group for effect algebras.

Some important basic classes of effect algebras, in particular orthoalgebras, orthomodular posets, orthomodular lattices, MV-algebras are studied. Also generalized effect algebras and D-posets as structures without unit are considered. We introduce many examples showing that these structures may appear in many different situations and fields of research.

Special attention is devoted to the important case of so-called interval effect algebras and their special subclasses, effect algebras with Riesz decomposition properties, and to convex effect algebras. We prove that they can be represented by the initial intervals of an interpolation group with order unit and of an ordered real linear space with order unit, respectively.

We consider also some basic constructions with D-posets and effect algebras, like subalgebras, direct products and central decompositions, and direct limits.

Borrowing the language of physics, we introduce the notions of compatibility, observables and commeasurability and show some of their basic properties. We show, e.g., that the notion of "blocks" with similar meaning as in orthomodular lattices can be generalized for lattice ordered effect algebras, with Boolean algebras replaced by MV-algebras. We also study some elements of functional calculus for observables on D-posets and conditions of their commeasurabilty.

1.1 D-posets

A partially ordered set is a set P endowed with the relation of a partial order "a is less or equal to b", written $a \leq b$, which satisfies the following properties for $a, b \in P$: a is less or equal to itself (reflexivity), if $a \leq b$ and $b \leq a$, then $a = b$ (antisymmetry) and if $a \leq b$ and $b \leq c$, then $a \leq c$ (transitivity). For example, the inclusion of sets is a partial order. A partially ordered set P in which the least upper bound (supremum) and the greatest lower bound (infimum) of any two elements exist, is a **lattice**.

Now we define a **difference** $\ominus$ on a partially ordered set P as follows:

Definition 1.1.1 A poset with difference *is a system* $(P; \leq, \ominus)$ *consisting of a partially ordered set* P *with a partially defined binary operation* $\ominus$ *satisfying the following conditions for all* $a, b, c \in P$.

(D1) $b \ominus a$ *is defined if and only if* $a \leq b$.

(D2) *If* $a \leq b$, *then* $b \ominus a \leq b$ *and* $b \ominus (b \ominus a) = a$.

(D3) *If* $a \leq b \leq c$, *then* $c \ominus b \leq c \ominus a$ *and* $(c \ominus a) \ominus (c \ominus b) = b \ominus a$.

These properties are very natural ones, it is easy to check that for example, subtraction of nonnegative numbers has all the above properties. If, in addition, P is a lattice, then P is called a **lattice with difference**.

Let us observe that the difference is not defined for an arbitrary pair of elements of P, but only for those pairs (a, b) where $a \leq b$. Therefore we say that the difference is a **partial binary operation**. For example, in the set of nonnegative numbers the result of a subtraction $b - a$ is again a nonnegative number if and only if the condition $a \leq b$ is satisfied.

In what follows, we will sometimes write $a \ominus b$, tacitly supposing that the latter expression is defined.

We note that a binary operation which is defined for any pair of elements is called **total**. For example, in the set of all real numbers subtraction is a total binary operation.

Let us consider some examples of posets with difference.

(a) A ring of sets partially ordered by inclusion, with the operation of difference of sets, is a lattice with difference.

(b) Let $(R; +, \cdot, \leq)$ be a partially ordered ring, let $R^+ = \{a \in R : a \geq 0\}$. Let us define a partial binary operation $\ominus$ on R^+ as follows:

$b \ominus a$ is defined iff $a \leq b$, and then $b \ominus a := b + (-a)$,
where $-a$ is the inverse element of a with respect to the operation $+$. Then R^+ is a poset with difference.

In particular, the set $\mathbb{R}^+$ of all nonnegative real numbers with respect to the usual difference of real numbers is a lattice with difference.

(c) Let $(G;+,\vee,\wedge)$ be an Abelian ℓ-group (= lattice ordered group) and $G^+ = \{a \in G : a \geq 0\}$. On G^+ we define a partial binary operation $\ominus$ in the same way as in the previous case. Then G^+ is a lattice with difference.

(d) Let $\mathcal{S} = \{x_0, x_1, x_2, \ldots, x_n, \ldots\}$ be a countable chain $x_0 < x_1 < x_2 < \cdots < x_{n-1} < x_n < \cdots, n \in \mathbb{N}$. Put

$x_k \ominus x_j := x_{k-j},\ k,j \in \{0,1,2,\ldots\}, k \geq j.$

Then $\mathcal{S}$ is a lattice with difference.

(e) The n-dimensional real vector space $\mathbb{R}^n$ with coordinatewise partial ordering and with the usual difference of vectors is a lattice with difference.

(f) Any system of fuzzy sets $\mathcal{F}(\Omega) = [0,1]^\Omega$, $\Omega \neq \emptyset$, with pointwise defined partial ordering, and with the partial binary operation $\ominus$ defined via $f \ominus g = f - g$ whenever $g \leq f$, is a poset with difference.

(g) A commutative BCK-algebra (see Chapter 5) is an algebra $(X; *, 0)$ of type (2,0) ([MeJu]) such that, for all $x, y, z \in X$ we have (i) $x * (x * y) = y * (y * x)$; (ii) $(x * y) * z = (x * z) * y$; (iii) $x * x = 0$; and (iv) $x * 0 = x$. If we define $x \leq y$ iff $x * y = 0$, then $\leq$ is a partial order, and we define $y \ominus x = y * x$ whenever $x \leq y$. Then $(X; \leq, \ominus)$ is a poset with difference with the least element 0 [DvKi] which is always a lower semilattice; it need not necessarily be a lattice. BCK-algebras have been introduced by Imai and Iséki [ImIs] as an algebraic structure for propositional calculi. Commutative BCK-algebras generalize MV-algebras (see Chapter 5), introduced by Chang [Cha], which play an important role in many-valued logic.

In what follows, we introduce more examples of posets and lattices with difference. In the following proposition, we list basic properties of posets with difference.

Proposition 1.1.2 *Let $(P; \leq, \ominus)$ be a poset with difference and $a, b, c, d \in P$.*

(i) *If $a \leq b \leq c$, then $b \ominus a \leq c \ominus a$ and $(c \ominus a) \ominus (b \ominus a) = c \ominus b$.*

(ii) *If $b \leq c$ and $a \leq c \ominus b$, then $b \leq c \ominus a$ and $(c \ominus b) \ominus a = (c \ominus a) \ominus b$.*

(iii) *If $a \leq b \leq c$, then $a \leq c \ominus (b \ominus a)$ and $(c \ominus (b \ominus a)) \ominus a = c \ominus b$.*

(iv) *If $a \leq c$ and $b \leq c$, then $c \ominus a = c \ominus b$ if and only if $a = b$.*

(v) *If $d \in P$, $d \leq a, b$ and $a, b \leq c$, then $c \ominus a = b \ominus d$ if and only if $c \ominus b = a \ominus d$.*

Proof. (i) From (D3) and (D1) we get that $(c \ominus a) \ominus (c \ominus b) = b \ominus a \leq c \ominus a$ and by (D2),

$$(c \ominus a) \ominus (b \ominus a) = (c \ominus a) \ominus ((c \ominus a) \ominus (c \ominus b)) = c \ominus b.$$

(ii) From the hypotheses it follows that $a \leq c \ominus b \leq c$, and from (D3) we obtain

$$c \ominus (c \ominus b) \leq c \ominus a, \text{ i.e., } b \leq c \ominus a.$$

Since by (i), $(c \ominus b) \ominus a \leq c \ominus a$, we get from (i) $(c \ominus a) \ominus ((c \ominus b) \ominus a) = c \ominus (c \ominus b) = b$, therefore

$$(c \ominus a) \ominus b = (c \ominus a) \ominus ((c \ominus a) \ominus ((c \ominus b) \ominus a)) = (c \ominus b) \ominus a.$$

(iii) According to (i), we have $b \ominus a \leq c \ominus a \leq c$ and, by (D3), we obtain

$$c \ominus (c \ominus a) \leq c \ominus (b \ominus a), \text{ i.e., } a \leq c \ominus (b \ominus a) \leq c.$$

Using (ii) and (i), we get

$$(c \ominus (b \ominus a)) \ominus a = (c \ominus a) \ominus (b \ominus a) = c \ominus b.$$

(iv) If $c \ominus a = c \ominus b$, then $b = c \ominus (c \ominus b) = c \ominus (c \ominus a) = a$. The converse assertion is evident.

(v) If $c \ominus a = b \ominus d$, then by (i) and (D3), $c \ominus b = (c \ominus d) \ominus (b \ominus d) = (c \ominus d) \ominus (c \ominus a) = a \ominus d$. The converse assertion can be proved analogously. □

In the definition of a poset with difference, the existence of a smallest and a greatest element is not required. Observe that if $a \leq b$, then $a \ominus a = b \ominus b$. Indeed, from $a \leq a \leq b$ we have $b \ominus a = (b \ominus a) \ominus (a \ominus a)$, and from $a \leq b \leq b$ we have $b \ominus a = (b \ominus a) \ominus (b \ominus b)$. Therefore (iv) in Proposition 1.1.2 yields the desired result. Using the transitive closure of the binary relation $\sim_z$, where $a \sim_z b$ iff $a \leq b$ or $b \leq a$, in [HePu], it has been observed that every poset with difference can be written as a set-theoretical union of posets with difference, each having a smallest element.

In particular, if there is a greatest element, usually denoted by 1, in a poset with difference, then $1 \ominus 1$ is the smallest element, usually denoted by 0.

Definition 1.1.3 *A poset with difference which contains a greatest element* 1 *and, consequently, a smallest element* 0, *is called a* **difference poset** *(a* **D-poset**, *in short).*

For any element a in a D-poset, the element $1 \ominus a$ is called the **orthosupplement** of a and is denoted by a'.

Recall that a unary operation $' : P \to P$ on a poset P with 0 and 1 as the smallest and the greatest elements, respectively, which is an involution, i.e., $a'' = (a')' = a$, order reversing, i.e., $a \leq b \Rightarrow b' \leq a'$ and a complementation, i.e., $a \vee a' = 1$ (dually $a \wedge a' = 0$) is called an **orthocomplementation**. A poset P with orthocomplementation is an **orthoposet**.

The operation $a \mapsto a'$ on a D-poset satisfies the following properties: $a \leq b \Rightarrow b' \leq a'$; $a'' = a$; but we need not have $a \wedge a' = 0$, in general. Therefore a D-poset need not be an orthoposet.

Let us observe that the following condition need not be satisfied in a poset with a difference $(P; \leq, \ominus)$ even if it contains a smallest element.

(C) If $a \leq b, c$ and $b \ominus a = c \ominus a$, then $b = c$ (**cancellation law**).

The simplest example where (C) is not satisfied is shown in Figure 1.1, where $b \ominus a = c \ominus a = a$, $x \ominus x = 0$ and $x \ominus 0 = x$ for all x.

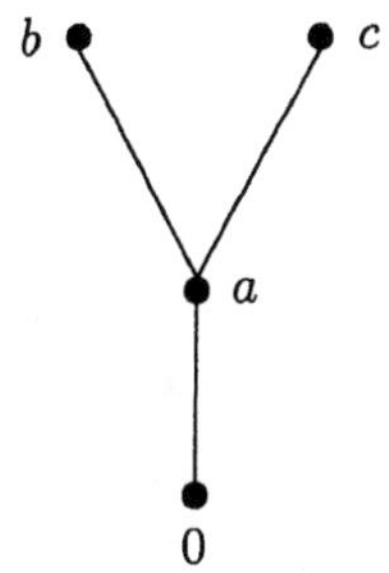

Fig. 1.1

Definition 1.1.4 *Let P be a poset with difference. We call P a* **generalized difference poset** *if P contains a smallest element* 0 *and condition (C) is satisfied.*

A generalized difference poset is a generalization of a generalized orthomodular lattice [Jan], and a generalized orthomodular poset [Malp]. Our examples (a)–(f) (and also example (g) whenever X has the relative cancellation property (see also Chapter 5) are examples of generalized difference posets. Some other examples, based on associative and alternative rings, tripotents in Jordan and alternative triples and JBW*-triples are introduced in [HePu] (see Section 1.6).

In every generalized D-poset, a dual operation $\oplus$ to the operation $\ominus$ can be defined as follows.

(S) $a \oplus b$ exists and equals $c \Leftrightarrow c \ominus b = a$.

Condition (S) reminds us of the usual transition from subtraction to addition of numbers. Condition (C) guarantees that the operation $\oplus$ is well-defined. Recall that a partially ordered set P is **upward directed** if, for all $a, b \in P$, there exists $d \in P$ such that $a \leq d$, $b \leq d$.

Lemma 1.1.5 *If P is an upward directed poset with difference, then condition (C) is satisfied. In particular, in every D-poset condition (C) is satisfied.*

Proof. Let $b \ominus a = c \ominus a$, $a, b, c \in P$. Since P is upward directed, there is an element $d \in P$ with $b, c \leq d$. Using Proposition 1.1.2(i), we obtain

$$d \ominus b = (d \ominus a) \ominus (b \ominus a) = (d \ominus a) \ominus (c \ominus a) = d \ominus c,$$

and so

$$b = d \ominus (d \ominus b) = d \ominus (d \ominus c) = c.$$

□

The relationships between the operations $\ominus$ and $\oplus$ are investigated in the following proposition.

Proposition 1.1.6 *Let P be a D-poset and $a, b, c \in P$. Then*

(i) *$a \oplus b$ exists in P iff $a \leq b'$, and $a \leq b'$ implies $a \leq a \oplus b$ and $(a \oplus b) \ominus a = b$.*

(ii) *$a \leq b'$ and $a \oplus b \leq c$ imply $c \ominus (a \oplus b) = (c \ominus a) \ominus b = (c \ominus b) \ominus a$.*

(iii) *$a \leq b \leq c'$ implies $a \oplus c \leq b \oplus c$ and $(b \oplus c) \ominus (a \oplus c) = b \ominus a$.*

(iv) *$a \leq b \leq c$ implies $a \oplus (c \ominus b) = c \ominus (b \ominus a)$ and also $(c \ominus b) \oplus (b \ominus a) = c \ominus a$.*

(v) *$a \leq b$ implies $b = a \oplus (b \ominus a)$;*

(vi) *$a \leq b'$ and $a \leq c'$ imply that $a \oplus b = a \oplus c$ if and only if $b = c$.*

(vii) *$a \leq b \leq c'$ and $c \leq d \leq a'$ imply that $a \oplus d = b \oplus c$ if and only if $b \ominus a = d \ominus c$.*

(viii) *$a \leq b' \leq c'$ implies $a \oplus (b \ominus c) = (a \oplus b) \ominus c$.*

(ix) *$c \leq a \leq b'$ and $c \leq b$ imply $(a \ominus c) \oplus (b \ominus c) = ((a \oplus b) \ominus c) \ominus c$.*

Proof. The proof requires routine verifications with the help of the axioms and other properties of the difference $\ominus$. □

We note that a (generalized) difference poset can also be introduced in another way, as an algebra with a partial binary operation $\ominus$ and a special element 0 as follows [DvPu 3], [Gud 5].

Proposition 1.1.7 *Let P be a set equipped with a special element 0 and endowed with a partial binary operation $\ominus$. Let $\leq$ be a binary relation on P given by*

$$a \leq b \text{ if and only if } b \ominus a \text{ is defined.}$$

Then $(P; \leq, \ominus, 0)$ is a generalized difference poset if and only if the following conditions are satisfied for all $a, b, c \in P$.

(d1) *$a \ominus 0$ is defined, and $a \ominus 0 = a$.*

(d2) *$a \ominus a$ is defined.*

(d3) *If $b \ominus a$ and $c \ominus b$ are defined, then $c \ominus a$ and $(c \ominus a) \ominus (c \ominus b)$ are defined, and $(c \ominus a) \ominus (c \ominus b) = b \ominus a$.*

(d4) *If $0 \ominus a$ is defined, then $a = 0$.*

(d5) *If $b \ominus a$ is defined and $c \ominus a$ is defined and $c \ominus a = b \ominus a$, then $b = c$.*

P is a difference poset if and only if it contains an element 1 *such that the following condition is satisfied:*

(d5') *For any* $a \in P$, $1 \ominus a$ *is defined.*

Proof. It is clear that every generalized D-poset satisfies (d1)–(d5), and a D-poset satisfies also (d5').

To prove the converse implication, observe that from (d1)–(d4), the following properties can be deduced.

(i) If $b \ominus a$ is defined, then $b \ominus (b \ominus a)$ is defined and $b \ominus (b \ominus a) = a$.

(ii) If $a \ominus b = a \ominus c$, then $b = c$.

(iii) $a \ominus a = 0$ for any $a \in P$.

Indeed, from (d1) and (d3) we conclude that if $b \ominus a$ is defined, then $(b \ominus 0) \ominus (b \ominus a) = a \ominus 0 = a$, which proves (i). By (i), we have $b = a \ominus (a \ominus b) = a \ominus (a \ominus c) = c$, which proves (ii). Since $a \ominus 0$ and $a \ominus a$ exist in P, from (d3) we obtain $(a \ominus 0) \ominus (a \ominus a) = a \ominus 0$, so that $a \ominus (a \ominus a) = a \ominus 0$, which, by (ii), means that $a \ominus a = 0$, i.e., (iii) holds.

Let us define an ordering $\leq$ on P by setting $a \leq b$ iff $b \ominus a$ is defined. From (d2) we have $a \leq a$, and from (d3) we have the transitivity of $\leq$. To prove antisymmetry, assume $a \leq b$ and $b \leq a$. Then $b \ominus a$ and $a \ominus b$ exist in P, and, by (d3), we conclude $(a \ominus a) \ominus (a \ominus b) = b \ominus a$, so that $0 \ominus (a \ominus b)$ is defined, and (d4) entails that $a \ominus b = 0$, and, by symmetry, $b \ominus a = 0$. Hence by (i), $a = b \ominus (b \ominus a) = b \ominus 0 = b$. This proves that $\ominus$ is a difference. Now (d5) implies condition (C), so that $(P; \leq, \ominus)$ is a generalized difference poset. Now assume that (d5') is satisfied. Conditions (d1)—(d4) and (d5') imply that $0 \leq a \leq 1$ for arbitrary $a \in P$. This proves that $(P; \leq, \ominus, 0)$ (resp. $(P; \leq, \ominus, 0, 1)$) is a generalized D-poset (a D-poset). Observe that by Lemma 1.1.5, (d5') implies (d5), so that in this case (d5) is redundant. □

1.2 Effect Algebras

An alternative structure to a D-poset is that of an **effect algebra** ([FoBe 1], or [GiGr], where it is called an "unsharp orthoalgebra"). Our definition of an effect algebra follows the first mentioned paper.

An effect algebra is based on a partial binary operation $\oplus$. The $\oplus$ operation goes back to the original ideas of G. Boole, [Boo], who supposed that $a + b$ will denote the logical disjunction and ab the logical conjunction of a, b, respectively. In fact, Boole only wrote $a + b$ when $ab = 0$, and this is all that is needed for probability theory: If $ab = 0$, then $P(a + b) = P(a) + P(b)$, where P is a probability measure. Therefore, $+$ can be introduced as a partially defined binary operation.

The axioms below give a natural algebraic structure, that of an effect algebra, using Boole's idea of $a + b$.

Definition 1.2.1 *An* **effect algebra** *is a system* $(E; \oplus, 0, 1)$ *consisting of a set* E *with two special elements* $0, 1 \in E$, *called the* **zero** *and the* **unit**, *and with a partially defined binary operation* $\oplus$ *satisfying the following conditions for all* $p, q, r \in E$.

(E1) (Commutative law) *If* $p \oplus q$ *is defined, then* $q \oplus p$ *is defined and* $p \oplus q = q \oplus p$.

(E2) (Associative law) *If* $q \oplus r$ *is defined and* $p \oplus (q \oplus r)$ *is defined, then* $p \oplus q$ *and* $(p \oplus q) \oplus r$ *are defined and* $p \oplus (q \oplus r) = (p \oplus q) \oplus r$.

(E3) (Orthosupplement law) *For every* $p \in E$ *there exists a unique* $q \in E$ *such that* $p \oplus q$ *is defined and* $p \oplus q = 1$.

(E4) (Zero-one law) *If* $p \oplus 1$ *is defined, then* $p = 0$.

A **generalized effect algebra** *(or* **GEA**, *for short) is a system* $(E; \oplus, 0)$ *satisfying conditions* (E1) *and* (E2) *and*

(E3') (Cancellation law) *If* $a \oplus b = a \oplus c$, *then* $b = c$.

(E4') (Positivity law) *If* $a \oplus b = 0$, *then* $a = b = 0$.

(E5) $a \oplus 0 = a$, $a \in E$.

The element q *in* (E3) *will be denoted by* p' *and called the* **orthosupplement** *of* p.

In every generalized effect algebra, a dual operation $\ominus$ to the operation $\oplus$ can be defined as follows

(D) $a \ominus c$ exists and equals b if and only if $b \oplus c$ exists and equals a.

Definition 1.2.2 *Let* E *be a generalized effect algebra.*

(1) *Define a binary relation on* E *by*

$$a \leq b \text{ if, for some } c \in E, \ c \oplus a = b. \tag{1.1}$$

(2) *Define a binary relation* $\perp$ *on* E *by*

$$a \perp b \ \Leftrightarrow \ \exists \ a \oplus b. \tag{1.2}$$

Proposition 1.2.3 *Let* E *be a generalized effect algebra. The binary relation* $\leq$ *defined by* (1.1) *is a partial ordering in* E. *Moreover,* 0 *is the smallest element of* E.

Proof. Reflexivity follows from $a \oplus 0 = a$.

To prove antisymmetry, assume $a \leq b$ and $b \leq a$. By definition, there exist elements u, v such that $b = a \oplus u$, $a = b \oplus v$. Hence $b = (b \oplus v) \oplus u$, and by (E2) and cancellativity, $v \oplus u = 0$, hence by positivity, $u = v = 0$. It follows $a = b$.

To prove transitivity, let $a \leq b$ and $b \leq c$. By definition, there are x, y such that $a \oplus x = b$, $b \oplus y = c$. Hence $(a \oplus x) \oplus y = c = a \oplus (x \oplus y)$, so that $a \leq c$.

From the fact that $a \oplus 0 = a$ for arbitrary $a \in E$ it follows easily that 0 is the smallest element. □

Lemma 1.2.4 *Let $(E;\oplus,0)$ be a generalized effect algebra. E is an effect algebra if and only if it has a greatest element* 1.

Proof. Let $(E;\oplus,0,1)$ be an effect algebra. We have to prove conditions (E3') and (E4'). Let $a\oplus b = a\oplus c$. By (E3), there is a unique element u such that $a\oplus b\oplus u = a\oplus c\oplus u = 1$. It follows, by uniqueness of the orthosupplement, that $(a\oplus u)' = b = c$. To prove (E4'), let $a\oplus b = 0$. It follows that $a' = a'\oplus a\oplus b = 1\oplus b$, so that by (E4), $b=0$. It is clear by (E3) that 1 is the greatest element of E.

Conversely, assume that E is a generalized effect algebra with greatest element 1. We have to prove conditions (E3) and (E4). Since $a\leq 1$ for any $a\in E$, there exits an element $b\in E$ such that $a\oplus b=1$. Now b is unique by (E3'). This proves (E3). To prove (E4), let $a\oplus 1 =: v$ be defined. Then $1\leq v$, hence $1=v$, which yields $a\oplus 1 = 1$ and so by (E3'), $a=0$. □

Lemma 1.2.5 *Let $(E;\oplus,0,1)$ be an effect algebra.* (i) *$a\perp b$ if and only if $a\leq b'$.*
(ii)

$$(a\oplus b)' = a'\ominus b = b'\ominus a. \tag{1.3}$$

Proof. (i) Assume $a\perp b$, i.e., $a\oplus b$ is defined. We have $a\oplus b\oplus(a\oplus b)'=1$, so by (E3), $a\oplus(a\oplus b)' = b'$, and hence $a\leq b'$. Conversely, if $a\leq b'$, there is a $c\in E$ such that $a\oplus c = b'$. Hence, $(a\oplus c)\oplus b = 1$. By (E2), $a\oplus b$ exists.

(ii) Using (D), we may write $a' = 1\ominus a$. Using (E3), we get $1\ominus(a\oplus b) = (1\ominus a)\ominus b$. The rest follows by symmetry. □

The operation $\oplus$ can be extended to any finite set of elements in a recurrent way: We say that $a_1,\dots,a_n$ are **summable** if $d := a_1\oplus\cdots\oplus a_{n-1}$ exists, and $d\oplus a_n$ exists, and put $a_1\oplus\cdots\oplus a_n = d\oplus a_n$. Owing to (E2), we can omit the parentheses. For an arbitrary subset $D\subseteq E$, we say that D is summable if every finite subset is summable. We define a **sum** of a system D, $\bigoplus D$, as the supremum of all partial finite sums, if it exists in E. We say that a (generalized) effect algebra is a **(generalized) σ-effect algebra**, if every countable summable subset has a sum. Similarly, a (generalized) effect algebra is called **complete** if an arbitrary summable subset has a sum. Clearly, E is a (generalized) σ-effect algebra iff for any nondecreasing chain $a_1\leq a_2\leq\cdots\leq a_i\leq\cdots$ the supremum $\bigvee_i a_i$ exists. For every element $a\in E$, we define its **isotropic index** $\iota(a)$, as the maximal nonnegative number n such that $na := \underbrace{a\oplus\cdots\oplus a}_{n-times}$ exists. If na exists for every integer n, we say that $\iota(a)=\infty$. We say that E is **Archimedean** if for every $a\neq 0$, $\iota(a)$ is finite. Observe that in any generalized effect algebra, $\bigoplus_{i=1}^{\infty} a$ exists iff $a=0$. Indeed, $\bigoplus_{i=1}^{\infty} a = a\oplus\bigoplus_{i=1}^{\infty} a$, which by cancellativity implies $a=0$.

Our aim is to show that every generalized effect algebra is an order ideal (with special properties) of an effect algebra. This result has been obtained in [Wil 2], and an alternative result for generalized D-posets in [HePu]. Similar results have been also obtained for particular structures: generalized orthomodular lattices, which are order ideals of orthomodular lattices [Jan] and (weak) generalized orthomodular posets, which are order ideals of orthomodular posets [MaIp].

Recall that a nonvoid subset I of a partially ordered set L is an **order ideal** if $a \in L$, $b \in I$ and $a \leq b$, then $a \in I$.

Let $(P; \leq, \ominus, 0)$ be a generalized effect algebra. Let $P^\sharp$ be a set disjoint from P with the same cardinality. Consider a bijection $a \mapsto a^\sharp$ from P onto $P^\sharp$ and let us denote $P \cup P^\sharp = \hat{P}$. Define a partial binary operation $\oplus^*$ on $\hat{P}$ by the following rules. For $a, b \in P$,

(i) $a \oplus^* b$ is defined if and only if $a \oplus b$ is defined, and $a \oplus^* b = a \oplus b$.

(ii) $b^\sharp \oplus^* a$ and $a \oplus^* b^\sharp$ are defined iff $b \ominus a$ is defined, and then $b^\sharp \oplus^* a = (b \ominus a)^\sharp = a \oplus^* b^\sharp$.

Theorem 1.2.6 *The structure $(\hat{P}; \oplus^*, 0, 0^\sharp)$ is an effect algebra. Moreover, P is an order ideal in $\hat{P}$ closed under $\oplus^*$, and the partial order induced by $\oplus^*$ when restricted to P, coincides with the partial order induced by $\oplus$.*

Proof. We have to check axioms (E1)–(E4). (E1) follows directly by definition.

To prove (E2), observe that if $a \oplus^* b$ and $(a \oplus^* b) \oplus^* c$ are defined by (i), then we have $(a \oplus^* b) \oplus^* c = (a \oplus b) \oplus c = a \oplus (b \oplus c) = a \oplus^* (b \oplus^* c)$. Now consider $a^\sharp \oplus^* (b \oplus c)$. Then by (ii), $a \ominus (b \oplus c)$ is defined in P, and since $a \ominus (b \oplus c) = (a \ominus b) \ominus c$, we obtain that

$$\begin{aligned} a^\sharp \oplus^* (b \oplus^* c) &= (a \ominus (b \oplus c))^\sharp \\ &= ((a \ominus b) \ominus c)^\sharp \\ &= (a \ominus b)^\sharp \oplus^* c \\ &= (a^\sharp \oplus^* b) \oplus^* c. \end{aligned}$$

The only remaining case is $a \oplus^* (b^\sharp \oplus^* c)$. It then follows

$$\begin{aligned} a \oplus^* (b^\sharp \oplus^* c) &= a \oplus^* (b \ominus c)^\sharp \\ &= ((b \ominus c) \ominus a)^\sharp \\ &= ((b \ominus a) \ominus c)^\sharp \\ &= (b \ominus a)^\sharp \oplus^* c \\ &= (b^\sharp \oplus^* a) \oplus^* c. \end{aligned}$$

To prove (E3), observe that $a^\sharp \oplus^* a$ exists for any $a \in P$ and equals $0^\sharp$, since $a \ominus a = 0$ for any $a \in P$. Let $x \in \hat{P}$ be such that $a \oplus^* x = 0^\sharp$. Then $x \in P$ is excluded by (i), hence $x \in P^\sharp$, $x = b^\sharp$, say. Then, by (ii), $a \oplus^* b^\sharp = (b \ominus a)^\sharp = 0^\sharp$, hence $b \ominus a = 0$, i.e., $b = a$.

To prove (E4), assume that $a \oplus 0^\sharp$ is defined. Then $0 \ominus a$ is defined, hence $a = 0$. Let $\leq^*$ be the order relation in $\hat{P}$ defined by $\oplus^*$. Observe that $a^\sharp \leq^* b$ is excluded by the definition of $\oplus^*$. This implies that P is an order ideal. If $a, b \in P$ and $a \oplus^* b$ exists, then by (i), $a \oplus b$ exists in P. It follows that P is closed under $\oplus^*$. Finally, let $a \leq^* b$, $a, b \in P$. Then there exists an element $x \in \hat{P}$ such that $a \oplus^* x = b$. Since P is an order ideal in $\hat{P}$, $x \leq^* b$ implies $x \in P$, hence $a \leq b$. The implication $\leq \Rightarrow \leq^*$ follows by (i). □

Proposition 1.2.7 *Let D be an effect algebra and let P be a proper order ideal in D closed under $\oplus$ and such that for every $a \in D$, $a \in P$ or $1 \ominus a \in P$. Denote $a^\sharp = 1 \ominus a$, $a \in P$. Then the effect algebra $(\hat{P}; \oplus^*, 0, 0^\sharp)$ coincides with $(D; \oplus, 0, 1)$.*

Proof. Since P is proper, for every $a \in D$ we have either that $a \in P$ or $1 \ominus a \in P$. It is now clear that $P^\sharp = D \setminus P$. Let $'$ be the following unary operation on D: $a' = 1 \ominus a$, $a \in D$. By Lemma 1.2.5, for any $a, b \in D$, $a \oplus b$ is defined if and only if $a \leq b'$, and moreover, $a \oplus b = (b' \ominus a)'$. Hence from $a' \leq b'$, which is equivalent to $b \leq a$, it follows $b' \ominus a' = (a' \oplus b)' = (b \oplus a')' = a \ominus b$. The proof is now clear from the definition of $(\hat{P}; \oplus^*, 0, 0^\sharp)$. □

Example 1.2.8 The set $P = \mathbb{R}^+$ of all nonnegative real numbers with the natural ordering and with the usual difference of numbers is a GEA (GDP). Since for every $a, b \in \mathbb{R}^+$ also $a + b \in \mathbb{R}^+$, we have $a \perp b$. Therefore, in $\hat{P} = P \cup P^\sharp$ every element in $P^\sharp$ is greater than any element in P. It follows that $\hat{P}$ is linearly (totally) ordered. Since for any $a \in \mathbb{R}^+$, na exists in $\mathbb{R}^+$, it is also defined in $\hat{P}$. Hence every $a \in P$ has an infinite isotropic index.

More general, the positive cone $P = G^+$ of any partially ordered abelian group $(G; +, 0, \leq)$ with the usual difference of group elements is a GEA/GDP. For every $a, b \in G^+$ also $a + b \in G^+$, hence $a \perp b$. Thus, in $\hat{P} = P \cup P^\sharp$ every element in $P^\sharp$ is greater than any element in P.

In the category of generalized difference posets, the following result can be obtained. Let $(P; \leq, \ominus, 0)$ be a generalized difference poset. Let $P^\sharp$ be a set disjoint from P with the same cardinality. Consider a bijection $a \mapsto a^\sharp$ from P onto $P^\sharp$ and let us denote $P \cup P^\sharp = \hat{P}$. Define a partial binary operation $\ominus^*$ on $\hat{P}$ by the following rules ($a, b \in P$):

(i) $b \ominus^* a$ is defined if and only if $b \ominus a$ is defined and $b \ominus^* a = b \ominus a$.

(ii) $b^\sharp \ominus^* a$ is defined if and only if $a \oplus b$ is defined and $b^\sharp \ominus^* a = (a \oplus b)^\sharp$.

(iii) $b^\sharp \ominus^* a^\sharp$ is defined if and only if $a \ominus b$ is defined and $b^\sharp \ominus^* a^\sharp = a \ominus b$.

Define a binary relation $\leq^*$ on $\hat{P}$ as follows: $x \leq^* y$ if and only if $y \ominus^* x$ is defined.

The proof of the next theorem is left to the readers as an exercise (see Theorem 2.4 in [HePu] for the details, or Exercise 24, Section 1.1).

Theorem 1.2.9 $(\hat{P}; \leq^*, \ominus^*, 0, 0^\sharp)$ *is a difference poset.*

1.3 Morphisms and Categorical Equivalence of D-posets and Effect Algebras

We have already observed that conditions (S) (for GDP) and (D) (for GEA) enable us to transfer from $\oplus$ to $\ominus$ and vice versa. The aim of this section is to find the exact relations between (generalized) effect algebras and D-posets.

Definition 1.3.1 *Let P, Q be two D-posets. A mapping $v : P \to Q$ is a* **morphism** *(of D-posets) if the following conditions are satisfied:*

(DM1) $v(1_P) = 1_Q$.

(DM2) *If $a, b \in P$, $a \leq b$, then $v(a) \leq v(b)$ and $v(b \ominus a) = v(b) \ominus v(a)$.*

A morphism $v : P \to Q$ is called a **σ-morphism** *if the following condition holds:*

(DM3) *If $\{a_n\}_{n=1}^{\infty} \subseteq P$, $a_n \uparrow a$ (i.e., $a_n \leq a_{n+1}$ for any $n \in \mathbb{N}$ and $a = \bigvee_{n=1}^{\infty} a_n$), then $v(a_n) \uparrow v(a)$.*

Definition 1.3.2 *Let E, F be effect algebras. A mapping $w : E \to F$ is a* **morphism** *(of effect algebras) if the following conditions are satisfied:*

(EM1) $w(1_E) = 1_F$.

(EM2) *If $a, b \in E$, $a \perp b$, then $w(a) \perp w(b)$ and $w(b \oplus a) = w(b) \oplus w(a)$.*

A morphism $w : P \to Q$ is called a **σ-morphism** *if the following condition holds:*

(EM3) *If $\{a_n\}_{n=1}^{\infty} \subseteq P$, $a_n \uparrow a$ (i.e., $a_n \leq a_{n+1}$ for any $n \in \mathbb{N}$ and $a = \bigvee_{n=1}^{\infty} a_n$), then $w(a_n) \uparrow w(a)$.*

A morphism h of difference posets or effect algebras is called a $\wedge$-**morphism** if

$$h(a \wedge b) = h(a) \wedge h(b)$$

whenever $a \wedge b$ exists. A morphism h such that $a \perp b$ iff $h(a) \perp h(b)$ is called a **monomorphism**. A bijective morphism h such that h^{-1} is also a morphism is an **isomorphism**. Observe that an isomorphism is necessarily a $\wedge$-morphism, and a surjective monomorphism is an isomorphism.

Definition 1.3.3 *A morphism from an effect algebra/D-poset L to the unit interval $[0,1]$ of reals is called a* **state** *on L.*

A σ-morphism from L to $[0,1]$ is a **σ-additive state**.

Theorem 1.3.4 (1) *Let $(P;\leq,\ominus,0,1)$ be a D-poset. Define a partial binary relation $\oplus$ on P by condition* (S). *Then $E(P):=(P;\oplus,0,1)$ is an effect algebra, and the partial ordering induced by $\oplus$ coincides with the initial partial ordering in P. Moreover, to every morphism of D-posets h from P to a D-poset Q there is a unique morphism $g: E(P)\to E(Q)$ such that $h(x)=g(x)$ for any $x\in P$.*

(2) *Let $(E;\oplus,0,1)$ be an effect algebra. Define a partial binary operation on E by condition* (D). *Then $D(E):=(E;\leq,\ominus,0,1)$ is a D-poset, where the partial ordering $\leq$ on E coincides with the effect algebra ordering. Moreover, to every morphism of effect algebras h from E to an effect algebra F there is a unique morphism of D-posets $g: D(E)\to D(F)$ such that $h(x)=g(x)$ for any $x\in E$.*

Proof. (1) We have to check the axioms of effect algebras.

(E1) Let $a\oplus b$ exist and $a\oplus b=c$. By condition (S), $c\ominus b=a$. It follows that $a\leq c$, and $c\ominus a=c\ominus(c\ominus b)=b$. By condition (S), $b\oplus a$ exists, and $b\oplus a=c$.

(E2) Assume $u:=a\oplus b$ and $v:=(a\oplus b)\oplus c$ exist. By (S), this means that $a=u\ominus b$ and $a\oplus b=u=v\ominus c$. Therefore $a\leq u\leq v$, and hence by (S), $v\ominus a=(v\ominus u)\oplus(u\ominus a)$. By (E1), $u\ominus a=b$, and so $v\ominus a=c\oplus b$. By (S), $v=a\oplus(c\oplus b)=a\oplus(b\oplus c)$.

(E3) For any a, $a\oplus(1\ominus a)=1$. Assume $a\oplus b=1$. By (E1), $b\oplus a=1$ and by (S), $b=1\ominus a$. Hence b such that $a\oplus b=1$ exists and is unique.

(E4) Assume $a\oplus 1$ exists. By (S), there exists u such that $a=u\ominus 1$. Since 1 is the greatest element, it follows $u=1$, hence $a=0$.

The proof of part (2) is dual. □

We note that a similar equivalence holds between generalized difference posets and generalized effect algebras. The proof is left to the reader as an exercise. Owing to the categorical equivalence, notions like summability, isotropic index, σ-structure can be transferred from (generalized) effect algebras to (generalized) D-posets.

Note that in spite of the above categorical equivalence, (generalized) effect algebras and difference posets may be different from the universal algebra point of view. Let us consider, for example, the unit interval $[0,1]$ of reals. The interval $[0,1)$ is a subalgebra of $[0,1]$ when it is considered as a generalized D-poset, but it is not a subalgebra when it is considered as a generalized effect algebra.

1.4 Group Representations of Effect Algebras. Interval Effect Algebras

In what follows, Abelian groups will be written additively. Recall that an **ordered Abelian group** is an Abelian group $(G; +, 0)$ equipped with a specified partial order $\leq$ which is translation-invariant, that is, given any $x, y, z \in G$ with $x \leq y$, it follows that $x + z \leq y + z$.

A **positive element** in a partially ordered Abelian group G is an element $x \in G$ such that $x \geq 0$.

A **cone** in an Abelian group G is any subset C of G such that $0 \in C$, and C is closed under $+$. A cone C is **strict** provided that 0 is the only element x in G for which both $x \in C$ and $-x \in C$. Given a cone C in G, we may define a relation $\leq_C$ on G such that $x \leq_C y$ if and only if $y - x \in C$ (for arbitrary $x, y \in G$). The relation $\leq_C$ is a translation-invariant pre-order on G, and $\leq_C$ is a partial order if and only if C is a strict cone.

The **positive cone** of a partially ordered Abelian group G is the set G^+ of all positive elements in G.

Given a partially ordered Abelian group G, its positive cone is denoted by $G^+ := \{g \in G : 0 \leq g\}$. Note that it is a strict cone and that the given partial order on G coincides with $\leq_{G^+}$.

In particular, $\mathbb{R}^+$ and $\mathbb{Z}^+$ denote the positive cones in the additive group $\mathbb{R}$ of real numbers and the additive group $\mathbb{Z}$ of integers. If G is a partially ordered Abelian group and $u \in G^+$, we define the interval

$$G^+[0, u] := \{g \in G : 0 \leq g \leq u\}.$$

The proof of the following theorem is straightforward.

Theorem 1.4.1 *If G is a partially ordered Abelian group and $u \in G^+$, then the interval $G^+[0, u]$ can be organized into an effect algebra $(G^+[0, u]; \oplus, 0, u)$ such that $p \oplus q$ is defined if and only if $p + q \leq u$, in which case $p \oplus q = p + q$. Moreover, the effect-algebra partial order coincides with the restriction to $G^+[0, u]$ of the partial order on G.*

Definition 1.4.2 *An effect algebra of the form $G^+[0, u]$ as in Theorem 1.4.1, or isomorphic to such an effect algebra, is called an* **interval effect algebra** *(or, for short, an* **interval algebra***) with unit u of the group G.*

The interval $[0, 1]$ of real numbers with the unit 1 is an interval effect algebra with unit 1 in the totally ordered Abelian group of real numbers.

The standard effect algebra $\mathcal{E}(H)$ of operators on a Hilbert space H is an interval algebra with unit I in the partially ordered Abelian group $\mathcal{B}_{sa}(H)$ of all self-adjoint operators on H.

Observe that if $p_1, \ldots, p_n$ is a finite sequence of elements in an interval algebra $G^+[0,u]$, then the orthogonal sum $\bigoplus_i p_i$ exists in $G^+[0,u]$ iff $\sum_i p_i \leq u$ in G, in which case $\bigoplus_i p_i = \sum_i p_i$. (The proof can be obtained by a simple mathematical induction.)

If G is a partially ordered Abelian group and $M \subseteq G$, we denote by $\langle M \rangle$ the subgroup generated by M and by $ssg(M)$ the subsemigroup of G consisting of 0 and all finite sums of elements of G. Clearly $\langle G^+ \rangle = G^+ - G^+$. Recall that a partially ordered Abelian group G is **directed** if, for arbitrary $a, b \in G$, there exists $c \in G$ with $a, b \leq c$. We say that the positive cone G^+ is **generating** if $G = G^+ - G^+$. The statement of the following lemma is well-known [Goo 1].

Lemma 1.4.3 *Let G be a partially ordered Abelian group. Then $G = \langle G^+ \rangle$ (in other words, G^+ is generating) if and only if G is directed.*

A subset S of an effect algebra E is a sub-effect-algebra if (i) $1 \in S$, (ii) $a \in S$ implies $a' \in S$, and (iii) if $a, b \in S$, $a \perp b$, then $a \oplus b \in S$ (see Definition 1.9.1).

Theorem 1.4.4 [BeFo 1] *Let K be a partially ordered Abelian group, $u \in K^+$ and let S be a sub-effect algebra of $K^+[0,u]$. Let $G := \langle S \rangle$ and $G^+ := ssg(S)$. Then G is a partially ordered Abelian group with the generating cone G^+ and $S = G^+[0,u]$.*

Proof. Evidently, $0, u \in S \subseteq G^+ \subseteq G \cap K^+$, $G^+ + G^+ \subseteq G^+$, $G^+ \cap -G^+ = \{0\}$. Hence G^+ is the generating cone in G, and G is partially ordered with respect to it. If $p \in S$, then $p' = u - p \in S$, so $p, u-p \in G^+[0,u]$, hence $p, u-p \in K^+[0,u]$. Conversely, suppose that $p \in G^+[0,u]$. Then $p, u-p \in G^+ \subseteq K^+$, and there exist $p_1, \ldots, p_n \in S \subseteq K^+[0,u]$ such that $p = \sum_i p_i$ and $p \leq u$ in K. Therefore $p = \bigoplus_i p_i$ in $K^+[0,u]$ and, since S is a sub-effect algebra of $K^+[0,u]$, it follows that $p \in S$. □

Corollary 1.4.5 *A sub-effect algebra of an interval algebra is an interval algebra.*

If G is a partially ordered Abelian group, an element $u \in G^+$ is an **order unit** or a **strong unit** for G if, for every $g \in G$, there exists $n \in \mathbb{Z}^+$ such that $g \leq nu$. If G admits an order unit, then G^+ is a generating cone, i.e., G is directed. Indeed, if $x, y \in G$, then there exist $n, m \in \mathbb{Z}^+$ with $x \leq nu$ and $y \leq mu$, whence $x, y \leq (n+m)u$.

Lemma 1.4.6 *If G is a partially ordered Abelian group with a generating cone G^+ and $G^+ = ssg(G^+[0,u])$, $u \in G^+$, then u is an order unit for G.*

Proof. Since G^+ generates G, for any $g \in G$ there exist elements $a, b \in G^+$ with $g = a - b$. Since $G^+ = ssg(G^+[0,u])$, there exist elements $p_1, \ldots, p_n \in G^+[0,u]$ with $a = \sum_i p_i$. Thus, $a = \sum_i p_i \leq nu \leq nu + b$, and it follows that $g = a - b \leq nu$. □

In what follows, when we refer to a family of morphisms on L, we do not necessarily mean to imply that all morphisms in the family map L into the same effect algebra. The conditions given in the following definition are suggested by analogous conditions that have been studied for families of states on quantum logics.

Definition 1.4.7 *Let Δ be a family of morphisms on an effect algebra L.*

(i) *Δ is* **positive** *if, for any $p \in L$ with $p \neq 0$, there is a $\phi \in \Delta$, $\phi(p) \neq 0$.*

(ii) *Δ is* **separating** *if, for $p, q \in L$ with $p \neq q$, there is a $\phi \in \Delta$, $\phi(p) \neq \phi(q)$.*

(iii) *Δ is* **unital** *if, for $p \in L$ with $p \neq 0$, there is a $\phi \in \Delta$, $\phi(p) = 1$.*

(iv) *Δ is* **ordering** *if, for $p, q \in L$, the condition $\phi(p) \leq \phi(q)$ for all $\phi \in \Delta$ implies that $p \leq q$.*

(v) *Δ is* **rich** *if, for $p, q \in L$ with $p \not\leq q$, there is a $\phi \in \Delta$, $\phi(p) = 1$, $\phi(q) \neq 1$.*

Group-valued measures, which are somewhat analogous to morphisms, play an important role in the theory of effect algebras.

Definition 1.4.8 (i) *Let E be an effect algebra, and G an Abelian group. A mapping $\phi : E \to G$ such that, whenever $p, q \in E$ with $p \perp q$ then $\phi(p \oplus q) = \phi(p) + \phi(q)$, is called a* **$G$-valued measure** *(or a* **group-valued measure** *or a* **charge** *if G is not specified) on E.*

Such a measure ϕ is said to be **positive** *if $\phi(p) = 0$ implies $p = 0$ for all $p \in E$, and it is said to be* **nondegenerate** *if $\phi(1) \neq 0$.*

(ii) *If $G = V$ is a vector space, over a field F, then a V-valued measure on E is called a* **vector-valued measure** *over F.*

Let Δ be a family of group-valued measures on E.

(i) *Δ is* **positive** *if, for $p \in L$ with $p \neq 0$, there exists a $\phi \in \Delta$, $\phi(p) \neq 0$.*

(ii) *Δ is* **separating** *if, for $p, q \in E$ with $p \neq q$, there exists a $\phi \in \Delta$, $\phi(p) \neq \phi(q)$.*

Definition 1.4.9 *Let G be an ordered Abelian group with the positive cone G^+.*

(i) *A G-valued measure $\phi : E \to G$ such that $\phi(E) \subseteq G^+$ is called a* **G-representation** *(or a* **group representation** *if G is not specified) of E. A representation $\phi : E \to G^+$ such that $\phi(p) \leq \phi(q) \Rightarrow p \leq q$, for all $p, q \in E$, is said to be* **faithful**.

(ii) *If V^+ is the positive cone in an ordered vector space V over $\mathbb{R}$, then a representation $\phi : E \to V^+$ is called a* **V-representation** *(or a* **vector representation** *if V is not specified) of E.*

A family Δ of representations of E is said to be **ordering** *if, for $p, q \in E$, the condition $\phi(p) \leq \phi(q)$ for all $\phi \in \Delta$ implies that $p \leq q$.*

Lemma 1.4.10 *A mapping $\phi : E \to G^+$ is a G-representation of E if and only if ϕ is a morphism of E into the effect algebra $G^+[0, \phi(1)]$. The representation is faithful if and only if ϕ is a monomorphism.*

Proof. Let $e \in G^+$ and let $\phi : E \to G^+[0,e]$ be a morphism. Then $\phi(1) = e$, and, for all $a \in E$, $\phi(a) \leq \phi(1)$ implies that $\phi(E) \leq G^+$. Hence ϕ is a G-representation of E. Conversely, let $\phi : E \to G^+$ be a G-representation of E. Let $\phi(1) = e$. Clearly, $e \in G^+$. For any $a \in E$, $a \oplus a' = 1$ implies $\phi(a) + \phi(a') = \phi(1) = e$, hence $\phi(E) \subseteq G^+[0,e]$, so ϕ is a morphism from E to the interval effect algebra $G^+[0,e]$. □

Definition 1.4.11 *Let E be an effect algebra. By a* **universal group** *(***unigroup***, for short) for E, we mean a pair $(\mathcal{G}, \gamma)$ consisting of an additive Abelian group $\mathcal{G}$ and a $\mathcal{G}$-valued measure $\gamma : E \to \mathcal{G}$ such that the following conditions hold:* (i) *$\gamma(E)$ generates $\mathcal{G}$.* (ii) *If G is an additive Abelian group and $\phi : E \to G$ is a G-valued measure, then there is a group homomorphism $\phi^* : \mathcal{G} \to G$ such that $\phi = \phi^* \circ \gamma$.*

Suppose X is a set, G is an additive Abelian group, and $\psi : X \to G$. We define the **support** of ψ by $supp(\psi) := \{x \in X : \psi(x) \neq 0\}$ and we define $G^{[X]}$ to be the set of all functions $\psi : X \to G$ such that $supp(\psi)$ is a finite subset of X. Under pointwise operations $G^{[X]}$ forms an additive Abelian group. The next theorem provides an intimate connection between effect algebras and Abelian groups.

Theorem 1.4.12 *If E is an effect algebra, then there is a universal group $(\mathcal{G}, \gamma)$ for E. The universal group is unique up to isomorphism.*

Proof. For any $p \in E$, define $\delta_p(q) \in \mathbb{Z}^{[E]}$ by $\delta_p(q) = 1$ if $q = p$ and $\delta_p(q) = 0$ if $q \neq p$, $q \in E$. Let H be the subgroup of $\mathbb{Z}^{[E]}$ generated by all elements of the form $\delta_p + \delta_q - \delta_r$ such that $p, q, r \in E$, $p \perp q$, and $p \oplus q = r$. Let $\mathcal{G} = \mathbb{Z}^{[E]}/H$, and let $\eta : \mathbb{Z}^{[E]} \to \mathcal{G}$ be the canonical group epimorphism with kernel H. Define $\gamma : E \to \mathcal{G}$ by $\gamma(p) := \eta(\delta_p)$ for all $p \in E$. Then, for $p, q, r \in E$ with $p \perp q$ and $p \oplus q = r$, we have $\eta(\delta_p + \delta_q - \delta_r) = 0$, so that

$$\gamma(p) + \gamma(q) - \gamma(r) = \eta(\delta_p) + \eta(\delta_q) - \eta(\delta_r) = 0,$$

and it follows that $\gamma : E \to \mathcal{G}$ is a $\mathcal{G}$-valued measure on E.

Now let G be an additive Abelian group and let $\phi : E \to G$ be a G-valued measure. Define the group homomorphism $\phi' : \mathbb{Z}^{[E]} \to G$ by

$$\phi'(\psi) := \sum_{p \in supp(\psi)} \psi(p)\phi(p), \ \psi \in \mathbb{Z}^{[E]}.$$

Notice that $\phi'(\delta_p) = \phi(p)$ for all $p \in E$. If $p, q, r \in E$ with $p \perp q$ and $p \oplus q = r$, then

$$\phi'(\delta_p + \delta_q - \delta_r) = \phi(p) + \phi(q) - \phi(r) = 0,$$

so $H = ker(\eta) \subseteq ker(\phi')$. Hence there is a group homomorphism $\phi^* : \mathcal{G} \to G$ such that $\phi' = \phi^* \circ \eta$. Thus, for $p \in E$,

$$\phi^*(\gamma(p)) = \phi^*(\eta(\delta_p)) = \phi'(\delta_p) = \phi(p),$$

so $\phi^* \circ \gamma = \phi$.

If both $(\mathcal{G}, \gamma)$ and $(\mathcal{F}, \chi)$ are universal groups for E, it is clear that there exists a group isomorphism $\chi^* : \mathcal{G} \to \mathcal{F}$ such that $\chi = \chi^* \circ \gamma$ and so the universal group for E is unique up to isomorphism. □

The next theorem shows that every morphism of effect algebras can be uniquely extended to a homomorphism of the corresponding universal groups.

Theorem 1.4.13 *If P and Q are effect algebras with universal groups $(\mathcal{G}, \alpha)$ and $(\mathcal{H}, \beta)$, respectively, and if $\phi : P \to Q$ is a morphism, then there is a unique group homomorphism $\phi' : \mathcal{G} \to \mathcal{H}$ such that $\phi' \circ \alpha = \beta \circ \phi$.*

Proof. Since $\beta \circ \phi : P \to \mathcal{H}$ is an $\mathcal{H}$-valued measure, there is a group homomorphism $\phi' : \mathcal{G} \to \mathcal{H}$ such that $\phi' \circ \alpha = \beta \circ \phi$. The uniqueness of ϕ' is a consequence of the fact that $\alpha(P)$ generates $\mathcal{G}$. □

It may happen that the unigroup-valued measure γ for an effect algebra E is trivial, i.e., $ker(\gamma) = E$, [Nav 5]. On the other hand, in the case when E is an interval effect algebra, γ is a faithful representation. In the next theorem, several possibilities are treated.

Theorem 1.4.14 *Let $(\mathcal{G}, \gamma)$ be the universal group for the effect algebra $(E; \oplus, 0, 1)$.*

(i) *γ is nondegenerate if and only if E admits a nondegenerate group-valued measure.*

(ii) *γ is positive if and only if E admits a positive family of group-valued measures.*

(iii) *γ is injective if and only if E admits a separating family of group-valued measures.*

(iv) *$\mathcal{G}$ can be organized into an ordered group in such a way that γ is a positive $\mathcal{G}$-representation of E if and only if E admits a positive family of representations.*

(v) *$\mathcal{G}$ can be organized into an ordered group in such a way that γ is a faithful $\mathcal{G}$-representation of E if and only if E admits an ordering family of representations.*

Proof. (i), (ii) and (iii) are obvious.

(iv) If $\mathcal{G}$ can be organized in such a way that $\gamma(E) \subseteq \mathcal{G}^+$ and γ is positive, then $\Delta := \{\gamma\}$ is a positive family of representations of E.

Conversely, suppose that Δ is a positive family of representations of E and define $P := ssg(\gamma(E))$. Evidently, $P + P \subseteq P$. Let $g \in P \cap -P$. Then there exists a finite nonempty subset A of $E \setminus \{0\}$ and there exist functions $\rho, \sigma : A \to \mathbb{Z}^+$ such that $g = \sum_{a \in A} \rho(a)\gamma(a)$ and $-g = \sum_{a \in A} \sigma(a)\gamma(a)$; hence, $0 = \sum_{a \in A}(\rho(a) + \sigma(a))\gamma(a)$.

If $\phi \in \Delta$, then $\phi : E \to G^+$ for an ordered group G and there exists a group homomorphism $\phi^* : \mathcal{G} \to G$ such that $\phi = \phi^* \circ \gamma$. We obtain

$$0 = \sum_{a \in A}(\rho(a) + \sigma(a))\phi^*(\gamma(a)) = \sum_{a \in A}(\rho(a) + \sigma(a))\phi(a) \in G^+,$$

and, since each $\phi(a) \in G^+$ and $\rho(a) + \sigma(a)$ is a nonnegative integer, it follows that $0 = (\rho(a) + \sigma(a))\phi(a) \in G^+$ for all $a \in A$. For each $a \in A \subseteq E \setminus \{0\}$, we can chose $\phi \in \Delta$ with $\phi(a) \neq 0$, and it follows that $\rho(a) + \sigma(a) = 0$ in $\mathbb{Z}^+$; hence, that $\rho = 0$. Consequently, $g = 0$, and so $P \cap -P = \{0\}$. Thus $\mathcal{G}$ can be organized into an ordered group with positive cone $\mathcal{G}^+ := P$. Evidently, $\gamma(E) \subseteq \mathcal{G}^+$ so γ is a representation of E. That γ is positive follows from part (ii).

(v) This can be proved in the same way as (iv). □

In the following theorems, we show the existence of universal vector spaces for effect algebras.

Definition 1.4.15 *Let F be a field. By a* **universal vector space over** *F for an effect algebra E, we mean a pair $(\mathcal{U}, \zeta)$ consisting of a vector space $\mathcal{U}$ over F and a vector measure $\zeta : E \to \mathcal{U}$ such that the following conditions hold:* (i) *$\zeta(E)$ spans $\mathcal{U}$.* (ii) *If U is a vector space over F and $\phi : E \to U$ is a vector measure, there is a linear transformation $\phi^\sharp : \mathcal{U} \to U$ such that $\phi = \phi^\sharp \circ \zeta$.*

Theorem 1.4.16 *If E is an effect algebra and F is a field, there exists a universal vector space over F for the effect algebra E.*

Proof. Let $(\mathcal{G}, \gamma)$ be the universal group for E and, regarding F as an Abelian group, let $\mathcal{U} := F \otimes \mathcal{G}$. Define the $\mathcal{U}$-valued measure $\zeta : E \to \mathcal{U}$ by $\zeta(p) = 1 \otimes \gamma(p)$ for all $p \in E$. The Abelian group $\mathcal{U}$ can be organized into a vector space over F in such a way that, for $\alpha, \beta \in F$ and $g \in \mathcal{G}$, $\alpha(\beta \otimes g) = (\alpha\beta) \otimes g$. Since $\gamma(E)$ generates $\mathcal{G}$, it is clear that $\zeta(E)$ spans the vector space $\mathcal{U}$ over F. Let U be a vector space over F, let $\phi : E \to U$ be a measure, and let $\phi^* : \mathcal{G} \to U$ be the corresponding homomorphism into the Abelian group U such that $\phi = \phi^* \circ \gamma$. Define the group bihomomorphism $\phi' : F \times \mathcal{G} \to U$ by $\phi'(\beta, g) := \beta\phi^*(g)$ for all $\beta \in F$, $g \in \mathcal{G}$ and let $\phi^\sharp : \mathcal{U} \to U$ be the corresponding group homomorphism such that $\phi'(\beta, g) = \phi^\sharp(\beta \otimes g) = \beta\phi^*(g)$. Evidently, $\phi^\sharp$ is a linear transformation and, for $p \in E$, we have $\phi^\sharp(\zeta(p)) = \phi^\sharp(1 \otimes \gamma(p)) = \phi^*(\gamma(p)) = \phi(p)$. □

For a field F, we denote by $V(E, F)$ the set of all F-valued measures $\mu : E \to F$. Under pointwise operations, it is clear that $V(E, F)$ is a vector space over F. The proof of the following theorem is straightforward.

Theorem 1.4.17 *If F is a field, $(\mathcal{U},\zeta)$ is the universal vector space over F for the effect algebra E, and $\mathcal{U}'$ is the algebraic dual space of $\mathcal{U}$ over F, then the mapping $\natural : V(E,F) \to \mathcal{U}'$ given by $\mu \mapsto \mu^\natural$ is a vector-space isomorphism of $V(E,F)$ onto $\mathcal{U}'$.*

If $V(E,F)$ is finite dimensional (which happens, for example, when E is finite), then Theorem 1.4.17 implies that $\mathcal{U}$ is finite dimensional and that $\mathcal{U}$ can be identified with the dual space $V(E,F)'$. Under this identification, the vector measure ζ is identified with the mapping $p \mapsto f_p$, where f_p is the linear functional on $V(E,F)$ obtained by evaluation $f_p(\mu) := \mu(p)$.

Definition 1.4.18 *If G is a partially ordered Abelian group, then an element $u \in G^+$ such that $G^+ = ssg(G^+[0,u])$ and $G = \langle G^+\rangle$ is called a* **generative order unit for** *G. If u is a generative order unit for G and an effect algebra E is isomorphic to $G^+[0,u]$, we say that G is an* **ambient group with order unit** *u for E.*

As a direct consequence of Theorem 1.4.4 we obtain the following result.

Theorem 1.4.19 *Every interval algebra has an ambient group G.*

An ambient group for an interval algebra E need not necessarily be uniquely determined by E. For example, Let $G := \mathbb{Z}$ as an additive group with the non-standard positive cone $G^+ := \{3n + 4m : n,m \in \mathbb{Z}^+\} = \{0,3,4\} \cup (6+\mathbb{Z}^+)$. Then G is an ambient group with order unit 7 for $E := G^+[0,7] = \{0,3,4,7\}$. The interval algebra E, which is isomorphic to the Boolean algebra 2^2 also has $\mathbb{Z}\times\mathbb{Z}$ with positive cone $\mathbb{Z}^+ \times \mathbb{Z}^+$ and order unit $(1,1)$ as an ambient group.

The next theorem tells us that every interval algebra has an ambient unigroup.

Theorem 1.4.20 *Let $(\mathcal{G},\gamma)$ be the universal group for the effect algebra E and let $\mathcal{G}^+ := ssg(\gamma(E))$. Then a necessary and sufficient condition that E is an interval effect algebra is that $\mathcal{G}^+ \cap -\mathcal{G}^+ \subseteq \{0\}$, so that $\mathcal{G}$ is partially ordered with $\mathcal{G}^+$ as its positive cone, and $\gamma : E \to \mathcal{G}^+[0,\gamma(u)]$ is a surjective faithful representation.*

Proof. The sufficiency of the condition is evident. Conversely, suppose that E is an interval effect algebra. By Theorem 1.4.19, there is an ambient group for E and we can identify E with $G^+[0,u]$. The injection mapping $\iota : E \to G$ is a G-valued measure on E, so there is a group homomorphism $\iota^* : \mathcal{G} \to G$ such that $\iota^*(\gamma(p)) = p$ for all $p \in E$.

Let $x \in \mathcal{G}^+$. Then there exist elements $p_1, p_2, \ldots, p_n \in E$ with $x = \sum_i \gamma(p_i)$ and it follows that $\iota^*(x) = \sum_i p_i \in G^+$. Thus, $\iota^*(\mathcal{G}^+) \subseteq G^+$. If $x, -x \in \mathcal{G}^+$, then $\sum_i p_i, -\sum_i p_i \in G^+$, so $\sum_i p_i = 0$; hence, since $p_i \in G^+, i = 1,2,\ldots,n$, $p_1 = \cdots = p_n = 0$, and so $x = 0$. Thus, $\mathcal{G}$ is partially ordered with $\mathcal{G}^+$ as its positive cone.

Let $p \in E$. Then $p' = u - p \in E$, $\gamma(u) = \gamma(p \oplus p') = \gamma(p) + \gamma(p')$, and it follows that $\gamma(p), \gamma(u) - \gamma(p) = \gamma(p') \in \gamma(E) \subseteq \mathcal{G}^+$; hence, $\gamma(p) \in \mathcal{G}^+[0, \gamma(u)]$. Consequently, $\gamma(E) \subseteq \mathcal{G}^+[0, \gamma(u)]$. Conversely, suppose that $x \in \mathcal{G}^+[0, \gamma(u)]$, so that $x, \gamma(u) - x \in \mathcal{G}^+$. Thus, $\iota^*(x), \iota^*(\gamma(u) - x) = u - \iota^*(x) \in \iota^*(\mathcal{G}^+) \subseteq G^+$, and it follows that $\iota^*(x) \in G^+[0, u] = E$. Since $x \in \mathcal{G}^+$, there exist $p_1, \ldots, p_n \in E$ with $x = \sum_i \gamma(p_i)$, so $\sum_i p_i = \iota^*(x) \in E$, hence, $\iota^*(x) = \bigoplus_i p_i \in E$, and it follows that $x = \sum_i \gamma(p_i) = \gamma(\iota^*(x)) \in \gamma(E)$. Therefore, $\mathcal{G}^+[0, \gamma(u)] \in \gamma(E)$, and we have $\gamma(E) = \mathcal{G}^+[0, \gamma(u)]$.

Suppose that $p, q \in E$ with $\gamma(p) \leq \gamma(q)$ in $\mathcal{G}^+[0, \gamma(u)]$. Then $\gamma(q) - \gamma(p) \in \mathcal{G}^+$, so $q - p = \iota^*(\gamma(q) - \gamma(p)) \in \iota^*(\mathcal{G}^+) = G^+$, and it follows that $p \leq q$ in E. □

The following Corollary is obvious.

Corollary 1.4.21 *If E is an interval algebra with unit u, there exists an ambient group G with order unit u for E such that $E = G^+[0, u]$ and every group-valued measure $\phi : E \to H$ can be extended to a group homomorphism $\phi^* : G \to H$.*

We refer to the group G in Corollary 1.4.21 as the **universal ambient group** for the interval algebra E and the order unit u in G is called an **universal** order unit for E.

From the point of view of the category theory, the last two theorems yield the following statement.

Corollary 1.4.22 *There is a categorical equivalence between the category of interval effect algebras with their morphisms and partially ordered Abelian groups with group homomorphisms.*

Lemma 1.4.23 *Let G be a partially ordered Abelian group. The following two conditions are equivalent:*

(i) *Every generative order unit in G is universal.*

(ii) *If u is a generative order unit for G and $\phi : G^+[0, u] \to H$ is a group-valued measure, then ϕ can be extended to a group-valued measure $\phi^\sharp : G^+[0, 2u] \to H$.*

Proof. (i) ⇒ (ii). By (i), $\phi : G^+[0, u] \to H$ can be extended to a group homomorphism $\phi^* : G \to H$, and the restriction $\phi^\sharp$ of ϕ^* to $G^+[0, 2u]$ is an H-valued measure.

(ii) ⇒ (i). Let u be a generative order unit for G, and let $\phi : G^+[0, u] \to H$ be a group-valued measure. Let $1 \leq n \in \mathbb{Z}$. By mathematical induction and the fact that $ssg(G^+[0, u]) = G^+$, we see that there exists a unique H-valued measure $\phi_n : G^+[0, 2^n u] \to H$ that extends ϕ. Since u is an order unit, G^+ is the set-theoretical union of the intervals $G^+[0, 2^n u]$, so there is a uniquely determined mapping $\phi^+ :$

$G^+ \to H$ that extends each of the H-valued measures ϕ_k. Evidently, $\phi^+(x+y) = \phi^+(x) + \phi^+(y)$ for all $x, y \in G^+$. Since $G = G^+ - G^+$, ϕ^+ can be extended uniquely to a group homomorphism $\phi^* : G \to H$. □

Theorem 1.4.24 *Let G be a partially ordered Abelian group with no nonzero element of order two such that, for every $x \in G^+$, there exists an element $y \in G^+$ with $2y = x$. Then every order unit in G is universal.*

Proof. Let u be an order unit in G. Then G^+ is a generating cone. If $g \in G$, there are elements $a, b \in G^+$ with $g = a - b$ and there are elements $c, d \in G^+$ with $a = 2c$, $b = 2d$, hence there is an element $h := c - d \in G$ with $g = 2h$. Since G contains no nonzero elements of order two, h is uniquely determined by g, and we define $\frac{1}{2}g := h$. Evidently, $x \mapsto \frac{1}{2}x$ is an order preserving automorphism of G. If $a \in G^+$, there is a positive integer n with $a \le 2^n u$ and, by 2^n applications of the latter automorphism, we can find $b \in G^+$ with $b \le u$ such that $2^n b = a$. Consequently, $G^+ = ssg(G^+[0,u])$, so u is generative.

Let $\phi : G^+[0,u] \to H$ be an H-valued measure. Then $\phi^\sharp : G^+[0,2u] \to H$ defined by $\phi^\sharp(x) = 2\phi(\frac{1}{2}x)$, for all $x \in G^+[0,2u]$, is an H-valued measure that extends ϕ, so u is universal by Lemma 1.4.23. □

Corollary 1.4.25 *Let F be a subfield of $\mathbb{R}$, let V be a vector space over F ordered by the positive cone V^+ in such a way that $F^+V^+ \subseteq V^+$ (i.e., V^+ partially orders V as a vector space). Then every order unit in the partially ordered Abelian group V is universal.*

In particular, if $\mathcal{B}_{sa}(H)$ is the Abelian group of self-adjoint operators on a Hilbert space H, ordered in the usual way, then $(\mathcal{B}_{sa}(H), I)$ is the universal group for the standard effect algebra $\mathcal{E}(H) := \mathcal{B}^+_{sa}[O, I]$.

Recall that a **probability measure** (or a **state**) on the effect algebra E is an $\mathbb{R}$-valued measure $\omega : E \to \mathbb{R}$ such that $\omega(E) \subseteq \mathbb{R}^+$ and $\omega(u) = 1$. In other words, a state is a morphism from E to the interval algebra $[0,1]$. We denote by $\Omega(E)$ the set of all probability measures on E. In agreement with Definition 1.4.7, we say that a subset $\Delta \subseteq \Omega(E)$ is **positive** if, for every $p \in E$, $p \ne 0$, there is a state $\omega \in \Delta$ such that $\omega(p) \ne 0$, and Δ is **ordering** if, for all $p, q \in E$, $\omega(p) \le \omega(q)$, for all $\omega \in \Delta$ we have $p \le q$.

Let $\mathcal{G}$ be the universal group for E, and $\gamma : E \to \mathcal{G}$ be the corresponding $\mathcal{G}$-valued measure. Then there is a one-to-one correspondence $\omega :\to \omega^*$, determined by $\omega = \omega^* \circ \gamma$, between probability measures $\omega \in \Omega(E)$ and group homomorphisms $\omega^* : \mathcal{G} \to \mathbb{R}$ such that $\omega^*(\gamma(E)) \subseteq \mathbb{R}^+$ and $\omega^*(\gamma(u)) = 1$.

If G is the universal group for an interval effect algebra $E = G^+[0,u]$, then the probability measures on E are precisely the restrictions to E of order-preserving group homomorphisms $\omega^* : G \to \mathbb{R}$ that are normalized in the sense that $\omega^*(u) = 1$ (note that the latter group homomorphisms are called the **states** on the partially ordered Abelian group G with order unit). It is also clear that every state ω on E is a $\mathbb{R}$-representation of E.

Proposition 1.4.26 *Let E be an effect algebra admitting a positive set $\Omega(E)$ of states. Let $(\mathcal{G},\gamma)$ be the universal group for E and the corresponding $\mathcal{G}$-valued measure. Then $\mathcal{G}^+ := ssg(\gamma(E))$ is a positive cone for $\mathcal{G}$, and $\Omega(E)$ is the set of mappings of the form $\omega^* \circ \gamma$, where $\omega^* : \mathcal{G} \to \mathbb{R}$ is an order-preserving group homomorphism such that $\omega^*(\gamma(u)) = 1$. Futhermore, $\gamma(E) \subseteq \mathcal{G}^+[0,\gamma(u)]$.*

Proof. By Theorem 1.4.14, $\mathcal{G}$ is partially ordered by the positive cone $\mathcal{G}^+$. For every $\omega \in \Omega(E)$, the group homomorphism $\omega^* : \mathcal{G}^+ \to \mathbb{R}$ satisfies

$$\omega^*(\mathcal{G}^+) = \omega^*(ssg(\gamma(E))) \subseteq ssg(\omega^*(\gamma(E))) \subseteq ssg(\mathbb{R}^+) \subseteq \mathbb{R}^+,$$

so ω^* is order preserving and $\omega^*(\gamma(u)) = \omega(u) = 1$. Conversely, if $\omega^* : \mathcal{G} \to \mathbb{R}$ is an order-preserving group homomorphism such that $\omega^*(\gamma(u)) = 1$, it is clear that $\omega^\sharp := \omega^* \circ \gamma \in \Omega(E)$.

If $p \in E$ and p' is its orthosupplement in E, then $\gamma(u) = \gamma(p \oplus p') = \gamma(p) + \gamma(p')$ and, since $\gamma(p), \gamma(p') \in \mathcal{G}^+$, we have $0 \leq \gamma(p) = \gamma(u) - \gamma(p') \leq \gamma(u)$. Therefore, $\gamma(E) \subseteq \mathcal{G}^+[0,\gamma(u)]$. □

Theorem 1.4.27 *If an effect algebra E admits an ordering set of states, then it is an interval algebra.*

Proof. Since an ordering set of states is positive, by Proposition 1.4.26, we can organize the universal group $\mathcal{G}$ of E into a partially ordered Abelian group with positive cone $\mathcal{G}^+ = ssg(\gamma(E))$, where $\gamma : E \to \mathcal{G}$ is the canonical $\mathcal{G}$-valued measure. If $\omega \in \Omega(E)$, denote by $\omega^* : \mathcal{G} \to \mathbb{R}$ the corresponding order-preserving group homomorphism such that $\omega = \omega^* \circ \gamma$.

Suppose $p, q \in E$ with $\gamma(p) \leq \gamma(q)$. Then for every state ω on E we have $\omega^*(\gamma(p)) \leq \omega^*(\gamma(q))$, so $\omega(p) \leq \omega(q)$, and since $\Omega(E)$ is ordering, it follows $p \leq q$. So $\gamma : E \to \gamma(E)$ is an order isomorphism.

From $\gamma(p) + \gamma(p') = \gamma(u)$, we get that $\gamma(E)$ is closed under the orthosupplementation in $\mathcal{G}^+[0,\gamma(u)]$. Suppose $p, q \in E$, $\gamma(p) \perp \gamma(q)$ in $\mathcal{G}^+[0,u]$. Then $\gamma(p) \leq \gamma(q')$, so $p \leq q'$, i.e., $p \perp q$ in E, and $\gamma(p \oplus q) = \gamma(p) + \gamma(q) = \gamma(p) \oplus \gamma(q)$ in $\mathcal{G}^+[0,u]$. This shows that $\gamma(E)$ is a sub-effect algebra of $\mathcal{G}^+[0,u]$, hence by Corollary 1.4.5, $\gamma(E)$ is an interval algebra. Since $\gamma : E \to \gamma(E)$ is an effect-algebra isomorphism, it follows that E is an interval algebra. □

Note that the converse need not hold, for example, there are linearly ordered interval effect algebras which do not have an ordering set of states. But according the following result, every interval effect algebra possesses at least one state.

Theorem 1.4.28 *Every interval algebra* $(E; \oplus, 0, 1)$ *admits at least one state.*

Proof. We may assume that $E = \mathcal{G}^+[0, u]$ for the universal group $\mathcal{G}$ with generative order unit u. By [GoHa, Corollary 3.3], there is a group homomorphism $\phi : \mathcal{G} \to \mathbb{R}$ such that $\phi(\mathcal{G}^+) \subseteq \mathbb{R}^+$ and $\phi(u) = 1$. The restriction ω of ϕ to E is a state on E. □

1.4.1 Examples of Interval Algebras

In this subsection, we exhibit some examples of interval algebras [FoBe]. We recall that an interval algebra is a **scale algebra** if it is linearly ordered.

Example 1.4.29 The simplest possible interval algebra is $\mathbf{2} := \mathbb{Z}^+[0, 1] = \{0, 1\}$. Note that $\mathbf{2}$ is the only orthoalgebra that is also a scale algebra and, as a poset, $\mathbf{2}$ is the two-element Boolean algebra.

Example 1.4.30 The **standard scale algebra** $\mathbb{R}^+[0, 1]$ has $\mathbb{R}$, ordered in the usual way, as its universal group. A scale algebra is isomorphic to a sub-effect algebra of $\mathbb{R}^+[0, 1]$ if and only if it has no nonzero elements of infinite isotropic index (see Proposition 2.2.33).

Example 1.4.31 If n is a positive integer, we define the n-**chain** $C_n := n\mathbf{2} = \mathbb{Z}^+[0, n]$. Evidently, C_n is a scale algebra and $\mathbb{Z}$ with its usual order is its universal group.

If r is a positive integer, we define $\mathbb{Z}^r$ to be the r-fold direct product $\mathbb{Z} \times \mathbb{Z} \times \cdots \times \mathbb{Z}$ of the additive Abelian group $\mathbb{Z}$ with itself. The standard positive cone $(\mathbb{Z}^+)^r$ in $\mathbb{Z}^r$ is understood to be the r-fold cartesian product $\mathbb{Z}^+ \times \mathbb{Z}^+ \times \cdots \times \mathbb{Z}^+$.

Example 1.4.32 If $n_1, n_2, \ldots, n_r$ is a finite sequence of positive integers, we define the **rectangular treillis**

$$RT(n_1, n_2, \ldots, n_r) := (\mathbb{Z}^+)^r[(0, 0, \ldots, 0), (n_1, n_2, \ldots, n_r)]$$

As a poset, $RT(n_1, n_2, \ldots, n_r)$ forms a finite distributive lattice. We define the interval algebra $\mathbf{2}^r := RT(n_1, n_2, \ldots, n_r)$ for $n_1 = n_2 = \cdots = n_r = 1$. As a poset, 2^r is isomorphic to the Boolean algebra with 2^r elements.

If X is a set, we denote by $\mathbb{Z}^X$ the set of all functions $f : X \to \mathbb{Z}$ organized into an additive Abelian group under pointwise operations. The *standard positive cone* in $\mathbb{Z}^X$ is understood to be the subset $(\mathbb{Z}^+)^X$ consisting of all functions $f \in \mathbb{Z}^X$ such that $f(x) \in \mathbb{Z}^+$ for all $x \in X$.

Example 1.4.33 Let X be a Stone space, i.e., a totally disconnected, compact, Hausdorff topological space. Let G be the subgroup of $\mathbb{Z}^X$ consisting of all functions $f : X \to \mathbb{Z}$ that are continuous when $\mathbb{Z}$ is given the discrete topology. Partially order G by the positive cone $G^+ := G \cap (\mathbb{Z}^+)^X$ and let u be the constant function $u(x) = 1$ for all $x \in X$. Then G is the universal group for the interval algebra $G^+[0, u]$, and the Boolean algebra of all compact open subsets M of X is isomorphic as a sublattice to $G^+[0, u]$ under the mapping $M \mapsto \chi_M$ that carries M into the characteristic set function χ_M of M. Thus, by Stone's theorem, every Boolean algebra can be organized into an interval algebra. (Regarding a Boolean algebra as an MV-algebra, or even as an effect algebra with the Riesz decomposition property, provides another approach).

If G and H are partially ordered Abelian groups, the group $P := G \times H$, partially ordered by the positive cone

$$P^+ := \{(g, h) \in G \times H : 0 \neq g \in G^+ \text{ or } (g = 0 \text{ and } h \in H^+)\}$$

is called the **lexicographic product** of G and H, and we will denote it by $G \times_{lex} H$. If G and H are totally ordered, so is their lexicographic product. For examples using the lexicographic product, see Exercises 2 and 3, Section 1.12.

Example 1.4.34 An effect algebra P is **convex** if for every $a \in P$ and $\lambda \in [0, 1] \subseteq \mathbb{R}$ there exists an element $\lambda a \in P$ such that the following conditions hold.

(C1) If $\alpha, \beta \in [0, 1]$ and $a \in P$, then $\alpha(\beta)a = (\alpha\beta)a$.

(C2) If $\alpha, \beta \in [0, 1]$ with $\alpha + \beta \leq 1$ and $a \in P$, then $\alpha a \perp \beta a$ and $(\alpha + \beta)a = \alpha a \oplus \beta a$.

(C3) If $a, b \in P$ with $a \perp b$ and $\lambda \in [0, 1]$, then $\lambda a \perp \lambda b$ and $\lambda(a \oplus b) = \lambda a \oplus \lambda b$.

(C4) If $a \in P$, then $1a = a$.

If P and Q are convex effect algebras, we say that they are **affinely isomorphic** if there is an effect algebra isomorphism $\phi : P \to Q$ which is an **affine morphism**, i.e., $\phi(\lambda a) = \lambda\phi(a)$ for every $\lambda \in [0, 1]$, $a \in P$.

A map $(\lambda, a) \mapsto \lambda a$ that satisfies (C1)–(C4) is an example of a bimorphism from $[0, 1] \times P$ into P (see Chapter 4, p. 236) and we call this map a **convex structure** on P. The effect algebras that arise in practice are usually convex. For example, the unit interval $[0, 1]$ and the Hilbert space effect algebra $\mathcal{E}(H)$ are examples of convex effect algebras.

More generally, let V be a real linear space with zero θ. A subset K of V is a positive cone if $\mathbb{R}^+K \subseteq K$, $K + K \subseteq K$ and $K \cap (-K) = \{\theta\}$. Then K defines a partial order $\leq_K$ on V in the usual way. Two ordered linear spaces (V_1, K_1) and (V_2, K_2) are **order isomorphic** if there exists a linear bijection $T : V_1 \to V_2$ such that

$T(K_1) = K_2$. The interval effect algebra $K[\theta, u]$, $u \in K$, is a convex effect algebra (with the convex structure $(\lambda, x) \mapsto \lambda x$), which we call a **linear effect algebra**. In [GuPu 1], the following representation theorem has been proved, which shows that every convex effect algebra is a linear algebra.

Theorem 1.4.35 *If $(P; \oplus, 0, 1)$ is a convex effect algebra, then P is affinely isomorphic to a linear effect algebra $K[\theta, u]$ that generates an ordered linear space (V, K) and the effect algebra order $\leq$ on $K[\theta, u]$ coincides with the linear space order $\leq_K$ restricted to $K[\theta, u]$. Moreover, (V, K) is unique in the sense that if P is affinely isomorphic to a linear effect algebra $K_1[\theta_1, u_1]$ that generates (V_1, K_1), then (V_1, K_1) is order isomorphic with (V, K).*

Proof. We give a sketch of the proof here, the details are left to the reader (see [GuPu 1]). We need the following steps.

(1) Define the set $\hat{K} \subseteq \mathbb{R} \times P$ by

$$\hat{K} = \{(\alpha, a) : \alpha \geq 1, a \in P\}.$$

For $(\alpha, a), (\beta, b) \in \hat{K}$, define the relation $\sim$ on $\hat{K}$ by $(\alpha, a) \sim (\beta, b)$ if $\beta^{-1}a = \alpha^{-1}b$. Prove that $\sim$ is an equivalence.

Denote the equivalence class containing (α, a) by $[(\alpha, a)]$ and let $\tilde{K} = \{[(\alpha, a)] : (\alpha, a) \in \hat{K}\}$. For $\beta \geq 0$, define

$$\beta[(\alpha, a)] = \begin{cases} [(\beta\alpha, a)] & \text{if } \beta \geq 1, \\ [(\alpha, \beta a)] & \text{if } \beta < 1. \end{cases}$$

Show that this operation is well defined.

Define also an operation $+$ on $\tilde{K}$ by

$$[(\alpha, a)] + [(\beta, b)] = \left[\left(\alpha + \beta, \frac{\alpha}{\alpha + \beta} a \oplus \frac{\beta}{\alpha + \beta} b\right)\right].$$

Show that the operation $+$ is well defined.

(2) Show that $\tilde{K}$ forms an abstract cone with a zero $(= \tilde{\theta} = [(1, 0)])$. That is, we must verify that $\tilde{K}$ with the operations $+$ defined above is commutative, associative, and cancellative (i.e., $X + Y = X + Z$ implies $Y = Z$), and $X + \tilde{\theta} = X$, for all $X, Y, Z \in \tilde{K}$. Moreover, the usual distributivity conditions for $+$ and multiplication by constants holds, and $\alpha(\beta X) = (\alpha\beta)X$, $1X = X$ for all $\alpha, \beta \in [0, 1]$ and $X \in \tilde{K}$.

Show that $\tilde{K}$ can be extended to a real linear space. To do this, put

$$V_0 = \left\{(X, Y) : X, Y \in \tilde{K}\right\}.$$

Define the relation $\approx$ on V_0 by $(X_1, Y_1) \approx (X, Y)$ if $X_1 + Y = X + Y_1$, and show that $\approx$ is an equivalence. Put

$$V = \{[(X, Y)] : (X, Y) \in V_0\}.$$

Define the addition $+$ on V componentwise, and show that it is well defined. Then verify that $(V;+)$ is an Abelian group with zero $\theta = [(\tilde{\theta}, \tilde{\theta})]$. Define a scalar multiplication by real numbers as follows. If $\lambda \geq 0$, then $\lambda[(X,Y)] = [(\lambda X, \lambda Y)]$ and if $\lambda < 0$, then $\lambda[(X,Y)] = [((-\lambda)Y, (-\lambda)X)]$. It is easy to show that V is a real linear space.

(3) Define $K \subseteq V$ by

$$K = \left\{ \left[(X, \tilde{\theta}) \right] : X \in \tilde{K} \right\}.$$

Show that K is a positive cone. Moreover, $V = K - K$.

(4) Define $u \in K$ by $u = \left[\left([(1,1)], \tilde{\theta} \right) \right]$ and form the interval $K[\theta, u]$. Show that $\mathbb{R}^+ K[\theta, u] = K$, so that $K[\theta, u]$ generates (V, K).

(5) Show that P is affinely isomorphic to $K[\theta, u]$ by defining $\phi : P \to K[\theta, u]$ by $\phi(a) = \left[\left([1, a)], \tilde{\theta} \right) \right]$ and showing that ϕ is an affine isomorphism.

(6) To prove uniqueness, suppose that $\phi_1 : P \to K_1[\theta_1, u_1]$ is an affine isomorphism, where $K_1[\theta_1, u_1]$ is a linear effect algebra that generates (V_1, K_1). It is easy to check that $\psi = \phi_1 \circ \phi^{-1}$ is an affine bijection from $K[\theta, u]$ onto $K_1[\theta_1, u_1]$. Define $T : V \to V_1$ as follows. If $x \in V$, then $x = \alpha y - \beta z$, where $\alpha, \beta \geq 0$ and $y, z \in K[\theta, u]$. We define $T(x)$ by

$$T(x) = \alpha\psi(y) - \beta\psi(z).$$

Show that T is a well defined order isomorphism. □

1.5 Generalized Orthoalgebras and Orthomodular Posets as Orthomodular Structures

In what follows, we will mean by an "orthomodular structure" a nonempty poset P with two partially defined binary operations $\oplus$ and $\ominus$, with a neutral element 0, both satisfying the corresponding properties ((E1), (E2), (E3'), (E4') and (D1), (D2), (D3), (C), respectively) and a generalized orthomodular law

$$a \leq b \;\Rightarrow\; b = a \oplus c, \tag{1.4}$$

where $c \in P$ is uniquely defined, namely $c = b \ominus a$. Clearly, D-posets, effect algebras, generalized D-posets and generalized effect algebras are orthomodular structures. In this Chapter, we will consider some more specific substructures of them.

Orthomodular posets (OMPs) and lattices (OMLs) play an important role in the mathematical foundations of quantum theories as so-called quantum logics. From the probabilistic point of view, a quantum logic is considered as a mathematical model of the set of random events of a quantum mechanical experiment. In comparison with the classical model, a Boolean algebra, the distributive law is replaced by the orthomodular law. Orthomodularity can be considered as the smallest weakening of distributivity which is satisfied by both the classical model - a Boolean algebra and the standard model of quantum mechanics - the projection lattice of a Hilbert space. In effect algebras, a generalized form of the orthomodular law is satisfied, namely $a \leq b$ implies there exists c such that $a \oplus c = b$. So effect algebras (or D-posets, equivalently) can be considered as a generalized form of quantum logics. Since the logical excluding middle law $a \vee a' = 1$ need not be satisfied, they are sometimes called "unsharp quantum logics". Orthoalgebras play an important role in the Foulis and Randall theory of test spaces (for test spaces see Chapter 4).

Other examples of orthomodular structure - MV-algebras, QMV-algebras, BCK-algebras - are treated in separate chapters.

The notion of a **generalized OML** (GOML) has been introduced by Janowitz [Jan], and it has been shown that every GOML can be embedded into an OML as an orthomodular ideal (see also [Ber], [Kal]). A generalization of the latter result has been proved by Mayet-Ippolito [MaIp] for a so-called weak generalized OMP (WGOMP).

Definition 1.5.1 *An* **orthoalgebra** *(OA) is a set A containing two special elements $0, 1$ and equipped with a partial binary operation $\oplus$ satisfying, for all $a, b, c \in A$, the following conditions:*

(OA1) *If $a \oplus b$ is defined, then $b \oplus a$ is defined and $a \oplus b = b \oplus a$ (commutativity).*

(OA2) *If $a \oplus b$ and $(a \oplus b) \oplus c$ are defined, then $b \oplus c$ and $a \oplus (b \oplus c)$ are defined, and $(a \oplus b) \oplus c = a \oplus (b \oplus c)$ (associativity).*

(OA3) *For every $a \in A$ there is a unique $b \in A$ such that $a \oplus b$ is defined and $a \oplus b = 1$ (orthocomplementation).*

(OA4) *If $a \oplus a$ is defined, then $a = 0$ (consistency).*

For an element $a \in A$, the unique element $b \in A$ satisfying condition (OA3) is denoted by a' and is called the **orthosupplement** of a.

Observe that axioms (OA1)–(OA3) coincide with axioms (E1)–(E3), respectively. Observe that axiom (E4) is weaker than (OA4). Indeed, if (OA4) holds, then $a \oplus 1$ is defined implies $a \oplus (a \oplus a')$ is defined, and by associativity, $a \oplus a$ is defined, whence $a = 0$. We obtain the following result:

Lemma 1.5.2 *An effect algebra (or a D-poset) is an orthoalgebra if and only if $a \perp a$ implies $a = 0$.*

Note that an orthoalgebra can be also characterized as an effect algebra in which all nonzero elements have isotropic index 1.

Proposition 1.5.3 *For an effect algebra E, the following conditions are mutually equivalent:*

(i) *E is an orthoalgebra.*

(ii) *For $p, q \in L$ with $p \perp q$ we have that $p \oplus q$ is a minimal upper bound for p and q.*

(iii) *For any $p \in L$, $p \wedge p' = 0$.*

(iv) *E is orthocomplemented by the mapping $p \mapsto p'$.*

Proof. (i) $\Rightarrow$ (ii). Assume (i). Clearly, $p, q \leq p \oplus q$. Suppose $r \in L$, $p, q \leq r$, $r \leq p \oplus q$. Then there are p_1, q_1, r_1 such that $r = p \oplus p_1 = q \oplus q_1$, $p \oplus q = r \oplus r_1$, so that $p \oplus q = r \oplus r_1 = p \oplus p_1 \oplus r_1$. By cancellativity, $q = p_1 \oplus r_1$. So $r_1 \leq q \leq r$, but also $r_1 \perp r$, which implies $r_1 \perp r_1$, and by (OA4), $r_1 = 0$.

(ii) $\Rightarrow$ (iii). Assume (ii) and suppose $r \in L$ with $r \leq p, p'$. Then $p, p' \leq r' \leq 1 = p \oplus p'$, so $r' = 1$, $r = 0$.

(iii) $\Rightarrow$ (iv). It can be easily checked that $p \mapsto p'$ is an order-inverting involution on the bounded poset L, so that the de Morgan law is effective. Therefore, since $p \wedge p' = 0$ for all $p \in L$, it follows that $p \vee p' = 0' = 1$ for all $p \in L$.

(iv) $\Rightarrow$ (i). Suppose L is orthocomplemented by $p \mapsto p'$ and let $q \in L$ with $q \perp q$. Then $q \leq q'$, so $0 = q \wedge q' = q$. □

Definition 1.5.4 *An* **orthomodular poset** *(OMP, in short) is a bounded poset $(L; \leq, ', 0, 1)$ with a unary operation $' : L \to L$ (an orthocomplementation) such that the following conditions are satisfied for all $a, b, c \in L$:*

(OMP1) $a \leq b \Rightarrow b' \leq a'$.

(OMP2) $(a')' = a$.

(OMP3) $a \vee a' = 1$.

(OMP4) *If $a \leq b'$ then $a \vee b$ exists in L.*

(OMP5) *(***Orthomodular law***) If $a \leq b$, then there is a $c \in L$ such that $c \leq a'$ and $a \vee c = b$.*

We will say that an effect algebra P satisfies the **coherence law** if

$$p, q, r \in P,\ p \perp q, q \perp r, r \perp p \ \Rightarrow\ \exists\, p \oplus q \oplus r.$$

Two elements a, b in an OMP L are **orthogonal** if $a \leq b'$. Two elements $a, b \in L$ are **compatible** if there are mutually orthogonal elements a_1, b_1, c such that $a = a_1 \vee c$, $b = b_1 \vee c$.

It is a well known fact that an OMP in which any two elements a, b are compatible is a Boolean algebra.

Theorem 1.5.5 *An effect algebra $(L; \oplus, 0, 1)$ can be organized into an OMP such that $a \oplus b = a \vee b$ provided $a \perp b$ if and only if it satisfies the coherence law.*

Proof. Let L be an OMP. Define a partial operation $\oplus$ on L by

$$a \oplus b \text{ is defined iff } a \perp b \text{ and then } a \oplus b = a \vee b.$$

We have to check that $(L; \oplus, 0, 1)$ is an effect algebra satisfying the coherence law. The proof of properties (E1)–(E4) is easy, we leave it to the readers. To prove the coherence law, assume that a, b, c are pairwise orthogonal. This yields that $a \leq c'$, $b \leq c'$ entail $a \vee b \leq c'$, so that $a \oplus b \oplus c = a \vee b \vee c$ exists.

Conversely, assume that L is an effect algebra satisfying the coherence law. We will prove that $a \oplus b = a \vee b$ whenever $a \perp b$. Clearly, $a, b \leq a \oplus b$. Suppose $a, b \leq r$. It means that $a, b \perp r'$, so that a, b, r' are pairwise orthogonal. By the coherence law, $a \oplus b \oplus r'$ exists, so that $a \oplus b \perp r'$. In other words, $a, b \leq r$ implies $a \oplus b \leq r$, hence $a \oplus b = a \vee b$. Then we have $a \vee a' = a \oplus a' = 1$ for all $a \in L$, and it follows that L is an OMP. □

The following proposition gives some characterizations of OMPs among orthoalgebras.

Proposition 1.5.6 [FGR] *The following conditions are equivalent for an orthoalgebra A:*

(i) *A is an OMP.*

(ii) *If $a \oplus b$ is defined, then $a \vee b$ exists.*

(iii) *If $a \oplus b$ is defined, then $a \vee b$ exists, and $a \oplus b = a \vee b$.*

(iv) *If $a \oplus b$, $a \oplus c$ and $b \oplus c$ exist, then $(a \oplus b) \oplus c$ exists.*

Proof. Left to the readers as an exercise. □

Any OMP may be regarded as a DP by defining $b \ominus a = b \wedge a'$ precisely when $a \leq b$. A relation between difference posets and orthomodular posets is as follows (cf. [NaPt]).

Proposition 1.5.7 *A difference poset P is an orthomodular poset if and only if the following two conditions are satisfied:* (i) *If $a \leq 1 \ominus a$, then $a = 0$.* (ii) *For all $a, b \in P$, $a \leq 1 \ominus b$ implies $a \vee b$ exists.*

Proof. Left to the readers as an exercise. □

An **orthomodular lattice** (OML) can be defined as a lattice ordered OMP. Next proposition shows that an OML can be defined as a lattice ordered orthoalgebra.

Proposition 1.5.8 *A lattice ordered orthoalgebra is an OML.*

Proof. Let $(L; \oplus, 0, 1)$ be a lattice ordered orthoalgebra. For any $p, q \in L$, $p \perp q$, we have, by $p \vee q \leq p \oplus q$ and Proposition 1.5.3 (ii), that $p \oplus q = p \vee q$. Let a, b, c be pairwise orthogonal, from $a, b \leq c'$ we have $a \vee b \leq c'$, so that the coherence law is satisfied. By Theorem 1.5.5 L is an OMP. □

So far we considered examples having the greatest element. Now we will consider orthomodular structures without it.

Definition 1.5.9 *A* **generalized orthoalgebra** *(generalized* OA, GOA*) is a set A containing a special element* 0 *and endowed with a partial binary operation $\oplus$ satisfying conditions* (OA1), (OA2), (OA4) *from Definition* 1.5.1 *and*

(GOA1) *If $a \oplus b$ and $a \oplus c$ are defined and $a \oplus b = a \oplus c$, then $b = c$ (cancellativity);*

(GOA2) $a \oplus 0 = a$.

Proposition 1.5.10 *A generalized orthoalgebra is a generalized effect algebra satisfying* (OA4).

Proof. We have only to prove that (OA4) in a GOA implies (E4'). Assume $a \oplus b = 0$. Then $(a \oplus b) \oplus (a \oplus b)$ is defined, hence by (E1) and (E2), $a \oplus a$ is defined, hence $a = 0$. □

Theorem 1.5.11 *Let $(P; \oplus, 0)$ be a generalized effect algebra and let $(\hat{P}; \oplus^*, 0, 0^\sharp)$ be the effect algebra constructed in Theorem* 1.2.6. *Then $\hat{P}$ is an orthoalgebra if and only if P is a generalized orthoalgebra.*

Proof. We have to prove that $\hat{P}$ satisfies (OA4) iff P satisfies (OA4). This is clear from the fact that $a^\sharp \oplus^* a^\sharp$ $(a \in P)$ is never defined. □

Definition 1.5.12 [MaIp] *Let $(P; \leq)$ be a poset with a smallest element* 0*, such that every interval $[0, a]$ of P is equipped with a unary operation $x \mapsto x^{\sharp_a}$. P is called a* **weak generalized orthomodular poset** *(WGOMP) if it satisfies the following conditions:*

(G1) *For any $a \in P$, $([0, a]; \leq, ^{\sharp_a}, 0, a)$ is an OMP.*

(G2) *If $a \leq b \leq c$, then $a^{\sharp_b} = b \wedge a^{\sharp_c}$.*

Elements $a, b \in P$ are said to be **orthogonal** *(in notation $a \perp b$) if $a, b \leq c$ and $a \leq b^{\sharp_c}$ for some $c \in P$.*

(G3) *If $a \perp b$, then $a \vee b$ exists.*

(G4) *If* $a \perp b$, $a \perp c$ *and* $b \perp c$, *then* $a \vee b \perp c$.

According to (G1) and (G2) every WGOMP can be regarded as a GDP by defining $b \ominus a = a^{\sharp_b}$ precisely when $a \leq b$. The following result shows a relation between generalized difference posets and weak generalized orthomodular posets.

Theorem 1.5.13 *Let* $(P; \leq, \ominus, 0)$ *be a generalized D-poset. For every* $a \in P$ *define* $x^{\sharp_a} = a \ominus x$ $(x \in P,\ x \leq a)$. *Then* P *is a weak generalized OMP if and only if the following conditions are satisfied:*

(W1) *If* $a, b \in P$ *and* $a \perp b$, *then* $a \oplus b$ *is the supremum of* a, b.

(W2) *If* $a, b, c \in P$, $a \perp b$, $a \perp c$, *and* $b \perp c$, *then* $a \oplus b \perp c$.

Proof. If P is a WGOMP and $a, b \in P$ are such that $a \perp b$, then we get $(a \vee b) \ominus b = (a \vee b) \wedge ((a \oplus b) \ominus b) = (a \vee b) \wedge a = a$, hence $a \vee b = a \oplus b$. From this observation it is clear that (W1) and (W2) are satisfied.

Conversely, let (W1) and (W2) be satisfied. We have to check properties (G1)–(G4) of the preceding definition. (G3) and (G4) are straightforward. To prove (G1) consider the D-poset $[0, a]$ $(a \in P)$. According to Proposition 1.5.6, for $[0, a]$ to be an OMP, it suffices to show that $[0, a]$ is an orthoalgebra, which will follow when (i) in Proposition 1.5.7 is satisfied. So, if $b \in P$, $b \leq a$ and $b \leq a \ominus b$, then $b \perp b$, hence by (W1), $b \oplus b = b$, i.e., $b = b \ominus b = 0$.

To prove (G2), let $a, b, c \in P$ be such that $a \leq b \leq c$. Since $[0, c]$ is an OMP we get $c \ominus (b \ominus a) = c \ominus ((c \ominus a) \ominus (c \ominus b)) = a \oplus (c \ominus b) = a \vee b^{\sharp_c} = (b \wedge a^{\sharp_c})^{\sharp_c}$, hence $b \ominus a = b \wedge a^{\sharp_c}$ which proves (G2). □

Lemma 1.5.14 *Let* $(P; \leq, ', 0, 1)$ *be an OMP and let* J *be an order ideal of* P *such that, for all* $a, b \in J$ *with* $a \leq b'$, *also* $a \vee b \in J$. *Equip every interval* $[0, a]$ *of* J *with a unary operation* $\sharp_a$ *given by* $x \mapsto x^{\sharp_a} = a \wedge x'$. *Then* J *is a WGOMP.*

Proof. Conditions (G1)–(G2) of a WGOMP are clear; (G3) and (G4) follow from the following observation: For all $a, b \in J$, $a \leq b'$ if and only if $a \leq b^{\sharp_c}$ for some $c \in J$ with $a, b \leq c$. □

Theorem 1.5.15 *Let* $(P; \oplus, 0)$ *be a generalized effect algebra and let* $(\hat{P}; \oplus^*, 0, 0^\sharp)$ *be the effect algebra constructed in Theorem 1.2.6. Let* $'^*$ *be a unary operation on* $\hat{P}$ *given by* $x'^* = 0^\sharp \ominus^* x$ *(this means that if* $a \in P$, *then* $a'^* = a^\sharp$ *and* $(a^\sharp)'^* = a$*). Then* $(\hat{P}; \leq, '^*, 0, 0^\sharp)$ *is an OMP if and only if* P *is a WGOMP.*

Proof. If P is a WGOMP then P satisfies (OA4), hence by Theorem 1.5.11, $\hat{P}$ satisfies (OA4). Owing to Proposition 1.5.7, it suffices to show that, for all $x, y \in \hat{P}$, $x \vee^* y$ exists whenever $x \leq^* y'^*$. So, let $a, b \in P$. Since $a \leq^* (b^\sharp)'^*$ is equivalent to $b^\sharp \leq^* a'^*$, we have to check the following two possibilities.

1) If $a \leq^* (b^\sharp)'^*$, i.e., $a \leq b$, then clearly $a, b^\sharp \leq^* (b \ominus a)^\sharp$. If $a, b^\sharp \leq^* c^\sharp$ for some $c \in P$, then $a \perp c$ and $c \leq b$, hence by (W1), $a \oplus c \leq b$ and thus $c \leq b \ominus a$ which means that $(b \ominus a)^\sharp \leq^* c^\sharp$. Therefore $a \vee^* b^\sharp = (b \ominus a)^\sharp$.

2) If $a \leq^* b'^*$, i.e., $a \leq^* b^\sharp$, then $a \perp b$ and by (W1), $a \oplus b = a \vee b$. Clearly, $a, b \leq^* a \oplus b$. If $a, b \leq^* c^\sharp$ for some $c \in P$, then $a \perp c$ and $b \perp c$, hence by (W2), $a \oplus b \perp c$, i.e., $a \oplus b \leq^* c^\sharp$. Therefore $a \vee^* b = a \vee b$.

Conversely, let $\hat{P}$ be an OMP. Since $\hat{P}$ satisfies (OA4), P satisfies (OA4) by Theorem 1.5.11.

Let $a, b \in P$ with $a \leq^* b'^*$. This means that $a \leq^* b^\sharp$, i.e. $a \oplus b$ is defined, and $a \vee^* b$ exists. From $a, b \leq a \oplus b$ it follows $a, b \leq^* a \oplus b$, hence $a \vee^* b \leq^* a \oplus b$ which with $a \oplus b \in P$ gives $a \vee^* b \in P$ and thus $a \vee^* b = a \vee b$. Denote $c = a \vee b$ and $d = a \oplus b$. $b \leq c \leq d$ implies $d \ominus c \leq d \ominus b = a \leq c$, hence $d \ominus c \leq d \ominus (d \ominus c)$ which by (OA4) gives $d \ominus c = 0$, i.e., $d = c$. We have $a \oplus b = a \vee b$.

Since, by the preceding considerations, P, as an order ideal of $\hat{P}$, satisfies the condition of Lemma 1.5.14, it suffices now to show that for all $a, b \in P$ with $a \leq b$ it holds that $b \ominus a = b \wedge^* a^\sharp$. $b \ominus a$ is a lower bound of b and $a^\sharp$ in $(\hat{P}; \leq^*)$ because $b \ominus a \leq b$ and $(b \ominus a) \oplus a = b$. If $c \leq^* b$, $a^\sharp$ for some $c \in P$, then $c \leq b$ and $a \oplus c$ exists, hence $a \oplus c = a \vee c \leq b$ which implies $c \leq b \ominus a$, i.e., $c \leq^* b \ominus a$. Therefore $b \ominus a$ is the greatest lower bound of b and $a^\sharp$. □

Remark 1.5.16 By the preceding considerations, the embedding of a WGOMP P into an orthomodular poset $\hat{P}$ from Theorem 1.5.15 coincides with that of [MaIp] and, as observed there, this embedding preserves the infimum but not generally the supremum whenever they exist in P. If $a, b \in P$ with $a \perp b$, then, by (G4), the supremum of a and b in P is also the supremum of a and b in $\hat{P}$.

According to [MaIp], for a WGOMP P the following conditions are equivalent:

(i) The embedding of P into an OMP $\hat{P}$ preserves all existing suprema of two elements.

(ii) If $a, b, c \in P$ with $a \perp c$ and $b \perp c$ and if $a \vee b$ exists in P, then $a \vee b \perp c$.

(There is no difficulty to see that a similar statement is true for a GDP P and a corresponding D-poset $\hat{P}$.)

A **generalized orthomodular poset** (GOMP) is defined in [MaIp] as a poset $(P; \leq)$ with a smallest element 0 such that every interval $[0, a]$ of P is equipped with a unary operation $x \mapsto x^{\sharp_a}$, satisfying the axioms (G1), (G2), (G3) and

(G4)' If $a, b, c \in P$ are such that $a \perp c$, $b \perp c$ and if $a \vee b$ exists, then $a \vee b \perp c$.

Since (G3) and (G4)' imply (G4), every GOMP is a WGOMP (see Definition 1.5.12). Theorem 2 in [MaIp] and Theorem 1.5.15 imply the following.

Theorem 1.5.17 *Let $(P;\oplus,0)$, $(\hat{P};\oplus^*,0,0^\sharp)$ and $'^*$ be as in Theorem 1.5.15. Then $(\hat{P};\leq,'^*,0,0^\sharp)$ is an OMP for which the supremum of $a,b\in P$ in P, if it exists, is also the supremum of a,b in $\hat{P}$ if and only if P is a GOMP.*

1.6 Examples and Applications

There is an abundance of various examples of difference structures mentioned in Section 1.5. In this Section, we follow [HePu]. In particular, we will study Abelian groups with a special partial order introduced in [Che] under the name "orthomodular groups". Actually, we study a generalization of orthomodular groups. We prove that an orthomodular group is always a generalized orthomodular poset (GOMP).

As a further generalization of an orthomodular group we study subsets of Abelian groups with a special order. We prove for example, that sets of idempotents (projections) in rings (*-rings) satisfying special conditions form WGOMPs. In another concrete example motivated by a theory of triple systems (alternative and Jordan triples), we introduce a "triple group" as an Abelian group endowed with a ternary operation, and prove that the set of all tripotents in it forms a WGOMP. As a special case, the set of all tripotents of a JBW*-triple ([Bat], [Rut 2]) forms a GOMP. Using triple groups, known partial orders of idempotents (projections) in rings (*-rings) are extended to tripotents and it is shown that they remain to form WGOMPs.

Example 1.6.1 Interval algebras studied in Section 1.4 yield a very important class of effect algebras/difference posets. Since every partially ordered vector space V is at the same time a partially ordered Abelian group, the ordering cone $P=V^+$ with the naturally defined difference is a GDP. The set of all positive operators on a complex Hilbert space, positive operators in a von Neumann algebra, positive elements in a C*-algebra or a Jordan algebra are such examples of generalized difference posets.

Example 1.6.2 Let X be a nonempty set and $\mathcal{F}\subseteq[0,1]^X$ satisfy

(i) $1\in\mathcal{F}$,

(ii) $f,g\in\mathcal{F}$ and $f\leq g$ implies $g-f\in\mathcal{F}$,

where $\leq$ and $-$ are componentwise partial order and difference of real functions, respectively. Then $\mathcal{F}$ with the partial binary operation $\ominus$ given by: $g\ominus f$ is defined if and only if $f\leq g$ and $g\ominus f=g-f$, is a D-poset.

$\mathcal{F}$ is an orthoalgebra if and only if

(iii) $0\neq f\in\mathcal{F}$ implies $2f\notin\mathcal{F}$.

$\mathcal{F}$ is an OMP if and only if

(iv) $f, g, h \in \mathcal{F},\ f + g \leq 1,\ f + h \leq 1,\ g + h \leq 1$ imply $f + g + h \in \mathcal{F}$.

As concerns the latter example, see [MaTr], [BuMa] and [Pul 1].

Example 1.6.3 Let X be a nonempty set and let $\mathcal{F} \subseteq (\mathbb{R}^+)^X$ satisfy

(i) $0 \in \mathcal{F}$,

(ii) $f, g \in \mathcal{F}$ and $f \leq g$ implies $g - f \in \mathcal{F}$,

where $\leq$ and $-$ are as in Example 1.6.2. Then $\mathcal{F}$ with $\ominus$ as in Example 1.6.2 is a GDP. Observe that for $f, g \in \mathcal{F}$, $f \perp g$ if and only if $f + g \in \mathcal{F}$, where $+$ is pointwise sum of real functions.

$\mathcal{F}$ is a GOA if and only if

(iii) $0 \neq f \in \mathcal{F}$ implies $2f \notin \mathcal{F}$.

$\mathcal{F}$ is a WGOMP if and only if

(iv) $f, g, h, f + g \in \mathcal{F}$ and $f, g \leq h$ imply $f + g \leq h$,

(v) $f, g, h, f + g, f + h, g + h \in \mathcal{F}$ imply $f + g + h \in \mathcal{F}$.

As a concrete example we present the following set of real functions which is a GOA but which is not a WGOMP. Let $\mathcal{F} = \{0, f, g, h, f + g, h - f, h - g\} \subset (\mathbb{R}^+)^{\mathbb{R}^+}$ be a set of seven different functions $\mathbb{R}^+ \to \mathbb{R}^+$ described in Figure 1.2. All functions on Figure 1.2 are linear on the intervals [0,3], [3,6] and [6,∞), and $f(0) = 1$, $g(0) = 4$, $h(0) = 7$. $\mathcal{F}$ under pointwise partial order of real functions forms a poset which is on Figure 1.3. Conditions (i)–(iii) and (v) are satisfied, but (iv) is not.

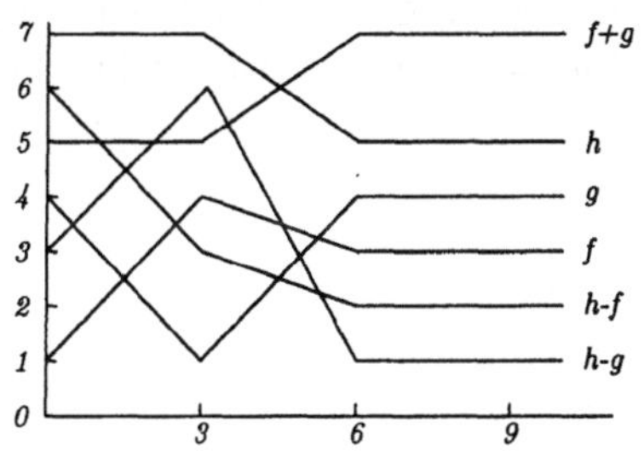

Fig. 1.2

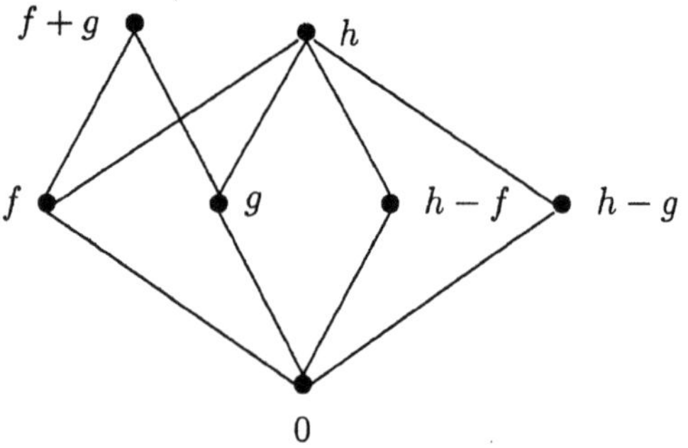

Fig. 1.3

Example 1.6.4 Let X be a nonempty set and let $\mu : \mathbb{R}^+ \to \mathbb{R}^+$ be a strictly increasing continuous function such that $\mu(0) = 0$. Define a partial binary operation $\ominus$ on $(\mathbb{R}^+)^X$ as follows: If $f, g \in (\mathbb{R}^+)^X$, then $g \ominus f$ is defined if and only if $f \leq g$ and in this case $(g \ominus f)(x) = \mu^{-1}(\mu(g(x)) - \mu(f(x)))$ for all $x \in X$, where $\leq$ is a pointwise partial order of real functions. Then $(\mathbb{R}^+)^X$ is a GDP (cf. [KoCh]).

Example 1.6.5 Since every interval $[0, a]$ in a generalized D-poset is itself a D-poset, every interval $[0, a]$, $a \geq 0$, in a partially ordered Abelian group is a D-poset and hence we have the following examples of D-posets: the set of all effects, that is, selfadjoint operators A on a complex Hilbert space such that $O \leq A \leq I$ (which plays an important role in quantum axiomatic [BLM]), the interval $[0, e]$ in an Archimedean order-unit space (A, e) with the order unit e ([Alf]), the interval $[O, I]$ in a JB-algebra.

Example 1.6.6 Let $(G; +, 0)$ be an Abelian group and let $\leq$ be a partial order on G such that:

(i) If $a, b, c \in G$ and $a \leq b \leq c$, then $c - b \leq c - a$.

Define a partial binary operation $\ominus$ on G by: If $a, b \in G$, then $b \ominus a$ is defined if and only if $a \leq b$, and let $b \ominus a := b - a$. Then the following three conditions are equivalent:

(1) $(G; \leq, \ominus, 0)$ is a GDP.

(2) 0 is a smallest element in $(G; \leq)$.

(3) If $a, b \in G$ and $a \leq b$, then $b - a \leq b$.

(1) $\Rightarrow$ (2). This is clear.

(2) $\Rightarrow$ (3). If $a, b \in G$ are such that $a \leq b$, then from $0 \leq a \leq b$ by (i) it follows that $b - a \leq b - 0 = b$.

(3) ⇒ (1). This follows from group properties.

Assume that $(G; \leq, \ominus, 0)$ is a GDP. Recall that for $a, b \in G$, $a \perp b$ if and only if $a = c \ominus b$ for some $c \in G$. From (3) and the fact that, for all $a, b \in G$, $a = (a+b) - b$ and $b = (a+b) - a$ it follows the following:

(4) If $a, b \in G$, then $a \perp b$ if and only if $a \leq a + b$.

The sum $\oplus$ is then given by $(a, b \in G)$: $a \oplus b$ is defined if and only if $a \perp b$, and $a \oplus b := a + b$. And $(G; \oplus, 0)$ is a GOA if and only if the following condition is satisfied:

(5) If $a \in G$ and $a \leq a + a$, then $a = 0$.

For every $a \in G$ define $x^{\sharp a} = a \ominus x$ $(x \in G,\ x \leq a)$. Then G is a WGOMP if and only if the following two conditions are satisfied for all $a, b, c \in G$:

(ii) If $a \leq a + b$, then $a \vee b$ exists and $a \vee b = a + b$.

(iii) If $a \leq a + b,\ a + c$ and $b \leq b + c$, then $a \leq a + b + c$.

To prove this, it suffices to observe that (2) is satisfied and then to use Theorem 1.5.13. Indeed, if $a \in G$ is arbitrary, then from $a \leq a = a + 0$ it follows by (ii) that $a \vee 0$ exists and $a \vee 0 = a + 0 = a$ which means that $0 \leq a$.

Consequently, G is a GOMP if and only if (ii) and the following condition are satisfied:

(iv) If $a \leq a + b,\ a + c$ and $b \vee c$ exists, then $a \leq a + (b \vee c)$.

In [Che] the notion of an **orthomodular group** is introduced as an Abelian group G equipped with a partial order $\leq$ satisfying the following conditions for all $a, b, c \in G$:

(OG1) $a \leq b \leq b + c$ implies $a \leq a + c$.

(OG2) $a \leq b \leq c$ implies $c - b \leq c - a$.

(OG3) $a \leq a + b$ implies that $a \vee b$ exists and $a \vee b = a + b$.

(OG4) $a \leq a + b,\ a + c$ implies $a \leq a + b + c$.

We show that conditions (OG2)–(OG4) imply conditions (i), (ii) and (iv), hence every orthomodular group is a GOMP (in [Che], it was proved that an orthomodular group is a WGOMP). So, (OG2) is (i) and (OG3) is (ii). Since (OG3) implies that 0 is a smallest element, G is a GDP and hence (4) is satisfied. To prove (iv), let $a, b, c, d \in G$ be such that $a \leq a + b,\ a + c$ and $d = b \vee c$. From $b \leq b + a,\ b + (d - b)$ it follows by (OG4) that $b \leq a + b + (d - b) = a + d$. Similarly we get $c \leq a + d$ and thus $a + d$ is an upper bound of b, c, hence $d \leq a + d$, which implies $a \leq a + d$.

Let us note that from the preceding considerations it follows that, in the above definition of an orthomodular group, condition (OG1) can be omitted. Namely, (OG1) means that if $a, b, c \in G$ are such that $a \leq b$ and $b \perp c$, then $a \perp c$, which is true in every GDP.

Let X be a nonempty set and let $S \subseteq 2^X$ be such that $\emptyset \in S$ and $a \Delta b \in S$, for all $a, b \in S$, where $a \Delta b = (a \cap b') \cup (a' \cap b)$. Then $(S; \Delta, \emptyset)$ is an Abelian group and with respect to $\leq$ defined by set inclusion S is an orthomodular group.

As shown in [Che] there is an abundance of orthomodular groups:

(a) Let A be an alternative ring with no nonzero nilpotent elements. Define a binary relation $\leq$ on A by $a \leq b$ if and only if $ab = a^2$ [MyJi], then $\leq$ is a partial order.

(b) Let A be an associative *-ring with a proper involution [Berb], that is, $a^*a = 0$ implies $a = 0$. Define a binary relation $\leq$ on A by $a \leq b$ if and only if $aa^* = ba^*$ and $a^*a = a^*b$. Then $\leq$ is a partial order (called the *-order) [Dra] and A with $\leq$ is a WGOMP [MaIp]. In particular, a commutative ring A without nonzero nilpotent elements, a Rickart *-ring A [Berb] (in [MaIp], using results from [Jan 1], it is shown that A is a GOMP), a C*-algebra.

(c) Let A be a Jordan algebra ([Top], [HOSt]) without nonzero nilpotent elements and satisfying the following condition:

$$[x, x, y] = 0 \text{ implies } [xy, x, y] = 0,$$

where $[a, b, c] = (ab)c - a(bc)$ is the associator of a, b, c. Define a binary relation $\leq$ on A by $a \leq b$ if and only if $ab = a^2$ and $a^2b = a^3$ [Che], then $\leq$ is a partial order.

(d) Let A be a JB-algebra ([HOSt], [MaLo]). Define a binary relation $\leq$ on A by $a \leq b$ if and only if $a^2b = a^3$ [Che], then $\leq$ is a partial order.

Since A is always an Abelian group, (a), (b), (c) and (d) with the order relations defined above are examples of orthomodular groups and thus examples of GOMPs.

We present yet an example of an Abelian group which is not an orthomodular group (even which is not a WGOMP) but which is a GOA. Consider the Abelian group $(\mathbb{Z}_7; +, 0)$ of integers modulo 7 partially ordered as in Figure 1.4.

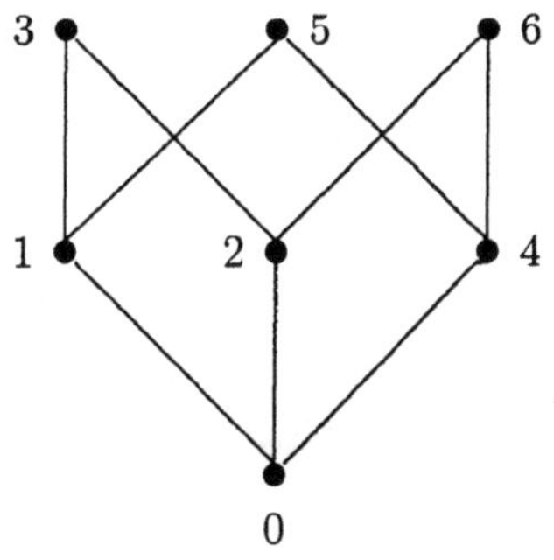

Fig. 1.4

Then conditions (i), (2), (5) and (ii) are satisfied, hence $\mathbb{Z}_7$ is a GDP which is a GOA (let us note that, in general, (ii) need not be satisfied in a GOA). $\mathbb{Z}_7$ is not a WGOMP since (iii) is not satisfied: We have $1 \leq 1+2,\ 1+4$ and $2 \leq 2+4$, but $1 \not\leq 1+2+4$. Thus $\mathbb{Z}_7$ is not a GOMP, and hence $\mathbb{Z}_7$ is not an orthomodular group.

A simple example of an Abelian group which is not an orthomodular group, but which is a GOMP, is the group $(\mathbb{Z}_4; +, 0)$ of integers modulo 4 partially ordered as in Figure 1.5. Conditions (i), (2), (ii) and (iv) are satisfied, hence $\mathbb{Z}_4$ is a GOMP, but condition (OG4) is not satisfied, since $2 \leq 1+2$ but $2 \not\leq 1+1+2$.

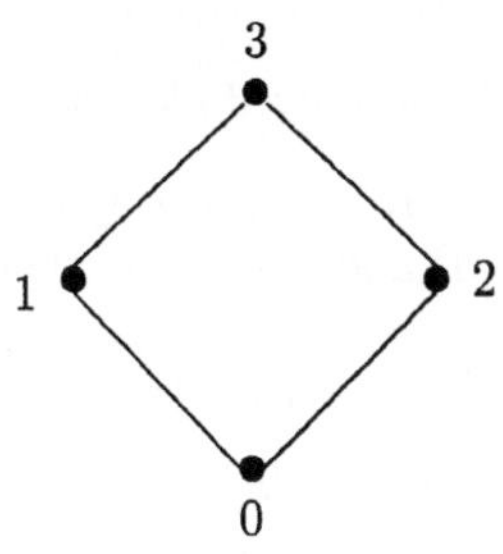

Fig. 1.5

There is also an example of an Abelian group which is a WGOMP but which is not a GOMP (hence, which is not an orthomodular group). Consider the Abelian group $(\mathbb{Z}_9; +, 0)$ of integers modulo 9 partially ordered as in Figure 1.6.

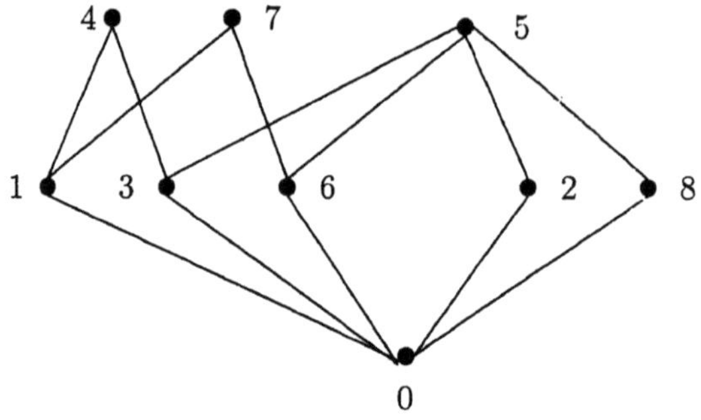

Fig. 1.6

Then conditions (i)–(iii) and (2) are satisfied (this means that $\mathbb{Z}_9$ is a WGOMP) but condition (iv) is not satisfied, since $1 \leq 1+3, 1+6$ and $3 \vee 6$ exists but $1 \not\leq 1+(3 \vee 6)$ (thus $\mathbb{Z}_9$ is not a GOMP). Let us note that the poset in Figure 1.6 considered as an abstract poset is an example of Roddy from [MaIp].

Example 1.6.7 We present generalizations (modifications) of examples (a), (b) and (c) in Example 1.6.6.

(a) Let R be a ring such that the following binary relation $\leq$ on R is a partial order:

$$a \leq b \text{ if and only if } ab = ba = a^2.$$

Clearly, 0 is a smallest element in $(R; \leq)$. Since $\leq$ is antisymmetric, $a^2 = 0$ implies $a = 0$ for every $a \in R$. If $a, b, c \in R$ and $a \leq b \leq c$, then

$$\begin{gathered}(c-b)(c-a) = c^2 - bc - ca + ba = c^2 - b^2, \text{ similarly}\\ (c-a)(c-b) = c^2 - b^2,\\ (c-b)^2 = c^2 - bc - cb + b^2 = c^2 - b^2,\end{gathered}$$

which means that $c - b \leq c - a$. Hence R is a GDP and by (4) in Example 1.6.6, for all $a, b \in R$, $a \perp b$ if and only if $a \leq a + b$. It is easy to see that for all $a, b \in R$,

$$a \perp b \text{ if and only if } ab = ba = 0.$$

We prove that $(R; +, 0)$ is an orthomodular group. It remains to show conditions (OG3) and (OG4).

(OG3) If $a, b \in R$ and $a \leq a+b$, then $b \leq a+b$ (since $\perp$ is symmetric), hence $a+b$ is an upper bound of a, b. If $c \in R$ and $a, b \leq c$, then

$$(a+b)c = ac + bc = a^2 + b^2 = ca + cb = c(a+b),$$
$$(a+b)^2 = a^2 + ba + ab + b^2 = a^2 + b^2,$$

hence $a+b \leq c$. Thus $a+b$ is the join of a, b in $(R; \leq)$.

(OG4) If $a, b, c \in R$ and $a \leq a+b, a+c$, then $a \perp b$ and $a \perp c$, hence $a(b+c) = ab + ac = 0 + 0 = 0$ and $(b+c)a = ba + ca = 0 + 0 = 0$, which means that $a \perp b+c$ and thus $a \leq a+b+c$.

For example, $\leq$ as defined above is a partial order on every member R of the class of rings studied by Abian in [Abi], hence every R is an orthomodular group. As shown by Hentzel in [Hen, Lemma 1], these rings generalize alternative rings without nonzero nilpotent elements. Let us note that the partial order $\leq$ on R reduces to:

$$a \leq b \text{ if and only if } ab = a^2.$$

Cf. Example 1.6.6 (a).

(b) Let R be a *-ring (i.e., a ring R with an involution *) such that the following binary relation $\leq$ on R is a partial order:

$$a \leq b \text{ if and only if } aa^* = ba^* \text{ and } a^*a = a^*b.$$

Observe that 0 is a smallest element in $(R; \leq)$ and that, for all $a, b \in R$, $a \leq b$ implies $aa^* = ab^*$ and $a^*a = b^*a$. If $a, b, c \in R$ and $a \leq b \leq c$, then

$$((c-b) - (c-a))(c-b)^* = (a-b)(c^* - b^*) = ac^* - bc^* - ab^* + bb^* = 0,$$

and similarly we get $(c-b)^*((c-b) - (c-a)) = 0$, hence $c - b \leq c - a$. Thus R is a GDP and by (4) in Example 1.6.6, for all $a, b \in R$, $a \perp b$ if and only if $a \leq a+b$. It is easy to show that, for all $a, b \in R$,

$$a \perp b \text{ if and only if } a^*b = ba^* = 0.$$

To prove that $(R; +, 0)$ is an orthomodular group, it remains to show conditions (OG3) and (OG4).

(OG3) If $a, b \in R$ and $a \leq a+b$, then $b \leq a+b$. If $c \in R$ and $a, b \leq c$, then

$$(a+b)(a+b)^* = aa^* + ba^* + ab^* + bb^* = aa^* + bb^*,$$
$$c(a+b)^* = ca^* + cb^* = aa^* + bb^*,$$

and similarly we get $(a+b)^*(a+b) = a^*a + b^*b = (a+b)^*c$, hence $a+b \leq c$. Thus $a+b$ is the join of a, b in $(R; \leq)$.

(OG4) If $a, b, c \in R$ and $a \le a+b, a+c$, then $a \perp b$, $a \perp c$, hence $a^*(b+c) = a^*b + a^*c = 0 + 0 = 0$ and $(b+c)a^* = ba^* + ca^* = 0 + 0 = 0$, which means that $a \perp b+c$ and thus $a \le a+b+c$.

(c) Let R be a ring in which, for all $a \in R$, $a^2a = aa^2$. Consider the following binary relation $\le$ on R:

$$a \le b \text{ if and only if } ab = ba = a^2 \text{ and } a^2b = ba^2 = ab^2 = b^2a = a^3.$$

Then $\le$ is reflexive and $0 \le a$ for all $a \in R$. The relation $\le$ is antisymmetric if and only if, for all $a \in R$, $a^2 = 0$ implies $a = 0$. Assume that $\le$ is a partial order on R. Using Example 1.6.6 we show that the Abelian group $(R; +, 0)$ is a WGOMP, where for every $a \in R$ we define $x^{\sharp_a} = a - x$ $(x \in R, x \le a)$. So, we prove that conditions (i)–(iii) of Example 1.6.6 are satisfied.

(i) If $a, b, c \in R$ and $a \le b \le c$, then $(c-a)^2 = c^2 - a^2$, $(c-b)^2 = c^2 - b^2$ and $(c-b)^3 = c^3 - b^3$. Then we get

$$\begin{aligned}
(c-b)(c-a) &= c^2 - bc - ca + ba = c^2 - b^2,\\
(c-b)^2(c-a) &= c^3 - b^2c - c^2a + b^2a = c^3 - b^3,\\
(c-b)(c-a)^2 &= c^3 - bc^2 - ca^2 + ba^2 = c^3 - b^3,
\end{aligned}$$

and similarly we get $(c-a)(c-b) = c^2 - b^2$ and $(c-a)(c-b)^2 = c^3 - b^3 = (c-a)^2(c-b)$, which means that $c - b \le c - a$.

Hence R is a GDP and by (4) in Example 1.6.6, for all $a, b \in R$, $a \perp b$ if and only if $a \le a + b$. It is easy to see that for all $a, b \in R$,

$$a \perp b \text{ if and only if } ab = ba = a^2b = ba^2 = ab^2 = b^2a = 0.$$

(ii) We show that if $a, b \in R$ with $a \le a + b$, then $a + b$ is the join of a and b in $(R; \le)$. Since $\perp$ is symmetric, from $a \le a + b$ it follows $b \le a + b$, hence $a + b$ is an upper bound of a and b. From $a \perp b$ we get $(a+b)^2 = a^2 + b^2$ and $(a+b)^3 = a^3 + b^3$. If $c \in R$ is an upper bound of a and b, then we obtain

$$\begin{aligned}
(a+b)c &= ac + bc = a^2 + b^2 = (a+b)^2,\\
(a+b)^2c &= (a^2+b^2)c = a^2c + b^2c = a^3 + b^3 = (a+b)^3,\\
(a+b)c^2 &= ac^2 + bc^2 = a^3 + b^3 = (a+b)^3,
\end{aligned}$$

and similarly $c(a+b) = (a+b)^2$ and $c(a+b)^2 = (a+b)^3 = c^2(a+b)$, which means that $a + b \le c$.

(iii) If $a, b, c \in R$, $a \le a+b$, $a+c$ and $b \le b+c$, then $a \le a+b+c$ since $a(b+c) = ab+ac = 0$, $a^2(b+c) = a^2b+a^2c = 0$, $a(b+c)^2 = a(b^2+c^2) = ab^2+ac^2 = 0$ and similarly $0 = (b+c)a = (b+c)a^2 = (b+c)^2a$.

Let us note that if R is zero commutative (i.e., if, for all $x, y \in R$, $xy = 0$ implies $yx = 0$) or commutative, then the partial order $\leq$ on R as defined above reduces to:

$$a \leq b \text{ if and only if } ab = a^2 \text{ and } a^2b = ab^2 = a^3.$$

And the orthogonality relation $\perp$ on R reduces to:

$$a \perp b \text{ if and only if } ab = a^2b = ab^2 = 0.$$

A Jordan ring is a commutative ring R satisfying $(xy)x^2 = x(yx^2)$ for all $x, y \in R$, this is to say $[x, y, x^2] = 0$. Let R be a Jordan ring satisfying the condition $2x = 0$ implies $x = 0$ for all $x \in R$, without nonzero nilpotent elements and satisfying the condition in Example 1.6.6 (c). In [GoMa] it is proved that the binary relation $\leq$ on R defined above is a partial order, hence R is a WGOMP.

Example 1.6.8 A convenient generalization of Example 1.6.6 which includes various known examples is as follows. Let $(G; +, 0)$ be an Abelian group and let P be a nonempty subset of G. Assume that there is a partial order $\leq$ on P such that the following two conditions are satisfied for all $a, b, c \in P$:

(o) $a \leq b$ implies $b - a \in P$,

(i) $a \leq b \leq c$ implies $c - b \leq c - a$.

By (o), $0 \in P$. Define a partial binary operation $\ominus$ on P by $(a, b \in P)$: $b \ominus a$ is defined if and only if $a \leq b$ and let $b \ominus a = b - a$. Then the following three conditions are equivalent:

(1) $(P; \leq, \ominus, 0)$ is a GDP.

(2) 0 is a smallest element in $(P; \leq)$.

(3) If $a, b \in P$ and $a \leq b$, then $b - a \leq b$.

Assume that $(P; \leq, \ominus, 0)$ is a GDP. Recall that for $a, b \in P$, $a \perp b$ if and only if $a = c \ominus b$ for some $c \in P$. Then the following condition is satisfied:

(4) If $a, b \in P$, then $a \perp b$ if and only if $a + b \in P$ and $a \leq a + b$.

Under the conditions (o) and (i), for every $a \in P$ define $x^{\sharp_a} = a \ominus x$ $(x \in P, x \leq a)$. Then P is a WGOMP if and only if the following two conditions are satisfied for all $a, b, c \in P$:

(ii) $a + b \in P$ and $a \leq a + b$ implies $a \vee b$ exists, and $a \vee b = a + b$,

(iii) $a + b, a + c, b + c \in P$, $a \leq a + b, a + c$ and $b \leq b + c$ implies $a + b + c \in P$, and $a \leq a + b + c$.

And P is a GOMP if and only if (ii) and the following condition are satisfied for all $a, b, c \in P$:

(iv) If $a + b, a + c \in P$, $a \leq a + b, a + c$ and $b \vee c$ exists, then $a + (b \vee c) \in P$ and $a \leq a + (b \vee c)$.

Proofs of all mentioned statements can be carried on analogously as in Example 1.6.6. We present some more concrete examples.

(a) Let R be a ring and let P be the set of all idempotents in R (i.e., elements $a \in R$ with $a = a^2$) such that the following binary relation $\leq$ on P is a partial order:

$$a \leq b \text{ if and only if } ab = ba = a.$$

P is nonempty since $0 \in P$. Clearly, $0 \leq a$ for all $a \in P$. If $a, b \in P$ and $a \leq b$, then $(b - a)^2 = b^2 - ab - ba + a^2 = b - a - a + a = b - a$, hence $b - a \in P$. If $a, b, c \in P$ and $a \leq b \leq c$, then $(c - b)(c - a) = c^2 - bc - ca + ba = c - b - a + a = c - b$ and $(c-a)(c-b) = c^2 - ac - cb + ab = c - a - b + a = c - b$, which means that $c - b \leq c - a$. Thus conditions (o) and (i) are satisfied and therefore P is a GDP. According to (4), if $a, b \in P$, then $a \perp b$ if and only if $ab = ba = 0$. It is easy to show that conditions (ii) and (iii) are satisfied (cf. Example 1.6.7(a)), hence P is a WGOMP, where for every $a \in P$ we define $x^{\sharp_a} = a - x$ $(x \in P, x \leq a)$.

As an example of such a WGOMP we can take the set P of all idempotents of a ring R in Example 1.6.7(a) (which is an orthomodular group, hence a GOMP) with the partial order $\leq$ on R restricted to the set P. Similarly, the set P of all idempotents of a ring R in Example 1.6.7 (c) (which is a WGOMP) with the partial order $\leq$ on R restricted to the set P is also such an example of a WGOMP.

(b) Let R be a *-ring and let P be the set of all projections in R (i.e., elements $a \in R$ such that $a^2 = a^* = a$) such that the following binary relation $\leq$ on P is a partial order:

$$a \leq b \text{ if and only if } ab = a.$$

Let us observe that $a, b \in P$ and $a \leq b$ implies $ba = a$. Then P is a GDP since $0 \in P$ and $0 \leq a$ for all $a \in P$, and conditions (o) and (i) are satisfied. By (4), for all $a, b \in P$, $a \perp b$ if and only if $ab = 0$ if and only if $ba = 0$. Conditions (ii) and (iii) are satisfied, too (cf. Example 1.6.7(b)), and therefore P is a WGOMP.

The set P of all projections of a *-ring R in Example 1.6.7(b) (which is an orthomodular group, hence a GOMP) with the partial order $\leq$ on R restricted to the set P is such an example of a WGOMP.

The set of all idempotents (projections) of an associative ring (*-ring) with 1 is an OMP ([Kal], [Bir]) and the set of all idempotents (projections) of an associative ring (*-ring) which need not have 1 is a WGOMP, see [Malp]. Let R be an associative ring (*-ring) without 1 and let $\tilde{R}$ denote a unification of R, see [Berb]. Let P denote the set of all idempotents (projections) of R. Then the OMP $\hat{P}$ (cf. Theorem 1.5.15) is isomorphic with the OMP $\tilde{P}$ of all idempotents (projections) in $\tilde{R}$. Namely, $\tilde{R} = R \times A$, where A is an auxiliary ring (*-ring) with a unit, and for all $(a, \alpha), (b, \beta) \in \tilde{R}$, $(a,\alpha)+(b,\beta) = (a+b, \alpha+\beta)$, $(a,\alpha)(b,\beta) = (ab+\beta a+\alpha b, \alpha\beta)$ (and $(a,\alpha)^* = (a^*, \alpha^*)$). If $(a,\alpha)^2 = (a,\alpha)$, then $(a^2 + 2\alpha a, \alpha^2) = (a, \alpha)$, hence $\alpha^2 = \alpha$ and $a^2 + 2\alpha a = a$ and thus $\alpha = 0$ and $a^2 = a$ or $\alpha = 1$ and $a^2 = -a$. Therefore $\tilde{P} = \{(a,0) : a \in P\} \cup \{(-a,1) : a \in P\}$.

Example 1.6.9 Another concrete example, which is motivated by a theory of triple systems - alternative and Jordan triples ([Bat], [EdRu], [Loo], [Loo 1], [Loo 2], [Mey], [Mey 1]) and which is still very general of the preceding general example, is as follows. Let $(A; +, 0)$ be an additive Abelian group endowed with a ternary operation $(x, y, z) \mapsto (xyz)$ on A which is additive in all three variables such that the following conditions are satisfied for all $a, b, c \in A$:

(1) $((aba)c(aba)) = (a(b(aca)b)a)$,

(2) $((aba)bc) = (a(bab)c)$, $(cb(aba)) = (c(bab)a)$.

A is called **a triple group**. Elementary consequences of the additivity are the following two properties $(a, b, c \in A)$:

$$(0ab) = (a0b) = (ab0) = 0,$$
$$-(abc) = (-a\ bc) = (a\ -b\ c) = (ab\ -c).$$

An element $a \in A$ is called a **tripotent** if $(aaa) = a$. Let P denote the collection of all tripotents of A. Clearly, $0 \in P$. Consider the following binary relation $\leq$ on the set P:

$$a \leq b \text{ if and only if } a = (aba) = (bab).$$

Then $\leq$ is reflexive, antisymmetric and $0 \leq a$ for all $a \in P$. If $a, b, c \in P$, $a \leq b$ and $b \leq c$, then $a \leq c$ since by (1),

$$\begin{aligned} (aca) &= ((bab)c(bab)) = (b(a(bcb)a)b) = (b(aba)b) = (bab) = a, \\ (cac) &= (c(bab)c) = (c((bcb)a(bcb))c) = (c(b(c(bab)c)b)c) \\ &= (c(b(cac)b)c) = ((cbc)a(cbc)) = (bab) = a. \end{aligned}$$

Hence $\leq$ is transitive and thus a partial order with a smallest element 0. Let us observe that the following condition is satisfied for all $a, b \in P$ and $c \in A$:

(3) $a \leq b$ implies $(abc) = (bac) = (aac)$ and $(cab) = (cba) = (caa)$.

Namely, if $a, b \in P, a \leq b$ and $c \in A$, then according to (2), $(abc) = ((aba)bc) = (a(bab)c) = (aac)$ and $(bac) = (b(aba)c) = ((bab)ac) = (aac)$. Similarly, $(cab) = (caa) = (cba)$. In particular, the following condition is satisfied for all $a, b \in P$:

(3*) $a \leq b$ implies $a = (aab) = (baa) = (abb) = (bba)$.

If $a, b \in P$ and $a \leq b$, then by (3*), $(b-a\ b-a\ b-a) = (bbb) - (bba) - (bab) + (baa) - (abb) + (aba) + (aab) - (aaa) = b - a$, which means that $b - a \in P$.

We show that if $a, b, c \in P$ and $a \leq b \leq c$, then $c - b \leq c - a$. Indeed, using conditions (3) and (3*), we obtain $(c-b\ c-a\ c-b) = (ccc) - (ccb) - (cac) + (cab) - (bcc) + (bcb) + (bac) - (bab) = c - b$ and $(c-a\ c-b\ c-a) = (ccc) - (cca) - (cbc) + (cba) - (acc) + (aca) + (abc) - (aba) = c - b$.

Thus conditions (o) and (i) in Example 1.6.8 are satisfied, hence P is a GDP. According to condition (4) in Example 1.6.8, for $a, b \in P$, $a \perp b$ if and only if $a + b \in P$ and $a \leq a + b$. Hence, using (3), we obtain for all $a, b \in P$ and $c \in A$:

(4) $a \perp b$ implies $(abc) = (bac) = (cab) = (cba) = 0$.

By the additivity we get the following. For all $a, b \in P$,

$$a + b \in P \text{ if and only if } (aab) + (aba) + (abb) + (baa) + (bab) + (bba) = 0.$$

If $a, b, a + b \in P$, then

$$a \leq a + b \text{ if and only if } (aba) = 0 = (aab) + (baa) + (bab).$$

According to condition (4), it is now clear that for all $a, b \in P$,

(5) $a \perp b$ if and only if $(aba) = (bab) = (aab) = (baa) = (abb) = (bba) = 0$.

Our aim is to prove that the collection of all tripotents of a triple group, partially ordered as above, forms a WGOMP. So, it remains to show that conditions (ii) and (iii) in Example 1.6.8 are satisfied.

(ii) If $a, b, a + b \in P$ and $a \leq a + b$, i.e., $a \perp b$, then $b \perp a$, hence $b \leq a + b$ and thus $a + b$ is an upper bound of a and b. If $c \in P$ and $a, b \leq c$, then by (2),

$$(acb) = ((aca)cb) = (a(cac)b) = (aab) = 0$$

and similarly $(bca) = 0$. Hence we get $(a+b\ c\ a+b) = (aca) + (acb) + (bca) + (bcb) = a + b$ and $(c(a+b)c) = (cac) + (cbc) = a + b$, which means that $a + b \leq c$. Therefore $a + b$ is the join of a and b in $(P; \leq)$.

(iii) Using the additivity and conditions (4) and (5) it is easy to see that if $a, b, c \in P$, $a \perp b, a \perp c$ and $b \perp c$, then $a \perp b + c$.

An Abelian group $(A; +, 0)$ together with a mapping $A^3 \to A$, $(x, y, z) \mapsto (xyz)$ which is symmetric in the first and third variables and additive in all three variables is called a **Jordan triple group** if the following identities are satisfied:

(J) $(ab(cde)) - (cd(abe)) = ((abc)de) - (c(dab)e)$,

(j) $((aba)ca) = (ab(aca))$.

(J) is known as the **Jordan triple identity**. Let us note that if A satisfies the condition $3x = 0$ implies $x = 0$ for all $x \in A$, then the identity (j) follows from (J) and the symmetry (this can be shown similarly as in [Mey 1]). Every Jordan triple group is a triple group. Indeed, replacing (c, d) by (a, b) in (J), we get

$$(ab(abe)) - (ab(abe)) = ((aba)be) - (a(bab)e),$$

which, using the symmetry, implies (2). The identity (1) is known as the fundamental-formula and can be derived similarly as in [Mey 1, Satz 2.2].

By a **Jordan triple** we mean a module A over a ring R with a unit endowed with an operation $A^3 \to A$, $(x, y, z) \mapsto (xyz)$ which is linear in all three variables such that A is a Jordan triple group (cf. [Mey 1]). The mapping $A^3 \to A$, $(x, y, z) \mapsto (xyz)$ is called the **Jordan triple product**.

In [Bat], a Jordan triple is a complex vector space $\mathcal{A}$ equipped with a mapping $\mathcal{A}^3 \to \mathcal{A}$, $(x, y, z) \mapsto (xyz)$ which is symmetric and linear in the first and third variables, conjugate linear in the second variable and satisfies the Jordan triple identity (J). In [Loo 2], this system is called a **Hermitean Jordan triple**. Every Jordan triple is a Jordan triple group.

A **JB*-triple** $\mathcal{A}$ is a Hermitean Jordan triple which is a Banach space such that the mapping D from $\mathcal{A}^2$ to the Banach space of bounded linear operators on $\mathcal{A}$, defined on all pairs (a, b) of elements in $\mathcal{A}$ and all c in $\mathcal{A}$, by $D(a, b)c := (abc)$, is continuous and such that, for all elements a in $\mathcal{A}$, $D(a, a)$ is Hermitean and with nonnegative spectrum and $\|D(a, a)\|$ is equal to $\|a\|^2$. For all elements a, b and c in a JB*-triple $\mathcal{A}$, $\|(abc)\| \leq \|a\|\|b\|\|c\|$ and $\|(aaa)\| = \|a\|^3$. A JB*-triple which is a Banach space dual is called a **JBW*-triple**. In a JBW*-triple $\mathcal{A}$ the Jordan triple product is separately w*-continuous.

The collection of all tripotents in a JBW*-triple together with the partial order defined by Loos has been studied in [Bat], [Rut 2], where it was shown that the collection of all tripotents in a JBW*-triple is a GOMP.

Indeed, (G1) follows by [Bat, Lem 3.5(iii)]. In fact, the interval $[0, w]$ is a complete orthomodular lattice with an orthocomplementation defined by $v'^w = w - v$.

To prove (G2), observe that if $u \leq v \leq w$, then u, v are idempotents in a JBW-algebra $\mathcal{A}_2(w)^J_{sa}$,and $v - u$ is the relative complement in the subinterval $[0, v]$ of the orthomodular lattice of the idempotents in $\mathcal{A}_2(w)^J_{sa}$. This gives $v - u = (w - u) \wedge v$.

(G3) follows by [Bat, Cor 3.7(ii)].

Condition (G4') follows by [Bat, Cor 3.10(i)].

We note that an alternative proof of properties (G1), (G2) and (G3) can be obtained by observing that the partial order introduced by Loos coincides with that we introduced in a triple group, therefore the collection of all tripotents in a JBW*-triple is a WGOMP.

An Abelian group $(A;+,0)$ equipped with a mapping $A^3 \to A$, $(x,y,z) \mapsto (xyz)$ which is additive in all three variables is called an **alternative triple group** if the following identities are satisfied:

(A) $(ab(cde)) + (cd(abe)) = ((abc)de) + (c(bad)e)$,

(a1) $((abc)dc) = (ab(cdc))$,

(a2) $(ab(abc)) = ((aba)bc)$.

Putting $a+d$ instead of a in the identity (a2) and then using (a2) and (A), we obtain the following identity (cf. the proof of (1.5) in [Loo]):

(i) $((abc)bd) = (a(bcb)d)$.

Every alternative triple group is a triple group. Indeed, using the identity (a1) twice and then using (i), we get

$$((aba)c(aba)) = (((aba)ca)ba) = ((ab(aca))ba) = (a(b(aca)b)a),$$

which proves the fundamental-formula (1). Putting a instead of c in (i), we get the first identity in (2). The second one is obtained as follows:

$$(cb(aba)) =^{(a1)} ((cba)ba) =^{(i)} (c(bab)a).$$

By an **alternative triple** we mean a module A over a ring R with a unit endowed with a mapping $A^3 \to A$, $(x,y,z) \mapsto (xyz)$ which is linear in all three variables such that A is an alternative triple group (cf. [Loo 2]). Alternative triples are investigated in [Loo], where an equivalent system of axioms is dealt with, instead of (a2), the following identity is used:

(ii) $((aba)cd) = (a(cab)d)$.

Example 1.6.10 Using the partial order $\leq$ on the set of all tripotents in a triple group A from Example 1.6.9, we can obtain generalizations of known results about idempotents (projections) in rings (*-rings). Known partial orders on idempotents (projections) in rings (*-rings) are extended to tripotents. It follows that the tripotents in special classes of rings and *-rings form a WGOMP (as to the examples (a) and (c) below cf. [Che, Proposition 10]).

(a) Let R be an alternative ring, i.e., a ring R satisfying $x^2y = x(xy)$ and $yx^2 = (yx)x$ for all $x, y \in R$. Artin's theorem in R says that any subring of R generated by two elements is associative. Other well-known properties are Moufang identities ([Scha], [ZSSS]): $(aba)c = a(b(ac))$, $c(aba) = ((ca)b)a$ and $a(bc)a = (ab)(ca)$ for all $a, b, c \in R$.

R together with the mapping $f : R^3 \to R$, $(x,y,z) \mapsto (xy)z$ is a triple group. Clearly, f is additive in all three variables. The first identity in (2) from Example 1.6.9 is clear for f by Artin's theorem. The second one can be shown as follows. If $a,b,c \in R$, then

$$cb.aba = (cb.a)b.a = (c.bab)a.$$

Using Moufang identities we get for all $a,b,c \in R$,

$$aba.c.aba = a(b.ac).aba = a.(b.ac)(ab).a = a(b.aca.b)a,$$

which proves the fundamental-formula (1) in Example 1.6.9.

The Cayley numbers [ZSSS] form an alternative ring which is an example of a triple group which is not an alternative triple group in the sense of the definition from the preceding example. Indeed, conditions (a1), (a2) are satisfied, but (A) is not satisfied (even condition (ii) is not satisfied).

An element $a \in R$ is a tripotent if $f(a,a,a) = a$, i.e., if $a^3 = a$. Every idempotent of R is a tripotent. Let T denote the set of all tripotents in R. The partial order $\leq$ on T from Example 1.6.9 has then the following form:

$$a \leq b \text{ if and only if } a = aba = bab.$$

By Example 1.6.9, $(T; \leq)$ is a WGOMP. Let us note that the binary relation $\leq$ on T has also the following form (cf. Example 1.6.8 (a)):

$$a \leq b \text{ if and only if } a^2 = ab = ba.$$

Indeed, if $a,b \in T$ and $a \leq b$, then $ab = aba.b = a.bab = a^2$ and $ba = b.aba = bab.a = a^2$. Conversely, if $a,b \in T$ and $ab = ba = a^2$, then $aba = a^2a = a$ and $bab = a^2b = aa^2 = a$, hence $a \leq b$.

(b) Let R be an alternative ring with an involution *. Then R together with the operation $f : R^3 \to R$, $(x,y,z) \mapsto (xy^*)z$ is an alternative triple group (cf. [Loo]). An element $a \in R$ is a tripotent if $f(a,a,a) = a$, i.e., if $aa^*a = a$. Every projection of R is a tripotent. Let T denote the collection of all tripotents in R. Then the partial order $\leq$ on T from Example 1.6.9 has the following form:

$$a \leq b \text{ if and only if } a = ab^*a = ba^*b.$$

By Example 1.6.9, since R is a triple group, $(T; \leq)$ is a WGOMP. Let us note that the binary relation $\leq$ on T when restricted to the set P of all projections of R has the following form (cf. Example 1.6.8 (b)):

$$a \leq b \text{ if and only if } ab = a.$$

If R is associative, then the relation $\leq$ on T has the following form (cf. Examples 1.6.6 (b), 1.6.7 (b)):

$$a \leq b \text{ if and only if } aa^* = ba^* \text{ and } aa^* = a^*b.$$

Namely, if $a, b \in T$ and $a \leq b$, then $aa^* = ba^*b.a^* = b.a^*ba^* = b(ab^*a)^* = ba^*$ and similarly $a^*a = a^*b$. Conversely, if $a, b \in T$, $aa^* = ba^*$ and $a^*a = a^*b$, then $aa^* = ab^*$, hence $ab^*a = aa^*.a = a$ and $ba^*b = aa^*.b = a.a^*b = aa^*a = a$, and thus $a \leq b$.

(c) Let R be a Jordan ring equipped with the mapping $f : R^3 \to R$, $(x, y, z) \mapsto x(yz) - y(xz) + z(xy)$ As known, R is a Jordan triple group (cf. [Mey 1]). An element $a \in R$ is a tripotent if $f(a, a, a) = a$, i.e., if $a^3 = a$. Every idempotent of R is a tripotent. Let T denote the set of all tripotents of R. The partial order $\leq$ on T from Example 1.6.9 has the following form:

$$a \leq b \text{ if and only if } a = 2a(ab) - a^2b = 2b(ab) - ab^2.$$

If $a, b \in T$ and $a \leq b$, then by (3*) in Example 1.6.9, $a = f(a, a, a) = a^2b$ and $a = f(a, b, b) = ab^2$, and by (3) in Example 1.6.9, $f(a, b, a^2) = f(a, a, a^2)$ from which it follows $ab = a^2$. Thus the relation $\leq$ on T has the following form (cf. Example 1.6.7 (c)):

$$a \leq b \text{ if and only if } ab = a^2 \text{ and } a^2b = ab^2 = a.$$

1.7 Effect Algebras with the Riesz Decomposition Property

We will start with some well-known properties in partially ordered Abelian groups, and we will define analogous notions in effect algebras. For example, we will show that an effect algebra with the Riesz decomposition property is always an interval effect algebra ([Rav]).

Proposition 1.7.1 *For a partially ordered Abelian group G, the following conditions are equivalent:*

(a) *Given x_1, x_2, y_1, y_2 in G such that $x_i \leq y_j$ for all i, j, there exists z in G such that $x_i \leq z \leq y_j$ for all i, j.*

(b) *Given x, y_1, y_2 in G^+ such that $x \leq y_1 + y_2$, there exist x_1, x_2 in G^+ such that $x = x_1 + x_2$ and $x_j \leq y_j$ for each j.*

(c) *Given x_1, x_2, y_1, y_2 in G^+ such that $x_1 + x_2 = y_1 + y_2$, there exist $z_{11}, z_{12}, z_{21}, z_{22}$ in G^+ such that $x_i = z_{i1} + z_{i2}$ for each i and $y_j = z_{1j} + z_{2j}$ for each j.*

Proof. (a) $\Rightarrow$ (b). Let $x, y_1, y_2 \in G^+$, $x \leq y_1 + y_2$. Then $0 \leq x$ and $0 \leq y_1$, $x - y_2 \leq y_1$. Since also $y_2 \geq 0$, we have $x - y_2 \leq x$. By (a), there exists $x_1 \in G$ such that $0 \leq x_1 \leq x$ and $x - y_2 \leq x_1 \leq y_1$. Set $x_2 = x - x_1$. Then x_1 and x_2 are elements of G^+ such that $x = x_1 + x_2$ and $x_1 \leq y_1$. Since $x - y_2 \leq x_1$ we also have $x_2 = x - x_1 \leq y_2$.

(b) $\Rightarrow$ (a). Let x_1, x_2, y_1, y_2 satisfy the assumptions of (a). Then $y_j - x_i \in G^+$ for all i, j and

$$y_2 - x_1 \leq (y_2 - x_1) + (y_1 - x_2) = (y_1 - x_1) + (y_2 - x_2).$$

By (b), there exist z_1, z_2 in G^+ such that $y_2 - x_1 = z_1 + z_2$ and each $z_j \leq y_j - x_j$. Set $z = x_1 + z_1$, and note that $x_1 \leq z$. Since $z_1 \leq y_1 - x_1$, we have $z \leq y_1$. Moreover, $z = y_2 - z_2 \leq y_2$. Finally, since $z_2 \leq y_2 - x_2$, we have $x_2 \leq y_2 - z_2 = z$. Therefore $x_i \leq z \leq y_j$ for all i, j.

(b) $\Rightarrow$ (c). Let $x_1, x_2, y_1, y_2 \in G^+$ and $x_1 + x_2 = y_1 + y_2$. Since $x_2 \geq 0$, we have $x_1 \leq y_1 + y_2$. By (b), there exist z_{11}, z_{12} in G^+ such that $x_1 = z_{11} + z_{12}$ and each $z_{1j} \leq y_j$. Set $z_{2j} = y_j - z_{1j}$ for each j, so that $z_{2j} \in G^+$ and $y_j = z_{1j} + z_{2j}$. Since

$$x_1 + x_2 = y_1 + y_2 = z_{11} + z_{21} + z_{12} + z_{22},$$

we also have $x_2 = z_{21} + z_{22}$.

(c) $\Rightarrow$ (b). Let $x, y_1, y_2 \in G^+$. $x \leq y_1 + y_2$. Set $w_1 = x$ and $w_2 = y_1 + y_2 - x$, so that w_1, w_2 are elements of G^+ satisfying $w_1 + w_2 = y_1 + y_2$. By (c), there exist $z_{11}, z_{12}, z_{21}, z_{22}$ in G^+ such that each $w_i = z_{i1} + z_{i2}$ and each $y_j = z_{1j} + z_{2j}$. Then z_{11} and z_{12} are elements of G^+ such that $x = w_1 = z_{11} + z_{12}$. Since each $z_{2j} \geq 0$, we also have $z_{1j} \leq z_{1j} + z_{2j} = y_j$. □

Definition 1.7.2 *A partially ordered set X is said to have the* **Riesz interpolation property** *(or just interpolation, for short) provided X satisfies the condition* (a) *of Proposition* 1.7.1 *(given any x_1, x_2, y_1, y_2 in X such that $x_i \leq y_j$ for all i, j, there exists $z \in X$ such that $x_i \leq z \leq y_j$ for all i, j).*

Definition 1.7.3 *A partially ordered Abelian group G is said to satisfy the* **Riesz decomposition property** *provided conditions (b) or, equivalently, (c) of Proposition* 1.7.1 *hold in G. An* **interpolation group** *is a partially ordered group that satisfies the Riesz interpolation property, and, consequently, the Riesz decomposition properties as well.*

Note that property (b) above can be alternatively written as

$$[0, a] + [0, b] = [0, a + b], \ a, b \geq 0.$$

Roughly speaking, a partially ordered Abelian group G is an interpolation group if and only if sums of positive intervals are intervals.

For example, any lattice ordered Abelian group is an interpolation group. An example, where interpolation does not hold, is the Abelian group $G = \mathbb{Z}^2$ with the positive cone

$$G^+ = \{(a,b) \in G : 2a \geq b \geq 0\}.$$

Set $x_1 = (0,0)$ and $x_2 = (0,1)$, while $y_1 = (1,1)$ and $y_2 = (1,2)$. Then $x_i \leq y_j$ for all i,j, but there is no element $z \in G$ such that $x_i \leq z \leq y_j$ for all i,j.

The class of interpolation groups is closed under the formation of direct products, direct sums and direct limits.

In what follows, we adapt the notion of the Riesz decomposition property for effect algebras. We prove that this property is a necessary and sufficient condition for embedding an effect algebra into an interpolation group. The latter result has been obtained by Ravindran [Rav].

Definition 1.7.4 *A (generalized) effect algebra E has the* **Riesz decomposition property** *if, for $a_1, a_2, b_1, b_2 \in E$, $a_1 \oplus a_2 = b_1 \oplus b_2$ implies the existence of $c_{ij} \in E$ such that $a_i = c_{i1} \oplus c_{i2}$ and $b_j = c_{1j} \oplus c_{2j}$ for all $i, j \in \{1,2\}$.*

Lemma 1.7.5 *Let E be a generalized effect algebra. The following conditions are equivalent:*

(i) *For $a, b_1, b_2 \in E$ with $a \leq b_1 \oplus b_2$, there exist $a_1, a_2 \in E$ such that $a = a_1 \oplus a_2$ and $a_i \leq b_i$ $i = 1,2$.*

(ii) *E has the Riesz decomposition property.*

Proof. Essentially the same as the proof of the equivalence of (b) and (c) in Proposition 1.7.1, one must only take care that the desired $\oplus$-sums exist. □

Notice that property (i) in the preceding theorem can be reformulated as follows: For all $a, b \geq 0$, $a \perp b$, $[0,a] \oplus [0,b] = [0, a \oplus b]$.

The technique used in the following construction goes back to R. Baer and O. Wyler. For the purpose of effect algebras, it has been modified by K. Ravindran.

Let E be a generalized effect algebra with the Riesz decomposition property. Following Baer, let us first construct sequences $W = (a_1, a_2, \ldots, a_n)$ of elements of E, called **words**. The **length** of a word W, denoted by $|W|$, is defined to be n. The elements a_i $i = 1, 2, \ldots, n$ are referred as the **components** of W. For each $a \in E$, the word (a) is of length 1, and to simplify notation, we write simply a. Denote by $\mathcal{W}(E)$ the collection of all words on E.

Next introduce a binary operation $+$ on the collection $\mathcal{W}(E)$ of all words as follows: For two words $W_1 = (a_1, \ldots, a_n)$, $W_2 = (b_1, \ldots, b_m)$, the word $W_1 + W_2 = (a_1, \ldots, a_n, b_1, \ldots, b_m)$. It is easily verified that $\mathcal{W}(E)$ is a semigroup (with no zero element).

We will say that two words W_1 and W_2 are **directly similar** or that W_2 is obtained from W_1 by an **elementary transformation**, and write $W_1 \to W_2$, if

$W_1 = (a_1, \ldots, a_n)$, $a_k \perp a_{k+1}$, $W_2 = (a_1, \ldots, a_{k-1}, a_k \oplus a_{k+1}, a_{k+2}, \ldots, a_n)$. Clearly, $|W_2| = |W_1| - 1$.

For two words W^1, W^2, define $W^1 \sim W^2$ if there is a finite sequence of words $W_0 := W^1, W_1, W_2, \ldots, W_{m-1}, W_m := W^2$ such that, for $0 \le i \le m-1$ we have either $W_i = W_{i+1}$, or $W_i \to W_{i+1}$, or $W_{i+1} \to W_i$. In this case, W^1 and W^2 are said to be **related by an elementary chain of length** m. The relation $\sim$ is clearly an equivalence, and we will denote by $\mathcal{G}^+$ the collection of the equivalence classes on $\mathcal{W}(E)$.

Define the binary operation $+$ in $\mathcal{G}^+$ by $[W^1]+[W^2] = [W^1+W^2]$. This operation is well-defined; if $W^1 \sim W^2$ and $W^3 \sim W^4$, then $W^1+W^3 \sim W^2+W^4$, as can be seen by a successive application of elementary transformations in W^1 and W^3. Thus $\mathcal{G}^+$ is the quotient semigroup $\mathcal{W}(E)/\!\sim$ consisting of all equivalence classes with the binary operation $+$. Since $\mathcal{W}(E)$ is associative, so is $\mathcal{G}^+$. Given $(a_1, \ldots, a_n) \in \mathcal{W}(E)$, we have $(a_1, \ldots, a_n) = (a_1)+(a_2)+\cdots+(a_n)$, consequently $[(a_1, \ldots, a_n)] = [a_1]+\cdots+[a_n]$ in $\mathcal{G}^+$.

In what follows, we need some definitions and lemmas from [Bae].

Given a word W, we inductively define a subset of E, written $S(W)$, as follows.

(i) If $n = 1$, define $S(W) = \{a_1\}$.

(ii) If $n > 1$, then $a \in S(W)$ if and only if there are $i \in \mathbb{N}$ and $b, c \in E$ such that $1 \le i \le n-1$, $b \in S(a_1, \ldots, a_i)$, $c \in S(a_{i+1}, \ldots, a_n)$, and $a = b \oplus c$.

Clearly, that $S(W)$ is finite for any word W and may be void. Note that in [Bae], Baer defined an **add** A to be a nonempty set with a partial binary operation $+$. He formed **vectors** $(a_1, \ldots, a_n)$, $a_1, \ldots, a_n \in A$, and introduced an equivalence relation $\sim$ on the collection $V(A)$ of vectors using the $+$. A (generalized) effect algebra is an add and thus Baer's statements on adds are true if translated to the case of a generalized effect algebra (our "words" are just "vectors" in Baer's terminology). We omit the straightforward but lengthy proof of the following statements. The details can be found in [Rav].

Lemma 1.7.6 [Bae] *Let $W \in \mathcal{W}(E)$ with $|W| = n > 1$.*

(i) *If $a \in S(W)$, then there exists i with $1 \le i \le n-1$ and a pair a_i, a_{i+1} in W such that $a_i \oplus a_{i+1}$ exists and $a \in S(W_1)$, where W_1 is the word obtained from W by replacing a_i, a_{i+1} with $a_i \oplus a_{i+1}$.*

(ii) *If there exists a pair a_j, a_{j+1} in W such that $a_j \oplus a_{j+1}$ exists for some $j \in \mathbb{N}$ with $1 \le j \le n-1$, then $S(W_2) \subseteq S(W)$, where W_2 is the word obtained from W by replacing a_j, a_{j+1} with $a_j \oplus a_{j+1}$.*

Proposition 1.7.7 [Bae] *The following are equivalent in an add A.*

(i) *For $a \in A$ and $W \in V(A)$, $(a) \sim W$ implies $a \in S(W)$.*

(ii) *For* $a \in A$ *and* $W \in V(A)$, $(a) \sim W$ *implies either* $(a) = W$ *or that there exists a pair* a_i, a_{i+1} *in* W *such that* $a_i \oplus a_{i+1}$ *exists.*

(iii) *Given* W_1, $W_2 \in V(A)$, $W_1 \sim W_2$ *and* $S(W_1) \neq \emptyset$ *imply that* $S(W_2)$ *consists of single element.*

(iv) *For any* $W \in V(A)$, $S(W)$ *contains at most one element, and for* $W_1, W_2 \in V(A)$ *such that* $W_1 \to W_2$, *if* $S(W_1) \neq \emptyset$, *then* $S(W_2) \neq \emptyset$.

(v) *For* $W_1, W_2 \in V(A)$, $W_1 \sim W_2$ *implies* $S(W_1) = S(W_2)$.

Condition (v) is what Baer called "the strong associative law". It turns out that the associative law in generalized effect algebras is even stronger and enables us to prove that all conditions in Proposition 1.7.7 are fulfilled for them.

Proposition 1.7.8 *Let* E *be a generalized effect algebra and* $W \in \mathcal{W}(E)$.

(i) *If the orthogonal sum* a *of the components of* W *exists, then* $a \in S(W)$.

(ii) *If* $S(W) \neq \emptyset$, *then the orthogonal sum* a *of the components of* W *exists, and* $S(W) = \{a\}$.

(iii) *Conditions* (i)–(v) *from Proposition* 1.7.7 *are satisfied.*

Proof. (i) We will proceed by induction on the length of words.

If $|W| = 1$, then $S(W)$ contains the one-element orthogonal sum a, where $W = (a)$.

Suppose that (i) holds for all words of length less that k. Let $W = (a_1, a_2, \dots, a_k)$ be a word of length $k \geq 2$ such that $a_1 \oplus \cdots \oplus a_k$ exists. Then $a_1 \oplus \cdots \oplus a_{k-1}$ exists. By the induction hypothesis, $a_1 \oplus \cdots \oplus a_{k-1} \in S(a_1, \dots, a_{k-1})$. Since $a_k \in S(a_k)$, by the definition of $S(a_1, \dots, a_k)$, we have $a_1 \oplus \cdots \oplus a_k \in S(a_1, \dots, a_k)$. This completes the proof of (i).

(ii) Assume $S(W) \neq \emptyset$. We use again the induction on the length of words. If $|W| = 1$, there is nothing to prove. Suppose (ii) has been proved for all words of length less than k. Let $W = (c_1, \dots, c_k)$ be a word of length $k \geq 2$ such that $S(W) \neq \emptyset$. Then by Lemma 1.7.6, there exists $1 \leq i \leq k-1$ such that $c_i \oplus c_{i+1}$ exists, and there is $g_1 \in S(W_1)$, where $W_1 = (c_1, \dots, c_{i-1}, c_i \oplus c_{i+1}, c_{i+2}, \dots, c_k)$. Since W_1 has length $k-1 < k$, we get by the induction hypothesis, that $g_1 = c_1 \oplus \cdots \oplus c_{i-1} \oplus (c_i \oplus c_{i+1}) \oplus \cdots \oplus c_k$. It follows from the associativity of $\oplus$, that $g_1 = c_1 \oplus \cdots \oplus c_k$. This completes the proof of (ii).

(iii) It suffices to prove condition (iv) of Proposition 1.7.7. Let $W \in \mathcal{W}(E)$. By (ii), either $S(W) = \emptyset$, or $S(W)$ contains a single element. Suppose W_1, W_2 are two words such that $W_1 \to W_2$ and $S(W_1) \neq \emptyset$. Write $W_1 = (a_1, \dots, a_n)$. It then follows that $|W_1| \geq 2$, $W_2 = (a_1, \dots, a_i \oplus a_{i+1}, \dots, a_n)$, for some $1 \leq i \leq n-1$, and $|W_2| = |W_1| - 1$. By (ii), $S(W_1) = a_1 \oplus \cdots \oplus a_n$. We have $g := a_1 \oplus \cdots \oplus a_n = a_1 \oplus \cdots \oplus a_{i-1} \oplus (a_i \oplus a_{i+1}) \oplus a_{i+2} \oplus \cdots \oplus a_n = g_1 \oplus g_2$, where $g_1 = a_1 \oplus \cdots \oplus a_i \oplus a_{i+1}$,

$g_2 = a_{i+2} \oplus \cdots \oplus a_n$. By (ii), $g_1 \in S(a_1, \ldots, (a_i \oplus a_{i+1}))$ and $g_2 \in S(a_{i+2}, \ldots, a_n)$. Therefore, by definition of $S(W_2)$, $g \in S(W_2)$ so that $S(W_2) \neq \emptyset$. This completes the proof. □

It is useful to reformulate condition (ii) in Proposition 1.7.7 in the following form.

Corollary 1.7.9 *$(a_1, \ldots, a_n) \sim (b)$ if and only if $a_1 \oplus \cdots \oplus a_n$ exists and equals b. In particular,* (i) *if $b = 0$, then $a_1 = a_2 = \cdots = a_n = 0$, and* (ii) *if $[a] = [b]$, then $a = b$.*

In the next statement, the Riesz decomposition property can be replaced by a weaker Mackey compatibility property: Two elements x, y in a generalized effect algebra E are Mackey compatible, if there are elements $x_1, y_1, z \in E$ such that $x_1 \oplus y_1 \oplus z$ exists and $x = x_1 \oplus z$, $y = y_1 \oplus z$. E is said to satisfy **the Mackey property** if any two elements $x, y \in E$ are Mackey compatible. It is easy to verify that in any upward directed generalized effect algebra, the Riesz decomposition property implies the Mackey property. Namely, by the upward directedness, for any $a, b \in E$, there is $c \in E$ with $a, b \leq c$, hence $c = a \oplus a_1 = b \oplus b_1$. By the Riesz decomposition property, there is $\{w_{ij} : i, j = 1, 2\}$ in E such that $a = w_{11} \oplus w_{12}$, $a_1 = w_{21} \oplus w_{22}$ and $b = w_{11} \oplus w_{21}$, $b_1 = w_{12} \oplus w_{22}$. Since $w_{11} \oplus w_{12} = a$, $w_{21} \leq a_1$, it follows that $w_{11} \oplus w_{12} \oplus w_{21}$ exists, and gives the Mackey decomposition of a, b.

Lemma 1.7.10 *Let E be a generalized effect algebra satisfying the Mackey property. Then $[a] + [b] = [b] + [a]$. Thus, given a word W, if W' is any word obtained by permuting the components of W, then $W \sim W'$.*

Proof. Let $a, b \in E$. Owing to Mackey property, there are a_1, b_1, c, such that $a_1 \oplus b_1 \oplus c$ exists, and $a = a_1 \oplus c$, $b = b_1 \oplus c$. Now $[a] + [b] = [(a, b)]$. Clearly, $(a_1, c) \to (a)$, $(b_1, c) \to (b)$. By commutativity of $\oplus$, we also have $(c, a_1) \to (a)$, $(c, b_1) \to (b)$. Therefore $(a, b) \sim (c, a_1, b_1, c)$. Further, $(c, a_1, b_1, c) \to (c, a_1 \oplus b_1, c) = (c, b_1 \oplus a_1, c)$. Hence $(a, b) \sim (c, b_1 \oplus a_1, c) \sim (c, b_1, a_1, c) \sim (c \oplus b_1, a_1 \oplus c) = (b, a)$. Finally, $[a] + [b] = [(a, b)] = [(b, a)] = [b] + [a]$. □

Lemma 1.7.11 *Let E be a generalized effect algebra with the Riesz decomposition property. Suppose there exist words $(a_1, \ldots, a_n), (b_1, \ldots, b_m), W \in \mathcal{W}(E)$ such that*

$$(a_1, \ldots, a_n) + (b_1, \ldots, b_m) \sim W, \ |W| = k.$$

Then there exist elements $d_1, \ldots, d_k$, $e_1, \ldots, e_k \in E$ such that $W = (d_1 \oplus e_1, \ldots, d_k \oplus e_k)$, $(a_1, \ldots, a_n) \sim (d_1, \ldots, d_k)$, and $(b_1, \ldots, b_m) \sim (e_1, \ldots, e_k)$.

Proof. Put $W_0 := (a_1, \ldots, a_n) + (b_1, \ldots, b_m)$. Since $W_0 \sim W$, there exists an elementary chain linking W_0 and W. The proof is by induction on the minimal length l of all elementary chains linking W_0 and W.

If $l = 1$, then $W_0 = W$. That is, $(a_1, \dots, a_n) + (b_1, \dots, b_m) = W$, and $m + n = k$. Take

$$\begin{aligned} d_i &= a_i,\ 1 \le i \le n,\ d_{n+1} = d_{n+2} = 0, \dots, d_k = 0, \\ e_i &= 0,\ 1 \le i \le n,\ e_{n+1} = b_1, e_{n+2} = b_2, \dots, e_k = b_m. \end{aligned}$$

Assume the statement to be true for all words W^1 such that $W_0 \sim W^1$ and such that an elementary chain of length less than l links W_0 and W^1.

Let $W = (c_1, \dots, c_k)$ be a word linked with W_0 by an elementary chain $W_0, W_1, \dots,$ $W_l := W$ of length l. We may assume that $W_{l-1} \ne W$ so that either $W_{l-1} \to W$ or $W \to W_{l-1}$. These two cases have to be considered.

(1) $W_{l-1} \to W$. In this case, there exist elements c_{1j}, c_{2j} such that $c_{1j} \oplus c_{2j}$ exists and equals c_j for some j, and W_{l-1} is of the form

$$(c_1, \dots, c_{j-1}, c_{1j}, c_{2j}, c_{j+1}, \dots, c_k).$$

Since W_0 and W_{l-1} are connected by an elementary chain of length $l - 1$, by the induction hypothesis, there are

$$d_1, \dots, d_{j-1}, d_{1j}, d_{2j}, d_{j+1}, \dots, d_k$$

and

$$e_1, \dots, e_{j-1}, e_{1j}, e_{2j}, e_{j+1}, \dots, e_k$$

such that $c_i = d_i \oplus e_i$ for $1 \le i \le j - 1$ and $j + 1 \le i \le k$, and $c_{1j} = d_{1j} \oplus e_{1j}$, $c_{2j} = d_{2j} \oplus e_{2j}$, where

$$\begin{aligned} (a_1, \dots, a_n) &\sim (d_1, \dots, d_{j-1}, d_{1j}, d_{2j}, d_{j+1}, \dots, d_k) \\ (b_1, \dots, b_m) &\sim (e_1, \dots, e_{j-1}, e_{1j}, e_{2j}, e_{j+1}, \dots, e_k). \end{aligned}$$

Since $c_{1j} \oplus c_{2j}$ exists, $(d_{1j} \oplus e_{1j}) \oplus (d_{2j} \oplus e_{2j})$ exists. By the associativity and commutativity of $\oplus$, we get $c_j = c_{1j} \oplus c_{2j} = (d_{1j} \oplus d_{2j}) \oplus (e_{1j} \oplus e_{2j})$. Write $d_j := d_{1j} \oplus d_{2j}$, $e_j := e_{1j} \oplus e_{2j}$. Then $c_j = d_j \oplus e_j$. Moreover,

$$\begin{aligned} (a_1, \dots, a_n) &\sim (d_1, \dots, d_{j-1}, d_{1j}, d_{2j}, d_{j+1}, \dots, d_k) \\ &\sim (d_1, \dots, d_{j-1}, d_{1j} \oplus d_{2j}, d_{j+j}, \dots, d_k) \\ &= (d_1, \dots, d_{j-1}, d_j, d_{j+1}, \dots, d_k) \end{aligned}$$

and, similarly, $(b_1, \dots, b_m) \sim (e_1, \dots, e_k)$. This completes the proof of case (1).

(2) $W \to W_{l-1}$. In this case, there exists i such that $c_i \oplus c_{i+1}$ exists and

$$W_{l-1} = (c_1, c_2, \dots, c_{i-1}, c_i \oplus c_{i+1}, c_{i+2}, \dots, c_k).$$

Applying the induction hypothesis to the elementary chain of length $l-1$ linking W_0 and W_{l-1}, we get elements

$$d_1, \ldots, d_{i-1}, d, d_{i+2}, \ldots, d_k$$

and

$$e_1, \ldots, e_{i-1}, e, e_{i+2}, \ldots, e_k$$

such that $c_j = d_j \oplus e_j$ for $1 \leq j \leq i-1$ and $i+2 \leq j \leq k$, and $c_i \oplus c_{i+1} = d \oplus e$,

$$(a_1, \ldots, a_n) \sim (d_1, \ldots, d_{i-1}, d, d_{i+2}, \ldots, d_k)$$

and

$$(b_1, \ldots, b_m) \sim (e_1, \ldots, e_{i-1}, e, e_{i+2}, \ldots, e_k).$$

Since $c_i \oplus c_{i+1} = d \oplus e$, by the Riesz decomposition property, there are elements $f_{11}, f_{12}, f_{21}, f_{22}$ such that $c_i = f_{11} \oplus f_{12}$, $c_{i+1} = f_{21} \oplus f_{22}$, $d = f_{11} \oplus f_{21}$, and $e = f_{12} \oplus f_{22}$. Write $d_i := f_{11}$, $d_{i+1} := f_{21}$, $e_i := f_{12}$, $e_{i+1} := f_{22}$. Then

$$\begin{aligned} W &= (c_1, \ldots, c_{i-1}, c_i, c_{i+1}, \ldots, c_k) \\ &= (d_1 \oplus e_1, \ldots, d_{i-1} \oplus e_{i-1}, d_i \oplus e_i, d_{i+1} \oplus e_{i+1}, \ldots, d_k \oplus e_k). \end{aligned}$$

Moreover, it is clear that $(a_1, \ldots, a_n) \sim (d_1, \ldots, d_k)$ and $(b_1, \ldots, b_m) \sim (e_1, \ldots, e_k)$. This completes the proof of case (2). □

By Birkhoff [Bir], a cancellative, positive Abelian monoid M is the generating positive cone of a partially ordered Abelian group. In the next theorem, we will prove that the collection of the equivalence classes $\mathcal{G}^+$ of words for an upward directed generalized effect algebra with the Riesz decomposition property is a generating positive cone of a partially ordered Abelian group $\mathcal{G}$.

Theorem 1.7.12 *Let E be an upward directed generalized effect algebra with the Riesz decomposition property. Then $\mathcal{G}^+$ is a cancellative, positive Abelian monoid. Consequently, $\mathcal{G}^+$ is the generating positive cone of a partially ordered Abelian group $\mathcal{G}$.*

Proof. Since $\mathcal{G}^+$ is an Abelian (Lemma 1.7.11) semigroup, we have only to check the existence of the identity, positivity and cancellativity. It is straightforward that $[0]$ is the additive identity in $\mathcal{G}^+$. Suppose $[W_1] + [W_2] = [0]$, $W_1 = (a_1, \ldots, a_n)$, $W_2 = (b_1, \ldots, b_m)$. Then $[W_1 + W_2] = [0]$, i.e. $(a_1, \ldots, a_n, b_1, \ldots, b_m) \sim 0$, so by Corollary 1.7.9, $a_i = b_j = 0$ for all $i = 1, \ldots, n$, $j = 1, \ldots, m$. This shows positivity of $\mathcal{G}^+$.

It remains to prove cancellativity. So let $[W], [W_1], [W_2] \in \mathcal{G}^+$ and $[W] + [W_1] = [W] + [W_2]$. We may assume that $W = (a)$. For, if $W = (a_1, \ldots, a_n)$, then $[W] =$

$[a_1] + \cdots + [a_n]$, and the $[a_i]$'s can be cancelled successively. Write $W_1 = (b_1, \ldots, b_m)$ and $W_2 = (c_1, \ldots, c_k)$. Then $[W] + [W_1] = [W] + [W_2]$ implies that $(a, b_1, \ldots, b_m) \sim (a, c_1, \ldots, c_k)$. By Lemma 1.7.11, there exist elements $d, e, d_1, \ldots d_k, e_1, \ldots, e_k$ with $a = d \oplus e$ and $c_i = d_i \oplus e_i$, $1 \leq i \leq k$, such that $a \sim (d, d_1, \ldots, d_k)$ and $(b_1, \ldots, b_m) \sim (e, e_1, \ldots, e_k)$. By Corollary 1.7.9, we have $a = d \oplus d_1 \oplus \cdots \oplus d_k$ so since $a = d \oplus e$ as well, we have $e = d_1 \oplus \cdots \oplus d_k$ by the cancellativity of $\oplus$. This implies that

$$\begin{aligned}(b_1, \ldots, b_m) &\sim (e, e_1, \ldots, e_k)\\ &= (d_1 \oplus \cdots \oplus d_k, e_1, \ldots, e_k) \sim (d_1, \ldots, d_k, e_1, \ldots, e_k),\end{aligned}$$

so using Lemma 1.7.10, we conclude that

$$\begin{aligned}(b_1, \ldots, b_m) &\sim (d_1, e_1, d_2, e_2, \ldots, d_k, e_k)\\ &\sim (d_1 \oplus e_1, \ldots, d_k \oplus e_k) = (c_1, \ldots, c_k).\end{aligned}$$

Consequently, $[W_1] = [W_2]$. □

Recall that a subset S of a partially ordered group G is **convex** if $x, y \in S$, $x \leq z \leq y$ imply $z \in S$.

Theorem 1.7.13 *Let E be an upward directed generalized effect algebra with the Riesz decomposition property. Then there is a partially ordered Abelian group $(\mathcal{G}, \mathcal{G}^+)$ and a monomorphism $\Phi : E \to \mathcal{G}^+$ such that $\Phi(E)$ is a convex subset of $\mathcal{G}^+$ that generates $\mathcal{G}^+$ (in the sense that to every element $g \in \mathcal{G}^+$, there are $x_1, \ldots, x_n \in \Phi(E)$ such that $g = x_1 + \cdots + x_n$, i.e., $\mathcal{G}^+ = ssg(\Phi(E))$). In particular, every effect algebra with the Riesz decomposition property is an interval effect algebra.*

Proof. Let $\mathcal{G}$ be the partially ordered Abelian group as given in Theorem 1.7.12. Define the map $\Phi : E \to \mathcal{G}^+$ by $\Phi(a) = [a]$. By Corollary 1.7.9 (ii), Φ is an injection. Moreover, $\Phi(E)$ generates $\mathcal{G}^+$.

Assume that $a \leq b$. Then there is $c \in E$ such that $b = a \oplus c$. This implies $(a, c) \sim (b)$, i.e., $[a] + [c] = [b]$, hence $[a] \leq [b]$. So Φ preserves order.

Let $v \in E$ and let $[(a_1, \ldots, a_n)] \in \mathcal{G}^+$ be such that $[(a_1, \ldots, a_n)] \leq [v]$. Then there exists $[(b_1, \ldots, b_m)] \in \mathcal{G}^+$ such that $[a_1, \ldots, a_n] + [b_1, \ldots, b_m] = [v]$. By Corollary 1.7.9, $a_1 \oplus \cdots \oplus a_n \oplus b_1 \oplus \cdots \oplus b_m$ exists in E and equals v. It follows that Φ is a monomorphism. In particular, $a_1 \oplus \cdots \oplus a_n$ exists in E. Thus $\Phi(E)$ is a convex subset of $\mathcal{G}^+$, and Φ is an isomorphism of E onto $\Phi(E)$.

In particular, if E is an effect algebra with unit u, then $0 \leq a \leq u$ for any $a \in E$ implies that $\Phi(E) \subseteq \mathcal{G}^+[[0], [u]]$. Observe that if $u \neq 0$, then $[u] \neq [0]$ (Corollary 1.7.9 (i)). Repeating the arguments in the preceding paragraph with v replaced by u, we get that $\Phi(E) = \mathcal{G}^+[[0], [u]]$. □

Theorem 1.7.14 *Let E be an upward directed generalized effect algebra with the Riesz decomposition property. Then the group $\mathcal{G}$ constructed in Theorem* 1.7.12 *has the interpolation property.*

Proof. Assume that $[W],[W_1],[W_2] \in \mathcal{G}^+$ are such that $[W] \leq [W_1]+[W_2]$. We intend to prove that there are $[W_1^1],[W_2^2] \in \mathcal{G}^+$ such that $[W_1^1] \leq [W_1]$, $[W_2^2] \leq [W_2]$ and $[W] = [W_1^1]+[W_2^2]$.

The proof is by induction on the length of W. If the length is 1, i.e., $W = a$ for some $a \in E$, then $[W] = [a]$. Write $W_1 = (b_1,\ldots,b_m)$, $W_2 = (c_1,\ldots,c_n)$. Now there exists a word $(d_1,\ldots,d_k)$ such that $[a]+[(d_1,\ldots,d_k)] = [(b_1,\ldots,b_m)] + [(c_1,\ldots,c_n)]$, that is, $[(a,d_1,\ldots,d_k)] = [(b_1,\ldots,b_m,c_1,\ldots,c_n)]$. Using Lemma 1.7.11, we get elements $e, f, e_1,\ldots,e_l, f_1,\ldots,f_l$ such that

$$a = e \oplus f, d_i = e_i \oplus f_i,\ 1 \leq i \leq l,$$
$$(e,e_1,\ldots,e_l) \sim (b_1,\ldots,b_m),\ (f,f_1,\ldots,f_l) \sim (c_1,\ldots,c_n).$$

This implies that $[e] \leq [(b_1,\ldots,b_m)] = [W_1]$ and $[f] \leq [(c_1,\ldots,c_n)] = [W_2]$. Since $[W] = [a] = [e \oplus f] = [e]+[f]$, the desired decomposition is $W_1^1 := e$, and $W_2^2 = f$.

Assume that we have the desired decomposition for all words W of length less that k. Let $W_3 = (e_1,\ldots,e_k)$ be a word of length k such that $[W_3] \leq [W_1]+[W_2]$. Then $[W_3] = [W_4]+[e_k]$, where $W_4 = (e_1,\ldots,e_{k-1})$ is a word of length $k-1$ and we have $[W_4] \leq [W_1]+[W_2]$. By induction hypothesis, there are $[W_4^1],[W_4^2] \in \mathcal{G}^+$ such that $[W_4] = [W_4^1]+[W_4^2]$ with $[W_4^1] \leq [W_1]$, $[W_4^2] \leq [W_2]$. This implies that both $[W_1]-[W_4^1]$ and $[W_2]-[W_4^2]$ are in $\mathcal{G}^+$. Since $[e_k] \leq ([W_1]-[W_4^1])+([W_2]-[W_4^2])$, the step for length 1 implies that $[e_k] = [p_k]+[q_k]$ for some $[p_k] \leq [W_1]-[W_4^1]$, $[q_k] \leq [W_2]-[W_4^2]$. Therefore $[W_4^1]+[p_k] \leq [W_1]$ and $[W_4^2]+[q_k] \leq [W_2]$. Thus $[W_3] = ([W_4^1]+[p_k])+([W_4^2]+[q_k])$ gives the desired decomposition. The proof is complete. □

Observe that a generalized effect algebra E with the Riesz property is embedded into $\mathcal{G}^+$ as a convex subset. Indeed, $y \in E$, $z \leq y$ imply $z \leq y \oplus 0$. By the Riesz property, there is $y_1 \in E$, $y_1 \leq y$ such that $z = y_1 \oplus 0$, hence $z = y_1 \in E$.

Conversely, every convex subset $K \subseteq G^+$ in a partially ordered interpolation group G containing 0 can be organized into a generalized effect algebra with the Riesz decomposition property if we define for $a, b \in K$, $a \oplus b$ to be defined if and only if $a+b \in K$ in which case $a \oplus b = a+b$.

Proposition 1.7.15 *Let G be a directed partially ordered Abelian group that has the Riesz decomposition property. Let $H \subseteq G^+$ be such that $0 \in H$, H is upward directed and convex, and $G^+ = ssg(H)$. Let K be any Abelian group. Then every K-valued measure $\psi : H \to K$ can be uniquely extended to a group homomorphism $\psi^* : G \to K$. If, moreover, K is partially ordered and ψ maps H to K^+, then $\psi^*(G^+) \subseteq K^+$, hence ψ^* is a positive group homomorphism.*

Proof. Let $x \in G^+$. Then there are $x_1, \dots, x_n \in H$ with $x = \sum_{i=1}^n x_i$. Define $\psi^*(x) := \sum_{i=1}^n \psi(x)$. Then $\psi^* : G^+ \to K$ is a well defined additive map. Indeed, let $y_1, \dots, y_m$ be any elements of H such that $x = \sum_{j=1}^m y_j$. Owing to the Riesz decomposition property of G, there are w_{ij} in G^+ such that $x_i = \sum_{j=1}^m w_{ij}$, $1 \le i \le n$, and $y_j = \sum_{i=1}^n w_{ij}$, $0 \le j \le m$. Since H is convex and $0 \le w_{ij} \le x_i, y_j$, $x_i, y_j \in H$, we have $w_{ij} \in H$ for all i, j. Since $\psi : H \to K$ is a K-valued measure, we obtain

$$\psi^*(x) = \sum_{i=1}^n \psi(x_i) = \sum_{i=1}^n \sum_{j=1}^m \psi(w_{ij}) = \sum_{j=1}^m \sum_{i=1}^n \psi(w_{ij}) = \sum_{j=1}^m \psi(y_j).$$

Additivity of ψ^* on G^+ is clear. Since G is directed, any $z \in G$ is of the form $z = x - y$ with $x, y \in G^+$. Define $\psi^*(z) = \psi^*(x) - \psi^*(y)$. Now if x_1, y_1 are another pair of elements in G^+ such that $z = x_1 - y_1$, then $x + y_1 = x_1 + y \in G^+$, hence $\psi^*(x) + \psi^*(y_1) = \psi^*(x_1) + \psi^*(y)$, whence $\psi^*(x) - \psi^*(y) = \psi^*(x_1) - \psi^*(y_1)$. So $\psi^* : G \to K$ is well defined, and additivity of ψ^* is clear.

Finally, assume that K is partially ordered and $\psi(G^+[0,u]) \subseteq K^+$. For every $x \in G^+$, there are $x_1, \dots, x_n$ in H such that $g = x_1 + \dots + x_n$. Therefore $\psi^*(g) = \psi(x_1) + \dots + \psi(x_n) \in K^+$. Hence $\psi^*(G^+) \subseteq K^+$. □

Proposition 1.7.16 *Let G be a partially ordered Abelian group that has the Riesz decomposition property. Let u be an order unit for G. Then G is the universal group for the effect algebra $G^+[0,u]$.*

Proof. Since G has an order unit, G is directed. Moreover, $G^+[0,u]$ is a convex upward directed subset of G^+. It remains to prove that $G^+ = ssg(G^+[0,u])$, then we may apply Proposition 1.7.15. Let $x \in G^+$; since u is an order unit, there is $n \ge 1$ such that $x \le nu$. By the Riesz decomposition property, there are $x_i, i = 1, \dots, n$ with $x_i \le u$, and $x = x_1 + \dots + x_n$, hence $x_i \in G^+[0,u]$, and $x \in ssg(G^+[0,u])$.

The proof is concluded. □

Theorem 1.7.17 (a) *Let E be an effect algebra that has the Riesz decomposition property. Then there exists an order unit u in the universal group G for E such that E is isomorphic to the interval $G^+[0,u]$, and G is an interpolation group.*

(b) *Conversely, if G is an interpolation group with an order unit u, then $G^+[0,u]$ is an interval affect algebra that has the Riesz decomposition property and G is its universal group.*

(c) *There is a categorical isomorphism between effect algebras with the Riesz decomposition property and interpolation groups with order unit.*

Proof. (a) It follows from Theorems 1.7.13 and 1.7.14.

(b) It follows by Proposition 1.7.16.

(c) Let P, Q be effect algebras with the Riesz decomposition property. Let $(\mathcal{G}_P, u)$, $(\mathcal{G}_Q, v)$ be their universal groups, respectively, and $\psi : P \to Q$ be an effect algebras morphism. Then ψ can be considered as a $\mathcal{G}_Q$-measure on P, therefore by Proposition 1.7.15, it can be uniquely extended to a positive group homomorphism $\psi^* : \mathcal{G}_P \to \mathcal{G}_Q$. Clearly, $\psi^*(u) = \psi(u) = v$, therefore ψ^* is a positive homomorphism of order unit groups.

Conversely, let (G, u) and (H, v) be interpolation groups with order units. Assume that $\kappa : G \to H$ is a positive homomorphism such that $\kappa(u) = v$. Clearly, the restriction of κ to $G^+[0, u]$ is an effect algebra morphism to $H^+[0, v]$. □

1.8 D-lattices and Lattice-ordered Effect Algebras

Recall that a difference poset which is a lattice is called a **difference lattice**. A **generalized difference lattice (GDL)** is a lattice ordered generalized difference poset. Fortunately, we have the following.

Proposition 1.8.1 *Any lattice with a difference is a generalized difference lattice.*

Proof. We have to prove that (1) L contains 0 and (2) condition (C) is satisfied. (1) follows by the observation after Proposition 1.1.2. (2) follows by Lemma 1.1.5. □

We say that a (generalized) effect algebra is lattice ordered if it is a lattice with respect to the partial ordering induced by the operation $\oplus$.

Using (S) and (D), it is easily seen that there is a one-to-one correspondence between lattices with difference and lattice ordered generalized effect algebras, and between D-lattices and lattice ordered effect algebras.

The following statements have been proved in [ChKo 1].

Proposition 1.8.2 (1) *Let $(P; \leq, \ominus)$ be a poset with a difference, $a, b, c \in P$, $a, b \leq c$. If $a \vee b$ exists in P, then $(c \ominus a) \wedge (c \ominus b)$ exists in P, and*

$$c \ominus (a \vee b) = (c \ominus a) \wedge (c \ominus b).$$

In particular, if $a \perp b$, then

$$(a \oplus b) \ominus (a \vee b) = a \wedge b.$$

(2) *Let $(P; \leq, \ominus)$ be a lattice with a difference, $a, b, c \in P$, $a \leq c$, $b \leq c$. Then*

$$c \ominus (a \wedge b) = (c \ominus a) \vee (c \ominus b).$$

In particular, if we put $c = a \vee b$,

$$(a \vee b) \ominus (a \wedge b) = ((a \vee b) \ominus a) \vee ((a \vee b) \ominus b).$$

(3) *In a lattice with a difference for* $c \leq a, b$,

$$(a \wedge b) \ominus c = (a \ominus c) \wedge (b \ominus c).$$

Proof. (1) From the inequalities $a \leq a \vee b \leq c$, $b \leq a \vee b \leq c$ and (D3) we have $c \ominus (a \vee b) \leq c \ominus a$ and $c \ominus (a \vee b) \leq c \ominus b$. For any other $w \in P$ with $w \leq c \ominus a$, $w \leq c \ominus b$, $a = c \ominus (c \ominus a) \leq c \ominus w$, $b \leq c \ominus (c \ominus b) \leq c \ominus w$, therefore, $a \vee b \leq c \ominus w \leq c$, and so $w = c \ominus (c \ominus w) \leq c \ominus (a \vee b)$, which implies that $c \ominus (a \vee b)$ is the greatest lower bound of the set $\{c \ominus a, c \ominus b\}$.

(2) From the inequalities $a \wedge b \leq a \leq c$ and $a \wedge b \leq b \leq c$ it follows that $c \ominus a \leq c \ominus (a \wedge b)$ and $c \ominus b \leq c \ominus (a \wedge b)$. For $w \in P$ with $c \ominus a \leq w$, $c \ominus b \leq w$, then $c \ominus a = (c \ominus a) \wedge c \leq w \wedge c \leq c$, which gives $c \ominus (w \wedge c) \leq a$, and similarly $c \ominus (w \wedge c) \leq b$, therefore, $c \ominus (w \wedge c) \leq a \wedge b$. Then we obtain $c \ominus (a \wedge b) \leq w \wedge c \leq w$ which implies that $c \ominus (a \wedge b)$ is the least upper bound of the set $\{c \ominus a, c \ominus b\}$.

(3) From $c \leq a \wedge b \leq a, b$ it follows that

$$(a \wedge b) \ominus c \leq (a \ominus c) \wedge (b \ominus c) \leq a \wedge b.$$

If $w \in P$ is such that $w \leq a \ominus c, b \ominus c$, then $w \oplus c \leq a, b$, hence $w \leq (a \wedge b) \ominus c$. Hence $(a \wedge b) \ominus c$ is the greatest lower bound of $\{a \ominus c, b \ominus c\}$, which concludes the proof. □

The following two propositions were proved in [ChKo 1] for D-posets (D-lattices), but they hold true also in more general situations.

Proposition 1.8.3 *Let P be a poset with difference having a smallest element 0. If* $a \vee b \in P$, *then* $((a \vee b) \ominus a) \wedge ((a \vee b) \ominus b) \in P$ *and* $((a \vee b) \ominus a) \wedge ((a \vee b) \ominus b) = 0$.

Proof. In Proposition 1.8.2 (1) put $c = a \vee b$. □

Proposition 1.8.4 *Let P be a lattice with difference,* $c \leq a$, $c \leq b$. *Then* $(a \ominus c) \vee (b \ominus c) = (a \vee b) \ominus c$.

Proof. From $c \leq a \leq a \vee b$, $c \leq b \leq a \vee b$ we get $a \ominus c \leq (a \vee b) \ominus c$, $b \ominus c \leq (a \vee b) \ominus c$. Let $w \in P$ be such that $a \ominus c$, $b \ominus c \leq w$. Then $a \ominus c \leq w \wedge ((a \vee b) \ominus c) \leq (a \vee b) \ominus c$ imply $((a \vee b) \ominus c) \ominus (w \wedge ((a \vee b) \ominus c)) \leq ((a \vee b) \ominus c) \ominus (a \ominus c) = ((a \vee b) \ominus a)$; and similarly, $((a \vee b) \ominus c) \ominus (w \wedge ((a \vee b) \ominus c)) \leq ((a \vee b) \ominus b)$. Therefore $((a \vee b) \ominus c) \ominus (w \wedge ((a \vee b) \ominus c)) \leq ((a \vee b) \ominus a) \wedge ((a \vee b) \ominus b) = 0$ by Proposition 1.8.3. Hence $(a \vee b) \ominus c \leq w$, which gives the desired result. □

Corollary 1.8.5 *Let P be a lattice with difference. Then* $(a \ominus (a \wedge b)) \wedge (b \ominus (a \wedge b)) = 0$.

Proof. In Proposition 1.8.2 (3) put $c = a \wedge b$. □

From the above properties, the following statement can be derived.

Proposition 1.8.6 *Let P be a lattice with difference. If $x \oplus y$ and $x \oplus z$ exist, then*

(1) $x \oplus (y \wedge z) = (x \oplus y) \wedge (x \oplus z)$.

(2) $x \oplus (y \vee z) = (x \oplus y) \vee (x \oplus z)$.

Proof. (1) By Proposition 1.8.2 (3), $((x \oplus y) \wedge (x \oplus z)) \ominus x = ((x \oplus y) \ominus x) \wedge ((x \oplus z) \ominus x) = y \wedge z$, whence by (S), $(x \oplus y) \wedge (x \oplus z) = x \oplus (y \wedge z)$.

(2) By Proposition 1.8.4 $((x \oplus y) \vee (x \oplus z)) \ominus x = ((x \oplus y) \ominus x) \vee ((x \oplus z) \ominus x) = y \vee z$, whence by (S), $(x \oplus y) \vee (x \oplus z) = x \oplus (y \vee z)$. □

According to the next proposition, $\oplus$ distributes over existing suprema [BeFo].

Proposition 1.8.7 *Let $\{b_\alpha\}_\alpha$ be a family of elements in an effect algebra E, suppose the supremum $\bigvee_\alpha b_\alpha$ exists in E, and let $a \in E$ with $a \perp \bigvee_\alpha b_\alpha$. Then $\bigvee_\alpha (a \oplus b_\alpha)$ exists in E and $a \oplus \bigvee_\alpha b_\alpha = \bigvee_\alpha (a \oplus b_\alpha)$.*

Proof. Let $b := \bigvee_\alpha b_\alpha$ Then $b_\alpha \leq b \leq a'$ and $a \oplus b_\alpha \leq a \oplus b$ for all α. Suppose $a \oplus b_\alpha \leq q$ for all α. Then $a \leq q$ and $a \oplus b_\alpha \leq q = a \oplus (q \ominus a)$ for all α, whence $b_\alpha \leq q \ominus a$ for all α by the cancellation law. Consequently, $b \leq q \ominus a$, so $a \oplus b \leq q$. □

Corollary 1.8.8 (i) *Let $\{b_\alpha\}_\alpha$ be a family of elements in E, suppose the infimum $\bigwedge_\alpha b_\alpha$ exists in E, and let $b \in E$ with $b \leq b_\alpha$ for all α. Then $\bigwedge_\alpha (b_\alpha \ominus b)$ exists in E and $\bigwedge_\alpha (b_\alpha \ominus b) = (\bigwedge_\alpha b_\alpha) \ominus b$.*

(ii) *If $\{b_\alpha\}_\alpha$ is a family of elements in E and $b := \bigwedge_\alpha b_\alpha$ exists in E, then $\bigwedge_\alpha (b_\alpha \ominus b) = 0$.*

Proposition 1.8.9 *Let P be a D-lattice. Then the following assertions are true:*

(i) $(a \vee b) \ominus (a \wedge b) = ((a \vee b) \ominus a) \oplus (a \ominus (a \wedge b))$.

(ii) *If $a \leq b'$, then* $(a \ominus (a \wedge b)) \oplus (b \ominus (a \wedge b)) = ((a \oplus b) \ominus (a \wedge b)) \ominus (a \wedge b)$.

(iii) *If $a \leq b'$, then $a \vee b \leq a \oplus b$.*

(iv) *If $a \leq b'$ and $a \wedge b = 0$, then $a \oplus b = a \vee b$.*

(v) *If $a \leq b'$, then* $(a \ominus (a \wedge b)) \oplus (b \ominus (a \wedge b)) = (a \ominus (a \wedge b)) \vee (b \ominus (a \wedge b))$.

Proof. (i) and (ii) follow directly from (iv) and (ix) of Proposition 1.1.5, respectively.

(iii) The inequality $b' \ominus a \leq b'$ implies the inequality $b \leq (b' \ominus a)' = a \oplus b$. Similarly $a \leq a \oplus b$, therefore $a \vee b \leq a \oplus b$.

(iv) Follows from (iii) and from $a \ominus b = 0$ iff $a = b$;

(v) It holds

$$a \ominus (a \wedge b) \leq a \leq b' \leq (b \ominus (a \wedge b))',$$

$$(a \ominus (a \wedge b)) \wedge (b \ominus (a \wedge b)) = 0,$$

and by (iv), the assertion holds. □

In the following proposition we show that in a lattice ordered effect algebra, $\oplus$ distributes over existing infima [BeFo].

Proposition 1.8.10 *Let $\{b_\alpha\}_\alpha$ be a family of elements in the lattice ordered effect algebra E, suppose the infimum $\bigwedge_\alpha b_\alpha$ exists in E, and let $a \in E$ with $a \perp b_\alpha$ for all α. Then $\bigwedge_\alpha (a \oplus b_\alpha)$ exists in E, and $a \oplus \bigwedge_\alpha b_\alpha = \bigwedge_\alpha (a \oplus b_\alpha)$.*

Proof. Let $b := \bigwedge_\alpha b_\alpha$. Then $a \perp b$ and $a \oplus b \leq a \oplus b_\alpha$ for all α. Suppose $q \in E$ with $q \leq a \oplus b_\alpha$ for all α. Then $a \leq q \vee a \leq a \oplus b_\alpha$ for all α, so by (D3), we have $(q \vee a) \ominus a \leq (a \oplus b_\alpha) \ominus a = b_\alpha$ for all α. Consequently, $(q \vee a) \ominus a \leq b$, and since $a \perp b$, we get $q \vee a \leq a \oplus b$, so $q \leq a \oplus b$. □

Theorem 1.8.11 *Let P be a D-lattice. Then there is a total binary operation $-$ on P such that the following properties are satisfied for all $a, b, c \in P$:*

(i) $a - 0 = a$.

(ii) $a - (a - b) = b - (b - a)$.

(iii) $a \leq b$ *implies* $c - b \leq c - a$.

(iv) $a \leq b \leq c$ *implies* $(c - a) - (c - b) = b - a$.

Conversely, let P be a poset with the least element 0 and greatest element 1. Let $-$ be a binary operation on P with the properties (i)–(v). Then P is a D-lattice.

Proof. Let $(P; \wedge, \vee, \ominus, 1)$ be a D-lattice. We define a binary operation $-$ on P as follows:

$$b - a = b \ominus (a \wedge b).$$

It is easy to check that the binary operation $-$ has, for all $a, b, c \in P$, the following properties:

(1) If $a \leq b$, then $b - a = b \ominus a$.

(2) $b - a \leq b$.

(3) $a - (a - b) = a \wedge b$.

(4) If $b \leq a$, then $b - a = 0$.

(5) $a \wedge b = 0$ if and only if $b - a = b$.

(6) $c \ominus (a \wedge b) = (c - a) \vee (c - b)$.

Now the properties (i)–(iv) are evident.

Conversely, suppose that $(P; \leq, -, 0, 1)$ is a poset with a binary operation $-$ such that the properties (i)–(iv) hold. We show that the operation $-$ has the properties (D1)–(D3) and that P is a lattice.

From (i) and (iii) it follows that $b - a \leq b - 0 = b$, which proves (D1).

By the above and (ii) we have $a - (a - b) \leq a$ and $a - (a - b) = b - (b - a) \leq b$. If $w \in P$ and $w \leq a$, $w \leq b$, then

$$\begin{aligned} w &= w - 0 = w - (w - b) = b - (b - w) \\ &\leq b - (b - a) = a - (a - b), \end{aligned}$$

which shows that the element $a - (a - b)$ is the greatest lower bound of a and b, therefore, $b - (b - a) = a - (a - b) = a \wedge b$. Now it is easy to see that if $a \leq b$, then $b - (b - a) = a$, which proves (D2).

Property (D3) follows directly from (iii) and (iv).

Let us denote

$$a' = 1 - a.$$

The unary operation $'$ is an involution and order reversing operation. Then the element $(a' \wedge b')'$ $(= (a' - (a' - b'))')$ is the least upper bound of a and b, that is $(a' \wedge b')' = a \vee b$, therefore P is a lattice. □

A very important special example of a D-lattice is an MV-algebra. MV-algebras are treated in Chapter 2 of this book. For the convenience of the readers, we bring MV-algebras axioms and basic properties here. Proofs can be found in Section 2.2. We will use here the signs $+$ and $-$ for the total MV-algebra operations, to discern them from the partial operation $\oplus$ and $\ominus$ in effect algebras and D-posets. We note that in the original papers by Chang, the sign $+$ was used, but in more recent literature about MV-algebras the sign $\oplus$ is used. Therefore, to keep the tradition, in Chapter 2 devoted especially to MV-algebras and QMV-algebras, we will use the sign $\oplus$ for the total MV-algebra and QMV-algebra operations, while the partial operations, if necessary, will be denoted by $\boxplus$ and $\boxminus$.

In the present section, we will introduce two more classes of D-lattices (or lattice ordered effect algebras), namely Boolean D-lattices introduced in [ChKo 2] and ϕ-symmetric effect algebras introduced in [BeFo]. We will prove that there is a one to one correspondence between Boolean D-lattices and MV-algebras and between ϕ-symmetric effect algebras and MV-algebras. With an abuse of a language, we sometimes say that Boolean D-lattices (or ϕ-symmetric effect algebras) are MV-algebras.

An **MV-algebra** is an algebra $\mathcal{M} = (M; +, ^*, 0, 1)$, where M is a nonempty set, 0 and 1 are constants, $+$ is a total binary operation, and $*$ is a unary operation, satisfying the following axioms (cf. [Man], this definition is equivalent to the definition introduced by Chang [Cha]).

(MV1) $(a+b)+c = a+(b+c)$.

(MV2) $a+0 = a$.

(MV3) $a+b = b+a$.

(MV4) $a+1 = 1$.

(MV5) $(a^*)^* = a$.

(MV6) $0^* = 1$.

(MV7) $a+a^* = 1$.

(MV8) $(a^*+b)^*+b = (a+b^*)^*+a$.

Given an MV-algebra, we can define

(1) $a \circ b = (a^*+b^*)^*$.

(2) $a \wedge b = (a+b^*) \circ b$.

(3) $a \vee b = (a \circ b^*) + b$.

Then $a \le b$ iff $a = a \wedge b$ is a partial order in M, and $(M; \wedge, \vee)$ is a distributive lattice. We note that in every MV-algebra, the following cancellation property holds (Proposition 2.1.2)

$$a \le c^*,\ b \le c^*,\ a+c = b+c \ \Rightarrow\ a = b.$$

The following theorem shows relations between MV-algebras and D-lattices.

Theorem 1.8.12 [ChKo 1] *Let $\mathcal{M} = (M; +, ^*, 0, 1)$ be an MV-algebra. If we put $b \ominus a := (a+b^*)^*$ for $a \le b$, then we obtain that M is a distributive D-lattice (equivalently, a distributive lattice ordered effect algebra). Moreover, $b - a := b \ominus (a \wedge b) = b \circ a^*$ for any $a, b \in M$.*

Conversely, in any D-lattice $(P; \le, \ominus, 0, 1)$ we can define a binary operation $b - a := b \ominus (a \wedge b)$ for any $a, b \in M$. A D-lattice P is an MV-algebra (with $a^* = a'$, $a + b = a \oplus a' \wedge b$) if and only if

$$a \ominus (a \wedge b) = (a \vee b) \ominus b,\ a, b \in P. \tag{1.5}$$

Proof. We prove that every MV-algebra is a distributive lattice ordered effect algebra.

For $a, b \in M$, define $a \perp b$ (a is orthogonal to b) iff $a \le b^*$. Now introduce a partial binary operation $\oplus$ on M by restricting the operation $+$ to orthogonal pairs. Since $a \le b^*$ iff $b \le a^*$ and $+$ is commutative, $\oplus$ is commutative (in the sense of axiom (E1)). To prove (E2), assume that $(a \oplus b) \oplus c$ is defined. Then $a \perp b$, and $a + b \perp c$. By Proposition 2.2.3 (ii), $a + b \le c^*$ iff $a + b + c \le 1$. In the same way, the latter inequality implies $b + c \le a^*$, and since $a + b \perp c$ implies $b \perp c$, $(b \oplus c) \oplus a$

is defined, and equals $a + b + c$. To prove (E3), define $a' := a^*$, and observe that $a \oplus a'$ exists and equals $a + a^* = 1$. Clearly, $a \perp b$ iff $a \leq b'$. For any $b \in M$ with $b \perp a$ and $a \oplus b = 1$, by Proposition 2.1.2 (xiv), $b = a^* = a'$. Finally, if $1 \oplus a$ is defined, then $1 \leq a^*$, hence $a^* = 1$, hence $a = 0$. So $(M; \oplus, 0, 1)$ is an effect algebra, and so for $a \leq b$, $b \ominus a = (b' \oplus a)'$. By the definition of ordering in MV-algebras (Proposition 2.2.3) we can easily derive that the partial order in M as an MV-algebra coincides with the effect algebra order, and with respect to this order, $(M; \oplus, 0, 1)$ is a distributive lattice.

Now assume that $(P; \leq, \ominus, 0, 1)$ is a D-lattice. Using the interplay between $\ominus$ and $\oplus$, we define $a^* := 1 \ominus a$, $a + b := a \oplus (a^* \wedge b)$. Clearly,

$$\begin{aligned} (a^*)^* &= a, \\ a \leq b &\Rightarrow b^* \leq a^*, \\ a \perp b &\Rightarrow a + b = a \oplus b, \\ a \leq c &\Rightarrow a + b \leq c + b. \end{aligned}$$

Commutativity of the operation + requires

$$a \oplus (a^* \wedge b) = b \oplus (b^* \wedge a),$$

equivalently (since $x \oplus y = (x' \ominus y)'$ whenever $x \perp y$), $a^* \ominus (a^* \wedge b) = (b \vee a^*) \ominus b$. This shows that the condition is necessary. To show sufficiency, assume that the condition is satisfied. We have to check the MV-algebra axioms. (MV2), (MV3), (MV4), (MV5), (MV6), and (MV7) are immediate.

(MV1) $(a+b)+c = (a \oplus (a^* \wedge b)) \oplus [(a \oplus (a^* \wedge b))^* \wedge c] \leq (a \oplus (a^* \wedge c)) \oplus (a^* \wedge b) \leq (a+c)+b$; by symmetry we get the equality.

(MV8) $a + (a + b^*)^* = a \oplus [a^* \wedge (a \oplus (a^* \wedge b^*))^*] = a \oplus (a^* \wedge (a^* \ominus (a^* \wedge b^*))) = a \oplus (a^* \ominus (a^* \wedge b^*)) = (a^* \ominus (a^* \ominus (a^* \wedge b^*)))^* = (a^* \wedge b^*)^* = a \vee b$. By symmetry also $b + (b + a^*)^* = a \vee b$. □

Definition 1.8.13 *A poset P with the least element 0 and the greatest element 1 is said to be a* **Boolean D-poset**, *if there is a binary operation $-$ on P satisfying the following conditions:*

(BD1) $a - 0 = a$ *for any* $a \in P$.

(BD2) $a - (a - b) = b - (b - a)$ *for any* $a, b \in P$.

(BD3) $a, b \in P$, $a \leq b$ *implies* $c - b \leq c - a$ *for any* $c \in P$.

(BD4) $(a - b) - c = (a - c) - b$ *for any* $a, b, c \in P$.

Proposition 1.8.14 *Let P be a Boolean D-poset. Then the following assertions are true for every $a, b, c \in P$:*

(i) $b - a \leq b$.

(ii) $a - a = 0$.

(iii) *If* $a \leq b$, *then* $a - b = 0$.

(iv) $(c - a) - (c - b) = (b - a) - (b - c)$.

(v) *If* $a \leq b \leq c$, *then* $(c - a) - (c - b) = b - a$.

(vi) *If* $a \leq b$, *then* $b - (b - a) = a$.

Proof. (i) Since $0 \leq a$, then $b - a \leq b - 0 = b$.
(ii)

$$\begin{aligned} a - a &= (a - 0) - (a - 0) = (a - (a - 0)) - 0 \\ &= (0 - (0 - a)) - 0 = 0 - (0 - a) \leq 0, \end{aligned}$$

which gives $a - a = 0$.
(iii) If $a \leq b$, then $a - b \leq a - a = 0$.
(iv) $(c - a) - (c - b) = (c - (c - b)) - a = (b - (b - c)) - a = (b - a) - (b - c)$.
(v) If $a \leq b \leq c$, then $c - b \leq c - a$ and $(c - a) - (c - b) = (b - a) - (b - c) = (b - a) - 0 = b - a$.
(vi) If $a \leq b$, then $b - (b - a) = a - (a - b) = a - 0 = a$. □

Corollary 1.8.15 *A Boolean D-poset is a D-lattice.*

Proof. Axioms (BD1), (BD2) and (BD3) are in fact properties (i),(ii) and (iii) in Theorem 1.8.11. Axiom (iv) in Theorem 1.8.11 follows from (v) in Proposition 1.8.14. □

Theorem 1.8.16 *Every MV-algebra is a Boolean D-poset and, conversely, a Boolean D-poset is an MV-algebra.*

Proof. Let $(P; +, ^*, 0, 1)$ be an MV-algebra. Let us define

$$b - a = (a + b^*)^* \text{ for all } a, b \in P.$$

Then $a - 0 = (0 + a^*)^* = (a^*)^* = a$, which proves axiom (BD1). Using (MV8), we have

$$a - (a - b) = ((b + a^*)^* + a)^* = ((a + b^*)^* + b^*)^* = b - (b - a),$$

which proves (BD2).

The operation * is order reversing, and from the monotony of $+$ we have $a + c^* \leq b + c^*$ whenever $a \leq b$, hence $c - b = (b + c^*)^* \leq (a + c^*)^* = c - a$, and (BD3) is proved.

To prove (BD4), let us calculate

$$\begin{aligned}(a-b)-c &= (c+(a-b)^*)^* = (c+(b+a^*))^* \\ &= (b+(c+a^*))^* = (b+(a-c)^*)^* = (a-c)-b.\end{aligned}$$

Conversely, let $(P;-,0,1)$ be a Boolean D-poset. Let us put

$$a^* = 1-a \text{ for any } a \in P.$$

It is evident that $*$ is an involution (axiom (MV5)) and $0^* = 1$ (axiom (MV6)). From (BD3) it follows that $*$ is an order reversing operation. Now we define a binary operation $+$ as follows:

$$a+b = (a^* - b)^* \text{ for every } a,b \in P.$$

Using (BD4) we have

$$\begin{aligned}a+b &= (a^*-b)^* = ((1-a)-b)^* \\ &= ((1-b)-a)^* = (b^*-a)^* = b+a,\end{aligned}$$

therefore $+$ is a commutative operation (axiom (MV1)). Similarly

$$\begin{aligned}(a+b)+c &= ((b+a)^*-c)^* = ((b^*-a)^{**}-c)^* = ((b^*-a)-c)^* \\ &= ((b^*-c)-a)^* = (b+c)+a = a+(b+c),\end{aligned}$$

therefore $+$ is associative (axiom (MV2)). Let us calculate

$$\begin{aligned}a+0 &= (a^*-0)^* = (a^*)^* = a, \\ a+1 &= (a^*-1)^* = 0^* = 1, \\ a+a^* &= (a^*-a^*)^* = 0^* = 1,\end{aligned}$$

which proves the validity of axioms (MV3), (MV4) and (MV7). Moreover,

$$a+(a+b^*)^* = (a^*-(a^*-b^*))^* = (b^*-(b^*-a^*))^* = b+(b+a^*)^*,$$

which gives axiom (MV8). □

If E is an orthomodular lattice and $p \in E$, the **Sasaki projection** $\phi_p : E \to E$ is defined by $\phi_p(q) := p \wedge (p' \vee q)$ for all $q \in E$ ([Kal], [Ber]). Thus, defining $a \oplus b := a \vee b$ exactly when $a \perp b$, $(\phi_p(q))' = p' \vee (p \wedge q') = p' \oplus (p \wedge q')$, so $\phi_p(q) = (p' \oplus (p \wedge q'))' = p \ominus (p \wedge q')$. This suggests the following definition (cf. [BeFo]).

Definition 1.8.17 *Let E be a lattice ordered effect algebra. The mapping $\phi : E \times E \to E$ defined by $\phi(p,q) := p \ominus (p \wedge q') = [p' \oplus (p \wedge q')]'$ is called the* **Sasaki mapping** *on E.*

The relationship between $\phi(p,q)$ and $\phi_p(q)$ is as follows:

Lemma 1.8.18 *For $p,q \in E$, $\phi(p,q) \leq p \wedge (p' \vee q)$, and equality holds if and only if $p \wedge p' \wedge q' = 0$.*

Proof. By 1.8.2, we have $[\phi(p,q)]' = p' \oplus (p \wedge q') = (p' \wedge (p \wedge q')) \oplus [p' \vee (p \wedge q')]$, so $\phi(p,q) = [p \wedge (p' \vee q)] \ominus (p \wedge p' \wedge q')$. □

It is well-known that in an orthomodular lattice, the Sasaki projection is a self-adjoint and idempotent residuated mapping. Recall that a mapping $\alpha : E \to E$ is residuated if there is a mapping $\beta : E \to E$, called the residual of α, such that, for all $x, y \in E$, $\alpha(x) \leq y \Leftrightarrow x \leq \beta(y)$ [BeFo]. A residuated mapping $\alpha : E \to E$ is called self-adjoint if its residual has the form $\beta(y) = [\alpha(y')]'$ for all $y \in E$. Evidently, α is self-adjoint iff $\alpha(x) \perp y \Leftrightarrow x \perp \alpha(y)$ for all $x, y \in E$. Next lemma shows that the mapping $\phi(p, .)$ is self-adjoint.

Lemma 1.8.19 *Let $p, q, r, s \in E$. Then*

(i) $\phi(p,q) \perp r \Leftrightarrow \phi(p,r) \perp q$.

(ii) $p \ominus (p \wedge r) \leq s \Leftrightarrow p \ominus (p \wedge s) \leq r$.

Proof. (i) Suppose $\phi(p,q) \perp r$. Then $r \leq p' \oplus (p \wedge q')$, whence $p' \leq p' \vee r \leq p' \oplus (p \wedge q')$, so that $(p' \vee r) \ominus p' \leq p \wedge q'$. Therefore $p' \vee q \leq p' \oplus (p \wedge r')$, whence $q \leq p' \vee q \leq [\phi(p,r)]'$, and it follows that $\phi(p,r) \perp q$. The converse implication follows by symmetry.

(ii) Replace q by r' and r by s' in (i) and use the fact that $\phi(x,y') = x \ominus (x \wedge y)$, for $x, y \in E$. □

Since residuated mappings preserve existing suprema, we have the following corollary.

Corollary 1.8.20 *If $p \in E$ and $\{b_\alpha\}$ is a family of elements of E such that $\bigvee_\alpha b_\alpha$ exists, then $\bigvee_\alpha \phi(p, b_\alpha)$ exists in E and $\bigvee_\alpha \phi(p, b_\alpha) = \phi(p, \bigvee_\alpha b_\alpha)$.*

In the following theorem, a "parallelogram rule" in terms of Sasaki mappings is studied.

Theorem 1.8.21 (Parallelogram Theorem) [BeFo] *Let E be a lattice ordered effect algebra and $p, q \in E$. Then*

(i) $p = (p \wedge q) \oplus \phi(p,q')$ *and* $q = (p \wedge q) \oplus \phi(q,p')$;

(ii)

$$\begin{aligned} p \vee q &= p \oplus \phi(p',q) = q \oplus \phi(q',p) \\ &= (p \wedge q) \oplus \phi(p,q') \oplus \phi(p',q); \end{aligned}$$

(iii)

$$\begin{aligned}\phi(p,q') \oplus \phi(p',q) &= \phi(q,p') \oplus \phi(q',p) \\ &= (p \vee q) \ominus (p \wedge q) = \phi(p',q) \vee \phi(q',p);\end{aligned}$$

(iv) $\phi(p,q') \wedge \phi(q,p') = 0$ *and* $\phi(p',q) \wedge \phi(q',p) = 0$.

Proof. (i) It follows from the fact that $\phi(p,q') = p\ominus(p\wedge q)$ and $\phi(q,p') = q\ominus(p\wedge q)$.

(ii) There exists $k \in E$ with $p \vee q = p \oplus k$. Thus $p \oplus k \oplus (p \vee q)' = 1$, so $p \oplus (p' \wedge q') \oplus k = 1$, and $k = \phi(p',q)$. By symmetry, $p \vee q = q \oplus \phi(q',p)$, and the remaining parts of (ii) follow from (i).

(iii) All but the last equality follows from (ii). To prove the last equality, observe that

$$\begin{aligned}p \vee q &= (p \wedge q) \oplus [(p \wedge q) \oplus (p \vee q)']' \\ &= (p \wedge q) \oplus [(p \wedge q) \oplus (p' \wedge q')]'.\end{aligned}$$

Also, by Proposition 1.8.6

$$\begin{aligned}(p \wedge q) \oplus (p' \wedge q') &= [p \oplus (p' \wedge q')] \wedge [q \oplus (p' \wedge q')] \\ &= [\phi(p',q)]' \wedge [\phi(q',p)]',\end{aligned}$$

so

$$(p \vee q) \ominus (p \wedge q) = \phi(p',q) \vee \phi(q',p).$$

(iv) The first part follows from Corollary 1.8.5, the second by symmetry. □

Definition 1.8.22 *An effect algebra E is* **ϕ-symmetric** *if it is lattice ordered and $\phi(p,q) = \phi(q,p)$.*

Theorem 1.8.23 *A lattice ordered effect algebra is ϕ-symmetric if and only if it is an MV-algebra.*

Proof. It follows from Theorem 1.8.12 and the fact that $a \ominus (a \wedge b) = \phi(a,b')$ and $\phi(b',a) = (a \vee b) \ominus b$ and the symmetry condition $\phi(a,b') = \phi(b',a)$. □

Corollary 1.8.24 [BeFo] *If E is an interval effect algebra with a lattice ordered ambient group, then E is ϕ-symmetric (hence an MV-algebra).*

Proof. If G is a lattice ordered group and $a,b,u \in G^+$, then $a+b = (a\wedge b)+(a\vee b)$, so $a - (a \wedge b) = (a \vee b) - b$, hence in $G^+[0,u]$, the condition (1.5) of Theorem 1.8.12 is satisfied. □

The following theorem gives some characterizations of ϕ-symmetric effect algebras among lattice ordered effect algebras ([BeFo]). The proof is left to the readers as an exercise.

Theorem 1.8.25 *For the lattice ordered effect algebra E, the following conditions are mutually equivalent:*

(i) *E is ϕ-symmetric.*

(ii) *E has the Riesz decomposition property.*

(iii) *For $x, y, z \in E$ with $y \perp z$, $x \wedge (y \oplus z) \leq (x \wedge y) \oplus (x \wedge z)$.*

(iv) *For $x, y, z \in E$ with $y \perp z$, $x \wedge (y \oplus z) \leq (x \wedge y) \oplus z$.*

(v) *$a \leq (a \wedge b) \oplus (a \wedge b')$ for all $a, b \in E$.*

(vi) *For $a, b \in E$ with $a \wedge b = 0$, $a \perp b$.*

(vii) *For $a, b \in E$, there exist $x, y \in E$ with $x \leq b$, $y \leq b'$, $a = x \oplus y$.*

(ix) *$\phi(a, b) \leq b$ for all $a, b \in E$.*

1.9 Some Basic Constructions with D-posets/Effect Algebras

In this Section, we study constructions that enable us to obtain new effect algebra/D-posets from the given ones. In particular, we study subalgebras, direct products, horizontal sums and direct limits. Direct products are closely related with so-called central elements. A generalization of the notion of direct products to generalized effect algebras by replacing central elements by central ideals can be found in [Pul 17], [RiZ 11].

If effect algebras and their morphisms are organized into a category, the direct product is the categorical product and the horizontal sum is the categorical coproduct. Owing to the categorical equivalence of effect algebras and D-posets, there is also an equivalence between the corresponding constructions.

1.9.1 Subalgebras and Ideals

Recall that on an effect algebra $(E; \oplus, 0, 1)$, the relation $\leq$, defined by $a \leq b$ if there is a c such that $b = a \oplus c$, is a partial order on E, and with this order, E becomes a difference poset: $b \ominus a$ is defined if and only if $a \leq b$, and $b \ominus a = c$ if and only if $a \oplus c = b$. Furthermore, E satisfies the following cancellation law: If $a \oplus b \leq a \oplus c$, then $b \leq c$.

If $e \in E$, we define the **interval** $E[0, e] := \{p \in E : p \leq e\}$.

Definition 1.9.1 *A subset F is called a* **sub-effect algebra** *of an effect algebra E if* (i) $0, 1 \in F$, (ii) *for $p \in F$, $p' \in F$, and* (iii) *for $p, q \in F$ with $p \perp q$, $p \oplus q \in F$.*

A subset Q of a D-poset P is called a **sub-D-poset** *of P if* (i) $0, 1 \in P$, (ii) *for $p \in Q$, $p' \in Q$, and* (iii) *for $p, q \in Q$, with $p \leq q$, $q \ominus p \in Q$.*

Evidently, a sub-effect algebra F is an effect algebra in its own right under the restriction of $\oplus$ to F. Similarly, a sub D-poset Q of a D-poset is a D-poset in its own right under the restriction of $\ominus$ to Q. Owing to the equivalence between effect algebras and D-posets, it is easy to see that a sub-effect algebra is a sub-D-poset, and vice versa. Indeed, assume that F is a sub-effect algebra of an effect algebra E. Let $a, b \in E$ with $a \leq b$. It then follows that $b' \in F$, $a \perp b'$, so that $a \oplus b' \in F$, and hence $b \ominus a = (a \oplus b')' \in F$. The proof of the converse is similar. The family $\mathcal{S}$ of all sub-effect algebras of E is closed under set-theoretic intersections. Therefore it forms a complete lattice under set-theoretic inclusion. In particular, to every subset $X \subseteq E$, there is the smallest sub-effect algebra F_X such that $X \subseteq F_X$. We call F_X a subalgebra **generated** by X.

Lemma 1.9.2 *A subalgebra generated by a subset X of an effect algebra can be constructed in the following way: Let $Y' := \{x' : x \in Y\}$, for any $Y \subseteq E$, and define*

$$F_0 := X, \ F_{n+1} = \{x \oplus y : x, y \in F_n \cup F_n', \ x \perp y\}.$$

Then $F_X = \bigcup_{n=0}^{\infty} F_n$.

Proof. Straightforward. □

Recall that nonempty subset I of an effect algebra E is an **order ideal** if for $a, b \in E$, $a \in I$ and $b \leq a$ imply $b \in I$.

Definition 1.9.3 *A nonempty subset I of a generalized effect algebra E is called an* **ideal** *in E if, for all $p, q \in E$ with $p \perp q$, $p \oplus q \in I$ if and only if $p \in I$ and $q \in I$.*

In other words, an ideal is an order ideal closed under $\oplus$. The collection $\mathcal{I}(E)$ of all ideals in E is closed under the formation of arbitrary set-theoretic intersections, so it forms a complete lattice under set-theoretic inclusion. Consequently, if $X \subseteq E$, there will be a smallest ideal I_X with $X \subseteq I_X$. We call I_X the ideal **generated** by X. In particular, the ideal I_e generated by a single element $e \in E$ is called a **principal ideal**.

Lemma 1.9.4 *An ideal I_X generated by a subset X of a generalized effect algebra can be constructed in the following way: define*

$$I_0 = X, \ I_{n+1} := \bigcup_{x \in I_n} E[0, x], \ I_{n+2} := \{x \oplus y : x, y \in I_{n+1}, x \perp y\}, \ n = 0, 1, \ldots.$$

Clearly, $I_n \subseteq I_{n+1} \subseteq I_{n+2}$, $n = 0, 1, \ldots$. Then $I_X = \bigcup_{n=0}^{\infty} I_n$.

Proof. It is a routine verification that $\bigcup_{n=0}^{\infty} I_n$ is an ideal. □

1.9.2 Center and Direct Products

If E is an orthomodular poset and $e \in E$, then the principal ideal I_e is the interval $E[0, e]$. If E is a more general effect algebra, then the interval $E[0, e]$ may be a proper subset of I_e. Indeed, the necessary and sufficient condition that $I_e = E[0, e]$ is that e is principal in the sense of the following definition.

Definition 1.9.5 *An element e of a GEA/GDP E, is said to be* **principal** *if, for $p, q \in E$, $p \perp q$ and $p, q \leq e$, $p \oplus q \leq e$.*

Lemma 1.9.6 *If $q \in E$ is principal, and $q \leq u$, $u \in E$, then $q \wedge (u \ominus q) = 0$. In particular, if E is an effect algebra, then $q \wedge q' = 0$.*

Proof. If q is principal and $p \leq q, u \ominus q$, then $p \perp q$ and $p, q \leq q$, hence $p \oplus q \leq q = 0 \oplus q$, hence $p = 0$ by the cancellation law. □

In the next lemma, we do not necessarily assume that e is principal. Thus the interval $E[0, e]$ is not necessarily an ideal in E although it is a poset under the partial order inherited from E.

Lemma 1.9.7 [GFP] *Let E be a GEA/GDP, and $0 \neq e \in E$. Define $\tau : E[0, e] \to E[0, e]$ by $\tau(p) := e \ominus p$ for all $p \in E[0, e]$. Then $\tau(0) = e$, $\tau(e) = 0$, and the following de Morgan-type laws hold:*

(i) *If $p, q \in E[0, e]$ and $p \vee q$ exists in E, then $\tau(p) \wedge \tau(q)$ exists in E and $\tau(p \vee q) = \tau(p) \wedge \tau(q)$.*

(ii) *If $p, q \in E[0, e]$, and $\tau(p) \vee \tau(q)$ exist in E, then $p \wedge q$ exists in E and $\tau(p \wedge q) = \tau(p) \vee \tau(q)$.*

Proof. It is a routine verification. □

If $\{E_\alpha\}_{\alpha \in J}$ is a family of effect algebras/D-posets, the cartesian product $E := \prod_{\alpha \in J} E_\alpha$ is organized into an effect algebra in the obvious way using coordinatewise operations and relations. The projection mappings $\pi_\alpha : E \to E_\alpha$ are surjective effect algebra/D-poset morphisms for all $\alpha \in J$. The proof of the following theorem is a routine verification.

Theorem 1.9.8 *For $j = 1, 2, \ldots,$ let E_j be an effect algebra with unit 1_j, let $E := E_1 \times E_2 \times \cdots \times E_n$, and define $e_j := (0, 0, \ldots, 0, 1_j, 0, \ldots, 0)$ with 1_j in the jth coordinate position. Then:*

(i) *Each e_j is principal in E.*

(ii) *The principal ideal $E[0, e_j]$ in E is isomorphic to E_j under the restriction to $E[0, e_j]$ of the projection homomorphism $\pi_j : E \to E_j$.*

(iii) *$e_1, e_2, \ldots, e_n$ is an orthogonal sequence in E and $\bigoplus e_j =: 1$ is the unit element in E.*

(iv) *Every element $p \in E$ can be represented uniquely in the form $p = \bigoplus_j p_j$ with $p_j \in E[0, e_j]$; indeed, $p_j = p \wedge e_j$ for $j = 1, 2, \ldots, n$.*

The following theorem deals with factoring an effect algebra E as an "internal direct product" of ideals generated by orthogonal principal elements. A similar result can be obtained for D-posets.

Theorem 1.9.9 [GFP, Thm 4.2] *Let $e_1, e_2, \ldots, e_n$ be a finite orthogonal sequence of nonzero, principal elements of E. Define $\phi : \prod_j E[0, e_j] \to E$ by $\phi(p_1, p_2, \ldots, p_n) := \bigoplus_j p_j$ for all $(p_1, p_2, \ldots, p_n) \in \prod_j E_E[0, e_j]$. Then ϕ is an isomorphism if and only if it is surjective. Furthermore, if ϕ is an isomorphism, then, for $p \in E$, the infima $p \wedge e_1, p \wedge e_2, \ldots, p \wedge e_n$ exist in E and $p = \bigoplus_j (p \wedge e_j)$.*

Proof. Since $e_1, e_2, \ldots, e_n$ is an orthogonal sequence of nonzero principal elements of E, it follows that $\phi : \prod_j E[0, e_j] \to E$ is well-defined and it preserves inequalities and orthogonal sums. If ϕ is an isomorphism, it is surjective by definition.

Conversely, suppose $\phi : \prod_j E[0, e_j] \to E$ is surjective and let u be the unit in E. Then there exists $(u_1, u_2, \ldots, u_n) \in \phi^{-1}(u)$, and we have $u = \bigoplus_j u_j$ and $u_j \le e_j$ for all $j = 1, 2, \ldots, n$. Therefore,

$$e_1' \le u_1' = \bigoplus_{j=2}^{n} u_j \le \bigoplus_{j=2}^{n} e_j \le e_1'.$$

(Notice that the last inequality follows from the orthogonality of the sequence $e_1, e_2, \ldots, e_n$, and does not require that e_1' be principal.) Hence, $e_1' = u_1'$, so $u_1 = e_1$, and by symmetry, $u_j = e_j$ for all j. Consequently, $\phi(e_1, e_2, \ldots, e_n) = u$, so $\phi : \prod_j E[0, e_j] \to E$ preserves the units, and so it is a surjective morphism.

Suppose $(p_1, p_2, \ldots, p_n), (q_1, q_2, \ldots, q_n) \in \prod_j E[0, e_j]$ with

$$p := \phi(p_1, p_2, \ldots, p_n) \le q := \phi(q_1, q_2, \ldots, q_n)$$

in E. To prove that ϕ is an isomorphism, we have only to show that

$$(p_1, p_2, \ldots, p_n) \le (q_1, q_2, \ldots, q_n)$$

in $\prod_j E[0, e_j]$. Let $(r_1, r_2, \ldots, r_n)$ be the orthosupplement of $(q_1, q_2, \ldots, q_n)$ in $\prod_j E[0, e_j]$, so that $r := \phi(r_1, r_2, \ldots, r_n) = q'$. Then

$$r_1 \le \bigoplus_j r_j = r = q',$$

so $q \le r_1'$ and, therefore, $p_1 \le \bigoplus_j p_j = p \le q \le r_1'$. Since e_1 is principal and $p_1, r_1 \le e_1$, it follows that $p_1 \oplus r_1 \le e_1 = q_1 \oplus r_1$; hence, $p_1 \le q_1$ by the cancellation law. By symmetry, $p_j \le q_j$ for all j, so ϕ is an isomorphism.

Again suppose that $(p_1, p_2, \ldots, p_n) \in \prod_j E[0, e_j]$ with $p := \phi(p_1, p_2, \ldots, p_n) = \bigoplus_j p_j$. We note that $p_1 \le p, e_1$ and claim that, in fact, $p_1 = p \wedge e_1$. To prove this, suppose that $x \le p, e_1$. We have to prove that $x \le p_1$. Then $(x, 0, 0, \ldots, 0) \in \prod_j E[0, e_j]$ with $\phi(x, 0, 0, \ldots, 0) = x \le p = \phi(p_1, p_2, \ldots, p_n)$ Because ϕ is an isomorphism, we conclude that $x \le p_1$ as desired. □

Corollary 1.9.10 *Suppose that $\{e_1, e_2, \ldots, e_n\}$ is an orthogonal sequence of principal elements in E such that, for all $p \in E$, there are elements $p_j \le e_j$ for $j = 1, 2, \ldots, e_n$ with $p = \bigoplus_j p_j$. Then all orthogonal sums of subsequences of $e_1, e_2, \ldots, e_n$ are principal elements of E.*

With Theorem 1.9.9 in mind, we make the following definition ([GFP]).

Definition 1.9.11 *An element $a \in E$ is* **central** *if*

(i) *for all $p \in E$, there exist $q, r \in E$ such that $q \le a, r \le a'$, and $p = q \oplus r$;*

(ii) *a and a' are principal.*

The **center** *$C(E)$ is the set of all central elements in E.*

Lemma 1.9.12 *If $a \in C(E)$ and $p \in E$, then:*

(i) *$a \wedge p$ and $a' \wedge p$ exist in E.*

(ii) *$p = (a \wedge p) \oplus (a' \wedge p)$.*

(iii) *$a \perp p$ if and only if $a \wedge p = 0$.*

(iv) *The isotropic index of a, $\iota(a)$, is 1.*

Proof. We can assume $a \ne 0, 1$, so a, a' form an orthogonal sequence of nonzero principal elements of E. Applying Theorem 1.9.9 to $e_1 := a$ and $e_2 = a'$, we obtain (i) and (ii). Part (iii) is a consequence of (ii). If $a \perp a$, then $a = a \oplus a$ by (ii), and $a = 0$ by the cancellation law. □

Lemma 1.9.13 *If $a \in E$, then $a \in C(E)$ if and only if there is an orthogonal sequence $e_1, e_2, \ldots, e_n$ of principal elements in E such that, for all $p \in E$, there are elements $p_j \le e_j$ for $j = 1, 2, \ldots, n$ with $p = \bigoplus_j p_j$; and a is an orthogonal sum of a subsequence of $e_1, e_2, \ldots, e_n$.*

Proof. If $a \in C(E)$, let $e_1 := a$ and $e_2 := a'$. Conversely, suppose $e_1, e_2, \ldots, e_n$ is an orthogonal sequence of principal elements in E such that, for all $p \in E$, there are elements $p_j \leq e_j$ for $j = 1, 2, \ldots, n$ with $p = \bigoplus_j p_j$, and suppose that a is a sum of a subsequence of $e_1, e_2, \ldots, e_n$. Without loss of generality, we can assume that $a = e_1 \oplus e_2 \oplus \cdots \oplus e_k$. By Theorem 1.9.9, $a' = e_{k+1} \oplus e_{k+2} \oplus \cdots \oplus e_n$ and, by Corollary 1.9.10, both a and a' are principal. If $p \in E$ there exist $p_j \leq e_j$ for $j = 1, 2, \ldots, n$ with $p = \bigoplus p_j$. Then, with $q := p_1 \oplus p_2 \oplus \cdots \oplus p_k \leq a$ and $r := p_{k+1} \oplus p_{k+2} \oplus \cdots \oplus p_n \leq a'$, we have $p = q \oplus r$. □

Theorem 1.9.14 [GFP, Thm 5.4] *The center $C(E)$ is a sub-effect algebra of E, and (as an effect algebra in its own right) $C(E)$ forms a Boolean algebra. Furthermore, if $a, b \in C(E)$, then $a \wedge b$ and $a \vee b$ as calculated in $C(E)$ are also the infimum and supremum of a and b as calculated in E.*

Proof. Clearly, $0, 1 \in C(E)$ and $a \in C(E) \Rightarrow a' \in C(E)$. Suppose $a, b \in C(E)$. By Lemma 1.9.12, applied first to $p = b$ then to $p = b'$,

$$e_1 := a \wedge b,\ e_2 := a \wedge b',\ e_3 := a' \wedge b,\ \text{and } e_4 := a' \wedge b'$$

all exist in E; $b = e_1 \oplus e_3$, and $b' = e_2 \oplus e_4$, so

$$1 = b \oplus b' = e_1 \oplus e_3 \oplus e_2 \oplus e_4,$$

and e_1, e_2, e_3, e_4 form an orthogonal sequence in E. Since $b \in C(E)$, Lemma 1.9.12 also implies that $a = e_1 \oplus e_2$. Evidently, if two principal elements have an infimum in E, that infimum is again principal, so e_1, e_2, e_3 and e_4 are principal elements of E.

If $p \in E$ we have $p = q \oplus r$ with $q \leq a$ and $r \leq a'$. Also, since $b \in C(E)$, we have $q = p_1 \oplus p_2$ and $r = p_3 \oplus p_4$ with $p_1, p_3 \leq b$ and $p_2, p_4 \leq b'$. Therefore, $p = q \oplus r = p_1 \oplus p_2 \oplus p_3 \oplus p_4$ with

$$p_1 \leq q, b,\ p_2 \leq q, b',\ p_3 \leq r, b,\ p_4 \leq r, b'.$$

Since $q \leq a$ and $r \leq a'$, it follows that

$$p_1 \leq a \wedge b = e_1, \qquad p_2 \leq a \wedge b' = e_2,$$
$$p_3 \leq a' \wedge b = e_3, \qquad p_4 \leq a' \wedge b' = e_4,$$

so e_1, e_2, e_3, e_4 satisfy the conditions in Corollary 1.9.10; hence e_1, e_2, e_3, e_4 and all orthogonal sums of them belong to $C(E)$. In particular, $a \wedge b = e_1 \in C(E)$.

If $a \perp b$, then $a = a \wedge b' = e_2$, $b = a' \wedge b = e_3$, so $a \oplus b = e_2 \oplus e_3 \in C(E)$, and it follows that $C(E)$ is a sub-effect algebra of E. Since every element of $C(E)$ is principal in E, it is principal in $C(E)$, hence $C(E)$ is an orthomodular poset (cf. Exercise 13, Section 1.12). Furthermore, if $a, b \in C(E)$, then $a \wedge b$ exists in E and belongs to $C(E)$; hence, it is also the infimum of a and b as calculated in $C(E)$. Therefore, $C(E)$ is an orthomodular lattice. By part (iii) of Lemma 1.9.12, disjoint elements of $C(E)$ are orthogonal, so $C(E)$ is a Boolean algebra. □

Definition 1.9.15 *An element a in an effect algebra E is* **sharp** *if $a \wedge a' = 0$ (or equivalently, if $a \vee a' = 1$).*

We have the following implications: central $\Rightarrow$ principal $\Rightarrow$ sharp. As we shall see in the next section, the converse implications need not hold in general.

1.9.3 Principal and Sharp Elements in Effect Rings

Definition 1.9.16 [GFP] *By an* **effect ring** *we mean a ring R with unit 1 ordered by a positive cone R^+, such that the following conditions hold for all $a, b \in R^+$:*

(i) $ab = ba \Rightarrow ab \in R^+$.

(ii) $aba \in R^+$.

(iii) $aba = 0 \Rightarrow ab = ba = 0$.

(iv) $a = a^2 \Rightarrow 1 - a \in R^+$.

For example, a unital C*-algebra R can be organized into an effect ring by taking $R^+ := \{aa^* : a \in R\}$.

Clearly, if R is an effect ring, the interval $R^+[0,1]$ can be organized into an interval effect algebra. In what follows, we assume that R is an effect ring. Observe that for any $a, b \in R^+[0,1]$, $aba, ab'a \in R^+$ by Definition 1.9.16 (ii), and $ab'a = a^2 - aba$, $(aba)' = (a^2)' + ab'a$. So $aba \leq a^2 \leq a$.

By a **projection** we will mean an idempotent element belonging to R^+. We denote by P the set of all projections in R. Clearly, $e \in P$ implies $e' = 1 - e \in P \subseteq R^+$. It then follows that $e, e' \leq 1$.

Lemma 1.9.17 *If $e \in P$, $a \in R^+[0,1]$, then $a \leq e$ if and only if $a = ea = ae$ if and only if $a = eae$.*

Proof. If $a \leq e$, then $b := e - a \in R^+$, and $e'ae' + e'be' = e'ee' = 0$. Since both $e'ae'$ and $e'be'$ belong to R^+, it follows that $e'ae' = 0$, hence $e'a = ae' = 0$ by part (iii) of Definition 1.9.16. Therefore, $a = ae = ea$. That $a = ae = ea$ iff $a = eae$ is clear. Finally, if $a = ae = ea$, then $a'e = ea' = e - a$ belongs to R^+ by Definition 1.9.16 (i), and hence $a \leq e$. □

The proof of the following theorem is left to the reader (see [GFP] for the details).

Theorem 1.9.18 [GFP] *P is a subeffect algebra of $R^+[0,1]$ and, as an effect algebra in its own right, P forms an orthomodular poset.*

Theorem 1.9.19 [GFP, Thm 6.8] *If $e \in R^+[0,1]$, then the following conditions are mutually equivalent:*

(i) *e is principal.*

(ii) $e \wedge e' = 0$.

(iii) $e \vee e' = 1$.

(iv) $ee' = 0$.

(v) $e \in P$.

(vi) *e' is principal.*

Proof. (i) $\Rightarrow$ (ii) $\Leftrightarrow$ (iii) is clear.

(ii) $\Rightarrow$ (iv). We have $ee' = e(1-e) = e - e^2 = (1-e)e = e'e$. Hence, by Definition 1.9.16 (i), $0 \leq ee' = e - e^2 = e' - e'^2 \leq e, e'$, so the implication holds.

(iv) $\Rightarrow$ (v) is obvious.

(v) $\Rightarrow$ (i). Suppose $e \in P$ and $a, b \in R^+[0,1]$ with $a \perp b$ and $a, b \leq e$. By Lemma 1.9.17, $a = ae = ea$ and $b = eb = be$, whence, $a+b = (a+b)e = e(a+b)$, so $a+b \leq e$. Thus $e \in P$ is principal.

Since $e \in P$ iff $e' \in P$, we have (i) $\Leftrightarrow$ (vi). □

We note that in MV-algebras, sharp and central elements coincide (cf. Exercise 19 Section 1.12). Theorem 1.9.19 applied to the Hilbert space effects $\mathcal{E}(H)$ shows that the set of all sharp elements as well as that of principal elements coincides with the projections. But a central element in $\mathcal{E}(H)$ is central also in the projection lattice $\mathcal{P}(H)$, but the center of the latter is $\{0,1\}$.

1.9.4 Finite Distributive Lattice Ordered Effect Algebras

In what follows, we give a complete characterization of effect algebras which are finite distributive lattices. We follow the ideas of [GFP]. Recall that $\iota(a)$ denotes the isotropic index of a, that is, $\iota(a)$ is the largest positive integer n with property na exists. The element $smt(a) := \iota(a)a$ is called the **isotropic summit** of a.

Recall that the effect algebra E is a scale algebra, if it is totally ordered, and it is **distributive** if, as a poset, forms a distributive lattice.

As usual, a nonzero element $a \in E$ is called an **atom** if $E[0,a] = \{0,a\}$. We say that E is **atomic** if, for $0 \neq p \in E$, there is an atom $a \in E$ with $a \leq p$. An effect algebra E satisfies the **chain condition** if every chain (i.e., totally ordered subset) has finite length.

Proposition 1.9.20 [FoBe 1, Thm 4.11] *If E satisfies the chain condition, then every nonzero element in E is a finite orthogonal sum of atoms.*

Proof. Let E satisfy the chain condition. Since every descending chain is finite, L is atomic. Suppose $0 \neq p \in E$ and p cannot be written as a finite sum of orthogonal atoms in E. Choose an atom $a_1 \leq p$, noting that $p \ominus a_1 \neq 0$. Choose an atom $a_2 \leq p \ominus a_1$, noting that $(p \ominus a_1) \ominus a_2 \neq 0$. Continuing in this way, we produce an infinite sequence of (not necessarily distinct) atoms $a_1, a_2, a_3, \ldots$ in E such that $(a_1, a_2, \ldots, a_n)$ is an orthogonal sequence for every $n \geq 1$. Let $p_n := \bigoplus_{i=1}^{n} a_i$ for $n \geq 1$. Then $p_1 < p_2 < \cdots < p_n < \cdots$ contradicts the ascending chain condition. □

Lemma 1.9.21 *Let E be distributive and suppose $e_1, e_2, \ldots, e_n$ are a pairwise disjoint and pairwise orthogonal sequence of elements of E. Then $e_1, e_2, \ldots, e_n$ is a summable sequence in E and $e_1 \oplus e_2 \oplus \cdots \oplus e_n = e_1 \vee e_2 \vee \cdots \vee e_n$.*

Proof. If $n = 2$, using the fact that $e_1 \wedge e_2 = 0$, we get from Proposition 1.8.2(1) that $e_1 \oplus e_2 = e_1 \vee e_2$. The rest follows by induction on n. □

Definition 1.9.22 *An atom $a \in E$ is of* **type** 1 *if whenever m is a positive integer and ma is defined, the interval $E[0, ma]$ consists only of* 0 *and positive integer multiples of a; otherwise it is of* **type** 2.

An atom with isotropic index 1 is of type 1. If a is a type 1 atom in E and the isotropic index of a is n, it is clear that $smt(a) = na$ is principal and that $E[0, na]$ is a scale algebra with $n+1$ elements. Up to isomorphism, there is only one such scale algebra; the n-chain $C_n = \{0, a, 2a, \ldots, na\}$.

Example 1.9.23 The four-element effect algebra D, called the **diamond**, consists of 0, two atoms a, a^* with $a \neq a^*$, $a = a'$, $a^* = a^{*\prime}$, and the unit $2a = 2a^* = smt(a) = smt(a^*) = a \vee a^*$. The diamond D is a distributive effect algebra and both a and a^* are type 2 atoms in D.

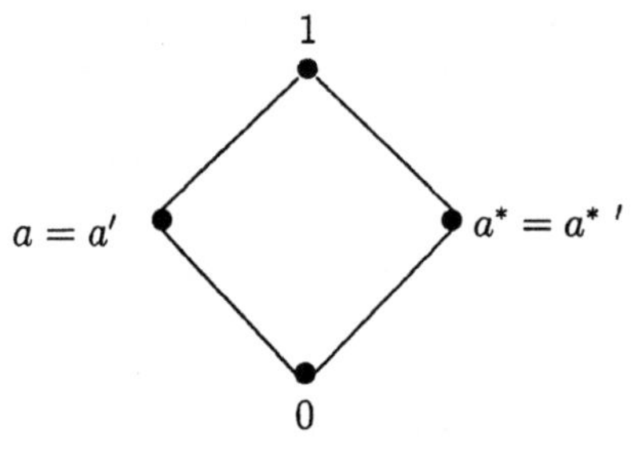

Fig. 1.7 Diamond

If a and a^* are the two atoms in the diamond D, then $a \not\perp a^*$, else $a \oplus a^* = 2a$, so $a^* = a$ by the cancellation law, contradicting $a \neq a^*$.

Theorem 1.9.24 [GFP, Thm 7.5] *Let E be a distributive effect algebra and let a, a^* be atoms in E with $a \neq a^*$ and $a \not\perp a^*$. Then a and a^* have isotropic index 2, $smt(a) = smt(a^*) = 2a = 2a^* = a \vee a^*$, a and a^* are type 2 atoms, $a \vee a^*$ is principal, and $E[0, a \vee a^*]$ is the diamond.*

Proof. By the distributive law and the fact that a and a^* are atoms, the only elements in $E[0, a \vee a^*]$ are $0, a, a^*$ and $a \vee a^*$.

Since $a \not\perp a^*$, we have $a \vee a^* \not\perp a$ and $a \vee a^* \not\perp a^*$. Since $a < a \vee a^*$, there is an $r \in E$ with $r \neq 0$, $a \perp r$ and $a \oplus r = a \vee a^*$. Thus $0 \neq r \in E[0, a \vee a^*]$, so $r = a$ or $r = a^*$ or $r = a \vee a^*$. Since $a \perp r$, $a \not\perp a^*$, and $a \not\perp a \vee a^*$, it follows that $r = a$, so $2a = a \vee a^*$. By symmetry, $2a^* = a \vee a^* = 2a$.

If the isotropic index of a exceeds 2, then $3a = 2a \oplus a = 2a^* \oplus a$, contradicting the fact that $a \not\perp a^*$. Therefore, $smt(a) = 2a = a \vee a^*$ and by symmetry, $\iota(a^*)a^* = 2a^* = a \vee a^*$.

Since $E[0, a \vee a^*] = \{0, a, a^*, a \vee a^*\}$, and the only orthogonalities in $E[0, a \vee a^*]$ that do not involve 0 are $a \perp a$ and $a^* \perp a^*$, it follows that $a \vee a^*$ is principal in E. So $E[0, a \vee a^*]$ is the diamond. □

Lemma 1.9.25 (i) *Let a be an atom in an effect algebra E, let m be a positive integer and suppose that ma is defined. Then, if $0 \neq q$, $q \perp a$, $q \vee a$ exists, and $q \leq ma$, it follows that $a \leq q$.*

(ii) *If E is lattice ordered, a is an atom, m is a positive integer such that ma is defined, $0 \neq q \in E$, $q \perp a$ and $q \leq ma$, then $q = na$ for some positive integer $n \leq m$.*

Proof. (i) Since a is an atom, $a \wedge q$ is either a or 0. Assume $0 \neq q$, $q \perp a$, then $q \wedge a = 0$ implies $q \oplus a = q \vee a \leq ma$, whence by cancellation, $q \leq (m-1)a$. By induction, we get $q = a$, contradicting $q \wedge a = 0$. Therefore $a \leq q$.

(ii) By (i), we may assume that $a \leq q$. Let n be the largest positive integer for which $na \leq q$. Then there exists $r \in E$ with $na \oplus r = q$. Consequently, $na \oplus r \leq ma$, hence $r \leq ma$, $r \perp a$. If $r \neq 0$, then by the above part of proof for q replaced by r, $a \leq r$, which contradicts the choice of n. Therefore $r = 0$, and $q = na$. □

Theorem 1.9.26 [GFP, Thm 7.6] *Let E be a distributive and atomic effect algebra. Let a be an atom in E with finite isotropic index n. Then a is of type 1 if and only if $E[0, smt(a)]$ is the n-chain C_n, and a is of type 2 if and only if $n = 2$ and $E[0, smt(a)]$ is the diamond D. In either case, $smt(a)$ is principal.*

Proof. Clearly, a is of type 1 iff $E[0, smt(a)]$ is the n-chain C_n, in which case $smt(a)$ is principal. If $[0, smt(a)]$ is the diamond D, then a is obviously type 2 and again $smt(a)$ is principal. Conversely, suppose a is of type 2. If there is an atom $b \in E$ such that $b \neq a$ and $b \not\perp a$, the proof is complete by Theorem 1.9.24.

Assume that every atom $b \neq a$ satisfies $b \perp a$. Since a is of type 2, there is a $q \neq 0, q < smt(a) = na$, but q is not a positive integer multiple of a. By Lemma 1.9.25 (ii), $q \not\perp a$. Since E is atomic and $q \leq na$, there is an atom b with $b \perp q$ and $q \oplus b \leq na$. Since $q \not\perp a$, it follows that $b \neq a$, so $b \perp a$. Also, $b \leq q \oplus b \leq na$; hence, $a \leq b$ by Lemma 1.9.25 (i). But b is an atom, so $a = b$, contradicting $a \neq b$. □

Lemma 1.9.27 *Let E be a distributive and atomic effect algebra, suppose a and b are atoms in E both of which have finite isotropic index, a is of type 1, and $a \neq b$. Then $smt(a) \wedge smt(b) = 0$ and $smt(a) \perp smt(b)$.*

Proof. If $smt(a) \wedge smt(b) \neq 0$, there is an atom $c \leq smt(a), smt(b)$. Since a is of type 1, $c = a$, so $a \leq smt(b)$. As $b \neq a$, b is type 2, so $E[0, smt(b)] = \{0, a, b, a \vee b\}$, contradicting the fact that a is of type 1. Thus, $smt(a) \wedge smt(b) = 0$.

Since a is of type 1 and $b \neq a$, we have $b \neq smt(a)$, so $smt(a) < smt(a) \vee b$. Therefore, there is an atom c in E with $smt(a) \perp c$ and $smt(a) \oplus c \leq smt(a) \vee b$. By the definition of $smt(a)$, we cannot have $c = a$. By the distributive law, $c \leq smt(a)$ or else $c = b$, the first possibility is ruled out because a is of type 1 and $c \neq a$, and therefore $c = b$. Thus, $b \perp smt(a)$.

Let the positive integer n be maximal with respect to the properties that nb is defined and $nb \perp smt(a)$. To complete the proof, it is sufficient to show that n is the isotropic index of b. Assume the contrary, so that $nb < (n+1)b \leq smt(b)$. Since $smt(b)$ is disjoint from $smt(a)$, it follows that nb and $(n+1)b$ are disjoint from $smt(a)$, so $nb \vee smt(a) = nb \oplus smt(a)$; moreover, $(n+1)b \not\leq nb \vee smt(a)$ by the distributive law.

Thus, $nb \oplus smt(a) < (n+1)b \vee smt(a)$ so there is an atom c with $c \perp (nb \oplus smt(a))$ and $nb \oplus smt(a) \oplus c \leq (n+1)b \vee smt(a)$. Note $c \perp nb$, so $c \perp b$. Also, $c \perp smt(a)$, so $c \neq a$. If $c = b$, then$(n+1)b = nb \oplus b \perp smt(a)$, contradicting the maximality of n, so $c \neq b$. By the distributive law, $c \leq (n+1)b$ or else $c \leq smt(a)$. Since a is of type 1 and $c \neq a$, $c \leq (n+1)b$ is the only possibility; hence, since $c \neq b$, b is of type 2, $n = 1$, $c \leq 2b$, and $c \not\perp b$, contradicting $c \perp b$. □

Lemma 1.9.28 *Let E be a distributive and atomic effect algebra and suppose a and b are atoms in E both of which have finite isotropic index in E. Then, either $smt(a) = smt(b)$ or else $smt(a)$ and $smt(b)$ are disjoint and orthogonal.*

Proof. Suppose $smt(a) \neq smt(b)$, so $a \neq b$. By Lemma 1.9.27, we can assume that both a and b are type 2. By Theorem 1.9.24, $a \perp b$, both a and b have isotropic index 2, and both $E[0, 2a]$ and $E[0, 2b]$ are diamonds with $2a = smt(a) \neq smt(b) = 2b$. thus, there are atoms $a^* \neq a$, $b^* \neq b$ in E such that $E[0, 2a] = \{0, a, a^*, 2a\}$, $2a = a \vee a^*$, $E[0, 2b] = \{0, b, b^*, 2b\}$, $a \perp b, b^*$ and $b \perp a, a^*$. Obviously, $2a \wedge 2b = 0$. Also, $smt(a) = a \vee a^* \perp b \vee b^* = smt(b)$. □

Theorem 1.9.29 [GFP, Thm 7.9] *A finite distributive effect algebra E is a direct product of finite chains and diamonds.*

Proof. By Lemma 1.9.21, Theorem 1.9.26, and Lemma 1.9.28, the disjoint isotropic summits of atoms in E form an orthogonal family $e_1, e_2, \ldots, e_n$ of nonzero principal elements. Since every nonzero element of E is a finite orthogonal sum of atoms, and each atom is dominated by its own isotropic summit, Theorem 1.9.9 implies that E is isomorphic to the direct product $\prod_j E[0, e_j]$. By Theorem 1.9.26, each $E[0, e_j]$ is either a finite chain or a diamond. □

As a consequence of Theorem 1.9.29, the center $C(E)$ of a finite distributive effect algebra E is the sub-effect algebra of E generated by the isotropic summits of the atoms in E, and these isotropic summits are precisely the atoms in the Boolean algebra $C(E)$.

Corollary 1.9.30 [GFP, Cor. 7.10] *A finite distributive effect algebra is an interval algebra.*

Proof. Every finite chain is an MV-algebra, therefore an interval algebra (cf. Corollary 5.3.8, Mundici's theorem). The diamond D is an interval algebra (see Exercise 3, Section 1.12). We will see (Theorem 1.9.31), that a direct product of interval algebras is an interval algebra. Therefore, the result follows from Theorem 1.9.29. □

As a poset (not as an effect algebra) the diamond D is isomorphic to the 4-element Boolean algebra $\mathbf{2}^2$. However, there are finite distributive lattices - for instance the free distributive lattice on three generators - that cannot be factored into chains and copies of $\mathbf{2}^2$; hence not every distributive lattice can be organized into an effect algebra.

1.9.5 Direct Products of Interval Algebras

In the next theorem, we prove that the direct product of interval effect algebras is an interval effect algebra.

Theorem 1.9.31 [FGB, Thm 7.1] *Let A and B be interval algebras with units u and v and with the universal groups G and H, respectively. With $G^+ \times H^+$ as the positive cone, $G \times H$ is the universal group with unit (u, v) for $A \times B$.*

Proof. Condition (i) of Definition 1.4.11 is obviously satisfied. To verify condition (ii), assume that $\phi : A \times B \to K$ is a K-valued measure. The mappings $\alpha : A \to K$ and $\beta : B \to K$ defined by $\alpha(a) := \phi(a, 0)$ and $\beta(b) := \phi(0, b)$ for $a \in A$, $b \in B$ are K-valued measures; hence they can be extended to group homomorphisms $\alpha^* : G \to K$

and $\beta^* : H \to K$, respectively. Therefore the mapping $\phi^* : G \times H \to K$ defined by $\phi^*(g,h) := \alpha^*(g) + \beta^*(h)$ for $(g,h) \in G \times H$ is a group homomorphism that extends ϕ. □

For example, the rectangular treillis $RT(n_1, n_2, \ldots, n_r)$ (Example 1.4.32) is an interval algebra and its universal group is $\mathbb{Z}^r$ partially ordered by the standard positive cone $(\mathbb{Z}^+)^r$ and with unit $(n_1, n_2, \ldots, n_r)$.

For an effect algebra E, let $\Omega(E)$ denote the set of all probability measures (states) on E, and let $\delta_e\Omega(E)$ denote the set of all extremal points in the convex set $\Omega(E)$.

Let $A \times B$ be the direct product of effect algebras A and B with units u and v, respectively. Let $\Omega_A := \{\omega \in \Omega(A \times B) : \omega(0,v) = 0\}$, $\Omega_B := \{\omega \in \Omega(A \times B) : \omega(u,0) = 0\}$. The mapping $\mu \mapsto \mu_A$ from $\Omega(A)$ to $\Omega(A \times B)$ given by $\mu_A(a,b) := \mu(a)$ for $(a,b) \in A \times B$ is an affine isomorphism of $\Omega(A)$ onto $\Omega_A \subseteq \Omega(A \times B)$. Likewise, the mapping $\nu \mapsto \nu_A$ given by $\nu_A(a,b) := \nu(b)$ for $(a,b) \in A \times B$ is an affine isomorphism of $\Omega(B)$ onto $\Omega_B \subseteq \Omega(A \times B)$. In the sense of the following theorem, $\Omega(A \times B)$ can be regarded as the "convex hull" of $\Omega(A)$ and $\Omega(B)$.

Theorem 1.9.32 [FGB, Thm 7.3] *Let $A \times B$ be the direct product of effect algebras A and B. If $\Omega(A), \Omega(B) \neq \emptyset$ and $\omega \in \Omega(A \times B)$, there is a unique $t \in \mathbb{R}^+[0,1]$ and there are states $\mu \in \Omega(A)$ and $\nu \in \Omega(B)$ such that $\omega = t\mu_A + (1-t)\nu_B$.*

Proof. We may assume that $\omega \notin \Omega_A \cup \Omega_B$, so that $\omega(0,v), \omega(u,0) \neq 0$. Let $t := \omega(u,0)$. Since $\omega(0,v) + \omega(u,0) = \omega(u,v) = 1$, we have $1 - t = \omega(0,v)$. Evidently, $\mu : A \to \mathbb{R}^+[0,1]$ defined by $\mu(a) := \omega(a,0)/t$ for $a \in A$ is a state on A. Likewise, $\nu : B \to \mathbb{R}^+[0,1]$ defined by $\nu(b) := \omega(0,b)/(1-t)$ for $b \in B$ is a state on B. Furthermore,

$$\begin{aligned}(t\mu_A + (1-t)\nu_B)(a,b) &= t\mu(a) + (1-t)\nu(b)\\ &= \omega(a,0) + \omega(0,b)\\ &= \omega(a,b).\end{aligned}$$

Conversely, if $\omega = t\mu_A + (1-t)\nu_B$, then $\omega(u,0) = t\mu_A(u) = t$, so t is uniquely determined. □

As a consequence of Theorem 1.9.32 we obtain that

$$\delta_e\Omega(A \times B) = \delta_e\Omega_A \dot{\cup} \delta_e\Omega_B.$$

1.9.6 Horizontal Sums

Let A and B be effect algebras with units u and v, respectively. To form the horizontal sum $A \dot{+} B$ of A and B, we begin with relabeling the elements of A and B, if necessary, so that $A \cap B = \{0, w\}$, where $w = u = v$. The **horizontal sum** is then defined to be $A\dot{+}B := A \cup B$, organized into an effect algebra in such a way that, for $x, y \in A\dot{+}B$, $x \oplus y$ is defined iff $x, y \in A$ or $x, y \in B$, in which case $x \oplus y$ is defined as in A or B, respectively.

For example, the diamond D is isomorphic to the horizontal sum $C_2\dot{+}C_2$ of two 2-chains.

An n-fold horizontal sum $A_1\dot{+}A_2\dot{+}\cdots\dot{+}A_n$ is defined in the obvious way. For example, if n is a positive integer, the effect algebra $MO(n)$ is defined to be the horizontal sum $\mathbf{2}^2\dot{+}\mathbf{2}^2\dot{+}\cdots\dot{+}\mathbf{2}^2$ of n-copies of the interval algebra $\mathbf{2}^2$. As a poset, $MO(n)$ forms a finite modular orthocomplemented lattice [Kal, p. 29].

The following statement is obvious.

Lemma 1.9.33 *An effect algebra C with unit 1 is isomorphic to the horizontal sum $A\dot{+}B$ if and only if there are morphisms $\alpha : A \to C$, $\beta : B \to C$ such that*

(i) *α is an isomorphism of A onto the sub-effect algebra $\alpha(A)$ of C.*

(ii) *β is an isomorphism of B onto the sub-effect algebra $\beta(B)$ of C.*

(iii) $\alpha(A) \cap \beta(B) = \{0, 1\}$.

(iv) $\alpha(A) \cup \beta(B) = C$.

(v) *If $x \in \alpha(A)$ and $y \in \beta(B)$ and $x \oplus y$ is defined, then $x = 0$ or $y = 0$.*

The universal group of a horizontal sum of interval algebras is constructed from the quotient group of a direct product. Note that if U is a subgroup of the partially ordered Abelian group G and $U \cap G^+ = \{0\}$, then G/U can be organized into a partially ordered Abelian group with $(G/U)^+ = G^+/U$. Indeed, G^+/U is closed under addition, and if $g_1 + U = -g_2 + U$ with $g_1, g_2 \in G^+$, then $g_1 + g_2 \in U \cap G^+ = \{0\}$; thus $g_1 = -g_2$, so that $(G^+/U) \cap (-G^+/U)$ is the zero element of G/U.

Assume that A and B are interval algebras with units u and v and with universal groups G and H, respectively. Let U be the cyclic subgroup of $G \times H$ generated by $(u, -v)$, let $Q = (G \times H)/U$, and let $\eta : G \times H \to Q$ be the canonical epimorphism. Thus, for $n \in \mathbb{Z}$, $g \in G$, $h \in H$, we have

$$\eta(g, h) = \eta(g + nu, h - nv).$$

Because $U \cap (G^+ \times H^+) = \{(0, 0)\}$, it follows that Q can be organized into a partially ordered group with $Q^+ := \eta(G^+ \times H^+)$ as a generating positive cone. Note that $w := \eta(u, 0) = \eta(0, v)$ is a nonzero element of Q^+, every element of Q^+ is a sum of a sequence of elements in the interval $Q^+[0, w]$, and Q^+ generates Q.

Define $\alpha : A \to Q^+[0, w]$ and $\beta : B \to Q^+[0, w]$ by $\alpha(a) = \eta(a, 0)$ and $\beta(b) = \eta(0, b)$ for all $a \in A$, $b \in B$. Thus, α and β are effect-algebra isomorphisms of A and B onto sub-effect algebras $\alpha(A)$ and $\beta(B)$, respectively, of $Q^+[0, w]$.

Theorem 1.9.34 [FGB, Thm 8.1] *With the notation above, $Q^+[0,w] = \alpha(A)\dot{+}\beta(B)$ and, with w as the unit, Q is the universal group for the horizontal sum $\alpha(A)\dot{+}\beta(B)$.*

Proof. Suppose $\eta(g,h) \in \alpha(A) \cap \beta(B)$. Then there are elements $a \in A$, $b \in B$. and $m, n \in \mathbb{Z}$ such that $a = g + nu$, $h = nv$, $g = -mu$, and $b = h - mv$. Therefore, $a = (n-m)u$ and $b = (n-m)v$. Since $0 \leq a \leq u$, it follows that $n - m = 0$ or $n - m = 1$ and so $\eta(g,h) = 0$ or $\eta(g,h) = w$. Therefore, $\alpha(A) \cap \beta(B) = \{0, w\}$.

Suppose $q \in Q^+[0,w]$. Since $q \in Q^+$, there exist $g \in G^+$ and $h \in H^+$ with $q = \eta(g,h)$. Since $w - \eta(g,h) \in Q^+$, there is an integer n such that $0 \leq g \leq (n+1)u$ in G and $0 \leq h \leq -nv$ in H. Therefore, $n = 0$ and $q = \eta(g,h) \in \alpha(A)$ or else $n = -1$ and $q = \eta(g,h) \in \beta(b)$. Consequently, $Q^+[0,w] = \alpha(A) \cup \beta(B)$.

Suppose $a \in A$, $b \in B$ and $\alpha(a) + \beta(b) \in Q^+[0,w]$, that is, $w - \eta(a,b) \in Q^+$. Then there is an integer n such that $0 \leq a \leq (n+1)u$ in G and $0 \leq b \leq -nv$ in H, and it follows that $n \in \{0,-1\}$, so that $b = 0$ or $a = 0$. Therefore, $Q^+[0,w]$ is the horizontal sum of its sub-effect algebras $\alpha(A)$ and $\beta(B)$.

Let $\phi : Q^+[0,w] \to K$ be a K-valued measure. To complete the proof, we only have to show that ϕ can be extended to a group homomorphism $\phi^* : Q \to K$. The K-valued measures $\phi \circ \alpha : A \to K$ and $\phi \circ \beta : B \to K$ can be extended to group homomorphisms $(\phi\circ\alpha)^* : G \to K$ and $(\phi\circ\beta)^* : H \to K$. The group homomorphism $\xi : G \times H \to K$ defined by $\xi(g,h) = (\phi \circ \alpha)^*(g) + (\phi \circ \beta)^*(h)$ satisfies the condition $\xi(u,-v) = 0$, so there exists a group homomorphism $\phi^* : Q \to K$ such that $\phi^* \circ \eta = \xi$. For $a \in A$,

$$\phi^*(\alpha(a)) = \phi^*(\eta(a,0)) = \xi(a,0) = (\phi \circ \alpha)^*(a) = \phi(\alpha(a))$$

and likewise, $\phi^*(\beta(b)) = \phi(\beta(b))$ for $b \in B$. Since $Q^+[0,w] = \alpha(A) \cup \beta(B)$, it follows that ϕ^* is an extension of ϕ. □

Proof of the following theorem is straightforward.

Theorem 1.9.35 [FGB, Thm 8.2] *If $\Omega(A) \neq \emptyset$, $\Omega(B) \neq \emptyset$, the mapping $\Phi : \Omega(A) \times \Omega(B) \to \Omega(A\dot{+}B)$ given by $\Phi(\mu,\nu)(x) = \mu(x)$ for $x \in A$ and $\Phi(\mu,\nu)(x) = \nu(x)$ for $x \in B$ is an affine isomorphism of $\Omega(A) \times \Omega(B)$ onto $\Omega(A\dot{+}B)$ and Φ maps $\delta_e\Omega(A) \times \delta_e\Omega(B)$ onto $\delta_e\Omega(A\dot{+}B)$.*

1.9.7 Direct Limits of D-posets/Effect Algebras

In this subsection, we study direct limits in the category with D-posets as objects and D-poset-morphisms as morphisms. Owing to the categorical equivalence of D-posets and effect algebras, it is easy to transfer this construction to the category with effect algebras as objects and effect-algebra morphisms as morphisms.

The existence of direct limits can be obtained from a more general consideration (e.g., [Bur]). We bring a more specialized proof here. Notice that the construction of direct limits can be used in the construction of infinite tensor products, and applied in the so-called histories approach to quantum mechanics (see, e.g., [GeHa 1], [GeHa 2] for general approach, [Ish], [Ish 1], [Ish 2], [IsLi], [IsLi 1], [ILS], [Pul 2], [Rud], [Rud 1] [Rud 2], [Wri], [Wri 1], [Wri 2]) for applications to quantum logics and for decoherence functionals.

Definition 1.9.36 (i) *A* **directed system of D-posets** *is a family* $A_{\underline{I}} := (A_i; (f_{ij} : A_i \to A_j), i, j \in I, i \leq j)$ *where* $(I, \leq)$ *is a directed set,* A_i *is a D-poset for each* $i \in I$*, and each* f_{ij} *is a morphism* $(i \leq j)$*, such that*

(i1) $f_{ii} = id_{A_i}$ *for every* $i \in I$.

(i2) *If* $i \leq j \leq m$ *in* I*, then* $f_{jm} f_{ij} = f_{im}$.

(ii) *Let* $A_{\underline{I}}$ *be a directed system of D-posets; then* $\underline{f} := (A; (f_i : A_i \to A; i \in I))$ *is called its* **direct limit** *if* (ii1) *through* (ii3) *below hold:*

(ii1) *A is a D-poset;* f_i *is a morphism for each* $i \in I$.

(ii2) *If* $i \leq j$ *in* I*, then* $f_j f_{ij} = f_i$ *(i.e.,* $\underline{f}$ *is compatible* with $A_{\underline{I}}$*).*

(ii3) If $\underline{g} := (B; (g_i : A_i \to B, i \in I))$ is any system compatible with $A_{\underline{I}}$ (i.e., $g_j f_{ij} = g_i$ for all $i \leq j$ in I), then there exists exactly one morphism $g : A \to B$ such that $g f_i = g_i$ for every $i \in I$.

The system $\underline{f}$ *above (or sometimes only the object A) is denoted as* $\lim_{\to} A_{\underline{I}}$.

It is easy to see that every direct limit is unique up to isomorphism.

Theorem 1.9.37 [Pul 2, Thm 2.7] *Let* $A_{\underline{I}} = (A_i; (f_{ij} : A_i \to A_j), i, j \in I, i \leq j)$ *be a directed system of D-posets, where* f_{ij} *is a morphism for every* $i, j \in I, i \leq j$*. Then a direct limit* $\lim A_{\to \underline{I}}$ *exists.*

Proof. Put $A = \dot{\bigcup}_{i \in I} A_i$ and define a relation $\equiv$ on A as follows: We put $a \equiv b$ $(a \in A_i,\ b \in A_j)$ if there is a $k \in I$ with $i, j \leq k$ such that $f_{ik}(a) = f_{jk}(b)$ in A_k. Now, if we choose an arbitrary $k_1 \in I$ with $i, j \leq k_1$, then for any $\ell \in I$ with $k, k_1 \leq \ell$, the following equalities hold:

$$
\begin{aligned}
f_{i\ell}(a) &= f_{k_1\ell} f_{ik_1}(a);\\
f_{i\ell}(a) &= f_{k\ell} f_{ik}(a) = f_{k\ell} f_{jk}(b) = f_{j\ell}(b);\\
f_{j\ell}(b) &= f_{k_1\ell} f_{jk_1}(b);
\end{aligned}
$$

hence

$$f_{ik_1}(a) \equiv f_{jk_1}(b).$$

Reflexivity and symmetry of the relation $\equiv$ are evident. To prove transitivity, let $a \equiv b$ and $b \equiv c$, where $a \in A_i$, $b \in A_j$, $c \in A_k$. Then there are $\ell_1, \ell_2 \in I$ with $i, j \leq \ell_1$ and $j, k \leq \ell_2$ and $f_{i\ell_1}(a) = f_{j\ell_1}(b)$ and $f_{j\ell_2}(b) = f_{k\ell_2}(c)$. Let $\ell_1, \ell_2 \leq \ell$, then $f_{j\ell}(b) = f_{\ell_1\ell}f_{j\ell_1}(b) = f_{\ell_1\ell}f_{i\ell_1}(a) = f_{i\ell}(a)$, and $f_{j\ell}(b) = f_{\ell_2\ell}f_{j\ell_2}(b) = f_{\ell_2\ell}f_{k\ell_2}(c) = f_{k\ell}(c)$, hence $f_{i\ell}(a) = f_{k\ell}(c)$, so that $a \equiv c$. Put $\tilde{A} := A/\equiv$. We prove that $\tilde{A}$ can be endowed with a structure of a D-poset. Let $a, b \in A$, $a \in A_i$, $b \in A_j$ and let $\tilde{a} = a/\equiv$, $\tilde{b} = b/\equiv$ be the corresponding equivalence classes in $\tilde{A}$. We define a partial binary operation $\ominus$ on $\tilde{A}$ as follows:

$$\begin{gathered}\tilde{b} \ominus \tilde{a} \text{ exists iff there is } k \in I,\ i, j \leq k, \text{ with}\\ f_{jk}(b) \ominus f_{ik}(a) \text{ being defined in } A_k,\\ \text{and then } \tilde{b} \ominus \tilde{a} = (f_{jk}(b) \ominus f_{ik}(a))^{\sim}.\end{gathered}$$

To prove that the operation $\ominus$ is well-defined, let a_1, b_1 be any other representants of $\tilde{a}$, $\tilde{b}$ with $a_1 \in A_{i_1}$, $b_1 \in A_{j_1}$. There is an $\ell \in I$ with $i_1, j_1, k \leq \ell$, and

$$\begin{aligned} f_{j_1\ell}(b_1) &= f_{j\ell}(b) = f_{k\ell}f_{jk}(b),\\ f_{i_1\ell}(a_1) &= f_{i\ell}(a) = f_{k\ell}f_{ik}(a). \end{aligned}$$

Now that $f_{jk}(b) \ominus f_{ik}(a)$ exists in A_k implies $f_{j_1\ell}(b_1) \ominus f_{i_1\ell}(a_1)$ exists in A_ℓ, and $f_{j_1\ell}(b_1) \ominus f_{i_1\ell}(a_1) = f_{k\ell}(f_{jk}(b) \ominus f_{ik}(a))$, hence $f_{j_1\ell}(b_1) \ominus f_{i_1\ell}(a_1) \equiv f_{jk}(b) \ominus f_{ik}(a)$.

Clearly, $\tilde{1} = \bigcup\{a \in A_i : \exists\ j \geq i, f_{ij}(a) = 1_j\}$, $\tilde{0} = \bigcup\{a \in A_i : \exists\ j \geq i, f_{ij}(a) = 0_j\}$, where 1_i and 0_i are the greatest and the smallest elements in A_i, respectively.

Define $\tilde{a} \leq \tilde{b}$ iff $\tilde{b} \ominus \tilde{a}$ exists. Let $\tilde{a} \leq \tilde{b} \leq \tilde{c}$, and $a \in A_i$, $b \in A_j$, $c \in A_k$ be any representants. There is $\ell \in I$ with $i, j, k \leq \ell$, and $f_{i\ell}(a) \leq f_{j\ell}(b) \leq f_{k\ell}(c)$ hold in A_ℓ. The fact that A_ℓ is a D-poset implies that $\ominus$ is a difference operation on $\tilde{A}$.

For every $i \in I$, let $f_i : A_i \to \tilde{A}$ be the natural projection $f_i(a) = \tilde{a}$. If $a \in A_i$ and $i \leq j$, then for any k, $i \leq j \leq k$, $f_{ik}(a) = f_{jk}f_{ij}(a)$, so that $a \equiv f_{ij}(a)$, and hence $f_j f_{ij} = f_i$.

Let $\underline{g} := (B; (g_i : A_i \to B, i \in I))$ be any system compatible with $A_{\underline{I}}$. Define $g : \tilde{A} \to B$ via $g(f_i(a)) = g_i(a)$. Now $f_i(a) = f_i(b)$, $a, b \in A_i$ iff there is $j \in I$, $i \leq j$ with $f_{ij}(a) = f_{ij}(b)$. But then $g_i(a) = g_j f_{ij}(a) = g_j f_{ij}(b) = g_i(b)$, hence g is a well-defined morphism. Clearly, g is unique. This concludes the proof that $\tilde{A} = \lim_{\to} A_{\underline{I}}$. □

Recall that a D-poset is an orthoalgebra iff (i) $a \leq 1 \ominus a \Rightarrow a = 0$; and a D-poset is an orthomodular poset iff it is an orthoalgebra and the following condition is satisfied: (ii) $a \perp b$, $b \perp c$, $c \perp a$ imply $a \leq (1 \ominus c) \ominus b$. Indeed, if A is an OMP, then $b \ominus a = b \wedge a'$ for $a \leq b$, and (ii) becomes $a \leq b' \wedge c'$. Conversely, recall that an orthoalgebra is an OMP iff $a \perp b$, $b \perp c$, $c \perp a$ imply $a \oplus b \perp c$. Now if (ii) holds, then $a = ((1 \ominus c) \ominus b) \ominus d = ((1 \ominus c) \ominus d) \ominus b$, which implies $a \oplus b = (1 \ominus c) \ominus d \leq 1 \ominus c$.

Corollary 1.9.38 *If $A_i, i \in I$, in a directed system $A_{\underline{I}}$ are orthoalgebras (orthomodular posets), then the direct limit* $\lim_{\rightarrow} A_{\underline{I}}$ *is also an orthoalgebra (orthomodular poset).*

Proof. From the proof of Theorem 1.9.37 it can be derived that conditions (i), (ii) for the difference $\ominus$ are satisfied in $\lim_{\rightarrow} A_{\underline{I}}$ if they are satisfied in every A_i, $i \in I$. □

It is sometimes more convenient to work in the category with monomorphisms as morphisms, since then A_i is embedded into A_j for $i \leq j$. For direct limits in the category with effect algebras as objects and monomorphisms as morphisms, see [Hab 1] (see also Exercise 15, Section 1.12).

1.10 Compatibility and Observables

Recall that a state has been defined as a morphism from a D-poset (effect algebra) to the unit interval $[0,1]$. A state is σ- (or completely) additive if the corresponding morphism is such. In this Section, we define an observable as a σ-morphism from a σ-field of sets (most usually, the Borel σ-field $\mathcal{B}(\mathbb{R}^n)$) to an effect algebra (or to a D-poset). These definitions are generalizations of the definitions of states and observables in quantum logics, where the quantum logic - in usual sense an orthomodular lattice or poset - is replaced by a D-poset (effect algebra). In the Hilbert space approach, states are represented by density operators, as usual, but "sharp" observables (projection-valued measures, PV-measures) are replaced by "unsharp" observables (positive-operator-valued measures, POV-measures). In turns out that the latter play an important role in the theory of quantum measurements (cf. [BLM], [BGL], [Lud]).

From the physical point of view, an observable corresponds to a measuring procedure, and effects contained in the range of an observable are coexistent, in the sense they are simultaneously measurable. Therefore, from the physical point of view, a privileged notion of compatibility is represented by the relation of commeasurability (called also simultaneous measurability or coexistence).

It would be desirable to have an intrinsic definition of commeasurability. As it is well known, in orthomodular lattices, such a condition is represented by a decomposition property ([Var], [Mack]). We will extend this definition of compatibility to effect algebras and more general orthomodular structures.

1.10.1 Compatibility in Orthomodular Structures

Definition 1.10.1 *Let P be a D-poset (effect algebra), $a, b \in P$. Elements a and b are called* **(Mackey) compatible** *($a \leftrightarrow b$) if there are $a_1, b_1, c \in P$ such that*

(C1) $a_1 \oplus b_1 \oplus c$ *exists in* P.

(C2) $a = a_1 \oplus c$, $b = b_1 \oplus c$.

Elements a_1, b_1, c *satisfying conditions* (C1), (C2) *are called a* **Mackey decomposition** *of* a, b.

Theorem 1.10.2 [Kop 1] *Let* P *be a D-poset,* $a, b \in P$. *The following statements are equivalent:*

(i) *There are* $u, v \in P$ *such that* $v \le a \le u$, $v \le b \le u$ *and* $u \ominus a = b \ominus v$.

(ii) *There exists a* $u \in P$ *such that* $a \le u$, $b \le u$, $u \ominus a \le b$ *and* $u \ominus b \le a$.

(iii) *There exists a* $v \in P$ *such that* $v \le a$, $v \le b$, $b \ominus v \le 1 \ominus a$ *and* $a \ominus v \le 1 \ominus b$.

(iv) $a \leftrightarrow b$.

Proof. (i) $\Rightarrow$(ii) follows immediately.
(ii) $\Leftrightarrow$ (iii) follows by the transition $x \mapsto 1 \ominus x$.
(iii) $\Rightarrow$(iv). $a = v \oplus (a \ominus v)$, $b = v \oplus (b \ominus v)$. It remains to prove $\oplus$-orthogonality of $v, a \ominus v, b \ominus v$. Recall that in a D-poset, elements x, y are orthogonal iff $x \le 1 \ominus y$. Then $v \oplus (a \ominus v) = a$, and $b \ominus v \le 1 \ominus a$ imply the desired result.
(iv) $\Rightarrow$ (i). Let a_1, b_1, c be the Mackey decomposition of a, b. Put $v = c$, $u = c \oplus a_1 \oplus b_1$. □

We note that in [Kop 1], compatibility was defined by property (i) in the above theorem.

The definition of compatibility makes sense also in a generalized difference poset, and the equivalence of (i), (ii), and (iv) of Theorem 1.10.2 holds true.

Observe that $a \le b$ implies $a \leftrightarrow b$. Indeed, by the generalized orthomodular law, $b = a \oplus (b \ominus a)$, hence $a, 0, b \ominus a$, is a Mackey decomposition of a, b.

Observe that (i) in Theorem 1.10.2 holds for a, b if and only if it holds for $1\ominus a$, $1\ominus b$. Indeed, $v \le a \le u$, $v \le b \le u$ are equivalent with $1 \ominus u \le 1 \ominus a \le 1 \ominus v$, $1 \ominus u \le 1 \ominus b \le 1 \ominus v$, and $(1 \ominus v) \ominus (1 \ominus a) = a \ominus v = u \ominus b = (1 \ominus b) \ominus (1 \ominus u)$. It follows that $a \leftrightarrow b$ if and only if $1 \ominus a \leftrightarrow 1 \ominus b$.

Proposition 1.10.3 *Let* P *be a GDP. The following statements are equivalent for all* $a, b \in P$:

(i) $a \leftrightarrow b$;

(ii) *There are* $u, d, e \in P$ *such that* $a \le u$ *and* $b = d \oplus e$ *with* $d \le a$, *and* $e \le u \ominus a$.

Corollary: (1) $a \perp b$ *implies* $a \leftrightarrow b$.
(2) *If* P *is an effect algebra, then* $a \leftrightarrow b$ *implies* $a \leftrightarrow b'$.

Proof. (i) $\Rightarrow$ (ii). Let $a = a_1 \oplus c$, $b = b_1 \oplus c$ be a Mackey decomposition of a, b. Put $u = a_1 \oplus b_1 \oplus c$, $d = c$, $e = b_1$.

(ii) $\Rightarrow$ (i). We have $d \leq a$, hence $a = d \oplus (a \ominus d)$. Now $d \oplus (a \ominus d) \oplus e = a \oplus e \leq u$, therefore $e, d, a \ominus d$ is a Mackey decomposition of a, b.

The remaining statements are clear. □

We introduce the following strengthening of the definition of compatibility [Pul 5].

Definition 1.10.4 *Let P be a GDP. We say that $a, b \in P$ are* **strongly compatible** *if there exists a $\oplus$-orthogonal triple $\{a_1, b_1, c\}$ in P such that $a = a_1 \oplus c$, $b = b_1 \oplus c$, and $a_1 \wedge b_1 = 0$.*

Proposition 1.10.5 *If a, b are strongly compatible elements of a GDL P, then $a \wedge b = c$, $a \vee b = a_1 \oplus b_1 \oplus c$.*

Proof. We have $c \leq a, b$. If $d \in P$ is such that $d \leq a, b$, then $c \leq c \vee d \leq a, b$ implies $(c \vee d) \ominus c \leq a \ominus c = a_1, b \ominus c = b_1$, hence $(c \vee d) \ominus c \leq a_1 \wedge b_1 = 0$, whence $d \leq c$.

Similarly we prove the second statement. □

We note that in orthomodular posets, compatibility coincides with strong compatibility, and Mackey decompositions are uniquely defined as in Proposition 1.10.5. On the other hand, Mackey decompositions in effect algebras need not be unique. For example, consider the chain $0, \frac{1}{4}, \frac{1}{2}, \frac{3}{4}, 1$ and the elements $\frac{1}{4}, \frac{3}{4}$. Then both $0, \frac{1}{4}, \frac{3}{4}$ and $\frac{1}{4}, 0, \frac{1}{2}$ are Mackey decompositions of $\frac{1}{4}$ and $\frac{3}{4}$. In orthoalgebras compatibility coincides with strong compatibility too, but several Mackey decompositions may occur. For a counterexample in orthoalgebras see the **Frazer cube** [LH2]. [1]

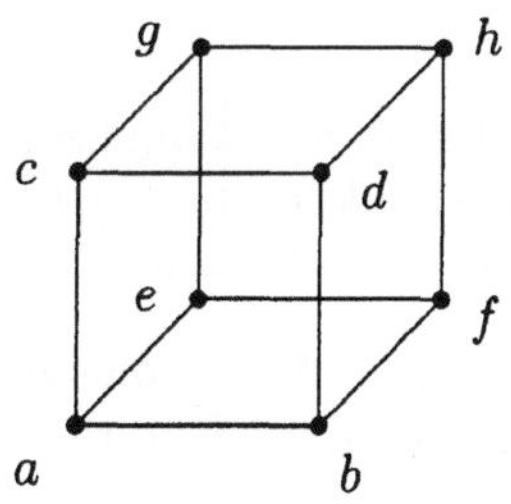

Fig 1.8 Frazer cube

[1] On the figure, every face corresponds to a Boolean algebra of four atoms (the "blocks" of the orthoalgebra represented by Frazer cube). Hence, we have for example, $c \oplus d = e \oplus f = (a \oplus b)'$ and $c \oplus g = b \oplus f = (a \oplus e)'$. Hence, (c, d, g) and (b, e, f) are two different Mackey decompositions of $(a \oplus b)'$ and $(a \oplus e)'$.

In more general D-posets strong compatibility and compatibility need not coincide, as we can see from the following simple example. Let P be a chain $0 \leq a_1 \leq a_2 \leq a_3 \leq a_4 \leq a_5$, where $a_i \oplus a_j = a_{i+j}$, iff it is defined. For a_2, a_4 we have two Mackey decompositions, $a_2 = a_2 \oplus 0$, $a_4 = a_2 \oplus a_2$ (strong compatibility), and $a_2 = a_1 \oplus a_1$, $a_4 = a_1 \oplus a_3$.

Theorem 1.10.6 [ChKo 2] *Let P be a GDL. Then*

$$a \leftrightarrow b \text{ if and only if } (a \vee b) \ominus b = a \ominus (a \wedge b). \tag{1.6}$$

Proof. If $a \leftrightarrow b$, due to Theorem 1.10.2 (i), there exist elements $c, d \in P$ such that $d \leq a \leq c$, $d \leq b \leq c$ and $c \ominus a = b \ominus d$. From the inequalities $a \leq a \vee b \leq c$ and $b \leq a \vee b \leq c$ it follows that $(a \vee b) \ominus b \leq c \ominus b = a \ominus d \leq a$, and

$$a \ominus ((a \vee b) \ominus b) = [(a \vee b) \ominus ((a \vee b) \ominus a] \ominus ((a \vee b) \ominus b) = b \ominus ((a \vee b) \ominus a).$$

Denote $w = a \ominus ((a \vee b) \ominus b)$. By the above equality, $w \leq a \wedge b$ and

$$a \ominus w = (a \vee b) \ominus b, \;\; b \ominus w = (a \vee b) \ominus a.$$

Let us calculate

$$(a \wedge b) \ominus w = (a \ominus w) \wedge (b \ominus w) = ((a \vee b) \ominus b) \wedge ((a \vee b) \ominus a) = 0$$

hence $w = a \wedge b$, which gives $a \ominus (a \wedge b) = (a \vee b) \ominus b$. The converse implication follows from Theorem 1.10.2 (i) with $u = a \vee b$, $v = a \wedge b$. □

Corollary 1.10.7 *In a GDL P, elements $a, b \in P$ are compatible if and only if they are strongly compatible.*

Proof. By Theorem 1.10.6, $a \leftrightarrow b$ implies that $a \ominus (a \wedge b) = (a \vee b) \ominus b$. It follows that $a \ominus (a \wedge b) \perp b$, hence $a \wedge b, a \ominus (a \wedge b), b \ominus (a \wedge b)$ is a Mackey decomposition of a, b, and $(a \ominus (a \wedge b)) \wedge (b \ominus (a \wedge b)) = 0$ (see Corollary 1.8.5). □

Some additional properties of the compatibility in D-lattices are stated in the following proposition. The proof is left to the readers (recall that $a - b = a \ominus (a \wedge b)$, see Theorem 1.8.11).

Proposition 1.10.8 [RiNe, A.3.7] *Let P be a D-lattice, $a, b \in P$, and $a \leftrightarrow b$. Then:*

(i) $a' - b' = b - a$.

(ii) $a \vee b = a \oplus (b - a)$.

(iv) *If $c \leq a$ and $c \leq b$, then $(a \ominus c) \leftrightarrow (b \ominus c)$.*

(v) *If* $a \leq c'$ *and* $b \leq c'$, *then* $(a \oplus c) \leftrightarrow (b \oplus c)$.

We have already observed (see remarks before Lemma 1.7.10) that in a generalized effect algebra E with the Riesz decomposition property, every two elements are compatible if and only if E is upward directed. In particular, if E is an effect algebra, or if E is lattice ordered, then $a \leftrightarrow b$ holds for all $a, b \in E$. There are examples of effect algebras, where for all a, b, $a \leftrightarrow b$ holds, but the Riesz property is not satisfied (e.g., Fano plane, see below).[2] In the next theorem, we give a characterization of effect algebras with the Riesz property by means of a strengthened compatibility condition.

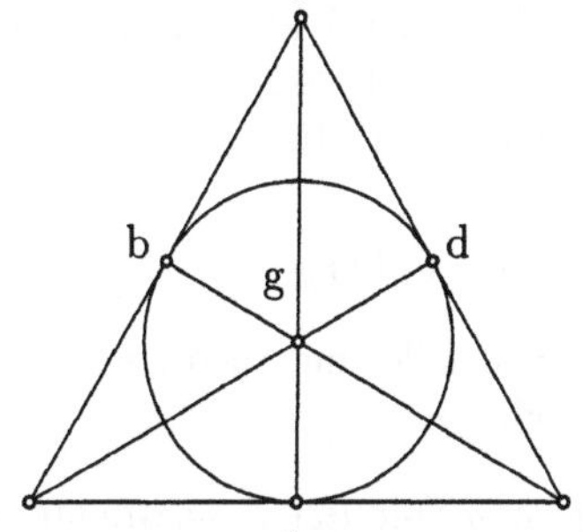

Fig. 1.9 Fano plane

Definition 1.10.9 *Let* x *be any element of* E. *Two elements* a, b *are called* x-**(Mackey) compatible** *(*$a \leftrightarrow_x b$*) if and only if* $a, b \leq x$ *and there is a Mackey decomposition* a_1, b_1, c *of* a, b *such that* $a_1 \oplus b_1 \oplus c \leq x$.

Theorem 1.10.10 *An effect algebra* E *satisfies the Riesz decomposition property if and only if, for any* $a, b, x \in E$, $a, b \leq x$ *implies* $a \leftrightarrow_x b$.

Proof. Assume the Riesz decomposition property. Let $a, b \leq x$. Then there is y such that $a \leq b \oplus y = x$. By the Riesz decomposition property, there are $c \leq b$, $a_1 \leq y$ such that $a = c \oplus a_1$. Let b_1 be such that $c \oplus b_1 = b$. Now $c \oplus b_1 = b$, and $a_1 \leq y$ imply $a_1 \oplus b_1 \oplus c \leq b \oplus y = x$. Hence $a \leftrightarrow_x b$.

To prove the converse, let $x \leq y \oplus z$. Put $w := y \oplus z$. By hypothesis, $x \leftrightarrow_w y$, hence there are x_1, y_1, u such that $x_1 \oplus y_1 \oplus u \leq w$ and $x = x_1 \oplus u$, $y = y_1 \oplus u$. Thus

[2]On the figures like Fano plane, Wright triangle, the Greechie diagrams are used. That is, points represent atoms in the corresponding effect algebra, and smooth lines connect maximal sets of orthogonal atoms. Every smooth line represents the effect subalgebra (called a block) generated by these atoms in the whole effect algebra. Angles in the diagram represent atoms belonging to the intersection of two blocks.

$u \leq y$. To prove $x_1 \leq z$, let v be such that $x_1 \oplus y_1 \oplus u \oplus v = w = y \oplus z$. By the cancellation property, $x_1 \oplus v = z$, hence $x_1 \leq z$. □

For a characterization of MV-algebras by means of compatibility, see Exercise 12, Section 1.12.

Corollary 1.10.11 *A GDL P satisfies the Riesz decomposition property if and only if any two elements in P are compatible.*

Proof. Since a GDL is upward directed, if the Riesz decomposition property is satisfied, then any two elements are compatible. Conversely, if any two elements are compatible, then by Corollary 1.10.7 and Proposition 1.10.5, there is a Mackey decomposition a_1, b_1, c such that $a_1 \oplus b_1 \oplus c = a \vee b$. Hence $a \leftrightarrow_x b$ for any $x \geq a, b$. The proof now follows by Theorem 1.10.10. □

1.10.2 Compatibility in GDLs

The method of the proof of the following theorem is a refinement of that used in Abelian ℓ-groups, see e.g. [Fuc].

Theorem 1.10.12 [Pul 10] *Let a D-lattice A satisfy the equality in* (1.5). *Let $(a_1, a_2, 0 \ldots, a_m)$ and $(b_1, b_2, \ldots, b_n)$ be two finite sequences of orthogonal elements of A such that*

$$a_1 \oplus a_2 \oplus \cdots \oplus a_m = b_1 \oplus b_2 \oplus \cdots \oplus b_n.$$

Then there are orthogonal elements $\{w_{ij}\}_{i \leq m, j \leq n}$ in A such that

$$a_i = \bigoplus_{j=1}^{n} w_{ij}, \; i = 1, 2, \ldots, m,$$

$$b_j = \bigoplus_{i=1}^{m} w_{ij}, \; j = 1, 2, \ldots, n,$$

and

$$\bigoplus_{s>j} w_{is} \wedge \bigoplus_{p>i} w_{pj} = 0, \; i = 1, 2, \ldots, m-1, \; j = 1, 2, \ldots, n-1.$$

Proof. We proceed by induction with respect to m, n.
Let $m = n = 2$. Define

$$w_{11} := a_1 \wedge b_1, \; w_{12} := a_1 \ominus (a_1 \wedge b_1), \; w_{21} := b_1 \ominus (a_1 \wedge b_1).$$

Then we have $a_1 = w_{11} \oplus w_{12}$, $b_1 = w_{11} \oplus w_{21}$. Using (1.5), we obtain

$$w_{21} = (a_1 \vee b_1) \ominus a_1, \; w_{12} = (a_1 \vee b_1) \ominus b_1.$$

Then we have

$$\begin{aligned} a_1 \vee b_1 &= a_1 \oplus w_{21} = b_1 \oplus w_{12} \\ &\leq a_1 \oplus a_2 = b_1 \oplus b_2 =: c. \end{aligned}$$

Define

$$w_{22} := c \ominus (a_1 \oplus w_{21}) = c \ominus (b_1 \oplus w_{12}).$$

Using the cancellation property,

$$a_1 \oplus a_2 = a_1 \oplus w_{21} \oplus w_{22} \Rightarrow a_2 = w_{21} \oplus w_{22},$$

and similarly,

$$b_1 \oplus b_2 = b_1 \oplus w_{12} \oplus w_{22} \Rightarrow b_2 = w_{12} \oplus w_{22}.$$

To prove the last statement, observe that by Proposition 1.8.3

$$w_{12} \wedge w_{21} = ((a_1 \vee b_1) \ominus a_1) \wedge ((a_1 \vee b_1) \ominus b_1) = 0.$$

Now suppose the statement true for m', n' with $m' \leq m, n' \leq n$ $(n \geq 3)$, and write the given equality in the form

$$a_1 \oplus a_2 \oplus \cdots \oplus a_m = b_1 \oplus b_2 \oplus \cdots \oplus (b_{n-1} \oplus b_n).$$

By the induction hypothesis, there exist elements

$$w_{ij},\ i = 1, 2, \ldots m;\ \ j = 1, 2, \ldots, n-2,$$

and

$$u_i,\ i = 1, 2, \ldots, m$$

such that

$$a_i = w_{i1} \oplus \cdots \oplus w_{in-2} \oplus u_i,\ i = 1, 2, \ldots, m,$$

and

$$\begin{aligned} b_j &= w_{1j} \oplus \cdots \oplus w_{mj}\ (j \leq n-2), \\ b_{n-1} \oplus b_n &= u_1 \oplus \cdots \oplus u_m. \end{aligned}$$

We again apply the induction hypothesis, now to the last equation, to conclude the existence of elements w_{ij}, $i = 1, 2, \ldots, m; j = n-1, n$ with the property

$$u_i = w_{in-1} \oplus w_{in};\ b_j = w_{ij} \oplus \cdots \oplus w_{mj}.$$

In addition, by the induction hypothesis,

$$(w_{i+1j} \oplus \cdots \oplus w_{mj}) \wedge (w_{ij+1} \oplus \cdots \oplus w_{i,n-2} \oplus u_i) = 0$$

for all $i < m, j \le n-2$, and also

$$(w_{i+1n-1} \oplus \cdots \oplus w_{mn-1}) \wedge w_{in} = 0.$$

It remains to prove the uniqueness assumption. We have

$$w_{ij} = (w_{ij} \oplus (w_{i+1j} \oplus \cdots \oplus w_{mj})) \wedge (w_{ij} \oplus (w_{ij+1} \oplus \cdots \oplus w_{in}))$$

where

$$w_{ij} \oplus \cdots \oplus w_{mj} = b_j \ominus (w_{1j} \oplus \cdots \oplus w_{i-1j}),$$

and

$$w_{ij} \oplus \cdots \oplus w_{in} = a_i \ominus (w_{i1} \oplus \cdots \oplus w_{ij-1}).$$

This shows that the w_{ij} may be computed recursively, completing the proof. □

The notion of a commutative minimal clan as a common abstraction of Boolean rings and commutative lattice ordered groups was introduced by Schmidt [Sch]. A similar notion appears also in [Wyl]. In what follows, we show relations between commutative minimal clans and GDLs.

A **minimal clan** is a set E with a relation $S \subseteq E \times E$, a map $\oplus : S \to E$, and an order relation $\le$ such that

(MC-1) there exists an element 0 in E satisfying $(0,x) \in S$, $(x,0) \in S$, and $0 \oplus x = x = x \oplus 0$ for all $x \in E$;

(MC-2) for all $x, y, z \in E$, $(x,y) \in S$ and $(x \oplus y, z) \in S$ if and only if $(y,z) \in S$ and $(x, y \oplus z) \in S$, and in this case $(x \oplus y) \oplus z = x \oplus (y \oplus z)$;

(MC-3) $x = y$ holds for all $x, y \in E$ satisfying $u \oplus x \oplus v = u \oplus y \oplus v$ for some $u, v \in E$ satisfying $(u,x) \in S$, $(u \oplus x, v) \in S$, and $(u \oplus y, v) \in S$;

(MC-4) $u \oplus x \oplus v \le u \oplus y \oplus v$ holds for all $x, y \in E$ satisfying $x \le y$ and for all $u, v \in E$ satisfying $(u,x) \in S$, $(u \oplus x, v) \in S$, $(u,y) \in S$, and $u \oplus y, v) \in S$;

(MC-5) $x \vee y$ and $x \wedge y$ exist for all $x, y \in E$;

(MC-6) for all $x, y \in E$, there exist $u, v \in E$ satisfying $0 \le u, 0 \le v$, $(u,v) \in S$, $(x,v) \in S$, $(u, x \wedge y) \in S$, $(x \wedge y, v) \in S$, $u \oplus x = x \vee y = x \oplus v$, and $u \oplus (x \wedge y) = y = (x \wedge y) \oplus v$.

Throughout this section, let $(E; S, \oplus, \le)$ be a minimal clan.

Two elements $x, y \in E$ are **summable** if $(x,y) \in S$, and the map $\oplus : S \to E$ is called **(partial) addition**, and the unique element $0 \in E$ satisfying (MC-1) is said to be the **zero element** of E. Axiom (MC-3) is the **cancellation property**, axiom (MC-6) is referred as the **difference property**; it may be used to define partial left and right subtractions.

An element $x \in E$ is called **invertible** if there exist $u, v \in E$ satisfying $u \oplus x = 0 = x \oplus v$, this is equivalent with the existence of some $w \in E$ satisfying $w \oplus x = 0$ or $0 = x \oplus w$, and this is in turn equivalent with the existence of a (unique) element $x^* \in E$ satisfying $x^* \oplus x = 0 = x \oplus x^*$, which is said to be the **inverse** of x. The set of all invertible elements will be denoted by $E_\circ$. An element $x \in E$ is **positive** if $0 \leq x$. The set of all positive elements of E will be denoted by E_+. A minimal clan is **positive** if $E = E_+$. Two elements $x, y \in E$ are **disjoint** if $x \wedge y = 0$. The set of all pairs of disjoint elements of E will be denoted by $\mathcal{D}$. A minimal clan is **commutative** if $(x, y) \in S$ if and only if $(y, x) \in S$ and $x \oplus y = y \oplus x$. The following lemma collects some simple results that are satisfied in minimal clans.

Lemma 1.10.13 [Sch] *Let $(E; S, \oplus, \leq)$ be a minimal clan.*

(i) *If $u \leq x, v \leq y$, and $(x, y) \in S$, then $(u, v) \in S$.*

(ii) *If $u \leq x$ and $x \in E_\circ$, then $u \in E_\circ$.*

(iii) *For all $x \in E$ $x \vee 0 \oplus x \wedge 0 = x = x \wedge 0 \oplus x \vee 0$.*

(iv) *$E_\circ \times E \subseteq S$ and $E \times E_\circ \subseteq S$.*

(v) *If $E_\circ = E$, then $S = E \times E$.*

(vi) *$(E_\circ; E_\circ \times E_\circ, \oplus, \leq)$ is a minimal clan.*

(vii) *The following are equivalent: $E_\circ = \{0\}$ and $E_+ = E$.*

(viii) *$(E_+; (E_+ \times E_+) \cap S, \oplus, \leq)$ is a positive minimal clan.*

(ix) *If $x \wedge y = 0$, then $x \oplus y = x \vee y = y \oplus x$.*

(x) *$\mathcal{D} \subseteq S$, and if $\mathcal{D} = S$, then $E_+ = E$.*

(xi) *$E = E_+$ if and only if the following modularity property holds: For all $x, y \in E$, $(x, y) \in S$ if and only if $(x \vee y, x \wedge y) \in S$, and in this case $x \oplus y = (x \vee y) \oplus (x \wedge y)$.*

(xii) *$u \oplus (x \wedge y) = y = (x \wedge y) \oplus v$ if and only if $u \oplus x = x \vee y = x \oplus v$, and in this case $u \in E_+$, $v \in E_+$. Moreover, u, v are uniquely defined.*

(xiii) *If $u \oplus x \oplus v \leq u \oplus y \oplus v$, then $x \leq y$ (the order cancellation property).*

It is easy to see that every commutative, positive minimal clan is a GEA. Indeed, (E1) is satisfied by commutativity, (E2) by (MC-2), (E3') by (MC-3), (S4) existence of a zero follows by (MC-1) and (E4') by positivity and Lemma 1.10.13 (vii), the lattice structure is guaranteed by (MC-5). Moreover, every minimal clan is a distributive lattice ([Sch, Thm 1.1.14]).

Theorem 1.10.14 [Pul 5] *Let P be a GDL. The following statements are equivalent.*

(i) *P is a positive, commutative minimal clan.*

(ii) *For any* $x, y \in P$ *there is a* $u \in P$ *such that* $u \oplus x = x \vee y$ *and* $u \oplus (x \wedge y) = y$. *Equivalently,*

$$(x \vee y) \ominus x = u = y \ominus (x \wedge y).$$

(iii) *If* $x, y \in P$ *and* $x \wedge y = 0$, *then* $x \oplus y$ *exists and* $x \oplus y = x \vee y$.

(iv) *Any* $x, y \in P$ *are compatible.*

(v) *Any* $x, y \in P$ *are strongly compatible.*

(vi) *Every two orthogonal sets* $(u_1, u_2, \ldots, u_m)$ *and* $(v_1, v_2, \ldots, v_n)$ *such that*

$$\bigoplus_{i=1}^{m} u_i = \bigoplus_{j=1}^{n} v_j$$

have a common refinement, i.e., there is an orthogonal system $\{w_{ij}\}_{ij}$ *such that* $u_i = \bigoplus_{j=1}^{n} w_{ij}$, $i = 1, 2, \ldots, m$ *and* $v_j = \bigoplus_{i=1}^{m} w_{ij}$, $j = 1, 2, \ldots, n$, *and*

$$\bigoplus_{k=i+1}^{m} w_{kj} \wedge \bigoplus_{l=j+1}^{n} w_{il} = 0$$

for all $i \in \{1, 2, \ldots, m-1\}$ *and* $j \in \{1, 2, \ldots, n-1\}$.

Proof. (i) $\Rightarrow$ (ii) follows by (MC-6) in the definition of the minimal clan.

(ii) $\Rightarrow$ (iii). $x \wedge y = 0 \Rightarrow (x \vee y) \ominus y = x \ominus (x \wedge y) = x \Rightarrow x \vee y = x \oplus y$.

(iii) $\Rightarrow$ (iv). For any $x, y \in P$, $x = (x \wedge y) \oplus (x \ominus (x \wedge y))$, $y = (x \wedge y) \oplus (y \ominus (x \wedge y))$. Since $(x \ominus (x \wedge y)) \wedge (y \ominus (x \wedge y)) = 0$, it follows by (iii) that $x \ominus (x \wedge y)$ and $y \ominus (x \wedge y)$ are orthogonal. Moreover, by Proposition 1.8.4 with $c = a \wedge b$ and Proposition 1.8.5,

$$\begin{aligned} x \vee y &= (x \wedge y) \oplus ((x \vee y) \ominus (x \wedge y)) = (x \wedge y) \oplus ((x \ominus (x \wedge y)) \vee (y \ominus (x \wedge y))) \\ &= (x \wedge y) \oplus (x \ominus (x \wedge y)) \oplus (y \ominus (x \wedge y)). \end{aligned}$$

Hence, $x \wedge y$, $x \ominus x \wedge y$, $y \ominus x \wedge y$ are jointly orthogonal.

(iv)$\Leftrightarrow$(v). Corollary 1.10.7.

(v) $\Rightarrow$ (vi). Theorem 1.10.12

(vi) $\Rightarrow$ (i). (MC-1)–(MC-5) and positivity follow from properties of a GDL. It remains to prove (MC-6). For any $x, y \in P$, we have $x \oplus u = x \vee y = y \oplus v$ for some $u, v \in P$. By (vi), $x = w_{11} \oplus w_{12}$, $y = w_{11} \oplus w_{21}$, and $w_{11} \oplus w_{12} \oplus w_{21} \oplus w_{22} = x \vee y$, hence x and y are compatible. The desired result follows from Theorem 1.10.6. □

In orthomodular lattices and posets, a very important role is played by blocks, that is, maximal Boolean subalgebras. In orthomodular lattices, as well as in so-called regular orthomodular posets, blocks coincide with the maximal subsets of pairwise compatible elements. In more general orthomodular posets, pairwise compatibility must be replaced by the stronger notion of f-compatibility (cf. [PtPu], [PtPu 1] and citations therein).

The definition of blocks as maximal subsets of pairwise compatible elements can be extended also to D-lattices ([RiZ 10]). We will end this subsection with this interesting result. We will start with a proposition which is itself of interest.

Proposition 1.10.15 [JeRi] *Let E be a lattice ordered effect algebra. Assume $b \in E$, $A \subseteq E$ are such that $\bigvee A$ exists in E and $b \leftrightarrow a$ for all $a \in A$. Then*

(a) $b \leftrightarrow \bigvee A$.

(b) $\bigvee\{b \wedge a : a \in A\}$ *exists in E and equals* $b \wedge (\bigvee A)$.

Proof.
(a) For every $a \in A$,

$$a \leq (b \ominus (a \wedge b))' \leq (b \ominus ((\bigvee A) \wedge b))'.$$

Therefore,

$$\bigvee A \leq (b \ominus ((\bigvee A) \wedge b))'$$

and this is equivalent to $\bigvee A \leftrightarrow b$.

(b) First, let us prove

$$\bigwedge\{b \ominus (b \wedge a) : a \in A\} = b \ominus (b \wedge (\bigvee A)). \tag{1.7}$$

Assume d is a lower bound of $\{b \ominus (b \wedge a) : a \in A\}$. Since $b \leftrightarrow a$, we have $b \ominus (b \wedge a) = (b \vee a) \ominus a$. Evidently, $d \leq (b \vee (\bigvee A)) \ominus a$ for all $a \in A$. This implies $a \leq (b \vee (\bigvee A)) \ominus d$ and we have $\bigvee A \leq (b \vee (\bigvee A)) \ominus d$ which is equivalent to $d \leq (b \vee (\bigvee A)) \ominus (\bigvee A)$. By part (a), $b \leftrightarrow \bigvee A$. Therefore,

$$d \leq (b \vee (\bigvee A)) \ominus (\bigvee A) = b \ominus (b \wedge (\bigvee A)),$$

which completes the proof of (1.7).

Now, let e be an upper bound of $\{b \wedge a : a \in A\}$. For all $a \in A$, $b \wedge a \leq b \wedge e$ and $b \ominus (b \wedge e) \leq b \ominus (b \wedge a)$. Therefore, $b \ominus (b \wedge e) \leq \bigwedge\{b \ominus (b \wedge a) : a \in A\}$. Applying (1.7), $b \ominus (b \wedge e) \leq b \ominus (b \wedge (\bigvee A))$. Thus, $b \wedge (\bigvee A) \leq (b \wedge e) \leq e$. Moreover, $b \wedge (\bigvee A)$ is an upper bound of $\{b \wedge a : a \in A\}$. This completes the proof. □

Corollary 1.10.16 *Let E be a lattice ordered effect algebra. Assume $b \leftrightarrow a_1$, $b \leftrightarrow a_2$. Then $b \leftrightarrow (a_1 \vee a_2)$, and $b \wedge (a_1 \vee a_2) = (b \wedge a_1) \vee (b \wedge a_2)$.*

Theorem 1.10.17 [RiZ 10] *Let $(P; \leq, \ominus, 0, 1)$ be a D-lattice and let $x, y, z \in P$ be such that $x \leftrightarrow z$ and $y \leftrightarrow z$. Then*

(i) $x \vee y \leftrightarrow z$.

(ii) *If* $x \leq y$, *then* $y \ominus x \leftrightarrow z$,

(iii) $x' = 1 \ominus x \leftrightarrow z$.

(iv) $x \wedge y \leftrightarrow z$.

(v) $x \oplus y \leftrightarrow z$ *whenever* $x \perp y$.

Proof. By assumptions there exist $x \oplus (z \ominus (x \wedge z))$ and $y \oplus (z \ominus (y \wedge z))$.

(i) It follows by Corollary 1.10.16.

(ii) If $x \leq y$, then $x \wedge z \leq y \wedge z$ and $x \vee z \leq y \vee z$. It follows that there exists $w \in P$ such that $(x \wedge z) \oplus w = y \wedge z$ and $x \vee z = x \oplus (z \ominus (x \wedge z)) \leq y \vee z = y \oplus (z \ominus (y \wedge z)) = (y \wedge z) \oplus (y \ominus (y \wedge z)) \oplus (z \ominus (y \wedge z))$, thus $(x \wedge z) \oplus (x \ominus (x \wedge z)) \oplus (z \ominus (x \wedge z)) \leq (y \wedge z) \oplus (y \ominus (y \wedge z)) \oplus (z \ominus (y \wedge z))$ and since $z = (x \wedge z) \oplus (z \ominus (x \wedge z)) = (y \wedge z) \oplus (z \ominus (y \wedge z))$ we obtain $x \ominus (x \wedge z) \leq y \ominus (y \wedge z)$. The last inequality implies that there is $e \in P$ such that $(x \ominus (x \wedge z)) \oplus e = y \ominus (y \wedge z)$. We obtain $y = (x \wedge z) \oplus w \oplus e \oplus (x \ominus (x \wedge z))$ and $y \oplus (z \ominus (y \wedge z)) = (x \wedge z) \oplus w \oplus e \oplus (x \ominus (x \wedge z)) \oplus (z \ominus (y \wedge z))$. These equalities imply that $y \ominus x = w \oplus e$ and $z = w \oplus [(x \wedge z) \oplus (z \ominus (y \wedge z))]$, since $(x \wedge z) \oplus w = y \wedge z$. We conclude that $y \ominus x \leftrightarrow z$ since $w \oplus e \oplus [(x \wedge z) \oplus (z \ominus (y \wedge z))]$ exists.

(iii) follows by Proposition 1.10.3.

(iv) By (iii) $x' \leftrightarrow z$ and $y' \leftrightarrow z$ which by (i) implies that $x' \vee y' \leftrightarrow z$ and by (iii) $x \wedge y = (x' \vee y')' \leftrightarrow z$.

(v) $x \oplus y = 1 \ominus (x' \ominus y) \leftrightarrow z$ by conditions (ii) and (iii). □

Corollary 1.10.18 (a) *Every maximal subset M of mutually compatible elements of a D-lattice $(P; \leq, \ominus, 0, 1)$ is a sub-D-lattice of P.*

(b) *Every maximal subset M of mutually compatible elements of a lattice effect algebra $(P; \oplus, 0, 1)$ is a sub-effect-algebra of P.*

Definition 1.10.19 *A maximal subset M of mutually compatible elements of a D-lattice (lattice effect algebra) P is called a* **block of** P.

Theorem 1.10.20 *Every D-lattice (lattice effect algebra) P is a set-theoretical union of its blocks. Every subset $A \subseteq P$ of mutually compatible elements is contained in a block.*

Proof. Let $\emptyset \neq A \subseteq P$ be a set of mutually compatible elements of P and $\mathcal{A} = \{B \subseteq P : A \subseteq B, B$ is a set of mutually compatible elements$\}$. Then for every chain $\mathcal{B} \subseteq \mathcal{A}$ (i.e., for $X, Y \in \mathcal{B}$ we have $X \subseteq Y$ or $Y \subseteq X$) the set $\bigcup \mathcal{B} \in \mathcal{A}$. By Zorn lemma, there exists a maximal element $M \in \mathcal{A}$. Moreover, for $a \in P$, the set $A = \{0, a, a', 1\}$ is mutually compatible. □

From Corollary 1.10.18, we derive the following result.

Theorem 1.10.21 [RiZ 10] *Every D-lattice/lattice ordered effect algebra can be written as a set-theoretical union of MV-algebras.*

1.10.3 Observables

Let P and Q be D-posets (effect algebras). Recall that a morphism $w : P \to Q$ is a σ-morphism if $a_n \in P$, $n = 1, 2, \ldots$, and $a_n \uparrow a$ (that is, $a_n \leq a_{n+1}$, $n = 1, 2, \ldots$, and $\bigvee a_n = a$), $a \in P$, imply $w(a_n) \uparrow w(a)$.

Recall that a (σ-additive) state on P is a (σ-morphism from P) to the unit interval $[0, 1]$ of real numbers. Observe that a state $s : P \to [0, 1]$ is a σ-state if and only if s is σ-additive, that is, if $\{a_n\}_{n \in \mathbb{N}}$ is a $\oplus$-orthogonal system with the sum $a = \bigoplus_{n=1}^{\infty} a_n$, then $s(a) = \sum_{n=1}^{\infty} s(a_n)$. Note that also a completely additive state can be defined in an obvious way.

Definition 1.10.22 *Let (Ω, Σ) be measurable space, and P be a D-poset (effect algebra). A σ-morphism x from Σ to P is called a (Ω, Σ)-observable on P. If $(\Omega, \Sigma) = (\mathbb{R}, \mathcal{B}(\mathbb{R}))$, x is called a* **real observable.**

If we replace σ-morphism by a morphism, we obtain a **finitely additive observable.**

In what follows, if not stated otherwise, by an **observable** we will mean a real observable.

If x is an observable and s is a σ-additive state on P, then $s \circ x : \mathcal{B}(\mathbb{R}) \to [0, 1]$, $s \circ x(E) = s(x(E))$, is a probability measure on $\mathcal{B}(\mathbb{R})$, which is called the **probability distribution** of the observable x in the state s. Now a mean value of the observable x in the state s can be defined by the integral

$$E_s(x) = \int_{\mathbb{R}} t(s \circ x)(dt),$$

if it exists (and is finite).

The set $\mathcal{R}(x) = \{x(E) : E \in \Sigma\}$ is said to be the **range** of x. It is well known that on an orthomodular poset the range of every observable is a Boolean algebra [PtPu 1]. In particular, each countably generated Boolean algebra is the range of a real observable [Var]. Moreover, every orthomodular poset can be represented as the pasting of Boolean algebras [Rog].

Differently from the sharp case, the range of an unsharp observable not only fails to be a Boolean algebra, but is not even closed under the orthogonal sum.

Lemma 1.10.23 *There exists an observable α such that*

(1) *$\mathcal{R}(\alpha)$ is not closed under $\oplus$;*

(2) *$\mathcal{R}(\alpha)$ does not satisfy the associativity condition* (E2) *of Definition* 1.2.1*).*

(3) *Let $a, b \in \mathcal{R}(\alpha)$. The relation $\lessdot$ (where $a \lessdot b$ if and only if there exists $c \in \mathcal{R}(\alpha)$ such that $(a \oplus c)$ exists and $b = a \oplus c$) does not define a partial order in $\mathcal{R}(\alpha)$.*

Proof. Let $\mathcal{A} = [0,1]$ be the unit-interval effect algebra. Let $\Omega := \{1,2,3,4\}$ and let $\mathcal{B}(\Omega)$ be the power-set of Ω. Let α be the observable defined as follows:

$$\alpha(\{1\}) = \tfrac{3}{12}, \qquad \alpha(\{2\}) = \tfrac{4}{12},$$
$$\alpha(\{3\}) = \tfrac{1}{12}, \qquad \alpha(\{4\}) = \tfrac{4}{12}.$$

It turns out that $\mathcal{R}(\alpha) = \{0, 1, \frac{3}{12}, \frac{4}{12}, \frac{1}{12}, \frac{7}{12}, \frac{5}{12}, \frac{8}{12}, \frac{9}{12}, \frac{11}{12}\}$.

(1) $\mathcal{R}(\alpha)$ is not closed under $\oplus$. For example, $\alpha(\{1\}) \oplus \alpha(\{1\}) = \frac{6}{12} \notin \mathcal{R}(\alpha)$.

(2) Associativity fails. For example, there exist $(\alpha(\{1\}) \oplus \alpha(\{3\}))$, and $(\alpha(\{1\}) \oplus (\alpha(\{1\}) \oplus \alpha(\{3\}))) = \frac{7}{12}$; however, $\alpha(\{1\}) \oplus \alpha(\{1\}) = \frac{6}{12} \notin \mathcal{R}(\alpha)$.

(3) The relation $\lessdot$ is not transitive. For example, $\frac{1}{12} \lessdot \frac{4}{12}$ since $\frac{1}{12} \oplus \frac{3}{12} = \frac{4}{12}$ and $\frac{3}{12} \in \mathcal{R}(\alpha)$. Further, $\frac{4}{12} \lessdot \frac{7}{12}$ since $\frac{4}{12} \oplus \frac{3}{12} = \frac{7}{12}$. But, $\frac{1}{12} \not\lessdot \frac{7}{12}$ since $\frac{1}{12} \oplus \frac{6}{12} = \frac{7}{12}$ and $\frac{6}{12} \notin \mathcal{R}(\alpha)$.

As a consequence $\mathcal{R}(\alpha)$ is not a subeffect algebra of $\mathcal{A}$. □

Definition 1.10.24 *Let P be an effect algebra, $a, b \in P$. Elements a and b are called* **commeasurable** *($a \, C \, b$) if there is an observable x on P with $a, b \in \mathcal{R}(x)$.*

Proposition 1.10.25 *Let x be an observable on a D-poset P. The range $\mathcal{R}(x)$ consists of pairwise compatible elements.*

Proof. Let $a, b \in \mathcal{R}(x)$, $a = x(A)$, $b = x(B)$ where $A, B \in \mathcal{B}(\mathbb{R})$. Put $c = x(A \cup B)$, and $d = x(A \cap B)$. Evidently $d \le a \le c$, $d \le b \le c$ and

$$\begin{aligned} c \ominus a &= x(A \cup B) \ominus x(A) = x((A \cup B) \setminus A) \\ &= x(B \setminus (A \cap B)) = x(B) \ominus x(A \cap B) = b \ominus d. \end{aligned}$$

Therefore, $a \leftrightarrow b$. □

From Proposition 1.10.25 we see that commeasurable elements are pairwise compatible. The converse need not be true, even in an orthomodular poset, see e.g., [PtPu 1]. To characterize commeasurable elements in orthomodular posets, finite partitions of unity are needed. In agreement with the definition in OMPs, we give the following definitions.

A **finite partition of unity** in an effect algebra $(A; \oplus, 0, 1)$ is a finite sequence $a_1, \ldots, a_n$ of elements of A such that $\bigoplus_{i \le n} a_i = 1$ (i.e., the latter sum exists and equals 1).

Definition 1.10.26 *Let $(A; \oplus, 0, 1)$ be an effect algebra. An observable $\alpha : \mathcal{B}(\Omega) \to A$ is* **simple** *if and only if there is a finite subset $\Omega_0 = \{\omega_1, \ldots, \omega_n\}$ of Ω such that $\{\omega_i\} \in \mathcal{B}(\Omega)$ for $i = 1, 2, \ldots, n$, and $\alpha(\{\omega_1\}) \oplus \cdots \oplus \alpha(\{\omega_n\}) = 1$.*

Definition 1.10.27 *A finite subset $F = \{a_1, \ldots, a_n\}$ of a D-poset A is* **f-compatible** *if there is a finite partition of unity G such that every a_i is a $\oplus$-sum of a subset G_i of G.*

An arbitrary subset of A is **f-compatible** *if every finite subset is f-compatible.*

Proposition 1.10.28 *There is a one-to-one correspondence between simple observables and finite partitions of unity in A.*

Proof. If $\alpha : \mathcal{B}(\Omega) \to A$ is a simple observable, there is a finite set $\Omega_0 := \{\omega_1, \ldots, \omega_n\}$ of points in Ω such that each singleton $\{\omega_i\}$, for $i = 1, 2, \ldots, n$, belongs to $\mathcal{B}(\Omega)$ and $R := \{\alpha(\{\omega_1\}), \ldots, \alpha(\{\omega_n\})\}$ is a partition of unity. Let us notice that Ω_0 and any of its subsets are elements of $\mathcal{B}(\Omega)$, i.e., the whole power set 2^{Ω_0} of Ω_0 is contained in $\mathcal{B}(\Omega)$. Therefore, we can introduce $\alpha^R : 2^{\Omega_0} \to A$ defined as $\alpha^R(X) = \bigoplus_{\{i : \omega_i \in X\}} \alpha(\{\omega_i\})$, $X \in 2^{\Omega_0}$. Then α^R is an observable which coincides with α on 2^{Ω_0}, this latter being such that $\alpha(\Omega_0) = 1$ [whence $\alpha(\Omega \setminus \Omega_0) = 0$], i.e., the support of α is Ω_0. It follows that we can identify the original A-valued observable α on $\mathcal{B}(\Omega)$ with the A-valued observable α^R on 2^{Ω_0}.

Conversely, let $R := \{a_1, \ldots, a_n\}$ be a finite partition of unity in A. Let $\Omega_0 = \{\omega_1, \ldots, \omega_n\}$ be a finite set equipped with the algebra of subsets 2^{Ω_0}, and define, for all $X \in 2^{\Omega_0}$, $\alpha(X) = \bigoplus_{\{i : \omega_i \in X\}} a_i$. It is obvious that $\alpha : 2^{\Omega_0} \to A$ is a simple A-valued observable on 2^{Ω_0}. □

Corollary 1.10.29 *A finite subset of elements of a D-poset is f-compatible if and only if it is contained in a range of a simple observable.*

The functional calculus for compatible observables on quantum logics is based on a representation of these observables by Borel measurable functions. In a similar way, we will try to construct a functional calculus for observables in D-posets [ChKo], [RiNe, Thm A.4.12].

Theorem 1.10.30 (Representation Theorem) *Let x, y be two observables on a D-poset P. The following conditions are equivalent:*

(i) *There is a Borel measurable function $f : \mathbb{R} \to \mathbb{R}$ such that $x(E) = y(f^{-1}(E))$ for any $E \in \mathcal{B}(\mathbb{R})$.*

(ii) *There is a chain M, $M \subseteq \mathcal{B}(\mathbb{R})$, such that $\{x((-\infty, r)) : r \in \mathbb{Q}\} \subseteq \{y(A) : A \in M\}$, where $\mathbb{Q}$ is the set of all rationals.*

Proof. Let M be a linearly ordered set of the Borel subsets such that

$$\{x((-\infty, r)) : r \in \mathbb{Q}\} \subseteq \{y(A) : A \in M\}.$$

Then for every $r \in \mathbb{Q}$ there is a Borel subset $A_r \in M$ such that $x((-\infty, r)) = y(A_r)$. We note that if $y(A) \leq y(B)$ for $A, B \in M$, then there are $C, D \in M$ such that $A \subseteq C$ and $y(B) = y(C)$, $D \subseteq B$ and $y(A) = y(D)$. Indeed, it suffices to put $C = A \cup B$, $D = A \cap B$. Similarly, if $A, B, C \in M$, $A \subseteq C$ and $y(A) \leq y(B) \leq y(C)$, then there is $D \in M$ such that $A \subseteq D \subseteq C$ and $y(D) = y(B)$. It suffices to put $D = A \cup (B \cap C)$.

Now we can construct, by induction, a sequence $\{B_n\}_{n=1}^{\infty}$ in M such that

$$x((-\infty, r_n)) = y(B_n) \text{ for any } r_n \in \mathbb{Q}; \text{ and if } r_i < r_j, \text{ then } B_i \subseteq B_j.$$

Let $B = \bigcap_{n=1}^{\infty} B_n$. Put $A_n = B_n \setminus B$. Since $y(B) = x(\emptyset) = 0$, we have

$$y(A_n) = y(B_n \setminus B) = y(B_n) \ominus y(B) = y(B_n) = x((-\infty, r_n)).$$

The sequence $\{A_n\}_{n=1}^{\infty}$ is constructed such that:

(i) $x((-\infty, r_n)) = y(A_n)$ for any $r_n \in \mathbb{Q}$, $n = 1, 2, \ldots$;

(ii) $A_i \subseteq A_j$ if $r_i < r_j$;

(iii) $\bigcap_{n=1}^{\infty} A_n = \emptyset$.

We define an $\mathcal{B}(\mathbb{R})$-measurable function $f : \mathbb{R} \to \mathbb{R}$ as follows:

$$f(t) = \begin{cases} 0, & \text{if } t \in \mathbb{R} \setminus \bigcup_{n=1}^{\infty} A_n, \\ \inf \{r_i \in \mathbb{Q} : t \in A_i\}, & \text{if } t \in \bigcup_{n=1}^{\infty} A_n. \end{cases}$$

The function f is everywhere well-defined and finite. Moreover,

$$f^{-1}((-\infty, r_k)) = \begin{cases} \bigcup_{r_i < r_k} A_i, & \text{if } r_k \leq 0, \\ \bigcup_{r_i < r_k} A_i \cup (R \setminus \bigcup_{n=1}^{\infty} A_n), & \text{if } r_k > 0, \end{cases}$$

hence f is $\mathcal{B}(\mathbb{R})$-measurable. Let $r \in \mathbb{Q}$, $r \leq 0$. Then

$$y(f^{-1}((-\infty, r))) = y(\bigcup_{r_i < r} A_i) = y(\bigcup_{i=1}^{\infty} A_{j_i}) = \bigvee_{n=1}^{\infty} y(\bigcup_{i=1}^{n} A_{j_i}) =$$

$$= \bigvee_{n=1}^{\infty} y(A_{K_n}) = \bigvee_{n=1}^{\infty} x((-\infty, r_{K_n})) = x(\bigcup_{n=1}^{\infty} (-\infty, r_{K_n})) = x((-\infty, r)),$$

where $\{r_{j_i}\}_{n=1}^{\infty} = \{r_i \in \mathbb{Q} : r_i < r\}$, $r_{K_n} = \max \{r_{j_1}, r_{j_2}, \ldots, r_{j_n}\}$.

Similarly, if $r > 0$.

It is clear that $y(f^{-1}(\mathbb{R})) = x(\mathbb{R})$, because $y(f^{-1}(\mathbb{R})) = 1$.

Let $[a, b)$ be an interval, $a, b \in \mathbb{Q}$, $a < b$. Then $[a, b) = (-\infty, b) \setminus (-\infty, a)$, therefore, $y(f^{-1}([a, b))) = x([a, b))$. Let us denote $\mathcal{S} = \{[a, b) : a, b \in \mathbb{Q}, a < b\}$. It is not difficult to show that

$$y(f^{-1}([a,b)\cup[c,d)))=x([a,b)\cup[c,d))$$

and

$$y(f^{-1}([a,b)\setminus[c,d)))=x([a,b)\setminus[c,d)).$$

Now we put

$$\mathcal{K}=\{A\in\mathcal{B}(\mathbb{R}) : y(f^{-1}(A))=x(A)\}.$$

The system $\mathcal{K}$ contains the algebra $a(\mathcal{S})$ over the system $\mathcal{S}$. We show that $\mathcal{K}$ is a monotone system.

Let $\{E_n\}_{n=1}^{\infty}\subset\mathcal{K}$, $E_n\subseteq E_{n+1}$ for any $n\in N$. Then

$$y(f^{-1}(\bigcup_{n=1}^{\infty}E_n))=y(\bigcup_{n=1}^{\infty}f^{-1}(E_n))=\bigvee_{n=1}^{\infty}y(f^{-1}(E_n))=\bigvee_{n=1}^{\infty}x(E_n)=x(\bigcup_{n=1}^{\infty}E_n).$$

Then $\mathcal{B}(\mathbb{R})=\sigma(\mathcal{S})\subseteq\mathcal{M}(a(\mathcal{S}))\subseteq\mathcal{K}$, where $\sigma(\mathcal{S})$ denotes the least σ-algebra over $\mathcal{S}$ and $\mathcal{M}(a(\mathcal{S}))$ denotes the least monotone system over $a(\mathcal{S})$, which implies that $\mathcal{K}=\mathcal{B}(\mathbb{R})$.

Conversely, let $f:\mathbb{R}\to\mathbb{R}$ be a Borel measurable function with $y(f^{-1}(E))=x(E)$ for every $E\in\mathcal{B}(\mathbb{R})$. Then the system $M=\{f^{-1}((-\infty,r)) : r\in\mathbb{Q}\}$ is a chain such that

$$\{x((-\infty,r)) : r\in\mathbb{Q}\}\subseteq\{y(A) : A\in M\}.$$

□

Definition 1.10.31 [ChKo] *Let x be an observable on a D-poset P and let Q be a subset of the range $\mathcal{R}(x)$. We say that the observable x has a* **V-property** *on Q, if for every two Borel sets A,B, $A\subseteq B$, and for every element $c\in Q$ such that $x(A)\le c\le x(B)$, there exists a Borel set C such that $x(C)=c$ and $A\subseteq C\subseteq B$.*

Lemma 1.10.32 [ChKo] *Let x be an observable such that the following implication holds: If $x(A)\le x(B)$, then $x(A\cup B)=x(B)$ (equivalently $x(A\cap B)=x(A)$) for every $A,B\in\mathcal{B}(\mathbb{R})$. Then the observable x has the V-property on the range $\mathcal{R}(x)$.*

Proof. If $A,B\in\mathcal{B}(\mathbb{R})$, $A\subseteq B$ and $c\in\mathcal{R}(x)$ such that $x(A)\le c\le X(B)$, then there exists a set $C_1\in\mathcal{B}(\mathbb{R})$ such that $c=x(C_1)$. We put $C=A\cup(B\cap C_1)$. Then $c=x(C_1)=x(B\cap C_1)=x(A\cup(B\cap C_1))=x(C)$. It is evident that $A\subseteq C\subseteq B$. □

Lemma 1.10.33 [ChKo] *Let x be an observable on a D-poset P with the V-property on $\mathcal{R}(x)$. If $x(A)\le x(B)$, then $x(B)\ominus x(A)\in\mathcal{R}(x)$.*

Proof. It holds $x(A) \leq x(B) \leq x(A \cup B)$. From the V-property of the observable x it follows that there exists a set $B_1 \in \mathcal{B}(\mathbb{R})$ such that $A \subseteq B_1 \subseteq A \cup B$, $x(B) = x(B_1)$ and, moreover, $x(B) \ominus x(A) = x(B_1) \ominus x(A) = x(B_1 \setminus A) \in \mathcal{R}(x)$. □

Theorem 1.10.34 [ChKo] *Let x and y be two observables such that the following conditions hold:*

(i) $\mathcal{R}(x) \subseteq \mathcal{R}(y)$.

(ii) *The observable y has the V-property on $\mathcal{R}(x)$.*

Then there exists a Borel measurable function $f : \mathbb{R} \to \mathbb{R}$ such that $x(E) = y(f^{-1}(E))$ for every $E \in \mathcal{B}(\mathbb{R})$.

Proof. This proof may be realized in the same way as the proof of Theorem 1.10.30. □

In the end of this section, we will discuss the following question: Under what conditions a subset F of elements of an effect algebra $(A; \oplus, 0, 1)$ can be embedded into the range of a (finitely additive) observable? We will answer the question by using projective limits of simple observables (equivalently, partitions of unity in A) in a similar way as in [LaPu 1] (see also [LaPu] for the special case of the Hilbert space effects).

Let Ω be a nonempty set and $\mathcal{B}(\Omega)$ an algebra of subsets of Ω. We will consider $(\Omega, \mathcal{B}(\Omega))$-observables on an effect algebra A.

Definition 1.10.35 (a) *Let $(D; \leq)$ be a directed set and Ω_i a finite set for each $i \in D$. Whenever $i, j \in D$ and $i \leq j$, let there be given a mapping $g_{i,j} : \Omega_j \to \Omega_i$ and denote by $\mathcal{G} := \{g_{i,j} : i, j \in D$ and $i \leq j\}$ the collection of all such mappings. The pair $((\Omega_i)_{i \in D}, \mathcal{G})$ is called a* **projective system of finite sets** *if the following conditions hold:*

(i) *$g_{i,i}$ is the identity map on Ω_i for each $i \in D$.*

(ii) *$g_{i,j} \circ g_{j,k} = g_{i,k}$, whenever $i \leq j \leq k$.*

(b) *We say that $((\Omega_i, \alpha_i)_{i \in D}, \mathcal{G})$ is a* **projective system of simple observables** *if $((\Omega_i)_{i \in D}, \mathcal{G})$ is a projective system of finite sets and for each $i \in D$, $\alpha_i : 2^{\Omega_i} \to A$ is a simple observable such that the following* compatibility condition *holds:*

(iii) *For all $X \in 2^{\Omega_i}$, $\alpha_i(X) = \alpha_j(g_{i,j}^{-1}(X))$, whenever $i \leq j$.*

Let $((\Omega_i)_{i \in D}, \mathcal{G})$ be a projective system of finite sets with directed set D. If $\Omega_D := \prod_{i \in D} \Omega_i$ is the cartesian product of the Ω_i, let Ω be its subset consisting of those elements, called **threads**, $\Omega = \{(\omega_i)_{i \in D} \in \Omega_D$ such that, for all $i, j \in D$ satisfying

$i \le j$ we have $g_{i,j}(\omega_j) = \omega_i\}$. Then Ω is called the **projective limit of the system** $((\Omega_i)_{i\in D}, \mathcal{G})$ and denote $\Omega =: \lim_{\leftarrow}((\Omega_i)_{i\in D}, \mathcal{G})$ (cf. [Rao]). For this space we consider the i-projection $g_i : \Omega \to \Omega_i$, $\omega \mapsto \omega_i$, $i \in D$. Trivially, $g_i = g_{i,j} \circ g_j$ whenever $i \le j$.

In general, Ω can be very small (or empty) even if each $g_{i,j}$ is an onto mapping. In order to overcome this difficulty, the following (sufficient) condition (due to Bochner) is useful.

Definition 1.10.36 *A projective system of finite sets $((\Omega_i, g_{i,j}) : i, j \in D)$ is said to satisfy the* **sequential maximality condition (s.m. condition)** *if, for each sequence $i_1 \le i_2 \le \cdots$ in D, and any point $\omega = (\omega_i)_{i\in D} \in \Omega_D$ such that*

$$g_{i_n, i_{n+1}}(\omega_{i_{n+1}}) = \omega_{i_n},\ n \ge 1,$$

we have $\omega \in \Omega$.

It is not hard to verify that the s.m. condition holds if each Ω_i is a nonempty compact Hausdorff space (in particular, if it is finite) and all $g_{i,j}$'s are onto mappings ([Rao]). In the latter case $\emptyset \neq \Omega \subset \Omega_D$ is also compact.

Theorem 1.10.37 *Let A be an effect algebra. A subset K of A can be embedded into a range of an observable α if and only if there is a projective system $((\Omega_i, \alpha_i)_{i\in D}, \mathcal{G})$ of simple observables such that for each $k \in K$ there is an $i \in D$ such that $k \in \mathcal{R}(\alpha_i)$.*

Proof (i) Assume that $K \subseteq \mathcal{R}(\alpha)$ for an observable $\alpha : \mathcal{B}(\Omega) \to A$, where $\mathcal{B}(\Omega)$ is a Boolean algebra of subsets of a nonempty set Ω [i.e., for every $k \in K$, there exists $X_k \in \mathcal{B}(\Omega)$ such that $\alpha(X_k) = k$]. Let $\mathcal{P}$ be the set of all finite disjoint partitions of Ω with elements in $\mathcal{B}(\Omega)$ [in particular, for every $k \in K$ the pair $R_k := \{X_k, \Omega \setminus X_k\}$ is a finite partition of Ω, i.e., an element of $\mathcal{P}$]. For each $R = \{X_i\}_{i \le n} \in \mathcal{P}$, the mapping $R \ni X_i \mapsto \alpha(X_i) \in A$ defines uniquely a simple observable $\alpha^R : 2^R \to A$ by the law for all $\{X_1, X_2, \ldots, X_j\} \subseteq R$, $\alpha^R(\{X_1, X_2, \ldots, X_j\}) = \alpha^R(X_1) \oplus \alpha^R(X_2) \oplus \cdots \oplus \alpha^R(X_j)$ [in particular, $\alpha^{R_k}(X_k) = k$].

Let $R_1, R_2 \in \mathcal{P}$ and assume that R_1 is a refinement of R_2, that is, for every $X_i \in R_1$ there is (exactly one) $Y_j \in R_2$, such that $X_i \subseteq Y_j$. We then write $R_2 \le R_1$. For $R_2 \le R_1$ the function $g_{R_1 R_2} : R_1 \to R_2$, $X_i \mapsto Y_j$ (with $X_i \subseteq Y_j$) is then a uniquely defined onto mapping. If $R_2 \le R_1$ and $R_3 \le R_2$, then also $R_3 \le R_1$ and $g_{R_3 R_1} = g_{R_3 R_2} \circ g_{R_2 R_1}$. Finally, for any two partitions $R_1, R_2 \in \mathcal{P}$ the partition $R = \{X_i \cap Y_j : X_i \in R_1, Y_j \in R_2\}$ belongs to $\mathcal{P}$, and is a common refinement of R_1 and R_2, that is, $R_1 \le R$, $R_2 \le R$, and so $(\mathcal{P}; \le)$ is a directed set. Let $\mathcal{G} := \{g_{RS} : R, S \in \mathcal{P}, R \le S\}$. It is now immediate to check that $((R; \alpha^R)_{R\in\mathcal{P}}, \mathcal{G})$ is a projective system of simple observables. By construction, $K \subseteq \bigcup_{R\in\mathcal{P}} \mathcal{R}(\alpha^R)$.

Conversely, let $((\Omega_i, \alpha_i)_{i\in D}, \mathcal{G})$ be a projective system of simple observables. Then the Bochner s.m. condition is satisfied and, by [Rao, Prop 4, p. 120], the projective limit $\Omega = \lim_{i\leftarrow}$ is a compact Hausdorff space. Let $g_i : \Omega \to \Omega_i$ be the natural coordinate projection, $i \in D$. The s.m. condition implies that $g_i(\Omega) = \Omega_i$. Define $\mathcal{B}(\Omega)_0 := \bigcup_i g_i^{-1}(\mathcal{B}(\Omega)_i)$. Then $\mathcal{B}(\Omega)_0$ is an algebra of subsets of Ω, which is generated by the cylinder sets of Ω with bases in $\mathcal{B}(\Omega)_i, i \in D$, where $\mathcal{B}(\Omega)_i := 2^{\Omega_i}$. Let $X \in \mathcal{B}(\Omega)_0$. If $X \in g_i^{-1}(\mathcal{B}(\Omega)_i) \cap g_j^{-1}(\mathcal{B}(\Omega)_j)$, then there exist $Y_1 \in \mathcal{B}(\Omega)_i$, $Y_2 \in \mathcal{B}(\Omega)_j$ such that $X = g_i^{-1}(Y_1) = g_j^{-1}(Y_2)$. Since D is directed, there exists an $l \in D$ such that $l \geq i$, $l \geq j$, and by compatibility of the mappings, we have $g_i = g_{il} \circ g_l$, $g_j = g_{jl} \circ g_l$. Hence

$$g_l^{-1}(g_{il}^{-1}(Y_1)) = g_i^{-1}(Y_1) = X = g_j^{-1}(Y_2) = g_l^{-1}(g_{jl}^{-1}(Y_2)).$$

Since $g_l : \Omega \to \Omega_l$ is onto, it follows that $g_l^{-1} : \mathcal{B}(\Omega)_l \to \mathcal{B}(\Omega)_0$ is one-to-one so that the latter equations imply that

$$g_{il}^{-1}(Y_1) = g_{jl}^{-1}(Y_2).$$

Thus

$$\alpha_i(Y_1) = \alpha_l(g_{il}^{-1}(Y_1)) = \alpha_l(g_{jl}^{-1}(Y_2)) = \alpha_j(Y_2),$$

by the compatibility of observables. If we set $\alpha(X) := \alpha_i(Y_1) = \alpha_j(Y_2)$, $X \in \mathcal{B}(\Omega)_0$, then α is unambiguously defined on $\mathcal{B}(\Omega)_0$. Clearly, $\alpha(\Omega) = \alpha_i(\Omega_i) = 1$, and α is finitely additive on $\mathcal{B}(\Omega)_0$. Moreover, the prescription also implies that $\alpha_i = \alpha \circ g_i^{-1}$, $i \in D$. Hence, for all $i \in D$, $\mathcal{R}(\alpha_i) \subset \mathcal{R}(\alpha)$. □

Applying the above characterization and using Theorem 1.10.12, we obtain the following result ([Pul 10], for finite subsets see also [ChKo 5]).

Theorem 1.10.38 *Any at most countable subset of an MV-algebra A is contained in a range of an observable.*

Proof. Let us consider a sequence $\{a_i\}_i$ of elements of an MV-algebra A. Consider the sequence of equalities

$$\begin{aligned} 1 &= a_1 \oplus a_1^* = w_{11} \oplus w_{12} \oplus w_{21} \oplus w_{22} \\ &= a_2 \oplus a_2^* = \cdots, \end{aligned}$$

where (w_{ij}) are obtained applying Theorem 1.10.12 to the second equality. Now we order w_{ij}'s lexicographically, and apply Theorem 1.10.12 to the fourth equality, etc. Now we define observables $x_1, x_2, \ldots$ as follows:

$$x_1(\{\omega_1^1\}) = a_1, \ x_1(\{\omega_2^1\}) = a_1^*,$$
$$x_2(\{\omega_1^2\}) = w_{11}, \ x_2(\{\omega_2^2\}) = w_{12}, \ x_2(\{\omega_3^2\}) = w_{21}, \ x_2(\{\omega_4^2\}) = w_{22},$$
$$\text{etc.,}$$

so that $\mathcal{R}(x_1) \subseteq \mathcal{R}(x_2) \subseteq \mathcal{R}(x_3) \subseteq \cdots$. We obtain a sequence of sets

$$\Omega_1 = \{\omega_1^1, \omega_2^1\},\ \Omega_2 = \{\omega_1^2, \omega_2^2, \omega_3^2, \omega_4^2\}, \ldots$$

and define mappings

$$g_{n,n+1} : \Omega_{n+1} \to \Omega_n,$$
$$g_{n,n+1}\{\omega_i^{n+1}\} = \omega_j^n,$$

where j is such that $x^{n+1}(\{\omega_i^{n+1}\})$ is contained in $x_n(\{\omega_j^n\})$. Then $(\Omega_i, g_{i,i+1})_i$ is a projective system of sets, and $(\Omega_i, x_i, g_{i,i+1})$ is a projective system of simple observables. By Theorem 1.10.37, there is an observable (Ω, x) such that the range of x contains the ranges of all x_i, and hence the elements $a_i, i = 1, 2, \ldots$. □

1.11 Order Properties of $\mathcal{E}(H)$

The well known Kadison antilattice theorem, [LuZa, Thm 58.4], says that for bounded self-adjoint operators A, B the infimum $A \wedge B$ exists in $\mathcal{B}_{sa}(H)$ with respect to the order induced by positive operators only when $A \leq B$ or $B \leq A$. The situation if different if we consider infima of operators in the positive cone $\mathcal{B}_{sa}^+(H)$, or the infima in the set of effects $\mathcal{E}(H)$. For instance, it is well known that for any two (orthogonal) projections the infimum always exists and is equal to the projection onto the intersection of their ranges. In this Section, we present a necessary and sufficient condition that given $A, B \geq O$, the infimum $A \wedge B$ exists in the cone $\mathcal{B}_{sa}^+(H)$. Clearly, if A, B are effects, $A \wedge B$, if it exists, is an effect as well. The problem of the existence of the infima of effects has been studied by several authors (see [LaMa], [Gud 6], [Gud 8], [Dvu 18], also Exercise 17. of Section 1.12). Recently, in [MoGu], the problem has been solved in the finite dimensional case using methods of linear algebra. The general case has been solved by Ando [And], and we give a sketch of the methods used there. For the details see [And].

To discuss the existence of the infimum $A \wedge B$ in $\mathcal{E}(H)$, we may assume that $ker(A + B) = \{0\}$ (here ker and ran denote the kernel and the range, respectively). In fact, let P be the orthoprojection to the orthocomplement of $ker(A+B)$. Then the map $X \mapsto \Phi(X) = PXP$, $X \in \mathcal{B}_{sa}^+(H)$, gives an affine order isomorphism from $\{X \in \mathcal{B}_{sa}^+(H) : 0 \leq X \leq A+B\}$ onto a subset Q of $\mathcal{B}_{sa}^+(H)$ such that if $Y \in Q, O \leq Z \leq Y$, then $Z \in Q$ and $ker(\Phi(A)+\Phi(B)) = \{0\}$. When $ker(A+B) = \{0\}$, there is an affine order isomorphism $\phi : \{X \in \mathcal{B}_{sa}^+(H) : O \leq X \leq A+B\} \to \{Y \in \mathcal{E}(H) : O \leq Y \leq I\}$ given by

$$X = (A+B)^{1/2}\phi(X)(A+B)^{1/2}, \tag{1.8}$$

where $\phi(A) + \phi(B) = I$, hence $\phi(A)$ and $\phi(B)$ commute. Therefore we first discuss existence of $A \wedge B$ in $\mathcal{E}(H)$ under the assumptions

$$A + B = I. \tag{1.9}$$

We may write, according to the spectral theorem,

$$A = \int_0^1 \lambda dE(\lambda), \text{ and } B = \int_0^1 (1-\lambda) dE(\lambda). \tag{1.10}$$

Lemma 1.11.1 *Let $A, B \in \mathcal{E}(H)$ and $A + B = I$. If the infimum $A \wedge B$ in $\mathcal{E}(H)$ exists, then $A \wedge B = \int_0^1 \min(\lambda, 1-\lambda) dE(\lambda)$.*

Proof. Let $C := \int_0^1 \min(\lambda, 1-\lambda) dE(\lambda)$. Then obviously $O \leq C \leq A, B$. Let D be such that $C \leq D \leq A, B$. Then $O \leq D - C \leq A - C, B - C$, so that there exist contractions X, Y such that $(D-C)^{1/2} = (A-C)^{1/2} X$ and $(D-C)^{1/2} = (B-C)^{1/2} Y$, which implies

$$ran(D-C)^{1/2} \subseteq ran(A-C)^{1/2} \cap ran(B-C)^{1/2} = \{0\},$$

the latter equality is obtained by functional calculus. It follows $D - C = O$, that is, $D = C$. So C is a maximal lower bound of A, B, therefore if $A \wedge B$ exists, it is equal to C. □

Theorem 1.11.2 *If $A + B = I$, $A, B \in \mathcal{E}(H)$, the infimum of $A \wedge B$ exists in $\mathcal{E}(H)$ if and only if $U := \int_{0+}^{1-} \lambda dE(\lambda)$ and $V := \int_{0+}^{1-} (1-\lambda) dE(\lambda)$ are comparable. Then the infimum coincides with the smaller of the above too, and is equal to $\int_0^1 \min(\lambda, 1-\lambda) dE(\lambda)$.*

Proof. Observe that

$$U = \int_{0+}^{1-} \lambda dE(\lambda) = A[I - E(\{0\}) - E(\{1\})], \tag{1.11}$$

$$V = \int_{0+}^{1-} (1-\lambda) dE(\lambda) = B[I - E(\{0\}) - E(\{1\})]. \tag{1.12}$$

The proof of the "if" part follows by direct calculation.

To prove the "only if" part by contradiction, suppose that U and V are incomparable but A, B admit the infimum. Then by Lemma 1.11.1, the infimum coincides with $C = \int_0^1 \min(\lambda, 1-\lambda) dE(\lambda)$. By equalities (1.11) and (1.12), supposing $U \not\leq V$ and $V \not\leq U$, there exist disjoint subsets $\Delta_1, \Delta_2 \subseteq (0,1)$ and $\epsilon > 0$ such that $E(\Delta_1) \neq O, E(\Delta_2) \neq O$, and

$$1 - 3\epsilon \geq \lambda \geq (1-\lambda) + \epsilon, \quad \lambda \in \Delta_1, \tag{1.13}$$
$$1 - 3\epsilon \geq 1 - \lambda \geq \lambda + \epsilon, \quad \lambda \in \Delta_2. \tag{1.14}$$

We may assume, without loss of generality, that $\dim(ranE(\Delta_1)) \geq \dim(ranE(\Delta_2))$. Then there is a partial isometry $V : ranE(\Delta_2) \to ranE(\Delta_1)$. Define an operator D by

$$D = B\,E(\Delta_1) - \epsilon E(\Delta_1) + A\,E(\Delta_2) - \epsilon E(\Delta_2) + 2^{1/2}\epsilon V\,E(\Delta_2) + 2^{1/2}\epsilon E(\Delta_2)V^*.$$

We claim that $A, B \geq D \geq O$. From (1.13) and (1.10) we get

$$B\,E(\Delta_1) - \epsilon E(\Delta_1) \geq 2\epsilon E(\Delta_1)$$

and from (1.14) and (1.10)

$$A\,E(\Delta_2) - \epsilon E(\Delta_2) \geq 2\epsilon E(\Delta_2).$$

By definition of V,

$$E(\Delta_2)\,V^*V\,E(\Delta_2) = E(\Delta_2) \text{ and } E(\Delta_1)\,V\,E(\Delta_2) = V\,E(\Delta_2)$$

we can conclude that

$$\begin{aligned} D &\geq 2^{1/2}\epsilon\{E(\Delta_1) + E(\Delta_2) + V\,E(\Delta_2) + E(\Delta_2)V^*\} \\ &= 2^{1/2}\epsilon\{E(\Delta_1) + V\,E(\Delta_2)\}^*\{E(\Delta_1) + V\,E(\Delta_2)\} \geq O. \end{aligned}$$

Since by (1.10) and (1.13),

$$(A - B)\,E(\Delta_1) \geq \epsilon E(\Delta_1),$$

it follows by direct computation using the definition of D and properties of V that

$$A - D = \epsilon\{2^{1/2}E(\Delta_1) - V\,E(\Delta_2)\}^*\{2^{1/2}E(\Delta_1) - V\,E(\Delta_2)\} \geq O.$$

This proves that $A \geq D$. The inequality $B \geq D$ is proved similarly. By Lemma 1.11.1 we can conclude $C \geq D$. To see that this causes a contradiction, take a unit vector $x \in ranE(\Delta_2)$ and let $y := Vx$. Then y is a unit vector in $E(\Delta_1)$. From

$$C - D = C\,E((\Delta_1 \cup \Delta_2)^c) + \epsilon\{E(\Delta_1) + E(\Delta_2) - 2^{1/2}V\,E(\Delta_2) - 2^{1/2}E(\Delta_2), V^*\},$$

we get $\langle (C - D)(x + y), x + y\rangle < 0$, which contradicts $C - D \geq O$. □

To transfer to the general case, we need to get rid of the spectral theorem. To this aim, we need the definition of **parallel sum** operation $X : Y$, and that of the **short** (operation).

As a positive quadratic sum, $X : Y$ for $X, Y \geq O$ is defined as

$$\langle (X : Y)a, a\rangle := \inf\{\langle Xb, b\rangle + \langle Yc, c\rangle : b + c = a\}. \tag{1.15}$$

When $X + Y$ is invertible, this definition takes the form

$$X : Y = X - X(X + Y)^{-1}X = Y - Y(X + Y)^{-1}Y. \tag{1.16}$$

Let P be a projection. The **short** $[P]X$ of a positive operator X with respect to P is defined by as a maximal element in the set

$$\{Z : O \leq Z \leq X, \gamma P, \text{ for some } \gamma \in \mathbb{R}^+\}.$$

If P commutes with X, then $[P]X = XP$. In general case it can be checked that $[P]X = \lim_{n\to\infty}(nP) : X$ (strong convergence).

A **generalized short** with $Y \geq O$ is defined by

$$[Y]X := \lim_{n\to\infty} (nY) : X. \tag{1.17}$$

Lemma 1.11.3 *For $A, B \geq O$ with $A + B = I$, we have $\int_{0+}^{1-} \lambda dE(\lambda) = [B]A$, and $\int_{0+}^{1-}(1 - \lambda)dE(\lambda) = [A]B$.*

Proof. By (1.16),

$$[B]A = \lim_{n\to\infty} ((nB) : A) = \lim_{n\to\infty} \int_0^1 f_n(\lambda)dE(\lambda) = \int_{0+}^{1-} \lambda dE(\lambda),$$

where

$$f_n(\lambda) := \lambda - \frac{\lambda^2}{\lambda + n(1 - \lambda)}.$$

As $n \to \infty$, $f_n(\lambda)$ converges to λ for $\lambda \in (0, 1)$, while $f_n(1) = f_n(0) = 0$ $(n = 1, 2, \ldots)$. Similarly, $[A]B = \int_{0+}^{1-}(1 - \lambda)dE(\lambda)$. □

Now the context of Theorem 1.11.2 can be stated without the use of spectral representation.

Theorem 1.11.4 *When $A, B \geq O$ and $A+B = I$, the infimum $A \wedge B$ exists in $\mathcal{E}(H)$ if and only if $[A]B$ and $B[A]$ are comparable, in which case $A \wedge B = \min([A]B, [B]A)$.*

Now fix any $A, B \in \mathcal{E}(H)$ with $ker(A+B) = \{0\}$. By (1.8), there is an affine order isomorphism ϕ from the set $\{X : O \leq X \leq A+B\}$ onto the set $\{Z : O \leq Z \leq I\}$, for which $\phi(A) + \phi(B) = I$.

Lemma 1.11.5 *The mapping* ϕ *satisfies the condition*

$$\phi([Y]X) = [\phi(Y)]\phi(X), \ O \leq X, Y \leq A+B.$$

Proof. First we establish the identity

$$\phi(X : Y) = \phi(X) : \phi(Y). \tag{1.18}$$

We use the fact that, since $ker(A+B) = \{0\}$, $ran(A+B)^{1/2}$ is dense in H, and prove (1.18) for vectors in $ran(A+B)^{1/2}$, by a direct computation. Then we use the definition of a generalized short and prove the statement of the theorem. Again, it suffices to prove it for vectors in $ran(A+B)^{1/2}$. □

Now we are in the position to transform Theorem 1.11.4 to the general case.

Theorem 1.11.6 *Given* $A, B \in \mathcal{E}(H)$, *the infimum* $A \wedge B$ *in* $\mathcal{E}(H)$ *exists if and only if* $[B]A$ *and* $[A]B$ *are comparable. In this case* $A \wedge B = \min([B]A, [A]B)$.

Proof. We may assume that $ker(A+B) = \{0\}$. The proof then follows by Theorem 1.11.4 and Lemma 1.11.5. □

We note that Gudder and Moreland [MoGu] using methods of linear algebra in the finite dimensional case, proved the following (we denote by $P_{A,B}$ the projection onto $ranA \cap ranB$).

Theorem 1.11.7 *For* $A, B \in \mathcal{E}(\mathbb{C}^n)$, $A \wedge B$ *exists in* $\mathcal{E}(\mathbb{C}^n)$ *if and only if* $A \wedge P_{A,B}$ *and* $B \wedge P_{A,B}$ *are comparable, in which case* $A \wedge B$ *is the smaller of* $A \wedge P_{A,B}$ *and* $B \wedge P_{A,B}$.

In [And], there is proved that the content of Theorem 1.11.7 is the same as that of Theorem 1.11.6. It follows from the fact that in the finite dimensional case, $P_{A,B} = P_{A:B}$ and that $[A]B = [P_A]B$, where P_A is the projection onto the (closed) range of A.

1.12 Exercises

1. Prove that $MO(n)$ can be realized as $G^+[0,u]$, where $G = \mathbb{Z}^{n+1}$, $u := (1, 1, \ldots, 1)$, the $2n+2$ elements of $MO(n)$ are $0, u$, the n elements $a_1 := (1, 0, 0, \ldots, 0)$, $a_2 := (1, 1, 0, 0, \ldots, 0)$, $a_3 := (1, 1, 1, 0, 0, \ldots, 0)$, …, $a_n := (1, 1, \ldots, 1, 0)$, and n

more elements of the form $b_i := u - a_i$ for $i = 1, 2, \dots, n$. Here G^+ is the subcone of the standard positive cone $(\mathbb{Z}^+)^{n+1}$ consisting of all nonnegative-integer linear combinations of $a_1, a_2, \dots, a_n, b_1, b_2, \dots, b_n$.

(Hint: Use Theorem 1.9.34 and mathematical induction ([FGB]).)

2. Let $\mathbb{Z}$ be ordered by the standard positive cone $\mathbb{Z}^+$ and let $P := \mathbb{Z} \times_{lex} \mathbb{Z}$ be the lexicographic product. Prove that in the scale algebra $A := P^+[(0,0),(1,0)]$ every element of the form $(0,k)$, $k \in \mathbb{Z}^+$, has infinite isotropic index. Moreover, P is the universal group for A. ([FGB, Ex 4.6])

3. Let $\mathbb{Z}$ be ordered by the standard positive cone $\mathbb{Z}^+$, let $0 \neq m, n \in \mathbb{Z}^+$, let $\mathbb{Z}_m$ denote the additive group of integers modulo m partially ordered by the trivial cone $(\mathbb{Z}_m)^+ = \{0\}$, and let $P := \mathbb{Z} \times_{lex} \mathbb{Z}_m$ be the lexicographic product. We define the **polychain of height n and width m** by $C_{n,m} := P^+[(0,0),(n,m-1)]$. For $m = 2$, we define the **diamond** by $D := P^+[(0,0),(2,0)]$. Show that (1) The diamond D and $C_{2,2}$ are isomorphic as posets, but not as effect algebras, whereas $C_{2,2}$ and 2^2 are isomorphic as effect algebras. (2) For $n \geq 3$, lexicographically ordered group $\mathbb{Z} \times_{lex} \mathbb{Z}_m$ is the universal group for $C_{n,m}$, and $\mathbb{Z} \times_{lex} \mathbb{Z}_2$ is the universal group for the diamond D; however, $\mathbb{Z} \times_{lex} \mathbb{Z}_2$ is not the universal group for $C_{2,2}$.

4. Prove that a horizontal sum of polychains (cf. the previous Exercise) of height three or more admits only one state.

5. The *multiple of an interval algebra* $A = G^+[0,u]$ is defined by $nA := G^+[0,nu]$ for a positive integer n. Prove that if G with unit u is the unigroup for A, then G with unit nu is the unigroup for nA.

6. A finite effect algebra F with the Riesz decomposition property is an MV-algebra.

   (Hint: Observe that F is lattice ordered because it has the interpolation property and is finite). ([BeFo, Cor. 3.12])

7. Prove that an orthomodular lattice is ϕ-symmetric if and only if it is a Boolean algebra ([Kal], [Ber]).

8. Prove that an effect algebra E is an OMP iff every element in E is principal (see Definition 1.9.5, [FoBe 1, Theorem 5.3]).

9. Prove that for a lattice ordered effect algebra E the following conditions are mutually equivalent:

   (i) E is an orthomodular lattice.
   (ii) For all $p \in E$, $\phi(p,.)$ is an idempotent.
   (iii) For all $p, q \in E$, $\phi(p,q) = p \wedge (p' \vee q)$.

   ([BeFo])

10. By transferring the usual notion of compatibility from orthomodular posets to effect algebras, we obtain the following weaker relation. a and b are called *weakly-Mackey compatible* ($a \leftrightarrow_{WM} b$) iff there exist a_1, b_1, c [$a_1 \perp b_1$, $a_1 \perp c$, $b_1 \perp c$ and $a = a_1 \oplus c$, and $b = b_1 \oplus c$]. In such a case, we will say that a_1, b_1, c represent a *Mackey weak decomposition* of a, b. Show that compatibility implies weakly-Mackey compatibility, but not the other way round.

    (Hint: Use Wright triangle Figure 4.3.3 on p. 273 as a counterexample.)

11. Prove that an effect algebra A is an orthomodular poset if and only if the following condition is satisfied: For all $a, b \in E$, $a = a_1 \oplus c$, $b = b_1 \oplus c$ is a Mackey weak decomposition iff it is a Mackey decomposition of a, b.

12. Prove that an effect algebra E is an MV-algebra iff $\forall\ a, b \in E$, $a \leftrightarrow_x b$, with the same Mackey decomposition a_1, b_1, c for any upper bound x of a, b.

13. Let P be an effect algebra and let a be a sharp element of P. Prove that if $b \in P$ with $a \perp b$, then $a \oplus b$ is a minimal upper bound for a and b.

14. Let E be a lattice ordered effect algebra. Recall that a block in E is any maximal pairwise compatible subset M of E. Denote by $C(M)$ the set of all central elements in M. Since M is an MV-algebra, $C(M)$ is a Boolean algebra. Prove that the following statements are equivalent for $a \in E$:

    (i) a is sharp.
    (ii) a is central in every block containing a.
    (iii) a is central in some block.

    Moreover, denote $E_S := \bigcup\{C(M) : M \text{ is a block of } E\}$. Then E_S is a subeffect algebra of E closed under (all existing) lattice operations, and as an effect algebra in its own right, it is an OML. ([JeRi])

15. Prove that direct limits in the category with effect algebras as objects and effect-algebra monomorphisms as morphisms exists. We will say that a morphism $\phi : E \to F$, where E and F are effect algebras, *preserves disjointness* if $a \wedge b = 0 \Rightarrow \phi(a) \wedge \phi(b) = 0$. Prove that the direct limit of a directed system $(L_\alpha, (\phi_{\alpha,\beta})_{\alpha \leq \beta})$ where all L_α is an OML, and $\phi_{\alpha,\beta}$ is a disjointness preserving monomorphism, is an OML. ([Hab 1])

16. Let $\mathcal{F} = [0,1]^X$ be a D-poset of functions. Let x, y be observables defined via

$$x(E) = \begin{cases} 1_x, & \text{if } 1,2,3 \in E, \\ (\frac{3}{4})_X, & \text{if } 1,3 \in E \text{ and } 2 \notin E \text{ or } 2,3 \in E \text{ and } 1 \notin E, \\ (\frac{1}{2})_X, & \text{if } 1,2 \in E \text{ and } 3 \notin E \text{ or } 3 \in E \text{ and } 1,2 \notin E, \\ (\frac{1}{4})_X, & \text{if } 1 \in E \text{ and } 2,3 \notin E \text{ or } 2 \in E \text{ and } 1,3 \notin E, \\ 0_X, & \text{if } 1,2,3 \notin E, \end{cases}$$

$$y(F) = \begin{cases} 1_x, & \text{if } a,b,c,d \in F, \\ (\frac{3}{4})_x, & \text{if } a,b,c \in F \text{ and } d \notin F, \\ (\frac{1}{2})x, & \text{if } a,b \in F \text{ and } c,d \notin F, \\ (\frac{1}{4})x, & \text{if } a \in F \text{ and } b,c,d \notin F, \\ 0_x, & \text{if } a,b,c,d \notin F, \end{cases}$$

where a, b, c, d are mutually different elements from the set $\{4, 5, 6, 7\}$.

Prove that the observable x has not the V-property on the range $\mathcal{R}(x)$ and the observable y has the V-property on the range $\mathcal{R}(y)$.

17. Let H be a complex Hilbert space. Denote the set of all bounded self-adjoint operators on H by $\mathcal{B}_{sa}(H)$ and the set of all (orthogonal) projections on H by $\mathcal{P}(H)$. We say that an element $A \in \mathcal{B}_{sa}(H)$ is **positive** if $\langle Ax, x\rangle \geq 0$ for all $x \in H$. Let $\mathcal{B}_{sa}^+(H)$ denote the set of all bounded positive operators on H. Then $\mathcal{B}_{sa}^+(H)$ becomes a positive cone in the Abelian group $\mathcal{B}_{sa}(H)$. Recall that an element $A \in \mathcal{B}_{sa}(H)$ that satisfies $0 \leq A \leq I$ is called an **effect** and the set of effects on H is denoted by $\mathcal{E}(H)$. For $A, B \in \mathcal{E}(H)$ we write $A \wedge B$ ($A \vee B$) for the infimum (supremum) of A and B within $\mathcal{E}(H)$, if it exists. It is well-known that $\mathcal{E}(H)$ is not lattice ordered (unless $\dim(H) \leq 1$).

For $A \in \mathcal{B}_{sa}(H)$, let $ranA$ be the closure of the range AH of A and $kerA$ be the null space of A. For $A, B \in \mathcal{S}(H)$ we denote the projection onto $ranA$ by P_A and the projection onto $ranA \cap ranB$ by $P_{A,B}$.

For $A \in \mathcal{B}_{sa}(H)$, define $|A| = (A^2)^{1/2}$ where $(A^2)^{1/2}$ is the unique positive square root of A^2. For $A, B \in \mathcal{B}_{sa}(H)$, we define the **generalized infimum** $A \sqcap B$ by

$$A \sqcap B = \frac{1}{2}(A + B - |A - B|).$$

Similarly, the **generalized supremum** $A \sqcup B$ is defined by

$$A \sqcup B = \frac{1}{2}(A + B + |A - B|).$$

Prove the following statements:

(i) If $A, B \in \mathcal{B}_{sa}(H)$, then $A \sqcap B \leq A, B$ and

$$(A - A \sqcap B)(B - A \sqcap B) = O.$$

(ii) If $A, B \in \mathcal{E}(H)$ and $A \sqcap B \geq O$, then

$$(A - A \sqcap B) \wedge (B - A \sqcap B) = O.$$

(iii) If $A, B \in \mathcal{B}_{sa}(H)$, then $A, B \leq A \sqcup B$ and

$$(A \sqcup B - A)(A \sqcup B - B) = O.$$

(iv) Let $A, B \in \mathcal{E}(H)$ with $A \sqcup B \leq I$, then

$$(A \sqcup B - A) \wedge (A \sqcup B - B) = O.$$

Moreover, $A \sqcup B$ is a minimal upper bound for A and B in $\mathcal{E}(H)$. If $A \vee B$ exists, then $A \vee B = A \sqcup B$.

(v) For $A, B \in \mathcal{E}(H)$, if $\dim(P_{A,B}) \leq 1$, then $A \wedge B$ exists.

(vi) If A and B are commuting effects with pure point spectrum, then $A \wedge B$ exists if and only if $AP_{A,B}$ and $BP_{A,B}$ are comparable. Moreover, in this case $A \wedge B$ is the smaller of $AP_{A,B}$ and $BP_{A,B}$.

([LaMa], [Gud 8])

18. A **sum Brouwer-Zadeh algebra (SBZ)** is a structure $(E; \oplus, ', ^{\sim}, 0, 1)$, where $(E; \oplus, ', 0, 1)$ is a regular effect algebra, $a \mapsto a' = 1 \ominus a$ is the **Kleene orthocomplement**, satisfying

(K1) $a = a''$.
(K2) $a \leq b$ implies $b' \leq a'$.
(K3) $a \leq a'$ and $b' \leq b$ imply $a \leq b$ (regularity condition).

The mapping $^{\sim} : E \to E$ is the **Brouwer complementation** satisfying

(B1) $\forall a, b \in E, \exists r : a \oplus r = b^{\sim}$ implies $\exists s : b \oplus s = a^{\sim}$.
(B2) $\forall a \in E, a \perp a^{\sim}$ (i.e., $\exists\, a \oplus a^{\sim}$).
(B3) Let $a, c \in E$. Then $\exists r : a^{\sim} \oplus r = c$ and $\exists s : a^{\sim\sim} \oplus s = c$ imply $c = 1$.

Prove that the standard effect algebra $\mathcal{E}(H)$ of the Hilbert space effects is an SBZ algebra, if we define $A^{\sim} = N_A$, where N_A is the projection onto the null space of A. ([Cat])

19. Prove that in MV-algebras, sharp and central elements coincide.

20. Prove that a convex effect algebra P is "convex" in the usual way, that is, for $\alpha \in [0, 1]$, $a, b \in P$, $\alpha a \oplus \beta b$ is defined in P. ([GuPu 1])

21. Prove that the standard effect algebra of Hilbert space effects $\mathcal{E}(H)$ is the norm-closure of the convex envelope of the projection lattice $\mathcal{P}(H)$.

(Hint: Show that every effect A in $\mathcal{E}(H)$ of finite rank can be expressed as a convex combination of projections. Then use the fact that operators of finite rank are norm dense in the algebra of all bonded operators.)

22. Prove that every maximal subset of pairwise commuting effects in $\mathcal{E}(H)$, H separable, can be organized into an MV-algebra.

    (Hint: Use the functional calculus for commuting self-adjoint operators.)

23. (Naimark Theorem) Let $x : \mathcal{B}(\mathbb{R}) \to \mathcal{E}(H)$ be a POV observable. Then there is a Hilbert space H_0 such that $H \subseteq H_0$ and a PV observable $y : x : \mathcal{B}(\mathbb{R}) \to \mathcal{P}(H_0)$ such that for any $E \in \mathcal{B}(\mathbb{R})$, $x(E) = Py(E)P$, where P is the projection from H_0 onto H. (Hint: See, e.g., [Hol].)

24. Prove Theorem 1.2.9. ([HePu])

25. Prove: Let $A \in \mathcal{E}(H)$. Then $A \vee_{\mathcal{E}(H)} (I - A) = I$ iff $A \in \mathcal{P}(H)$. ([Dvu 18, Thm 4.4])

26. Let $\mathcal{S} \neq \emptyset$ be a convex system of σ-additive states on L. A mapping $f : \mathcal{S} \times \mathcal{B}(\mathbb{R}) \to [0,1]$ such that

    (i) given $s \in \mathcal{S}$, $f(s, \cdot)$ is a finitely additive measure on $\mathcal{B}(\mathbb{R})$;
    (ii) for any $M \in \mathcal{B}(\mathbb{R})$, $f(\lambda\, s_1 + (1-\lambda)\, s_2, M) = \lambda\, f(s_1, M) + (1-\lambda)\, f(s_2, M)$ whenever $\lambda \in [0,1]$ and $s_1, s_2 \in \mathcal{S}$,

    is said to be the **effect function** on $\mathcal{S}$.

    We say that a convex system $\mathcal{S}$ of σ-additive states on L has the **E-property** on L (E as for existence), if, given an effect function f on $\mathcal{S}$, for any $M \in \mathcal{B}(\mathbb{R})$, there exists an element $x(M) \in L$ such that $f(s, M) = s(x(M))$, $s \in \mathcal{S}$.

    Prove: (1) If $\mathcal{S} = \mathcal{S}(H) := \{m_T : T \in Tr_1(H)\}$, then $\mathcal{S}$ has the E-property on $\mathcal{E}(H)$.

    (2) Show that the E-property fails on $\mathcal{P}(H)$. ([DDdL, Thm 4.1])

27. Prove: Let $L = [0,1]$ be ordered by the natural way. Then the set of all states $\mathcal{S} = \{s_0\}$, where $s_0(t) = t$, $t \in [0,1]$, is order determining. This state is also completely additive. Then $\mathcal{S}$ has the E-property in $[0,1]$. ([DDdL, Ex 5.1])

28. Prove: Let $[0,1]^2 = [0,1] \times [0,1]$ be the product of two copies of $[0,1]$. Then $[0,1]^2$ is a complete effect algebra, where $(u_1, v_1) \oplus (u_2, v_2)$ is defined iff $u_1 \oplus u_2$ and $v_1 \oplus v_2$ are defined in $[0,1]$; then $(u_1, v_1) \oplus (u_2, v_2) := (u_1 \oplus u_2, v_1 \oplus v_2)$. The space of all states on $[0,1]^2$ is the set $\mathcal{S} = \{s_\alpha : \alpha \in [0,1]\}$, where $s_\alpha(u,v) := \alpha u + (1-\alpha)v$, $(u,v) \in [0,1]^2$, which is order determining. Any state is completely additive, and $\mathcal{S}$ has the E-property in $[0,1]^2$. Generalize this result for product of n copies of $[0,1]$. ([DDdL, Ex 5.2])

29. Let (G, G^+, u) be an interpolation group with order unit u. Prove that G is lattice ordered if and only if $G^+[0,u]$ is lattice ordered. Derive as a corollary that a unigroup for an MV-algebra is lattice ordered.

(Hint: [MuPa]. By [Fuc 1], G is a subdirect product of antilattice ordered groups (recall that a partially ordered set is an antilattice if, for any elements a, b, $a \wedge b$ is defined if and only if $a \leq b$, or $b \leq a$, dually for $a \vee b$). Let $G \subseteq \prod G_i$, and let π_j be the projection onto j-th coordinate. If $G^+[0, u]$ is a lattice, then $\pi_j(G^+[0, u]) \subseteq G_i^+[0, u_j]$ is linearly ordered, and hence $\pi_j(G^+)$ is linearly ordered. It follows that G^+ is a lattice. By [Fuc], G is a lattice.)

30. Prove that a convex effect algebra E can be represented by an interval $V^+[0, u]$ of a lattice ordered linear space (V, V^+) with order unit u if and only if E is a lattice. ([BBGP])
31. Prove that an extreme element in a convex effect algebra must be sharp. ([BBGP])
32. A convex effect algebra is **Archimedean** if whenever $a, b, c \in P$ and $c \leq a \oplus n^{-1}b$ for every $n \in \mathbb{N}$, then $c \leq a$. Prove that the following statements are equivalent:

    (i) P is Archimedean.
    (ii) The representing linear space (V, V^+) is Archimedean.
    (iii) P has an order determining set of states.

    ([BBGP])

Chapter 2

MV-algebras and QMV-algebras

In this chapter, we study supplement algebras, MV-algebras and QMV-algebras. MV-algebras were introduced by C. Chang [Cha] as algebraic models of many-valued logics. The prototypical model of MV-algebras is based on the real interval $[0, 1]$. Nowadays, there is a huge literature devoted to MV-algebras. In this chapter, we introduce only their basic properties. For more details see, e.g., the recent monograph [CDM].

Quantum MV-algebras (or QMV-algebras) were introduced by R. Giuntini in [Giu] as a non-lattice theoretic generalization of MV-algebras and as a non-idempotent generalization of orthomodular lattices. The introduction of QMV-algebras was mainly motivated by the search for an adequate algebraic structure for the quantum counterpart of the real interval $[0, 1]$, which is the class $\mathcal{E}(H)$ of all effects of a Hilbert space H. Recall that an effect of H is a positive linear operator with spectrum contained in $[0, 1]$. Effects can be considered as mathematical representatives of "unsharp properties" of a quantum physical system in that their possible values are contained in $[0, 1]$, in the contrast to the "sharp properties" represented by projection operators, whose spectrum is contained in $\{0, 1\}$.

From an algebraic point of view, MV and QMV-algebras share a common set of axioms, which S. Gudder [Gud 4] has called supplement algebra (S-algebra). An S-algebra is essentially an Abelian $\oplus$ monoid[1] (with an absorbing element 0) equipped with a complement (*). What makes an S-algebra an MV-algebra is the addition of the Łukasziewicz axiom (see axiom (MV) in Section 2.2 or (MV8) in Section 1.8). This axiom is precisely what makes an MV-algebra a lattice ordered structure.

Effects of a Hilbert space do not determine a lattice (see Section 1.11). Therefore for an algebraic structure that is to be an abstraction of the Hilbert space effects, the Łukasziewicz axiom has to be weakened. Accordingly, QMV-algebras are obtained from S-algebras by adding axiom (QMV), which is weaker than (MV). The interesting property of this axiomatization is that both MV-algebras and orthomodular

[1]Throughout this chapter, the operation $\oplus$ is **totally defined**.

lattices are obtained from QMV-algebras by adding the Łukazsiewicz axiom and the idempotence condition ($a \oplus a = a$), respectively.

2.1 S-algebras

Definition 2.1.1 *A* **supplement algebra** (**S-algebra** for short) *is an algebraic structure* $\mathcal{M} = (M; \oplus, ^*, 0, 1)$ *consisting of a nonempty set* M*, two constant elements* $0, 1$ *in* M*, a unary operation* * *and a binary operation* $\oplus$ *on* M *satisfying the following axioms:*

(S1) $a \oplus b = b \oplus a$;

(S2) $a \oplus (b \oplus c) = (a \oplus b) \oplus c$;

(S3) $a \oplus a^* = 1$;

(S4) $a \oplus 0 = a$;

(S5) $a^{**} = a$;

(S6) $0^* = 1$;

(S7) $a \oplus 1 = 1$.

On every supplement algebra, the following additional operations can be introduced:

$$a \odot b := (a^* \oplus b^*)^*, \tag{2.1}$$

$$a \Cap b := (a \oplus b^*) \odot b, \tag{2.2}$$

$$a \Cup b := (a \odot b^*) \oplus b. \tag{2.3}$$

We will consider $\odot$ as more binding than $\oplus$. For all $a, b \in M$, define $\leq$ by [2]

$$a \leq b \quad \text{iff} \quad a = a \Cap b.$$

Proposition 2.1.2 [Giu 1] Let $\mathcal{M} = (M; \oplus, ^*, 0, 1)$ be an S-algebra. The following properties hold.

(i) $a \odot b = b \odot a$.

(ii) $a \odot (b \odot c) = (a \odot b) \odot c$.

(iii) $a \odot a^* = 0$.

(iv) $a \odot 0 = 0$.

[2]Sometimes we write $\leq_Q$ for $\leq$, see p. 424.

(v) $a \odot 1 = a$.

(vi) $a \Cap 1 = a = 1 \Cap a$.

(vii) $a \Cap 0 = 0 \Cap a = 0$.

(viii) $a = a \Cap a$.

(ix) $(a \Cup b)^* = a^* \Cap b^*$.

(x) $(a \Cap b)^* = a^* \Cup b^*$.

(xi) If $a \leq b$, then $a = b \Cap a$.

(xii) $a \oplus b = 0$ implies $a = b = 0$, dually $a \odot b = 1$ implies $a = b = 1$.

(xiii) $a \Cup b = 0$ implies $a = b = 0$, dually $a \Cap b = 1$ implies $a = b = 1$.

(xiv) If $a \oplus c = b \oplus c$, $a \leq c^*$, $b \leq c^* \Rightarrow a = b$.

(xv) $a \leq b \Rightarrow a^* \oplus b = 1$.

Proof. The proof of (i)–(viii) is straightforward.

(ix) $(a \Cup b)^* = [(a \odot b^*) \oplus b]^* = (a \odot b^*)^* \odot b^* = (a^* \oplus b) \odot b^* = a^* \Cap b^*$.

(x) $(a \Cap b)^* = [(a \oplus b^*) \odot b]^* = (a \oplus b^*)^* \oplus b^* = (a^* \odot b) \oplus b^* = a^* \Cup b^*$.

(xi) $b \Cap a = b \Cap (a \Cap b) = [b \oplus (a^* \Cup b^*)] \odot (a \Cap b) = [b \oplus (a^* \odot b) \oplus b^*] \odot (a \Cap b) = [b \oplus b^* \oplus (a^* \odot b)] \odot (a \Cap b) = [1 \oplus (a^* \odot b)] \odot (a \Cap b) = a \Cap b = a$.

Proofs of the remaining statements are similar, and are left to the readers as an exercise. □

In general, $a = b \Cap a$ does not imply $a = a \Cap b$, and $a^* \oplus b = 1$ does not imply $a \leq b$. But we have the following result.

Corollary 2.1.3 *Relation $\leq$ is reflexive and antisymmetric and $0 \leq a \leq 1$ for every a. In general, it can be non-transitive as the example of an S-algebra below shows.*

Proof. Reflexivity follows by Proposition 2.1.2 (viii). To prove antisymmetry, let $a \leq b$ and $b \leq a$, then $a = a \Cap b$ and $b = b \Cap a$, and by Proposition 2.1.2 (xi), $a \Cap b = b \Cap a$, hence $a = b$.

Proposition 2.1.2 (vi) and (vii) imply that for any a, $0 \leq a \leq 1$.

The following S-algebra $(\{0,1,2,3,4,5,6\}; \oplus, {}^*, 0, 1)$, where $\oplus$ and * are defined in the table below, has the order $\leq$ non-transitive[3], because $4 \leq 6$, $6 \leq 3$, but $4 \not\leq 3$.

[3]This example was found by R. Giuntini using the programme MACE.

$\oplus$	0	1	2	3	4	5	6
0	0	1	2	3	4	5	6
1	1	1	1	1	1	1	1
2	2	1	1	1	1	1	3
3	3	1	1	1	1	1	1
4	4	1	1	1	2	3	5
5	5	1	1	1	3	1	1
6	6	1	3	1	5	1	1
*	0	1	2	3	4	5	6
	1	0	2	4	3	6	5

$\Cap$	0	1	2	3	4	5	6
0	0	0	0	0	0	0	0
1	0	1	2	3	4	5	6
2	0	2	2	3	4	6	6
3	0	3	2	3	4	5	6
4	0	4	2	0	4	0	4
5	0	5	2	2	4	5	6
6	0	6	0	6	4	5	6

Tab 2.0 S-algebra with non-transitive order

□

2.2 MV-algebras

In this Section, we introduce axioms and basic properties of MV-algebras, their algebraic properties, ideals and congruences. More details about MV-algebras, in particular their representation theorem in Abelian ℓ-groups with order unit, can be found in Chapter 5, Corollary 5.3.8 (Mundici's theorem).

For a systematic treatment of MV-algebras see, e.g., the monograph [CDM].

2.2.1 Basic Properties

Combining [Man] and supplement algebras, we can define MV-algebras as follows (compare with the definition on p. 73).

Definition 2.2.1 *An* **MV-algebra** *is an S-algebra* $\mathcal{M} = (M; \oplus, ^*, 0, 1)$ *satisfying the condition*

(MV) $$(a^* \oplus b)^* \oplus b = (a \oplus b^*)^* \oplus a, \; a, b \in M.$$

The condition (MV) is a so-called **Łukasiewicz axiom**. Owing to (2.1), and (2.3), (MV) is equivalent to

$$a \Cup b = b \Cup a, \; a, b \in M.$$

Hence operations $\Cup$ and its dual $\Cap$ are commutative. In this chapter, we replace $\Cap$ by $\wedge$ and $\Cup$ by $\vee$ whenever we will deal with MV-algebras. In Proposition 2.2.4 (vii) and (vii*), we will prove that they are the infimum and supremum of a, b, respectively.

Remark 2.2.2 Axioms (S1)–(S7) and (MV) are not independent. Indeed, as regards (S3), we may write

$$\begin{aligned} a \oplus a^* = (a^*)^* \oplus a^* &= (a^* \oplus 0)^* \oplus a^* \\ &= (0^* \oplus a)^* \oplus 0^* \quad \text{(MV), (S1)} \\ &= 0^* = 1 \quad \text{(S6)}. \end{aligned}$$

(S5) can be weakened to
(S5') $(0^*)^* = 0$.
Indeed,

$$\begin{aligned} (a^*)^* = (a^* \oplus 0)^* \oplus 0 &= (a \oplus 0^*)^* \oplus a \quad \text{(MV), (S1)} \\ &= a \oplus (0^*)^* = a \oplus 0 = a. \end{aligned}$$

(S7) and (S6) can be replaced by
(S7') $a \oplus 0^* = 0^*$, $0^* := 1$.
In [CaLo], independence of axioms (S1), (S2), (S4), (S5'), (S7') and (MV) has been proved.
Moreover, (S1) and (S2) can be reduced to one axiom
(S1,2) $(a \oplus b) \oplus c = b \oplus (c \oplus a)$.

The relation $\leq$ can be alternatively characterized as follows.

Proposition 2.2.3 [Cha], [CDM] *Let $\mathcal{M} = (M; \oplus, ^*, 0, 1)$ be an MV-algebra. The following conditions are equivalent:* (i) $x \leq y$, (ii) $y \oplus x^* = 1$, (iii) $x \odot y^* = 0$, *and* (iv) *there is $z \in M$ such that $y = x \oplus z$.*

Proof. Clearly (ii) and (iii) are equivalent.
(i) $\Rightarrow$ (ii). $x \leq y$ implies $x^* \oplus y = 1$ by Proposition 2.1.2 (xv).
(iii) $\Rightarrow$ (iv). Assume that $x \odot y^* = 0$, then $x \oplus (y \odot x^*) = (x \odot y^*) \oplus y = y$ (by (MV)), hence $z = y \odot x^*$.
(iv) $\Rightarrow$ (i). $x \wedge y = x \wedge (x \oplus z) = (x \oplus z) \wedge x = (x \oplus z \oplus x^*) \odot x = 1 \odot x = x$ by Proposition 2.1.2 (v), (i). □

From (S1)–(S7) and (MV), the following properties of MV-algebras can be derived.

Proposition 2.2.4 *Let $\mathcal{M} = (M; \oplus, ^*, 0, 1)$ be an MV-algebra. For all $a, b, c \in M$, the following properties hold.*

(i) $a \wedge b = b \wedge a$ *(commutativity of $\wedge$).*

(i*) $a \vee b = b \vee a$ *(commutativity of $\vee$).*

(ii) $a \leq b$ *iff* $a \vee b = b$.

(iii) $a \leq b$ *implies* $b^* \leq a^*$ *(order-reversibility of* **).*

(iv) $a \leq b, b \leq c$ *imply* $a \leq c$ *(transitivity of* $\leq$*).*

(v) $a \leq b$ *implies* $a \oplus c \leq b \oplus c$ $(c \in M)$ *(monotonicity of* $\oplus$*).*

(v*) $a \leq b$ *implies* $a \odot c \leq b \odot c$ $(c \in M)$ *(monotonicity of* $\odot$*).*

(vi) $a = (b \wedge a) \oplus (b^* \odot a)$.

(vii) $a \wedge b$ *is the infimum of* a, b *in* $(M; \leq)$.

(vii*) $a \vee b$ *is the supremum of* a, b *in* $(M; \leq)$.

(viii) $a \odot b \leq a \leq a \oplus b$.

(ix) $a \odot b \leq a \wedge b$.

(ix*) $a \vee b \leq a \oplus b$.

(x) $a \oplus b = a \oplus (a^* \wedge b)$.

(xi) $a \oplus (b \wedge c) = (a \oplus b) \wedge (a \oplus c)$ *(distributivity of* $\oplus$ *over* $\wedge$*).*

(xi*) $a \odot (b \vee c) = (a \odot b) \vee (a \odot c)$ *(distributivity of* $\odot$ *over* $\vee$*).*

(xii) $a \vee (b \wedge c) = (a \vee b) \wedge (a \vee c)$ *(distributivity of* $\vee$ *over* $\wedge$*).*

(xii*) $a \wedge (b \vee c) = (a \wedge b) \vee (a \wedge c)$ *(distributivity of* $\wedge$ *over* $\vee$*).*

Proof. (i)

$$\begin{aligned} a \wedge b = (a \oplus b^*) \odot b &= ((a \oplus b^*)^* \oplus b^*)^* \\ &= ((a^* \oplus b)^* \oplus a^*)^* \quad \text{(MV)} \\ &= (a^* \oplus b) \odot a \\ &= b \wedge a. \end{aligned}$$

(i*) $a \vee b = (a^* \wedge b^*)^* = (b^* \wedge a^*)^* = b \vee a$.
(ii) $a \leq b \ \Rightarrow \ a^* \oplus b = 1$ (Proposition 2.2.3 (ii)) implies

$$\begin{aligned} a \vee b &= a \odot b^* \oplus b \\ &= (a^* \oplus b)^* \oplus b \\ &= b. \end{aligned}$$

On the other hand,

$$\begin{aligned} a \vee b = b \quad &\Rightarrow \quad a \odot b^* \oplus b = b \\ &\Rightarrow \quad (a^* \oplus b)^* \oplus b = b \\ &\Rightarrow \quad (b^* \oplus a)^* \oplus a = b \quad \text{(MV)} \\ &\Rightarrow \quad a^* \oplus b = a^* \oplus a \oplus (b^* \oplus a)^* \\ &= \quad 1 \\ &\Rightarrow \quad a \leq b \quad \text{(Proposition 2.2.3 (ii))}. \end{aligned}$$

(iii) $a \leq b$ iff $a \wedge b = a$ iff $a^* \vee b^* = a^*$ iff $b^* \leq a^*$ by (ii).

(iv) $a \leq b$, $b \leq c$ imply by Proposition 2.2.3 (iv) that there are u, v such that $b = a \oplus u$, $c = b \oplus v$, hence $c = a \oplus (u \oplus v)$, and again by Proposition 2.2.3 (iv), $a \leq c$.

(v) $a \leq b \Rightarrow a^* \oplus b = 1$ by Proposition 2.2.3. Hence

$$\begin{aligned}
(a \oplus c)^* \oplus b \oplus c &= (a \oplus c)^* \oplus c \oplus b \\
&= (c^* \oplus a^*) \oplus a^* \oplus b \quad \text{(MV)} \\
&= 1 \text{ (hypothesis)},
\end{aligned}$$

hence $a \oplus c \leq b \oplus c$ by Proposition 2.2.3.

(v*) follows by (iii).

(vi)

$$\begin{aligned}
(b \wedge a) \oplus (b^* \odot a) &= ((b \oplus a^*) \odot a) \oplus (b^* \odot a) = ((b^* \odot a)^* \odot a) \oplus (b^* \odot a) \\
&= (a^* \odot b^* \odot a) \oplus a \quad \text{(MV)} \\
&= a.
\end{aligned}$$

(vii) $a \wedge b \leq a, b$ by (vi) and Proposition 2.2.3 (iv).

Assume that $c \in M$ is such that $c \leq a, b$. Then

$$\begin{aligned}
c^* \oplus (a \wedge b) &= c^* \oplus ((a \oplus b^*) \odot b) \\
&= (c^* \odot b) \oplus (b^* \oplus ((a \oplus b^*) \odot b)) \\
&= (c^* \odot b) \oplus ((a \oplus b^*) \oplus (a \oplus b^*)^* \odot b^*) \quad \text{(MV)} \\
&= a \oplus ((c^* \odot b) \oplus b^*) \\
&= a \oplus (c^* \vee b^*) \\
&= a \oplus c^* \quad \text{hypothesis } c \leq b, \text{ (iii) and (ii)} \\
&= 1.
\end{aligned}$$

By Proposition 2.2.3 (ii), $c \leq a \wedge b$.

(vii*) follows from (vii) by duality.

(viii) $a \leq a \oplus b$ follows directly from Proposition 2.2.3 (iv).

$a \odot b \leq a$ follows by (iii).

(ix) follows by (viii) and (vii).

(ix*) follows from (ix) by (iii).

(x)

$$\begin{aligned}
a \oplus (a^* \wedge b) &= a \oplus (b \oplus a) \odot a^* \quad \text{(i)} \\
&= a \oplus ((b \oplus a)^* \oplus a)^* \\
&= b \oplus a \oplus (b \oplus a \oplus a^*)^* \quad \text{(MV)} \\
&= b \oplus a.
\end{aligned}$$

(xi)

$$\begin{aligned}
a\oplus(b\wedge c) &= a\oplus[(b\wedge c)\wedge a^*] \quad \text{(i)}\\
&= a\oplus[b\wedge((c\oplus a)\odot a^*)] \quad \text{(vii)}\\
&= a\oplus[b\oplus(c\oplus a)^*\oplus a]\odot(c\oplus a)\odot a^*\\
&= a\oplus[(b\oplus a\oplus(c\oplus a)^*)^*\oplus(c\oplus a)^*]^*\odot a^*\\
&= a\oplus[(c\oplus a\oplus(b\oplus a)^*)^*\oplus(b\oplus a)^*]^*\odot a^* \quad \text{(MV)}\\
&= a\oplus[(c\oplus a)^*\odot(b\oplus a)\oplus(b\oplus a)^*]^*\odot a^*\\
&= a\oplus((c\oplus a)^*\vee(b\oplus a)^*)^*\odot a^*\\
&= a\oplus((c\oplus a)\wedge(b\oplus a))\odot a^*\\
&= ((c\oplus a)\wedge(b\oplus a))\vee a\\
&= (c\oplus a)\wedge(b\oplus a).
\end{aligned}$$

(xi*) follows by duality.
(xii*)

$$\begin{aligned}
a\wedge(b\vee c) &= (a\oplus(b\vee c)^*)\odot(b\vee c)\\
&= (a\oplus(b^*\wedge c^*))\odot(b\vee c)\\
&= ((a\oplus b^*)\wedge(a\oplus c^*))\odot(b\vee c) \quad \text{(xi)}\\
&= [((a\oplus b^*)\wedge(a\oplus c^*))\odot b]\vee[((a\oplus b^*)\wedge(a\oplus c^*))\odot c]\\
&= [((a\oplus c^*)\oplus(a\oplus b^*)^*)\odot(a\oplus b^*)\odot b]\\
&\vee\ [((a\oplus b^*)\oplus(a\oplus c^*)^*)\odot(a\oplus c^*)\odot c]\\
&= [((a\oplus c^*)\oplus(a\oplus b^*)^*)\odot(a\wedge b)]\vee[((a\oplus b^*)\oplus(a\oplus c^*)^*)\odot(a\wedge c)]\\
&\le (a\wedge b)\vee(a\wedge c).
\end{aligned}$$

On the other hand, $a\wedge b\le a$, $a\wedge c\le a$ imply

$$\begin{aligned}
(a\wedge b)\vee(a\wedge c) &\le a\vee(a\wedge c)\\
&= a \quad \text{(ii)}
\end{aligned}$$

and $a\wedge b\le b$, $a\wedge c\le c$ imply by (vii*), $(a\wedge b)\vee(a\wedge c)\le b\vee(a\wedge c)\le b\vee c$. Therefore $[(a\wedge b)\vee(a\wedge c)]\wedge[a\wedge(b\vee c)]=(a\wedge b)\vee(a\wedge c)$, and hence $(a\wedge b)\vee(a\wedge c)\le a\wedge(b\vee c)$.
(xii) by duality. □

Remark 2.2.5 MV-algebras were originally introduced by [Cha] as algebras $(M;\oplus,\odot,^*,0,1)$ satisfying (S1)–(S7) (and their duals), properties (i), (i*), (xi), (xi*) of Proposition 2.2.4, and $(a\vee b)\vee c=a\vee(b\vee c)$ and $(a\wedge b)\wedge c=a\wedge(b\wedge c)$. Equivalence of Chang's axioms with (S1)–(S7) and (MV) follows by Proposition 2.2.4. Originally, it has been proved by Mangani [Man] in a different way.

Putting together results obtained so far, we arrive at the following description of the algebraic structure of MV-algebras.

Theorem 2.2.6 [Cha] *Let $\mathcal{M} = (M; \oplus, ^*, 0, 1)$ be an MV-algebra. The relation $\leq$ is a partial order of M, and 0 and 1 are the least and the greatest elements, respectively. The elements $x \vee y$ and $x \wedge y$ are respectively the l.u.b. and the g.l.b. of the elements x and y with respect to the ordering $\leq$. With respect to these operations, M is a distributive lattice.*

Proposition 2.2.7 *In an MV-algebra $\mathcal{M} = (M; \oplus, ^*, 0, 1)$, the following four conditions are equivalent:*

(i) $x \oplus y = y$.

(ii) $x \odot y = x$.

(iii) $y \vee x^* = 1$.

(iv) $x \wedge y^* = 0$.

Proof. It suffices to prove the equivalence of (ii) and (iii). By duality, the equivalence of (i) and (iv) then follows, and, clearly, (iii) is equivalent with (iv).

If $x \odot y = x$, then $x^* \vee y = x^* \oplus (x \odot y) = x^* \oplus x = 1$. If $x^* \vee y = 1$ then $x = x \odot 1 = x \odot (x^* \vee y) = (x \odot x^*) \vee (x \odot y) = 0 \vee x \odot y = x \odot y$. □

The following proposition shows that in an MV-algebra the set of elements B which are idempotent with respect to the operations $\oplus$ and $\odot$ are precisely the elements which satisfy the law of excluded middle with respect to the operations $\vee$ and $\wedge$. The proof follows by Proposition 2.2.7.

Proposition 2.2.8 [Cha] *In an MV-algebra $\mathcal{M} = (M; \oplus, ^*, 0, 1)$, the following conditions are equivalent:*

(i) $x \oplus x = x$.

(ii) $x \odot x = x$.

(iii) $x^* \oplus x^* = x^*$.

(iv) $x^* \odot x^* = x^*$.

(v) $x \vee x^* = 1$.

(vi) $x \wedge x^* = 0$.

Proposition 2.2.8 leads to the following

Theorem 2.2.9 [Cha] *Let B be the set of elements x of M such that $x \oplus x = x$. Then B is closed under the operations $\oplus$, $\odot$ and * and $x \oplus y = x \vee y$ and $x \odot y = x \wedge y$ for $x, y \in B$. Furthermore, the system $(B; \oplus, \odot, ^*, 0, 1)$ is the largest subalgebra of M which is in the same time a Boolean algebra with respect to the same operations $\oplus$, $\odot$ and **

In the following propositions we collect some further properties that hold in MV-algebras. We will often use axiom (MV) in the form $x \odot y^* \oplus y = y \odot x^* \oplus x$.

Proposition 2.2.10 $y \odot (x \oplus z) \leq x \oplus y \odot z$.

Proof. We have

$$\begin{aligned}
(y \odot (x \oplus z))^* \oplus x \oplus y \odot z &= y^* \oplus x^* \odot z^* \oplus x \oplus y \odot z \\
&= (y^* \oplus y \odot z) \oplus (x \oplus x^* \odot z^*) \\
&= (z \oplus y^* \odot z^*) \oplus (z^* \oplus z \odot x) \\
&= (z \oplus z^*) \oplus y^* \odot z^* \oplus z \odot x \\
&= 1.
\end{aligned}$$

By Proposition 2.2.3, the inequality follows. □

Proposition 2.2.11 *If $x \wedge y^* = 0$, then $x \oplus y \odot z = y \odot (x \oplus z)$.*

Proof. Letting $w = x^* \odot (y^* \oplus z^*) \oplus y \odot (x \oplus z)$, we see by Proposition 2.2.3 that it is sufficient to prove that $w = 1$. Now

$$\begin{aligned}
x \oplus w &= (x \oplus x^* \odot (y^* \oplus z^*)) \oplus y \odot (x \oplus z) \\
&= y^* \oplus z^* \oplus x \odot y \odot z \oplus y \odot (x \oplus z) \quad \text{(MV)} \\
&= z^* \oplus x \odot y \odot z \oplus (y^* \oplus y \odot (x \oplus z)) \\
&= z^* \oplus x \odot y \odot z \oplus x \oplus z \oplus y^* \odot x^* \odot z^* \quad \text{(MV)} \\
&= 1,
\end{aligned}$$

and

$$\begin{aligned}
y^* \oplus w &= y^* \oplus y(\odot(x \oplus z)) \oplus x^* \odot (y^* \oplus z^*) \\
&= x \oplus z \oplus y^* \odot x^* \odot z^* \oplus x^* \odot (y^* \oplus z^*) \quad \text{(MV)} \\
&= z \oplus y^* \odot x^* \odot z^* \oplus (x \oplus x^* \odot (y^* \oplus z^*)) \\
&= z \oplus y^* \odot x^* \odot z^* \oplus (y^* \oplus z^* \oplus x \odot y \odot z) \quad \text{(MV)} \\
&= 1.
\end{aligned}$$

From the above equalities we obtain

$$(x \oplus w) \wedge (y^* \oplus w) = 1.$$

By Proposition 2.2.4 (xi) and the hypothesis

$$1 = (x \wedge y^*) \oplus w = w,$$

and the proof is complete. □

Proposition 2.2.12 $(x \oplus y^*) \vee (y \oplus x^*) = 1.$

Proof. Let us calculate

$$\begin{aligned}
(x \oplus y^*) \vee (y \oplus x^*) &= (x \oplus y^*) \odot [(y \oplus x^*)]^* \oplus y \oplus x^* \\
&= [(x \oplus y^*) \odot y^* \odot x \oplus x^*] \oplus y \\
&= ((x \oplus y^*) \odot y^* \oplus y) \oplus x^* \odot (x^* \odot y \oplus y) \quad \text{(MV)} \\
&= x \oplus y^* \oplus y \odot (x \oplus y^*)^* \oplus x^* \odot (x^* \odot y \oplus y) \quad \text{(MV)} \\
&= y^* \oplus y \odot x^* \odot y \oplus x^* \odot y \oplus y \oplus x \odot (x^* \odot y \oplus y)^* \quad \text{(MV)} \\
&= 1.
\end{aligned}$$

□

Proposition 2.2.13 *If* $x \vee y = 1$, *then* $x \odot x \vee y \odot y = 1$.

Proof. By using the hypothesis and Proposition 2.2.11 twice, we get

$$(x^* \oplus x^*) \odot y \odot y = (x^* \oplus x^* \odot y) \odot y = x^* \oplus x^* \odot y \odot y.$$

Using this and

$$x \odot x \vee y \odot y = x \odot x \oplus (x^* \oplus x^*) \odot y \odot y,$$

we get

$$\begin{aligned}
x \odot x \vee y \odot y &= (x \odot x \oplus x^*) \oplus x^* \odot y \odot y \\
&= (x^* \odot x^* \oplus x) \oplus x^* \odot y \odot y \quad \text{(MV)} \\
&= x^* \odot x^* \oplus (x \oplus x^* \odot y \odot y) \quad \text{(MV)} \\
&= x^* \odot x^* \oplus y \odot y \oplus x \odot (y^* \oplus y^*).
\end{aligned}$$

Again by the hypothesis and Proposition 2.2.11, $x \odot (y^* \oplus y^*) = x \odot y^* \oplus y^*$. Thus the latter equation leads to

$$\begin{aligned} x \odot x \vee y \odot y &= x^* \odot x^* \oplus y \odot y \oplus x \odot y^* \oplus y^* \\ &= x^* \odot x^* \oplus x \odot y^* \oplus (y \odot y \oplus y^*) \\ &= x^* \odot x^* \oplus x \odot y^* \oplus y^* \odot y^* \oplus y \\ &= x^* \odot x^* \oplus y^* \odot y^* \oplus (x \odot y^* \oplus y) \\ &= x^* \odot x^* \oplus y^* \odot y^* \oplus (x \vee y) \\ &= 1, \end{aligned}$$

as desired. □

We introduce the following definitions: $0 \odot x = 0$ and $(n+1) \odot x = n \odot x \oplus x$, $x^0 = x$ and $x^{n+1} = (x^n) \odot x$. The **order** of an element x, in symbols $ord(x)$, is the least integer m such that $m \odot x = 1$. If no such integer exists, then $ord(x) = \infty$. According to [Cha], an MV-algebra M is **locally finite** if every element of M different from 0 has a finite order.

We easily obtain by induction from the definition that

$$\begin{aligned} (n \odot x)^* &= (x^*)^n,\ (x^n)^* = n \odot x^*,\ m \odot (n \odot x) = (m.n)x, \\ x^{(m+n)} &= (x^n) \odot (x^m),\ x^{m.n} = (x^n)^m. \end{aligned}$$

We recall that sometimes without loss of precision, we shall use simply $n\,x$ instead of $n \odot x$.[4]

Proposition 2.2.14 *If $x \vee y = 1$, then $x^n \vee y^n = 1$ for each n.*

Proof. By an easy induction from Proposition 2.2.13 we obtain that $x^{2^m} \vee y^{2^m} = 1$ for each m. Since for each n there exists an m such that $n \leq 2^m$, we see that $x^{2^m} \leq x^n$ and $y^{2^m} \leq y^n$. By monotonicity of $\vee$, $x^n \vee y^n = 1$. □

Proposition 2.2.15 *If $ord(x \odot y) < \infty$, then $x \oplus y = 1$.*

Proof. By the hypothesis, for some $n \geq 1$, $n \odot (x \odot y) = 1$. Hence $(x^* \oplus y^*)^n = 0$. By Proposition 2.2.12, $(x \oplus y) \vee (x^* \oplus y^*) = 1$, and by Proposition 2.2.14, $(x \oplus y)^n \vee (x^* \oplus y^*)^n = 1$. Thus $(x \oplus y)^n = 1$. By induction with respect to n and using Proposition 2.2.4 (x), we obtain $x \oplus y = 1$. □

An MV-algebra $\mathcal{M} = (M; \oplus, ^*, 0, 1)$ is **linearly ordered** if, for every $x, y \in M$, either $x \leq y$ or $y \leq x$. It is easily seen that the only Boolean algebra that is linearly ordered is a two-element Boolean algebra consisting of 0 and 1.

[4]Sometimes in literature it is written also as $n.x$. Since we study also nx as sum of n-times of x supposing that $x \leq ((n-1)x)^*$, we shall prefer $n \odot x$ instead of $n.x$.

Theorem 2.2.16 [Cha] *Every locally finite MV-algebra is linearly ordered.*

Proof. Let $x, y \in M$. We wish to show that either $x \oplus y^* = 1$ or $x^* \oplus y = 1$. If $x \oplus y^* \neq 1$, then $x^* \odot y \neq 0$. Hence by the assumption on M, $ord(x^* \odot y) < \infty$. Now it follows from Proposition 2.2.15 that $x^* \oplus y = 1$. □

The following example furnishes a counterexample to the converse of Theorem 2.2.16. Consider the following set C of formal symbols:

$$0, c, c \oplus c, c \oplus c \oplus c, \ldots,$$
$$1, 1 \ominus c, 1 \ominus c \ominus c, 1 \ominus c \ominus c \ominus c, \ldots.$$

We define $0.c = 0$ and $n.c = \underbrace{c \oplus \cdots \oplus c}_{n\text{-times}}$ and $1 \ominus 0.c = 1$ and $1 \ominus n.c = 1 \ominus \underbrace{c \ominus \cdots \ominus c}_{n\text{-times}}$.
In terms of this convention the rules of addition, multiplication, and complementation are written down as follows:

(i) If $x = n.c$ and $y = m.c$ then $x \oplus y = (m+n).c$; if $x = n.c$ and $y = 1 \ominus m.c$, then $x \oplus y = 1$ if $m \leq n$ and $x \oplus y = 1 \ominus (m-n).c$ if $n < m$; if $x = 1 \ominus m.c$ and $y = n.c$, then $x \oplus y = 1$ if $m \leq n$ and $x \oplus y = 1 \ominus (m-n).c$ if $n < m$; if $x = 1 \ominus n.c$ and $y = 1 \ominus m.c$, then $x \oplus y = 1 \ominus (m+n)c$.

(ii) If $x = n.c$ then $x^* = 1 \ominus n.c$, if $x = 1 \ominus n.c$ then $x^* = n.c$.

One can easily check that the system $(C; \oplus, ^*, 0, 1)$ defined above is an MV-algebra by showing that each of the axioms (S1)–(S7) and (MV) is satisfied.

The inclusion relation $\leq$ in the algebra C can be described as follows: $x \leq y$ if and only if one of the conditions below is satisfied: (i) $x = n.c$ and $y = 1 \ominus m.c$. (ii) $x = n.c$ and $y = m.c$, where $n \leq m$. (iii) $x = 1 \ominus n.c$ and $y = 1 \ominus m.c$, where $m \leq n$. In what follows, there will be seen that this MV-algebra C provide a crucial counterexample; we call this MV-algebra often as a **Chang MV-algebra** (see also Example 5.2.39). We will also see that similar examples can be obtained using any Abelian linearly ordered non-Archimedean group.

The following function d plays the role of a distance function in MV-algebras. The function d is defined as follows:

$$d(x, y) = x^* \odot y \oplus y^* \odot x.$$

In Boolean algebras, the element $d(x, y)$ is simply the familiar symmetric difference of x and y. We note here that in Boolean algebras there is another way of defining the symmetric difference of x and y, namely, $d'(x, y) = (x \oplus y) \odot (x^* \oplus y^*)$. However, the operation d' thus defined for MV-algebras would not have the desired property that $d'(x, x) = 0$. For example, in the algebra $\{0, \frac{1}{2}, 1\}$ we easily see that $d'(\frac{1}{2}, \frac{1}{2}) \neq 0$.

Theorem 2.2.17 *The function d has the following properties:*

(i) $d(x,x) = 0$, $d(x,y) = d(y,x)$, $d(x,y) = d(x^*,y^*)$, $d(x,0) = x$, *and* $d(x,1) = x^*$.

(ii) *If* $d(x,y) = 0$, *then* $x = y$.

(iii) $d(x,z) \leq d(x,y) \oplus d(y,z)$.

(iv) $d(x \oplus u, y \oplus v) \leq d(x,y) \oplus d(u,v)$.

(v) $d(x \odot u, y \odot v) \leq d(x,y) \oplus d(u,v)$.

Proof. (i) This is obvious from the definition.
(ii) This follows from the fact that $x \oplus y = 0$ implies $x = 0 = y$.
In order to prove (iii), we have to show that

$$x \odot z^* \oplus x^* \odot z \leq x^* \odot y \oplus x \odot y^* \oplus z^* \odot y \oplus z \odot y^*. \tag{2.4}$$

Now

$$\begin{aligned} [z \odot x^*]^* \oplus x^* \odot y \oplus y^* \odot z &= z^* \oplus x \oplus x^* \odot y \oplus y^* \odot z \\ &= (x \oplus x^* \odot y) \oplus (z^* \oplus y^* \odot z) \\ &= y \oplus x \odot y^* \oplus y^* \oplus y \odot z^* \\ &= 1, \end{aligned}$$

hence

$$z \odot x^* \leq x^* \odot y \oplus y^* \odot z, \tag{2.5}$$

In an entirely similar fashion,

$$x \odot z^* \leq x \odot y^* \oplus y \odot z^*. \tag{2.6}$$

(2.5) and (2.6) yield (2.4) and (iii).
Condition (iv) is essentially

$$(x \oplus u)^* \odot (y \oplus v) \oplus (x \oplus u) \odot (y \oplus v)^* \leq x^* \odot y \oplus y^* \odot x \oplus u^* \odot v \oplus v^* \odot u. \tag{2.7}$$

Now

$$\begin{aligned} (x \oplus u \oplus y^* \odot v^*) \oplus x^* \odot y \oplus u^* \odot v &= (x \oplus x^* \odot y) \oplus y^* \odot v^* \oplus (u \oplus u^* \odot v) \\ &= (y \oplus x \odot y^*) \oplus y^* \odot v^* \oplus (v \oplus u \odot v^*) \\ &= x \odot y^* \oplus u \odot v^* \oplus v \oplus (y \oplus y^* \odot v^*) \\ &= x \odot y^* \oplus u \odot v^* \oplus v \oplus v^* \oplus y \odot v \\ &= 1 \end{aligned}$$

which implies

$$(x \oplus u)^* \odot (y \oplus v) \leq x^* \odot y \oplus u^* \odot v.$$

Similarly we derive

$$(x \oplus u) \odot (y \oplus v)^* \leq (y^* \odot x) \oplus (v^* \odot u).$$

The last two relations give (2.7) and (iv).

To prove condition (v), observe that

$$\begin{aligned} d(x \odot u, y \odot v) &= d([x^* \oplus u^*]^*, [y^* \oplus v^*]^*) \\ &= d(x^* \oplus u^*, y^* \oplus v^*) \\ &\leq d(x^*, y^*) \oplus d(u^*, v^*) \\ &= d(x, y) \oplus d(u, v). \end{aligned}$$

□

2.2.2 Congruences and Ideals in MV-algebras

Definition 2.2.18 *A subset $I \neq \emptyset$ of M is an* **ideal** *of M if and only if*

(i) *if $x, y \in I$ then $x \oplus y \in I$,*

(ii) *if $x \in I$ and $y \leq x$, then $y \in I$.*

We have the following easy characterizations of an ideal.

Lemma 2.2.19 (i) *$I \subseteq M$ is an ideal if and only if $I \neq \emptyset$, I is closed under $\oplus$, and $x \in I$, $y \in M$ imply $x \wedge y \in I$.*

(ii) *$I \subseteq M$ is an ideal if and only if $I \neq \emptyset$, I is closed under $\oplus$ and $x \in I$, $y \in M$ imply $x \odot y \in I$.*

Proof. (i) This is obvious because $a \leq b$ iff $a \wedge b = a$. (ii) One direction is clear because $x \odot y \leq x$. Assume that $I \neq \emptyset$, I is closed under $\oplus$ and $x \in I$, $y \in M$ imply $x \odot y \in M$. Now let $x \in I$, $y \leq x$. Then $x \odot (x^* \oplus y) \in I$, hence $x \wedge y \in I$, so $x \wedge y = y \in I$. □

An ideal is said to be **proper** if $I \neq M$. Clearly, I is proper if and only if $1 \notin I$. Let $\mathcal{I}(M)$ denote the set of all ideals of M. Clearly, the intersection of any set of ideals is also an ideal. Since M is the greatest element in $\mathcal{I}(M)$, for every subset W of M, there exists a smallest ideal $\langle W \rangle$ that contains W. In more details, if $W = \emptyset$, then $\langle W \rangle = \{0\}$. If $W \neq \emptyset$, then

$$\langle W \rangle = \{x \in M : x \leq w_1 \oplus \cdots \oplus w_k, \text{ for some } w_1, \ldots, w_k \in W\}. \tag{2.8}$$

In particular, for each element $z \in M$, the ideal $\langle z \rangle := \langle \{z\} \rangle$ is called the **principal ideal generated by** z, and we have

$$\langle z \rangle = \{x \in M : x \leq n \odot z \text{ for some integer } n \geq 0\}. \tag{2.9}$$

Further, for every ideal J of M and each $z \in M$ we have

$$\langle J \cup \{z\} \rangle = \{x \in M : x \leq n \odot z \oplus a \text{ for some } n \in \mathbb{N}, a \in J\}. \tag{2.10}$$

Definition 2.2.20 *R is a* **congruence** *over M if and only if*

(i) *R is an equivalence relation over M,*

(ii) *if xRy and uRv, then x^*Ry^*, $x \oplus uRy \oplus v$.*

The definition of a congruence is such that it automatically preserves the operations $\odot$, $\wedge$, and $\vee$.

The following theorem shows that there are one-to-one correspondences between ideals and congruence relations and homomorphisms. For a homomorphism f of M, let $Ker(f)$ denote the set of elements $x \in M$ such that $f(x) = 0$.

Theorem 2.2.21 *Let $\mathcal{M} = (M; \oplus, ^*, 0, 1)$ be an MV-algebra.*

(i) *If f is a homomorphism of M onto another MV-algebra, then $Ker(f)$ is an ideal, and the relation R defined by*

$$xRy \quad \text{if and only if} \quad f(x) = f(y)$$

is a congruence relation.

(ii) *If R is a congruence relation, then the set of elements of the zero class $[0]$ is an ideal and the set of all equivalence classes forms an MV-algebra*

$$(M/R; \oplus, ^*, [0], [1])$$

under the definitions $[x] \oplus [y] = [x \oplus y]$, $[x] \odot [y] = [x \odot y]$, $[x]^* = [x^*]$, where M/R is a homomorphic image of M under the mapping $f(x) = [x]$.

(iii) *If R is a congruence relation, then xRy if and only if $d(x, y)R0$.*

(iv) *If R and S are congruences, then $R = S$ if and only if $[0]_R = [0]_S$.*

(v) *If I is an ideal, then the relation R defined by xRy if and only if $d(x, y) \in I$ is a congruence relation.*

Proof. (i) and (ii) are obvious.

(iii) can be obtained as follows. Assume that xRy, then since x^*Rx^* and y^*Ry^*, we obtain $x \odot y^*R0$, $x^* \odot yR0$, and $(x \odot y^* \oplus y \odot x^*)R0$. Hence $d(x,y)R0$. Assume now $d(x,y)R0$. Since $(x \odot y^* \oplus 0)R(x \odot y^* \oplus 0)$,

$$[(x \odot y^* \oplus 0) \wedge (x \odot y^* \oplus y \odot x^*)]R[(x \odot y^* \oplus 0) \wedge 0]$$

and hence

$$[x \odot y^* \oplus (0 \wedge (y \odot x^*))]R0.$$

This implies $x \odot y^*R0$. Similarly $y \odot x^*R0$. Now, $(y \oplus 0)R(y \oplus x \odot y^*)$, $(x \oplus y \odot x^*)R(x \oplus 0)$, $y \oplus x \odot y^* = x \oplus y \odot x^*$, and xRy.

(iv) follows directly from (iii).

As for (v), it is a consequence of the definition of d. □

Theorem 2.2.22 *There exists a proper ideal I such that $x \in I$ if and only if $ord(x) = \infty$.*

Proof. Clearly, if x belongs to a proper ideal, then $ord(x) = \infty$. On the other hand, if $ord(x) = \infty$, then we take I to be the set of the elements y such that $y \leq m \odot x$ for some m. Obviously, I is a proper ideal. □

Thus we see that, in general, not every element $x \neq 1$ can be a member of a proper ideal.

Definition 2.2.23 (i) *J is a* **maximal ideal** *of M if and only if J is a proper ideal and whenever I is any ideal such that $J \subseteq I \subseteq M$, then either $J = I$ or $I = M$.*

(ii) *R is a* **maximal congruence relation** *of M if and only if $R \neq M \times M$ and whenever S is a congruence relation such that $R \subseteq S \subseteq M \times M$, then either $R = S$ or $S = M \times M$.*

It is easily seen from the one-to-one correspondence between ideals and congruence relations that maximal ideals correspond to maximal congruence relations and vice versa. It can be proved, by standard arguments using the Axiom of Choice and the fact that the element 1 never belongs to a proper ideal, that every proper ideal can be extended to a maximal ideal.

Theorem 2.2.24 *Let I be an ideal of M, then the following conditions are equivalent:*

(i) *I is a maximal ideal.*

(ii) *For every element $x \notin I$, $x^{*n} \in I$ for some $n \in \mathbb{N}$.*

(iii) *M/I is locally finite.*

Proof. Assume (i) and let x be such that $x \notin I$. Let J be defined as the set of all t's such that for some $y \in I$ and some n, $t \leq y \oplus n \odot x$. Clearly J is an ideal and $I \subseteq J$. Since $x \in J$ and $x \notin I$, we see that $J = M$ and $1 \in J$. This means that for some $y \in I$ and some n, $1 = y \oplus n \odot x$. This last leads to $(n \odot x)^* \leq y$, $x^{*n} \leq y$ and $x^{*n} \in I$.

Assume (ii) and let $[x]_I \in M/I$ be such that $[0]_I \neq [x]_I$, i.e., $x \notin I$. Thus, for some n, $x^{*n} \in I$ and $[0]_I = [x^{*n}]_I = [x^*]_I^n = ([x]_I^*)^n$. This last implies that $n \odot [x]_I = [1]_I$.

Assume (iii) and let J be an ideal for which $I \subseteq J$ and there exists an element x such that $x \in J$ and $x \notin I$. Thus, for some n, $n \odot [x]_I = [1]_I$. Since $x \in J$, $n \odot [x]_J = [0]_J$, and since $I \subseteq J$, $n \odot [x]_J = [1]_J$. This gives $[0]_J = [1]_J$ and $J = M$. □

Due to Theorem 2.2.24, we see that the class of locally finite MV-algebras corresponds precisely to the class of simple MV-algebras, i.e., those MV-algebras M in which the only ideals are the sets $\{0\}$ and M. This fact can also be easily seen directly from the definitions.

Definition 2.2.25 *A proper ideal I of M is* **prime** *if it satisfies the following condition:*

$$\forall\, x, y \in M, \text{ either } x \odot y^* \in I \text{ or } y \odot x^* \in I.$$

Remark 2.2.26 Observe that M is linearly ordered iff, for any $x, y \in M$, either $x \odot y^* = 0$ or $y \odot x^* = 0$. Hence every proper ideal in a totally ordered MV-algebra is prime. Moreover, the set $\mathcal{I}(M)$ is totally ordered by inclusion. Indeed, if there are two proper ideals I and J such that $I \nsubseteq J$, $J \nsubseteq I$, then there are $a \in I \setminus J$, $b \in J \setminus I$ which implies $a \nleq b$, $b \nleq a$, a contradiction.

Proposition 2.2.27 *The following properties hold in any MV-algebra M:*

(i) *Every proper ideal of M that contains a prime ideal is prime.*

(ii) *For any proper ideal I, the quotient M/I is linearly ordered if and only if I is prime.*

(iii) *For each prime ideal J of M, the set $\{I \in I(M) : J \subseteq I\}$ is totally ordered by inclusion. Consequently, every prime ideal is contained in a unique maximal ideal.*

Proof. (i) This follows immediately from the definition.

(ii) For the equivalence classes $[x], [y]$ we have $[x] \leq [y]$ iff $[x] \odot [y]^* = [0]$ iff $x \odot y^* \in I$. Hence M/I is linearly ordered iff I is prime.

(iii) By (ii), M/J is totally ordered. If $I \supseteq J$, then $I = \bigcup\{[a]_J : a \in I\}$, so to every I there corresponds an ideal in M/J. By Remark 2.2.26, these ideals are linearly ordered by inclusion, which entails that also the considered set of ideals is linearly ordered. □

Proposition 2.2.28 *Let J be an ideal of an MV-algebra M. For every $a \in M \setminus J$ there is a prime ideal P of M such that $J \subseteq P$ and $a \notin P$.*

Proof. A routine application of Zorn's lemma shows that there is an ideal I of M which is maximal with the property that $J \subseteq I$ and $a \notin I$. We shall show that I is a prime ideal. Let x and y be elements of M, and suppose that both $x \odot y^* \notin I$ and $y \odot x^* \notin I$. By the maximality assumption, the ideal generated by I and $x \odot y^*$ must contain the element a. By (2.10), $a \leq s \oplus p \odot (x \odot y^*)$ for some $s \in I$ and some integer $p \geq 1$. Similarly, there is an element $t \in I$ and an integer $q \geq 1$ such that $a \leq t \oplus q \odot (y \odot x^*)$. Let $u = s \oplus t$ and $n = \max(p, q)$. Then $u \in I$, $a \leq u \oplus n \odot (x \odot y^*)$ and $a \leq u \oplus n \odot (y \odot x^*)$. Hence we have $a \leq (u \oplus n \odot (x \odot y^*)) \wedge (u \oplus n \odot (y \odot x^*)) = u \oplus (n \odot (x \odot y^*) \wedge n \odot (y \odot x^*)) = u$ (Proposition 2.2.4 (xi) and dual of Propositions 2.2.12 and 2.2.14), whence $a \in I$, a contradiction. □

Corollary 2.2.29 (i) *Every proper ideal in an MV-algebra is an intersection of prime ideals.* (ii) *Every nontrivial MV-algebra has a maximal ideal.* (iii) *Every maximal ideal is prime.*

2.2.3 Subdirect Representation Theorem

Let Γ denote a nonempty set. The **direct product** of a family $\{M_i\}_{i\in\Gamma}$ of MV-algebras denoted by $\prod_{i\in\Gamma} M_i$, is the MV-algebra obtained by endowing the set-theoretical cartesian product of the family with the MV-operations defined pointwise. In other words, $\prod_{i\in\Gamma} M_i$ is the set of all functions $f : \Gamma \to \bigcup_{i\in\Gamma} M_i$ such that $f(i) \in M_i$, for all $i \in \Gamma$, with the operations * and $\oplus$ defined by

$$f^*(i) := f(i)^* \text{ and } (f \oplus g)(i) := f(i) \oplus g(i).$$

The zero element of $\prod_{i\in\Gamma} M_i$ is the function $\Gamma \ni i \mapsto 0_i \in M_i$. For each $j \in \Gamma$, the map $\pi_j : \prod_{i\in\Gamma} M_i \to M_j$ is defined by

$$\pi_j(f) := f(j).$$

Each π_j is a homomorphism onto M_j, called the j-th **projection function.** In particular, for each MV-algebra M and a nonempty set X, the MV-algebra M^X is the direct product of the family $\{M_x\}_{x\in X}$, where $M_x = M$ for all $x \in X$.

Definition 2.2.30 *An MV-algebra M is a* **subdirect product** *of a family $\{M_i\}_{i\in\Gamma}$ of MV-algebras if and only if there exists a one-one homomorphism $h : M \to \prod_{i\in\Gamma} M_i$ such that, for each $j \in \Gamma$, the composite map $\pi_j \circ h$ is a homomorphism onto M_j.*

The following result is a particular case of a theorem of Universal Algebra, due to Birkhoff [Bir].

Theorem 2.2.31 *An MV-algebra M is a subdirect product of a family $\{M_i\}_{i\in\Gamma}$ of MV-algebras if and only if there is a family $\{J_i\}_{i\in\Gamma}$ of ideals of M such that*

(i) $M_i \cong M/J_i$ *for each* $i \in \Gamma$,

(ii) $\bigcap_{i\in\Gamma} J_i = \{0\}$.

Proof. Supposing first that M is a subdirect product of a family $\{M_i\}_{i\in\Gamma}$ of MV-algebras, let $h : M \to \prod_{i\in\Gamma} M_i$ be the one-one homomorphism as given by Definition 2.2.30. For each $j \in \Gamma$, let $J_i = Ker(\pi_j \circ h)$. Since $\pi_j \circ h$ is surjective, $M_j \cong M/J_j$. If $x \in \bigcap_{i\in\Gamma} J_i$, then $\pi_j(h(x)) = 0$ for all $j \in \Gamma$. This implies that $h(x) = 0$, and since h is injective, $x = 0$. Therefore $\bigcap_{i\in\Gamma} J_i = \{0\}$, and conditions (i) and (ii) hold true.

Conversely, suppose $\{J_i\}_{i\in\Gamma}$ to be a family of ideals of M satisfying conditions (i) and (ii). Let ϵ_i be an isomorphism of M/J_i onto M_i, as given by condition (i). Let the function $h : \prod_{i\in\Gamma} M_i \to M_i$ be defined as follows: for each $x \in M$, $(h(x))(i) = \epsilon_i(x/J_i)$. It follows from (ii) that $Ker(h) = \{0\}$, whence h is injective. Since for each $i \in \Gamma$ the map $a(\in M) \mapsto a/J_i(\in M/J_i)$ is surjective, then $\pi_i \circ h$ maps M onto M_i. Thus, M is a subdirect product of the family $\{M_j\}_{j\in\Gamma}$, as required. □

The following result is fundamental:

Theorem 2.2.32 (Chang's Subdirect Representation Theorem) *Every nontrivial MV-algebra is a subdirect product of linearly ordered MV-algebras.*

Proof. By Theorem 2.2.31 and Proposition 2.2.27, an MV-algebra M is a subdirect product of a family of linearly ordered MV-algebras if and only if there is a family $\{P_i\}_{i\in\Gamma}$ of prime ideals of M such that $\bigcap_{i\in\Gamma} P_i = \{0\}$. Now apply Corollary 2.2.29 (i) to the ideal $\{0\}$. □

2.2.4 Semisimple MV-algebras and Bold Fuzzy Set Theory

One of the most important examples of an MV-algebra is the unit interval $[0,1]$ of real numbers. Here we define, for $x, y \in [0,1]$,

$$x \oplus y = \min\{1, x+y\}, \ \ x^* = 1 - x.$$

Then

$$x \odot y = \max\{0, x+y-1\}, \quad x \vee y = \max\{x,y\}, \quad x \wedge y = \min\{x,y\}.$$

Given a set X, a **fuzzy subset** of X is a function $f : X \to [0,1]$. Consequently, $M = [0,1]^X$ becomes an MV-algebra under pointwise induced operations. Any subalgebra of M is called a **Bold algebra** of fuzzy sets.

Let G be any lattice ordered Abelian group, written additively, with the positive cone G^+, and let $a \in G$, $a > 0$. Let

$$M = G^+[0,a] = \{x \in G : 0 \leq x \leq a\}.$$

Then M is an MV-algebra $\mathcal{M} = (M; \oplus, ^*, 0, a)$ under the operations

$$x \oplus y = a \wedge (x+y), \quad x^* = a - x, \quad x, y \in G^+[0,a].$$

In this MV-algebra, the role of 1 is played by a. It was shown that every MV-algebra arises in this manner from some lattice ordered Abelian group (see Corollary 5.3.8 below). This result was obtained in [Cha 1] for linearly ordered MV-algebras, and in [Mun] for general MV-algebras.

For maximal ideals we have the following result.

Proposition 2.2.33 *Let I be a maximal ideal in M. Then M/I is isomorphic to a subalgebra of* $[0,1]$.

Proof. Since M/I is locally finite and linearly ordered, it can be identified with an interval $G^+[0,u]$ of an Abelian linearly ordered Archimedean group, where u is an order unit. Every such group is isomorphic to a subgroup of $\mathbb{R}$ ([Goo 1]). □

Corollary 2.2.34 *Every locally finite and linearly ordered MV-algebra is isomorphic either to $C(M) := \{0, 1/m, 2/m, \ldots, 1\}$ or to a dense subset of $[0,1]$ (in the sense that whenever $x, y \in M$, $x < y$, there exists $z \in M$, $x \leq z \leq y$).*

Proof. An additive subgroup G of $\mathbb{R}$ is either cyclic or dense in $\mathbb{R}$ (cf. [Goo 1]). □

By Proposition 2.2.33, for each maximal ideal J there is a homomorphism

$$\phi_J : M \to [0,1].$$

Now let $\mathcal{M}(M)$ denote the set of all maximal ideals in M. With each $x \in M$ we associate a fuzzy subset of $\mathcal{M}(M)$, namely

$$\tilde{x} : \mathcal{M}(M) \to [0,1], \text{ where } \tilde{x}(J) = \phi_J(x).$$

The set

$$Rad(M) := \bigcap \{M \in \mathcal{M}(M)\}$$

is said to be the **radical** of M.

If $\mathcal{F}(M) := \{\tilde{x} : x \in M\}$, then $\mathcal{F}(M)$ is a Bold algebra of fuzzy sets, and the map $x \mapsto \tilde{x}$ is a homomorphism of M into $\mathcal{F}(M)$. In general, this map is not an injection.

Theorem 2.2.35 [Bel] *For each MV-algebra M there is a Bold algebra of fuzzy sets $\mathcal{F}(M)$ and a homomorphism of M onto $\mathcal{F}(M)$. The homomorphism $x \mapsto \tilde{x}$ is one-to-one if and only if $Rad(M) = \{0\}$.*

Proof. Clearly, the map $x \mapsto \tilde{x}$ is a homomorphism. $\tilde{x} = \tilde{y}$ if and only if for each maximal ideal J, $d(x, y) \in J$. Thus $Rad(M) = \{0\}$ implies $d(x, y) = 0$, hence $x = y$. If $Rad(M) \neq \{0\}$, choose $y \in Rad(M)$, $y \neq 0$. Then $\tilde{y} = \tilde{0}$, so $x \mapsto \tilde{x}$ is not one-to-one. □

The above theorem can be considered as a generalization of Stone's theorem for Boolean algebras.

We say that M is **semisimple** if $Rad(M) = \{0\}$. The above theorem then tells us that semisimple MV-algebras are Bold algebras of fuzzy sets.

For each $x \in M$ let

$$S(x) := \{J : J \in \mathcal{M}(M) \text{ and } x \in J\}. \tag{2.11}$$

It is obvious that $S(x \odot y) \subseteq S(x) \cap S(y)$, hence the sets $S(x)$ generate a topology $\mathcal{T}$ on $\mathcal{M}(M)$. It is also easy to prove the following theorem.

Theorem 2.2.36 *$\mathcal{T}$ is a compact T_1-space and if x is idempotent, then $S(x)$ is clopen.*

By Corollary 2.2.29 (iii), every maximal ideal is prime. There are examples of linearly ordered MV-algebra that are not contained in $[0, 1]$ (see the example after Theorem 2.2.16). Consequently, prime ideals need not be maximal.

Letting $\mathcal{P}(M)$ denote the set of all prime ideals of M and defining

$$R(x) := \{P \in \mathcal{P}(M) : x \in P\} \tag{2.12}$$

we have as before that the sets $R(x), x \in M$, generate a topology on $\mathcal{P}(M)$. This topology is also a compact T_1-space and for x idempotent $R(x)$ will be clopen.

Definition 2.2.37 *An MV-algebra $\mathcal{M} = (M; \oplus, ^*, 0, 1)$ is* **Archimedean**[5] *if, for each $x, y \in M$, if $n \odot x \leq y$ for all $n \geq 0$, then $x \odot y = x$.*

It is easy to see that a locally finite MV-algebra is Archimedean in the Belluce sense.

We have the following partial converse of Theorem 2.2.16.

Theorem 2.2.38 *A linearly ordered Archimedean MV-algebra $\mathcal{M} = (M; \oplus, ^*, 0, 1)$ is locally finite.*

[5]Or Archimedean in the Belluce sense, see also page 395.

Proof. Suppose the converse. Let x be such that $x \neq 0$, $n \odot x < 1$ for all n. Since M is linearly ordered, we must have $n \odot x \leq x^*$, for all n, as $x^* \leq n \odot x$ would imply $(n+1) \odot x = 1$. $\mathcal{M}$ is Archimedean in the Belluce sense, so $x \odot x^* = x$. But then $x = 0$ contrary to our choice of x. □

The following theorem gives a characterization of semisimple MV-algebras.

Theorem 2.2.39 [Bel] *An MV-algebra $\mathcal{M}$ is semisimple if and only if it is Archimedean in the Belluce sense.*

Proof. (i) Let $\mathcal{M}$ be semisimple. By Chang's Representation Theorem 2.2.32, there is a family $\{M_i\}_{i\in\Gamma}$ of linearly ordered MV-algebras, each M_i locally finite, such that

$$M \subseteq \prod_{i\in\Gamma} M_i$$

as a subdirect product. Suppose $x, y \in M$ are such that $n \odot x \leq y$ for all $n \geq 0$. So in the direct product, we would have have $n \odot x_i \leq y_i$ for each i. Since M_i are locally finite, in each M_i we have

$$(x \odot y)_i = x_i \odot y_i = \begin{cases} 0, & \text{if } x_i = 0_i, \\ x_i, & \text{if } x_i \neq 0_i. \end{cases}$$

Thus $x \odot y = x$.

(ii) Let $\mathcal{M}$ be Archimedean in the Belluce sense. Suppose $x \in Rad(M)$. The x has an infinite order, i.e., $n \odot x < 1$ for all n. Let P be a prime ideal. If $x \notin P$ then P is not maximal. In M/P, x/P must also have an infinite order, for P can be extended to a maximal ideal J and $x \in J$, so $x/P \in J/P$ and J/P is a proper ideal in M/P since $J \neq P$. Thus $n \odot (x/P) \leq x^*/P$ for all n, M/P being linearly ordered. Now M has a subdirect representation in the product of $(M/P), P \in \mathcal{P}(M)$. Consider the element $(n \odot x \wedge x^*)$. Its image in the P-th component M/P is

$$(n \odot x \wedge x^*)/P = \begin{cases} 0, & \text{if } x \in P, \\ n \odot x/P & \text{if } x \notin P. \end{cases}$$

so for each $P \in \mathcal{P}(M)$,

$$(n \odot x \wedge x^*)/P = n \odot x/P,$$

hence $n \odot x \wedge x^* = n \odot x$ for all n. Thus $n \odot x \leq x^*$ for all $n \geq 0$ and by the Archimedean property, $x \odot x^* = x$, therefore $x = 0$. □

Now consider the set $\mathcal{P}(M)$ of prime ideals of M. Let $\mathbf{S} \subseteq \mathcal{P}(M)$ be any nonempty subset. Define on M $x \equiv y (\mathrm{mod}\, \mathbf{S})$ if and only if for each $P \in \mathbf{S}$, $x \in P$ if and only if $y \in P$. In other words, two element of M are related $(\mathrm{mod}\, \mathbf{S})$ iff no prime ideal $P \in \mathbf{S}$ can separate them. Proofs of the statements in the following proposition are left to the readers as an exercise (it can be found in [Bel]).

Proposition 2.2.40 *Let $\mathcal{M} = (M; \oplus, ^*, 0, 1)$ be an MV-algebra.*

(i) $\equiv (mod\, \mathbf{S})$ *is a congruence relation on M with respect to $\oplus, \wedge$.*

(ii) *For each $x \in M$, let $[x]$ denote the equivalence class under $\equiv$ and let $[M]_{\mathbf{S}}$ denote the set of all such equivalence classes. Define on $[M]_{\mathbf{S}}$, the operations $+$ and $\cdot$ by $[x] + [y] = [x \oplus y]$, $[x] \cdot [y] = [x \odot y]$. Let $1 = [1], 0 = [0]$ and define $[x] \leq [y]$ iff $[x] \cdot [y] = [x]$. Then $([M]_{\mathbf{S}}, +, \cdot, \leq, 0, 1)$ is a distributive lattice with least element 0 and greatest element 1.*

(iii) *Let I be an ideal of M. Define $I^* := \{[y] \in [M]_{\mathbf{S}} : \text{for some } y \in [x], y \in I\}$. Then I^* is an ideal of $[M]_{\mathbf{S}}$.*

(iv) *Let L be an ideal of $[M]_{\mathbf{S}}$. Define $L_* := \bigcup\{[x] : [x] \in L\}$. Then L_* is an ideal of M, $(L_*)^* = L$, and $1 \in L_*$ iff $[1] \in L$.*

(v) *Let $\mathcal{P}(M)_{\mathbf{S}}$ denote the set of prime ideals of the lattice $[M]_{\mathbf{S}}$. Then $P \in \mathbf{S}$ implies $P^* \in \mathcal{P}(M)_{\mathbf{S}}$. Conversely, if $L \in \mathcal{P}(M)_{\mathbf{S}}$, then $L_* \in \mathcal{P}(M)$. Moreover, $(P^*)_* = P$.*

(vi) *Let us consider the case $\mathbf{S} = \mathcal{P}(M)$. Then $P \mapsto P_*$ is a one-one map of $\mathcal{P}(M)$ to $(\mathcal{P}(M))_*$. This mapping, moreover, carries $\mathcal{M}(M)$ onto $(\mathcal{M}(M))_*$, the set of maximal ideals of $[M]_{\mathcal{P}(M)}$.*

Summarizing, we get the following theorem.

Theorem 2.2.41 [Bel] *For each MV-algebra M, there is a distributive lattice $[M]_{\mathcal{P}(M)}$ with $0, 1$ such that the prime ideal spaces of M and $[M]_{\mathcal{P}(M)}$ are homeomorphic.*

2.3 QMV-algebras

In this Section, we derive some basic properties of QMV-algebras and introduce some important examples. For example, orthomodular lattices form a subvariety of QMV-algebras characterized by the idempotence of $\oplus$, the subvariety of MV-algebras is characterized by the commutativity of $\Cap$. Another important class of QMV-algebras are so-called quasilinear QMV-algebras, which include the standard QMV-algebra whose elements are the Hilbert space effects.

Then we study commutativity (with respect to the operation $\Cap$) relation in QMV-algebras. We prove that the commutative center of a QMV-algebra is an MV-subalgebra.

Then we introduce a hierarchy of ideals in QMV-algebras and study the relations among them. Every QMV-algebra gives rise to an effect algebra if we restrict the total $\oplus$ operation only to orthogonal pairs of elements. In correspondence with that, there are two basic classes of ideals in QMV-algebras: e-ideals (effect algebra ideals) and q-ideals (QMV-algebra ideals). In both cases, most important subclasses of ideals are those which produce congruences, namely Riesz e-ideals and p-ideals (i.e., q-ideals closed under perspectivity).[6] We prove that the latter two classes form distributive lattices with respect to the set-theoretical inclusion.

In the last section, we consider quasilinear and weakly linear QMV-algebras. In the case of MV-algebras, both these conditions coincide with linearity. We introduce the notion of a prime ideal, and prove that a quotient of a QMV-algebra with respect to a p-ideal is weakly linear if and only if the ideal is prime.

2.3.1 Basic Properties of QMV-algebras

Definition 2.3.1 *A* **QMV-algebra** *is an S-algebra* $\mathcal{M} = (M; \oplus, ^*, 0, 1)$ *satisfying, for all* $a, b, c \in M$, *the condition*[7]

(QMV) $$a \oplus ((a^* \Cap b) \Cap (c \Cap a^*)) = (a \oplus b) \Cap (a \oplus c).$$

Proposition 2.3.2 [Giu 5] Let $\mathcal{M} = (M; \oplus, ^*, 0, 1)$ be an QMV-algebra. For all $a, b, c \in M$, the following properties hold:

(i) $a \oplus (b \Cap a^*) = a \oplus b$.

(ii) $a \oplus (a^* \Cap b) = a \oplus b$.

(iii) $a \oplus [b \Cap (c \Cap a^*)] = (a \oplus b) \Cap (a \oplus c)$.

(iv) $a \odot b \leq a$.

(iv*) $a \leq a \oplus b$.

(v) $a \Cap b \leq b$.

(vi) $a \oplus [b \Cap (a \oplus c)^*] = (a \oplus b) \Cap (a \oplus a^* \odot c^*)$.

(vii) $a \Cup (b \Cap a) = a$.

(viii) $a \leq b$ implies $b = b \Cup a$.

(ix) $a \leq b$ entails $b^* \leq a^*$ (order-reversibility of *).

(x) $a \leq b$ implies $a \oplus c \leq b \oplus c$, (monotonicity of $\oplus$).

(x*) $a \leq b$ implies $a \odot c \leq b \odot c$ (monotonicity of $\odot$).

(xi) $(a \Cap b) \Cap c = (a \Cap b) \Cap (b \Cap c)$.

(xii) $a \oplus b = a \oplus b \oplus a \odot b$.

(xiii) $(a^* \oplus b) \Cup (b^* \oplus a) = 1$.

---

[6]In MV-algebras all these classes coincide, see Theorem 3.1.32.

[7]An example of an S-algebra which is neither a QMV-algebra nor an MV-algebra is in Exercise 6.6.3.1 or in Table 2.0 on p. 132.

Proof. (i)

$$\begin{aligned} a \oplus (b \Cap a^*) &= a \oplus 1 \odot (b \Cap a^*) \\ &= a \oplus (a \oplus a^* \oplus a^* \odot b^*) \odot (b \Cap a^*) \\ &= a \oplus (a^* \oplus (b^* \Cup a)) \odot (b \Cap a^*) \\ &= a \oplus (a^* \oplus (b \Cap a^*)^*) \odot (b \Cap a^*) \\ &= a \oplus (a^* \Cap (b \Cap a^*)) \\ &= a \oplus [(a^* \Cap 1) \Cap (b \Cap a^*)] \\ &= (a \oplus 1) \Cap (a \oplus b) \text{ (QMV)} \\ &= a \oplus b. \end{aligned}$$

(ii)

$$\begin{aligned} a \oplus b &= (a \oplus b) \Cap (a \oplus 1) \\ &= a \oplus [(a^* \Cap b) \Cap (1 \Cap a^*)] \text{ (QMV)} \\ &= a \oplus [(a^* \Cap b) \Cap a^*] \\ &= a \oplus (a^* \Cap b) \quad \text{(i)}. \end{aligned}$$

(iii)

$$\begin{aligned} a \oplus [b \Cap (c \Cap a^*)] &= a \oplus [b \oplus (c^* \Cup a)] \odot (c \Cap a^*) \\ &= a \oplus [b \oplus c^* \odot a^* \oplus a] \odot (c \Cap a^*) \\ &= a \oplus [(a^* \Cap b) \oplus a \oplus c^* \odot a^*] \odot (c \Cap a^*) \quad \text{(ii)} \\ &= a \oplus [(a^* \Cap b) \Cap (c \Cap a^*)] \\ &= (a \oplus b) \Cap (a \oplus c) \quad \text{(QMV)}. \end{aligned}$$

(iv)

$$\begin{aligned} (a \odot b) \Cap a &= (a \odot b \oplus a^*) \odot a \\ &= (b \Cup a^*) \odot a \\ &= [a^* \oplus (b^* \Cap a)]^* \\ &= (a^* \oplus b^*)^* \quad \text{(i)} \\ &= a \odot b. \end{aligned}$$

(iv*)

$$\begin{aligned} a \Cap (a \oplus b) &= (a \oplus 0) \Cap (a \oplus b) \\ &= a \oplus [(a^* \Cap 0) \Cap (b \Cap a^*)] \quad \text{(QMV)} \\ &= a \oplus 0 \\ &= a. \end{aligned}$$

(v) This follows from (iv) and the fact that $a \Cap b = (a \oplus b^*) \odot b$.
(vi)

$$\begin{aligned} a \oplus [b \Cap (a \oplus c)^*] &= a \oplus [b \Cap (a \oplus (c \Cap a^*))^*] \quad \text{(i)} \\ &= a \oplus [b \Cap (a \oplus (c \oplus a) \odot a^*)^*] \\ &= a \oplus [b \Cap ((a \oplus c) \Cup a)^*] \\ &= a \oplus [b \Cap ((a^* \odot c^*) \Cap a^*)] \\ &= (a \oplus b) \Cap (a \oplus a^* \odot c^*) \quad \text{(iii)}. \end{aligned}$$

(vii)

$$\begin{aligned} a \Cup (b \Cap a) &= a \odot (b^* \Cup a^*) \oplus (b \Cap a) \\ &= [a^* \oplus (b \Cap a)]^* \oplus (b \Cap a) \\ &= (a^* \oplus b)^* \oplus (b \Cap a) \quad \text{(i)} \\ &= a \odot b^* \oplus (b \oplus a^*) \odot a \\ &= a \Cup (a \odot b^*) \\ &= [a^* \Cap (a^* \oplus b)]^* \\ &= a^{**} \quad (\text{iv}^*) \\ &= a. \end{aligned}$$

(viii)
$b = b \Cup (a \Cap b)$ (vii) $= b \Cup a$ (hypothesis).
(ix) This follows from (viii).
(x)

$$\begin{aligned} (a \oplus c) \Cap (b \oplus c) &= [(a \oplus c) \oplus b^* \odot c^*] \odot (b \oplus c) \\ &= [(a \oplus c) \oplus (b^* \Cap a^*) \odot c^*] \odot (b \oplus c) \quad \text{(ix), hypothesis} \\ &= [(a \oplus c) \oplus (b^* \oplus a) \odot a^* \odot c^*] \odot ((b \Cup a) \oplus c) \ \text{(viii), hypothesis} \\ &= [(a \oplus c) \oplus (b^* \oplus a) \odot a^* \odot c^*] \odot (b \odot a^* \oplus a \oplus c) \\ &= (a \oplus c) \Cap (b \odot a^* \oplus a \oplus c) \\ &= (a \oplus c) \quad (\text{iv}^*). \end{aligned}$$

(x^*) This follows from (x) by duality
(xi)

$$\begin{aligned} (a \Cap b) \Cap (b \Cap c) &= [(a \Cap b) \oplus (b^* \Cup c^*)] \odot (b \Cap c) \\ &= [(a \Cap b) \oplus b^* \odot c \oplus c^*] \odot (b \oplus c^*) \odot c \\ &= [((a \Cap b) \oplus c^*) \Cap (b \oplus c^*)] \odot c \\ &= [(a \Cap b) \oplus c^*] \odot c \quad \text{(v) and (x)} \\ &= (a \Cap b) \Cap c. \end{aligned}$$

(xii)

$$\begin{aligned}
a \oplus b \oplus a \odot b &= b \oplus (b^* \Cap a) \oplus a \odot b \quad \text{(ii)}\\
&= b \oplus (b^* \oplus a^*) \odot a \oplus a \odot b\\
&= b \oplus (a \Cup (a \odot b))\\
&= b \oplus a \quad \text{(v) and (viii)}.
\end{aligned}$$

(xiii)

$$\begin{aligned}
(a^* \oplus b) \Cup (b^* \oplus a) &= (a^* \oplus b) \odot (b \odot a^*) \oplus b^* \oplus a\\
&= (a^* \oplus b) \odot (b \odot a^*) \oplus b^* \oplus a \oplus b^* \odot a \quad \text{(xii)}\\
&= (a^* \oplus b) \odot (b \odot a^*) \oplus [(b \odot a^*) \odot (b \oplus a^*)]^* = 1. \qquad \square
\end{aligned}$$

Remark 2.3.3 QMV-algebras were originally introduced by Giuntini [Giu 1] as algebraic structures $\mathcal{M} = (M; \oplus, ^*, 0, 1)$ satisfying axioms (S1)–(S7) and (vii), (xi), (vi), (ii), (xiii) of Proposition 2.3.2. Equivalence of the latter axioms with the axiom (QMV) was proved in [Giu 5]. We will prove it in the following proposition.

Recently in [Giu 6], independence of the following axioms has been proved: (S1,2), (S4), (S5'), (S7') and (QMV) (see Remark 2.2.2). Observe that the first four axioms in MV-algebras and QMV-algebras coincide.

Proposition 2.3.4 *Let* $\mathcal{M} = (M; \oplus, ^*, 0, 1)$ *be an S-algebra. Axiom* (QMV) *is equivalent to the axioms* (vii), (xi), (vi), (ii), (xiii) *of Proposition* 2.3.2.

Proof. In Proposition 2.3.2 we proved that (S1)–(S7) and (QMV) imply (vii), (xi), (vi), (ii), (xiii). To prove the converse, assume that (vii), (xi), (vi), (ii), (xiii) are satisfied. Then

$$\begin{aligned}
a \oplus [(a^* \Cap b) \Cap (c \Cap a^*)] &= a \oplus [(a^* \Cap b) \oplus (c \Cap a^*)^*] \odot (c \Cap a^*)\\
&= a \oplus [(a^* \Cap b) \oplus a \oplus c^* \odot a^*] \odot (c \Cap a^*)\\
&= a \oplus [b \oplus a \oplus c^* \odot a^*] \odot (c \Cap a^*) \quad \text{(ii)}\\
&= a \oplus [b \oplus (c^* \Cup a)] \odot (c \Cap a^*)\\
&= a \oplus [b \Cap c \Cap a^*)]\\
&= a \oplus [b \Cap ((c \oplus a) \odot a^*)]\\
&= a \oplus [b \Cap (a \oplus c^* \odot a^*)^*]\\
&= (a \oplus b) \Cap (a \oplus (a^* \odot (c \oplus a))) \quad \text{(vi)}\\
&= (a \oplus b) \Cap (a \oplus (c \Cap a^*))\\
&= (a \oplus b) \Cap (a \oplus c) \quad \text{(i)}.
\end{aligned}$$

$\square$

In what follows, we assume that $\mathcal{M} = (M; \oplus, ^*, 0, 1)$ is a QMV-algebra.

Proposition 2.3.5 *The following properties hold for every $a, b, c \in M$:*

(i) (Absorption law)

$$a \Cap ((a \oplus b) \Cap c) = a \Cap c.$$

(ii) (Cancellation law)

$$a \leq c^*,\ b \leq c^*,\ a \oplus c = b \oplus c \ \Rightarrow \quad a = b.$$

(iii) (Monotonicity of $\Cap$)

$$a \leq b \ \Rightarrow \ a \Cap c \leq b \Cap c.$$

(iii*) (Monotonicity of $\Cup$)

$$a \leq b \quad \Rightarrow \quad a \Cup c \leq b \Cup c.$$

(iv) (Transitivity of $\leq$) $a \leq b,\ b \leq c \Rightarrow a \leq c$.

Proof. (i)

$$\begin{aligned}
a \Cap ((a \oplus b) \Cap c) &= (a \oplus ((a \oplus b)^* \Cup c^*)) \odot ((a \oplus b) \Cap c) \\
&= (a \oplus (a \oplus b)^* \odot c \oplus c^*) \odot (a \oplus b \oplus c^*) \odot c \\
&= ((a \oplus c^*) \oplus (a \oplus b \oplus c^*)^*) \odot (a \oplus b \oplus c^*) \odot c \\
&= ((a \oplus c^*) \Cap (a \oplus b \oplus c^*)) \odot c \\
&= (a \oplus c^*) \odot c \\
&= (a \Cap c).
\end{aligned}$$

(ii) The hypotheses imply $a = a \Cap c^* = (a \oplus c) \odot c^* = (b \oplus c) \odot c^* = b \Cap c^* = b$.

(iii), (iii*) They follow from monotonicity of $\oplus$ and $\odot$, Proposition 2.3.2 (x), (x*).

(iv) By the definition of $\leq$, we have $a = a \Cap b$, $b = b \Cap c$, which yields $a = (a \Cap b) \Cap (b \Cap c)$. Using Proposition 2.3.2 (xi), we get $a = (a \Cap b) \Cap c = a \Cap c$, hence $a \leq c$. □

Notice that Proposition 2.3.5 (iv) together with Corollary 2.1.3 imply that $\leq$ is a partial ordering.

Proposition 2.3.6 *For every $a, b, c \in M$, $a \Cap c = 0$ and $b \Cap c = 0$ imply $(a \oplus b) \Cap c = 0$.*

Proof. Suppose $a \Cap c = 0$ and $b \Cap c = 0$. From $c^* \oplus (a \oplus c^*)^* = 1$ and orthogonality of c^* and $(a \oplus c^*)^*$ we conclude by Proposition 2.3.5 that $c^* = c^* \oplus a$ and similarly, $c^* \oplus b = c^*$. Thus

$$\begin{aligned}
(a \oplus b) \Cap c &= (a \oplus b \oplus c^*) \odot c \\
&= (c^* \oplus b) \odot c \\
&= c^* \odot c \\
&= 0.
\end{aligned}$$

□

Corollary 2.3.7 *Let $C := \{a_1, \ldots, a_n\}$ be any finite sequence of elements of M and let c be any element of M such that $c \Cap a_i = 0$ for all i such that $1 \leq i \leq n$. Then*

$$(a_1 \oplus \cdots \oplus a_n) \Cap c = 0.$$

Proof. We proceed by induction on n. If $n = 1$, the statement holds by hypothesis. Suppose that the statement holds for all m such that $m \leq n - 1$. Then, using $c^* = c^* \oplus a_n$,

$$\begin{aligned}
(a_1 \oplus \cdots \oplus a_n) \Cap c &= (a_1 \oplus \cdots \oplus a_{n-1} \oplus c^*) \odot c \\
&= (a_1 \oplus \cdots \oplus a_{n-1}) \Cap c \\
&= 0.
\end{aligned}$$

□

2.3.2 Examples of QMV-algebras

Example 2.3.8 The smallest QMV-algebra that is not an MV-algebra is the "diamond" $M_4 = \{0, 1, a, b\}$ with the operations $\oplus$ and * defined as follows: $a^* = a$, $b^* = b$, $0^* = 1$, $1^* = 0$, $a \oplus a = 1 = b \oplus b$, $0 \oplus 0 = 0$, $0 \oplus a = a$, $0 \oplus b = b$, all other $\oplus$-sums are equal to 1. (See also Figure 1.7 on p. 88.) It is easy to check that $a \Cap b = b \neq a = b \Cap a$. As this example shows, QMV-algebras, differently from MV-algebras, can contain more than one fixed point of the operation *.

Theorem 2.3.9 *Every orthomodular lattice $\mathcal{L} = (L; \vee, \wedge, ', 0, 1)$ determines a QMV-algebra by taking $\oplus$ as the supremum $\vee$ and * as the orthocomplement $'$. Conversely, if an ortholattice $\mathcal{L}$ determines a QMV-algebra taking $\oplus = \vee$ and $^* = '$, then $\mathcal{L}$ is orthomodular.*

Proof. Axioms (S1)–(S7) are easily verified. To verify (QMV), replace $\oplus$ by $\vee$ and * by $'$. Clearly, $\odot$ is to be replaced by $\wedge$, and $a \Cap b$ becomes $(a \vee b') \wedge b = \phi_b(a)$, the Sasaki projection. Thus we obtain the left-hand side of (QMV) in the form

$$\begin{aligned}
& a \vee [[[(a' \vee b') \wedge b] \vee [(c' \wedge a') \vee a]] \wedge [(c \vee a) \wedge a']] \\
&= a \vee [(((a' \vee b') \wedge b) \vee a \vee (c' \wedge a')) \wedge ((c \vee a) \wedge a')] \\
&= a \vee [(((a' \vee b' \vee a) \wedge (b \vee a)) \vee (c' \wedge a')) \wedge (c \vee a) \wedge a'] \ (a\,\mathrm{C}\,(a' \vee b'),\ b\,\mathrm{C}\,(a' \vee b')) \\
&= a \vee [(b \vee a \vee (c' \wedge a')) \wedge (c \vee a) \wedge a'] \\
&= (a \vee b \vee a \vee (c' \wedge a')) \wedge (a \vee ((c \vee a) \wedge a')) \ (a\,\mathrm{C}\,(b \vee a \vee c' \wedge a'),\ a\,\mathrm{C}\,(c \vee a) \wedge a') \\
&= (a \vee b \vee (c' \wedge a')) \wedge (a \vee c) \\
&= \phi_{(a \vee c)}(a \vee b),
\end{aligned}$$

where $a\,\mathrm{C}\,b$ means that a is compatible with b and we used the Foulis-Holland theorem ([Kal]).

On the other hand, if we take $a \leq b$, $b = c$, where a, b, c are elements of an ortholattice, we obtain from (QMV)

$$a \vee (a' \wedge b \vee b' \vee a) \wedge (b \wedge a') = a \vee (b \wedge a') = b,$$

the orthomodular law. □

Definition 2.3.10 Let $\mathcal{M} = (M; \oplus, ^*, 0, 1)$ be a QMV-algebra. Let

$$M_I := \{a \in M : a = a \oplus a\}.$$

We will show, in the next theorem, that M_I is an orthomodular lattice. First we need the following lemma.

Lemma 2.3.11 *Let $\mathcal{M} = (M; \oplus, ^*, 0, 1)$ be a QMV-algebra. The following conditions are equivalent:*

(i) $b = b \oplus a^*$.

(ii) $a \Cup b = 1$.

(iii) $a = a \oplus b^*$.

(iv) $b \Cup a = 1$.

Proof. (i) $\Rightarrow$ (ii). $a \Cup b = (a \odot b^*) \oplus b = (a^* \oplus b)^* \oplus b = b^* \oplus b = 1$.

(ii) $\Rightarrow$ (iii). $a^* = a^* \odot 1 = a^* \odot (a \Cup b)$, by Proposition 2.3.2 (ii) we may conclude that $a^* \odot (a \Cup b) = a^* \odot b$. Thus $a = a \oplus b^*$.

(iii) $\Rightarrow$ (iv). $b \Cup a = (b \odot a^*) \oplus a = (b^* \oplus a)^* \oplus a = a^* \oplus a = 1$.

(iv) $\Rightarrow$ (i). $b^* = b^* \odot 1 = b^* \odot (b \Cup a)$. By Proposition 2.3.2 (ii) we can conclude that $b^* \odot (b \Cup a) = b^* \odot a$. Thus $b = b \oplus a^*$. □

Theorem 2.3.12 $\mathcal{M}_I = (M_I; \oplus, \odot, ^*, 0, 1)$ *is an orthomodular lattice.*

Proof. M_I is clearly closed under $\oplus$. Suppose $a \in M_I$. Then $a = a \oplus (a^*)^*$. By Lemma 2.3.11 (i) and (iii), $a^* = a^* \oplus a^*$, hence $a^* \in M_I$.

Let a, b be any elements of M_I. We need to show that $a \oplus b$ is the supremum of a and b in M_I under the restriction to M_I of the partial ordering $\leq$. We have $a, b \leq a \oplus b$ (monotonicity of $\oplus$). Suppose $a, b \leq c$. Then $a \oplus b \leq c \oplus c = c$ (see Proposition 2.3.2 (iv*)).

We now show that $a \leq b$ iff $a \oplus b = b$. Suppose $a \leq b$. Then $a \oplus b \leq b \oplus b = b \leq a \oplus b$. Thus $a \oplus b = b$. Conversely, if $a \oplus b = b$, then $a \leq (a \oplus b) = b$.

Therefore the partial ordering determined by $\oplus$ is precisely $\leq$, i.e. the partial ordering determined by $\Cup$. It should be noticed that in general, $a \oplus b \neq a \Cup b$.

The operation * is clearly an orthocomplementation. It remains to prove orthomodularity. Suppose $a \leq b$, then $a \oplus b = b$, and hence $b = b \Cup a = b \odot a^* \oplus a = (b^* \oplus a)^* \oplus a$. □

Corollary 2.3.13 *The variety of orthomodular lattices is a subvariety of the variety of QMV-algebras. This subvariety satisfies the condition $a = a \oplus a$, or equivalently, $a \Cap a^* = 0$.*

Proof. Equivalence of conditions $a \oplus a = a$ and $a \Cap a^* = 0$ follows from Lemma 2.3.11 (i) and (ii).

Condition $a = a \oplus a$ is satisfied in every QMV-algebra which is an orthomodular lattice under the lattice operation $\vee' = \oplus$.

Conversely, if the latter condition is satisfied, then $\mathcal{M}$ coincides with $\mathcal{M}_I$, and hence by Theorem 2.3.12, it is an orthomodular lattice. □

Example 2.3.14 Every MV-algebra is a QMV-algebra. Both MV- and QMV-algebras are S-algebras. It suffices to show that (MV) implies (QMV). It is easy to see that (MV) is equivalent to $a \Cap b = b \Cap a$. Then (QMV) reduces to

$$a \oplus [(b \Cap a^*) \Cap (a^* \Cap c)] = (a \oplus b) \Cap (a \oplus c).$$

By Proposition 2.2.4 (vii), the left-hand side equals $a \oplus [(b \Cap a^*) \Cap c]$, and by Proposition 2.2.4 (i), (vii) and (x), we get $a \oplus (b \Cap c)$. Using Proposition 2.2.4 (xi) we get the desired equality.

Thus we have the following statement.

Theorem 2.3.15 *The variety of MV-algebras is a subvariety of the variety of QMV-algebras. This subvariery satisfies the condition $a \Cap b = b \Cap a$.*

Example 2.3.16 Standard QMV-algebras. Let us consider the standard effect algebra $\mathcal{E}(H)$ consisting of the set $\mathcal{E}(H)$ of all self-adjoint operators A such that $O \leq A \leq I$, where O is the null and I is the identity operator, partially ordered by the usual ordering inherited from the set of all self-adjoint operators, endowed with the partial operation $\oplus$ defined by $A \oplus B = A + B$ if and only if $A + B \in \mathcal{E}(H)$, where $+$ is the usual operator sum.

Let us extend the operation $\oplus$ to $\oplus^c$ as follows. For any $E, F \in \mathcal{E}(H)$,

$$E \oplus^c F := \begin{cases} E+F, & \text{if } E+F \in \mathcal{E}(H), \\ I, & \text{otherwise.} \end{cases}$$

Thus, the operation $\oplus^c$ can be regarded as a **quantum collapsed sum**. We define, as usual, $E^* := I - E$. It is easy to see that $E \oplus^c F = E + F$ iff $E + F \leq I$, where $\leq$ is the usual ordering of self-adjoint operators.

One can easily check that the structure $\mathcal{E}(H) = (\mathcal{E}(H); \oplus^c, ^*, O, I)$ satisfies the axioms of an S-algebra. We will check (QMV). We have, according to the general definitions,

$$E \Cap F = E, \text{ if } E \leq F, \; E \Cap F = F, \text{ otherwise.}$$

It turns out that the ordering defined by $A \leq B$ if $A = A \Cap B$ coincides with the usual ordering for $\mathcal{E}(H)$. It is easy to check that $X \leq Y \Rightarrow X \oplus^c Z \leq Y \oplus^c Z$ for any Z. Let $A, B, C \in \mathcal{E}(H)$. Recall that (QMV) states that $U = V$, where

$$U := A \oplus^c ((A^* \Cap B) \Cap (C \Cap A^*)), \; V := (A \oplus^c B) \Cap (A \oplus^c C).$$

The following cases can occur:

(1) $A^* \leq B, C \leq A^*$. It then follows $C \leq B$, which in turn implies $A \oplus^c C \leq A \oplus^c B$. Then $U = A \oplus^c A^* \Cap C = A \oplus^c C$, $V = A \oplus^c C$.

(2) $A^* \leq B$, $C \not\leq A^*$. Then $A \oplus^c C = I$, $I = A \oplus^c A^* \leq A \oplus^c B$, hence $U = A \oplus^c A^* = I$, $V = I \Cap I = I$.

(3) $A^* \not\leq B, C \leq A^*$. Then $U = A \oplus^c B \Cap C$. Two cases can occur.

(3.1) $B \leq C$, then $A \oplus^c B \leq A \oplus^c C$. So we get $U = A \oplus^c B$, $V = A \oplus^c B$.

(3.2) $B \not\leq C$. Then $U = A \oplus^c C$. Since $A \oplus^c C \leq I$, $A \oplus^c B \leq A \oplus^c C$ would imply $B \leq C$, a contradiction. Therefore $A \oplus^c B \not\leq A \oplus^c C$, hence $V = A \oplus^c C$.

(4) $A^* \not\leq B$, $C \not\leq A^*$. Then $U = A \oplus^c B \Cap A^*$. Since $C \not\leq A^*$, $A \oplus^c C = I$. Two cases can occur.

(4.1) $B \leq A^*$. Then $U = A \oplus^c B = V$.

(4.2) $B \not\leq A^*$. Then $U = I = V$.

The structure $\mathcal{E}(H) = (\mathcal{E}(H); \oplus^c, ^*, O, I)$ will be called the **standard QMV-algebra**.

Example 2.3.17 Let $\mathcal{E} = (E; \oplus, ^*, 0, 1)$ be any effect algebra. We write $a \perp b$ iff $a \leq b^*$. Let us extend the partial operation $\oplus$ to a total operation $\oplus^c$ as follows:

$$a \oplus^c b = \begin{cases} a \oplus b, & \text{if } a \perp b, \\ 1, & \text{otherwise.} \end{cases}$$

Then it can be checked, exactly as in Example 2.3.16, that $(E; \oplus^c, ^*, 0, 1)$ is a QMV-algebra. The operation $\oplus^c$ will be called the **collapsed sum extension** of $\oplus$. It turns out that the partial order in the latter QMV-algebra coincides with the partial order in E (see Proposition 2.3.2 (iv*) and Lemma 2.3.20 below).

QMV-algebras obtained this way belong to the class of quasi-linear QMV-algebras (see Subsection 2.3.7).

Suppose that $(M; \oplus, ^*, 0, 1)$ is a QMV-algebra. We define $a \boxplus b = a \oplus b$ iff $a \perp b$. Then $(M; \boxplus, 0, 1)$ is an effect algebra.

We note that starting from a QMV-algebra to an effect algebra and from this effect algebra again to a QMV-algebra, the final and starting structures are not always the same. They are the same only if the QMV-algebra is quasi-linear.

Remark 2.3.18 Let X be a nonempty set, and let $A := [0,1]^X$. If we define $f^* = \underline{1} - f$, $(f \oplus^t g)(x) = \sup\{f(x) + g(x), 1\}$, $\underline{1}$ and $\underline{0}$ as the identically 1 and 0 functions, then the structure $(A; \oplus^t, ^*, \underline{0}, \underline{1})$ is the standard MV-algebra.

A can also be considered as an effect algebra with partial operation $\oplus$ defined by $f \oplus g = f + g$ iff $f + g \leq \underline{1}$, and $f' = \underline{1} - f = f^*$. The partial order in this effect algebra coincides with the usual pointwise ordering of functions, and A is a lattice with respect to this order.

We can define the collapsed extension $\oplus^c$ of $\oplus$ as in Example 2.3.17. We obtain a quasilinear QMV-algebra, where $a \Cap b = a$ if $a \leq b$ and $a \Cap b = b$ otherwise. The ordering still coincides with the pointwise ordering of functions. So, in general, $a \Cap b \neq b \Cap a$, and the structure $(A; \oplus^c, ^*, \underline{0}, \underline{1})$ is not an MV-algebra. But it still has the lattice structure, and it is easy to check that every two elements $f, g \in A$ have a Mackey decomposition $f \wedge g, f - f \wedge g, g - f \wedge g$. So every two elements are compatible in the Mackey sense. Therefore by [ChKo 2], [Pul 5], the effect algebra A can be organized into an MV-algebra by defining the extension of $\oplus$ in the following way: $f \oplus' g = f \oplus (f^* \wedge g)$. It is easy to check that $\oplus' \equiv \oplus^t$.

We may conclude that there are QMV-algebras in which any two elements are Mackey compatible, which are not MV-algebras.

2.3.3 Commutativity in QMV-algebras

MV-algebras can be characterized as those QMV-algebras where the operation $\Cap$ is commutative. Therefore it is quite natural to define a relation of commutativity for QMV-algebras, which turns out to be universal for MV-algebras. We will define the

(commutative) center of a QMV-algebra in a usual way as the subset of those elements that commute with all its elements. In this Section we prove that the commutative center of a QMV-algebra is an MV-algebra. In this sequel, we mainly follow [Giu 4].

Definition 2.3.19 *Let* $\mathcal{M} = (M; \oplus, ^*, 0, 1)$ *be a QMV-algebra and let* $a, b \in M$. *We say that* a **commutes** *with* b *(and we write* $a \,\mathrm{C}\, b$*) if* $a \Cap b = b \Cap a$.

In what follows $\mathcal{M} = (M; \oplus, ^*, 0, 1)$ is a QMV-algebra.

Lemma 2.3.20 *For any* $a, b \in M$,

$$a = a \odot b^* \oplus (b \Cap a).$$

Proof. We have $a \odot b^* \oplus (b \Cap a) = a \odot b^* \oplus (b \oplus a^*) \odot a = a \odot b^* \oplus (b^* \odot a)^* \odot a = a \Cup (b^* \odot a) = a$, since $b^* \odot a \leq a$ (Proposition 2.3.2 (iv)). □

Lemma 2.3.21 *Let* $a, b \in M$. *The following conditions are equivalent:*

(i) $a \,\mathrm{C}\, b$.

(ii) $a = a \odot b^* \oplus (a \Cap b)$.

Proof. (i) $\Rightarrow$ (ii) follows from Lemma 2.3.20. (ii) $\Rightarrow$ (i). Suppose (ii). By Lemma 2.3.20, $a = a \odot b^* \oplus (b \Cap a)$. Now $a \odot b^* \leq b^* \leq a^* \odot b \oplus b^* = (a \Cap b)^*$, $a \odot b^* \leq b^* \odot a \oplus a^* = (b \Cap a)^*$. Thus, by cancellation law (Proposition 2.3.5 (ii), $a \Cap b = b \Cap a$. □

Lemma 2.3.22 *The following properties hold for any* $a, b \in M$:

(i) *The relation* C *is reflexive and symmetric.*

(ii) *If* $a \leq b$ *or* $b \leq a$, *then* $a \,\mathrm{C}\, b$.

(iii) *If* $a \,\mathrm{C}\, b$, *then* $a^* \,\mathrm{C}\, b^*$.

Proof. (i) This is straightforward.

(ii) This follows by Proposition 2.1.2 (xi).

(iii) By Lemma 2.3.20, $b = b \odot a^* \oplus (a \Cap b)$, $a = a \odot b^* \oplus (b \Cap a) = a \odot b^* \oplus (a \Cap b)$ (hypothesis). Thus, $a \Cup b = a \odot b^* \oplus b = a \odot b^* \oplus b \odot a^* \oplus (a \Cap b) = b \odot a^* \oplus a = b \Cup a$, hence $a^* \Cap b^* = b^* \Cap a^*$. □

Lemma 2.3.23 *For all* $a, b, c \in M$, $(c \Cap a)^* \oplus (b \Cap a) = (c \Cap a)^* \oplus b$.

Proof.

$$\begin{aligned}(c \Cap a)^* \oplus (b \Cap a) &= (c^* \Cup a^*) \oplus (b \Cap a)\\ &= (c^* \odot a) \oplus a^* \oplus (b \Cap a)\\ &= (c^* \odot a) \oplus a^* \oplus b \quad \text{(Proposition 2.3.2 (i))}\\ &= (c^* \Cup a^*) \oplus b\\ &= (c \Cap a)^* \oplus b.\end{aligned}$$

□

Lemma 2.3.24 *For all $a, b, c, d \in M$, if $a \leq b$, $c \leq d$, $b \,\mathrm{C}\, c$ and $b \,\mathrm{C}\, d$, then $a \Cup c \leq b \Cup d$.*

Proof. From monotonicity of $\odot$ (Proposition 2.3.2 (x*)) we can derive $a \Cup c \leq b \Cup c$ and $c \Cup b \leq d \Cup b$. By Lemma 2.3.22 (iii), $b^* \,\mathrm{C}\, c^*$ and $b^* \,\mathrm{C}\, d^*$. Hence $b \Cup c = c \Cup b$ and $d \Cup b = b \Cup d$. Thus, $a \Cup c \leq b \Cup d$. □

Lemma 2.3.25 *For all $a, b, c \in M$, if $a \,\mathrm{C}\, b$ and $a \,\mathrm{C}\, c$, then $(a \Cap b) \Cap c = b \Cap (a \Cap c)$.*

Proof.

$$\begin{aligned}(a \Cap b) \Cap c &= (b \Cap a) \Cap c \quad (a \,\mathrm{C}\, b)\\ &= (b \Cap a) \Cap (a \Cap c) \ \text{(Proposition 2.3.2 (xi))}\\ &= ((b \Cap a) \oplus (a \Cap c)^*) \odot (a \Cap c)\\ &= (b \Cap a) \oplus (c \Cap a)^*) \odot (c \Cap a) \quad (a \,\mathrm{C}\, c)\\ &= ((c \Cap a)^* \oplus b) \odot (c \Cap a) \ \text{(Lemma2.3.23)}\\ &= b \Cap (c \Cap a)\\ &= b \Cap (a \Cap c) \quad (a \,\mathrm{C}\, c).\end{aligned}$$

□

Lemma 2.3.26 *For all $a, b, c \in M$, if $a \,\mathrm{C}\, b$, $b \,\mathrm{C}\, c$, $a \,\mathrm{C}\, c$, then $(a \Cap b) \Cap c = c \Cap (a \Cap b)$.*

Proof. $(a \Cap b) \Cap c = b \Cap (a \Cap c)$ (Lemma 2.3.25) $= b \Cap (c \Cap a)$ $(a \,\mathrm{C}\, c) = c \Cap (b \Cap a)$ (Lemma 2.3.25) $= c \Cap (a \Cap b)$ $(a \,\mathrm{C}\, b)$. □

Lemma 2.3.27 *For all $a, b, c \in M$, if $a \,\mathrm{C}\, c^*$ and $b \,\mathrm{C}\, c$, then*

$$a \oplus (b \Cap c) = (a \oplus b) \Cap (a \oplus c).$$

Proof. First we prove that $a \oplus (b \Cap c) \leq (a \oplus b) \Cap (a \oplus c)$. Let $\alpha := (a \oplus (b \Cap c)) \Cap ((a \oplus b) \Cap (a \oplus c))$. We want to show that $\alpha = a \oplus (b \Cap c)$.

$$\begin{aligned}
\alpha &= (a \oplus (b \Cap c) \oplus ((a \oplus b)^* \Cup (a \oplus c)^*)) \odot (a \oplus b \oplus (a \oplus c)^*) \odot (a \oplus c) \\
&= (a \oplus (b \Cap c) \oplus (a \oplus b)^* \odot (a \oplus c) \oplus (a \oplus c)^*) \odot ((a \oplus b) \oplus (a \oplus c)^*) \odot (a \oplus c) \\
&= ((a \oplus (b \Cap c) \oplus (a \oplus c)^*) \Cap ((a \oplus b) \oplus (a \oplus c)^*)) \odot (a \oplus c) \\
&= ((a \oplus (c \Cap b) \oplus (a \oplus c)^*) \Cap ((a \oplus b) \oplus (a \oplus c)^*)) \odot (a \oplus c) \quad (b \, \mathrm{C} \, c) \\
&= (a \oplus (b \Cap c) \oplus (a \oplus c)^*) \odot (a \oplus c) \quad (c \Cap b \leq b) \\
&= (a \oplus (b \Cap c)) \Cap (a \oplus c) \\
&= a \oplus (b \Cap c).
\end{aligned}$$

This proves $a \oplus (b \Cap c) \leq (a \oplus b) \Cap (a \oplus c)$. In order to prove the opposite inequality, we have to show that $((a \oplus b) \Cap (a \oplus c))^* \oplus a \oplus (b \Cap c) = 1$. Let $\beta := ((a \oplus b) \Cap (a \oplus c))^* \oplus a \oplus (b \Cap c)$. Then

$$\begin{aligned}
\beta &= a \oplus (b \Cap c) \oplus ((a \oplus b)^* \Cup (a \oplus c)^*) \\
&= a \oplus (b \Cap c) \oplus (a^* \odot b^*) \odot (a \oplus c) \oplus (a^* \odot c^*) \\
&= (c^* \Cup a) \oplus (b \Cap c) \oplus (a^* \odot b^*) \odot (a \oplus c) \\
&= b^* \odot (a^* \Cap c) \oplus (a^* \Cap c)^* \oplus (b \Cap c) \quad (a \, \mathrm{C} \, c^*) \\
&= b^* \odot (a^* \Cap c) \oplus (a^* \Cap c)^* \oplus b \text{ (Lemma 2.3.23)} \\
&= 1.
\end{aligned}$$

□

Definition 2.3.28 *The* **centre** *of* $\mathcal{M} = (M; \oplus, ^*, 0, 1)$ *is the set* $\mathcal{Z}(M) := \{a \in M : a \Cap b = b \Cap a \ \ \forall\, b \in M\}$.

Proposition 2.3.29 *For any* $a, b \in \mathcal{Z}(M)$ *such that* $a \leq b^*$, $a \oplus b \in \mathcal{Z}(M)$.

Proof. Let $c \in M$. By Lemma 2.3.21 and hypothesis,

$$c = c \odot a^* \oplus (c \Cap a),$$

and similarly,

$$c \odot a^* = (c \odot a^*) \odot b^* \oplus ((c \odot a^*) \Cap b)$$

and

$$c \Cap a = (c \Cap a) \odot b \oplus ((c \Cap a) \Cap b^*).$$

By hypothesis, $a \leq b^*$, hence $a \odot b = 0$ and $a \Cap b^* = a$. Since $c \Cap a \leq a \leq b^*$, we get $(c \Cap a) \odot b = 0$ and $(c \Cap a) \Cap b^* = c \Cap a$. Thus,

$$\begin{aligned}
c &= (c \odot a^*) \odot b^* \oplus ((c \odot a^*) \Cap b) \oplus (c \Cap a) \\
&= (c \odot a^*) \odot b^* \oplus (((c \Cap a) \oplus (c \odot a^*)) \Cap ((c \Cap a) \oplus b)) \quad \text{(Lemma 2.3.27)} \\
&= (c \odot a^*) \odot b^* \oplus (((a \Cap c) \oplus (c \odot a^*)) \Cap ((c \Cap a) \oplus b)) \quad (a\,\mathrm{C}\,c) \\
&= (c \odot a^*) \odot b^* \oplus (((a \oplus c^*) \odot c \oplus (c^* \oplus a)^*) \Cap ((c \Cap a) \oplus b)) \\
&= (c \odot a^*) \odot b^* \oplus ((c \Cup (c \odot a^*)) \Cap ((c \Cap a) \oplus b)) \\
&= (c \odot a^*) \odot b^* \oplus (c \Cap ((c \Cap a) \oplus b)) \quad (c \odot a^* \leq c) \\
&= (c \odot a^*) \odot b^* \oplus (c \Cap (b \oplus c) \Cap (a \oplus b)) \quad \text{(Lemma 2.3.27)} \\
&= (c \odot a^*) \odot b^* \oplus (((b \oplus c) \Cap c) \Cap (a \oplus b)) \quad \text{(Lemma 2.3.25)} \\
&= (c \odot a^*) \odot b^* \oplus (c \Cap (a \oplus b)) \\
&= c \odot (a \oplus b)^* \oplus (c \Cap (a \oplus b)).
\end{aligned}$$

Thus, by Lemma 2.3.21, $(a \oplus b)\,\mathrm{C}\,c$. □

Theorem 2.3.30 *Let $\mathcal{M} = (M; \oplus, ^*, 0, 1)$ be a QMV-algebra. The structure*

$$\mathcal{Z}(\mathcal{M}) = (\mathcal{Z}(M); \oplus, ^*, 0, 1)$$

is an MV-subalgebra of $\mathcal{M}$. Moreover, for any $a, b \in M$, the inf *(*sup*) in $\mathcal{Z}(M)$ coincides with the* inf *(*sup*) of a, b in M, and is equal to $a \Cap b$ ($a \Cup b$). Thus $(\mathcal{Z}(M); \Cap, \Cup, ^*, 0, 1)$ is a de Morgan distributive sublattice of the involutive bounded poset $(M; \leq, ^*, 0, 1)$.*

Proof. Clearly, $\mathcal{Z}(M)$ contains 0 and 1. By Lemma 2.3.22 (iii), $\mathcal{Z}(M)$ is closed under the operation *. To prove that $\mathcal{Z}(M)$ is a subalgebra of $\mathcal{M}$, it remains to prove that $\mathcal{Z}(M)$ is closed under $\oplus$.

Suppose $a, b \in \mathcal{Z}(M)$. By Proposition 2.3.2 (i), $a \oplus b = a \oplus (b \Cap a^*)$. Since $\mathcal{Z}(M)$ is closed under $\Cap$ (Lemma 2.3.26) and * and $a \leq b^* \Cup a = (b \Cap a^*)^*$, it follows by Proposition 2.3.29 that $a \oplus b \in \mathcal{Z}(M)$. Thus $\mathcal{Z}(M)$ is an MV-subalgebra of $\mathcal{M}$.

Let $a, b \in \mathcal{Z}(M)$. By commutativity, $a \Cap b \leq a, b$. Let c be any element of M such that $c \leq a, b$. Then $c = (c \Cap a) \Cap b = c \Cap (a \Cap b)$ by Lemma 2.3.26, hence $c \leq a \Cap b$. Thus, $a \Cap b$ is the infimum of a, b in M, and consequently in $\mathcal{Z}(M)$.

The remaining statement follows by the properties of MV-algebras. □

Corollary 2.3.31 (1) *Let $\mathcal{L} = (L; \wedge, \vee, ', 0, 1)$ be an orthomodular lattice. Let $\oplus := \wedge$ and $^* :=\, '$. The center of $\mathcal{L}$ (understood as a QMV-algebra) is a Boolean subalgebra of $\mathcal{L}$ and coincides with the center of $\mathcal{L}$ (understood as a lattice).*

(2) *Let $\mathcal{M} = (M; \oplus, ^*, 0, 1)$ be a quasilinear QMV-algebra. The center of $\mathcal{M}$ is a totally ordered MV-algebra.*

(3) *Let $\mathcal{E}(H)$ be the standard QMV-algebra. The center of $\mathcal{E}(H)$ is $\{O, I\}$.*

2.3.4 Congruences and Ideals in Quantum MV-algebras

In what follows, $\mathcal{M} = (M; \oplus, ^*, 0, 1)$ is a QMV-algebra. We mainly follow [GiPu].

Definition 2.3.32 *An* **ideal** *of* $\mathcal{M} = (M; \oplus, ^*, 0, 1)$ *is a nonempty subset* $I \subseteq M$ *such that*

(i) $a, b \in I$ *entails* $a \oplus b \in I$,

(ii) $a \in I$ *entails* $a \odot b \in I$ *for any* $b \in M$.

Lemma 2.3.33 *A nonempty subset* I *of* M *is an ideal if and only if* I *is an order ideal closed under* $\oplus$.

Proof. Assume that I is an ideal. Let $a \in I$ and $b \leq a$. Then $b = b \Cap a$ $=(b \oplus a^*) \odot a \in I$.

Conversely, if I is an order ideal closed under $\oplus$, then $a \odot b \leq a$ entails $a \odot b \in I$ whenever $a \in I$. □

Definition 2.3.34 *Two elements* $a, b \in M$ *are* **perspective** *(*$a \sim b$ *in symbols) if they have a common complement, i.e., there is* $c \in M$ *such that* $a \odot c = b \odot c = 0$ *and* $a \oplus c = b \oplus c = 1$.

The relation $\sim$ is reflexive and symmetric, but in general not transitive. We denote its transitive closure by $\approx$. Thus, for any $a, b \in M$, $a \approx b$ iff there exist $x_1, \ldots, x_n \in M$ such that $x_1 = a, x_n = b$ and $x_i \sim x_{i+1}$, $i = 1, \ldots, n-1$.

Proposition 2.3.35 *The following properties hold for every* $a, b, c \in M$:

(i) $a \sim 0$ *implies* $a = 0$.

(ii) $a \sim b$ *implies* $a^* \sim b^*$.

(iii) $(a \Cap b) \sim (b \Cap a)$.

(iv) $c \leq a$ *and* $a \sim b$ *imply there exists* $d \in M$ *such that* $d \leq b$ *and* $c \sim d$.

Proof. (i) This is straightforward.

(ii) If c is a common complement of a, b, then c^* is a common complement of a^*, b^*.

(iii) The element $(a \Cap b)^*$ is a common complement of $a \Cap b$ and $b \Cap a$. Indeed,

$$\begin{aligned}
(b \Cap a) \oplus (a \Cap b)^* &= (b \Cap a) \oplus (a^* \odot b \oplus b^*) \\
&= (a^* \odot b) \oplus (b^* \oplus b \Cap a) \\
&= a^* \odot b \oplus (b^* \oplus a) \text{ (Proposition 2.3.2 (ii))} \\
&= 1.
\end{aligned}$$

(iv) Suppose $c \leq a$ and $a \sim b$. Let x be a common complement of a and b. Let $d := (c \Cap x^*) \Cap b$. Clearly, $d \leq b$. It remains to prove that $c \sim d$. Let $y := c^* \Cup x$. We show that y is a common complement of c and d.

Clearly, $c \oplus y = c \oplus (c^* \Cup x) = 1$.

$d \odot y = ((c \Cap x^*) \Cap b) \odot y = (y^* \Cap b) \odot y = (y^* \oplus b^*) \odot b \odot y = 0.$

$$\begin{aligned} c \odot y &= c \odot (c^* \Cup x) \\ &= c \odot x \text{ (Proposition2.3.2(ii))} \\ &\leq a \odot x \quad (c \leq a) \\ &= 0. \end{aligned}$$

$$\begin{aligned} d \oplus y &= ((c \Cap x^*) \Cap b) \oplus (c \Cap x^*)^* \\ &= (c \Cap x^*)^* \oplus b \quad \text{(Proposition 2.3.2 (ii))} \\ &= (c^* \Cup x) \oplus b \\ &= c^* \odot x^* \oplus x \oplus b \\ &= 1. \end{aligned}$$

□

Definition 2.3.36 *A* **p-ideal** *is an ideal I which is closed under perspectivity, i.e., for all $a, b \in M$, $a \in I$ and $a \sim b$ imply $b \in I$.*

Proposition 2.3.37 *Let I be an ideal in M. The following statements are equivalent:*

(i) *I is a p-ideal.*

(ii) *$a \in I$ implies $a \Cap b \in I$.*

(iii) *$b \in I$ and $a^* \oplus b = 1$ imply $a \in I$.*

Proof. (i) $\Rightarrow$ (ii). Let $a \in I$. Then $b \Cap a \leq a$ and $a \Cap b \sim b \Cap a$ (Proposition 2.3.35 (ii)) imply $a \Cap b \in I$.

(ii) $\Rightarrow$ (iii). By hypothesis, $b \Cap a = (b \oplus a^*) \odot a = a$, hence $a \in I$.

(iii) $\Rightarrow$ (i). Let $a \in I$ and $a \sim b$. Let c be a common complement of a and b. Then $a \oplus c = 1$ implies $c^* \in I$, and $b^* \oplus c^* = 1$ implies $b \in I$. □

Definition 2.3.38 *Let I be an ideal in M. For $a, b \in M$, $a \equiv_I b$ if and only if there exists $x \in M$ such that $x \leq a, b$ and $x^* \odot a, x^* \odot b \in I$.*

Proposition 2.3.39 *Let I be an ideal in M. The following conditions are equivalent:*

(i) $a \equiv_I b$.

(ii) *There exist* $i, j \in I$ *such that* $i \leq a, j \leq b$ *and* $a \odot i^* = b \odot j^*$.

(iii) *There exist* $i, j \in I$ *such that* $a \odot i^* \leq b$, $b \odot j^* \leq a$.

Proof. (i) ⇒ (ii). Suppose $a \equiv_I b$, then there is $x \leq a, b$ such that $x^* \odot a, x^* \odot b \in I$. Let $i := x^* \odot a$ and $j := x^* \odot b$. Clearly, $i \leq a$ and $j \leq b$. Moreover, $a \odot i^* = a \odot (x \oplus a^*) = x \Cap a = x = x \Cap b = (x \oplus b^*) \odot b = j^* \odot b$.

(ii) ⇒ (iii) is straightforward.

(iii) ⇒ (ii). Put $l := i \Cap a$ and $k := (i \oplus a^*) \odot b$. Clearly, $l \leq a$, $k \leq b$ and $l \in I$. We want to prove that $k \in I$. Since $i \oplus j \in I$, and I is a p-ideal, it suffices to prove that $k^* \oplus i \oplus j = 1$. Since $a \odot i^* = a \odot (i \Cap a)^*$, we may assume that $i \leq a$. Then

$$\begin{aligned} k^* \oplus i \oplus j &= i^* \odot a \oplus b^* \oplus i \oplus j \\ &= (a \Cup i) \oplus b^* \oplus j \\ &= a \oplus b^* \oplus j \\ &= a \oplus (b \odot j^*)^* \\ &\geq a \oplus a^* = 1. \end{aligned}$$

It remains to show that $a \odot l^* = b \odot k^*$. Since $(a \odot i^*) \leq b$, we have $a \odot l^* = a \odot (i \Cap a)^* = a \odot i^* = (a \odot i^*) \Cap b = (a \odot i^* \oplus b^*) \odot b = k^* \odot b$.

(ii) ⇒ (i). Suppose there exists $i, j \in I$ such that $i \leq a$, $j \leq b$ and $x := i^* \odot a = j^* \odot b$. Clearly, $x \leq a$ and $x \leq b$. Moreover, $a \odot x^* = a \odot (i \oplus a^*) = i \Cap a = i \in I$, and $b \odot x^* = b \odot (j \oplus b^*) = j \Cap b = j \in I$. □

Lemma 2.3.40 (i) *Let* I *be an ideal. Then, for all* $a, b \in M$, $a \equiv_I b$ *entails* $d(a, b) \in I$, *where* $d(a, b) := (a \odot b^*) \oplus (a^* \odot b)$. (ii) *Let* I *be a p-ideal. Then* $a \in I$ *and* $a \equiv_I b$ *entail* $b \in I$.

Proof. (i) Suppose $a \equiv_I b$. Then there is $x \leq a, b$ such that $a \odot x^* \in I$ and $b \odot x^* \in I$. Thus, $b \odot a^* \leq b \odot x^*$ and $a \odot b^* \leq a \odot x^*$. Since I is an ideal, we have $a \odot b^* \in I$ and $a^* \odot b \in I$. Hence $d(a, b) \in I$.

(ii) Suppose $a \in I$ and $a \equiv_I b$ By (i), $b \odot a^* \in I$. Thus $b \Cup a = (b \odot a^*) \oplus a \in I$. Since $b \Cup a \sim a \Cup b$, and $b \leq a \Cup b$, we have $b \in I$. □

Definition 2.3.41 *A binary relation* $\sim$ *is a* **congruence** *if it is an equivalence relation and preserves* * *and* $\oplus$.

Theorem 2.3.42 *Every p-ideal* I *determines a congruence relation* $\equiv_I$ *such that the set* $\{a : a \equiv_I 0\}$ *is equal to* I.

Conversely, given a congruence relation $\equiv$, *the set* $I := \{a : a \equiv 0\}$ *is a p-ideal and* $\equiv_I \leq \equiv$.

Proof. First we prove that the relation $\equiv_I$ is a congruence. The relation $\equiv_I$ is clearly reflexive and symmetric. So, we have to prove that $\equiv_I$ is (i) transitive, (ii) preserves *, and (iii) preserves $\oplus$.

(i) Suppose $a \equiv_I b$ and $b \equiv_I c$. Then there exist $x, y \in M$ such that $x \leq a, b$, $y \leq b, c$ and $x^* \odot a$, $x^* \odot b$, $y^* \odot b$, $y^* \odot c \in I$. We show that there exist $l, m \in I$ such that $a \odot l^* \leq c$ and $c \odot m^* \leq a$. Then we can conclude by (iii) of Proposition 2.3.39 that $a \equiv_I c$. Let

$$l := a \odot x^* \oplus b \odot y^*, \quad m := b \odot x^* \oplus c \odot y^*.$$

Clearly, $l, m \in I$. We have

$$\begin{aligned} a \odot l^* &= a \odot (x^* \odot a)^* \odot (b \odot y^*)^* \leq (a \odot (x^* \odot a)^* \oplus x^* \odot b) \odot (b \odot y^*)^* \\ &= (a \odot (x \oplus a^*) \oplus x^* \odot b) \odot (b^* \oplus y) = ((x \Cap a) \oplus x^* \odot b) \odot (b^* \oplus y) \\ &= (x \oplus x^* \odot b) \odot (b^* \oplus y) = (b \Cup x) \odot (b^* \oplus y) \\ &= b \odot (b^* \oplus y) = y \Cap b = y \leq c. \end{aligned}$$

Similarly we prove that $c \odot m^* \leq a$.

(ii) Let $a \equiv_I b$. We have to show that $a^* \equiv_I b^*$. Let $x \in M$ be such that $x \leq a, b$ and $x^* \odot a, x^* \odot b \in I$. Let $y := a^* \odot (x^* \odot b)^*$. Clearly, $y \leq a^*$. It suffices to prove that $y \leq b^*, y^* \odot a^*, y^* \odot b^* \in I$. We have $a^* \leq x^* = x^* \Cup b^*$. Therefore

$$\begin{aligned} y = a^* \odot (x^* \odot b)^* &\leq (x^* \odot b \oplus b^*) \odot (x^* \odot b)^* \\ &= b^* \Cap (x^* \odot b)^* \\ &= b^* \Cap (x \oplus b^*) \\ &= b^*. \end{aligned}$$

So $y \leq b^*$. Now $y^* \odot a^* = (a \oplus (x^* \odot b)) \odot a^* = (x^* \odot b) \Cap a^*$. Since $x^* \odot b \in I$ and I is a p-ideal, it follows that $(x^* \odot b) \Cap a^* \in I$. Moreover,

$$\begin{aligned} y^* \odot b^* &= (x^* \odot b \oplus a) \odot b^* \\ &= ((x^* \odot b) \oplus (a \odot x^*) \oplus x) \odot b^* \\ &\leq ((x^* \odot b) \oplus (a \odot x^*) \oplus b) \odot b^* = ((x^* \odot b) \oplus (x^* \odot a)) \Cap b^*. \end{aligned}$$

Now $(x^* \odot b) \oplus (x^* \odot a) \in I$, hence $((x^* \odot b) \oplus (x^* \odot a)) \Cap b^* \in I$, for I is a p-ideal. Thus $y^* \odot b^* \in I$.

(iii) Suppose $a \equiv_I b$ and $c \equiv_I d$. By hypothesis, there are $x, y \in M$ such that $x \leq a, b$, $y \leq c, d$ and $x^* \odot a, x^* \odot b \in I$, $y^* \odot c, y^* \odot d \in I$. Let $z := x \oplus y$. By monotonicity of $\oplus$, we get $z \leq a \oplus c$, $z \leq b \oplus d$. It remains to prove that $z^* \odot (a \oplus c), z^* \odot (b \oplus d) \in I$. It suffices to show that $(x^* \odot y^* \odot (a \oplus c))^* \oplus a \odot x^* \oplus c \odot y^* = 1$, and similarly for $a \oplus c$ replaced by $b \oplus d$. We have

$$\begin{aligned} x \oplus y \oplus a^* \odot c^* \oplus a \odot x^* \oplus c \odot y^* &= a \Cup x \oplus c \Cup y \oplus a^* \odot c^* \\ &= a \oplus c \oplus a^* \odot c^* \quad (x \leq a, y \leq c) \\ &= 1. \end{aligned}$$

The proof of $z^* \odot (b \oplus d) \in I$ is similar.

If $a \equiv_I 0$, there are $i, j \in I$ such that $i \leq a, j \leq 0$, $a \odot i^* = 0 \odot j^*$. But this yields $a \odot i^* = 0$, hence $i \Cap a = a$, but $i \leq a$ gives $a = i \in I$.

This proves the first part of Theorem.

Let $\equiv$ be a congruence, and let $I := \{a : a \equiv 0\}$. If $a, b \in I$, then $a \equiv 0, b \equiv 0$ imply $a \oplus b \equiv 0 \oplus 0$, hence $a \oplus b \in I$. If $a \in I$, $b \in M$, then $a \equiv 0$, $b \equiv b$ imply $a \odot b \equiv 0 \odot b = 0$, hence $a \odot b \in I$. If $a \in I$ and $b^* \oplus a = 1$, then $1 = b^* \oplus a \equiv b^*$, hence $b \equiv 0$. This proves that I is a p-ideal.

Assume $a \equiv_I b$. Then there are $i, j \in I$ such that $i \leq a$, $j \leq b$ and $i^* \odot a = j^* \odot b$. Since $i^* \equiv j^* \equiv 1$, we get $a \equiv b$. □

In general the equality $\equiv_I = \equiv$ does not hold, as can be seen from the following example. Let us consider the diamond M_4. Then the only proper p-ideal is the trivial ideal $\{0\}$. We will show that there is a congruence different from identity. Let us define $\equiv$ as follows: $a \equiv b$, $x \equiv x$, $x \in M_4$. It is easy to check that $\equiv$ is a congruence relation in the QMV sense. Then we get, e.g., $a \equiv b$, but $a \neq b$.

Let us consider the set $\mathcal{I}^p(M)$ of all p-ideals in M. It is clear that any set-theoretical intersection of p-ideals is a p-ideal, so the set $\mathcal{I}^p(M)$ is a complete lattice. Consequently, for any (nonempty) subset $X \subseteq M$, there exists a smallest ideal I_X that contains X. In particular, there is a smallest p-deal containing $x, x \in M$. In what follows, we describe the structure of these ideals in more details.

Let $\approx$ be the transitive closure of the perspectivity relation $\sim$. Then $\approx$ is an equivalence relation. Moreover, we have the following.

Lemma 2.3.43 (i) *$a \approx 0$ implies $a = 0$.*

(ii) *$a \approx b$ implies $a' \approx b'$.*

(iii) *$a \Cap b \approx b \Cap a$.*

(iv) *$c \leq a, a \approx b$ imply there exists $d \in M$ such that $d \leq b$ and $c \approx b$.*

Proof. Recall that $a \approx b$ iff there are $d_1, d_2, \ldots, d_n \in M$ such that $a = d_1$, $d_i \sim d_{i+1}, i = 1, 2, \ldots, n-1$, $d_n = b$. Then properties (i)–(iv) can be obtained from the properties of $\sim$ (see Proposition 2.3.35) by reccurence. □

Let A be any nonempty subset of M. Let us define

$$\overline{A} := \{a \in M : \exists\, C = \{b_1, \ldots, b_n\} \subset A, a = b_1 \oplus \cdots \oplus b_n\},$$

$$A^{\approx} := \{a \in M : \exists\, b \in M, \exists\, c \in A, \exists\, n \in \mathbb{N} \text{ such that } a \approx b \text{ and } b \leq n \odot c\}.$$

Proposition 2.3.44 [Giu 5] *Let A be a nonempty subset of M. Then $A^{\approx}$ is an order ideal closed under perspectivity.*

Proof. First we prove that $X^{\approx}$ is an order ideal. Let $a \in X^{\approx}$ and $d \leq a$. By definition, $a \approx b$ for some $b \in M$ such that $b \leq n \odot c$ for some $c \in X$ and $n \in \mathbb{N}$. By Lemma 2.3.43 (iv), there is $e \in M$ such that $e \leq b$ and $d \approx e$. Thus $e \leq n \odot c$ and therefore $d \in X^{\approx}$. Now we prove that $X^{\approx}$ is closed under perspectivity. Suppose $a \in X^{\approx}$ and $b \sim a$. By definition, $a \approx c$ for some $c \in M$ such that $c \leq n \odot d$ with $d \in X$ and $n \in \mathbb{N}$. By hypothesis, $b \sim a$, and hence $b \approx a$. By transitivity of $\approx$, $b \approx c$ and therefore $b \in X^{\approx}$. □

Lemma 2.3.45 [Giu 5] *Let I be an order ideal closed under perspectivity. The following property holds for any $a, b, x \in M$:*

$$b \in \overline{I}, \text{ and } x \odot (a \oplus b)^* = 0 \Rightarrow \exists\, y \in \overline{I} \text{ such that } x = y \oplus (a \Cap x).$$

Proof. We prove that our statement holds (i) for $b \in I$, and (ii) for $b \in \overline{I}$.

(i) Let $b \in I$ and $a, x \in M$ be such that $x \odot (a \oplus b)^* = 0$. Let $y := a^* \odot x$. We have $x = a^* \odot x \oplus (a \Cap x)$. Thus $x = y \oplus (a \Cap x)$. It remains to prove that $y \in \overline{I}$. Let $d := a^* \Cap b$. Now $a^* \Cap b \leq b$ and by hypothesis, $b \in I$. Hence $d \in I$ since I is an order ideal. Let $c := (b \Cap a^*) \Cup y$. Clearly, $y \leq c$. We want to show that $c \sim d$. Then we can conclude that $y \in I$, since I is an order ideal closed under perspectivity.

Let $z := b^* \Cup a$. We show that z is a common complement of c and d.

$$\begin{aligned} c \odot z &= ((b \Cap a^*) \Cup y) \odot (b^* \Cup a) \\ &= (b^* \Cup a) \odot y \quad \text{(Proposition 2.3.2 (ii))} \\ &= (b^* \Cup a) \odot a^* \odot x \quad \text{(definition of } y\text{)} \\ &= (a^* \odot b^*) \odot x \quad \text{(Proposition 2.3.2 (ii))} \\ &= 0 \quad \text{hypothesis.} \end{aligned}$$

Since $b^* \Cup a = (b \Cap a^*)^*$, we get $d \odot z = 0$ and $z \oplus d = 1$. Similarly $z \oplus c = 1$.

(ii) Suppose $b \in \overline{I}$ and $x \odot (a \oplus b)^* = 0$. By definition of $\overline{I}$, there exists $C = \{c_1, \ldots, c_n\} \subset I$ such that $b = c_1 \oplus \cdots \oplus c_n$. We will proceed by induction on the cardinality of C. If $|C| = 1$, then $b \in I$, and the statement follows by step (i). Suppose the statement holds for all C' such that $|C'| < |C|$. Let $c_i \in C$ and let $C_i := C \setminus \{c_i\}$. Let

$$d := c_1 \oplus \cdots \oplus c_{i-1} \oplus c_{i+1} \oplus \cdots \oplus c_n.$$

Thus, $b = d \oplus c_i$ and therefore

$$\begin{aligned} 0 &= x \odot (a \oplus b)^* \quad \text{hypothesis} \\ &= x \odot ((a \oplus c_i) \oplus d)^*. \end{aligned}$$

By induction hypothesis, there exists $z \in \overline{I}$ such that

$$x = z \oplus ((a \oplus c_i) \Cap x).$$

Now,

$$((a \oplus c_i) \Cap x) \odot (a \oplus c_i)^* = 0.$$

Thus, by (i), there exists $u \in I$ such that

$$(a \oplus c_i) \Cap x = u \oplus (a \Cap ((a \oplus c_i) \Cap x)).$$

By absorption law (Proposition 2.3.5 (i)),

$$a \Cap ((a \oplus c_i) \Cap x) = a \Cap x.$$

Hence,

$$x = z \oplus ((a \oplus c_i) \Cap x) = z \oplus u \oplus (a \Cap x).$$

Now, $z \in \overline{I}$ and $u \in I \subseteq \overline{I}$. Thus, $z \oplus u \in \overline{I}$, since $\overline{I}$ is closed under $\oplus$. □

Theorem 2.3.46 [Giu 5] *Let A be a nonempty subset of M. Then $\overline{A^{\approx}}$ is the smallest p-ideal containing A. Moreover, $C(X) := \overline{X^{\approx}}$ is a closure operator on the power set of M.*

Proof. Clearly, $\overline{A^{\approx}}$ is closed under $\oplus$. It remains to show
(i) $b \in \overline{A^{\approx}}$ and $a \leq b \Rightarrow a \in \overline{A^{\approx}}$, and
(ii) $\overline{A^{\approx}}$ is closed under perspectivity.

(i) Suppose $b \in \overline{A^{\approx}}$ and $a \leq b$. Then $a \odot b^* = 0$ and by Lemma 2.3.45, there exists $y \in \overline{A^{\approx}}$ such that $a = y \oplus (0 \Cap a) = y$. Hence $a \in \overline{A^{\approx}}$.

(ii) It suffices to prove that $a \in \overline{A^{\approx}}$ implies $a \Cap b \in \overline{A^{\approx}}$ for any $b \in M$. Assume $a \in \overline{A^{\approx}}$. Now $0 = (a \Cap b) \odot a^* = (a \Cap b) \odot (a \oplus 0)^*$. Thus by Lemma 2.3.45, there is $z \in \overline{A^{\approx}}$ such that $a \Cap b = z \oplus (0 \Cap a) = z$, hence $a \Cap b \in \overline{A^{\approx}}$.

To prove that $\overline{A^{\approx}}$ is the smallest p-ideal containing A, assume $A \subset J$ and J is a p-ideal. But then $A^{\approx} \subset J^{\approx} = J$, and $\overline{A^{\approx}} \subseteq \overline{J} = J$.

The rest of the proof is straightforward. □

If $\mathcal{M}$ is an MV-algebra, then, for all $a \in M$, $C(\{a\}) = \{b \in M : b \leq n \odot a, n \in \mathbb{N}\}$.

Lemma 2.3.47 *For all $a, b, c \in M$,*

$$a \perp c \Rightarrow a \oplus (b \Cap c) = (a \oplus b) \Cap (a \oplus c).$$

Proof. This is a consequence of Proposition 2.3.2 (iii). □

To every QMV-algebra $\mathcal{M} = (M; \oplus, ^*, 0, 1)$ there can be associated an effect algebra $(M; \boxplus, ^*, 0, 1)$, where $\boxplus$ is the partial binary operation on M obtained by the restriction of $\oplus$ to orthogonal pairs of elements. Therefore, we can consider the effect-algebra ideals of M (e-ideals, for short): a subset I of M is an **effect algebra ideal (e-ideal)** if

$$a, b \in M, \; a \perp b \Rightarrow (a \boxplus b \in I \Leftrightarrow a, b \in I).$$

In other words, an e-ideal is an order ideal closed under $\boxplus$. (Indeed, the partial order in the effect algebra $(M; \boxplus, ^*, 0, 1)$ coincides with the partial order in the QMV-algebra $\mathcal{M}$, cf. Proposition 2.3.2 and Lemma 2.3.20).

Definition 2.3.48 *We will say that a subset A of M has the* **Riesz property** *if and only if for $a \in A$, $b, c \in M$, $b \perp c$:*

$$a \leq b \oplus c \Rightarrow \exists\ b_1, c_1 \in A \text{ such that } b_1 \leq b,\ c_1 \leq c \text{ and } a \leq b_1 \oplus c_1.$$

An e-ideal with the Riesz property is called a **Riesz e-ideal** ([GuPu]).

A q-ideal with the Riesz property is called a **Riesz q-ideal** ([GiPu]).

Theorem 2.3.49 *Every p-ideal of M is a Riesz q-ideal.*

Proof. Assume I is a p-ideal. Let $i \in I$, $i \leq a \oplus b$ and $a \perp b$. By Proposition 2.3.2 (v), we have: $i \Cap a \leq a$, $i \Cap b \leq b$ and $i \Cap a$, $i \Cap b \in I$ since I is a p-ideal. To show that I is a Riesz q-ideal, it suffices to show

$$i \leq (i \Cap a) \oplus (i \Cap b). \tag{2.13}$$

By Lemma 2.3.47, since $i \Cap a \leq a \leq b^*$, we have

$$(i \Cap a) \oplus (i \Cap b) = ((i \Cap a) \oplus i) \Cap ((i \Cap a) \oplus b).$$

Since $b \perp a$, again by Lemma 2.3.47, we have

$$b \oplus (i \Cap a) = (b \oplus i) \Cap (b \oplus a).$$

As a consequence we have

$$\begin{aligned}(i \Cap a) \oplus (i \Cap b) &= ((i \Cap a) \oplus i) \Cap ((b \oplus i) \Cap (b \oplus a)) \\ &\geq ((i \Cap a) \oplus i) \Cap (i \Cap (b \oplus a)) \quad (b \oplus i \geq i) \\ &= ((i \Cap a) \oplus i) \Cap i \quad (i \leq a \oplus b) \\ &= i \quad ((i \Cap a) \oplus i \geq i).\end{aligned}$$

□

In what follows, we will present an example of a Riesz q-ideal, which is not a p-ideal [GiPu].

Let $\mathcal{M} = (M; \oplus, ^*, 0, 1)$ be an MV-algebra. We can define the following binary operation on M.

$$a \oplus^c b := \begin{cases} a \oplus b, & \text{if } a \perp b, \\ 1, & \text{otherwise.} \end{cases}$$

It turns out that the structure $\mathcal{M}^{\oplus^c} = (M; \oplus^c, {}^*, 0, 1)$ is a quasi-linear QMV-algebra such that the partial ordering $\leq$ of $\mathcal{M}^{\oplus^c}$ coincides with the partial ordering of $\mathcal{M}$.

Let us consider the direct product $\mathcal{M} \times \mathcal{M}$, which is an MV-algebra. Thus, $(\mathcal{M} \times \mathcal{M})^{\oplus^c}$ is a quasi-linear QMV-algebra. There follows:

$$(a, b) \oplus^c (c, d) = \begin{cases} (a, b) \oplus (c, d), & \text{if } a \perp c \text{ and } b \perp d, \\ (1, 1), & \text{otherwise.} \end{cases}$$

For all $a, b \in M$, and all $n \in \mathbb{N}$ we have

$$n \odot (a, b) := \underbrace{(a, b) \oplus \cdots \oplus (a, b)}_{n-\text{times}} = (n \odot a, n \odot b)$$

$$\overline{n} \odot (a, b) := \underbrace{(a, b) \oplus^c \cdots \oplus^c (a, b)}_{n-\text{times}}.$$

In general, $\overline{n} \odot (a, b) \neq n(a, b)$. Take for instance $\mathcal{M}$ as a Boolean algebra. $\mathcal{M}$ is an MV-algebra where $\oplus$ is supremum. Accordingly, for all $n \in \mathbb{N}$, $n \odot (a, b) = (a, b)$. Let $a \neq 0$ and $a \neq 1$. Then $\overline{2} \odot (a, b) = (a, b) \oplus^c (a, b) = (1, 1) \neq (a, b)$.

Lemma 2.3.50 *Let $\mathcal{M} = (M; \oplus, {}^*, 0, 1)$ be a linear MV-algebra. For all $a \in M$, $n \in \mathbb{N}$,*

$$n \odot (a, a) = \overline{n} \odot (a, a).$$

Proof. By induction on n. If $n = 1$, the proof is trivial. Suppose $(n-1) \odot (a, a) = \overline{n-1} \odot (a, a)$. We want to show $n \odot (a, a) = \overline{n} \odot (a, a)$. Two cases are possible: (i) $n \odot (a, a) = (1, 1)$; (ii) $n \odot (a, a) \neq (1, 1)$.

(i) Suppose $n \odot (a, a) = (1, 1)$. Then $na = 1$. By Proposition 2.2.3, $a^* \leq (n-1) \odot a$. Consequently, $(a^*, a^*) \leq ((n-1) \odot a, (n-1) \odot a) = (n-1) \odot (a, a)$. By QMV properties:

$$\begin{aligned} (1, 1) &= (a, a) \oplus^c (n-1) \odot (a, a) \\ &= (a, a) \oplus \overline{(n-1)} \odot (a, a) \quad \text{induction hypothesis} \\ &= \overline{n}(a, a). \end{aligned}$$

(ii) Suppose $n \odot (a, a) \neq (1, 1)$. Then $na \neq 1$. Since $\mathcal{M}$ is linear, $(n-1) \odot a < a^*$. Consequently, $((n-1) \odot a, (n-1) \odot a) \oplus^c (a, a) = (n \odot a, n \odot a)$.

$$\begin{aligned} \overline{n} \odot (a, a) &= \overline{(n-1)} \odot (a, a) \oplus^c (a, a) \\ &= (n-1) \odot (a, a) \oplus^c (a, a) \quad \text{induction hypothesis} \\ &= (n \odot a, n \odot a) \\ &= n \odot (a, a). \end{aligned}$$

□

Lemma 2.3.51 *Let $\mathcal{M} = (M; \oplus, ^*, 0, 1)$ be a linear MV-algebra and let $x \in M$. Let $I_x := \{(a,b) \in M \times M : (a,b) \leq n \odot (x,x), n \in \mathbb{N}\}$. Then, I_x is a Riesz q-ideal of $(\mathcal{M} \times \mathcal{M})^{\oplus^c}$ and I_x is proper if $C(\{x\})$ is proper, where $C(\{x\})$ is the q-ideal (in $\mathcal{M}$) generated by $\{x\}$.*

Proof. Clearly, I_x is an order ideal. We prove that I_x is closed under $\oplus^c$. Let $(b,c) \leq \overline{n} \odot (x,x)$ and $(e,f) \leq \overline{m} \odot (x,x)$, with $n, m \in \mathbb{N}$. Let $s := m + n$. By monotonicity,

$$(b,c) \oplus^c (e,f) \leq \overline{s} \odot (x,x).$$

By Lemma 2.3.50, $\overline{s} \odot (x,x) = s \odot (x,x)$ and therefore, I_x is closed under $\oplus^c$.

Suppose $C(\{x\})$ is proper. Then, for all $n \in \mathbb{N}$, $n \odot x \neq 1$ and therefore $n \odot (x,x) = (n \odot x, n \odot x) \neq (1,1)$.

The proof of the fact that I_x is a Riesz q-ideal follows from the following properties:

1. In any MV-algebra, every q-ideal is a Riesz q-ideal.
2. For all $(a,b), (c,d) \in M \times M$ such that $(a,b) \perp (c,d)$, there holds: $(a,b) \oplus^c (c,d) = (a,b) \oplus (c,d) = (a \oplus c, b \oplus d)$. □

Analogously as in MV-algebras, we introduce the definition of order of elements in a QMV-algebra.

Definition 2.3.52 *Let $\mathcal{M} = (M; \oplus, ^*, 0, 1)$ be a QMV-algebra. The* **order** *of an element a (shortly, $ord(a)$) is the least integer m, if it exists, such that $m \odot a = 1$. If such an integer does not exist, $ord(a) = \infty$.*

A QMV-algebra is said to be **locally finite** if and only if, for all $a \in M$, $ord(a) < \infty$. By 2.2.16, every locally finite MV-algebra is totally ordered. As shown by Chang [Cha], there are totally ordered MV-algebras, which are not locally finite (see example following Theorem 2.2.16).

Theorem 2.3.53 *Let $\mathcal{M}$ be a linear MV-algebra which is not locally finite. For any $x \in M$ such that $x \neq 0$ and $ord(x) = \infty$, the set*

$$I_x := \{(a,b) \in M \times M : (a,b) \leq n \odot (x,x), n \in \mathbb{N}\}$$

is a Riesz q-ideal, which is not a p-ideal.

Proof. Let us consider a totally ordered MV-algebra $\mathcal{M}$ which is not locally finite. Let x be any element of M such that $x \neq 0$ and $ord(x) = \infty$. Then, for all $n \in \mathbb{N}$, $n \odot x \neq 1$. Consequently, $C(\{x\})$ is proper and therefore, by Lemma 2.3.51, I_x is a proper Riesz q-ideal in $(\mathcal{M} \times M)^{\oplus^c}$. We will prove that I_x is not closed under perspectivity. Clearly, $(x,x) \in I_x$. Now $(0,1) \notin I_x$ since, otherwise, there exists

$n \in \mathbb{N}$ such that $(0,1) \leq n \odot (x,x)$ so that $n \odot x = 1$, which contradicts $ord(x) = \infty$. We prove that (x,x) is perspective to $(0,1)$ with common complement $(1,0)$. Clearly, $(1,0)$ is a complement of $(0,1)$. Now, $(x,x) \oplus^c (1,0) = (1,1)$ since $x \neq 0$ and therefore $(x,x) \not\leq (1,0)$. Further, $(x,x) \odot^c (1,0) := ((x^*,x^*) \oplus^c (0,1))^* = (1,1)^* = (0,0)$ since $(x^*,x^*) \not\leq (0,1)$. □

2.3.5 Lattices of Ideals in QMV-algebras

In this sequel, we mainly follow the ideas in [GiPu]. Consider the following sets:

- $\mathcal{I}^e(M)$ - the set of all effect algebra ideals (e-ideals) of M;
- $\mathcal{I}^q(M)$ - the set of all QMV-ideals (q-ideals) of M;
- $\mathcal{RI}^e(M)$ - the set of all Riesz e-ideals of M;
- $\mathcal{RI}^q(M)$ - the set of all Riesz q-ideals on M;
- $\mathcal{I}^p(M)$ - the set of all perspectivity closed q-ideals (p-ideals) of M.

By an **e-ideal** we mean an order ideal of M closed under the partial sum $\boxplus$. By a **q-ideal** we mean an order ideal of M closed under the total sum $\oplus$.

Recall that an ideal I is a Riesz ideal (see Definition 2.3.48) if it has the following property:

$$\forall\, i,a,b \in M,\ i \in I,\ a \leq b^*,\ i \leq a \oplus b\ \ (= a \boxplus b)$$
$$\Rightarrow\ \exists\, a_1, b_1 \in I,\ a_1 \leq a,\ b_1 \leq b,\ i \leq a_1 \oplus b_1.$$

We have the following diagram (Figure 2.1):

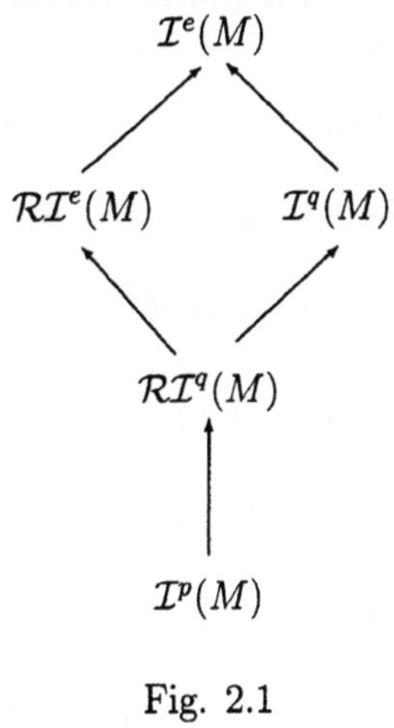

Fig. 2.1

Indeed, $\mathcal{I}^p(M) \subseteq \mathcal{RI}^q(M)$ is Theorem 2.3.49. All other inclusions are straightforward. Next theorem shows that all the inclusions are proper.

Theorem 2.3.54 [GiPu] *There are counterexample to the following statements:*

(i) $\mathcal{RI}^q(M) \subset \mathcal{I}^p(M)$.

(ii) $\mathcal{I}^e(M) \subset \mathcal{RI}^e(M)$.

(iii) $\mathcal{RI}^e(M) \subset \mathcal{I}^q(M)$.

(iv) $\mathcal{I}^e(M) \subset \mathcal{I}^q(M)$.

(v) $\mathcal{I}^q(M) \subset \mathcal{RI}^q(M)$.

Proof. (i) See Theorem 2.3.53.

(ii) Consider the standard effect algebra $\mathcal{E}(H)$ and let $\mathcal{P}(H)$ be the class of all projections in H. We recall that a **projection** is an effect E such that $E^2 = E$. Let $P \in \mathcal{P}(H)$ and $P \neq O$. Since P is principal (i.e., $A, B \leq P$, $A \perp B \Rightarrow A \oplus B \leq P$), the interval $[O, P] := \{A \in \mathcal{E}(H) : O \leq A \leq P\}$ is an e-ideal. But $[O, P]$ is not a Riesz e-ideal, unless P is a central projection (see Theorem 3.1.23), that is, $P = O$ or $P = I$.

(iii)–(iv) Let H be an infinite dimensional Hilbert space and let $J \subset \mathcal{P}(H)$ be the set of all finite dimensional projections on H. Then J is a Riesz e-ideal in $\mathcal{P}(H)$. Let $\mathcal{M} = (\mathcal{P}(H); \oplus^c, {}^*, O, I)$, defined as in Example 2.3.16. By definition of $\mathcal{M}$, J is a Riesz e-ideal in $\mathcal{M}$. But J cannot be a q-ideal since the only q-ideal of $\mathcal{M}$ is $\{O\}$.

This is also a counterexample to (iv).

(v) Let $\mathcal{M}$ be a QMV-algebra which is an OML (see Theorem 2.3.9). Since $\oplus = \vee$, every interval $[0, a]$, $a \in M$, is a q-ideal. But it is a Riesz ideal (and hence an orthomodular ideal) iff a is central. □

For any subsets A, B of M, define

$$A \boxplus B = \{a \boxplus b : a \in A, b \in B, a \leq b^*\}.$$

Whenever $a \perp b$, we will write $a \boxplus b$ instead of $a \oplus b$.

Lemma 2.3.55 *For any q-ideal J and any p-ideal I, we have*

$$I \vee^q J = I \boxplus J$$

(in the lattice $\mathcal{I}^q(M)$). If J is a Riesz ideal, then so are $I \vee^q J$ and $I \cap J$.

Proof. First we prove that $I \boxplus J$ is an e-ideal. Let $a, b \in I \boxplus J$ and $a \perp b$. Writing $a = i_1 \boxplus j_1$, $b = i_2 \boxplus j_2$, with $i_1, i_2 \in I$ and $j_1, j_2 \in J$, we have

$$a \boxplus b = (i_1 \boxplus j_1) \boxplus (i_2 \boxplus j_2) = (i_1 \boxplus i_2) \boxplus (j_1 \boxplus j_2) \in I \boxplus J.$$

Conversely, assume that $a \boxplus b \in I \boxplus J$. Then, there exists $i \in I$ and $j \in J$ such that $a \boxplus b = i \boxplus j$. Since I is a p-ideal (hence a Riesz e-ideal), there are $i_1, i_2 \in I$, $i_1 \leq a$, $i_2 \leq b$ and $i \leq i_1 \boxplus i_2$. Writing $a = i_1 \boxplus a_1$, $b = i_2 \boxplus b_2$, it follows

$$a \boxplus b = (a_1 \boxplus b_2) \boxplus (i_1 \boxplus i_2) = (a_1 \boxplus b_2) \boxplus (i \boxplus i_3),$$

since $i \leq i_1 \boxplus i_2$. As $(a_1 \boxplus b_2) \boxplus (i \boxplus i_3) = i \boxplus j$, it follows that $j = a_1 \boxplus b_2 \boxplus i_3$. Since J is an ideal and $j \in J$, we have $a_1 \in J$, $b_2 \in J$. Finally, $a \in I \boxplus J$, $b \in I \boxplus J$.

We will prove now that $I \boxplus J$ is closed under $\oplus$. Let $a, b \in I \boxplus J$. Then, $a = a_1 \boxplus a_2$, $b = b_1 \boxplus b_2$, $a_1, b_1 \in I$, $a_2, b_2 \in J$, so that $c := a_1 \oplus b_1 \in I$, $d := a_2 \oplus b_2 \in J$. Since $d \leq c \oplus d$, there is $d_1 \in M$ such that $d \boxplus d_1 = c \oplus d$. It then follows that $d \boxplus d_1 \boxplus (c \oplus d)^* = 1$ and $(c \oplus d) \oplus (c \oplus d)^* = 1$. Since I is a p-ideal and $c \in I$, the latter equality implies that $(d \oplus (c \oplus d)^*)^* \in I$. Since $d_1 \perp (d \boxplus (c \oplus d)^*)$, it follows that $d_1 = (d \oplus (c \oplus d)^*)^*$. Hence $d_1 \in I$. Consequently, $a \oplus b = c \oplus d = d_1 \boxplus d \in I \boxplus J$.

Now we prove that $I \vee^q J = I \boxplus J$. Clearly I, J are subsets of $I \boxplus J$. If K is any q-ideal containing I and J, then $I \boxplus J \subset K$. Hence, the equality holds.

Assume that J is a Riesz q-ideal. We have to prove that $I \vee^q J$ and $I \cap J$ are Riesz q-ideals. Let $a \in I \vee^q J$, $b, c \in M$, $a \leq b \boxplus c$. We can write $a = i \boxplus j$, $i \in I, j \in J$. Since I is a Riesz ideal, and $i \leq b \boxplus c$, there exist $i_1, i_2 \in I$, $i_1 \leq b$, $i_2 \leq c$ and $i \leq i_1 \boxplus i_2$. Now, $j = a \boxminus i \leq (b \boxplus c) \boxminus i = (b \boxminus i_1) \boxplus (c \boxminus i_2) \boxplus ((i_1 \boxplus i_2) \boxminus i)$. Since J is a Riesz ideal, there are $j_1, j_2, j_3 \in J$, $j_1 \leq b \boxminus i_1$, $j_2 \leq c \boxminus i_2$, $j_3 \leq (i_1 \boxplus i_2) \boxminus i$ and $j \leq j_1 \boxplus j_2 \boxplus j_3 \leq j_1 \boxplus j_2 \boxplus (i_1 \boxplus i_2 \boxminus i)$. Finally, $j_1 \boxplus i_1 \leq b$, $j_2 \boxplus i_2 \leq c$, $i_1 \boxplus j_1 \in I \boxplus J$, $i_2 \boxplus j_2 \in I \boxplus J$ and $a = i \boxplus j \leq i_1 \boxplus j_1 \boxplus i_2 \boxplus j_2$. Hence, $I \boxplus J$ has the Riesz property.

Let $x \in I \cap J$, $x \leq a \boxplus b$. Since $x \in I$, there are $i_1 \leq a$, $i_2 \leq b$, $i_1, i_2 \in I$, $x \leq i_1 \boxplus i_2$. But $x \in J$ implies that there are $j_1 \leq i_1 \leq a$, $j_2 \leq i_2 \leq b$, $j_1, j_2 \in J$ and $x \leq j_1 \boxplus j_2$. There follows that $j_1, j_2 \in I \cap J$. Hence $I \cap J$ has the Riesz property. □

In what follows, let $\vee^p$ and $\vee^q$ denote the supremum and the infimum in the set $\mathcal{I}^p(M)$ and $\mathcal{I}^q(M)$, respectively.

Lemma 2.3.56 *Let $I, J \in \mathcal{I}^p(M)$. Then*

$$C(I \oplus J) = C(I \cup J),$$

where $I \oplus J := \{a \oplus b \,:\, a \in I, b \in J\}$ and C is the the closure operator defined in Theorem 2.3.46.

Proof. Clearly $I \cup J \subseteq I \oplus J$. Therefore, $C(I \cup J) \subseteq C(I \oplus J)$. By Theorem 2.3.46, $C(I \cup J)$ is the smallest p-ideal containing $I \cup J$. Since every p-ideal containing I and J, contains also $I \oplus J$, we have $I \oplus J \subseteq C(I \cup J)$. Therefore, $C(I \oplus J) \subseteq C(C(I \cup J)) = C(I \cup J)$. □

Theorem 2.3.57 *The set $\mathcal{I}^p(M)$ forms a complete distributive lattice (with respect to the set-theoretical inclusion), where*

$$\bigvee_k I_k = C(\bigcup_k I_k), \qquad \bigwedge_k I_k = \bigcap_k I_k.$$

Proof. Clearly, the set-theoretical intersection of any set of p-ideals is a p-ideal. By Theorem 2.3.46, $C(\bigcup_k I_k)$ is the smallest p-ideal containing I_k for all k. It remains to prove distributivity.

Let I, J, K be p-ideals. Clearly,

$$(I \cap K) \vee^p (J \cap K) \subseteq (I \vee^p J) \cap K.$$

Assume $a \in (I \vee^p J) \cap K$, then $a \in K$, $a \in I \vee^p J = C(I \cup J)$. We have to prove that $a \in (I \cap K) \vee^p (J \cap K)$.

By definition of $C(I \cup J)$, $a = a_1 \oplus \cdots \oplus a_k$, $a_i \approx b_i$, $b_i \le n_i \odot c_i$, $c_i \in I \cup J$, $i = 1, 2, \ldots, k$. We can divide the set $c_1, \ldots, c_k$ into two parts

$$(c_1, \ldots, c_k) = (c_{ij})_j \cup (c_{il})_l,$$

where $c_{ij} \in I$, $c_{il} \in J$. Since $b_{ij} \le n_{ij} \odot c_{ij}$ and $a_{ij} \approx b_{ij}$, where $c_{ij} \in I$ and I is a p-ideal, we have $a_{ij} \in I$, and also $\bigoplus_j a_{ij} \in I$. Now since $\bigoplus_j a_{ij} \le a$ and $a \in K$, we have $\bigoplus_j a_{ij} \in I \cap K$. In the same way we prove that $\bigoplus_l a_{il} \in J \cap K$. Denote $u_1 := \bigoplus_j a_{ij}$, $u_2 := \bigoplus_l a_{il}$. So we have $a = u_1 \oplus u_2$, $u_1 \in I \cap K$, $u_2 \in J \cap K$, hence $a \in I \cap K \oplus J \cap K$. By Lemma 2.3.56

$$\begin{aligned} I \cap K \oplus J \cap K &\subseteq C(I \cap K \oplus J \cap K) \\ &= C((I \cap K) \cup (J \cap K)) \\ &= (I \cap K) \vee^p (J \cap K). \end{aligned}$$

Therefore $(I \vee^p J) \cap K = (I \cap K) \vee^p (J \cap K)$. □

Remark 2.3.58 (1) Since the set-theoretical intersection of any family of e-ideals is an e-ideal, $\mathcal{I}^e(M)$ is a complete lattice in the ordering by set-inclusion.

(2) By [ChPu], supremum of any family $\{I_k\}$ of Riesz e-ideals in $\mathcal{RI}^e(M)$ is

$$\bigvee_k^e I_k = \{i_{k1} \boxplus \cdots \boxplus i_{kp} : \ i_{kj} \in I_{kj}\}.$$

Since every ideal containing all I_k contains also $\bigvee^e I_k$, the suprema in $\mathcal{RI}^e(M)$ and $\mathcal{I}^e(M)$ are the same. Hence $\mathcal{RI}^e(M)$ is a distributive sublattice of $\mathcal{I}^e(M)$.

(3) Since the intersection of any family $\{I_k\}$ of q-ideals is a q-ideal, $\mathcal{I}^q(M)$ is a complete lattice with respect to ordering by the set-inclusion.

(4) From Lemma 2.3.55 it can be derived by induction that the supremum of any family $\{I_k\}$ of p-ideals in $\mathcal{I}^q(M)$ is

$$\bigvee_k^q I_k = \{i_{k1} \boxplus \cdots \boxplus i_{kp} : \ i_{kj} \in I_{kj}\},$$

and it coincides with the supremum in $\mathcal{RI}^e(M)$ (hence also in $\mathcal{RI}^q(M)$). By Theorem 2.3.57, supremum in $\mathcal{I}^p(M)$ may be larger.

2.3.6 Quotients of QMV-algebras

By Theorem 2.3.42, every p-ideal on a QMV-algebra $\mathcal{M}$ determines a (QMV) congruence. The quotient QMV-algebra induced by $\equiv_I$ will be denoted by $\mathcal{M}/I$. In this sequel, we follow mainly [GiPu].

Definition 2.3.59 *Let $I \in \mathcal{I}^q(M)$. I is* **maximal** *if and only if, for all $J \in \mathcal{I}^q(M)$,*

$$I \subseteq J \Rightarrow I = J \text{ or } J = M.$$

Definition 2.3.60 *Let $I \in \mathcal{I}^q(M)$. I is said to be*

(i) $\mathcal{I}^q(M)$**- maximal** *iff for all $J \in \mathcal{I}^q(M)$, $I \subseteq J \ \Rightarrow$ either $I = J$ or $J = M$;*

(ii) $\mathcal{I}^p(M)$**- maximal** *iff for all $J \in \mathcal{I}^p(M)$, $I \subseteq J \ \Rightarrow$ either $I = J$ or $J = M$;*

(iii) **strongly maximal** *iff, for all $a \in M$, $a \notin I \ \Rightarrow \ \exists\, n \in \mathbb{N}$ such that $(a^*)^n \in I$.*

Lemma 2.3.61 *Let $I \in \mathcal{I}^q(M)$. If I is strongly maximal then I is $\mathcal{I}^q(M)$-maximal.*

Proof. Suppose I is strongly maximal and assume $J \in \mathcal{I}^q(M)$ and $I \subseteq J$. Suppose $a \in J$ and $a \notin I$. Then there exists $n \in \mathbb{N}$ such that $(a^*)^n \in I$. Thus, $(a^*)^n \in J$. Since $a \in J$, $n \odot a \in J$. Therefore $1 = n \odot a \oplus (n \odot a)^* = n \odot a \oplus (a^*)^n \in J$. □

Lemma 2.3.62 *If every q-ideal is a p-ideal, then every maximal q-ideal is strongly maximal.*

Proof. Suppose I is a maximal q-ideal and suppose $a \notin I$. Let I^a be the q-ideal generated by $I \cup \{a\}$. Thus $I^a = \{b \in M : b \leq n \odot c,\, c \in I \cup \{a\},\, n \in \mathbb{N}\}$. Since I is maximal, $1 \in I^a$. Thus, $1 = b \oplus n \odot a$ for some $b \in I$. By hypothesis, I is a p-ideal. Thus, $(n \odot a)^* \in I$. □

Since in any MV-algebra every q-ideal is a p-ideal, the notions of $\mathcal{I}^q(M)$-maximality and $\mathcal{I}^p(M)$-maximality coincide. Accordingly, we will speak simply of maximal ideals.

Corollary 2.3.63 [Cha 1] *In any MV-algebra $\mathcal{M}$, every maximal q-ideal is strongly maximal.*

In a QMV-algebra, a maximal ideal need not be strongly maximal as the following counterexample shows. Let $\mathcal{M}_6$ be the modular ortholattice pictured in Figure 2.2.

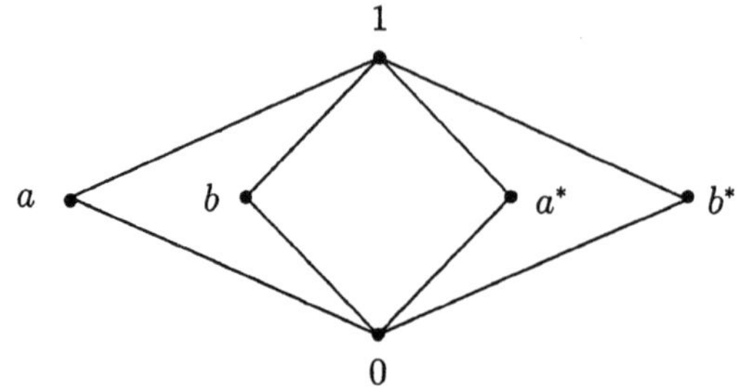

Fig. 2.2. $\mathcal{M}_6$

$\mathcal{M}$ can be thought of as a QMV-algebra, where $\oplus$ is the sup and * is the orthocomplement. Let $I = \{0, a\}$. Clearly I is a proper $\mathcal{I}^q(M_6)$-maximal ideal. However I is not strongly maximal since $b \notin I$ and $b^* \notin I$. It should be noted that even a maximal p-ideal need not be strongly maximal. Consider again $\mathcal{M}_6$. The set $I = \{0\}$ is a $\mathcal{I}^p(M)$-maximal p-ideal, which is not strongly maximal.

Lemma 2.3.64 [GiPu] *Let $I \in \mathcal{I}^p(M)$. The following conditions are equivalent:*

(i) *I is strongly maximal.*

(ii) *$\mathcal{M}/I$ is locally finite.*

Proof. (i) $\Rightarrow$ (ii). Suppose I is strongly maximal and $a \not\equiv_I 0$. Then $a \notin I$. Thus, there exists $n \in \mathbb{N}$ such that $(n \odot a)^* \in I$, i.e., $(n \odot a)^* \equiv_I 0$. Thus, $n \odot a \equiv_I 1$.

(ii) $\Rightarrow$ (i). Suppose $\mathcal{M}/I$ is locally finite and $a \notin I$. Then $a \not\equiv_I 0$ so that, by hypothesis, there exists $n \in \mathbb{N}$ such that $n \odot a \equiv_I 1$. Hence $(n \odot a)^* \in I$. □

Definition 2.3.65 $\mathcal{M} = (M; \oplus, ^*, 0, 1)$ *is* **perspectively finite** *if and only if, for all $a \neq 0$, there exists $\{x_1, \ldots, x_n\} \subseteq M$ such that $x_1 \oplus \cdots \oplus x_n = 1$, and there exists $\{y_1, \ldots, y_n\} \subseteq M$ such that, for all i with $1 \leq i \leq n$, $x_i \approx y_i$, and there exists $m \in \mathbb{N}$ such that $y_1 \oplus \cdots \oplus y_n \leq m \odot a$.*

Differently from MV-algebras, a perspectively finite QMV-algebra need not be locally finite. Let us consider the QMV-algebra $\mathcal{M}_6$ (see Figure 2.2). Clearly, $\mathcal{M}_6$ is not locally finite. We show that $\mathcal{M}_6$ is perspectively finite. We have $a \neq 0$. Let $x_1 := a$, $x_2 := b$. Then $x_1 \oplus x_2 = x_1 \vee x_2 = a \vee b = 1$. Let $y_1 := a$ and $y_2 := a$. Clearly, $x_1 \approx y_1$ and $x_2 \sim y_2$ with common complement a^*. Clearly, $y_1 \oplus y_2 = a \vee a = a$.

Theorem 2.3.66 [GiPu] *Let $I \in \mathcal{I}^p(M)$. The following conditions are equivalent:*

(i) *I is $\mathcal{I}^p(M)$-maximal.*

(ii) *$\mathcal{M}/I$ is perspectively finite.*

Proof. (i) $\Rightarrow$ (ii). Suppose I is $\mathcal{I}^p(M)$-maximal and $a \not\equiv_I 0$, i.e., $a \notin I$. Since $I \cup \underline{\{a\}^{\approx}}$ is an order ideal closed under perspectivity, it follows by Theorem 2.3.46 that $\overline{I \cup \{a\}^{\approx}}$ is a p-ideal. By $\mathcal{I}^p(M)$–maximality, $1 \in \overline{I \cup \{a\}^{\approx}}$. Then there exists $\{x_1, \ldots, x_n\} \subseteq I \cup \{a\}^{\approx}$ such that

$$1 = x_1 \oplus \cdots \oplus x_n.$$

Now, for any i, $(1 \leq i \leq n)$ such that $x_i \notin \{a\}^{\approx}$, we have $x_i \equiv_I 0$. Let $\{x_{i1}, \ldots, x_{ik}\}$ be the set of all x_i in $\{x_1, \ldots, x_n\}$ such that $x_i \in \{a\}^{\approx}$ and $x_i \not\equiv_I 0$. Thus

$$1 \equiv_I x_{i1} \oplus \cdots \oplus x_{ik}.$$

By definition of $\{a\}^{\approx}$, there are $u_{i1}, \ldots, u_{ik}$ and $n_{i1}, \ldots, n_{ik}$ such that

$$x_{ij} \approx u_{ij}, \text{ and } u_{ij} \leq n_{ij} \odot a.$$

Let $s = n_{i1} \oplus \cdots \oplus n_{ik}$. Then

$$u_{i1} \oplus \cdots \oplus u_{ik} \leq s \odot a.$$

Thus, $\mathcal{M}/I$ is perspectively finite.

(ii) $\Rightarrow$ (i). Suppose $\mathcal{M}/I$ is perspectively finite. Let J be a p-ideal such that $I \subsetneq J$. We want to show that J is not proper. Since $I \neq J$, there exists $a \in J$ such that $a \not\equiv_I 0$. Since $\mathcal{M}/I$ is perspectively finite, there exists $\{x_1, \ldots, x_m\} \subseteq M$ such that $x_1 \oplus \cdots \oplus x_m \equiv_I 1$ and there exists $\{y_1, \ldots, y_m\} \subseteq M$ such that $[x_i] \approx [y_i]$, for any i such that $1 \leq i \leq m$. Further, there exists $n \in \mathbb{N}$ such that $(y_1 \oplus \cdots \oplus y_m) \Cap (n \odot a) \equiv_I y_1 \oplus \cdots \oplus y_m$. Now $n \odot a \in J$ since $a \in J$. Then $y_1 \oplus \cdots \oplus y_m \in J$. Consequently, $[x_i] \subset J$ since $[x_i] \approx [y_i]$. Hence $x_1 \oplus \cdots \oplus x_m \in J$. Since $x_1 \oplus \cdots \oplus x_m \equiv_I 1$, we conclude that $1 \in J$. □

Definition 2.3.67 *An* **MV q-ideal** *is a q-ideal I of $\mathcal{M} = (M; \oplus, ^*, 0, 1)$ such that, for all $a, b \in M$,*

$$a \odot b^* \in I \;\Rightarrow\; \exists\, x \in M \text{ such that } x \leq a,\ x \leq a \Cap b, \text{ and } x^* \odot a \in I.$$

Theorem 2.3.68 *Let I be a p-ideal of $\mathcal{M} = (M; \oplus, ^*, 0, 1)$. The following are equivalent:*

(i) *I is an MV q-ideal.*

(ii) *$\mathcal{M}/I$ is an MV-algebra.*

Proof. (i) $\Rightarrow$ (ii). Suppose I is an MV q-ideal. To prove that $\mathcal{M}/I$ is an MV-algebra it suffices to show that, for all $a, b \in M$, if $a^* \oplus b \equiv_I 1$, then $a \Cap b \equiv_I a$.

Suppose $a \odot b^* \equiv_I 0$. Then, by hypothesis, there exists $x \leq a, x \leq a \Cap b, x^* \odot a \in I$. So, to prove that $a \Cap b \equiv_I a$, it remains to show that $x^* \odot (a \Cap b) \in I$. Now

$$\begin{aligned} x \oplus (a \Cap b)^* \oplus (a \odot x^*) &= (a \Cup x) \oplus (a \Cap b)^* \\ &= a \oplus (a^* \odot b) \oplus b^* \quad (x \leq a) \\ &= 1. \end{aligned}$$

Since $x^* \odot a \in I$, and I is a p-ideal, it follows $x^* \odot (a \Cap b) \in I$.

(ii) $\Rightarrow$ (i). Suppose $\mathcal{M}/I$ is an MV-algebra. Suppose $a \odot b^* \in I$, i.e., $a \odot b^* \equiv_I 0$. Then $a^* \oplus b \equiv_I 1$. Hence $a \Cap b \equiv_I a$ since $\mathcal{M}/I$ is an MV-algebra. The condition follows from the definition of $\equiv_I$. □

Definition 2.3.69 *A q-ideal I of $\mathcal{M} = (M; \oplus, ^*, 0, 1)$ is said to be* **idempotent** *if and only if, for all $a \in M$,*

$$a^n \in I \textit{ for some } n \in \mathbb{N} \Rightarrow a \in I.$$

Let

$$Ctr(M) := \{a \in M : a^2 = 0\}.$$

Theorem 2.3.70 *Let I be a p-ideal of $\mathcal{M} = (M; \oplus, ^*, 0, 1)$. The following are equivalent:*

(i) *I is idempotent.*

(ii) *$Ctr(M) \subseteq I$.*

Proof. (i) $\Rightarrow$ (ii). Straightforward.

(ii) $\Rightarrow$ (i). Suppose $Ctr(M) \subseteq I$. Suppose $a^n \in I$ for some $n \in \mathbb{N}$. We want to show that $a \in I$. Now, $(a \Cap a^*)^2 = (a \oplus a) \odot a^* \odot (a \oplus a) \odot a^* \leq (a \oplus a) \odot (a \oplus a)^* = 0$. Thus, by hypothesis, $a \Cap a^* \in I$ and therefore $a \Cap a^* \equiv_I 0$. By Corollary 2.3.7 and by (ii) of Proposition 2.3.35, $(n \odot a) \Cap (n \odot a^*) \equiv_I 0$. Hence, $a^n \oplus n \odot a = a^n \oplus ((n \odot a) \Cap (n \odot a^*)) \in I$. Now, $a^* \oplus a^n \oplus n \odot a = 1$. Since I is a p-ideal, it follows that $a \in I$. □

Theorem 2.3.71 *Let I be a p-ideal of $\mathcal{M} = (M; \oplus, ^*, 0, 1)$. The following conditions are equivalent:*

(i) *I is idempotent.*

(ii) *$\mathcal{M}/I$ is an orthomodular lattice.*

Proof. (i) $\Rightarrow$ (ii). Suppose I is idempotent. To prove that $\mathcal{M}/I$ is an OML, it suffices to show (by Theorem 2.3.9): $a \oplus a \equiv_I a$, $a \in M$. As proved in the previous Theorem, $a \Cap a^* \equiv_I 0$. Thus, since $a \leq a, a \leq (a \oplus a)$, $a^* \odot a = 0$, $(a \oplus a) \odot a^* \in I$, we have $a \oplus a \equiv_I a$.

(ii)$\Rightarrow$ (i). Suppose $\mathcal{M}/I$ is an OML. Then, for all $a \in M$, $a \oplus a \equiv_I a$ and therefore $a^* \oplus a^* \equiv_I a^*$. Thus $n \odot a \equiv_I a$, and $a^n \equiv_I a$, $n \in \mathbb{N}$. Suppose $a^n \in I$ for some $n \in \mathbb{N}$. Then $a \in I$. □

2.3.7 Quasilinearity and Weak Linearity

It is easily seen that a "genuine" QMV-algebra cannot be linearly ordered. However, there are some interesting QMV-algebras (e.g., the standard QMV-algebra) that satisfy some weaker conditions that turn out to be equivalent to linearity in any MV-algebra. In this sequel, we follow the ideas in [Giu 5].

Definition 2.3.72 *A QMV-algebra $\mathcal{M} = (M; \oplus, ^*, 0, 1)$ is said to be* **quasi-linear** *if, for all $a, b \in M$,*

$$a \not\leq b \;\Rightarrow\; a \Cap b = b. \tag{2.14}$$

In other words, a QMV-algebra $\mathcal{M}$ is quasilinear iff the following holds: $a \Cap b = a$ if $a \leq b$, $a \Cap b = b$ otherwise.

One can easily check that an MV-algebra is quasi-linear iff it is linear. As an example of a quasi-linear QMV-algebra which is not an MV-algebra we can consider the "diamond" $\mathcal{M}_4$. Both the standard MV-algebra $[0, 1]$ and the standard QMV-algebra $\mathcal{E}(H)$ are quasi-linear (see Examples 2.3.16, 2.3.17).

Proposition 2.3.73 *The following conditions are equivalent:*

(i) *$\mathcal{M} = (M; \oplus, ^*, 0, 1)$ is quasi-linear,*

(ii) *If $a \oplus b \neq 1$, then $a < b^*$, $a, b \in M$.*

(iii) *If $a \oplus c = b \oplus c \neq 1$, then $a = b$, $a, b, c \in M$.*

Proof. (i) $\Rightarrow$ (ii). Suppose $a < b^*$ does not hold. If $a = b^*$, then $a \oplus b = 1$. Assume $a \neq b^*$. By (i) then $a \Cap b^* = b^*$. Thus $a \oplus b = a \oplus (a \Cap b^*)^* = a \oplus (a^* \Cup b) = a \oplus [(a^* \odot b^*) \oplus b] = (a \oplus b) \oplus (a \oplus b)^* = 1$.

(ii) $\Rightarrow$ (iii). Suppose $a \oplus c = b \oplus c \neq 1$. By (ii), $a \leq c^*, b \leq c^*$. Then by the cancellation law $a = b$.

(iii) $\Rightarrow$ (i). Suppose $a \not\leq b$. Assume that $a \oplus b^* \neq 1$, then by Proposition 2.3.2 (i), $1 \neq b^* \oplus a = b^* \oplus (a \Cap b)$. By (iii) we obtain $a = a \Cap b$, a contradiction. Hence $a \oplus b^* = 1$, and so $a \Cap b = (a \oplus b^*) \odot b = b$. □

The notion of quasilinearity can be further weakened in the following way:

Definition 2.3.74 *A QMV-algebra* $\mathcal{M} = (M; \oplus, {}^*, 0, 1)$ *is* **weakly linear** *if, for all* $a, b \in M$,

$$a \oplus b^* = 1 \text{ or } b \oplus a^* = 1. \tag{2.15}$$

Proposition 2.3.75 *The following conditions are equivalent:*

(i) $\mathcal{M}$ *is weakly linear.*

(ii) *For all* $a, b \in M$, $a \Cap b = b$ *or* $b \Cap a = a$.

Proof. Straightforward. □

Let us consider the following condition.

$$a = (a \oplus b \odot c^*) \Cap (a \oplus c \odot b^*). \tag{2.16}$$

Lemma 2.3.76 *Let* $\mathcal{M} = (M; \oplus, {}^*, 0, 1)$ *be a weakly linear QMV-algebra. Then* (2.16) *holds.*

Proof. We have either $b \odot c^* = 0$ or $c \odot b^* = 0$. In both cases we obtain $(a \oplus b \odot c^*) \Cap (a \oplus c \odot b^*) = a$. □

Lemma 2.3.77 *Let* $\mathcal{M} = (M; \oplus, {}^*, 0, 1)$ *be a QMV-algebra satisfying* (2.16). *The following conditions hold, for all* $a, b, c \in M$:

(i) $b \Cap c = 0 \Rightarrow (a \oplus b) \Cap (a \oplus c) = a$.

(ii) $a \Cap (b \Cap c) = 0 \Rightarrow a \Cap (c \Cap b) = 0$.

(iii) $a \leq b$, $c \leq d$ *and* $b \Cap d = 0 \Rightarrow a \Cap c = 0$.

Proof. (i) Suppose $b \Cap c = 0$. From $c = c \odot b^* \oplus b \Cap c$ it follows that $c = c \odot b^*$ and, since $c \Cap b = 0$ by perspectivity, $b = b \odot c^*$. Thus, by (2.16), $(a \oplus b) \Cap (a \oplus c) = (a \oplus b \odot c^*) \Cap (a \oplus b^* \odot c) = a$.

(ii) Suppose $a \Cap (b \Cap c) = 0$. Then $(a^* \Cup (c \Cap b)^*) \oplus (b \Cap c) = 1$ and also $(a^* \Cup (c \Cap b)^*) \oplus a = 1$. Thus

$$\begin{aligned} 1 &= ((a^* \Cup (c \Cap b)^*) \oplus a) \Cap ((a^* \Cup (c \Cap b)^* \oplus (b \Cap c))) \\ &= a^* \Cup (c \Cap b)^* \quad \text{(i)}. \end{aligned}$$

(iii) Suppose $b \Cap d = 0$. Then $d \Cap b = 0$. Thus

$$\begin{aligned} 0 &= a \Cap (d \Cap b) \\ &= a \Cap (b \Cap c) \quad (c \leq d) \\ &= (a \Cap b) \Cap (b \Cap c) \quad (a \leq b) \\ &= (a \Cap b) \Cap c \quad \text{Proposition 2.3.2 (xi)} \\ &= a \Cap c \quad (a \leq b). \end{aligned}$$

□

Lemma 2.3.78 *Let $\mathcal{M} = (M; \oplus, ^*, 0, 1)$ be a QMV-algebra satisfying* (2.16). *The following conditions hold for any $a, b \in M$:*

(i) *$a \sim b$ and $a \Cap b = 0$ imply $a = b = 0$.*

(ii) *$a \approx b$ and $a \Cap b = 0$ imply $a = b = 0$.*

(iii) *$a \approx b$, $c \approx d$ and $b \Cap d = 0$ imply $a \Cap c = 0$.*

Proof. (i) Suppose $a \Cap b = 0$ and $a \sim b$. Let c be a common complement of a and b. Recall that, for any x, y, $x = y^* \odot x \oplus y \Cap x$. Since $a \Cap b = 0$, and $b \Cap a \sim a \Cap b = 0$ imply $b \Cap a = 0$, we get that $b = a^* \odot b$ and $a = a \odot b^*$. Thus, $1 = c \oplus a = c \oplus (a \odot b^*) = c \oplus (a^* \odot b)$. Therefore by (2.16), $c = (c \oplus (a^* \odot b)) \Cap (c \oplus (a \odot b^*)) = 1$. Hence $a = b = 0$.

(ii) By (i), the statement follows for $\sim$. By definition of $\approx$, there are $x_1, \ldots, x_n$ such that $a = x_1$, $x_i \sim x_{i+1}, i = 1, \ldots, n-1$, $x_n = b$. We prove (ii) by induction on n. Assume that the statement is true for all i such that $1 \leq i \leq n-1$. We have $a \Cap x_{n-1} \leq x_{n-1}$ and $x_{n-1} \sim x_n = b$. Therefore, by Proposition 2.3.35 (iv), there is $f \in M$, $f \leq x_n$ and $a \Cap x_{n-1} \sim f$. Since $x_{n-1} \Cap a \leq a = x_1$, $f \leq b$ and $a \Cap b = 0$, we conclude that $(x_{n-1} \Cap a) \Cap f = 0$. Therefore, $0 = (x_{n-1} \Cap a) \Cap f = f \Cap (x_{n-1} \Cap a)$ Since $f \sim (x_{n-1} \Cap a)$ we get by (i) that $a \Cap x_{n-1} = 0$. Now $a \Cap x_{n-1} = 0$ and $a \approx x_{n-1}$. Thus, by induction hypothesis, we can conclude that $a = x_{n-1} = 0$. Since $x_n \sim x_{n-1}$ and $b = x_n$, we have $b = 0$.

(iii) Assume $b \Cap d = 0$, $a \approx b$, $c \approx d$. Since $c \Cap a \leq a$, there is $e \in M$ such that $e \leq b$ and $c \Cap a \approx e$. Since $a \Cap c \leq c$ and $c \approx d$, there is $f \in M$ such that $f \leq d$ and $a \Cap c \approx f$. By transitivity of $\approx$ and perspectivity of $a \Cap c$ and $c \Cap a$, we conclude that $e \approx f$. By hypothesis, $b \Cap d = 0$. Hence by Lemma 2.3.77 (iii) we get $e \Cap f = 0$. Thus by (ii) of the present Lemma, $e = f = 0$. Since $a \Cap c \approx e$, we conclude that $a \Cap c = 0$. □

Definition 2.3.79 *An ideal I is* **prime** *if, for all $a, b \in M$, $a \odot b^* \in I$ or $a^* \odot b \in I$.*

Proposition 2.3.80 *A p-ideal I is prime if and only if $a \Cap b \in I$ implies $a \in I$ or $b \in I$.*

Proof. Let I be a prime p-ideal. Then $a \Cap b \in I$ iff $b \Cap a \in I$ together with the equalities $a = (b \Cap a) \oplus (b^* \odot a)$, $b = (a \Cap b) \oplus (a^* \odot b)$ imply the desired result.

On the other hand, if I is an ideal such that $a \Cap b \in I$ implies $a \in I$ or $b \in I$, then $(a \odot b^*) \Cap (b \odot a^*) = 0$ (Proposition 2.3.2 (xiii)) and $0 \in I$ imply that I is prime. □

Theorem 2.3.81 *Let $\mathcal{M} = (M; \oplus, ^*, 0, 1)$ be a QMV-algebra and let $I \in \mathcal{I}^p(M)$. The following conditions are equivalent:*

(i) *I is prime.*

(ii) *$\mathcal{M}/I$ is weakly linear.*

Proof. Suppose that $\mathcal{M}/I$ is weakly linear. Then, for all $a, b \in M$, $a \Cap b \equiv_I b$ or $b \Cap a \equiv_I a$. Then, $b^* \oplus (a \Cap b) \equiv_I 1$ or $a^* \oplus (b \Cap a) \equiv_I 1$. By Proposition 2.3.2 (i), $b^* \oplus a \equiv_I 1$ or $a^* \oplus b \equiv_I 1$. Thus $a^* \odot b \in I$ or $b^* \odot a \in I$.

Suppose that I is a prime ideal. We have to show that $a \Cap b \equiv_I b$ or $b \Cap a \equiv_I a$. Suppose $a \Cap b \not\equiv b$. Then, $a \oplus b^* \not\equiv 1$ and therefore $a^* \odot b \notin I$. Since I is prime, we have that $b^* \odot a \in I$. Thus $b^* \odot a \equiv_I 0$ hence, $b \Cap a \equiv_I a$. □

The main tool we will need in the sequel is the following lemma.

Lemma 2.3.82 *Let $\mathcal{M} = (M; \oplus, ^*, 0, 1)$ be a QMV-algebra satisfying* (2.16). *The following condition holds for all $x, y \in M$:*

$$x \Cap y = 0 \ \Rightarrow\ \overline{\{x\}^{\approx}} \cap \overline{\{y\}^{\approx}} = \{0\}.$$

Proof. Suppose $x \Cap y = 0$ and $b \in \overline{\{x\}^{\approx}} \cap \overline{\{y\}^{\approx}}$. By definition, there are $B := \{b_1, \ldots, b_n\} \subseteq \{x\}^{\approx}$ and $C := \{c_1, \ldots, c_m\} \subseteq \{y\}^{\approx}$ such that

$$b = b_1 \oplus \cdots \oplus b_n,\ b = c_1 \oplus \cdots \oplus c_m.$$

Let $b_i \in B$ and $c_j \in C$. We want to show that $b_i \Cap c_j = 0$. By definition of $\{x\}^{\approx}$, there is $e \in M$ such that $b_i \approx e$ and $e \leq r \odot x$ for some $r \in \mathbb{N}$. Similarly, there is $f \in M$ such that $c_j \approx f$ and $f \leq s \odot y$ for some $s \in \mathbb{N}$. Let $t := \max\{r, s\}$. By hypothesis, $x \Cap y = 0$, and therefore by Corollary 2.3.7 and Proposition 2.3.35 (iii), $(t \odot x) \Cap (t \odot y) = 0$. Thus by Lemma 2.3.78 (iii), $e \Cap f = 0$. Then by Lemma 2.3.78 (iii), $b_i \Cap c_j = 0$. By Corollary 2.3.7 and Proposition 2.3.35 (iii), $(b_1 \oplus \cdots \oplus b_n) \Cap (c_1 \oplus \cdots \oplus c_m) = 0$. □

Corollary 2.3.83 *Let $\mathcal{M} = (M; \oplus, ^*, 0, 1)$ be an MV-algebra and let $I \in \mathcal{I}^q(M)$. I is prime if and only if $\mathcal{M}/I$ is totally ordered.*

Theorem 2.3.84 *Let $\mathcal{M} = (M; \oplus, ^*, 0, 1)$ be a QMV-algebra which satisfies WL-axiom* (2.15). *Then, for all $a, b \in M$, $a \not\lesssim b$, there exists a prime p-ideal I on M such that $a \not\equiv_I a \Cap b$.*

Proof. Let $a \not\lesssim b$. Let I be the ideal maximal with respect the property "$a \not\equiv_I a \Cap b$". Such a p-ideal exists (by Zorn's lemma), since $\{0\}$ is a p-ideal and $a \neq a \Cap b$.

Suppose that I is not prime. Thus, there exist $x, y \in M$ such that $x \odot y^* \notin I$ and $y \odot x^* \notin I$. Let us consider the following sets

$$I_{x \odot y^*} := I \cup \overline{\{x \odot y^*\}^{\approx}} \text{ and } I_{x^* \odot y} := I \cup \overline{\{x^* \odot y\}^{\approx}}.$$

By Theorem 2.3.46, $\overline{\{x \odot y^*\}^{\approx}}$ and $\overline{\{x^* \odot y\}^{\approx}}$ are p-ideals. Then, clearly, $I_{x \odot y^*}$ and $I_{x^* \odot y}$ are order ideals closed under perspectivity. Thus, by Theorem 2.3.46, the sets

$$I^{\circ} := \overline{I_{x \odot y^*}} \text{ and } I^{\Box} := \overline{I_{x^* \odot y}}$$

are p-ideals both including I. Since I is maximal with respect the property "$a \not\equiv_I a \Cap b$", we have

$$a \equiv_{I^\circ} a \Cap b \text{ and } a \equiv_{I^\Box} a \Cap b. \tag{2.17}$$

Thus, there are two elements $r, s \in M$ such that $r, s \leq a, a \Cap b$ and $r^* \odot (a \Cap b) \in I^\circ$, $s^* \odot (a \Cap b) \in I^\Box$, $r^* \odot a \in I^\circ$, and $s^* \odot a \in I^\Box$.

Since I, $\overline{(x \odot y^*)^\approx}$ and $\overline{(x^* \odot y)^\approx}$ are p-ideals, every element in I° is of the form $b_1 \oplus b_2$ with $b_1 \in I$, $b_2 \in \overline{(x \odot y^*)^\approx}$ and similarly, every element in $I^\Box$ is of the form $c_1 \oplus c_2$ with $c_1 \in I$, $c_2 \in \overline{(x^* \odot y)^\approx}$.

By Proposition 2.3.2 (xii), we get $(x \odot y^*) \Cap (x^* \odot y) = 0$, hence by Lemma 2.3.82, $\overline{(x \odot y^*)^\approx} \cap \overline{(x^* \odot y)^\approx} = \{0\}$, so we have

$$I^\circ \cap I^\Box = I. \tag{2.18}$$

Now consider the element $t := r \Cup s$. By hypothesis, $r, s \leq a, a \Cap b$. Thus $t \leq a, a \Cap b$. If we prove $t^* \odot a \in I$, then we also have $t^* \odot (a \Cap b) \in I$ (indeed, observe that $(t^* \odot (a \Cap b))^* \oplus (t^* \odot a) = 1$) therefore $a \equiv_I a \Cap b$, which is a contradiction. By hypothesis, $r^* \odot a \in I^\circ$. To prove that $t^* \odot a \in I^\circ$, it suffices to prove that $(t^* \odot a)^* \oplus r^* \odot a = 1$.

$$\begin{aligned}
(t^* \odot a)^* \oplus r^* \odot a &= ((r^* \Cap s) \odot a^*)^* \oplus r^* \odot a \\
&= (r \Cup s) \oplus a^* \oplus r^* \odot a \\
&= (r \Cup s) \oplus (r^* \Cup a^*) \\
&= (r \Cup s) \oplus r^* \quad (r \leq a) \\
&= 1.
\end{aligned}$$

Since $t^* = r^* \Cap s^* \leq s^*$, and $s^* \odot a \in I^\Box$ by hypothesis, we also have $t^* \odot a \in I^\Box$. By (2.18), $t^* \odot a \in I$. Thus, we conclude $a \equiv_I a \Cap b$, a contradiction. □

2.3.8 Open Problems

The following open problems have been suggested by Giuntini.

1. Characterize the equational class generated by interval QMV-algebras (i.e., QMV-algebras whose corresponding effect algebra is an interval algebra). Is it finitely based ?

2. Characterize the equational class generated by standard QMV-algebras.

3. Characterize the class of QMV-algebras which are distributive lattices (the "diamond" is an example, which is not an MV-algebra).

Chapter 3

Quotients of Partial Abelian Monoids

The basic algebraic structure that is studied in this chapter is a partial Abelian monoid (PAM in short) (cf. [Wil 1], [Wil 2], [Pul 4], [GuPu]). A PAM is a structure $(P; 0, \oplus)$, where $\oplus$ is a commutative, associative partial binary operation on P and 0 is a neutral element. Beginning with a PAM at the lowest level, we shall consider a hierarchy of partial algebraic structures. The second level is a cancellative PAM (CPAM), the third level is a generalized effect algebra/generalized difference poset, which coincide with a cancellative, positive PAM. Commutative positive minimal clans and BCK-algebras are also included. An effect algebra is a unital generalized effect algebra/D-poset. On higher levels in the hierarchy we find orthoalgebras, orthomodular posets and lattices, MV-algebras, Boolean algebras.

If P is a PAM and $\sim$ is a binary relation on P, then the quotient of P relative to $\sim$ is a PAM, denoted by $P//\sim$ (for the definition of $P//\sim$ see below), which is the surjective image of a universal morphism on P that respects $\sim$. If the quotient exists, it is unique to within an isomorphism. Quotients are defined in a similar way for higher level categories in the hierarchy. We restrict our attention to relations $\sim$ that are congruences on P. Our definition of a congruence follows [Pul 4], which is more general then that given in [Wil 1].

Letting P be a PAM and $\sim$ be a congruence, we denote the set of congruence classes by $P/\sim$. Defining $\oplus$ on $P/\sim$ in a natural way, we show that $P/\sim$ is a PAM. Also if P is an effect algebra and $\sim \neq P \times P$ then $P/\sim$ is an effect algebra. For these two cases, it then follows that $P//\sim = P/\sim$. If P is a CPAM, we give an example which shows that $P/\sim$ need not be a CPAM.

We then introduce the concept of a Riesz ideal I on a CPAM P. We show that I generates a natural congruence $\sim_I$ on P and that $P/\sim_I$ is again a CPAM. Using the notation $P/I = P/\sim_I$, it follows that $P/I = P//\sim_I$ and we call P/I the quotient of P by I. Although a congruence need not be generated by a Riesz ideal, those that are so generated have stronger properties. For example, if P is an orthoalgebra, or an OMP or an OML, or an MV-algebra, or a Boolean algebra, then P/I belongs to the same class as P. We also show that a subset I of an OML is a Riesz ideal if and only if I is a p-ideal. On MV-algebras, every ideal is a Riesz ideal, and ideals in our sense coincide with MV-algebra ideals. Moreover, for effect algebras, congruences generated by Riesz ideals are precisely those that are given by a perspectivity [Wil 2], [Pul 4].

We also consider a more general type of a binary relation that we call a weak congruence, and we give an intrinsic characterization of weak congruences [Jen]. In Exercises 3.4, a large number of examples and counterexamples are included, that illustrate various concepts.

3.1 Congruences and Ideals on Partial Abelian Monoids

Definition 3.1.1 *Let P be a nonempty set and let $\perp \subseteq P \times P$ be a nonempty binary relation on P. A* **partial Abelian monoid** *(***PAM***) is an algebraic structure $(P; 0, \oplus)$ where $0 \in P$ and $\oplus$ is a partial binary operation on P that satisfies the following conditions.*

(P1) *If $a \perp b$, then $b \perp a$ and $a \oplus b = b \oplus a$.*

(P2) *If $a \perp b$ and $c \perp (a \oplus b)$, then $b \perp c$, $a \perp (b \oplus c)$ and $(a \oplus b) \oplus c = a \oplus (b \oplus c)$.*

(P3) *For all $a \in P$, $a \perp 0$ and $a \oplus 0 = a$.*

A PAM P is **cancellative** *(abbreviated* **CPAM***) if*

$$a \oplus c = b \oplus c \ \Rightarrow \ a = b.$$

We say that P is **positive**, *if*

$$a \oplus b = 0 \ \Rightarrow \ a = b = 0.$$

An element $1 \in P$ is called a **unit** *if*

$$\forall a \in P, \exists b \in P: \ a \perp b, a \oplus b = 1.$$

If P possesses a unit, we call P a **unital** *PAM.*

When we write and expression like $a \oplus b$ we are assuming that $a \perp b$. Moreover, due to (P2), we can omit parentheses and write $a \oplus b \oplus c$.

In a positive unital CPAM, the unit is unique. In a PAM P we define

$$a \leq b \text{ if } \exists c \in P : a \oplus c = b. \tag{3.1}$$

Then $\leq$ is reflexive and transitive, hence a preorder, and if P is a positive CPAM, then $\leq$ is a partial order (Exercise 1, Section 3.4).

It is easy to see that a PAM $(P; \oplus, 0, 1)$ is an effect algebra if and only if it is a positive CPAM with unit 1 (Exercise 2, Section 3.4).

Let P and Q be PAMs. A function $f : P \to Q$ is a **morphism** if $a \perp b$ implies $f(a) \perp f(b)$ and $f(a \oplus b) = f(a) \oplus f(b)$. If f is surjective, then $f(0) = 0$. This is because 0 is unique in any PAM and $f(0) \oplus f(a) = f(0 \oplus a) = f(a)$. If P is CPAM, then $f(0) = 0$ because $0 \oplus f(a) = f(0 \oplus a) = f(0) \oplus f(a)$. If P and Q are unital PAMs, we require that a morphism also satisfies $f(1) = 1$.

If $f : P \to Q$ is a bijective morphism and f^{-1} is a morphism, then f is an **isomorphism**. In this case, P and Q are isomorphic and we write $P \equiv Q$.

We will say that a morphism f is **strong** if

$$f(a) \perp f(b) \Rightarrow \exists\, c \in P \text{ such that } f(c) = f(b) \text{ and } c \perp a.$$

If $f : P \to Q$ is a morphism, we write $a \sim_f b$ for $a, b \in P$ if $f(a) = f(b)$.

Let P be a PAM and let $\sim$ be a nonempty binary relation on P. A **quotient** relative to $\sim$ is a pair (Q, q), where Q is a PAM and $q : P \to Q$ is a surjective morphism that satisfies

(Q1) $\sim \subseteq \sim_q$.

(Q2) If R is a PAM and $\phi : P \to R$ is a morphism such that $\sim \subseteq \sim_\phi$, then there exists a morphism $\psi : Q \to R$ such that $\psi \circ q = \phi$.

That is, we may regard $q : P \to Q$ as a universal morphism that respects $\sim$ and when it exists, we write $Q = P//\sim$. It is easy to see that if a quotient exists, then it is unique up to isomorphism. We can define quotients also in other categories, such as CPAMs, or effect algebras, etc. We restrict our attention to a special type of binary relations, which we call congruences.

3.1.1 Congruences and Quotients

In this subsection, we mainly follow the ideas in [GuPu].

Definition 3.1.2 *A binary relation $\sim$ on a PAM P is a* **weak congruence** *if it satisfies the following conditions.*

(C1) *$\sim$ is an equivalence relation.*

(C2) *$a \perp b$, $a_1 \perp b_1$, $a_1 \sim a$, $b_1 \sim b$ imply $a_1 \oplus b_1 \sim a \oplus b$.*

A weak congruence is a **congruence** *if*

(C3) *$a \perp b$, $c \sim a$ imply that there exists a $d \in P$ such that $d \sim b$, $d \perp c$.*

A congruence $\sim$ *is a c-***congruence** *if*

(C4) $a \perp b$, $a_1 \perp b_1$, $a_1 \sim a$, $a_1 \oplus b_1 \sim a \oplus b$ *imply* $b_1 \sim b$.

If $\sim$ is a congruence on a PAM P, we denote the equivalence class containing $a \in P$ by $[a]$ and denote the set of congruence classes $P/\sim$. We define $[a] \perp [b]$ if there exist $a_1, b_1 \in P$ such that $a_1 \sim a$, $b_1 \sim b$ and $a_1 \perp b_1$ and put $[a] \oplus [b] = [a_1 \oplus b_1]$. According to (C2), $[a] \oplus [b]$ is well defined.

Theorem 3.1.3 *Let* $\sim$ *be a congruence on a PAM. The following statements hold.*

(i) $P/\sim$ *is a PAM.*

(ii) $P/\sim$ *is a CPAM if and only if* $\sim$ *is a c-congruence.*

(iii) *If* P *is an effect algebra and* $\sim \neq P \times P$ *then* $P/\sim$ *is an effect algebra and* $a \sim b$ *implies* $a' \sim b'$.

Proof. Clearly, $\oplus$ is commutative and $[0]$ is the additive zero. To prove associativity, assume that $[a] \perp [b]$ and $([a] \oplus [b]) \perp [c]$. Then there exist $a_1, b_1 \in P$ such that $a_1 \sim a$, $b_1 \sim b$, $a_1 \perp b_1$ and $[a] + [b] = [a_1 \oplus b_1]$. By definition, $[c] \perp [a_1 \oplus b_1]$ so there exist $c_1, d \in P$ such that $c_1 \sim c$, $d \sim a_1 \oplus b_1$, $c_1 \perp d$ and

$$([a] \oplus [b]) \oplus [c] = [a_1 \oplus b_1] \oplus [c] = [d \oplus c_1].$$

Now $d \sim a_1 \oplus b_1, d \perp c_1$ imply by (C3) that there exists a $c_2 \in P$ such that $c_2 \sim c_1$ and $c_2 \perp a_1 \oplus b_1$. Since $b_1 \perp c_2$, $[b] \perp [c]$ and $[b] \oplus [c] = [b_1 \oplus c_2]$. Moreover, $a_1 \perp (b_1 \oplus c_2)$ so $[a] \perp ([b] \oplus [c])$ and

$$\begin{aligned} ([a] \oplus [b]) \oplus [c] &= [d \oplus c_1] = [(a_1 \oplus b_1) \oplus c_2] = [a_1 \oplus (b_1 \oplus c_2)] \\ &= [a_1] \oplus [b_1 \oplus c_2] = [a] \oplus ([b] \oplus [c]). \end{aligned}$$

(ii) Assume that $\sim$ is a c-congruence. To show that $P/\sim$ is cancellative, suppose that $[a] \oplus [b] = [a] \oplus [c]$. Then there exist $a_1, a_2, b_1, c_1 \in P$ such that $a_1 \sim a$, $b_1 \sim b$, $a_1 \perp b_1$, $a_2 \sim a$, $c_1 \sim c$, $a_2 \perp c_1$ and

$$[a_1 \oplus b_1] = [a] \oplus [b] = [a] \oplus [c] = [a_2 \oplus c_1].$$

Hence $a_1 \oplus b_1 \sim a_2 \oplus c_1$ and since $a_1 \sim a_2$ an application of (C4) gives $b_1 \sim c_1$. Hence, $b \sim c$ so $[b] = [c]$. Conversely, assume that $P/\sim$ is a CPAM. If $a \perp b$, $a_1 \perp b_1$, $a_1 \sim a$, $a_1 \oplus b_1 \sim a \oplus b$, then

$$[a] \oplus [b_1] = [a_1 \oplus b_1] = [a \oplus b] = [a] \oplus [b].$$

Since $P/\sim$ is cancellative, $[b_1] = [b]$ so $b_1 \sim b$. Hence, (C4) holds so $\sim$ is a c-congruence.

(iii) Assuming that P is an effect algebra, it follows from (i) that $P/\sim$ is a PAM. We now show that $a \sim b$ implies $a' \sim b'$. Indeed, $a \sim b$, $a \perp a'$ imply by (C3) that there exists a $c \in P$ such that $c \sim a'$, $c \perp b$. Also, there exists a $d \in P$ such that $c \oplus b \oplus d = 1$ so $b' = c \oplus d$. Then $a \sim b$, $a' \sim c$, $a \perp a'$, $b \perp c$ imply by (C2) that

$$b \oplus c \oplus d = 1 = a \oplus a' \sim b \oplus c.$$

Furthermore, $b \oplus c \sim b \oplus c \oplus d$, $d \perp b \oplus c$ imply by (C3) that there exists a $d_1 \in P$ such that $d_1 \sim d$, $d_1 \perp b \oplus c \oplus d$. Hence, $d \sim d_1 = 0$. Now $c \sim a'$, $d \sim 0$, $c \perp d$ imply by (C2) that

$$c \oplus d \sim a' \oplus 0 = a'.$$

Hence, $b' \sim a'$.

Clearly, $[a] \oplus [a'] = [a \oplus a'] = 1$ for every $a \in P$. To prove the uniqueness of $[a']$, assume that $[a] \oplus [b] = [1]$. Then there exist $a_1, b_1 \in P$ such that $a_1 \sim a$, $b_1 \sim b$ and $a_1 \oplus b_1 \sim 1$. Also, there exists a $c \in P$ such that $a_1 \oplus b_1 \oplus c = 1$ so $b_1 \oplus c = a_1'$. By (C3), $a_1 \oplus b_1 \sim a_1 \oplus b_1 \oplus c$, $c \perp a_1 \oplus b_1$ imply that there exists a $d \in P$ such that $d \sim c$, $d \perp a_1 \oplus b_1 \oplus c$. But then $d = 0$ and hence $c \sim 0$. Now $b_1 \sim b$, $c \sim 0$, $b_1 \perp c$ imply by (C2) and the previous paragraph that

$$b = b \oplus 0 \sim b_1 \oplus c = a_1' \sim a'.$$

Hence, $[b] = [a']$.

Now assume that $[a] \oplus [1]$ is defined. Then there exist $a_1, b \in P$ such that $a_1 \sim a$, $b \sim 1$, $a_1 \perp b$. Now by (C3), $b \sim 1$, $b \perp a_1$ imply that there exists an $a_2 \in P$ such that $a_2 \sim a_1$, $a_2 \perp 1$. Hence, $a_2 = 0$. Therefore, $a \sim a_1 \sim a_2 = 0$ so $[a] = [0]$.

Finally, if $1 \sim 0$ then for any $a \in P$, $a \perp 0$ so by (C3) there exists a $b \in P$ such that $b \sim a$ and $b \perp 1$. Hence, $b = 0$ so $a \sim 0$, a contradiction. Hence, $[0] \neq [1]$. □

Corollary 3.1.4 *A congruence $\sim$ on an effect algebra is a c-congruence.*

Corollary 3.1.5 *If $\sim$ is a congruence on a PAM P, then $(P/\sim, q)$ is a quotient of P relative to $\sim$, where $q(a) = [a]$.*

Proof. We have shown that $P/\sim$ is a PAM and that q is a surjective morphism satisfying (Q1). Suppose R is a PAM and $\phi : P \to R$ be such that $\sim \subseteq \sim_\phi$. Define $\psi : P/\sim \to R$ by $\psi([a]) := \phi(a)$. If $b \sim a$, then $\phi(b) = \phi(a)$, so ψ is well defined. If $[a] \perp [b]$, then there exist $a_1, b_1 \in P$ such that $a_1 \sim a$, $b_1 \sim b$, $a_1 \perp b_1$ and $[a] \oplus [b] = [a_1 \oplus b_1]$. Hence, $\psi([a]) = \phi(a_1) \perp \phi(b_1) = \phi(b)$, and

$$\psi([a] \oplus [b]) = \psi([a_1 \oplus b_1]) = \phi(a_1 \oplus b_1) = \phi(a_1) \oplus \phi(b_1) = \psi([a]) \oplus \psi([b]).$$

Clearly, $\phi = \psi \circ q$. □

Example 3.1.6 Let P be the set of nonnegative integers with the usual $+$ as a binary operation. Then P is a positive CPAM. Define $a \sim b$ if $a \equiv b (\bmod 3)$. Then $\sim$ is a congruence and, in the usual way, $P/\sim$ is isomorphic to $Q = \{0,1,2\}$ with addition $+$ modulo 3. Then Q is a CPAM that is not positive because $1 \oplus 2 = 0$. This shows that although $P/\sim$ is always cancellative when $\sim$ is a c-congruence, it need not be positive. Notice also that the preorder $\leq$ is not a partial order because $1 \leq 2$ and $2 \leq 1$, yet $1 \neq 2$.

Not every congruence is a c-congruence. Just let P be a PAM that is not a CPAM and define $a \sim b$ if $a = b$. Then $\sim$ is a congruence, but it is not a c-congruence. The next example shows that even on a CPAM, there can be congruences that are not c-congruences.

Example 3.1.7 Let P be the set of positive integers with the usual multiplication $\cdot$ as a binary operation and 1 as a neutral element. Then P is a positive CPAM. Again, define $a \sim b$ if and only if $a \equiv b$ (mod 3). Then $\sim$ is a congruence and in the usual way, $P/\sim$ is isomorphic to $Q = \{0,1,2\}$ with multiplication $\cdot$ modulo 3. Now Q is not a CPAM because $0 \cdot 1 = 0 \cdot 2 = 0$, yet $1 \neq 2$. Also, Q is not positive because $2 \cdot 2 = 1$. By Theorem 3.1.3 (ii), $\sim$ is not a c-congruence. This can also be shown directly since in P, $3 \sim 9$, $3 \cdot 2 \sim 9 \cdot 1$ and yet $2 \not\sim 1$.

Theorem 3.1.8 (i) *If $\sim$ is a congruence on a PAM P, then there exists a PAM Q and a surjective strong morphism f: $P \to Q$ such that $\sim = \sim_f$. Conversely, if P and Q are PAMs and f: $P \to Q$ is a strong morphism, then $\sim_f$ is a congruence.*

(ii) *If $\sim$ is a c-congruence on a PAM P, then there exists a CPAM Q and a surjective strong morphism f: $P \to Q$ such that $\sim = \sim_f$. Conversely, if P is a PAM, Q is a CPAM and $f : P \to Q$ is a strong morphism, then $\sim_f$ is a c-congruence.*

(iii) *If P is a PAM, Q is a CPAM and $f : P \to Q$ is a strong morphism, then $f(P) \equiv P/\sim_f$.*

Proof. (i) Applying Theorem 3.1.3 (i), $Q = P/\sim$ is a PAM. Let $f : P \to Q$ be the canonical surjection $f(a) = [a]$. Then $a \perp b$ implies $f(a) \perp f(b)$ and

$$f(a \oplus b) = [a \oplus b] = [a] \oplus [b] = f(a) \oplus f(b),$$

so f is a surjective morphism. To show that f is strong, assume that $f(a) \perp f(b)$. Then $[a] \perp [b]$ so there exists $a_1, b_1 \in P$ such that $a_1 \sim a$, $b_1 \sim b$ and $a_1 \perp b_1$. Since $a \sim a_1$, by (C3) there exists a $c \in P$ such that $c \sim b_1$ and $c \perp a$. Hence, $f(c) = f(b_1) = f(b)$ and $c \perp a$. Moreover, $a \sim b$ iff $a \sim_f b$. Conversely, if $f : P \to Q$ is a strong morphism, then clearly $\sim_f$ is an equivalence relation so (C1) holds. To show that (C2) holds, assume that $a \perp b$, $a_1 \perp b_1$, $a_1 \sim_f a$, $b_1 \sim_f b$. Then

$$f(a \oplus b) = f(a) \oplus f(b) = f(a_1) \oplus f(b_1) = f(a_1 \oplus b_1)$$

so $a \oplus b \sim_f a_1 \oplus b_1$. To show that (C3) holds, suppose that $a \perp b$ and $c \sim_f a$. Then $f(c) = f(a)$ and $f(a) \perp f(b)$ so $f(c) \perp f(b)$. Since f is strong, there exists a $d \in P$ such that $f(d) = f(b)$ and $d \perp c$. Hence, $d \sim_f b$ and $d \perp c$.

(ii) Applying Theorem 3.1.3 (ii), $Q = P/\sim$ is a CPAM. As in (i), $f : P \to Q$ given by $f(a) = [a]$ is a surjective strong morphism such that $\sim = \sim_f$. For the converse, we have shown in (i) that $\sim_f$ is a congruence. To show that $\sim_f$ is a c-congruence, assume that $a \perp b$, $a_1 \perp b_1$, $a_1 \sim_f a$ and $a \oplus b \sim_f a_1 \oplus b_1$. Then

$$f(a) \oplus f(b) = f(a \oplus b) = f(a_1 \oplus b_1) = f(a) \oplus f(b_1).$$

Since Q is cancellative, we have $f(b) = f(b_1)$ so $b_1 \sim_f b$.

(iii) Applying (ii), $\sim_f$ is a c-congruence so by Theorem 3.1.3 (ii), $P/\sim_f$ is a CPAM. To show that $f(P)$ is a PAM, suppose that $f(a), f(b) \in f(P)$ with $f(a) \perp f(b)$. Since f is strong, there exists a $c \in P$ such that $f(c) = f(b)$ and $c \perp a$. Hence,

$$f(a) \oplus f(b) = f(a) \oplus f(c) = f(a \oplus c) \in f(P).$$

Since $f(0) \oplus f(a) = f(0 \oplus a) = f(a)$, $f(0)$ is the additive zero on $f(P)$. Define $\phi : f(P) \to P/\sim_f$ by $\phi(f(a)) = [a]$. Notice that ϕ is well defined because $f(a) = f(b)$ implies that $[a] = [b]$. Clearly, ϕ is bijective. To show that ϕ is a morphism, suppose that $f(a) \perp f(b)$. Then as before, there exists a $c \in P$ such that $f(c) = f(b)$ and $c \perp a$. Hence, $[a] \perp [c]$ so $\phi(f(a)) \perp \phi(f(b))$ and

$$\phi(f(a) \oplus f(b)) = \phi(f(a \oplus c)) = [a \oplus c] = [a] \oplus [c] = \phi(f(a)) \oplus \phi(f(b)).$$

To show that ϕ^{-1} is a morphism, if $\phi(f(a)) \perp \phi(f(b))$ then $[a] \perp [b]$. Hence, there exists $a_1, b_1 \in P$ such that $a_1 \sim_f a$, $b_1 \sim_f b$ and $a_1 \perp b_1$. Hence, $f(a_1) \perp f(b_1)$ so $f(a) \perp f(b)$. Thus, ϕ is an isomorphism. □

Corollary 3.1.9 *If $\sim \neq P \times P$ is a congruence on an effect algebra P, then there exists an effect algebra Q and a surjective strong effect morphism $f : P \to Q$ such that $\sim = \sim_f$. Conversely, if P is an effect algebra, Q is a PAM and $f : P \to Q$ is a strong morphism, then $\sim_f$ is a congruence.*

Theorem 3.1.10 (i) *If P, Q, R are PAMs, $\phi : P \to Q$ is a surjective strong morphism and $\psi : Q \to R$ is a strong morphism, then $\psi \circ \phi : P \to R$ is a strong morphism.* (ii) *If P, Q are PAMs and $f : P \to Q$ is a strong morphism, then there exists a PAM R, a surjective strong morphism $\phi : P \to R$ and an injective strong morphism $\psi : R \to Q$ such that $f = \psi \circ \phi$.*

Proof. It is left to the readers as an exercise. □

3.1.2 Riesz Ideals on CPAMs

We will define an ideal on a PAM P in the same way as an ideal on an effect algebra. That is, a nonempty subset I of P is an **ideal** if, for all $a, b \in P$, $a \perp b$, we have $a, b \in I$ if and only if $a \oplus b \in I$.

In this subsection, we will consider a special class of ideals on a CPAM.

Definition 3.1.11 *An ideal I in a CPAM P is an* R_1**-ideal** *if the following condition holds.*

(R1) *If $i \in I$, $a, b \in P$ with $a \perp b$, $i \leq a \oplus b$, then there exists $j, k \in I$ such that $j \leq a$, $k \leq b$ and $i \leq j \oplus k$.*

An R_1-ideal I is a **Riesz ideal** *if the following condition holds.*

(R2) *If $i \in I$, $a, b \in P$ with $i \leq a$, $a \ominus i \perp b$, then there exists a $j \in I$ such that $j \leq b$ and $a \perp b \ominus j$.*

Although there is a formal analogy between the two concepts, the Riesz property is more general than the Riesz decomposition property (see Exercise 3 of Section 3.4). In what follows, we shall need the following lemma.

Lemma 3.1.12 *Let a, b, c, d be elements of a CPAM P.*

(i) *If $a \perp b$, $c \leq a$, $d \leq b$, then $c \perp d$, $c \oplus d \leq a \oplus b$ and*

$$(a \ominus c) \oplus (b \ominus d) = (a \oplus b) \ominus (c \oplus d).$$

(ii) *If $c \leq b \leq a$, $c \perp a \ominus b$, then $c \oplus (a \ominus b) = a \ominus (b \ominus c)$.*

(iii) *If $a \ominus b \leq a \ominus c$, then $c \leq b$.*

(iv) *If $c \leq b \leq a$, then $a \ominus b \leq a \ominus c$.*

(v) *If $b \leq a$ and $c \leq a \ominus b$, then $(a \ominus b) \ominus c = a \ominus (b \oplus c)$.*

Proof. It is left to the readers (cf. [GuPu, Lem 4.1]). □

Theorem 3.1.13 [GuPu], [ChPu] (i) *An ideal I in a CPAM P satisfies* (R1) *if and only if for every $i \in I$, $a, b \in P$ with $a \perp b$, $i \leq a \oplus b$ there exist $j, k \in I$ such that $j \leq a$, $k \leq b$ and*

$$(a \ominus j) \oplus (b \ominus k) \leq (a \oplus b) \ominus i.$$

(ii) *If an ideal I in an upwards directed CPAM P satisfies* (R1), *then I is a Riesz ideal. In particular, if an ideal in an effect algebra P satisfies* (R1), *it is a Riesz ideal.*

Proof. (i) Suppose I satisfies (R1) and $i \in I$ with $i \leq a \oplus b$. Then by (R1) there exist $j, k \in I$ such that $j \leq a$, $k \leq b$, $i \leq j \oplus k \leq a \oplus b$. Applying Lemma 3.1.12 (ii) we have

$$i \oplus [(a \oplus b) \ominus (j \oplus k)] = (a \oplus b) \ominus [(j \oplus k) \ominus i] \leq a \oplus b.$$

Hence, by Lemma 3.1.12 (i) and (iv)

$$(a \ominus j) \oplus (b \ominus k) = (a \oplus b) \ominus (j \oplus k) \leq (a \oplus b) \ominus i.$$

Conversely, suppose that I satisfies the condition of the theorem. If $i \leq a \oplus b$, then by this condition and Lemma 3.1.12 (i) there exist $j, k \in I$ such that $j \leq a$, $k \leq b$ and

$$(a \oplus b) \ominus (j \oplus k) = (a \ominus j) \oplus (b \ominus k) \leq (a \oplus b) \ominus i.$$

By Lemma 3.1.12 (iii) we have $i \leq j \oplus k$.

(ii) To show that (R2) holds, assume that $i \in I$, $i \leq a$ and $a \ominus i \perp b$. Since P is upwards directed, there is $u \in P$ such that $(a \ominus i) \oplus b, a \leq u$. Then we have $b \leq u \ominus (a \ominus i) = (u \ominus a) \oplus i$. Hence there is a $d \in P$ such that $(u \ominus a) \oplus i = b \oplus d$ and $u \ominus a = (b \oplus d) \ominus i$. By (R1) and (i) there exists $j, k \in I$ such that $j \leq b$, $k \leq d$ and

$$u \ominus a \geq (b \ominus j) \oplus (d \ominus k) \geq b \ominus j.$$

Hence, $a \perp b \ominus j$ so (R2) holds. □

Definition 3.1.14 *For an ideal I in a CPAM, we define $a \sim_I b$ if there exist $i, j \in I$ such that $i \leq a$, $j \leq b$ and $a \ominus i = b \ominus j$.*

It is easy to see that the above definition is equivalent to $a \sim_I b$ if there is $k \in P$ such that $k \leq a, b$ and $a \ominus k \in I$, $b \ominus k \in I$.

Lemma 3.1.15 *If I is an R_1-ideal in a CPAM P, then $a \sim_I b$ if and only if there exist $i, j \in I$ such that $i \leq a$, $j \leq b$, $a \ominus i \leq b$ and $b \ominus j \leq a$.*

Proof. Necessity is trivial. For sufficiency, suppose that $a \ominus i \leq b$ and $b \ominus j \leq a$. Then there exists a $c \in P$ such that $(a \ominus i) \oplus c = b$. Hence,

$$[(a \ominus i) \oplus c] \ominus j = b \ominus j \leq a.$$

Applying Theorem 3.1.13 (i), Lemma 3.1.12 (v) and the fact that I is an ideal, there exist $k, l \in I$ such that $(a \ominus k) \oplus (c \ominus l) \leq a$. Hence, there exists a $d \in P$ such that $(a \ominus k) \oplus (c \ominus l) \oplus d = a$. Thus, $(c \ominus l) \oplus d = k \in I$ so $c \ominus l \in I$ and $c = (c \ominus l) \oplus l \in I$. Since $a \ominus i = b \ominus c$, $a \sim_I b$. □

Theorem 3.1.16 *If I is a Riesz ideal in a CPAM P, then $\sim_I$ is a c-congruence. Moreover, $a \sim_I 0$ if and only if $a \in I$.*

Proof. Since $0 \in I$, $\sim_I$ is reflexive and $\sim_I$ is clearly symmetric. To prove transitivity, assume that $a \sim_I b$ and $b \sim_I c$. Then there exist $i, j, k, l \in I$ such that $a \ominus i = b \ominus j$ and $b \ominus k = c \ominus l$. Hence, $b = (a \ominus i) \oplus j$ so

$$[(a \ominus i) \oplus j] \ominus k = c \ominus l \leq c.$$

By Theorem 3.1.13 (i) and Lemma 3.1.12 (v), there exist $m, n \in I$ such that

$$[a \ominus (i \oplus m)] \oplus (j \ominus n) \leq c.$$

Hence, $a \ominus (i \oplus m) \leq c$ and $i \oplus m \in I$. Moreover, $b = (c \ominus l) \oplus k$ so

$$[(c \ominus l) \oplus k] \ominus j = a \ominus i \leq a.$$

The same argument shows that there exists a $p \in I$ such that $c \ominus (l \oplus p) \leq a$ and $l \oplus p \in I$. By Lemma 3.1.15, $a \sim_I c$ so (C1) holds.

To prove (C2), assume that $a \perp b$, $a_1 \perp b_1$, $a_1 \sim_I a$ and $b_1 \sim_I b$. Then there exist $i, j, k, l \in I$ such that $a_1 \ominus i = a \ominus j$, $b_1 \ominus k = b \ominus l$. Hence,

$$a_1 \oplus b_1 = [(a \ominus j) \oplus i] \oplus [(b \ominus l) \oplus k],$$

so by Lemma 3.1.12 (i) we have

$$(a_1 \oplus b_1) \ominus (i \oplus k) = (a \ominus j) \oplus (b \ominus l) = (a \oplus b) \ominus (j \oplus l).$$

Since $i \oplus k, j \oplus l \in I$ we have $a_1 \oplus b_1 \sim_I a \oplus b$ so (C2) holds.

To prove (C3), assume that $a \perp b$, $c \sim_I a$. Then there exist $i, j \in I$ such that $c \ominus i = a \ominus j$. Since $b \perp a$, we have $b \perp a \ominus j$ so $b \perp c \ominus i$. Applying (R2), there exists a $k \in I$ such that $b \ominus k \perp c$. Since $b \ominus k \sim_I b$, (C3) holds.

To prove (C4), assume that $a \perp b$, $a_1 \perp b_1$, $a_1 \sim_I a$ and $a_1 \oplus b_1 \sim_I a \oplus b$. Since $a_1 \perp b_1$ and $a \sim_I a_1$, by (C3) there exists a $d \in P$ such that $d \sim_I b_1$ and $d \perp a$. Hence, by (C2)

$$a \oplus b \sim_I a_1 \oplus b_1 \sim_I a \oplus d.$$

Thus, there exist $i, j \in I$ such that $(a \oplus b) \ominus i = (a \oplus d) \ominus j$. By Theorem 3.1.13 (i) there exist $k, l \in I$ such that

$$(a \ominus k) \oplus (b \ominus l) \leq (a \oplus d) \ominus j \leq a \oplus d.$$

Hence, there exists a $c \in P$ such that

$$(a \ominus k) \oplus (b \ominus l) \oplus c = a \oplus d = (a \ominus k) \oplus k \oplus d.$$

Since P is cancellative, we have $(b \ominus l) \oplus c = d \oplus k$. Hence,

$$[(b \ominus l) \oplus c] \ominus k = d.$$

Applying Theorem 3.1.13 (i) and Lemma 3.1.12 (v), there exists $m, n \in I$ such that

$$b \ominus (l \oplus m) \leq [b \ominus (l \oplus m)] \oplus (c \ominus n) \leq d.$$

Similarly, there exists a $p \in I$ such that $d \ominus p \leq b$. By Lemma 3.1.15, $b \sim_I d \sim_I b_1$ so (C4) holds.

Finally, if $a \sim_I 0$, then there exist $i, j \in I$ such that $a \ominus i = 0 \ominus j$. Hence, $a = (0 \ominus j) \oplus i$ and since $0 \ominus j \leq 0 \in I$, we have $0 \ominus j \in I$ so $a \in I$. Conversely, if $a \in I$, then $a \ominus a = 0 = 0 \ominus 0$ so $a \sim_I 0$. □

Let I be a Riesz ideal in a CPAM P. Since $\sim_I$ is a c-congruence, it follows from Theorem 3.1.3 (ii) that $P/\sim_I$ is again a CPAM. In contrast, for a general congruence $\sim$, $P/\sim$ need not be a CPAM. We use the notation $P/I = P/\sim_I$ and call P/I the **quotient** of P by I (cf. Theorem 3.1.3). We say that a congruence $\sim$ is **generated** by a Riesz ideal I if $\sim = \sim_I$. Theorem 3.1.17 states that every Riesz ideal on a CPAM generates a c-congruence.

Corollary 3.1.17 (i) *If I is a Riesz ideal in a CPAM P, then P/I is a positive CPAM.* (ii) *If $I \neq P$ is a Riesz ideal in an effect algebra P, then P/I is an effect algebra.*

Proof. (i) Applying Theorem 3.1.3 (ii) and Theorem 3.1.16, P/I is a CPAM. To show that P/I is positive, suppose that $[a] \oplus [b] = [0]$. Then there exist $a_1, b_1 \in P$ such that $a_1 \sim_I a$, $b_1 \sim_I b$, $a_1 \perp b_1$ and $[a] \oplus [b] = [a_1 \oplus b_1] = [0]$. Hence, $a_1 \oplus b_1 \sim_I 0$ so by Theorem 3.1.16, $a_1 \oplus b_1 \in I$. Since I is an ideal, $a_1, b_1 \in I$ so $[a] = [a_1] = [0]$ and $[b] = [b_1] = [0]$.

(ii) Since $I \neq P$, we have $1 \notin I$. Hence, by Theorem 3.1.16, $1 \not\sim_I 0$ so $\sim_I \neq P \times P$. The result now follows from Theorem 3.1.3 (iii) and Theorem 3.1.16. □

The following example shows that a c-congruence on a CPAM need not be generated by a Riesz ideal so in this sense, c-congruences are more general than Riesz ideals.

Example 3.1.18 In Example 3.1.6, P is a positive CPAM and $\sim$ is a c-congruence on P. We showed that $P/\sim$ is a CPAM that is not positive. It follows from Corollary 3.1.17 that $\sim$ is not generated by a Riesz ideal. This can also be seen as follows. If $\sim$ were generated by a Riesz ideal I, then by Theorem 3.1.16, $I = [0]$. Now $3 \in I$ and $3 \leq 1 + 2$, yet there do not exist $i, j \in I$ such that $i \leq 1$, $j \leq 2$ and $3 \leq i + j$.

Example 3.1.19 Let P_1 and P_2 be PAMs and let Q be their cartesian product $Q = P_1 \times P_2$. For $a = (a_1, a_2) \in Q$ and $b = (b_1, b_2) \in Q$ define $a \sim b$ if and only if $a_1 = b_1$. Then $\sim$ is a congruence on Q. Indeed, it is clear that $\sim$ is an equivalence relation so (C1) holds. To verify (C2), assume that $a \perp b$, $c \perp d$, $c \sim a$, $d \sim b$. Then $c_1 = a_1$, $d_1 = b_1$ so

$$c \oplus d = (c_1 \oplus d_1, c_2 \oplus d_2) = (a_1 \oplus b_1, c_2 \oplus d_2) \sim (a_1 \oplus b_1, a_2 \oplus b_2) = a \oplus b.$$

To verify (C3), assume that $a \perp b$ and $c \sim a$. Letting $d = (b_1, 0) \in Q$, we have $d \sim b$. Moreover, $d_1 = b_1 \perp a_1 = c_1$ and $d_2 = 0 \perp c_2$ so $d \perp c$. It follows from Theorem 3.1.3 (i) that $Q/\sim$ is a PAM. Define $\phi : Q/\sim \to P_1$ by $\phi([a]) = a_1$. Then ϕ is well defined and it is clear that ϕ is bijective. If $[a] \perp [b]$, then $a_1 \perp b_1$ and

$$\phi([a] \oplus [b]) = a_1 \oplus b_1 = \phi([a]) \oplus \phi([b])$$

so ϕ is a morphism. Moreover, if $\phi([a]) \perp \phi([b])$, then $a_1 \perp b_1$ so $[a] \perp [b]$. Hence, ϕ is an isomorphism so $Q/\sim = P_1$.

Now suppose that P_1 and P_2 are CPAMs. Then $Q = P_1 \times P_2$ and $Q/\sim$ are CPAMs. Letting $I = \{0\} \times P_2 \subseteq Q$, it is clear that I is an ideal. We now show that I is a Riesz ideal. To verify (R1), suppose that $i \in I$ and $i \leq a \oplus b$. Then $(0, a_2), (0, b_2) \in I$, $(0, a_2) \leq a$, $(0, b_2) \leq b$ and $i \leq (0, a_2) \oplus (0, b_2)$. To verify (R2), suppose that $i \in I$ and $a \ominus i \perp b$. Then $a_1 \perp b_1$ and letting $j = (0, b_2)$ we have $j \in I$ and $a \perp b \ominus j$. Hence, I is a Riesz ideal so $\sim_I$ is also a c-congruence and Q/I is a CPAM. If $a \sim_I b$ then there exist $i, j \in I$ such that $a \ominus i = b \ominus j$. Hence, $a_1 = b_1$ so $a \sim b$. Conversely, if $a \sim b$, then $a_1 = b_1$. Letting $i = (0, a_2)$, $j = (0, b_2)$, we have $i, j \in I$ and $a \ominus i = b \ominus j$ so $a \sim_I b$. We conclude that $\sim = \sim_I$ so $\sim$ is generated by I.

As a corollary we obtain that if p is a central element of an effect algebra P, then the interval $P[0, p]$ is a Riesz ideal in P.

Example 3.1.20 In Example 3.1.6, P is a positive CPAM, $\sim$ is a congruence on P but $\sim$ is not a c-congruence. It follows from Theorem 3.1.17 that $\sim$ is not generated by a Riesz ideal. This can also be seen as follows. If $\sim$ were generated by a Riesz ideal I, then by Theorem 3.1.16, $I = [1]$. Now $4 \in I$ and $4 \leq 2 \cdot 2$, yet there do not exist $i, j \in I$ such that $i, j \leq 2$ and $4 \leq i \cdot j$.

3.1.3 Congruences and Ideals in Effect Algebras

In this subsection, we study congruences and ideals in effect algebras and their subcategories, namely orthoalgebras, orthomodular posets and lattices. We will show that in all cases, quotients with respect to Riesz ideals belong to the same category. In orthomodular lattices, Riesz ideals coincide with p-ideals, and a congruence with respect to I as a Riesz ideal in the effect-algebra sense coincides with the congruence with respect to I as a p-ideal in the orthomodular-lattice sense. We also show that Riesz ideals are closed with respect to generalized Sasaki projections. We mainly follow [GuPu] and [ChPu]. We start with a characterization of congruences on effect algebras.

Proposition 3.1.21 *A weak congruence $\sim$ on an effect algebra is a congruence if and only if the following two conditions are satisfied.*

(C5) *If $a \sim b \oplus c$, then there exist a_1, a_2 such that $a_1 \sim b, a_2 \sim c$ and $a = a_1 \oplus a_2$.*

(C6) *$a \sim b$ entails $a' \sim b'$.*

Proof. Let $\sim$ be a congruence. We may assume that $\sim \neq P \times P$. Property (C6) is satisfied by Theorem 3.1.3 (iii). To prove (C5), assume $a \sim b \oplus c$. Then we have, by (C6), $a' \sim (b \oplus c)'$. Using (C3) and $b \perp (b \oplus c)'$, we obtain that there is a_1, $a_1 \sim b$ such that $a_1 \perp a'$, i.e., $a_1 \leq a$. Hence there is a_2 such that $a = a_1 \oplus a_2$. Now we have $a_1 \oplus a_2 \sim b \oplus c$, $a_1 \sim b$, and since a congruence on an effect algebra is cancellative, (C4) yields $a_2 \sim c$.

Assume that $\sim$ is a weak congruence satisfying (C5) and (C6). We have to prove (C3). Let $a \sim b$ and $b \perp c$. We have $b' = (b \oplus c)' \oplus c$, and $b' \sim a'$ by (C6). Therefore by (C5), $a' = u \oplus v$, $u \sim (b \oplus c)'$, $v \sim c$. Hence $v \sim c$ and $v \perp a$, and (C3) is satisfied. □

Weak congruences satisfying (C5) will be studied in the next section.

Theorem 3.1.22 [GiPu] *Let I be an ideal of an effect algebra P such that $\sim_I$ is a congruence. Then I is a Riesz ideal.*

Proof. Let $i \in I$ and $i \leq x \oplus y$. Then, since $i \sim_I 0$ and $\sim_I$ is a congruence, we have $(x \oplus y) \ominus i \sim_I x \oplus y$. Hence, by the preceding theorem, there are u, v such that $u \sim_I x$, $v \sim_I y$ and $(x \oplus y) \ominus i = u \oplus v$. But $u \sim_I x$, $v \sim_I y$ imply that there are elements $j_1, k_1, j_2, k_2 \in I$ such that $u \ominus j_1 = x \ominus k_1$ and $v \ominus j_2 = y \ominus k_2$. It then follows that

$$(x \oplus y) \ominus i = u \oplus v = (x \ominus k_1) \oplus (y \ominus k_2) \oplus (j_1 \oplus j_2),$$

denoting $j_1 \oplus j_2 = l$, we get

$$((x \oplus y) \ominus i) \ominus l = (x \ominus k_1) \oplus (y \ominus k_2) \leq (x \oplus y) \ominus i,$$

which yields $(x \oplus y) \ominus (k_1 \oplus k_2) \leq (x \oplus y) \ominus i$, hence $i \leq k_1 \oplus k_2$. □

Recall that an element p in an effect algebra is principal (Definition 1.9.4) if $a, b \leq p$, $a \perp b$ imply $a \oplus b \leq p$, and p is central (Definition 1.9.11) if (i) p and p' are principal, and (ii) for any $a \in P$, there is a unique decomposition $a = a_1 \oplus a_2$, where $a_1 \leq p$, $a_2 \leq p'$. If p is principal, then p is sharp, that is, $p \wedge p' = 0$ (Definition 1.9.15 and Lemma 1.9.6). The set of all central elements of an effect algebra E is denoted by $C(E)$.

Recall that an ideal I_e, generated by an element e, is called a **principal ideal**. It is clear that $I_e = E[0, e]$ if and only if e is a principal element.

Central elements allow us to characterize those intervals of an effect algebra which are Riesz ideals.

Theorem 3.1.23 *In an effect algebra E, an interval $E[0,p]$ is a Riesz ideal if and only if p is a central element of E.*

Proof. For sufficiency, see Example 3.1.19. To prove the necessity, assume that $E[0,p]$ is a Riesz ideal. Then p must be principal in order to make $E[0,p]$ an ideal. For every $a \le p$ and every $b \in E$, $p \le b \oplus b'$ implies, by the Riesz property, that there are $p_1 \le b$, $p_2 \le b'$, $p \le p_1 \oplus p_2 \le p$, hence $p = p_1 \oplus p_2$. Then there is also a decomposition $b = b_1 \oplus b_2$ such that $b_1 \le p, b_2 \le p'$ (see Proposition 1.10.3).

To prove that p' is principal, assume that $b, c \le p'$ and $b \perp c$. By the preceding part of the proof, there is a decomposition $b \oplus c = u_1 \oplus u_2$, $u_1 \le p$, $u_2 \le p'$. Now $(b \oplus c) \ominus u_2 = u_1 \le p$, and since $u_1 \le b \oplus c$, there are $b_1 \le b, c_1 \le c$ such that $u_1 \le b_1 \oplus c_1 \le p$. Since p is sharp, then $b_1 \le b \le p'$, $c_1 \le c \le p'$ imply that $b_1 = c_1 = 0$, and hence $u_1 = 0$. Therefore $b \oplus c = u_2 \le p'$.

To prove the uniqueness of the decomposition, assume that $b = b_1 \oplus b_2 = c_1 \oplus c_2$, where $b_1, c_1 \le p$ and $b_2, c_2 \le p'$. Then $(b_1 \oplus b_2) \ominus c = c_1 \le p$, and since $c_1 \le b_1 \oplus b_2$, there are $u_1 \le b_1, u_2 \le b_2$ such that $c_1 \le u_1 \oplus u_2 \le p$. But we also have $u_2 \le b_2 \le p'$, hence $u_2 = 0$. It follows $c_1 \le b_1$. By symmetry we obtain $b_1 \le c_1$. □

Let L be an orthomodular lattice (OML). We recall that a subset $I \subseteq L$ is a lattice ideal for $a, b \in L$ we have $a, b \in I$ iff $a \vee b \in I$. A lattice ideal $I \subseteq L$ is a p-ideal (or an orthomodular ideal) if $a \in I$, $b \in L$ implies that $(a \vee b') \wedge b \in I$. Here $\phi_b(a) = (a \vee b') \wedge b \in I$ is the Sasaki projection from a to b. A p-ideal is characterized also as a lattice ideal closed under perspectivity [Kal] ("p" stands for "perspectivity"). p-ideals in orthomodular lattices play an important role, because they are in a one-to-one correspondence with congruences.

In Definition 1.8.17, the notion of the Sasaki projection in lattice ordered effect algebras has been introduced. Now we will introduce a generalization of the Sasaki projection to any effect algebras ([BeFo 2]).

Definition 3.1.24 *Let E be an effect algebra. For two elements a and b of E define*

$$\nabla(a,b) = \{d \in E : d \le a \text{ and } b \le d \oplus a'\}.$$

Let $\Phi(a,b)$ be the set of all minimal elements of $\nabla(a,b)$ (note that $\Phi(a,b)$ may be the empty set). The set $\Phi(a,b)$ is called the **generalized Sasaki projection** *of b on a.*

In lattice ordered effect algebras, $\Phi(a,b) = \{\phi(a,b)\}$, where $\phi(a,b) = a \ominus (a \wedge b')$ is the Sasaki mapping introduced in Definition 1.8.17 (Exercise 4, Section 3.4). In particular, in an OML the set of generalized Sasaki projections reduces to the usual Sasaki projection $\phi_a(b)$.

Theorem 3.1.25 [ChPu] *In an effect algebra E, the Riesz ideals are closed for generalized Sasaki projections, that is,*

$$i \in I \text{ and } m \in \Phi(a,i) \Rightarrow m \in I.$$

Proof. Assume that I is a Riesz ideal of E, $i \in I$ and $m \in \Phi(a,i)$. We have $m \le a$ and $i \le m \oplus a'$. There exist i_1, $i_2 \in I$ such that $i_1 \le m$, $i_2 \le a'$ and $i \le i_1 \oplus i_2$. It follows that $i_1 \le a$ and $i \le i_1 \oplus a'$ and so $i_1 \in \nabla(a,i)$. As $i_1 \le m$, we have $i_1 = m$ and therefore $m \in I$. □

Proposition 3.1.26 *Let A be an OMP in which, for any $(a,b) \in A^2$, $\Phi(a,b) \neq \emptyset$. Any ideal of A closed for generalized Sasaki projections is a Riesz ideal.*

Proof. Consider an ideal I closed for generalized Sasaki projections and let $i \in I$ with $i \le a \oplus b$. Pick $i_1 \in \Phi(a,i)$ and $i_2 \in \Phi(b,i)$. By hypothesis, we have i_1, $i_2 \in I$ and $i_1 \le a$, $i_2 \le b$. Write $i_1 = a \ominus h_1$ with $h_1 \le a$, $h_1 \le i'$ and $i_2 = b \ominus h_2$ with $h_2 \le b$, $h_2 \le i'$. We have :

$$\begin{aligned} i_1 \oplus i_2 = (a \ominus h_1) \oplus (b \ominus h_2) &= (a \oplus b) \ominus (h_1 \oplus h_2) \\ &= (a \vee b) \wedge (h_1 \vee h_2)'. \end{aligned}$$

Since $i \le a \vee b$ and $h_1 \vee h_2 \le i'$, $(h_1 \vee h_2)' \ge i$, and therefore $i \le (a \vee b) \wedge (h_1 \vee h_2)' = i_1 \oplus i_2$. □

It is not known in which effect algebras the converse of Theorem 3.1.25 holds true (see Question 13 in Section 3.4). Proposition 3.1.26 gives only a partial answer.

Recall that an effect algebra E is an orthoalgebra if

$$\forall a \in E,\, a \perp a \Leftrightarrow a = 0.$$

An effect algebra E is an orthomodular poset (OMP) if

$$\forall a,b,c \in E, a \perp b, b \perp c, c \perp a \Rightarrow a \oplus b \perp c.$$

An OMP is an orthoalgebra. Indeed, $a \perp a$, $a \perp a'$, $a' \perp a$ imply $a \oplus a \perp a'$, and $a \oplus a \oplus a' = 1 \oplus a$, hence $a = 0$.

Let P be an orthomodular poset (OMP). Recall that, for $a, b \in P$, $a \oplus b$ is defined if $a \le b'$ and in this case $a \oplus b = a \vee b$. Two elements $a, b \in P$ are (Mackey) compatible $a \leftrightarrow b$ if there exists $a_1, b_1, c \in P$ such that $a_1 \oplus b_1 \oplus c$ is defined and $a = a_1 \oplus c$, $b = b_1 \oplus c$ (Definition 1.10.1). The compatibility center of P is

$$Z(P) = \{a \in P : a \leftrightarrow b \text{ for every } b \in P\}.$$

It is straightforward to show that this definition of compatibility is equivalent to the standard definition (e.g., [PtPu]).

Theorem 3.1.27 *Let P be an OMP. Then*

(i) $C(P) = Z(P)$.

(ii) *If $I \subseteq P$ is a Riesz ideal, then every maximal element of I is in $Z(P)$.*

(iii) *If $I \subseteq P$ is an ideal such that for every $i \in I$ there exists an $a \in I \cap Z(P)$ with $i \leq a$, then I is a Riesz ideal.*

(iv) *A principal ideal $I_e \subseteq P$ is a Riesz ideal if and only if $e \in Z(P)$.*

Proof. (i) It is clear that every element of P is principal. Suppose $a \in Z(P)$ and $b \in P$. Then there exist $a_1, b_1, c \in P$ such that $a_1 \oplus b_1 \oplus c$ is defined and $a = a_1 \oplus c$, $b = b_1 \oplus c$. Since $c \leq a$ and $b_1 \perp a$ so $b_1 \leq a'$, $a \in C(P)$. Conversely, let $a \in C(P)$ and $b \in P$. Then there exist $c, d \in P$ such that $c \leq a$, $d \leq a'$ and $b = c \oplus d$. Now $a = c \oplus (a \ominus c)$ and since $d \perp c \oplus (a \ominus c)$, $d \oplus c \oplus (a \ominus c)$ is defined. Hence, $a \in Z(P)$.

(ii) Let a be a maximal element of I. If $b \in P$, then $a \leq b \vee b'$. Since I is a Riesz ideal, there exist $i, j \in I$ such that $i \leq b$, $j \leq b'$ and $a \leq i \vee j$. Since $i \perp j$ we have $i \vee j \in I$ so $a = i \vee j$. Also, $b = i \vee (b \wedge i')$. But $a = j \vee i \leq b' \vee i$ so $b \wedge i' \leq a'$. Hence, $b = i \oplus (b \wedge i')$ where $i \leq a$ and $b \wedge i' \leq a'$ so $a \in Z(P)$.

(iii) Suppose $i \in I$ and $i \leq b \vee c$ where $b \perp c$. Let $a \in I \cap Z(P)$ with $i \leq a$. Then

$$i = i \wedge a \leq (b \vee c) \wedge a = (b \wedge a) \vee (c \wedge a) = b \wedge a \oplus c \wedge a.$$

But $b \wedge a \leq b$, $b \wedge a \in I$, $c \wedge a \leq c$, $c \wedge a \in I$ so I is a Riesz ideal.

(iv) Follows by Theorem 3.1.23 and (i). □

Theorem 3.1.28 *Let I be a Riesz ideal in an effect algebra P.* (i) *If P is an orthoalgebra, then so is P/I.* (ii) *If P is an OMP, then so is P/I.*

Proof. (i) Assume that $[a] \perp [a]$. Then there exist $a_1, b_1 \in P$ such that $a_1 \sim_I a$, $a_2 \sim_I a$ and $a_1 \perp a_2$. Since $a_1 \sim_I a_2$, there exist $i, j \in I$ such that $a_1 \ominus i = a_2 \ominus j$. But then

$$a_1 \ominus i \leq a_2 \leq a_1' \leq (a_1 \ominus i)',$$

which implies that $a_1 \ominus i = 0$. Therefore, $a_1 = i \in I$ so $[a] = [0]$.

(ii) Assume that $[a] \perp [b]$, $[b] \perp [c]$, $[c] \perp [a]$. Then there exist $a_1, b_1 \in P$ such that $a_1 \sim_I a$, $b_1 \sim_I b$ and $a_1 \perp b_1$. Similarly, there exist $b_2, c_2 \in P$ such that $b_2 \sim_I b$, $c_2 \sim_I c$, $b_2 \perp c_2$ and $c_3, a_3 \in P$ such that $c_3 \sim_I c$, $a_3 \sim_I a$ and $c_3 \perp a_3$. Since $\sim_I$ is a congruence, we have by (C3) that there exist $c_4, c_5 \in P$ such that $c_4 \sim_I c_2$, $c_4 \perp b_1$ and $c_5 \sim_I c_3$, $c_5 \perp a_1$. Moreover, since $c_4 \sim_I c_5$, there exist $i, j \in I$ such that $c_4 \ominus i = c_5 \ominus j$. Now $c_4 \ominus i \leq c_4 \leq b_1'$, $c_5 \ominus j \leq c_5 \leq a_1'$ and $a_1 \perp b_1$. Since P is an OMP, we have $c_4 \ominus i \perp a_1 \oplus b_1$. Since $c_4 \ominus i \sim_I c$ we conclude that $[c] \perp [a] \oplus [b]$. □

Theorem 3.1.29 *A subset I of an OML P is a Riesz ideal if and only if I is a p-ideal.*

Proof. Assume that I is a Riesz ideal, $a \in I$ and $b \in P$. Then $a \leq b \vee b'$ so there exist $i, j \in I$ such that $i \leq b$, $j \leq b'$ and $a \leq i \vee j$. Hence, $a \vee b' \leq i \vee j \vee b' = i \vee b'$ so $(a \vee b') \wedge b \leq (i \vee b') \wedge b$. But i, b, b' are mutually compatible, so by distributivity we have

$$(a \vee b') \wedge b \leq (i \wedge b) \vee (b' \wedge b) = i \wedge b \leq i.$$

Hence, $(a \vee b') \wedge b \in I$. To show that I is a lattice ideal, assume that $a, b \in I$. Since $b \in I$, by what we have just shown, $(b \vee a) \wedge a' \in I$. Moreover, a, $a \vee b$ and a' are mutually compatible so by orthomodularity we have $a \vee b = a \vee [(a \vee b) \wedge a']$. Since $a \perp (a \vee b) \wedge a'$ we conclude that $a \vee b \in I$. Hence, I is a p-ideal.

Conversely, assume that I is a p-ideal, $i \in I$, $a \perp b$ and $i \leq a \vee b$. Then $(i \vee a') \wedge a, (i \vee b') \wedge b \in I$. Since b is compatible with $i \vee b'$, $(i \vee a') \wedge a$, a and $i \vee a'$ we can apply the Foulis-Holland theorem [Kal] twice to obtain

$$\begin{aligned}
[(i \vee a') \wedge a] \vee [(i \vee b') \wedge b] &= \{[(i \vee a') \wedge a] \vee (i \vee b')\} \wedge \{[(i \vee a') \wedge a] \vee b\} \\
&= (i \vee b') \wedge [(i \vee a') \vee b] \wedge (a \vee b) \\
&= (i \vee b') \wedge (i \vee a') \wedge (a \vee b) \geq i.
\end{aligned}$$

It follows that I is a Riesz ideal. □

Proposition 3.1.30 *If I is an ideal in a lattice ordered effect algebra E, then $a \sim_I b$ if and only if $a \ominus (a \wedge b) \in I$ and $b \ominus (a \wedge b) \in I$. If I is also a lattice ideal, that is, $x, y \in I$ entails $x \vee y \in I$, then $a \sim_I b$ if and only if $(a \vee b) \ominus (a \wedge b) \in I$.*

Proof. Put $i := a \ominus (a \wedge b)$, $j := b \ominus (a \wedge b)$. Then $a \ominus i = a \wedge b = b \ominus j$, hence if $i, j \in I$, then $a \sim_I b$. Conversely, if there exist $u, v \in I$ such that $a \ominus u = b \ominus v$ then $a \ominus u \leq a, b$ and we have $a \ominus u \leq a \wedge b$, whence $a \ominus (a \wedge b) \leq u$. Similarly, $b \ominus (a \wedge b) \leq v$. Consequently, $a \ominus (a \wedge b), b \ominus (a \wedge b) \in I$. By Proposition 1.8.2 (2), $(a \ominus (a \wedge b)) \vee (b \ominus (a \wedge b)) = (a \vee b) \ominus (a \wedge b)$, so we have $(a \vee b) \ominus (a \wedge b) \in I$ whenever I is a lattice ideal. □

If I is a p-ideal in an OML P, the usual congruence definition is $a \approx_I b$ if $a \ominus a \wedge b, b \ominus a \wedge b \in I$ [Kal], [Che 2]. Since I is also a Riesz ideal, we have the congruence $a \sim_I b$.

Theorem 3.1.31 *Let I be a Riesz ideal in an OML P.*

(i) $\sim_I = \approx_I$.

(ii) *P/I is an OML.*

Proof. (i) This follows by Proposition 3.1.30.

(ii) It is well known that if I is a p-ideal in P, then $P/\approx_I$ is an OML. The result now follows from (i) and Theorem 3.1.29. □

Recall that an ideal in an MV-algebra $\mathcal{M} = (M; \oplus, ^*, 0, 1)$ is defined as a subset I of M such that $a, b \in I$ implies $a \oplus b \in I$, and $a \in I$, $b \leq a$ implies $b \in I$. If $\mathcal{M}$ happens to be a Boolean algebra, then $a \oplus b = a \vee b$, and so I is a lattice ideal. Recall also that a congruence with respect to an ideal is defined by $x \approx_I y$ iff $(x^* \odot y) \oplus (y^* \odot x) \in I$, where $x \odot y = (x^* \oplus y^*)^*$. Every MV-algebra becomes an effect algebra, if we restrict the total $\oplus$ operation to orthogonal pairs. Recall that for any $a, b \in M$, $a \oplus b = a \oplus a^* \wedge b$ (Theorem 2.2.4(x)). To discern an MV-algebra ideal from an effect algebra ideal, we call the former an **MV-ideal**, and the latter an **e-ideal**.

Theorem 3.1.32 (i) *A subset I of an MV-algebra M is an MV-ideal if and only if it is an e-ideal in the corresponding effect algebra.* (ii) *For every ideal I, $\approx_I = \sim_I$.* (iii) *Every ideal is a Riesz e-ideal.*

Proof. (i) Clearly, every MV-ideal is an e-ideal. Let I be an e-ideal. If $a, b \in I$, then $a \oplus b = a \oplus (a^* \wedge b) \in I$, because $a^* \wedge b \leq b \in I$ and $a \perp a^* \wedge b$. Hence I is closed under $\oplus$, so it is an MV-ideal.

(ii) Observe that $a \odot b^* = a \ominus (a \wedge b)$ and apply Proposition 3.1.30.

(iii) Owing to the Riesz property in MV-algebras, every e-ideal is a Riesz e-ideal. □

In what follows, we show an interesting example of a congruence on a standard effect algebra $\mathcal{E}(H)$ which is not generated by an ideal. We need the following lemma.

Lemma 3.1.33 *Let ω be a state on $\mathcal{E}(H)$.* (i) *For every $A \in \mathcal{E}(H)$ and $\alpha \in [0,1] \subseteq \mathbb{R}$, we have $\omega(\alpha A) = \alpha\omega(A)$.* (ii) *The morphism ω is strong.*

Proof. (i) Let $\mathbb{Q}^+[0,1]$ be the set of rational numbers in $[0,1]$. If $n \in \mathbb{Q}^+$, we have

$$\omega(A) = \omega(\tfrac{1}{n}A \oplus \cdots \oplus \tfrac{1}{n}A) = n\omega(\tfrac{1}{n}A), \qquad (n \text{ summands})$$

so $\omega(\frac{1}{n}A) = \frac{1}{n}\omega(A)$. If $m, n \in \mathbb{Q}^+$, $m \leq n$, then

$$\omega(\tfrac{m}{n}A) = \omega(\tfrac{1}{n}A \oplus \cdots \oplus \tfrac{1}{n}A) = m\omega(\tfrac{1}{n}A) = \tfrac{m}{n}\omega(A) \qquad (m \text{ summands}).$$

Hence, $\omega(rA) = r\omega(A)$ for every $r \in \mathbb{Q}^+[0,1]$. If $\alpha \in [0,1]$, then

$$\omega(\alpha A) \leq \inf\{\omega(rA) \colon r \in R, r \geq \alpha\} = \omega(A) \inf\{r \in R : r \geq \alpha\} = \alpha\omega(A)$$

Similarly, $\omega(\alpha A) \geq \alpha\omega(A)$ so the result holds.

(ii) To show that ω is strong, assume that $\omega(A) \perp \omega(B)$. If $\omega(A) = 1$, let $C = O$. Then $C \perp A$ and $\omega(C) = 0 = \omega(B)$. If $\omega(A) \neq 1$, let $C = \alpha A'$ where $\alpha = \omega(B)/[1 - \omega(A)]$. Notice that $\alpha \in [0,1]$, and $C \leq A'$ so $C \perp A$. Moreover, by (i) we have

$$\omega(C) = \omega(\alpha A') = \alpha\omega(A') = \alpha[1 - \omega(A)] = \omega(B).$$

□

Example 3.1.34 If ω is a state on $\mathcal{E}(H)$, then by Lemma 3.1.33 (ii) and Theorem 3.1.8 (i), $\sim_\omega$ is a congruence on $\mathcal{E}(H)$. Since $\omega(I) = 1 \neq 0 = \omega(O)$, we have that $\sim_\omega \neq \mathcal{E}(H) \times \mathcal{E}(H)$. Applying Theorem 3.1.3 (iii), we conclude that $\mathcal{E}(H)/\sim_\omega$ is an effect algebra. Define $\hat{\omega} : \mathcal{E}(H)/\sim_\omega \to [0,1]$ by $\hat{\omega}([A]) = \omega(A)$. Then $\hat{\omega}$ is well defined and $\hat{\omega}([I]) = \omega(I) = 1$. Also, if $[A] \perp [B]$ then there exist $A_1, B_1 \in \mathcal{E}(H)$ such that $A_1 \sim_\omega A$, $B_1 \sim_\omega B$, $A_1 \perp B_1$ and $[A] \oplus [B] = [A_1 \oplus B_1]$. Hence,

$$\hat{\omega}([A] \oplus [B]) = \hat{\omega}([A_1 \oplus B_1]) = \omega(A_1 \oplus B_1) = \omega(A_1) + \omega(B_1) = \hat{\omega}([A]) + \hat{\omega}([B]).$$

We conclude that $\hat{\omega}$ is a state on $\mathcal{E}(H)/\sim_\omega$. If $[A] \neq [O]$, then $\omega(A) \neq 0$ so $\hat{\omega}([A]) \neq 0$. Hence, $\hat{\omega}$ is positive on $\mathcal{E}(H)/\sim_\omega$.

Example 3.1.35 We showed in Lemma 3.1.33 (ii) that a state on $\mathcal{E}(H)$ is a strong morphism. We now show that this result need not hold for a state on $\mathcal{P}(H)$. Suppose $\dim H = 2$, let $\psi \in H$ with $\|\psi\| = 1$ and define the state ω on $\mathcal{P}(H)$ by $\omega(A) = \langle A\psi, \psi \rangle$. We now prove that $\sim_\omega$ is not a congruence on $\mathcal{P}(H)$ by showing that (C3) does not hold. It is clear that there exists a $B \in \mathcal{P}(H)$ such that $0 < \omega(B) < 1$. Letting $A = O$, we have $A \perp B$. Now let C be a one-dimensional projection satisfying $\omega(C) = 0$ (that is, $C\psi = 0$). Then $C \sim_\omega A$. If $D \perp C$, then $D = 0$ or $D = C'$. Hence, $\omega(D) = 0$ or $\omega(D) = 1$ so $\omega(D) \neq \omega(B)$ and $D \not\sim_\omega B$. Since $\sim_\omega$ is not a congruence, it follows from Theorem 3.1.8 (i) that ω is not a strong morphism on $\mathcal{P}(H)$.

Example 3.1.36 This example shows that a congruence in our sense on an OML need not be a lattice congruence. However, if I is a Riesz ideal on an OML, then by Theorem 3.1.31, $\sim_I$ is a lattice congruence. Also, if $\sim$ is a lattice congruence on an OML, then it is generated by a Riesz ideal so $\sim$ is a congruence. We conclude that our congruence is more general than a lattice congruence. Let H be a two-dimensional Hilbert space and consider the OML $\mathcal{P}(H)$. Let $P_1, P_2 \in \mathcal{P}(H)$ be one-dimensional projections with $P_1 \perp P_2$ and let W be the density operator $W = \frac{1}{2}P_1 + \frac{1}{2}P_2$. Then $\omega(A) = tr(WA)$ is a positive state on $\mathcal{P}(H)$. Defining $A \sim B$ iff $\omega(A) = \omega(B)$, it is clear that $\sim$ is a weak congruence on $\mathcal{P}(H)$. To show that (C3) holds, assume that $A \perp B$ and $C \sim A$. If A or B is 0 or 1, then we are finished. Otherwise, A and B are one-dimensional projections so $A \oplus B = 1$. Now $C' \perp C$ and

$$\omega(C') = 1 - \omega(C) = 1 - \omega(A) = \omega(B),$$

so $C' \sim B$. We conclude that $\sim$ is a congruence. Now $P_1 \sim P_2$, $P_1 \sim P_1$ but

$$P_1 \vee P_1 = P_1 \not\sim 1 = P_1 \vee P_2.$$

Hence, $\sim$ is not a lattice congruence. The zero class of $\sim$ is the Riesz ideal $I = \{O\}$ and $A \sim_I B$ iff $A = B$. Notice that $\sim$ is not generated by I.

In what follows, we will see many examples similar to the previous one (Section 3.3).

Example 3.1.37 Let M be a subset of a CPAM P. We say that $a, b \in P$ are **perspective with respect to** M and write $a \sim^M b$ if there exists a $c \in P$ such that $a \oplus c \in M$ and $b \oplus c \in M$. That is, $a \sim^M b$ if they share a **common relative complement with respect to** M (compare with the perspectivity in test spaces). We say that $M \subseteq P$ is **weakly algebraic** if $\sim^M$ is a congruence. And we say that M is **algebraic** if $\sim^M$ is a **faithful congruence**, that is, a binary relation satisfying (C1), (C2) and the following property

(C0) $\qquad a \sim a_1,\ a \perp b$ imply $a_1 \perp b$.

Clearly, a faithful congruence is a congruence. From the theory of D-test spaces (see Chapter 4) it follows that every effect algebra/D-poset is a quotient of a positive CPAM by the perspectivity with respect to an algebraic set (cf. [Wil 1], [Wil 2], [PuWi]).

In the next theorem, we show that weakly algebraic subsets of an effect algebra are in a one-to-one correspondence with Riesz ideals ([GuPu]).

For $M \subseteq P$, define $M' = \{a' : a \in M\}$.

Theorem 3.1.38 *Let P be an effect algebra.* (i) *If $I \subseteq P$ is a Riesz ideal, then $\sim_I = \sim^{I'}$.* (ii) *$M \subseteq P$ is weakly algebraic if and only if M' is a Riesz ideal.*

Proof. (i) If $a \sim_I b$, then there exist $i, j \in I$ such that $a \ominus i = b \ominus j$. Since $a' \oplus (a \ominus i) = i'$, we have $a \ominus i = i' \ominus a'$ so $i' \ominus a' = j' \ominus b'$. Hence, $a' \oplus (j' \ominus b') = i' \in I'$ and $b' \oplus (j' \ominus b') = j' \in I'$ so $a' \sim^{I'} b'$. Since $\sim_I$ is a congruence, we have $a \sim_I b$ iff $a' \sim_I b'$ so $a \sim^{I'} b$. Conversely, if $a \sim^{I'} b$, then there exists a $c \in P$ such that $a \oplus c = i'$, $b \oplus c = j'$ where $i, j \in I$. We then have $a = i' \ominus c = (i \oplus c)'$ and $b = j' \ominus c = (j \oplus c)'$. Hence, $a' \ominus i = c = b' \ominus j$ so $a' \sim_I b'$. We conclude that $a \sim_I b$ so $\sim_I = \sim^{I'}$.

(ii) If M' is a Riesz ideal, then $\sim_{M'}$ is a congruence on P. Applying (i), we have that $\sim^M$ is a congruence so M is weakly algebraic. Conversely, assume that $\sim^M$ is a congruence. We shall show that $I = \{a \in P\colon\ a \sim^M 0\}$ is a Riesz ideal. Since $\sim^M$ is a congruence, we get by reflexivity that for every $a \in P$ there is a $c \in P$ with $a \oplus c \in M$. It then follows that $1 \in M$, and $a \in M$ iff $a \sim^M 1$. Since $\sim^M$ is a congruence, $a \sim^M b$ implies $a' \sim^M b'$. If $a \in M$, then $a \sim^M 1$, hence $a' \in I$. If $a \sim^M 0$, then $a' \sim^M 1$, hence $a' \in M$. This shows that $I = M'$. If $a \in I$ and $b \leq a$, then $a' \leq b'$ and $a' \in M$. There is a v with $a' \oplus v = b'$. Now we have $a' \sim^M b \oplus b' = b \oplus a' \oplus v$. By (C4), $b \sim^M 0$, hence $b \in I$. Let $a, b \in I$, $a \perp b$. Then $[a \oplus b] = [a] \oplus [b] = [0]$ so $a \oplus b \in I$. This proves that I is an ideal. By Theorem 3.1.22, I is a Riesz ideal. □

3.2 Some Lattices of Ideals in Positive CPAMs

We have already observed that the set $\mathcal{I}(P)$ of all ideals of a CPAM P forms a complete lattice with respect to the set-theoretical ordering. Clearly, in a positive CPAM P, $\{0\}$ is the smallest and P is the greatest element of $\mathcal{I}(P)$. We will prove that R_1-ideals and Riesz ideals form distributive sublattices of $\mathcal{I}(P)$. Moreover, R_1-ideals are standard elements of $\mathcal{I}(P)$. We also show that the quotient with respect to an R_1-ideal is a positive CPAM. Moreover, the weak congruence with respect to an R_1-ideal I always satisfies property (C5), while property (C3) is satisfied if and only if I is a Riesz ideal.

3.2.1 R_1-ideals and Riesz Ideals

In this subsection, the symbol P is reserved for a positive CPAM. We will mainly follow the ideas on [ChPu].

Recall that a binary relation on P is called a weak congruence if it satisfies conditions (C1) and (C2) of Definition 3.1.2. We will say that an ideal I of P is a **weak** *c*-**ideal** if the relation $\sim_I$ is a weak congruence. We will say that an ideal I is a *c*-**ideal** if $\sim_I$ is a congruence. By Theorem 3.1.16, any Riesz ideal is a *c*-ideal.

In the following proposition, we give a characterization of weak *c*-ideals (cf. [Jen, Lemma 2]).

Proposition 3.2.1 *Let* $I \in \mathcal{I}(P)$. (i) $\sim_I$ *satisfies* (C2).

(ii) *Denote by* $[0, x]$ *the principal order ideal* $\{y \in P : y \leq x\}$. *I is a weak c-ideal if and only if, for every* $x \in P$, $I_x = ([0, x] \cap I; \leq)$ *is an upwards directed poset.*

Proof. (i) Assume $x_1 \sim x_2$, $y_1 \sim y_2$ and $x_1 \perp y_1, x_2 \perp y_2$. Then there are $i, j, k, l \in I$ such that $x_1 \ominus i = x_2 \ominus j$, $y_1 \ominus k = y_2 \ominus l$. It follows

$$\begin{aligned}(x_1 \oplus y_1) \ominus (i \oplus k) &= (x_1 \ominus i) \oplus (y_1 \ominus k) \\ &= (x_2 \ominus j) \oplus (y_2 \ominus l) = (x_2 \oplus y_2) \ominus (j \oplus l),\end{aligned}$$

hence $x_1 \oplus y_1 \sim_I x_2 \oplus y_2$.

(ii) Assume I is a weak *c*-ideal. Let $i_1, i_2 \in I$ be such that $i_1 \leq x$ and $i_2 \leq x$. There are $x_1, x_2 \in P$ such that $x = x_1 \oplus i_1 = x_2 \oplus i_2$. Then $(x_1 \oplus i_1) \ominus i_2 = x_2$ implies $x_1 \oplus i_1 \sim_I x_2$, and $(x_1 \oplus i_1) \ominus i_1 = x_1 \ominus 0$ implies $x_1 \sim_I x_1 \oplus i_1$, hence by (C1), $x_1 \sim_I x_2$. There are $j_1, j_2 \in I$ such that $x_1 \ominus j_1 = x_2 \ominus j_2$.

$$\begin{aligned}x &= (x_1 \ominus j_1) \oplus j_1 \oplus i_1 \\ &= (x_2 \ominus j_2) \oplus j_2 \oplus i_2.\end{aligned}$$

Therefore, $i = j_1 \oplus i_1 = j_2 \oplus i_2 \leq x$ is an upper bound of i_1 and i_2 in $([0, x] \cap I; \leq)$.

On the other hand, assume that for every $x \in P$, I_x is upwards directed. Let a, b, c be such that $a \sim_I b$ and $b \sim_I c$. There are $i_1, i_2, j_1, j_2 \in I$ such that $a \ominus i_1 = b \ominus i_2$ and $b \ominus j_1 = c \ominus j_2$. Since $j_1, j_2 \in I_b$ and I_b is upwards directed, there are $k_1, k_2 \in I$ such that $j_1 \oplus k_1 = i_2 \oplus k_2 \leq b$. So we get

$$\begin{aligned} b \ominus (i_2 \oplus k_2) &= b \ominus (j_1 \oplus k_1), \\ (b \ominus i_2) \ominus k_2 &= (b \ominus j_1) \ominus k_1, \\ (a \ominus i_1) \ominus k_2 &= (c \ominus j_2) \ominus k_1, \\ a \ominus (i_1 \oplus k_2) &= c \ominus (j_2 \oplus k_1). \end{aligned}$$

Thus, $\sim_I$ is transitive. □

Corollary 3.2.2 (i) *In an effect algebra, every weak c-ideal is upwards directed.*

(ii) *Let P be an effect algebra such that $(P; \leq)$ is a lattice. Assume I is an ideal of P. Then I is a weak c-ideal if and only if I is a lattice ideal of the lattice $(P; \leq)$.*

Proof. (i) Trivial. (ii) Let I be a weak c-ideal of P. Suppose $a, b \in I$. There is an upper bound of $\{a, b\}$ in I, say c. Since I is an order ideal, $a \vee b \leq c$ implies that $a \vee b \in I$.

Assume $I \in \mathcal{I}(P)$ is a lattice ideal of $(P; \leq)$. Let $a \in P$, $a \geq i_1, i_2 \in I$. Since I is a lattice ideal, $i = i_1 \vee i_2 \in I$. Evidently, $a \geq i \geq i_1, i_2$. □

Corollary 3.2.3 *Let I_1, I_2 be weak c-ideals of P. Then $I_1 \cap I_2$ is a weak c-ideal.*

Proof. Assume $i, j \in I_1 \cap I_2$, $i, j \leq a \in P$. It suffices to prove that there is an upper bound of $\{i, j\}$ in $[0, a] \cap I_1 \cap I_2$.

Since I_1 is a weak c-ideal, there is an upper bound of $\{i, j\}$ in $[0, a] \cap I_1$, say k. Since I_2 is a weak c-ideal, there is an upper bound of $\{i, j\}$ in $[0, k] \cap I_2$, say l. Evidently, $l \leq k \in I_1$ implies that $l \in I_1 \cap I_2$. Moreover, $l \leq a$ so l is an upper bound of $\{i, j\}$ in $[0, a] \cap I_1 \cap I_2$. □

Recall that an ideal I of P is an R_1-ideal if it satisfies condition (R1) in Definition 3.1.11.

Let us denote by $\mathcal{R}_1\mathcal{I}(P)$ the set, ordered by set-inclusion, of all R_1-ideals of the CPAM P. In a positive CPAM, the ideals $\{0\}$ and P are R_1-ideals.

The following proposition shows that there exists a simple form for the join of two elements of $\mathcal{I}(P)$ if one of them is an R_1-ideal.

Proposition 3.2.4 *For any ideal J and any R_1-ideal I of P,*

$$I \vee J = \{i \oplus j : i \in I, j \in J\}$$

(in the lattice $\mathcal{I}(P)$).

If J is an R_1-ideal, then so are $I \vee J$ and $I \cap J$ and $\mathcal{R}_1\mathcal{I}(P)$ is a sublattice of $\mathcal{I}(P)$.

Proof. Let $I \bigoplus J = \{i \oplus j : i \in I, j \in J\}$. Clearly, I and J are subsets of $I \bigoplus J$. We show first that $I \bigoplus J \in \mathcal{I}(P)$. For this end, let $a, b \in I \bigoplus J$ with $a \perp b$. Writing $a = i_1 \oplus j_1$, $b = i_2 \oplus j_2$ with $i_k \in I_k$ and $j_k \in J_k$, $k = 1, 2$, we have

$$a \oplus b = (i_1 \oplus j_1) \oplus (i_2 \oplus j_2) = (i_1 \oplus i_2) \oplus (j_1 \oplus j_2) \in I \bigoplus J.$$

Conversely, assume that $a \oplus b \in I \bigoplus J$. There exist $i \in I$, $j \in J$ such that $a \oplus b = i \oplus j$. As I is an R_1-ideal, there exist $i_1, i_2 \in I$ satisfying $i_1 \leq a$, $i_2 \leq b$ and $i \leq i_1 \oplus i_2$. Write $a = i_1 \oplus a_1$, $b = i_2 \oplus b_2$, we have

$$a \oplus b = (a_1 \oplus b_2) \oplus (i_1 \oplus i_2) = (a_1 \oplus b_2) \oplus (i \oplus i_3)$$

since $i \leq i_1 \oplus i_2$. As $(a_1 \oplus b_2) \oplus (i \oplus i_3) = i \oplus j$, $j = a_1 \oplus b_2 \oplus i_3$ holds. Since J is an ideal, $a_1 \oplus b_2 \in J$ and so $a_1 \in J$ and $b_2 \in J$. Finally, $a \in I \bigoplus J$ and $b \in I \bigoplus J$. The proof of $I \bigoplus J \in \mathcal{I}(P)$ is complete.

Now consider an ideal K such that $I \subset K$ and $J \subset K$. Let $x = i \oplus j \in I \bigoplus J$. We have $i \in I$ and so $i \in K$. Similarly, $j \in K$ and $x = i \oplus j \in K$. Thus, $I \bigoplus J \subset K$ and so

$$I \vee J = I \bigoplus J.$$

Assuming now that J is an R_1-ideal, we will prove that $I \cap J$ and $I \bigoplus J$ are also R_1-ideals.

Let $x \in I \cap J$ and assume that $x \leq a \oplus b$. As $x \in I$, there exist $i_1, i_2 \in I$ such that $i_1 \leq a$, $i_2 \leq b$ and $x \leq i_1 \oplus i_2$. By using $x \in J$ and $x \leq i_1 \oplus i_2$ there exist $j_1, j_2 \in J$ with $j_1 \leq i_1$, $j_2 \leq i_2$ and $x \leq j_1 \oplus j_2$. It follows that $j_1, j_2 \in I$ and so $j_1, j_2 \in I \cap J$. Thus $I \cap J \in \mathcal{R}_1\mathcal{I}(P)$.

Now if $x \in I \bigoplus J$ and $x \leq a \oplus b$, write $x = i \oplus j$ with $i \in I$ and $j \in J$. We have $i \leq a \oplus b$ and $i \in I$. Therefore, there exist $i_1, i_2 \in I$ satisfying $i \leq i_1 \oplus i_2$, $i_1 \leq a$ and $i_2 \leq b$. We have :

$$\begin{aligned} j \leq (a \oplus b) \ominus i &= [(a \oplus b) \ominus (i_1 \oplus i_2)] \oplus [(i_1 \oplus i_2) \ominus i] \\ &= [(a \ominus i_1) \oplus (b \ominus i_2)] \oplus [(i_1 \oplus i_2) \ominus i]. \end{aligned}$$

Since $j \in J$, there exist in J, $j_1 \leq a \ominus i_1$, $j_2 \leq b \ominus i_2$, $j_3 \leq (i_1 \oplus i_2) \ominus i$ with $j \leq j_1 \oplus j_2 \oplus j_3$. As $i \oplus j_3 \leq i_1 \oplus i_2$, we have

$$i \oplus j \leq j_1 \oplus j_2 \oplus i_1 \oplus i_2 = (i_1 \oplus j_1) \oplus (i_2 \oplus j_2).$$

We have $i_1 \oplus j_1 \in I \bigoplus J$, $i_2 \oplus j_2 \in I \oplus J$ and, as $j_1 \leq a \ominus i_1$ and $j_2 \leq b \ominus i_2$, $i_1 \oplus j_1 \leq a$ and $i_2 \oplus j_2 \leq b$ hold. Therefore, $I \bigoplus J$ is an R_1-ideal. □

Remarks. (1) For two R_1-ideals of a CPAM P,

$$I \vee J = \{r \in P : r \leq i \oplus j,\ i \in I,\ j \in J\}.$$

Proof. Let $I \triangle J = \{r \in P : r \leq i \oplus j,\ i \in I,\ j \in J\}$. Clearly, $I \vee J \subset I \triangle J$. Conversely, if $r \in I \triangle J$ then $r \leq i \oplus j$ with $i \in I$, $j \in J$. Since $i \oplus j \in I \vee J$ and $I \vee J$ is an ideal, it follows that $r \in I \vee J$.

(2) For any finite sequence $I_1, \ldots, I_n$ of R_1-ideals

$$I_1 \vee \cdots \vee I_n = \{i_1 \oplus \cdots \oplus i_n : i_k \in I_k\},$$

$$I_1 \wedge \cdots \wedge I_n = \bigcap_{k=1}^{n} I_k.$$

Theorem 3.2.5 *Let P be a positive CPAM. Ordered by the set-inclusion, $\mathcal{R}_1\mathcal{I}(P)$ is a complete lattice. Moreover, it is a distributive lattice and, for any family $\{I_k\}$ of R_1-ideals and for any ideal J,*

$$\bigvee I_k = \{i_{k_1} \oplus \cdots \oplus i_{k_p} : i_{k_j} \in I_{k_j}\},$$

$$J \wedge (\bigvee I_k) = \bigvee (J \wedge I_k).$$

Proof. Let $\bigoplus I_k = \{i_{k_1} \oplus \cdots \oplus i_{k_p} : i_{k_j} \in I_{k_j}\}$. Notice that $a \in \bigoplus I_k$ if and only if there exist $k_1, \ldots, k_p$ such that $a \in \bigvee_{m=1}^{p} I_{k_m}$.

Let $a, b \in \bigoplus I_k$ with $a \perp b$. We have $a \in I_{k_1} \vee \cdots \vee I_{k_j}$ and $b \in I_{i_1} \vee \cdots \vee I_{i_m}$ and so

$$a,\ b \in [I_{k_1} \vee \cdots \vee I_{k_j}] \vee [I_{i_1} \vee \cdots \vee I_{i_m}]$$

and

$$a \oplus b \in [I_{k_1} \vee \cdots \vee I_{k_j}] \vee [I_{i_1} \vee \cdots \vee I_{i_m}].$$

Therefore, $a \oplus b \in \bigoplus I_k$. Conversely if $a \oplus b \in \bigoplus I_k$, then $a \oplus b \in I_{k_1} \vee \cdots \vee I_{k_m}$ for some $k_1, \ldots, k_m$. Therefore, $a, b \in I_{k_1} \vee \cdots \vee I_{k_m}$ and $a, b \in \bigoplus I_k$. We have proved that $\bigoplus I_k$ is an ideal.

Let $i \in \bigoplus I_k$ with $i \leq a \oplus b$. We have $i \in I_{k_1} \vee \cdots \vee I_{k_m}$ and there exist $i_1, i_2 \in I_{k_1} \vee \cdots \vee I_{k_m}$ such that $i \leq i_1 \oplus i_2$, $i_1 \leq a$ and $i_2 \leq b$. Since $i_1 \in \bigoplus I_k$ and $i_2 \in \bigoplus I_k$, $\bigoplus I_k$ is an R_1-ideal.

Clearly, for any k, $I_k \subset \bigoplus I_k$. Let H be an ideal such that, for any k, $I_k \subset H$. If $i \in \bigoplus I_k$ write $i = i_{k_1} \oplus \cdots \oplus i_{k_m}$ with $i_{k_p} \in I_{k_p}$. As $I_{k_p} \subset H$, $i_{k_p} \in H$ and therefore $i \in H$. The set $\bigoplus I_k$ is the supremum of the family (I_k) in the ordered set $\mathcal{I}(P)$

and, a fortiori, in its subset $\mathcal{R}_1\mathcal{I}(P)$. As the smallest ideal $I_0 = \{0\}$ is an R_1- ideal, $\mathcal{R}_1\mathcal{I}(P)$ is a complete lattice and the supremum of a family of R_1-ideals is the same whether calculated in $\mathcal{I}(P)$ or in $\mathcal{R}_1\mathcal{I}(P)$.

Let J be an ideal and (I_k) be a family of R_1-ideals. We have $\bigvee(J \wedge I_k) \subset J \wedge (\bigvee I_k)$. Let $x \in J \wedge (\bigvee I_k)$. We have $x \in J$ and write $x = i_{k_1} \oplus \cdots \oplus i_{k_m}$ with $i_{k_p} \in I_{k_p}$. As $i_{k_p} \leq x$, $i_{k_p} \in J$ and therefore $i_{k_p} \in J \wedge I_{k_p}$. This follows that $x \in (J \wedge I_{k_1}) \vee \cdots \vee (J \wedge I_{k_p})$ and thus $x \in \bigvee(J \wedge I_k)$. In particular, if I, J and K are three R_1-ideals, $I \wedge (J \vee K) = (I \wedge J) \vee (I \wedge K)$ and $\mathcal{R}_1\mathcal{I}(P)$ is a distributive lattice. □

Remarks. (1) In general, it is not true that $J \vee (\bigwedge I_k) = \bigwedge(J \vee I_k)$. There exists a counterexample in Boolean algebras. (Obviously, Boolean algebras are CPAMs in which any ideal is an R_1-ideal.) Let J be a non principal maximal ideal in an infinite Boolean algebra B. Consider the family of ideals, $\{[0, x^\perp] : x \in J\}$. As J is a non principal maximal ideal, we have $\bigcap_{x \in J} [0, x^\perp] = \{0\}$ and, for any $x \in J$, $J \vee [0, x^\perp] = B$. Thus

$$J = J \vee (\bigwedge_{x \in J} [0, x^\perp]) \neq \bigwedge_{x \in J} (J \vee [0, x^\perp]) = B.$$

(2) By [Bir, Thm 24, p. 128] and the previous proposition, $\mathcal{R}_1\mathcal{I}(P)$ is a complete Brouwerian lattice. In particular, for any given R_1-ideals I and J the set of all R_1-ideals K such that $I \wedge K \subset J$ contains a greatest element $I : J$, the relative pseudo-complement of I in J. The lattice $\mathcal{R}_1\mathcal{I}(P)$ shares this property with the congruence lattice of a lattice. □

Recall, that in a lattice L, an element s is said to be **standard** if, for any x, $y \in L$, $(x \vee s) \wedge y = (x \wedge y) \vee (s \wedge y)$. An element d is called a **distributive** element if, for any x, $y \in L$, $(x \wedge y) \vee d = (x \vee d) \wedge (y \vee d)$. It is easy to see that any standard element is distributive. We note that these concepts are interesting for the study of lattices of congruences of a lattice. For example, kernels of congruences in a section complemented lattice L can be characterized as the standard or the distributive elements of the lattice of all ideals of L.

Theorem 3.2.6 *Any R_1-ideal of P is a standard element of the lattice $\mathcal{I}(P)$.*

Proof. Assume that G and H are two ideals of P and let I be an R_1-ideal. It will suffice to prove that $(G \vee I) \wedge H \subset (G \wedge H) \vee (I \wedge H)$. Let $x \in (G \vee I) \wedge H = (G \bigoplus I) \cap H$. We have $x \in H$ and write $x = g \oplus i$ with $g \in G$ and $i \in I$. As $i \leq x$, $i \in H$, therefore $i \in I \cap H$. Similarly, $g \in G \cap H$ and thus

$$x = g \oplus i \in (G \cap H) \vee (I \cap H).$$

□

According to Definition 3.1.11, an R_1-ideal of a CPAM is a Riesz ideal if it satisfies

If $i \in I$, $i \leq a$ and $a \ominus i \perp b$, then there exists $j \in I$ such that
$j \leq b$ and $a \perp b \ominus j$.

Denote by $\mathcal{RI}(P)$ the set of all Riesz ideals of P.

Proposition 3.2.7 *If I and J are two Riesz ideals of a positive CPAM P, then $I \bigoplus J = \{i \oplus j : i \in I, j \in J\}$ is an Riesz ideal.*

Proof. By Proposition 3.2.4, $I \bigoplus J$ is an R_1-ideal. Let $x \in I \bigoplus J$ and write $x = i \oplus j$, $i \in I$, $j \in J$. Assume $i \oplus j \leq a$ and $a \ominus (i \oplus j) \perp b$. Note that $a \ominus (i \oplus j) = (a \ominus i) \ominus j$. As $j \in J$, there exists $j_1 \in J$ such that $j_1 \leq b$ and $a \ominus i \perp b \ominus j_1$. Since $i \in I$, there exists $i_1 \in I$, $i_1 \leq b \ominus j_1$, such that $a \perp (b \ominus j_1) \ominus i_1 = b \ominus (i_1 \oplus j_1)$. We have $i_1 \oplus j_1 \in I \bigoplus J$ and therefore $I \vee J = I \bigoplus J$ is a Riesz ideal. □

The following reformulation of (R2) enables us to prove that the intersection of two Riesz ideals is a Riesz ideal. This result was obtained in [Jen], and answers an open question from [ChPu].

Proposition 3.2.8 *An ideal I of P satisfies* (R2) *if and only if the following condition holds:*

(R2a) *For all $a, b \in P$, $i \in I$ such that $a \geq b$ and $a \ominus b \perp i$, there is $j \in I$ such that $b \geq j$ and $(a \ominus j) \perp i$.*

Proof. Assume I satisfies (R2a). Let $a, b \in P$, $i \in I$ be such that $a \geq i$ and $a \ominus i \perp b$. Evidently,

$$((a \ominus i) \oplus b) \ominus b = (a \ominus i) \perp i.$$

By (R2a), there is $j \in I$ such that $b \geq j$ and

$$((a \ominus i) \oplus b) \ominus j \perp i.$$

Observe that $((a \ominus i) \oplus b) \ominus j = (a \ominus i) \oplus (b \ominus j) \perp i$. By associativity,

$$(b \ominus j) \perp (a \ominus i) \oplus i = a.$$

Thus, (R2) holds.

For the opposite direction, suppose that I satisfies (R2). Assume $a, b \in P$, $i \in I$, $a \geq b$, $a \ominus b \perp i$. Then,

$$((a \ominus b) \oplus i) \ominus i = a \ominus b \perp b.$$

By (R2), there is $j \in I$, $j \leq b$ such that $(a \ominus b) \oplus i \perp b \ominus j$. By associativity, this implies that $a \ominus b \perp b \ominus j$ and

$$(a \ominus b) \oplus (b \ominus j) = (a \ominus j) \perp i.$$

Thus, (R2a) holds. □

Proposition 3.2.9 *Assume I_1 and I_2 are ideals of P satisfying the* (R2a) *condition. Then $I_1 \cap I_2$ satisfies the* (R2a) *condition.*

Proof. Assume $a, b \in P$, $i \in I_1 \cap I_2$, $a \geq b$, $a \ominus b \perp i$. Since I_1 satisfies (R2a), $i \in I_1$ implies that there is $j_1 \in I_1$ with properties $j_1 \leq b$ and $a \ominus j_1 \perp i$. Further, $i \in I_2$ implies that there is $j \in I_2$ such that $j \leq j_1$ and $a \ominus j \perp i$. It remains to observe that $j \leq j_1 \in I_1$ implies that $j \in I_1 \cap I_2$ and that $j \leq b$. Thus, $I_1 \cap I_2$ satisfies (R2a). □

Theorem 3.2.10 [Jen] *Assume I_1, I_2 are Riesz ideals of P. Then $I_1 \cap I_2$ is a Riesz ideal of P.*

Proof. By Theorem 3.2.5, $I_1 \cap I_2$ is an R_1-ideal. By Proposition 3.2.8, I_1 and I_2 satisfy (R2a). By Proposition 3.2.9, $I_1 \cap I_2$ satisfies (R2a). By Proposition 3.2.8, $I_1 \cap I_2$ is a Riesz ideal. □

Corollary 3.2.11 *$\mathcal{RI}(P)$ is a complete distributive Browerian sublattice of $\mathcal{I}(P)$.*

Proof. By Theorem 3.2.4, I, J being Riesz ideals, $I \vee J = I \bigoplus J$ is a Riesz ideal. It then follows that the lattice of Riesz ideals ordered by $\subseteq$ is a complete lattice. Therefore, $\mathcal{RI}(P)$ is a sublattice of $\mathcal{R}_1\mathcal{I}(P)$, which is a distributive sublattice of $\mathcal{I}(P)$, hence $\mathcal{RI}(P)$ is a distributive sublattice of $\mathcal{I}(P)$. □

In the next subsection we will meet some interesting examples of weak congruences satisfying (C5). We will prove now that the quotient with respect to such congruence is a positive PAM. There are examples showing that cancellativity may fail.

Proposition 3.2.12 *Let $\sim$ be a binary relation in a positive CPAM P satisfying* (C1), (C2) *and* (C5). *Then the quotient $P/\!\sim$ is a positive PAM.*

Proof. We have to prove associativity of $\oplus$ in the quotient $P/\sim$. Assume that $([a] \oplus [b]) \oplus [c]$ is defined, $a, b, c \in P$. This means that there are $a_1, b_1, d, c_1 \in P$ such that $a_1 \sim a, b_1 \sim b, c_1 \sim c$, $d \sim a_1 \oplus b_1$ and $a_1 \perp b_1$, $c_1 \perp d$ so that

$$([a] \oplus [b]) \oplus [c] = [a_1 \oplus b_1] \oplus [c] = [d \oplus c_1].$$

But $d \sim a_1 \oplus b_1$ implies by (C5) that there are d_1, d_2, $d_1 \sim a_1$, $d_2 \sim b_2$ and $d = d_1 \oplus d_2$. Therefore

$$[d \oplus c_1] = [d_1 \oplus d_2 \oplus c_1] = [d_1 \oplus (d_2 \oplus c_1)] = [a] \oplus ([b] \oplus [c]) \text{ by(C1), (C2)}.$$

If $a \sim 0$, then $a \oplus b \sim 0 \oplus b = b$ for any $b \in P$, hence $[0]$ is the zero element in $P/\!\sim$. To prove positivity, let $[a] \oplus [b] = [0]$. Then there are $a_1 \sim a$, $b_1 \sim b$ with $a_1 \perp b_1$, and $a_1 \oplus b_1 \sim 0$. By (C5), there are $a_2 \sim a_1$, $b_2 \sim b_1$ with $0 = a_2 \oplus b_2$. By positivity of P, $a_2 = b_2 = 0$, hence by (C1) and (C2) $[a] = [b] = [0]$. □

Recall that if I is an R_1-ideal, then $a \sim_I b$ iff there exist $i, j \in I$ such that $i \leq a, j \leq b$ and $a \ominus i \leq b$ and $b \ominus j \leq a$ (Lemma 3.1.15).

Theorem 3.2.13 [ChPu], [Jen] (i) *If I is an R_1-ideal of P, then $\sim_I$ satisfies* (C1), (C2) *and* (C5). *Consequently, $P/\sim$ is a positive PAM. Moreover, $P/\sim_I$ is cancellative and $a \sim_I 0$ if and only if $a \in I$.* (ii) *An R_1-ideal I is a Riesz ideal if and only if $\sim_I$ satisfies* (C3).

Proof. (i) Let I be an R_1 ideal. To prove (C1), it suffices to prove transitivity of $\sim_I$. Assume $a \sim_I b$, $b \sim_I c$. Then there exist $i, j, k, l \in I$ such that $a \ominus i = b \ominus j$ and $b \ominus k = c \ominus l$. We may write $b = (b \ominus j) \oplus j$, and replace $b \ominus j$ by $a \ominus i$, so that

$$((a \ominus i) \oplus j) \ominus k = c \ominus l \leq c.$$

By R_1, there are $m, n \in I$ such that

$$(a \ominus (i \oplus m)) \oplus (j \ominus n) \leq c$$

and hence

$$a \ominus (i \oplus m) \leq c, i \oplus m \in I.$$

We also have $b = (c \ominus l) \oplus k$, whence

$$((c \ominus l) \oplus k) \ominus j = a \ominus i \leq a$$

and using R_1, we find $p \in I$ such that

$$c \ominus (l \oplus p) \leq a, l \oplus p \in I.$$

Owing to Lemma 3.1.15, we get $a \sim_I c$, and transitivity is proved.

(C2) follows by Proposition 3.2.1(i).

To prove (C5), assume $a \sim_I b \oplus c$. By the definition of $\sim_I$, there are $i, j \in I$ such that $a \ominus i = (b \oplus c) \ominus j$. Since $j \leq b \oplus c$, by R_1 there are $j_1, j_2 \in I$, such that $j_1 \leq b, j_2 \leq c$ and $j \leq j_1 \oplus j_2$. It follows that $(b \oplus c) \ominus j = (b \ominus j_1) \oplus (c \ominus j_2) \oplus ((j_1 \oplus j_2) \ominus j)$. Write $k := (j_1 \oplus j_2) \ominus j$. Clearly, $k \in I$. Then we obtain

$$\begin{aligned} a &= (b \ominus j_1) \oplus (c \ominus j_2) \oplus k \oplus i \\ &= [(b \ominus j_1) \oplus k] \oplus [(c \ominus j_2) \oplus i]. \end{aligned}$$

Now we have (by definition of $\sim_I$ and (C2)) $b \sim_I b \ominus j_1 \sim_I (b \ominus j_1) \oplus k$, and similarly $c \sim_I c \ominus j_2 \sim_I (c \ominus j_2) \oplus i$. Hence by (C1), $b \sim_I (b \ominus j_1) \oplus k$, $c \sim_I (c \ominus j_2) \oplus i$. Hence (C5) is satisfied.

We have $a \sim_I 0$ iff there are $j, k \in I$ such that $a \ominus j = 0 \ominus k$. Now $0 \ominus k \leq 0$ implies $0 \ominus k \in I$, hence $a = (0 \ominus k) \oplus j \in I$.

To prove cancellativity, assume $a \sim_I b$ and $a \oplus c \sim_I b \oplus d$. By (C5), $a \oplus c = u \oplus v$, $u \sim_I b$, $v \sim_I d$. By (C1), $u \sim_I a$, hence there are $i, j \in I$ such that $a \ominus i = u \ominus j$. It follows $a \oplus c = (a \ominus i) \oplus j \oplus v$, hence $v = (a \oplus c) \ominus [(a \ominus i) \oplus j] = [(a \ominus (a \ominus i)) \oplus c] \ominus j$, so that $v \oplus j = c \oplus i$. Now by (C2), $v \oplus 0 \sim_I v \oplus j = c \oplus i \sim_I c \oplus 0$, hence $v \sim_I c$. By (C1), $d \sim_I c$, which proves cancellativity.

Positivity follows by Proposition 3.2.12.

(ii) If I is a Riesz ideal, then $\sim_I$ satisfies (C3).

Assume $\sim_I$ satisfies (C3). We want to show that (R2) is satisfied. Let $i \in I$, $a, b \in P$, $a \ominus i \perp b$. Then $a \sim_I a \ominus i$ implies by (C3) that there exists $c \in P$, $c \sim_I b, c \perp a$. Hence there are $j, k \in I$ such that $c \ominus j = b \ominus k$, and $c \perp a$ implies $c \ominus j \perp a$, hence $b \ominus k \perp a$. Hence (R2) is satisfied. □

Example 3.2.14 Let $\mathbb{N}_0 = \mathbb{N} \cup \{0\}$, X a nonempty set and $\mathcal{T} \subset {\mathbb{N}_0}^X$. Recall that $(X, \mathcal{T})$ is a D-test space if the following conditions hold (see also Section 4.1).

(T1) For any $x \in X$ there exists a $t \in \mathcal{T}$ such that $t(x) \neq 0$.

(T2) If $s, t \in \mathcal{T}$ with $s \leq t$ (i.e., $s(x) \leq t(x)$ for all $x \in X$), then $s = t$.

We call $f \in {\mathbb{N}_0}^X$ an effect if $f \leq t$ for some $t \in \mathcal{T}$ and denote the set of effects by $\mathcal{E} = \mathcal{E}(X, \mathcal{T})$. The null effect is the function $f_0 \in \mathcal{E}$ that satisfies $f_0(x) = 0$ for all $x \in X$. For $f, g \in \mathcal{E}$, define $f \perp g$ if $f + g \in \mathcal{E}$ and, in this case, $f \oplus g = f + g$. Then $(\mathcal{E}; f_0, \oplus)$ becomes a positive CPAM which is not an effect algebra, in general.

Lemma 3.2.15 *If I is an ideal in $\mathcal{E}$, then I satisfies (R1) and $\sim_I$ is a weak congruence. Moreover, $\mathcal{E}/{\sim_I}$ is a positive CPAM, and $f \sim_I f_0$ if and only if $f \in I$.*

Proof. To show that I satisfies (R1), assume that $i \in I$ and $i \leq f \oplus g = f + g$. Then $i \wedge f \leq f, i \wedge g \leq g$ and $i \wedge f, i \wedge g \in I$. It remains to show that $i \leq (i \wedge f) \oplus (i \wedge g) = i \wedge f + i \wedge g$. If $i(x) \leq f(x)$, then

$$(f \wedge i)(x) + (g \wedge i)(x) = i(x) + (g \wedge i)(x) \geq i(x).$$

Similarly if $i(x) \leq g(x)$. If $i(x) > f(x)$, then since $i \leq f + g$ we have $i(x) < g(x)$ and this reduces to the previous case. The rest follows by Proposition 3.2.13. □

In what follows, we show that condition (R1) need not imply (R2). The same example also shows that (C4) and (C5) together need not imply (C3). The following simple example shows that (C3) need not imply (C5), in general. Let $\mathbb{N}_0$ be the set of all nonnegative integers with $\{0\}$ as zero and with $\oplus$ as the usual addition. Let $a \sim b$ iff $a = b \pmod 3$. Then $\sim$ satisfies (C3), but not (C5) (e.g., consider $1 \sim 2 \oplus 2$).

Let $(X, \mathcal{T})$ be a D-test space and let $Y \subseteq X$, $Y \neq \emptyset$. For $f, g \in \mathcal{E}$, define $f \sim g$ iff $f(y) = g(y)$ for all $y \in Y$. Thus $f \sim g$ iff $f|Y = g|Y$. Denoting the characteristic function of Y by χ_Y, we have $f \sim g$ iff $f\chi_Y = g\chi_Y$. Let $I = \{f \in \mathcal{E} : f(y) = 0 \text{ for all } y \in Y\}$ be the zero class for $\sim$. Then I is an ideal in $\mathcal{E}$, so Lemma 3.2.15 can be applied.

Theorem 3.2.16 *With the notation above, the following assertions hold.*

(i) $\sim = \sim_I$ *so* $\sim$ *is a weak congruence on* $\mathcal{E}$.

(ii) *In* $\mathcal{E}/\!\sim$ *we have* $[f] \perp [g]$ *if and only if* $f\chi_Y \perp g\chi_Y$.

(iii) $\mathcal{E}/\!\sim$ *is a positive CPAM.*

Proof. (i) If $f \sim_I g$, then there exist $i, j \in I$ such that $f \ominus i = g \ominus j$. It follows immediately that $f(y) = g(y)$ for all $y \in Y$ so $f \sim g$. Conversely, assume that $f \sim g$ so that $f\chi_Y = g\chi_Y$. Now $f\chi_{Y'}, g\chi_{Y'} \in I$ and $f = f\chi_Y \oplus f\chi_{Y'}$, $g = g\chi_Y \oplus g\chi_{Y'}$, so

$$f \ominus f\chi_{Y'} = f\chi_Y = g\chi_Y = g \ominus g\chi_{Y'}.$$

Hence, $f \sim_I g$.

The rest follows by Lemma 3.2.15 and Theorem 3.2.13. □

Although $\mathcal{E}/\!\sim$ is a positive CPAM, we now show that $\sim$ need not be a congruence and I need not be a Riesz ideal.

Example 3.2.17 Letting $X = \{x, y\}$, $s(x) = t(y) = 2$, $s(y) = t(x) = 1$ and $\mathcal{T} = \{s, t\}$ we see that $(X, \mathcal{T})$ is a D-test space. Let $Y = \{y\}$ so $f \sim g$ if and only if $f(y) = g(y)$. To show that $\sim$ does not satisfy (C3), define $f, g \in \mathcal{E}$ by $f(x) = f(y) = g(y) = 1$ and $g(x) = 0$. Then $f + g = t$ so $f \perp g$. Also, $s \sim f$. Suppose that $h \sim g$ and $h \perp s$. Then $h(y) = g(y) = 1$ so $h(y) + s(y) = 2$. Also, $h(x) + s(x) \geq s(x) = 2$. But then $h + s \notin \mathcal{E}$ which is a contradiction. We conclude that $\sim$ is not a congruence. It follows that $I = \{f \in \mathcal{E} : f(y) = 0\}$ is not a Riesz ideal. Indeed, if I were a Riesz ideal, then $\sim_I$ would be a congruence on $\mathcal{E}$. We would then conclude that $\sim$ is a congruence on $\mathcal{E}$ which is a contradiction. We can also show directly that I is not a Riesz ideal by demonstrating the failure of (R2). Letting $i(x) = 2$, $i(y) = 0$ we see that $i \in I$ and $s \ominus i + f = t$ so $s \ominus i \perp f$. Suppose there exists a $j \in I$ such that $s \perp f \ominus j$. Then $s(y) + (f \ominus j)(y) = s(y) + f(y) = 2$ and $s(x) + (f \ominus j)(x) \geq s(x) = 2$. But then $s + f \ominus j \notin \mathcal{E}$ which is a contradiction. Since I satisfies (R1) this also shows that (R1) does not imply (R2).

3.3 Applications to Dimension Theory

In this Section, we consider some applications of weak congruences in our sense to the dimension theory of orthomodular lattices, resp. more general structures. We briefly discuss some connections between interval effect algebras and K_0-theory for AF C*-algebras.

3.3.1 Congruences and Dimensional Equivalence

One of the interests of orthomodular lattices is that they give an abstract setting for the Murray-von Neumann dimension theory of operators algebras. In his pioneering paper, L. H. Loomis ([Loom]) defines a **dimension equivalence** on a complete OML L as an equivalence relation $\sim_d$ on L which satisfies :

(D1) If $a \sim_d 0$, then $a = 0$.

(D2) If $a \perp b$ and $c \sim_d a \vee b$, then $d \perp e$ exist such that $d \sim_d a$, $e \sim_d b$, $c = d \vee e$.

(D3) If $\{a_i\}$ and $\{b_i\}$ are families of pairwise orthogonal elements with the same indices i, and $a_i \sim_d b_i$ for all i, then $\bigvee a_i \sim_d \bigvee b_i$.

(D4) If a and b are perspective (i.e., have a common complement in L), then $a \sim_d b$.

In general, the quotient set $L/\sim_d$ is not naturally an OML but we will prove that it is a PAM and determine quotient sets which are effect algebras. A new characterization of congruences is necessary.

If $\sim_d$ is a dimension equivalence on a complete OML L, we can conclude, from (D2) and (D3), that $\sim_d$ satisfies (C1), (C2), and (C5). Therefore, by Proposition 3.2.12 the quotient with respect to $\sim_d$ is a positive PAM, which may be neither an effect algebra nor a CPAM, in general. In a complete modular orthomodular lattice, perspectivity relation is a dimension equivalence and in this case (C6) is also satisfied, so that this dimension equivalence is a congruence (cf. Proposition 3.1.21), and the quotient is an effect algebra.

Example 3.3.1 Consider a separable von Neumann factor A and let L be the complete lattice of its projections, endowed with the usual dimension equivalence $\sim$ (the Murray-von Neumann equivalence). The quotient $L/\sim$ is isomorphic to one of the following PAMs : $\{0, 1, \ldots, n\}$ (type I_n), $N \cup \{\infty\}$ (type I_∞), $[0, 1]$ (type II_{fin}), $R^+ \cup \{\infty\}$ (type II_∞), $\{0, \infty\}$ (type III). We remark that the quotient is an effect algebra if and only if L is a modular lattice. The following result generalizes this remark to dimensional OMLs.

Theorem 3.3.2 [ChPu] *Let L be a locally modular complete OML, endowed with a locally finite dimension equivalence $\sim$.*[1] *The following statements are equivalent*

(1) *L is modular.*

(2) *$L/\sim$ is a CPAM.*

(3) *$L/\sim$ is an effect algebra.*

[1]An OML L is **locally finite** if it contains a join-dense modular ideal I. A dimension equivalence $\sim_d$ on L is **locally finite** if it coincides with perspectivity on I. [Kal].

Proof. (3) $\Rightarrow$ (2). Obvious.

(2) $\Rightarrow$ (1). Assume that $L/\sim$ is a CPAM. If 1 is finite then L is a modular lattice ([Kal]). If $1 \sim a$ then, as $1 = a \vee a^{\perp}$, $[a] = [a] \oplus [a^{\perp}]$. Since $L/\sim$ is a CPAM, $[a^{\perp}] = 0$ and therefore $a^{\perp} = 0$. It follows that $a = 1$ and so 1 is finite.

(1) $\Rightarrow$ (3). If L is modular and $\sim$ a locally finite dimension equivalence, then $\sim$ coincides with perspectivity and satisfies condition (C6). Therefore $\sim$ is a congruence and $L/\sim$ is an effect algebra. □

Remark 3.3.3 (1) $\Rightarrow$ (3) is not true for any dimension equivalence. For example, consider a complete atomless Boolean algebra B (the regular open sets in $\mathbb{R}$ is a such algebra). There exists a dimensional equivalence $\sim$ on B defined by the quotient set $\{\{0\}, B \setminus \{0\}\}$. The algebra B is a complete modular lattice and $B/\sim$ is not a CPAM.

We will say that a positive (not necessarily cancellative) PAM P satisfies the **weak Riesz decomposition property** if

$$a \leq b \oplus c \;\Rightarrow\; a = a_1 \oplus a_2, a_1 \leq b, a_2 \leq c.$$

For the following results see [JePu].

Proposition 3.3.4 *Let P be a positive CPAM, and assume that $\sim$ is a weak congruence satisfying* (C5). *Let $[x]$ denote the equivalence class with respect to $\sim$ that contains x. Then $[a] \leq [b]$ if and only if there is $b_1 \leq b$ with $a \sim b_1$.*

Proof. If $b_1 \leq b$ and $a \sim b_1$, then $[a] \leq [b]$. Conversely, if $[a] \leq [b]$, then there is $c \in P$ with $[a] \oplus [c] = [b]$, and so there are a_1, c_1, $a_1 \sim a$, $c_1 \sim c$, $a_1 \perp c_1$, $a_1 \oplus c_1 \sim b$. By (C5), there are b_1, b_2 such that $b = b_1 \oplus b_2$ and $b_1 \sim a_1$, $b_2 \sim c_1$. Now $b_1 \leq b$ and $a \sim a_1$, $a_1 \sim b_1$ by (C1) implies $a \sim b_1$. □

Proposition 3.3.5 *Let P be a positive CPAM, and assume that $\sim$ is a weak congruence satisfying* (C5). *Then $P/\sim$ has the weak Riesz decomposition property if and only if the following condition is satisfied.*

(C7) $a, b, c \in P$, $b \perp c$, $a \leq b \oplus c \Rightarrow \exists a_1, a_2 \in P$, $a_1 \leq b$, $a_2 \leq c$ *and* $a \sim a_1 \oplus a_2$.

Proof. Assume that $P/\sim$ has the weak Riesz decomposition property. Let $x \leq y \oplus z$. Then $[x] \leq [y \oplus z] = [y] \oplus [z]$. By the weak Riesz decomposition property, there are x_1, x_2 with $[x_1] \leq [y]$, $[x_2] \leq [z]$ and $[x] = [x_1] \oplus [x_2]$. By Proposition 3.3.4, there are $y_1 \leq y$, $z_1 \leq z$ with $y_1 \sim x_1$, $z_1 \sim x_2$. Now $x \sim y_1 \oplus z_1$, hence (C7) holds.

Conversely, assume (C7). Let $[x] \leq [y] \oplus [z]$. There are $y_1 \sim y$, $z_1 \sim z$, $y_1 \perp z_1$. Moreover, there is $w \leq y_1 \oplus z_1$, $w \sim x$. By (C7), there are $y_2 \leq y_1$, $z_2 \leq z_1$ and

$w \sim y_2 \oplus z_2$. It follows $[w] = [x] = [y_2] \oplus [z_2]$, $[y_2] \leq [y]$, $[z_2] \leq [z]$. Hence the weak Riesz decomposition property is satisfied. □

As an application, we can prove the following theorem. We start with a lemma.

Lemma 3.3.6 [Ber, Ex. VIII.7] *Let L be a complete OML with dimension $\sim_d$, and $x, v, w \in L$ such that $x \wedge v = x \wedge w = 0$ and $v \sim_d w$. Then $x \vee v \sim_d x \vee w$.*

Proof. Put $e := (x \vee v) \wedge x'$, $f := (x \vee w) \wedge x'$. By orthomodularity, $e \vee x = v \vee x$. Since $e \wedge x = 0$, x is a common relative complement for v and e. Therefore e and v are perspective (cf. [Ber]) and, by (D4), $e \sim_d v$. Similarly $w \sim_d f$. Hence $e \sim_d f$, and $x \vee v = x \vee e \sim_d x \vee f = x \vee w$ by (D3). □

Theorem 3.3.7 *The quotient $L/\sim_d$ with respect to a dimension relation $\sim_d$ on an OML L has the weak Riesz decomposition property.*

Proof. It suffices to prove (C7). Let $a \leq b \vee c$, $b \perp c$. By orthomodularity, $a = (a \wedge b') \vee ((a' \vee b) \wedge a)$. Put $x = a \wedge b'$, $v = (b \vee a') \wedge a$, $w = (a \vee b') \wedge b$. Then $x \wedge v = x \wedge w = 0$. By the parallelogram law in OMLs, v and w are perspective, and hence $v \sim_d w$. Applying Lemma 3.3.6, we obtain $a \sim_d x \vee w$, $w \leq b$ and $x \leq (b \vee c) \leq b' = c$. This proves (C7). □

3.3.2 Relations to K_0 of AF C*-algebras

In the sequel we shall need some elements of the K_0 theory of C*-algebras. All C*-algebras that we shall consider here have a unit. We use [Bla] as the main reference book.

Let A be a C*-algebra. Recall that a **projection** is an element $p \in A$ satisfying $p = p^* = p^2$. An element u in A is a **partial isometry** if uu^* and u^*u are projections. Two projections p, q are **(Murray - von Neumann) equivalent** if there is a partial isometry u in A such that $u^*u = p$ and $uu^* = q$. The projections p, q are **orthogonal** (written $p \perp q$) if $pq = 0$. The equivalence has properties (C1), (C2) and (C5), moreover, $p \sim 0$ implies $p = 0$. Let $Proj(A)$ denote the set of all equivalence classes of projections of A.

The definition of $K_0(A)$ requires simultaneous consideration of all matrix algebras over A. Denote $M_n(A)$ the set of all $n \times n$ matrices with entries in A. Let $M_\infty(A)$ denote the algebraic direct limit of $M_n(A)$ under the embeddings $a \mapsto diag(a, 0)$. $M_\infty(A)$ can be thought as the algebra of all infinite matrices with only finitely many nonzero entries. The embeddings are isometries, so $M_\infty(A)$ has a natural norm. Let $V(A) := Proj(M_\infty(A))$. If A is separable, then $V(A)$ is countable. There is a binary operation on $V(A)$: If $[e], [f] \in V(A)$, choose $e' \in [e], f' \in [f]$ with $e' \perp f'$ (this is always possible by "moving down the diagonal"), and define $[e] + [f] = [e' + f']$.

This operation is well defined and makes $V(A)$ into an Abelian semigroup with the additive zero $[0]$.

For any Abelian semigroup H, there is a universal enveloping Abelian group $G(H)$ called the **Grothendieck group** of H. $G(H)$ can be thought of as the group of all equivalence classes of $H \times H$ under the equivalence relation $(x_1, y_1) \sim (x_2, y_2)$ iff there is z with $x_1 + y_2 + z = x_2 + y_1 + z$.

If A is a unital C*-algebra, then $K_0(A)$ is defined as the Grothendieck group of $V(A)$. The embedding of $V(A)$ into $K_0(A)$ is injective iff $V(A)$ has cancellation, i.e., whenever $e, f, g, h \in V(A)$ and $e \perp g, f \perp h, e \sim f$ and $e + g \sim f + h$ then $g \sim h$.

$K_0(A)$ can be preordered by taking the image of $V(A)$ in $K_0(A)$ as $K_0(A)^+$. We also define the **scale** $\Sigma(A)$ to be the image of $Proj(A)$ in $K_0(A)$. The triple $(K_0(A), K_0(A)^+, \Sigma(A))$ is called the **scaled group** of A. If A is unital C*-algebra with cancellation, then $\Sigma(A) = [0, [1_A]]$.

If $\phi : A \to B$ is a homomorphism of C*-algebras, then ϕ induces a map $\phi_* : Proj(A) \to Proj(B)$. ϕ_* extends to a homomorphism from $M_\infty(A)$ to $M_\infty(B)$, which induces a semigroup homomorphism, also denoted by ϕ_*, from $V(A)$ to $V(B)$. So V is a covariant functor from the category of C*-algebras to the category of Abelian semigroups. It has the following properties:

- If $A = A_1 \oplus A_2$, then $M_\infty(A) = M_\infty(A_1) \oplus M_\infty(A_2)$ and $V(A) = V(A_1) \oplus V(A_2)$.

- If $A = \lim_{\to}(A_i, \phi_{ij})$ is the inductive limit, then $V(A)$ is the algebraic direct limit of $(V(A_i), \phi_{ij}*)$.

According to [WeO], to every Abelian group G there is a C*-algebra A with $K_0(A) = G$. But this correspondence is not one-to-one; the C*-algebra A is not uniquely defined by $K_0(A)$. The situation is better in the simplest case of AF C*-algebras (AF stands for approximately finite). For the following facts see [Bla]. More details of the theory of AF C*-algebras can be found in [Goo]. A C*-algebra A is an **AF C*-algebra** (or AF algebra) if A is an inductive limit of a sequence of finite dimensional C*-algebras. Note that AF C*-algebras have a cancellation, and finite linear combinations with complex coefficients of projections form a norm-dense subset of A (in other words, the set of all projections is total in A). AF C*-algebras were first studied by Bratteli [Bra], following analysis of special cases by Glimm (UHF-algebras, [Gli]) and Dixmier (matroid C*-algebras, [Dix]). Examples of AF C*-algebras (not necessarily unital) are the CAR algebra, UHF and matroid C*-algebras, the algebra K of compact operators on a Hilbert space H (K is unital if and only if H is finite dimensional), $C_0(X)$, where X is locally compact, totally disconnected and second countable (it is unital iff X is compact). Note that $\Sigma(C_0(X))$ is a Boolean algebra.

Every finite dimensional C*-algebra is a direct sum of matrix algebras. If $A = A_1 \oplus A_2 \oplus \cdots \oplus A_r$, then $K_0(A) \equiv \mathbb{Z}^r$ with the usual ordering.

Let A, B be finite dimensional C*-algebras. The following facts are known:

(a) If $\sigma : K_0(A) \to K_0(B)$ is a homomorphism of scaled ordered groups, then there is a *-homomorphism $\phi : A \to B$ with $\phi_* = \sigma$.

(b) If $\phi, \psi : A \to B$ with $\phi_* = \psi_*$, then ϕ and ψ are unitarily equivalent. ϕ is injective iff $(Ker\phi_*) \cap K_0(A)^+ = 0$.

If $A = \lim_\to A_n$, then $K_0(A) = \lim_\to K_0(A_n)$ as a scaled ordered group. This is called the **dimension group** of A. So $K_0(A) = \lim_\to(\mathbb{Z}^{r_n}, \mathbb{R}_n)$ ($\mathbb{R}_n$=$n \times n$-matrices with nonnegative entries).

The fundamental theorem of AF C*-algebras theory is due to Elliot [Eli] and says that the dimension group of A is a complete isomorphism invariant for A (among AF C*-algebras).

Theorem 3.3.8 [Eli] *If A and B are AF C*-algebras and $\sigma : K_0(A) \to K_0(B)$ is an isomorphism of scaled ordered groups, then there is an isomorphism $\phi : A \to B$ with $\phi_* = \sigma$.*

The following theorem due to Effros, Handelman and Shen [EHS], gives an intrinsic characterization of dimension groups. (Recall that a partially ordered Abelian group G is **unperforated** if $na \geq 0$ for some $n = 1, 2, \ldots$ implies $a \geq 0$.)

Theorem 3.3.9 [EHS] *An ordered Abelian group is a dimension group if and only if it is countable, unperforated and has the interpolation property.*

From the relations between effect algebras with the Riesz decomposition property and their unigroups (see Section 1.7), the following results can be derived ([Pul 11]).

Proposition 3.3.10 *If A, B are AF C*-algebras and $\sigma_0 : \Sigma(A) \to \Sigma(B)$ is an isomorphism of effect algebras, then there is an isomorphism $\phi : A \to B$ with $\phi_* = \sigma$, where $\sigma : K_0(A) \to K_0(B)$ is the extension of σ_0.*

From Theorem 3.3.9 and Proposition 3.3.10, we can deduce the following result.

Theorem 3.3.11 [Pul 11] *There is a categorical equivalence between AF C*-algebras and effect algebras with the Riesz decomposition property whose unigroup is countable and unperforated.*

One can read much of the structure of an AF C*-algebra A from its dimension group. For example, the (closed two-sided) ideals of A are in a one-to-one correspondence with the order ideals of $K_0(A)$. In particular, A is simple iff $K_0(A)$ is simple (that is, every element of $K_0(A)^+$ is an order unit). The following result is of a special interest for us (cf. [Bla]). Recall that a state s on a C*-algebra A is a **tracial state** if $s(aa^*) = s(a^*a)$ for any $a \in A$. Let $T(A)$ denote the set of tracial states on A. A state s is **faithful** if $0 < a \in A$ implies $s(a) > 0$.

Theorem 3.3.12 [Bla] *The tracial states of an AF C*-algebra A correspond exactly to the states on $(K_0(A), K_0(A)^+, [1_A])$.*

Corollary 3.3.13 [Pul 11] *States on an effect algebra E with the Riesz decomposition property whose unigroup is countable and unperforated correspond exactly to the tracial states on the corresponding AF C^*-algebra A. The extremal (faithful) states on E correspond to pure (faithful) tracial states on A.*

Proof. Follows from Theorems 3.3.12 and 1.7.17. The correspondence between extremal and pure states follows by the fact that the correspondences in both latter theorems preserve convex combinations. Preservation of faithfulness follows from the fact that a state on A is faithful iff it is faithful on $Proj(A)$ (since $Proj(A)$ is total in A), and the fact that the zero class of equivalent projections contains only 0. □

The following result is well-known ([Mun]).

Theorem 3.3.14 *Up to isomorphism, the function $A \mapsto [0, [1_A]]$, where $\Sigma(A) = [0, [1_A]]$ is the scale in $(K_0(A), K_0(A)^+, \Sigma(A))$ provides a one-one correspondence between AF C^*-algebras whose Murray-von Neumann order is a lattice, and countable MV-algebras. Commutative AF C^*-algebras correspond to countable Boolean algebras.*

The correspondence between tracial states on AF C*-algebras with lattice Murray-von Neumann order and states on the corresponding MV-algebras has been studied e.g. in [Mun 3], [Mun 4].

Some interesting examples of the above correspondence have been collected in [Mun 2], Table 4.3. For example, Boolean algebras correspond to commutative AF C*-algebras, finite MV-algebras correspond to finite-dimensional AF C*-algebras, the standard MV-algebra $[0, 1]$ corresponds to $\mathbb{C}$, Łukasziewicz chain $[0, 1/n, \dots, 1]$ corresponds to $M_n(\mathbb{C})$, the $n \times n$ complex matrices, $\mathbb{Q} \cap [0, 1]$, $\mathbb{Q}$ (= the rationals), corresponds to the Glimm universal UHF-algebra, dyadic rationals in $[0, 1]$ correspond to the CAR algebra of the Fermi gas, etc.

A complete classification of all simple dimension groups is in [Goo 1, Thm 14.16, Prop 14.3].

Not all dimension groups are lattice ordered, and according to the result of [EHS], they are K_0-groups of some AF C*-algebras as well. In [Bla], the following four non-lattice ordered dimension groups are considered. Let $\mathbb{D} = \mathbb{Z}_{(2^\omega)}$, the dyadic rationals.

- $G_1 = \mathbb{Q}^2$ with strict ordering (i.e., for all $(x, y), (u, v) \in G_1$, $(x, y) < (u, v)$ iff $u - x > 0$ and $v - y > 0$).

- $G_2 = \mathbb{Q}^2$ with strict ordering from the first coordinate (i.e., for all $(x, y), (u, v) \in G_2$, $(x, y) < (u, v)$ iff $u - x > 0$).

- $G_3 = \mathbb{D}^2$ with strict ordering from the first coordinate.

- $G_4 = \mathbb{D} \oplus \mathbb{Z}$ with strict ordering from the first coordinate.

Choose order units $a_i \in G_i$ by $u_1 = (1,1)$, $u_2 = u_3 = u_4 = (1,0)$. Let A_i be the simple unital AF C*-algebras with dimension groups (G_i, u_i). The effect algebras $\Sigma(A_i)$ then are the following subsets of $\mathbb{R}^2$. $\Sigma(A_1) = \{(0,0)\} \cup \{(1,1)\} \cup \{(x,y) \in G_1 : 0 < x < 1, 0 < y < 1\}$, and for $i = 2,3,4$, $\Sigma(A_i) = \{(0,0)\} \cup \{(1,0)\} \cup \{(x,y) \in G_i : 0 < x < 1\}$. It is easy to see that $\Sigma(A_i)$ are antilattice ordered. G_1 has two extremal states τ_1 and τ_2 corresponding to the projections onto the coordinate axes; the others have a unique state corresponding to projection onto the x-axis. So A_1 has two extremal tracial states, and the others a unique tracial state.

3.4 Exercises

1. Prove that the relation $\leq$ defined by $a \leq b$ if there is c such that $b = a \oplus c$ in a positive CPAM is a partial order.
2. Prove that a PAM is an effect algebra if and only if it is a positive unital CPAM.
3. Find an example of a Riesz ideal I which does not have the Riesz decomposition property.

   (Hint: Consider the OML $\mathcal{P}(H)$ of projections on an infinite dimensional Hilbert space H. The set I of all finite dimensional projection is a p-ideal, hence a Riesz ideal. To show that the Riesz decomposition property is not satisfied, consider an atom $a \in I$.)
4. Prove that in a lattice ordered effect algebra the notion of Sasaki projections agrees with the notion of the Sasaki map from Definition 1.8.17.
5. Let P be a positive CPAM, $I \in \mathcal{I}(P)$. Prove that

   (i) I is an R_1-ideal if and only if I is a weak c-ideal and $\sim_I$ satisfies condition (C5). ([Jen, Prop 2])

   (ii) $\sim_I$ satisfies condition (C3) if and only if I is a weak c-ideal and $\sim_I$ satisfies condition (R2). (Hint: To prove (C3) implies (R2), apply (C3) to $a \sim_I a \ominus i$, $a \ominus i \perp b$.)
6. Let $\sim$ be a congruence on a positive CPAM P. Prove that condition (C3) is equivalent to

   (C3a) If $a, b, c \in P$ are such that $a \geq b$ and $a \ominus b \sim c$, then there are $a_1 \sim a$, $b_1 \sim b$ such that $a_1 \geq b_1$ and $a_1 \ominus b_1 = c$. ([Jen, Prop. 4])
7. Let P be a positive CPAM, $I \in \mathcal{RI}(P)$. Prove that, for every $a \in P$,

   $$I = \{a_1 \ominus a_2 : a_1 \geq a_2 \text{ and } a_1 \sim_I a_2 \sim_I a\}.$$

   ([Jen, Prop. 5])

8. Prove that in an effect algebra, conditions (C4) and (C6) for weak congruences are equivalent.

9. Prove the following statements.

   Let L be an effect algebra, and $a, b \in L$. Then $p \in \nabla(a, b)$ iff $\exists\, h \in a\downarrow \cap b'\downarrow$ such that $p = a \wedge h'$. If $h \in a\downarrow \cap b'\downarrow$ and $p = a \wedge h'$, then $p \in \Phi(a, b)$ if and only if h is a maximal element of $a\downarrow \cap b'\downarrow$. Here $x\downarrow = E[0, x]$.

   Moreover, if L is an OMP, then

   $$\Phi(a, b) = \{p \in \nabla(a, b) : p \wedge b' = 0\}.$$

   ([BeFo 2])

10. Let P be a positive CPAM, $I_1, I_2 \in \mathcal{R}_1\mathcal{I}(P)$, $I_1 \leq I_2$.

    (a) I_2 is an union of equivalence classes of $\sim_{I_1}$.

    (b) $I_2/I_1 = \{[i] \in P/I_1 : i \in I_2\}$ is an R_1-ideal of P/I_1.

    (c) For all $a, b \in P$, $a \sim_{I_2} b$ iff $[a]_{I_1} \sim_{I_2/I_1} [b]_{I_1}$.

    (d) P/I_2 is isomorphic to $(P/I_1)/(I_2/I_1)$.

    ([Jen])

11. (i) Let $\mathcal{P}$ be a CPAM, $I, J \in \mathcal{R}_1\mathcal{I}(P)$. For all $a, b \in P$, $a \sim_I b$ and $a \sim_J b$ iff $a \sim_{I \wedge J} b$.

    (ii) Let $\mathcal{P}$ be a CPAM, let $I, J \in \mathcal{R}_1\mathcal{I}(P)$. For every $a, b \in P$, $a \sim_{I \vee J} b$ iff $a(\sim_I \vee \sim_J)b$, where $\sim_I \vee \sim_J$ denotes the join of $\sim_I$ and $\sim_J$ in the lattice of all equivalence relations over P. ([Jen])

12. **Questions.** Let P be a positive CPAM, $I \in \mathcal{I}(P)$. Does condition (R2) on I imply (C1) and (C2) on $\sim_I$?

    Is the following implication true?

    I is a weak c-ideal and I satisfies (R2) $\Rightarrow$ $I \in \mathcal{RI}(P)$.

13. **Question.** What are the CPAMs in which the converse of the Theorem 3.1.25 is true? As any R_1-ideal of an OML is an orthomodular ideal the converse is true for OMLs ([Che 2, Prop 5]).

14. Prove that an Archimedean directed Abelian group G is unperforated. ([Goo 1, Prop. 1.24])

15. An orthomodular poset $(P; \leq, ', 0, 1)$ is called a **dimension poset** (in the sense of Sherstnev) if there is an equivalence relation $\sim_s$ on P such that

    (S1) If $a \sim_s 0$, then $a = 0$.

(S2) If $\{a_i : i \in I\}$ and $\{b_i : i \in I\}$ are two orthogonal subsets of P and if $a_i \sim_s b_i$ for every $i \in I$, then the elements $a = \bigvee a_i$ and $b = \bigvee b_i$ exist and $a \sim_s b$.

(S3) If $a \oplus b \sim_s c \oplus d$ then there exist elements $e, f, E, F \in P$ such that $a = e \oplus f$, $b = E \oplus F$, $e \oplus E \sim_s c$ and $f \oplus F \sim_s d$.

(S4) For any two elements $a, b \in P$ which are not orthogonal there exist nonzero elements c, d such that $c \leq a$, $d \leq b$ and $c \sim_s d$.

Prove:

(1) (C5) follows from (S3).

(2) Every dimension lattice in the sense of Loomis is a dimension lattice in the sense of Sherstnev.

(3) Find examples of a finite OML which are a dimension poset in the sense of Sherstnev but which are not dimension lattices in the sense of Loomis.

(Hint: See [Ber, VIII.3].)

Chapter 4

Tensor Product of D-Posets and Effect Algebras

The event structure of a quantum physical system is identified with a quantum logic [BLM] or an orthoalgebra [FoRa 1], [FoPt] in contrast to classical mechanics when it is assumed to be a Boolean algebra. One of important problems is a coupled system of two independent physical systems P and Q. The event structure L of this coupled system L, if it exists, is usually called a tensor product, and we write $L = P \otimes Q$.

Tensor products of quantum logics in various approaches have been studied in [Sik], [AeDa], [Mat], [Fou], [FGR], [KRF], [Loc], [Pul], [FoRa 1], [Wil], [Zec], [HuPu]. One of the factors motivating the study of orthoalgebras was the discovery [FGR] that there does not exist a satisfactory theory of tensor products of orthomodular posets and lattices. A tensor product of orthoalgebras has been investigated by Foulis and Bennett in [FoBe] via a universal mapping property, and a tensor product of an orthoalgebra and a Boolean algebra is given in [FoPt]. Unfortunately, Foulis and Bennett [FoBe] showed that the category of orthoalgebras is not closed with respect to the formation of tensor products. Therefore, more available structures have been expected.

The tensor product of D-posets, or equivalently of effect algebras, was firstly present by Dvurečenskij [Dvu 1], [Dvu 2], and we have shown that it depends on the notion of a D-test which was developed by the authors in [DvPu 2] and [DvPu 3]. Gudder and Greechie [GuGr] showed that the category of D-posets is also not closed under the formation of tensor products.

A chain tensor product was studied in [Gud 10], effect test spaces in [Gud 11], [Gud 12], [Gud 16]. We recall that E-test spaces (effect test spaces) are equivalent to D-test spaces.

4.1 D-test Spaces and Difference Posets

Foulis and Randall [FoRa], in 1972, (and in [RaFo 1]) gave a new mathematical foundation of an operational probability theory and statistics based upon a generalization of the conventional notion of a sample space in the sense of Kolmogorov [Kol], which was done in 1933.

Let us recall briefly the main notions of that approach according to [FoBe]:

Let X be a non-void set; elements of X are called **outcomes**. We say that a pair $(X, \mathcal{T})$ is a **test space** iff $\mathcal{T}$ is a non-empty family of subsets of X such that (i) for any $x \in X$, there is a $T \in \mathcal{T}$ containing x, and (ii) if $S, T \in \mathcal{T}$ and $S \subseteq T$, then $S = T$.

Any element of $\mathcal{T}$ is said to be a **test**. We say that a subset G of X is an **event** iff there is a test $T \in \mathcal{T}$ such that $G \subseteq T$. Let us denote the set of all effects in X by $\mathcal{E} = \mathcal{E}(X, \mathcal{T})$. We say that two events F and G are (i) **orthogonal** to each other, in symbols $F \perp G$, iff $F \cap G = \emptyset$ and there is a test $T \in \mathcal{T}$ such that $F \cup G \subseteq T$; (ii) **local complements** of each other, in symbols $F \, loc \, G$, iff $F \perp G$ and there is a test $T \in \mathcal{T}$ such that $F \cup G = T$; (iii) **perspective with axis** H iff they share a common local complement H. We write $F \approx_H G$ or $F \approx G$ if the axis is not emphasized.

The test space $(X, \mathcal{T})$ is **algebraic** iff, for $F, G, H \in \mathcal{E}$, $F \approx G$ and $F \, loc \, H$ entail $G \, loc \, H$. Then $\approx$ is the relation of an equivalence, and, for any $A \in \mathcal{E}(X, \mathcal{T})$, we put $\pi(A) := \{B \in \mathcal{E}(X, \mathcal{T}) : B \approx A\}$. Then $\Pi(X) := \{\pi(A) : A \in \mathcal{E}(X, \mathcal{T})\}$ is an orthoalgebra ([FoBe], [Gud 18]).

Conversely, for any orthoalgebra L, there is an algebraic test space $(X, \mathcal{T})$ such that $\Pi(X)$ is isomorphic with L, [FoBe], [Gud 18]. Orthoalgebras were used as a model for quantum measurements in [Pul 18].

4.1.1 D-test Spaces

We present now D-test spaces for D-posets (or equivalently for effect algebras). Another approach is due to [PuWi].

Let X be a nonempty set, elements of X are called **outcomes**. A function $F : T \to X$ is said to be of **finite multiplicity** if, for any $x \in X$, $F^{-1}(x)$ is of finite cardinality. For arbitrary sets I and J and functions $F \in X^I$ and $G \in X^J$ define $F \preccurlyeq G$ iff there is an injection $\sigma : I \to J$ such that $F = G \circ \sigma$, i.e. $F(i) = G(\sigma(i))$ for all $i \in I$. If $F \preccurlyeq G$ and $\sigma : I \to J$ is a bijection, we say that F and G are **equivalent**, in symbols $F \sim G$, and in what follows, we shall identify functions which are equivalent.[1] Therefore, we can define unambiguously a function $F \dot{\cup} G$ as follows: Let $K = I' \cup J'$, where $I' \cap J' = \emptyset$ and $\phi : I \to I'$, $\psi : J \to J'$ are bijections.

[1] Observe that if $I = \emptyset$, then $X^I = \{\emptyset\}$.

Then $F\dot{\cup}G$ is an element of X^K such that

$$F\dot{\cup}G(k) = \begin{cases} F(i), & \text{if } i \in I,\ k = \phi(i), \\ G(j), & \text{if } j \in J,\ k = \psi(j). \end{cases}$$

We note that the associativity $(F\dot{\cup}G)\dot{\cup}H = F\dot{\cup}(G\dot{\cup}H)$ holds, and therefore for both sides we write $F\dot{\cup}G\dot{\cup}H$. It can be easily checked that $\preccurlyeq$ is a preorder which becomes a partial order on equivalence classes with respect to $\sim$.

Let $\mathcal{R}(F) := \{F(i) : i \in I\}$ denote the range of $F \in X^I$.

Definition 4.1.1 *Let $\mathcal{T} = \{F \in X^I : I \in \mathcal{I}\}$, where $X \neq \emptyset$, and $\mathcal{I}$ is a nonvoid family of index sets. We say that the pair $(X, \mathcal{T})$ is a D-***test space** *if and only if the following two conditions are satisfied:*

(i) *Any $T \in \mathcal{T}$ is of finite multiplicity.*

(ii) *For every $x \in X$ there is a $T \in \mathcal{T}$ such that $x \in \mathcal{R}(T)$.*

(iii) *If $S, T \in \mathcal{T}$ and $S \preccurlyeq T$, then $S \sim T$.*

Any element of $\mathcal{T}$ is said to be a **D-test**.

Lemma 4.1.2 *If $F \in X^I$ is a D-test, then $I \neq \emptyset$.*

Proof. Let $F = \emptyset \in \mathcal{T}$. Then, for any $T \in \mathcal{T}$, $F \preccurlyeq T$ implies $F \sim T$ and, by (iii) of Definition 4.1.1, $T = \emptyset$ which is in contradiction with the condition (ii) of Definition 4.1.1. □

Definition 4.1.3 *Let $(X, \mathcal{T})$ be a D-test space. Let J be arbitrary and let $G \in X^J$. We say that G is an* **event** *if and only if there is a D-test $T \in \mathcal{T}$ such that $G \preccurlyeq T$. Let us denote the set of all events in $\mathcal{T}$ by $\mathcal{E} = \mathcal{E}(X, \mathcal{T})$.*

Clearly, $\emptyset \in \mathcal{E}$.

Definition 4.1.4 *Let $(X, \mathcal{T})$ be a D-test space. We say that two events F and G are*

(i) **orthogonal** *to each other, in symbols $F \perp G$, if and only if there is a D-test $T \in \mathcal{T}$ such that $F\dot{\cup}G \preccurlyeq T$;*

(ii) **local complements** *of each other, in symbols $F \operatorname{loc} G$, if and only if there is a D-test $T \in \mathcal{T}$ such that $F\dot{\cup}G \sim T$;*

(iii) **perspective with axis** *H if and only if they share a common local complement H.*

We write $F \approx_H G$ or $F \approx G$ if the axis is not emphasized.

Lemma 4.1.5 (i) *If $\emptyset \approx G$, then $\emptyset \sim G$.*

(ii) *If $T, S \in \mathcal{T}$ are D-tests, then $T \approx S$ with axis $\emptyset$.*

Proof. (i) Every local complement of $\emptyset$ is a D-test. Therefore, if $\emptyset \approx G$, there is a D-test T such that $T\dot{\cup}G$ is a D-test. But then $T \preccurlyeq T\dot{\cup}G$ implies $T \sim T\dot{\cup}G$, which implies $G \sim \emptyset$.

(ii) Observe that the unique local complement of every D-test is $\emptyset$. □

4.1.2 D-algebraic D-test Spaces

In the present part, D-algebraic D-test spaces are introduced. They will play an important role for representation of difference posets in the next section.

Definition 4.1.6 *A D-test space* $(X, \mathcal{T})$ *is D-***algebraic** *if and only if, for* $F, G, H \in \mathcal{E}$, $F \approx G$ *and* $F \perp H$ *entail* $G \perp H$.

We note that some examples of D-algebraic D-test spaces, but not yet under this name, are studied in [Dvu 1].

For simplicity, we usually refer to X, rather than to $(X, \mathcal{T})$, as a D-test space, similarly, we put $\mathcal{E} = \mathcal{E}(X, \mathcal{T})$. In what follows, let X be a D-test space.

Lemma 4.1.7 *A D-test space* X *is D-algebraic if and only if, for* $F, G, H \in \mathcal{E}$,

$$F \approx G,\ \ F\, loc\, H\ \Rightarrow G\, loc\, H.$$

Proof. Let X be D-algebraic and let $F, G, H \in \mathcal{E}$, $F \approx G$, $F\, loc\, H$. The algebraicity implies that $G \perp H$. Let Q be a local complement of $G \dot{\cup} H$. Then $H \dot{\cup} Q\, loc\, G$ implies $H \dot{\cup} Q \perp F$. But since $H\, loc\, F$, $H \dot{\cup} F \preccurlyeq H \dot{\cup} F \dot{\cup} Q$ implies $H \dot{\cup} F \sim H \dot{\cup} F \dot{\cup} Q$, we get $Q = \emptyset$.

The converse implication follows from the fact that if two events are orthogonal, then either one of them can be enlarged to a local complement of the other. □

Lemma 4.1.8 *If* X *is D-algebraic, then* $\approx$ *is transitive on* $\mathcal{E}$, *hence* $\approx$ *is an equivalence relation.*

Proof. Suppose $F, G, H \in \mathcal{E}$ with $F \approx G$ and $G \approx H$. Let K be the axis for $G \approx H$. By Lemma 4.1.7, K is a local complement of F, hence $F \approx H$ with axis K. □

Definition 4.1.9 *Let* X *be a D-algebraic D-test space. If* $F \in \mathcal{E}$, *we define* $\pi(F) := \{G \in \mathcal{E} : G \approx F\}$ *and we refer to* $\pi(F)$ *as the* **proposition affiliated with** F. *The set*

$$\Pi = \Pi(X) := \{\pi(F) : F \in \mathcal{E}\} \tag{4.1}$$

is called the **logic** *of the D-test space* X.

We define $0, 1 \in \Pi$ *by*

$$0 = \pi(\emptyset), \quad 1 = \pi(T), \tag{4.2}$$

where T *is any D-test.*

Suppose that $F, G, H \in \mathcal{E}$ and let both G, H be local complements of F. Then $G \approx H$ with axis F, and it follows that $\pi(G) = \pi(H)$. This observation leads to our next definition.

Definition 4.1.10 *Let X be a D-algebraic D-test space. If $F \in \mathcal{E}$, we define the negation $\pi(F)'$ of the proposition $\pi(F) \in \Pi(X)$ by $\pi(F)' = \pi(G)$, whenever G is any local complement of F in $\mathcal{E}$.*

Evidently, $(\pi(F)')' = \pi(F)$, $0' = 1$, and $1' = 0$.

For $a, b \in \Pi$, we shall write $a \perp b$ iff there are representants $F \in a$ and $G \in b$ with $F \perp G$.

Define a partial order $\leq$ on $\Pi(X)$ by putting $a \leq b$ iff $a \perp b'$.

Lemma 4.1.11 *For $a, b \in \Pi$, $a \leq b$ iff there are $F, G, H \in \mathcal{E}$ such that $a = \pi(F), b = \pi(G)$, $H \perp F$ and $G \approx F\dot{\cup}H$.*

Proof. If $a \leq b$, then there are $F, G, G' \in \mathcal{E}$ such that $G \, loc \, G'$, $a = \pi(F), b = \pi(G), b' = \pi(G')$ and $F \perp G'$. Then there is $H \in \mathcal{E}$ such that $F\dot{\cup}G'\dot{\cup}H$ is a D-test, which implies that $G \approx_{G'} F\dot{\cup}H$.

Conversely, let $a = \pi(F), b = \pi(G)$ and $G \approx F\dot{\cup}H$. Let G' be a local complement of G, then $G' \, loc \, F\dot{\cup}H$, which implies $F \perp G'$, hence $a \perp b'$.

Lemma 4.1.12 *Let X be a D-algebraic D-test space. Then, for $A, B, C, D \in \mathcal{E}$, we have that $A\dot{\cup}B \approx C$ and $A \approx D$ imply $D\dot{\cup}B \approx C$.*

Proof. Let C' be the axis for $A\dot{\cup}B \approx C$. Then $A\dot{\cup}B\dot{\cup}C' \approx C\dot{\cup}C'$, and $A \, loc \, B\dot{\cup}C'$, $A \approx D \Rightarrow D \, loc \, B\dot{\cup}C'$, which implies $D\dot{\cup}B \approx_{C'} C$. □

Lemma 4.1.13 *Let X be a D-algebraic D-test space. Then $\leq$ is a partial order on Π, and 0 and 1 from (4.2) are the smallest and greatest elements in Π.*

Proof. Clearly, $a \leq a$ for any $a \in \Pi$. Assume now that $a, b \in \Pi$, $a \leq b$ and $b \leq a$. By Lemma 4.1.11, $a \leq b$ implies that there are $F, G, F_1 \in \mathcal{E}$ with $a = \pi(F), b = \pi(G)$ and $F\dot{\cup}F_1 \approx G$. Similarly, $b \leq a$ implies that there are H, K, G_1 in $\mathcal{E}$ with $a = \pi(H), b = \pi(K)$ and $K\dot{\cup}G_1 \approx H$. Now $H \approx F$, $K \approx G$ and $K\dot{\cup}G_1 \approx H$ imply $G\dot{\cup}G_1 \approx F$. Further, $F\dot{\cup}F_1 \approx G$, $G\dot{\cup}G_1 \approx F$ imply $F\dot{\cup}F_1\dot{\cup}G_1 \approx F$. Let F' be the axis of the latter perspectivity. Then $F\dot{\cup}F_1\dot{\cup}G_1\dot{\cup}F'$ and $F\dot{\cup}F'$ are D-tests, which implies that $F_1\dot{\cup}G_1 = \emptyset$. Therefore, $F \approx G$, i.e., $a = b$.

Let now $a, b, c \in \Pi$, $a \leq b$, $b \leq c$. Then there are $F, G, H \in \mathcal{E}$ with $a = \pi(F), b = \pi(G), c = \pi(H)$, and $a \leq b$ entails that there is an $F_1 \in \mathcal{E}$ such that $F\dot{\cup}F_1 \approx G$, while $b \leq c$ entails that there is a $G_1 \in \mathcal{E}$ with $G\dot{\cup}G_1 \approx H$. But then $F\dot{\cup}F_1\dot{\cup}G_1 \approx G\dot{\cup}G_1 \approx H$, hence $\pi(F) \leq \pi(H)$, i.e., $a \leq c$.

The assertion on 0 and 1 is evident. □

4.1.3 D-Test Spaces versus Difference Posets

In the present Section, we show that the logic of any D-algebraic D-test space X can be organized into a difference poset $\Pi(X)$, and, conversely, for any difference poset L there is a D-algebraic D-test space X such that L and $\Pi(X)$ are isomorphic.

Theorem 4.1.14 *Let X be a D-algebraic D-test space. Then the logic $\Pi(X)$ of X can be organized into a D-poset.*

Proof. We have already proved that Π with $\leq$, 0, and 1, is a a partially ordered set with the smallest and greatest elements 0 and 1, respectively. Define now a partial binary operation $\ominus : \Pi \times \Pi \to \Pi$ as follows: If $a, b \in \Pi$ and $a \leq b$, and $F, G, H \in \mathcal{E}$ are such that $a = \pi(F), b = \pi(G)$, $G \approx F\dot{\cup}H$, then $b \ominus a := \pi(H)$. We claim that with this partial binary operation, Π becomes a D-poset.

The element $b \ominus a$ is well-defined. Indeed, $F\dot{\cup}H_1 \approx F\dot{\cup}H$ implies $H \approx H_1$ for any $F, H_1, H \in \mathcal{E}$. We have to verify two conditions: (i) $a \ominus 0 = a$; (ii) if $a \leq b \leq c$, then $c \ominus b \leq c \ominus a$ and $(c \ominus a) \ominus (c \ominus b) = b \ominus a$.

(i) is evident. To prove (ii), let $a \leq b \leq c$, $a = \pi(F), b = \pi(G), c = \pi(H)$, and $F_1, G_1 \in \mathcal{E}$ be such that $F\dot{\cup}F_1 \approx G$ and $G\dot{\cup}G_1 \approx H$. Then $F\dot{\cup}F_1\dot{\cup}G_1 \approx G\dot{\cup}G_1 \approx H$ imply $c \ominus b = \pi(G_1)$, and $c \ominus a = \pi(F_1\dot{\cup}G_1)$ which gives $c \ominus b \leq c \ominus a$. Moreover, $(c \ominus a) \ominus (c \ominus b) = \pi(F_1) = b \ominus a$.

In addition, if $a = \pi(F)$, then $a^\perp := 1 \ominus a = \pi(F)'$. □

Let P and L be two D-posets. We recall that a mapping $\phi : P \to L$ is

(i) a **morphism** iff $\phi(1) = 1$, and $p \perp q$, $p, q \in P$, implies $\phi(p) \perp \phi(q)$ and $\phi(p \oplus q) = \phi(p) \oplus \phi(q)$;

(ii) a **monomorphism** iff ϕ is a morphism and $\phi(p) \perp \phi(q)$ iff $p \perp q$;

(iii) an **isomorphism** iff ϕ is a surjective monomorphism, and we say that P is isomorphic to L.

Let P, Q, L be D-posets. A mapping $\beta : P \times Q \to L$ is called a **bimorphism** iff

(i) $a, b \in P$ with $a \perp b$, and $q \in Q$ imply $\beta(a, q) \perp \beta(b, q)$ and $\beta(a \oplus b, q) = \beta(a, q) \oplus \beta(b, q)$;

(ii) $c, d \in Q$ with $c \perp d$, and $p \in P$ imply $\beta(p, c) \perp \beta(p, d)$ and $\beta(p, c \oplus d) = \beta(p, c) \oplus \beta(p, d)$;

(iii) $\beta(1, 1) = 1$.

If $\beta : P \times Q \to L$ is a bimorphism, then $\beta(\cdot, 1) : P \to L$ and $\beta(1, \cdot) : Q \to L$ are morphisms. Therefore, for $p \in P$ and $q \in Q$, we have $\beta(p, 1)^\perp = \beta(p^\perp, 1)$, $\beta(1, q)^\perp = \beta(1, q^\perp)$ and $\beta(p, 0) = \beta(0, q) = 0$.

Also, if $a, b, p \in P$ and $c, d, q \in Q$, we have $a \leq b$ implies $\beta(a, q) \leq \beta(b, q)$ and $c \leq d$ implies $\beta(p, c) \leq \beta(p, d)$.

Theorem 4.1.15 *Let L be a D-poset. Then there exists a D-algebraic D-test space $(X, \mathcal{T})$ such that L is isomorphic to $\Pi(X)$.*

Proof. Let $X = L \setminus \{0\}$ and let $\mathcal{T}$ be the set of all finite partitions of 1 in L consisting of nonzero elements. We claim that $(X, \mathcal{T})$ is a D-algebraic D-test space. Indeed, we have $\{1\} \in \mathcal{T}$, and if $a \neq 0$, $a \neq 1$, then $\{a, a^\perp\} \in \mathcal{T}$ and $a \in \mathcal{R}(\{a, a^\perp\})$. This proves that (ii) of Definition 4.1.1 is satisfied. The conditions (i) and (iii) of Definition 4.1.1 are clear. Moreover, $(X, \mathcal{T})$ is D-algebraic. Indeed, let F, G, H be events such that $F \approx G$ and $F \, loc \, H$. Then $F = \{a_1, \dots, a_n\}$, $G = \{b_1, \dots, b_m\}$, $H = \{c_1, \dots, c_k\}$. Let $K = \{d_1, \dots, d_l\}$ be the axis for $F \approx G$. Now $\bigoplus F \oplus \bigoplus K = 1$, $\bigoplus G \oplus \bigoplus K = 1$, $\bigoplus F \oplus \bigoplus H = 1$, so that $\bigoplus H = \bigoplus K$ and $\bigoplus G \oplus \bigoplus H = 1$, hence $G \, loc \, H$.

Let $\Pi(X)$ be the logic of X. By Theorem 4.1.14, $\Pi(X)$ is a D-poset. Define a mapping $\phi : L \to \Pi(X)$ by

$$\phi(a) = \begin{cases} \pi(\{a\}), & \text{if } a \neq 0, \\ \pi(\emptyset), & \text{if } a = 0. \end{cases} \tag{4.3}$$

The mapping ϕ is a morphism from L onto $\Pi(X)$. Indeed, let $a, b \in L$, $a \perp b$. If one of a, b is 0, $\phi(a) \perp \phi(b)$ is evident. If $a \neq 0 \neq b$, then either $a = b^\perp$ or $a \neq b^\perp$, then $\{a, b\}$ and $\{a, b, (a \oplus b)^\perp\}$, respectively, are partitions of 1, which implies that $\phi(a) \perp \phi(b)$, and $\phi(a \oplus b) = \phi(a) \oplus \phi(b)$.

Let now $\pi(\{a\}) \perp \pi(\{b\})$. There is a partition of 1 containing a and b, hence $a \perp b$ in L. This proves that ϕ is a monomorphism. To prove that ϕ is onto, let $A = \{a_1, \dots, a_n\}$ be an event. Then A is $\bigoplus$-orthogonal, and put $a = \bigoplus_{i=1}^n a_i$. Then

$$\pi(A) = \pi(\{a_1, \dots, a_n\}) = \bigoplus_{i=1}^n \pi(\{a_i\}) = \pi(\{a\}) = \phi(a),$$

hence ϕ is onto. □

Remark 4.1.16 If we compare the notion of a (D-algebraic) D-test space with the notion of a usual (algebraic) test space, we see that a (D-algebraic) D-test space becomes an (algebraic) usual one if and only if all D-tests $F \in X^I$ are injective functions from I to X. In an algebraic case their logics are orthoalgebras.

4.1.4 Examples of D-Test Spaces

In this part, we show how the structure of σ-D-posets and complete D-posets is connected with D-algebraic D-test spaces. In addition, an example of a test space, which is related to inner product spaces, and which is not algebraic, is presented.

Let P and L be two σ-D-posets or complete D-posets. We recall that a mapping $\phi : P \to L$ is a **σ-morphism** or a **complete morphism** if, for any $\bigoplus$-orthogonal system $\{a_i\}_{i\in I}$ with the $\bigoplus$-join $a = \bigoplus_{i\in I} a_i$, we have that $\{\phi(a_i)\}_{i\in I}$ is $\bigoplus$-orthogonal with the $\bigoplus$-join $\bigoplus_{i\in I}\phi(a_i)$, and $\phi(\bigoplus_{i\in I} a_i) = \bigoplus_{i\in I}\phi(a_i)$, where I is always a countable or arbitrary index set, respectively. If, for ϕ, we have, in addition, $\phi(p) \perp \phi(q)$ iff $p \perp q$, ϕ is said to be a **σ-monomorphism** or a **complete monomorphism**. If, moreover, ϕ is surjective, ϕ is called a **σ-isomorphism** or a **complete isomorphism**, and P and L are called **σ-isomorphic** or **completely isomorphic**.

Let $F_t \in X^{I_t}$ for any $t \in \Sigma$. We define $\dot{\bigcup}_{t\in\Sigma} F_t$ as a unique equivalent function $F : \bigcup_{t\in\Sigma} I_t \to X$ such that $F(i) = F_t(i)$ whenever $i \in I_t$, supposing $I_t \cap I_s = \emptyset$, whenever $t \neq s$, $s,t \in \Sigma$.

We say that a D-test space $(X,\mathcal{T})$ has the **σ-property** or the **complete property** iff given a family of events, $\{E_t\}_{t\in\Sigma}$, Σ is countable or arbitrary, such that, for any finite $Q_0 \subseteq \Sigma$, $\dot{\bigcup}_{t\in Q_0} E_t \in \mathcal{E}(X)$, we have $\dot{\bigcup}_{t\in\Sigma} E_t \in \mathcal{E}(X)$.

Theorem 4.1.17 *Let $(X,\mathcal{T})$ be a D-algebraic D-test space with the σ-property or the complete property. Then the D-poset $\Pi(X)$ is a σ-D-poset or a complete D-poset.*

Conversely, if L is a σ-D-poset or a complete D-poset, then there exists a D-algebraic D-test space $(X,\mathcal{T})$ with the σ-property or the complete property, respectively, such that $\Pi(X)$ is σ-isomorphic or complete isomorphic to L.

Proof. Let $(X,\mathcal{T})$ be a D-algebraic D-test space with the σ-property. We assert that $\Pi(X)$ is a σ-D-poset. Let $\{a_i\}$ be a sequence of $\bigoplus$-orthogonal elements in $\Pi(X)$. Then any a_i is of the form $a_i = \pi(E_i)$. Then, for any $n \geq 1$, $\bigoplus_{i=1}^n a_i = \bigoplus_{i=1}^n \pi(E_i) = \pi(\dot{\bigcup}_{i=1}^n E_i)$, so that $E = \dot{\bigcup}_{i=1}^\infty E_i \in \mathcal{E}(X)$, and $\pi(E) \geq a_i$ for any i. Let now $\pi(F) \geq a_i$ for any i. Then $E_i \preccurlyeq F$ for any i. Let $E \in X^I$ and $F \in X^J$, there exists an injection $\sigma : I \to J$ such that $F = E\circ\sigma$. Put $K = J\setminus\sigma(I)$ and define $G \in X^K$ via $G(k) = F(k)$ iff $k \in K$. It is clear $G \in \mathcal{E}(X)$, $E \perp G$ and $E\dot{\cup}G = F$, so that $\pi(E) \leq \pi(F)$ which proves $\pi(E) = \bigoplus_{i=1}^\infty a_i$.

Conversely, let L be a σ-D-poset, put $X = L\setminus\{0\}$, and let $\mathcal{T}$ be a system of all $\bigoplus$-orthogonal sequences $\{a_i\}$ (finite or infinite) of nonzero elements such that $\bigoplus_i a_i = 1$. Then $(X,\mathcal{T})$ is a D-algebraic D-test space. An event of $\mathcal{E}(X)$ is any sequence of $\bigoplus$-orthogonal elements of L; they have, of course, the $\bigoplus$ -join in L.

Let $\{E_n\}_{n=1}^\infty$ be a sequence of events in $\mathcal{E}(X)$ such that $E_n = \{a_i^n\}_{i=1}^\infty \in \mathcal{E}(X)$, and let $F_m := \dot{\bigcup}_{j=1}^m E_j \in \mathcal{E}(X)$ for any integer m. This is possible iff F_m is $\bigoplus$-orthogonal (apply Exercise 1.4.3.10). Putting $E = \dot{\bigcup}_{n=1}^\infty E_n$, we conclude from applications of Exercise 1.4.3.10 that $\dot{\bigcup}_{n=1}^\infty E_n \in \mathcal{E}(X)$, and $(X,\mathcal{T})$ has the σ-property. The mapping ϕ defined via (4.3) is the σ-isomorphism in question.

The general case is similar to the "σ-case", when a test is any $\bigoplus$-orthogonal family $\{a_i\}_{i\in I}$ such that $\bigoplus_{i\in I} a_i = 1$ and $a_i \neq 0$, $i \in I$. $\square$

We now apply the notion of algebraic D-test spaces to a completeness criterion of inner product spaces.

Theorem 4.1.18 *Let S be a separable, real or complex inner product space. Let $X := \mathcal{S}(S) = \{x \in S : \|x\| = 1\}$ and let $\mathcal{T}$ consist of all maximal orthonormal systems (MONS in short). Then $(X, \mathcal{T})$ is an algebraic test space if and only if S is complete. In this case, $\Pi(X)$ is isomorphic to the system of all closed subspaces of S, which is a complete orthomodular lattice.*

Proof. It is clear that $(X, \mathcal{T})$ is always a test space. Suppose that S is complete, i.e., S is a Hilbert space, and any MONS in S is an orthonormal basis in S. Any event in $\mathcal{E}(X)$ is either the empty set or any orthonormal system. Then $\{a_i\} \approx \{b_j\}$ iff $\{a_i\}^{\perp\perp} = \{b_j\}^{\perp\perp}$ $(\emptyset^\perp = S)$.[2] The mapping $\phi : \pi(\{a_i\}) = \{a_i\}^{\perp\perp}$ is a σ-isomorphism which evidently implies the algebraicity of $(X, \mathcal{T})$.

Conversely, let $(X, \mathcal{T})$ be an algebraic test space. We have to show that S is Dacey, i.e., for any MONS $\{x_i\} \cup \{y_j\}$ in S, where $\{x_i\} \cap \{y_j\} = \emptyset$, we have $\{x_i\}^{\perp\perp} = \{y_j\}^\perp$, ([CFM]). Put $A = \{x_i\}$ and $B = \{y_j\}$. It is clear that $A \subseteq B^\perp$, so that $B^{\perp\perp} \subseteq A^\perp$. Let $x \in A^\perp$, $\|x\| = 1$, we have to show that $x \in B^{\perp\perp}$. If $x, y_1, y_2, \ldots$ are linearly dependent vectors, then $x \in B^{\perp\perp}$. If $x, y_1, y_2, \ldots$ are linearly independent vectors in S, applying the Gram-Schmidt orthogonalization process to $x, y_1, y_2 \ldots$ we obtain an orthonormal system $\{x, z_1, z_2, \ldots\}$. It is clear that $\{x, z_1, z_2, \ldots\} \cup \{x_1, x_2, \ldots\}$ is a MONS in S. The algebraicity of $(X, \mathcal{T})$ entails that $\pi(\{x, z_1, z_2, \ldots\}) = \pi(\{y_1, y_2, \ldots\})$.

It is clear that if M, N are two events in X, then $M \perp N$ as events iff $(x, y) = 0$ for all $x \in M$, $y \in N$. We define $M^{\perp_\mathcal{T}} := \{y \in \mathcal{S}(S) : (x, y) = 0 \text{ for all } x \in M\}$.

Denote $C = \{x, z_1, z_2, \ldots\}$. We assert that $\pi(C) = \pi(B)$ iff $C^{\perp_\mathcal{T}} = B^{\perp_\mathcal{T}}$. Indeed, let $\pi(C) = \pi(B)$ and let $\{z\} \perp C$. By algebraicity, $\{z\} \perp B$ so that $z \in B^{\perp_\mathcal{T}}$, and by symmetry, $C^{\perp_\mathcal{T}} = B^{\perp_\mathcal{T}}$. Conversely, let $C^{\perp_\mathcal{T}} = B^{\perp_\mathcal{T}}$. If C_1 is a local complement of C, then $C_1 \subseteq C^{\perp_\mathcal{T}}$ which yields $C_1 \subseteq B^{\perp_\mathcal{T}}$, hence $C_1 \perp B$, therefore $\pi(C)' \leq \pi(B)'$, so $\pi(B) \leq \pi(C)$, and by symmetry $\pi(C) \leq \pi(B)$.

Therefore, the equality $\pi(C) = \pi(B)$ implies $C^{\perp_\mathcal{T}} = B^{\perp_\mathcal{T}}$ and hence $C^\perp = B^\perp$ which is equivalent to $C^{\perp\perp} = B^{\perp\perp}$, so that $x \in B^{\perp\perp}$ which proves $A^\perp \subseteq B^{\perp\perp}$, and S is Dacey.

By [CFM], any separable Dacey inner product space is complete. □

Example 4.1.19 *Let H be an infinite-dimensional Hilbert space with an orthonormal basis $\{e_n\}_{n=1}^\infty$. Put $f = \sum_{n=1}^\infty e_n/2^n$, and let S be the span generated by the vectors in $\{f, e_2, e_3, \ldots\}$. Then $\mathcal{S}(S)$ with the system of all MONSs is a test space which is not algebraic. In general, any separable incomplete inner product space gives an example of a test space which is not algebraic.*

[2] If $M \subseteq S$, we define $M^\perp := \{x \in S : (x, y) = 0 \text{ for any } y \in M\}$.

Proof. The inner product space S is incomplete, and using Theorem 4.1.18 we obtain the assertion.

We note that $\{e_2, e_3, \ldots\}$ is a MONS in S for which the Dacey property holds, i.e., if $\{e_i\} \cup \{e_j\} = \{e_2, e_3, \ldots\}$ and $\{e_i\} \cap \{e_j\} = \emptyset$, then $\{e_i\}^{\perp\perp} = \{e_j\}^{\perp}$ ([CFM]). □

4.1.5 Weights on D-test spaces

Let $(X, \mathcal{T})$ be a D-test space. A **weight** on X is a function $\omega : X \to [0,1]$ such that for every $E \in \mathcal{T}$, $E \in X^I$,

$$\omega(E) := \sum_{i \in I} \omega(E(i)) = 1. \tag{4.4}$$

A set of weights Δ will be called **sufficient** if, for every $x \in X$, there is an $\omega \in \Delta$ with $\omega(x) \neq 0$.

Example 4.1.20 *Let S be a real or complex inner product space, not necessarily complete. Let X be the set of all unit vectors in S, and let $\mathcal{T}$ consist of all maximal orthogonal systems in S. Then $(X, \mathcal{T})$ is a D-test space (a test space, in fact), and there exists at least one weight on X iff S is complete* [Dvu, Thm 4.2.9].

Theorem 4.1.21 *Let $(X, \mathcal{T})$ be a D-test space and let Δ be a sufficient set of weights on X. Then there is a D-algebraic D-test space $(X, \mathcal{U})$ such that $\mathcal{T} \subseteq \mathcal{U}$.*

Proof. In what follows, I is an arbitrary set. Let $\mathcal{U} = \{E \in X^I : \sum_{i \in I} \omega(E(i)) = 1$ for all $\omega \in \Delta\}$. Then $\mathcal{T} \subseteq \mathcal{U}$ and $(X, \mathcal{U})$ is a D-test space. Indeed, (ii) of Definition 4.1.6 follows from $\mathcal{T} \subseteq \mathcal{U}$. To prove (iii) of Definition 4.1.1, let $E \preccurlyeq F$, $E, F \in \mathcal{U}$, $E \in X^I$, $F \in X^J$. Then there is a $G \in X^K$ such that $E \dot\cup G \sim F$, and hence

$$\begin{aligned} 1 &= \sum_{i \in I} \omega(E(i)) = \sum_{j \in J} \omega(F(j)) \\ &= \sum_{i \in I} \omega(E(i)) + \sum_{k \in K} \omega(G(k)) \end{aligned}$$

for every $\omega \in \Delta$, whence $K = \emptyset$.

To show that $(X, \mathcal{U})$ is D-algebraic, let $E, F, G \in \mathcal{E}(X, \mathcal{U})$ with $E \approx F$, $F \, loc \, G$. Let H be the axis for $E \approx F$. Then, for every $\omega \in \Delta$,

$$\begin{aligned} 1 &= \omega(E) + \omega(H) = \omega(F) + \omega(H) \\ &= \omega(F) + \omega(G), \end{aligned}$$

so that $\omega(H) = \omega(G)$ and $\omega(E) + \omega(G) = 1$, which gives $E \dot\cup G \in \mathcal{U}$, and hence $E \, loc \, G$. □

4.1.6 Morphisms and Bimorphisms

Let $(X,\mathcal{T})$, $(Y,\mathcal{S})$ be D-test spaces. Let $\phi : X \to Y$ be a mapping. For $E \in X^I$ we denote by $\phi(E)$ the function

$$\phi(E) := \phi \circ E \in Y^I,$$

that is,

$$\phi(E)(i) = \phi(E(i)),\ i \in I.$$

A mapping $\phi : X \to Y$ is called a **morphism** (of D-test spaces) if $\phi(T) \in \mathcal{S}$ for every $T \in \mathcal{T}$. Evidently, if $E, F \in \mathcal{E}(X)$ and $E \perp F$, then $\phi(E),\ \phi(F) \in \mathcal{E}(Y),\ \phi(E)\perp\phi(F)$ and $\phi(E\dot{\cup}F) = \phi(E)\dot{\cup}\phi(F)$ (we identify events related by $\sim$). Also, $E \approx F$ implies $\phi(E) \approx \phi(F)$.

Proposition 4.1.22 *Let $(X,\mathcal{T})$ and $(Y,\mathcal{S})$ be D-algebraic D-test spaces and let $\phi : X \to Y$ be a morphism. Then the mapping $\psi : \Pi(X) \to \Pi(Y)$ defined by*

$$\psi(\pi(E)) = \pi(\phi(E)), \quad E \in \mathcal{E}(X),$$

is a morphism of D-posets.

Proof. Clearly, $\psi(1) = 1$. Further, $\pi(F)\perp\pi(G)$ implies $F\perp G$ and $\phi(F)\perp\phi(G)$, so that $\psi(\pi(F))\perp\psi(\pi(G))$ and $\phi(F\dot{\cup}G) = \phi(F)\dot{\cup}\phi(G)$, which gives $\psi(\pi(F) \oplus \pi(G)) = \psi(\pi(F)) \oplus \psi(\pi(G))$. □

It is not difficult to see that if $(X,\mathcal{T})$ is a (D-algebraic) D-test space, and $\phi : X \to Y$, where $(Y,\mathcal{S})$ is any D-test space, is a morphism, then $(\phi(X), \phi(\mathcal{T}))$ (where $\phi(\mathcal{T})$ stands for $\{\phi(T) : T \in \mathcal{T}\}$), is a (D-algebraic) D-test space. Indeed, for any $y \in \phi(X)$, there is an $x \in X$ with $y = \phi(x)$. If $T \in \mathcal{T}$ is such that $x \in \mathcal{R}(T)$, then $\phi(x) \in \mathcal{R}(\phi(T))$. This proves that $(\phi(X), \phi(\mathcal{T}))$ is a D-test space. If now $E \in X^I$, $F \in X^J$ are such that $\phi(E)\perp\phi(F)$, then $I \cap J = \emptyset$, so that $E\perp F$, and moreover, $\phi(E\dot{\cup}F) = \phi(E)\dot{\cup}\phi(F)$. If $\phi(E)\, loc\, \phi(F)$, then $E\perp F$ implies that there is a $G \in \mathcal{E}(X)$ such that $E\dot{\cup}G\dot{\cup}F \in \mathcal{T}$. But then $\phi(E)\dot{\cup}\phi(G)\dot{\cup}\phi(F) \in \phi(\mathcal{T})$, and also $\phi(E)\dot{\cup}\phi(F) \in \phi(\mathcal{T})$, which yields $\phi(G) = \emptyset$, hence $G = \emptyset$, and $E\, loc\, F$. This proves that $E\, loc\, F$ iff $\phi(E)\, loc\, \phi(F)$, from which we easily derive that $(\phi(X), \phi(\mathcal{T}))$ is D-algebraic iff $(X,\mathcal{T})$ is D-algebraic.

A morphism $\phi : X \to Y$ of D-test spaces will be called an **isomorphism** iff ϕ is bijective.

Let $(X,\mathcal{T})$, $(Y,\mathcal{S})$ be D-test spaces and let $F \in X^I$, $G \in Y^J$. We define a function $F\dot{\times}G : I \times J \to X \times Y$ via

$$F\dot{\times}G(i,j) := (F(i), G(j)), \quad (i,j) \in I \times J.$$

Let $(X,\mathcal{T})$, $(Y,\mathcal{S})$ and $(Z,\mathcal{U})$ be D-test spaces. A mapping $\beta : X \times Y \to Z$ such that

(i) for any $T \in \mathcal{T}$, $S \in \mathcal{S}$, $\beta(T\dot{\times}S) \in \mathcal{U}$;

(ii) for any $T_1, T_2 \in \mathcal{T}$, $S_1, S_2 \in \mathcal{S}$, $E \in \mathcal{E}(X), F \in \mathcal{E}(Y)$,

$$\beta(T_1\dot{\times}F) \approx \beta(T_2\dot{\times}F),$$

and

$$\beta(E\dot{\times}S_1) \approx \beta(E\dot{\times}S_2),$$

will be called a **bimorphism**.

Clearly, if $E \in \mathcal{E}(X)$, $G \in \mathcal{E}(Y)$, then $\beta(E\dot{\times}G) \in \mathcal{E}(Z)$. Let $E, F \in \mathcal{E}(X)$, $E \perp F$, $E \in X^I$, $F \in X^J$ with $I \cap J = \emptyset$, and let $G \in \mathcal{E}(Y)$, $G \in Z^K$. Then, putting $E\dot{\cup}F = H \in X^{I\cup J}$, we get

$$\begin{aligned} \beta((E\dot{\cup}F)\dot{\times}G) &= \{\beta(H(i), G(j))\}_{(i,j)\in(I\cup J)\times K} \\ &= \{\beta(H(i), G(j))\}_{(i,j)\in I\times K}\dot{\cup}\{\beta(H(i), G(j))\}_{(i,j)\in J\times K} \\ &= \beta(E\dot{\times}G)\dot{\cup}\beta(F\dot{\times}G). \end{aligned}$$

Similarly, for any $E \in \mathcal{E}(X)$, $F, G \in \mathcal{E}(Y)$, $F \perp G$ we obtain

$$\beta(E\dot{\times}(F\dot{\cup}G)) = \beta(E\dot{\times}F)\dot{\cup}\beta(E\dot{\times}G).$$

From these observations, the following result can be derived.

Proposition 4.1.23 *Let $(X, \mathcal{T})$, $(Y, \mathcal{S})$ and $(Z, \mathcal{U})$ be D-algebraic D-test spaces. If $\beta : X \times Y \to Z$ is a bimorphism, then $\gamma : \Pi(X) \times \Pi(Y) \to \Pi(Z)$ defined by*

$$\gamma(\pi_X(E), \pi_Y(F)) = \pi_Z(\beta(E\dot{\times}F))$$

is a bimorphism of D-posets, where π_X, π_Y, π_Z are the mappings corresponding to π.

Proof. First we show that $E \approx_A E_1$, $F \approx_B F_1$ implies $\beta(E\dot{\times}F) \approx \beta(E_1\dot{\times}F_1)$. We have, by (ii) in the definition of bimorphism,

$$\beta((E\dot{\cup}A)\dot{\times}F) \approx \beta((E_1\dot{\cup}A)\dot{\times}F),$$

hence due to the above,

$$\beta(E\dot{\times}F) \approx \beta(E_1\dot{\times}F)$$

with the axis $\beta(A\dot{\times}F)\dot{\cup}D$, where D is the axis for $\beta((E\dot{\cup}A)\dot{\times}F) \approx \beta((E_1\dot{\cup}A)\dot{\times}F)$. Using similar arguments, replacing A with B, we obtain $\beta(E\dot{\times}F) \approx \beta(E\dot{\times}F_1)$ and $\beta(E_1\dot{\times}F) \approx \beta(E_1\dot{\times}F_1)$, which gives the desired result. This shows that γ is well-defined.

Assume now that $\pi_X(E) \perp \pi_X(E_1)$, $E, E_1 \in \mathcal{E}(X)$, $F \in \mathcal{E}(Y)$. Then $\beta(E\dot{\times}F) \perp \beta(E_1\dot{\times}F)$ and $\beta((E\dot{\cup}E_1)\dot{\times}F) = \beta(E\dot{\times}F)\dot{\cup}\beta(E_1\dot{\times}F)$ imply that

$$\gamma(\pi_X(E) \oplus \pi_X(E_1), \pi_Y(F)) = \gamma(\pi_X(E), \pi_Y(F)) \oplus \gamma(\pi_X(E_1), \pi_Y(F)).$$

Hence γ is a bimorphism of D-posets. □

4.1.7 Tensor Products of D-test Spaces

Definition 4.1.24 *Let $(X, \mathcal{T})$, $(Y, \mathcal{S})$, $(Z, \mathcal{U})$ be (D-algebraic) D-test spaces and let $\tau : X \times Y \to Z$ be a bimorphism. We say that the triple $(Z, \mathcal{U}, \tau)$ is a tensor product of $(X, \mathcal{T})$ and $(Y, \mathcal{S})$ in the category of (D-algebraic) D-test spaces if the following conditions are satisfied:*

(i) *For any (D-algebraic) D-test space $(V, \mathcal{W})$ and any bimorphism $\kappa : X \times Y \to V$, there is a morphism $\phi : Z \to V$ such that $\kappa = \phi \circ \tau$.*

(ii) *To every test $T \in \mathcal{U}$, there exist index sets I, J and a function $F \in (X \times Y)^{I \times J}$ such that $T = \tau(F)$.*

Theorem 4.1.25 *If a tensor product of (D-algebraic) D-test spaces exists, it is unique up to isomorphism.*

Proof. Let $(Z_1, \mathcal{U}_1, \tau_1)$ and $(Z_2, \mathcal{U}_2, \tau_2)$ be tensor products of D-test spaces $(X, \mathcal{T})$ and $(Y, \mathcal{S})$. By the definition, there are morphisms $\phi : Z_1 \to Z_2$ and $\psi : Z_2 \to Z_1$ such that $\tau_2 = \phi \circ \tau_1$ and $\tau_1 = \psi \circ \tau_2$. This gives

$$\tau_1 = \psi \circ \phi \circ \tau_1, \quad \tau_2 = \phi \circ \psi \circ \tau_2,$$

hence $\psi \circ \phi$ is the identity on the range of τ_1 and $\phi \circ \psi$ is the identity on the range of τ_2. Now, for every $z \in Z_1$, there is a test $U \in \mathcal{U}_1$ such that $z \in \mathcal{R}(U)$. By (ii) of Definition 4.1.24, there is $F \in (X \times Y)^{I \times J}$ such that $U = \tau_1(F)$. Hence, $z \in \mathcal{R}(\tau_1(F))$, which means that there are $(i, j) \in I \times J$ such that $z = \tau_1(F(i, j)) = \tau_1(x, y)$ for some $x \in X$, $y \in Y$. Therefore, $\psi \circ \phi$ is the identity on Z_1. Similarly, $\phi \circ \psi$ is the identity on Z_2. This yields $\phi = \psi^{-1}$ is a bijection. □

Theorem 4.1.26 *The tensor product of D-algebraic D-test spaces $(X, \mathcal{T})$ and $(Y, \mathcal{S})$ in the category of D-algebraic D-test spaces exists if and only if there is a D-algebraic D-test space $(Z, \mathcal{U})$ and a bimorphism $\beta : X \times Y \to Z$.*

Proof. Necessity is clear.

To prove sufficiency, for $(x, y) \in X \times Y$, $(x', y') \in X \times Y$ put $(x, y) \equiv (x', y')$ if $\beta(x, y) = \beta(x', y')$ for every bimorphism $\beta : X \times Y \to Z$, where $(Z, \mathcal{U})$ is a D-algebraic D-test space. Denote $V = (X \times Y)_{\equiv}$, and let $\mu : X \times Y \to V$ be the canonical mapping. Let $\mathcal{W}$ be the set of all $\mu(F)$, where $F \in (X \times Y)^{I \times J}$ (I, J are any index sets) such that $\beta(F)$ is a test for every bimorphism $\beta : X \times Y \to Z$. Then $(V, \mathcal{W})$ is a D-test space. Indeed, property (ii) of Definition 4.1.1 follows from the fact that $\mu(\mathcal{T} \times \mathcal{S}) \subseteq \mathcal{W}$. To prove property (iii) of Definition 4.1.1, let $A, B \in \mathcal{W}$ and $A \preccurlyeq B$. Then there are $E \in (X \times Y)^{I \times J}$, $F \in (X \times Y)^{I' \times J'}$ such that $\mu(E) = A$, $\mu(F) = B$. Now $A \preccurlyeq B$ implies $\beta(E) \preccurlyeq \beta(F)$, for every bimorphism β, but $\beta(E)$, $\beta(F)$ are tests, hence $\beta(E) = \beta(F)$ for every β, which implies $\mu(E) = \mu(F)$.

Let T_1, $T_2 \in \mathcal{T}$, $E \in \mathcal{E}(Y)$. Now $\beta(T_1\dot{\times}E) \approx \beta(T_2\dot{\times}E)$ for every bimorphism β, and, by D-algebraicity, $\beta((T_1\dot{\times}E)\dot{\cup}(T_2\dot{\times}E'))$ (where E' is any local complement of E) is a test. Therefore $\mu((T_1\dot{\times}E)\dot{\cup}(T_2\dot{\times}E'))$ is a test in $\mathcal{W}$, whence $\mu(T_1\dot{\times}E) \approx \mu(T_2\dot{\times}E)$ with axis $\mu(T_2\dot{\times}E')$. This proves that μ is a bimorphism.

We now prove that $(V, \mathcal{W})$ is D-algebraic. Let A, B, C, D be events such that $A \approx_D B$ and $B \, loc \, C$. Then there are A', B', C', D' with $\mu(A') = A, \mu(B') = B, \mu(C') = C, \mu(D') = D$, and E, F, G such that $\mu(E), \mu(F), \mu(G) \in \mathcal{W}$ and $\mu(E) = A\dot{\cup}D$, $\mu(F) = B\dot{\cup}D$, $\mu(G) = B\dot{\cup}C$. Then, for every β, $\beta(E) = \beta(A')\dot{\cup}\beta(D')$, $\beta(F) = \beta(B')\dot{\cup}\beta(D')$, $\beta(G) = \beta(B')\dot{\cup}\beta(C')$. The D-algebraicity of $(Z, \mathcal{U})$ implies that $\beta(A') \, loc \, \beta(C')$ for every β. Hence $\beta(A'\dot{\cup}C')$ is a test for every β, so that $\mu(A'\dot{\cup}C') = \mu(A')\dot{\cup}\mu(C') \in \mathcal{W}$. Therefore $A \, loc \, C$.

Let $(Z, \mathcal{U})$ be any D-algebraic D-test space and $\kappa : X \times Y \to \mathcal{U}$ be a bimorphism. Define $\phi : V \to Z$ by putting $\phi(a) = \kappa(x, y)$, where $a = \mu(x, y)$, $(x, y) \in X \times Y$. If $(u, v) \in a$ is any other representant, then $\kappa(u, v) = \kappa(x, y)$ by the definition of μ. Hence ϕ is well-defined, and $\kappa = \phi \circ \mu$. This proves that $(U, \mathcal{W}, \mu)$ is the tensor product of $(X, \mathcal{T})$ and $(Y, \mathcal{S})$ in the category of D-algebraic D-test spaces. □

Corollary 4.1.27 *Let $(X, \mathcal{T})$ and $(Y, \mathcal{S})$ be D-algebraic D-test spaces with sufficient sets of weights $\Delta(X)$ and $\Delta(Y)$, respectively. Then the tensor product of $(X, \mathcal{T})$ and $(Y, \mathcal{S})$ in the category of D-algebraic D-test spaces exists.*

Proof. For $m_1 \in \Delta(X)$ and $m_2 \in \Delta(Y)$ define $m_1 \otimes m_2 : X \times Y \to [0, 1]$ by $m_1 \otimes m_2(x, y) = m_1(x)m_2(y)$. It is easy to check that $m_1 \otimes m_2$ is a weight on the D-test space $(X \times Y, \mathcal{T} \times \mathcal{S})$, which then has a sufficient set of weights. According to Theorem 4.1.21, $(X \times Y, \mathcal{T} \times \mathcal{S})$ can be enlarged to a D-algebraic D-test space $(X \times Y, \mathcal{W})$. Therefore, by Theorem 4.1.26, the tensor product exists. □

4.2 Tensor Product of Difference Posets

4.2.1 Tensor Product

In the present Section, we define a tensor product of difference posets and a necessary and sufficient condition for it to exist. The following definition has been presented originally in [Dvu 1].

Definition 4.2.1 *Let P and Q be difference posets. We say that a pair (T, τ) consisting of a difference poset T and a bimorphism $\tau : P \times Q \to T$ is a* **tensor product** *of P and Q if and only if the following conditions are satisfied:*

(i) *If L is a D-poset and $\beta : P \times Q \to L$ is a bimorphism, there exists a morphism $\phi : T \to L$ such that $\beta = \phi \circ \tau$.*

(ii) *Every element of* T *is a finite orthogonal sum of elements of the form* $\tau(p,q)$ *with* $p \in P$, $q \in Q$.

It is not hard to show that if a tensor product (T,τ) of P and Q exists, it is unique up to isomorphism, i.e., if (T,τ) and (T^*,τ^*) are tensor products of D-posets P and Q, then there is a unique isomorphism $\phi : T \to T^*$ such that $\phi(\tau(p,q)) = \tau^*(p,q)$ for all $p \in P, q \in Q$.

We now present the main assertion (see also [Dvu 1]) of this section using the notion of D-test spaces.

Theorem 4.2.2 *The difference posets* P *and* Q *admit a tensor product if and only if there is at least one difference poset* L *for which there is a bimorphism* $\beta : P\times Q \to L$.

Proof. The necessary condition is evident.

For the sufficiency, suppose that N is the subset of $P \times Q$ consisting of all (p,q) such that $\beta(p,q) = 0$ for every bimorphism β on $P \times Q$. Define $X := (P \times Q) \setminus N$. If $A = \{(p_i,q_i)\}_{i=1}^n$ is a finite sequence of elements from $P \times Q$ and $\beta : P \times Q \to L$ is a bimorphism, it is clear that $\beta(A)$ is $\bigoplus$-orthogonal iff $\beta(\tilde{A})$ is $\bigoplus$-orthogonal, where $\tilde{A} = \{(p_i,q_i)\}_{i=1}^m, 0 \le m \le n$, and $(p_i,q_i) \in A, (p_i,q_i) \in X$; in this case $\bigoplus \beta(A) = \bigoplus \beta(\tilde{A})$ for every bimorphism β on $P \times Q$.

Denote by $\mathcal{H}$ the set of all finite sequences H of elements from X such that for every bimorphism β, $\beta(H)$ is a finite decomposition of 1. It is clear that $\mathcal{H}$ is nonempty, since $\{(1,1)\} \in \mathcal{H}$. We assert that $(X,\mathcal{H})$ is an algebraic D-test space. Then the set of all events, $\mathcal{E}(X,\mathcal{H})$, is the set of all finite sequences $A = \{(p_i,q_i)\}_{i=1}^n$ (may be also empty) such that there is a system $\{(a_j,b_j)\}_{j=1}^m$ of elements from X such that $((p_1,q_1),\ldots,(p_n,q_n),(a_1,b_1),\ldots,(a_m,b_m)) \in \mathcal{H}$. According to Theorem 4.1.14, $\Pi(X) := \{\pi(A) : A \in \mathcal{E}(X,\mathcal{H})\}$ can be organized into a difference poset.

Put now $P \otimes Q := \Pi(X)$ and define a mapping $\otimes : P \times Q \to P \otimes Q$ via

$$\otimes(p,q) = \begin{cases} \pi(\{(p,q)\}), & \text{if } (p,q) \in X, \\ 0, & \text{if } (p,q) \notin X. \end{cases}$$

For simplicity, we often write $p \otimes q$ rather than $\otimes(p,q)$.

We assert that $\otimes : P\times Q \to P\otimes Q$ is a bimorphism. Indeed, since $\{(1,1)\} \in \mathcal{H}$, we have $\otimes(1,1) = \pi(\{(1,1)\}) = 1$. Suppose that $a,b \in P$ with $a \perp b$ and $q \in Q$. We have to show that $a\otimes q \perp b\otimes q$ and $(a\oplus b)\otimes q = (a\otimes q)\oplus(b\otimes q)$. If $(a,q) \in N$ or $(b,q) \in N$, this is clear, so we may assume that $(a,q),(b,q) \in X$. If β is any bimorphism on $P\times Q$, we have $\beta(a \oplus b,q) = \beta(a,q) \oplus \beta(b,q)$. Hence $\{(a \oplus b,q)\} \sim \{(a,q),(b,q)\}$, so that $(a \oplus b) \otimes q = (a \otimes q) \oplus (b \otimes q)$.

A similar argument shows that $p\otimes(c\oplus d) = (p\otimes c)\oplus(p\otimes d)$ holds for $p \in P$ and $c,d \in Q$ with $c \perp d$.

It remains to prove that $(P \otimes Q, \otimes)$ is a tensor product of P and Q. Since every element of $P \otimes Q = \Pi(X)$ can be written in the form $\pi(A) = \bigoplus\{\pi(\{(p,q)\}) : (p,q) \in A\} = \bigoplus\{p \otimes q : (p,q) \in A\}$, every element of $P \otimes Q$ is a $\bigoplus$-sum of finitely many elements $p \otimes q$.

Finally, suppose that $\beta : P \times Q \to L$ is a bimorphism. If $A, B \in \mathcal{E}(X, \mathcal{H})$ and $A \approx B$, then $\bigoplus \beta(A) = \bigoplus \beta(B)$, hence we can define a mapping $\phi : P \otimes Q \to L$ by $\phi(\pi(A)) = \bigoplus \beta(A)$ for every $\pi(A) \in \Pi(X)$. Obviously, ϕ is a morphism and we have $\beta(p,q) = \phi(p \otimes q)$ for all $p \in P, q \in Q$. □

Unless confusion threatens, we usually refer to $P \otimes Q$ rather than to $(P \otimes Q, \otimes)$, as being a tensor product.

Theorem 4.2.3 *Let both D-posets P and Q possess at least one state. Then the tensor product of P and Q exists in the category of D-posets. In addition, for any state μ on P and any state ν on Q there is a unique state $\mu \otimes \nu$ on $P \otimes Q$ such that*

$$\mu \otimes \nu(p \otimes q) = \mu(p)\, \nu(q), \quad p \in P,\, q \in Q. \tag{4.5}$$

Proof. Let $L = [0, 1]$ be endowed with the natural ordering and the difference $b \ominus a := b - a$, $a, b \in [0, 1]$, $a \leq b$. Then L is a D-poset. Choose two states μ and ν on P and Q, respectively, and define a mapping $\beta_{\mu\nu} : P \times Q \to L$ such that

$$\beta_{\mu\nu}(p,q) = \mu(p) \cdot \nu(q), \quad p \in P,\, q \in Q. \tag{4.6}$$

Then $\beta_{\mu\nu}$ is a bimorphism, which, by Theorem 4.2.2, is a necessary and sufficient condition for P and Q to admit a tensor product.

Since $\beta_{\mu\nu}$ is a bimorphism, from the definition of $P \otimes Q$ it follows that there is a morphism $\phi : P \otimes Q \to [0, 1]$ such that $\phi(p \otimes q) = \beta_{\mu\nu}(p,q)$, $p \in P, q \in Q$. But it means that ϕ is a state on $P \times Q$ with the desired property (4.5). The uniqueness of ϕ is clear due to the property of $P \otimes Q$ that any element $t \in P \otimes Q$ is of the form $t = \bigoplus_{i=1}^{n} p_i \otimes q_i$. □

In [FoBe], it has been shown that the orthoalgebra, the Fano plane, illustrated by the Greechie diagram on page 101 or 279 has no tensor product $F \otimes F$ in the category of orthoalgebras. Since on F there is unique state, say μ, defined by $\mu(x) = 1/3$, $x \in \{a, b, c, d, e, f, g\}$, using Theorem 4.2.3, we see that $F \otimes F$ as a D-poset exists.

We now show that also the category of D-posets, or equivalently of effect algebras, is not closed with respect to formation of tensor product. We will follow the example of Gudder and Greechie [GuGr, Thm 4]. We recall that by an n-chain D-poset (or an n-chain effect algebra) we mean a chain generated by a non-zero element a, i.e., $\{0, a, 2a, \ldots, na\}$.

Example 4.2.4 *If Q is a 4-chain generated by d and P the horizontal sum of three 4-chains generated by a, b, c with the additional requirement that $a \oplus b \oplus c = 1$, then the tensor product of P and Q does not exists.*

Proof. Suppose a bimorphism $\beta: P \times Q \to R$ exists. Then

$$\begin{aligned} 1 &= \beta(1,1) = \beta(1,4d) = 4\beta(1,d) = 4\beta(a \oplus b \oplus c, d) \\ &= 4[\beta(a,d) \oplus \beta(b,d) \oplus \beta(c,d)] \\ &= 4\beta(a,d) \oplus 4\beta(b,d) \oplus 4\beta(c,d) \\ &= \beta(4a,d) \oplus \beta(4b,d) \oplus \beta(4c,d) = 3\beta(1,d). \end{aligned}$$

Since $3\beta(1,d) \oplus \beta(1,d) = 1$ we conclude that $\beta(1,d) \oplus 1$ is defined. Hence, $\beta(1,d) = 0$. It follows that $0 = 3\beta(1,d) = 1$, which is a contradiction. □

Corollary 4.2.5 *Let $\mathcal{E}(H_1)$ and $\mathcal{E}(H_2)$ be two sets of effects in Hilbert spaces H_1 and H_2 (not necessarily over the same field). Then $\mathcal{E}(H_1) \otimes \mathcal{E}(H_2)$ exists.*[3]

Proof. This follows from the fact that, for any von Neumann operator T_i on H_i, the mapping

$$\mu_i(A) = \mathrm{tr}(T_i A), \ A \in \mathcal{E}(H_i), \tag{4.7}$$

is a probability measure on $\mathcal{E}(H_i)$, and $\mathcal{S}(\mathcal{E}(H_i))$ is, in addition, sufficient for $i = 1, 2$. The assertion of the corollary follows now from Theorem 4.2.3. □

4.2.2 Bounded Boolean Powers and Tensor Products

A special kind of a tensor product is needed if we wish to describe a coupled system consisting of one quantum system and one classical one. This situation arises for example by quantum measurements, where we wish to measure a quantum observable by a measuring device [BLM]. A quantum logic approach to quantum measurements has been developed in [Pul 13], [Pul 19], [Pul 21]. The main tool used there is a Boolean power, defined below (for OMP see also [Pta 3] and for OA [FoPt]).

In the rest of the section, we shall study a tensor product of a difference poset L (i.e., a logic of a quantum system) and a Boolean algebra B (i.e., a logic of a classical system). It is well-known that $\mathcal{S}(B)$ is sufficient, but on the other hand it can happen that $\mathcal{S}(L) = \emptyset$, so that Theorem 4.2.3 is not effective. Nevertheless we show that the tensor product $L \otimes B$ exists and, in addition, it is isomorphic to a bounded Boolean power.

So let L be a difference poset and B a Boolean algebra with the smallest and greatest elements 0_B and 1_B, respectively. According to [DvPu 1], we define

$$L[B]^* = \{f \in B^L : \ a \neq b \Rightarrow f(a) \wedge f(b) = 0_B, f(L) \text{ is finite}, \bigvee_{a \in L} f(a) = 1_B\}. \tag{4.8}$$

[3]The same assertion holds if we use the systems of all effects in von Neumann algebras.

The set $L[B]^*$ is said to be a **bounded Boolean power** of L. In the paper [DvPu 1], there has been introduced also a Boolean power of L for the case when B is a complete Boolean algebra.

Define, for any $a \in L$, a mapping $\hat{a} : L \to B$ via

$$\hat{a}(x) = \begin{cases} 1_B, & \text{if } x = a, \\ 0_B, & \text{if } x \neq a, \end{cases} \quad x \in L. \tag{4.9}$$

Theorem 4.2.6 *Let $(L; \leq, \ominus, 0, 1)$ be a difference poset and let B be a Boolean algebra. For $f, g \in L[B]^*$ we define a partial binary relation $\preceq$ via*

$$f \preceq g \quad \textit{iff} \quad \bigvee_{\substack{x,y \in L \\ x \leq y}} (f(x) \wedge g(y)) = 1_B, \tag{4.10}$$

and for $f, g \in L[B]^*$ with $f \preceq g$ we put $g \ominus f : L \to B$ via

$$(g \ominus f)(x) = \bigvee_{\substack{a,b \in L \\ x = b \ominus a}} (g(b) \wedge f(a)), \; x \in L. \tag{4.11}$$

The Boolean power $L[B]^*$ with $\preceq, \ominus$ defined via (4.8), (4.10) *and* (4.11) *is a difference poset with the smallest and greatest elements $\hat{0}$ and $\hat{1}$, respectively.*

Proof. This follows the same ideas as that in [DvPu 1] for Boolean powers, and to illustrate it, we present a typical step of the proof: the antisymmetry of $\preceq$.

Let now $f \preceq g$, $g \preceq f$. Then

$$\begin{aligned} 1_B &= \left(\bigvee_{\substack{x,y \in L \\ x \leq y}} [f(x) \wedge g(y)] \right) \wedge \left(\bigvee_{\substack{u,v \in L \\ u \leq v}} [g(u) \wedge f(v)] \right) \\ &= \bigvee_{\substack{x,y \in L \\ x \leq y}} \bigvee_{\substack{u,v \in L \\ u \leq v}} [f(x) \wedge g(y) \wedge g(u) \wedge f(v)] \\ &= \bigvee_{\substack{x,y,v \in L \\ x \leq y \leq v}} [f(x) \wedge g(y) \wedge f(v)] = \bigvee_{x \in L} [f(x) \wedge g(x)] \end{aligned}$$

so that, for any $y \in L$, $f(y) = (\bigvee_{x \in L} f(x) \wedge g(x)) \wedge f(y) = \bigvee_{x \in L} [f(x) \wedge f(y) \wedge g(x)] = f(y) \wedge g(y)$, hence $f(y) \leq g(y)$. By symmetry, $g(y) \leq f(y)$, hence $f(y) = g(y)$, $y \in L$, and $f = g$.

For more details, see [DvPu 1]. □

Let B be a Boolean algebra and L a difference poset. Let $T = \{t_i : i \in I\}$ be a finite decomposition of 1_B, i.e., $t_i \wedge t_j = 0_B$ if $i \neq j$ and $\bigvee_{i \in I} t_i = 1_B$. If $\{f_i : i \in I\} \subseteq L[B]^*$, then

$$f(x) = \bigvee_{i \in I} (f_i(x) \wedge t_i), \quad x \in L, \tag{4.12}$$

is an element of $L[B]^*$. For (4.12) we can use $f = \bigvee_{i \in I} f_i \wedge t_i$ or the "sum" notation

$$f = \sum_{i \in I} f_i \cdot t_i. \tag{4.13}$$

In particular, if $\{a_i : i \in I\} \subseteq L$ and $\{t_i : i \in I\}$ is a finite decomposition of 1_B, then

$$\sum_{i \in I} \hat{a}_i \cdot t_i \tag{4.14}$$

belongs to $L[B]^*$. Conversely, any element $f \in L[B]^*$ can be written in the form (4.14) for appropriate pairwise different a_i's in L and a finite decomposition $T = \{t_i : i \in I\}$. Indeed, given $f \in L[B]^*$ we put $I = L$ and $T = \{f(a) : a \in L\}$. Then

$$f = \sum_{a \in L} \hat{a} \cdot t_a,$$

where $t_a = f(a)$, $a \in L$.

In addition, we may assume that the finite decomposition of 1_B is strictly positive (i.e., $t_i \neq 0_B$ for each i). This form is called the **reduced representation** of f by its values. We recall that in this case f has a unique reduced representation. Indeed, if $f = \sum_i \hat{a}_i \cdot t_i = \sum_j \hat{b}_j \cdot s_j$, $t_i, s_j > 0_B$, and $\{a_i\}$ and $\{b_j\}$ consist of pairwise different elements, then $f(x) = 0_B$ iff $x \neq a_i$ for any i, and $f(x) = t_i$ iff $x = a_i$, so that $a_i = b_j$ and $t_i = s_j$ for some i and j.

Theorem 4.2.7 *Let L be a difference poset and let B be a Boolean algebra. Then the mappings $\lambda :\ L \to L[B]^*$, defined via $\lambda(a) = \hat{a}$, where $\hat{a}$ is defined via (4.9) and $\beta :\ B \to L[B]^*$ defined for $b \in B$ via*

$$\beta(b)(x) \begin{cases} b, & \text{if } x = 1_L, \\ b^c, & \text{if } x = 0_L, \\ 0_B, & \text{otherwise}, \end{cases} \quad x \in L, \tag{4.15}$$

are monomorphisms preserving all existing suprema (infima) in L and $L[B]^*$, respectively. In particular,

$$\lambda(\bigoplus_i a_i) = \bigoplus_i \lambda(a_i),$$

whenever $\bigoplus_i a_i$ exists in L.

Proof. We recall that the partial binary operation $\oplus$ on $L[B]^*$ can be defined via

$$(f \oplus g)(x) = \bigvee_{\substack{u,v \in L \\ u \oplus v = x}} (f(u) \wedge g(v)),\ x \in L,$$

and $f^\perp(x) = f(x^\perp),\ x \in L$.

Since $\lambda(a) \leq \lambda(b)$ iff $a \leq b$, we argue that if $a = \bigvee_i a_i$, then $\lambda(a) \geq \lambda(a_i)$ for any i. If for $g \in L[B]^*$ we have $g \geq \lambda(a_i)$ for any i, we have

$$1_B = \bigvee_{\substack{x,y \in L \\ x \leq y}} (g(y) \wedge \hat{a}_i(x)) = \bigvee_{\substack{y \in L \\ y \geq a_i}} g(y),$$

which gives

$$1_B = \bigvee_{\substack{y \in L \\ y \geq a}} g(y) = \bigvee_{\substack{x,y \in L \\ x \leq y}} (g(y) \wedge \hat{a}(x)),$$

so that $\lambda(a) \leq g$.

For β we conclude as follows. We have $\beta(b)(x) = (\hat{1}(x) \wedge b) \vee (\hat{0}(x) \wedge b^c)$ or $\beta(b) = \hat{1} \cdot b + \hat{0} \cdot b^c$. Hence $\beta(1)(x) = \hat{1}(x)$ and $\beta(b^c)(x) = (\hat{1}(x) \wedge b^c) \vee (\hat{0}(x) \wedge b) = \beta(b)(x^\perp)$, since $\hat{1}(x^\perp) = \hat{0}(x)$ and $\hat{0}(x^\perp) = \hat{1}(x)$. Let $f \in L[B]^*$. Then

$$\beta(b)(x) \wedge f(y) = \begin{cases} b \wedge f(y), & \text{if } x = 1_L, \\ b^c \wedge f(y), & \text{if } x = 0_L, \\ 0_B, & \text{if } x \neq 1_L,\ x \neq 0_L, \end{cases}$$

and therefore

$$\bigvee_{\substack{x,y \in L \\ x \leq y}} [\beta(b)(x) \wedge f(y)] = (b \wedge f(1)) \vee \bigvee_{y \in L} [b^c \wedge f(y)] = b^c \vee (b \wedge f(1)),$$

which entails that $\beta(b) \leq f$ if and only if $b \leq f(1)$. For $b_1, b_2 \in B$ this gives that $\beta(b_1) \leq \beta(b_2)$ if and only if $b_1 \leq b_2$. Assume now that $b = \bigvee_i b_i$. Let f be any upper bound of $\beta(b_i)$, for any i. From $b_i \leq f(1)$, for any i we get $\bigvee_i b_i \leq f(1)$, hence $\beta(b) \leq f$. That is, $\beta(b) = \bigvee_i \beta(b_i)$. This proves that $\beta : B \to L[B]^*$ is an embedding. □

Theorem 4.2.8 *Let L be a difference poset and B a Boolean algebra. Then the tensor product $L \otimes B$ exists and is isomorphic to the bounded Boolean power $L[B]^*$.*

Proof. Define a mapping $\gamma_0 : L \times B \to L[B]^*$ as follows

$$\gamma_0(a, b) = \hat{a} \cdot b + \hat{0} \cdot b^c,\ (a, b) \in L \times B. \tag{4.16}$$

Then from Theorem 4.2.7 we conclude that γ_0 is a bimorphism on $L \times B$ and, by Theorem 4.2.2 $L \otimes B$ exists. Therefore, there is a morphism $\gamma : L \otimes B \to L[B]^*$ such that $\gamma(a \otimes b) = \gamma_0(a, b)$. If we use a reduced representation $f = \sum_{i=1}^n \hat{a}_i \cdot t_i = \bigoplus_{i=1}^n \gamma_0(a_i, t_i)$, we see from

(i) $\gamma_0(a_1, b_1) \perp \gamma_0(a_2, b_2)$ iff $a_1 \perp a_2$ or $b_1 \perp b_2$;

(ii) $t = \bigoplus_{i=1}^n a_i \otimes t_i \in L \otimes B$ and $\gamma(t) = f$,

that γ is surjective onto $L[B]^*$. Since the reduced representation is unique, we conclude that γ is a monomorphism, and so γ is an isomorphism, which proves that $L[B]^*$ is isomorphic with $L \otimes B$. □

Let, for any $i \in I$, L_i with $\leq_i$, 1_i, $\ominus_i$ be a D-poset. We recall that $L := \prod_{i\in I} L_i$ is a D-poset, called a **product D-poset** of $\{L_i : i \in I\}$, when $\leq$, 1, and $\ominus$ are defined on L as follows: $\{a_i\} \leq \{b_i\}$ iff $a_i \leq_i b_i$, $i \in I$, $1 = \{1_i\}$, $\{b_i\} \ominus \{a_i\} = \{b_i \ominus_i a_i\}$.

Theorem 4.2.9 *Let $B = 2^n$ and let L be a difference poset. Put $L_0 = \prod_{j=1}^n L_j$, where $L_j = L$ for $j = 1, \ldots, n$. Then $L_0, L[B]^*$ and $L \otimes B$ are isomorphic.*

Proof. We can assume that $N = \{1, \ldots, n\}$, and $B = 2^N$. Then $\{\{j\} : j = 1, \ldots, n\}$ is the set of atoms of B. For $\{a_j\}_{j=1}^n \in L_0$ we define an element of $L[B]^*$ via $\sum_{j=1}^n \hat{a}_j \cdot \{j\}$. It is not hard to show that the mapping $h : L_0 \to L[B]^*$ such that $\{a_j\}_{j=1}^n \mapsto \sum_{j=1}^n \hat{a}_j \cdot \{j\}$. is a monomorphism from L_0 into $L[B]^*$.

On the other hand, let $f \in L[B]^*$ and let it have the reduced representation $f = \sum_i \hat{a}_i \cdot t_i$. Then, for any $x \in L$, we have

$$\begin{aligned} f(x) &= \bigvee_i (\hat{a}_i(x) \wedge t_i) = \bigvee_i \bigvee_{j=1}^n \hat{a}_i(x) \wedge t_i \wedge \{j\} \\ &= \left(\bigvee_{j=1}^n \bigvee_{\substack{i \\ \{j\}\leq t_i}} \hat{a}_i(x) \wedge t_i \wedge \{j\} \right) \wedge \left(\bigvee_{j=1}^n \bigvee_{\substack{i \\ \{j\}\perp t_i}} \hat{a}_i(x) \wedge t_i \wedge \{j\} \right) \\ &= \bigvee_{j=1}^n \bigvee_{\substack{i \\ \{j\}\leq t_i}} (\hat{a}_i(x) \wedge t_i \wedge \{j\}). \end{aligned}$$

If we put $c_j = a_i$ whenever $\{j\} \leq t_i$, then $f = \sum_{j=1}^n \hat{c}_j \cdot \{j\}$, which proves that h is surjective. □

Finally, we note that if L is a Boolean algebra, so is $L[B]^*$, and there exists a tensor product of Boolean algebras, which in view of Theorem 4.2.8 is always a Boolean algebra.

4.2.3 Tensor Product of D-posets and D-test Spaces

Let L be a difference poset. The D-test space $(X, \mathcal{T})$, where $X = L \setminus \{0\}$ and $\mathcal{T}$ is the family of all finite partitions of unity, will be called the **standard** D-test space for L. Recall that for $E, F \in \mathcal{E}(X)$, $E \approx F$ iff $\bigoplus \pi(E) = \bigoplus \pi(F)$.

Proposition 4.2.10 *Let $(X,\mathcal{T})$, $(Y,\mathcal{S})$, $(Z,\mathcal{U})$ be standard D-test spaces for the D-posets P,Q,R, respectively.*

(i) *If $\psi : P \to Q$ is a morphism of D-posets, then $\phi : X \to Y$ defined by*

$$\phi(x) = \psi(\pi(\{x\}))$$

is a morphism of D-test spaces.

(ii) *If $\rho : P \times Q \to R$ is a bimorphism of D-posets, then $\sigma : X \times Y \to Z$ defined by*

$$\sigma(x,y) = \rho(\pi_X(\{x\}), \pi_Y(\{y\}))$$

is a bimorphism of D-test spaces, where π_X and π_Y are the corresponding mappings from $\mathcal{E}(X)$ and $\mathcal{E}(Y)$ onto P and Q, respectively.

Proof. (i) Let $T = \{T(i)\}_{i\le n} \in \mathcal{T}$. Then $\phi(T) = \{\psi(\pi_X(\{T(i)\}))\}_{i\le n}$, and since T is a finite partition of unity in P, $\phi(T)$ is a finite partition of unity in Q, hence ϕ is a morphism.

(ii) If $T = \{T(i)\}_{i\le n}$, $S = \{S(j)\}_{j\le m}$ are tests in $\mathcal{T}$ and $\mathcal{S}$, respectively, then $\sigma(T\dot{\times}S) = \{\rho(\pi_X(\{T(i)\}), \pi_Y(\{S(j)\}))\}_{i\le n, j\le m}$, and $\bigoplus \sigma(T\dot{\times}S) = 1$, hence $\sigma(T\dot{\times}S)$ is a test in $\mathcal{U}$. For $T_1, T_2 \in \mathcal{T}$ and $F \in \mathcal{E}(Y)$, we get that $\sigma(T_1\dot{\times}E)$, $\sigma(T_2\dot{\times}E)$ are finite sets of orthogonal elements, and $\bigoplus \sigma(T_1\dot{\times}E) = \rho(1,\pi(E)) = \bigoplus \sigma(T_2\dot{\times}E)$, whence $\sigma(T_1\dot{\times}E) \approx \sigma(T_2\dot{\times}E)$. This proves that σ is a bimorphism of D-test spaces. □

We now will study relations between tensor products of D-test spaces and tensor products of the corresponding D-posets.

Theorem 4.2.11 *Let $(Z,\mathcal{U},\kappa)$ be the tensor product of D-algebraic D-test spaces $(X,\mathcal{T})$ and $(Y,\mathcal{S})$ in the category of D-algebraic D-test spaces. Denote $P := \Pi_X(X)$, $Q := \Pi_Y(Y)$, where $\Pi_X(X)$ and $\Pi_Y(Y)$ are defined by (4.1). Then the tensor product of P and Q in the category of D-posets exists.*

Proof. Denote $R := \Pi_Z(Z)$. Define the mapping $\tau : P \times Q \to R$ by $\tau(a,b) = \pi_Z(\kappa(E\dot{\times}F))$, where $a = \pi_X(E)$, $b = \pi_Y(F)$. By Proposition 4.1.23, τ is a bimorphism. By Theorem 4.2.2, the tensor product of the D-posets P and Q exists. □

Corollary 4.2.12 *Let P and Q be D-posets and $(X,\mathcal{T})$ and $(Y,\mathcal{S})$ be the corresponding standard D-algebraic D-test spaces. Then the tensor product of P and Q exists in the category of D-posets iff there exists the tensor product of $(X,\mathcal{T})$ and $(Y,\mathcal{S})$ in the category of D-algebraic D-test spaces.*

Proof. This follows from Proposition 4.2.10 and Theorems 4.2.11 and 4.1.26. □

Theorem 4.2.13 *Let P, Q and R be D-posets and let $(X,\mathcal{T}),(Y,\mathcal{S})$ and $(Z,\mathcal{U})$ be the corresponding standard D-test spaces. Let (R,γ) be the tensor product of P and Q in the category of D-posets, $(V,\mathcal{W},\delta)$ be the tensor product of D-algebraic D-test spaces $(X,\mathcal{T})$, and $(Y,\mathcal{S})$, and D be the corresponding D-poset. Then there exist morphisms $\alpha : D \to R$ and $\beta : R \to D$ such that $\alpha \circ \beta = id_R$. Moreover, if every test in $\mathcal{W}$ is finite, then D and R are isomorphic.*

Proof. Since $(V,\mathcal{W},\delta)$ is a tensor product, there is a morphism $\phi : V \to Z$ such that $\phi\circ\delta = \kappa$, where $\kappa(x,y) = \gamma(\pi_X(\{x\}),\pi_Y(\{y\}))$. Let $\tilde{\phi} : D \to R$ be the morphism defined by $\tilde{\phi}(\pi_V(E)) = \pi_Z(\phi(E))$. Let $\tilde{\delta} : P\times Q \to D$ be the bimorphism defined by $\tilde{\delta}(\pi_X(E),\pi_Y(F)) = \pi_V(\delta(E\dot{\times}F))$. Since (R,γ) is a tensor product, there is a morphism $\psi : R \to D$ such that $\tilde{\delta} = \psi\circ\gamma$. Further, $\tilde{\kappa}(\pi_X(E),\pi_Y(F)) := \pi_Z(\kappa(E\dot{\times}F)) = \gamma(\pi_X(E)\times\pi_Y(F))$ implies $\tilde{\phi}\circ\tilde{\delta} = \tilde{\kappa} = \gamma$, whence $\gamma = \tilde{\phi}\circ\psi\circ\gamma$. Since every element in R is a finite orthogonal sum of elements of the form $\gamma(p,q)$, $p\in P$, $q \in Q$, we conclude $\tilde{\phi}\circ\psi = id_R$. Putting $\alpha = \tilde{\phi}$, $\beta = \psi$, we obtain the desired statement.

To prove the remaining statement, assume that every test in $\mathcal{W}$ is finite. Let $a \in D$ and let $E \in \mathcal{E}(V)$ be such that $a = \pi_V(E)$. Now $E \preccurlyeq T$ for some $T \in \mathcal{W}$. By (ii) of the definition of tensor product, there is a finite $E_0 \preccurlyeq F$, $F \in (X\times Y)^{I\times J}$ such that $E = \delta(E_0)$ and $T = \delta(F)$. Then

$$\begin{aligned} a &= \pi_V(\delta(E_0)) = \bigoplus_{(x,y)\in E_0} \pi_V(\{\delta(x,y)\}) \\ &= \bigoplus_{(x,y)\in E_0} \{\tilde{\delta}(\pi_X(\{x\}),\pi_Y(\{y\}))\}, \end{aligned}$$

which entails $\psi\circ\tilde{\phi}(a) = a$ (observe that $\tilde{\delta} = \psi\circ\tilde{\phi}\circ\tilde{\delta}$). This shows that ψ is surjective, and hence it is an isomorphism. □

4.2.4 State Tensor Product

Let P and Q be D-posets. Let $\mathcal{K} = (\mathcal{L},\mathcal{B})$ be a couple, where $\mathcal{L}$ is a nonvoid family of D-posets and $\mathcal{B}$ is a nonvoid family of bimorphisms on $P\times Q$ such that (i) for any $\beta\in\mathcal{B}$ there exists a D-poset $L\in\mathcal{L}$ such that $\beta : P\times Q\to L$, and (ii) for any $L\in\mathcal{L}$ there is a bimorphisms $\beta\in\mathcal{B}$ with $\beta : P\times Q \to L$; $\mathcal{K}$ is said to be a **consistent class** for P and Q.

Definition 4.2.14 *Let $\mathcal{K} = (\mathcal{L},\mathcal{B})$ be a consistent class for the D-posets P and Q. We say that a pair (T,τ) consisting of a difference poset T and a bimorphism $\tau : P\times Q\to T$ is a* **tensor product** *of P and Q* **in the class** *$\mathcal{K} = (\mathcal{L},\mathcal{B})$ iff the following conditions are satisfied:*

(i) $T\in\mathcal{L}$, $\tau\in\mathcal{B}$.

(ii) *If L is a D-poset in $\mathcal{L}$, and β is a bimorphism in $\mathcal{B}$, $\beta : P \times Q \to L$, there exists a morphism $\phi : T \to L$ such that $\beta = \phi \circ \tau$.*

(iii) *Every element of T is a finite $\bigoplus$-orthogonal sum of elements of the form $\tau(p,q)$ with $p \in P, q \in Q$.*

Similarly as above, if a tensor product (T, τ) of P and Q exists in the class $\mathcal{K}$, it is unique up to isomorphism. It is clear that if $\mathcal{L}$ consists of all D-posets L for which there exists a bimorphism $\beta : P \times Q \to L$ and $\mathcal{B}$ is the family of all those bimorphisms, then tensor products from Definition 4.2.1 and Definition 4.2.14 coincide. Moreover, let $\mathcal{K}_i = (\mathcal{L}_i, \mathcal{B}_i), i = 1,2$, be two consistent classes for P and Q such that $\mathcal{L}_1 \subseteq \mathcal{L}_2$ and $\mathcal{B}_1 \subseteq \mathcal{B}_2$. If $(T_i, \tau_i), i = 1,2$, is a tensor product of P and Q, then there exists a morphism $\phi : T_2 \to T_1$ such that $\tau_1 = \phi \circ \tau_2$.

Suppose that $\mathcal{P}_1$ and $\mathcal{P}_2$ are nonempty families of states on difference posets P and Q. We set $\mathcal{P} := \mathcal{P}_1 \times \mathcal{P}_2$ and, if $\lambda = (\mu, \nu) \in \mathcal{P}$ and $(p,q) \in P \times Q$, then $\lambda(p,q) := \mu(p) \cdot \nu(q)$.

We recall that a family of states $\mathcal{P}$ on a D-poset L is

(i) **separating** iff $\mu(a) = \mu(b)$ for any $\mu \in \mathcal{P}$ implies $a = b$;

(ii) **strong** iff, for $a, b \in L$, the statement "if $\mu(a) = 1$, then $\mu(b) = 1$, $\mu \in \mathcal{P}$," implies $a \leq b$;

(iii) **unital** iff, for any $a \in L$, $a \neq 0$, there is $\mu \in \mathcal{P}$ with $\mu(a) = 1$;

(iv) **order determining** iff the condition $\mu(a) \leq \mu(b)$ for all $\mu \in \mathcal{P}$ implies $a \leq b$;

(v) **full** iff $\mathcal{P}$ is separating, and if $\sum_{i=1}^n \mu(a_i) = 1$ for any $\mu \in \mathcal{P}$, then $\bigoplus_{i=1}^n a_i$ exists in L and $\bigoplus_{i=1}^n a_i = 1$.

If $\mathcal{P}$ is a strong system of states, then $\mathcal{P}$ is (i) unital; (ii) separating, and (iii) order determining. If $\mathcal{P}$ is order determining, so is separating.

It is not hard to show that if a D-poset L has a unital system of states, then L is an orthoalgebra: Indeed, let $a \perp a$. If $a \neq 0$, there is $\mu \in \mathcal{P}$ with $\mu(a) = 1$ which gives a contradiction $1 \geq \mu(a \oplus a) = \mu(a) + \mu(a) = 2$. Now, if an orthoalgebra has a strong system of states, then L is an OMP: $a \perp b$ and $a, b \leq c$, then $\mu(c) = 0$ implies $\mu(a) = \mu(b) = 0 = \mu(a \oplus b)$. Hence, $c^\perp \leq (a \oplus b)^\perp$, which gives $a \oplus b \leq c$ and $a \oplus b = a \vee b$.

If L is a D-poset and $\mathcal{P}$ is order determining, then $\mathcal{P}$ is full: Let $\sum_{i=1}^n \mu(a_i) = 1$. Then, for any $i \neq j$, $\mu(a_i) \leq \mu(a_j^\perp)$, so that $a_i \perp a_j$ and $a_i \oplus a_j \in L$. By induction we can show that $\bigoplus_{i=1}^n a_i \in L$, and the separateness of L implies $\bigoplus_{i=1}^n a_i = 1$.

As a corollary we have that $\mathcal{E}(H)$ has no strong system of states whenever $\dim H > 1$, but it has an order determining system of states and full (see the formula (4.7)). Similarly, the D-poset $L_g = ([0,1]; \leq, \ominus_g, 0, 1)$ where

$$b \ominus_g a := g^{-1}(g(b) - g(a)),\ a \leq b, \tag{4.17}$$

which is determined by a generator g, has the unique order determining and full system of states, namely $\mathcal{P} = \{\mu\}$, where $\mu(t) = g(t)$, $t \in [0, 1]$.

Theorem 4.2.15 *Let $\mathcal{P}_1$ and $\mathcal{P}_2$ be nonvoid systems of states on the D-posets P and Q, respectively. Let $\mathcal{L}_\mathcal{P}$ be the set of all D-posets L such that there is a bimorphism $\kappa : P \times Q \to L$, and the set $\mathcal{P}_\kappa := \{\mu \otimes_\kappa \nu : \ \mu \in \mathcal{P}_1, \nu \in \mathcal{P}_2\}$ is a full system of states on L, where $\mu \otimes_\kappa \nu(\kappa(p,q)) := \mu(p) \cdot \nu(q)$, $p \in P, q \in Q$, and let $\mathcal{B}_\mathcal{P}$ be the set of all these bimorphisms κ's. Then $\mathcal{K}_\mathcal{P} = (\mathcal{L}_\mathcal{P}, \mathcal{B}_\mathcal{P})$ is a consistent class for P and Q, and there exists a tensor product of P and Q in the class $\mathcal{K}$.*

Proof. Let X be the subset of $P \times Q$ consisting of all pairs (p, q) with $p \neq 0, q \neq 0$. If $M = \{(p_1, q_1), \ldots, (p_n, q_n)\}$ is a finite sequence of elements from X and $\lambda \in \mathcal{P} := \mathcal{P}_1 \times \mathcal{P}_2$, we put

$$\lambda(M) = \sum_{i=1}^{n} \lambda(p_i, q_i)$$

with the understanding that if $M = \emptyset$, then $\lambda(M) = 0$.

Define now the set $\mathcal{F}$ of all finite sequences $T = \{(p_i, q_i)\}_{i=1}^{n}$ of elements in X such that $\lambda(T) = 1$ for any $\lambda \in \mathcal{P}$. Since $\lambda(1, 1) = 1$, $\mathcal{F}$ is nonvoid. It is clear that if $(p, q) \in X$, then from the set $\{(p, q), (p^\perp, q), (p, q^\perp), (p^\perp, q^\perp)\} \cap X$ we can choose a finite sequence containing (p, q) and belonging to $\mathcal{F}$. Denote by $\mathcal{E}(\mathcal{F})$ the set of all finite sequences $\{(p_j, q_j) : j \in J\}$ such that $J \subseteq I$ and $\{(p_i, q_i) : i \in I\} \in \mathcal{F}$. We put $\{(p_i, q_i) : i \in \emptyset\} = 0$. It is evident that $(X, \mathcal{F})$ is a D-algebraic D-test space.

For two events $A, B \in \mathcal{E}(\mathcal{F})$ we define $A \approx B$ iff $\lambda(A) = \lambda(B)$ for any $\lambda \in \mathcal{P}$. Then $\approx$ is an equivalence on $\mathcal{E}(\mathcal{F})$, and let $\pi(A) := \{B \in \mathcal{E}(\mathcal{F}) : \ B \approx A\}$. Let $\Pi(X) = \{\pi(A) : \ A \in \mathcal{E}(\mathcal{F})\}$. We organize $\Pi(X)$ into a poset by defining a partial ordering $\leq$ on $\Pi(X)$ as follows: $\pi(A) \leq \pi(B)$, where $A = \{(p_1, q_1), \ldots, (p_n, q_n)\}, B = \{(r_1, s_1), \ldots, (r_m, s_m)\}$, iff there is $C = \{(p_1', q_1'), \ldots, (p_s', q_s')\} \in \mathcal{E}(\mathcal{F})$ such that $M := \{(p_1, q_1), \ldots, (p_n, q_n), (p_1', q_1'), \ldots, (p_s', q_s')\} \in \mathcal{E}(\mathcal{F})$ and $\lambda(M) = \lambda(B)$ for any $\lambda \in \mathcal{P}$. Then $\pi(\emptyset)$ and $\pi(T)$, where $T \in \mathcal{F}$, are the smallest and greatest elements in $\Pi(X)$.

The difference operation $\ominus$ on $\Pi(X)$ is defined whenever $\pi(A) \leq \pi(B)$, and $\pi(B) \ominus \pi(A) = \pi(C)$, where A, B, C satisfy the above mentioned conditions for the partial ordering $\leq$. Then $\ominus$ is defined correctly, and $\Pi(X)$ is a difference poset.

Define a mapping $\kappa_0 : P \times Q \to \Pi(X)$ via

$$\kappa_0(p, q) = \begin{cases} \pi(\{(p, q)\}), & \text{if } (p, q) \in X, \\ 0, & \text{if } (p, q) \notin X. \end{cases}$$

Then κ_0 is, evidently, a bimorphism. From the construction of $\Pi(X)$ we see that any $t \in \Pi(X)$ is of the form $t = \bigoplus_{i=1}^{n} \kappa_0(p_i, q_i)$, and the mapping $\mu \otimes_{\kappa_0} \nu$ on $\Pi(X)$ defined by $\mu \otimes_{\kappa_0} \nu(\kappa_0(p, q)) = \mu(p) \cdot \nu(q)$, $p \in P, q \in Q$, is a state on $\Pi(X)$. In addition, $\mathcal{P}_{\kappa_0}$ is a full system of states on $\Pi(X)$. Therefore, $\mathcal{L}_\mathcal{P} \neq \emptyset$, and let $\mathcal{B}_\mathcal{P}$ be the set of all bimorphisms κ such that κ maps $P \times Q$ into some $L \in \mathcal{L}_\mathcal{P}$ and $\mathcal{P}_\kappa$ be a full system of states on L. Then $\mathcal{K}_\mathcal{P} = (\mathcal{L}_\mathcal{P}, \mathcal{B}_\mathcal{P})$ is a consistent class for P and Q.

We assert that $(\Pi(X), \kappa_0)$ is a tensor product of P and Q in the class $\mathcal{K}_\mathcal{P}$. Choose $L \in \mathcal{L}_\mathcal{P}$ and a bimorphism $\kappa : P \times Q \to L$. Since $\mathcal{P}_\kappa$ is full for L, it follows that if $\kappa_0(p,q) = \kappa_0(p',q')$, then $\kappa(p,q) = \kappa(p',q')$ and we can define a mapping $\phi : \Pi(X) \to L$ such that $\phi(\kappa_0(p,q)) = \kappa(p,q)$, $p \in P$, $q \in Q$. We claim that if we extend ϕ to whole $\Pi(X)$ via $\phi(t) = \bigoplus_{i=1}^n \kappa(a_i,b_i)$, whenever $t = \bigoplus_{i=1}^n \kappa_0(a_i,b_i)$, then ϕ is a well-defined morphism. Indeed, let $t = \bigoplus_{i=1}^n \kappa_0(a_i,b_i) = \bigoplus_{j=1}^m \kappa_0(c_j,d_j)$. Then $t^\perp$ has a form $y^\perp = \bigoplus_{k=1}^s \kappa_0(u_k,v_k)$, and, for all $\mu \in \mathcal{P}_1$ and $\nu \in \mathcal{P}_2$, we have

$$\begin{aligned} 1 &= \mu \otimes_{\kappa_0} \nu(t \oplus t^\perp) = \mu \otimes_{\kappa_0} \nu(t) + \mu \otimes_{\kappa_0} \nu(t^\perp) \\ &= \sum_{i=1}^n \mu(a_i)\nu(b_i) + \sum_{k=1}^s \mu(u_k)\nu(v_k) \\ &= \sum_{j=1}^m \mu(c_j)\nu(d_j) + \sum_{k=1}^s \mu(u_k)\nu(v_k) \\ &= \sum_{i=1}^n \mu \otimes_\kappa \nu(\kappa(a_i,b_i)) + \sum_{k=1}^s \mu \otimes_\kappa \nu(\kappa(u_k,v_k)) \\ &= \sum_{j=1}^m \mu \otimes_\kappa \nu(\kappa(c_j,d_j)) + \sum_{k=1}^s \mu \otimes_\kappa \nu(\kappa(u_k,v_k)), \end{aligned}$$

hence, $\bigoplus_{i=1}^n \kappa(a_i,b_i) \oplus \bigoplus_{k=1}^s \kappa(u_k,v_k) = 1 = \bigoplus_{j=1}^m \kappa(c_j,d_j) \oplus \bigoplus_{k=1}^s \kappa(u_k,v_k)$, so that $\bigoplus_{i=1}^n \kappa(a_i,b_i) = \bigoplus_{j=1}^m \kappa(c_j,d_j)$.

It is easy to check that ϕ is a morphism in question, which proves the assertion of Theorem. □

The tensor product, $(P \otimes_\mathcal{P} Q, \otimes_\mathcal{P}) := (\Pi(X), \kappa_0)$, of the D-posets P and Q in the class $\mathcal{K}_\mathcal{P}$ is said to be a **state tensor product** of P and Q **with respect to the state system** $\mathcal{P} = \mathcal{P}_1 \times \mathcal{P}_2$. Unless confusion threatens, we usually refer to $P \otimes_\mathcal{P} Q$ rather than to $(P \otimes_\mathcal{P} Q, \otimes_\mathcal{P})$, and we shall write $p \otimes_\mathcal{P} q := \otimes_\mathcal{P}(p,q)$. Analogically we shall write $\mu \otimes_\mathcal{P} \nu$, for the unique state on $P \otimes_\mathcal{P} Q$, where $\mu \in \mathcal{P}_1, \nu \in \mathcal{P}_2$, defined by

$$\mu \otimes_\mathcal{P} \nu(p \otimes_\mathcal{P} q) = \mu(p) \cdot \nu(q), \quad p \in P,\ q \in Q. \tag{4.18}$$

Proposition 4.2.16 *Let $\mathcal{P}_1$ and $\mathcal{P}_2$ be order determining systems of states on the D-posets P and Q. Then the mappings $\beta_1 : P \to P \otimes_\mathcal{P} Q$, $\beta_2 : Q \to P \otimes_\mathcal{P} Q$, defined*

$$\begin{aligned} \beta_1(p) &= p \otimes_\mathcal{P} 1,\ p \in P, \\ \beta_2(q) &= 1 \otimes_\mathcal{P} q,\ q \in Q, \end{aligned}$$

where $\mathcal{P} = \mathcal{P}_1 \times \mathcal{P}_2$, are monomorphisms.

Proof. It is clear that β_1 is a morphism. Suppose that $\beta_1(p_1) \perp \beta_1(p_2)$. Then, for any state $\mu \otimes_{\mathcal{P}} \nu$, we have, by (4.18), $\mu \otimes_{\mathcal{P}} \nu(\beta_1(p_1) \oplus \beta_1(p_2)) = \mu \otimes_{\mathcal{P}} \nu(\beta_1(p_1)) + \mu \otimes_{\mathcal{P}} \nu(\beta_1(p_2)) = \mu(p_1) + \mu(p_2) \leq 1$, hence, $\mu(p_1) \leq \mu(p_2^\perp)$, which means $p_1 \leq p_2^\perp$. Similarly we deal with β_2. □

Proposition 4.2.17 *Let $\mathcal{P}_1$ and $\mathcal{P}_2$ be nonvoid systems of states on the D-posets P and Q, and let $\mathcal{P} = \mathcal{P}_1 \times \mathcal{P}_2$, and let $P \otimes_{\mathcal{P}} Q$ be the state tensor product of P and Q taken in the category of D-posets. Then the following is true:*

(i) *If $\mathcal{P}_1$ and $\mathcal{P}_2$ are unital, then P and Q are orthoalgebras, and $p \otimes_{\mathcal{P}} q = 0$ iff $p = 0$ or $q = 0$.*

(ii) *If $\mathcal{P}_1$ and $\mathcal{P}_2$ are strong systems of states, then P and Q are OMPs, and if $p \otimes_{\mathcal{P}} q = p_1 \otimes_{\mathcal{P}} q_1 \neq 0$, then $p = p_1$ and $q = q_1$.*

(iii) *If $\mathcal{P}_1'$ and $\mathcal{P}_2'$ are the σ-convex hulls generated by $\mathcal{P}_1$ and $\mathcal{P}_2$, then $P \otimes_{\mathcal{P}} Q = P \otimes_{\mathcal{P}'} Q$, where $\mathcal{P}' = \mathcal{P}_1' \times \mathcal{P}_2'$.*

Proof. (i) If, for example, $p \neq 0$, there exists a state $\mu \in \mathcal{P}_1$ such that $\mu(p) = 1$. Then, for any state $\nu \in \mathcal{P}_2$, we have $0 = \mu \otimes_{\mathcal{P}} \nu(p \otimes_{\mathcal{P}} q) = \nu(q)$, so that, $q = 0$.

(ii) Since by (i) $p \neq 0, q \neq 0$, we can find states $\mu \in \mathcal{P}_1$ and $\nu \in \mathcal{P}_2$ such that $\mu(p) = 1$ and $\nu(q) = 1$. Hence, $\mu(p_1) = 1$ and $\nu(q_1) = 1$, so that $p \leq p_1$ and $q \leq q_1$. By symmetry we have $p = p_1$ and $q = q_1$.

(iii) It is clear that both $\mathcal{P}_1'$ and $\mathcal{P}_2'$ are strong systems of states. Using the construction of the D-algebraic D-test spaces from the proof of Theorem 4.2.15 for $\mathcal{P}'$ and $\mathcal{P}$, we see that any D-test $T = \{(p_i, q_i)\}_{i=1}^n, p_i \in P, q_i \in Q, i = 1, \ldots, n$, corresponding to the system $\mathcal{P}'$ is a D-test corresponding to the system $\mathcal{P}$.

Conversely, let $T = \{(p_i, q_i)\}_{i=1}^n$ be a D-test corresponding to the system $\mathcal{P}$, i.e., $\sum_{i=1}^n \mu(p_i)\nu(q_i) = 1$ for any $\mu \in \mathcal{P}_1$ and any $\nu \in \mathcal{P}_2$. We choose any state $\mu \in \mathcal{P}_1'$ and $\nu \in \mathcal{P}_2'$. Then $\mu = \sum_j c_j \mu_j$, $\nu = \sum_k d_k \nu_k$, where $c_j > 0, d_k > 0$ and $\sum_j c_j = 1, \sum_k d_k = 1, \mu_j \in \mathcal{P}_1, \nu_k \in \mathcal{P}_2$. Then

$$\sum_{i=1}^n \mu(p_i)\nu(q_i) = \sum_j \sum_k c_j d_k \Big(\sum_{i=1}^n \mu_j(p_i)\nu_k(q_i) \Big) = \sum_j \sum_k c_j d_k = 1,$$

which proves $T = \{(p_i, q_i)\}_{i=1}^n$ is a D-test corresponding to the system $\mathcal{P}' = \mathcal{P}_1' \times \mathcal{P}_2'$. □

Remark 4.2.18 Similarly as for D-posets, we can define a tensor product of orthoalgebras (see, e.g., [FoBe]) as well as a state tensor product of ones in the category of orthoalgebras. This is possible, for example, if $\mathcal{P}_1$ and $\mathcal{P}_2$ are unital systems of states, since then, for any $0 \neq p \in P$, $0 \neq q \in Q$, there are $\mu \in \mathcal{P}_1$ and $\nu \in \mathcal{P}_2$ such that

$\mu(p)\nu(q) > 1/2$, (*) and in this case, by [FoBe, Thm 6.1], it is possible to show that $\Pi(X)$ from the proof of Theorem 4.2.15 is an orthoalgebra. In addition, all above results of the present section can be reformulated for orthoalgebras (with the condition (*)).

Let L_g be the D-poset from above which is determined by a generator g via (4.17), and let L be any D-poset. A mapping $\mu : L \to L_g$ is said to be a g-**state** iff (i) $\mu(1) = 1$, and (ii) $\mu(a \oplus b) = \mu(a) \oplus_g \mu(b)$, $a, b \in L$. We say that a family of g-states $\mathcal{P}$ on a D-poset L is **full** iff $\mu(a) = \mu(b)$ for any $\mu \in \mathcal{P}$ implies $a = b$, and if $\sum_{i=1}^n \mu(a_i) = 1$ for any $\mu \in \mathcal{P}$, then $\bigoplus_{i=1}^n {}_g a_i$ exists in L and $\bigoplus_{i=1}^n {}_g a_i = 1$. We now generalize Theorem 4.2.15.

Suppose that g is a generator. We define a binary operation $\odot_g : [0, 1] \times [0, 1] \to [0, 1]$ via

$$a \odot_g b = g^{-1}(g(a) \cdot g(b)), \; a, b \in [0, 1]. \tag{4.19}$$

Then $\odot_g$ is commutative and associative, and $[0, 1]$ is a semigroup with respect to the "multiplication" $\odot_g$ with the neutral element 1 and, for $a_1, a_2, b \in [0, 1]$, we have

$$(a_1 \oplus_g a_2) \odot_g b = (a_1 \odot_g b) \oplus_g (a_2 \odot_g b), \tag{4.20}$$

whenever one side of (4.20) exists in $[0, 1]$.

Theorem 4.2.19 *Let g be a generator on $[0, 1]$. Let $\mathcal{P}_1$ and $\mathcal{P}_2$ be nonvoid systems of g-states on the D-posets P and Q, respectively. Let $\mathcal{L}_{\mathcal{P}}^g$ be the set of all D-posets L such that there is a bimorphism $\kappa : P \times Q \to L$, and the set $\mathcal{P}_\kappa^g := \{\mu \otimes_\kappa^g \nu : \mu \in \mathcal{P}_1, \nu \in \mathcal{P}_2\}$ is a full system of g-states on L, where $\mu \otimes_\kappa^g \nu(\kappa(p, q)) := \mu(p) \odot_g \nu(q)$, $p \in P, q \in Q$, and let $\mathcal{B}_{\mathcal{P}}^g$ be the set of all these bimorphisms κ's. Then $\mathcal{K}^g = (\mathcal{L}_{\mathcal{P}}^g, \mathcal{B}_{\mathcal{P}}^g)$ is a consistent class for P and Q, and there exists a tensor product of P and Q in the class $\mathcal{K}^g$.*

Proof. This is similar to the proof of Theorem 4.2.15, therefore, we outline here only the main steps. Let X be the subset of $P \times Q$ consisting of all pairs (p, q) with $p \neq 0, q \neq 0$. If $M = \{(p_1, q_1), \dots, (p_n, q_n)\}$ is a finite sequence of elements from X and $\lambda = (\mu, \nu), \mu \in \mathcal{P}_1, \nu \in \mathcal{P}_2$, we put

$$\lambda(M) = \bigoplus_{i=1}^n {}_g \mu(p_i) \odot_g \nu(q_i)$$

with the understanding that if $M = \emptyset$, then $\lambda(M) = 0$.

Define now the set $\mathcal{F}_g$ of all finite sequences $T = \{(p_i, q_i)\}_{i=1}^n$ of elements in X such that $\lambda(T) = 1$ for any $\lambda = (\mu, \nu), \mu \in \mathcal{P}_1, \nu \in \mathcal{P}_2$. Since $\lambda(1, 1) = 1$, $\mathcal{F}_g$ is nonvoid.

Denote by $\mathcal{E}(\mathcal{F}_g)$ the set of all finite sequences $\{(p_j, q_j) : j \in J\}$ such that $J \subseteq I$ and $\{(p_i, q_i) : i \in I\} \in \mathcal{F}_g$. We put $\{(p_i, q_i) : i \in \emptyset\} = 0$. It is evident that $(X, \mathcal{F}_g)$ is a D-algebraic D-test space.

For two events $A, B \in \mathcal{E}(\mathcal{F}_g)$ we define $A \approx B$ iff $\lambda(A) = \lambda(B)$ for any $\lambda \in \mathcal{P}$. Then $\approx$ is an equivalence on $\mathcal{E}(\mathcal{F}_g)$, and let $\pi_g(A) := \{B \in \mathcal{E}(\mathcal{F}_g) : B \approx A\}$. Let $\Pi_g(X) = \{\pi_g(A) : A \in \mathcal{E}(\mathcal{F}_g)\}$. We organize $\Pi_g(X)$ into a D-poset as in Theorem 4.1.14.

Define a mapping $\kappa_0 : P \times Q \to \Pi_g(X)$ via

$$\kappa_0(p,q) = \begin{cases} \pi_g(\{(p,q)\}), & \text{if } (p,q) \in X, \\ 0, & \text{if } (p,q) \notin X. \end{cases}$$

Then κ_0 is, evidently, a bimorphism. From the construction of $\Pi_g(X)$ we see that any $t \in \Pi_g(X)$ is of the form $t = \bigoplus_{i=1}^{n}{}_g \kappa_0(p_i, q_i)$, and the mapping $\mu \otimes_{\kappa_0}^g \nu$ on $\Pi_g(X)$ defined by $\mu \otimes_{\kappa_0}^g \nu(\kappa_0(p,q)) = \mu(p) \odot_g \nu(q)$, $p \in P, q \in Q$, is a g-state on $\Pi_g(X)$. In addition, $\mathcal{P}_{\kappa_0}^g$ is a full system of g-states on $\Pi_g(X)$. Therefore, $\mathcal{L}_{\mathcal{P}}^g \neq \emptyset$, and let $\mathcal{B}_{\mathcal{P}}^g$ be the set of all bimorphisms κ such that κ maps $P \times Q$ into some $L \in \mathcal{L}_{\mathcal{P}}^g$ and $\mathcal{P}_\kappa^g$ be a full system of states on L. Then $\mathcal{K}_{\mathcal{P}}^g = (\mathcal{L}_{\mathcal{P}}^g, \mathcal{B}_{\mathcal{P}}^g)$ is a consistent class for P and Q.

We assert that $(\Pi_g(X), \kappa_0)$ is a tensor product of P and Q in the class $\mathcal{K}_{\mathcal{P}}^g$. Choose $L \in \mathcal{L}_{\mathcal{P}}^g$ and a bimorphism $\kappa : P \times Q \to L$. Since $\mathcal{P}_\kappa$ is full for L, it follows that if $\kappa_0(p,q) = \kappa_0(p',q')$, then $\kappa(p,q) = \kappa(p',q')$ and we can define a mapping $\phi : \Pi_g(X) \to L$ such that $\phi(\kappa_0(p,q)) = \kappa(p,q)$, $p \in P$, $q \in Q$. It is easy to check that ϕ is a well-defined mapping in question, which proves the assertion of Theorem. □

Proposition 4.2.20 *Let the conditions of Theorem* 4.2.19 *be satisfied. Define* $\tilde{\mathcal{P}}_i = \{g \circ \mu : \mu \in \mathcal{P}_i\}$, $i = 1, 2$. *Then* $\tilde{\mathcal{P}}_1$ *and* $\tilde{\mathcal{P}}_2$ *are nonempty systems of states on* P *and* Q, *respectively. In addition, the state tensor product of* P *and* Q *with respect to* $\tilde{\mathcal{P}}_1 \times \tilde{\mathcal{P}}_2$ *and the tensor product from Theorem* 4.2.19 *are the same.*

Proof. It is evident, that if μ is a g-state on P, then $g \circ \mu$ is a state on P.

If $M = \{(p_i, q_i)\}_{i=1}^n$ is a finite sequence of elements from X, then $\bigoplus_{i=1}^{n}{}_g \mu(p_i) \odot_g \nu(q_i) = 1$ iff $\sum_{i=1}^{n} \tilde{\mu}(p_i) \cdot \tilde{\nu}(g_i) = 1$, we conclude $\mathcal{F} = \mathcal{F}_g$, and similarly $\mathcal{E}(\mathcal{F}) = \mathcal{E}(\mathcal{F}_g)$ and $\pi(M) = \pi_g(M)$, $\Pi(X) = \Pi_g(X)$. □

4.2.5 Weight Tensor Product of D-test Spaces

We recall that any weight on a D-test space is any morphism on $(X, \mathcal{T})$ with values in the interval $[0, 1]$.

Theorem 4.2.21 *Let the systems of all weights on D-algebraic D-test spaces* $(X,\mathcal{T})$ *and* $(Y,\mathcal{S})$ *be nonvoid. Then there is a tensor product* $(X\otimes Y,\mathcal{T}\otimes\mathcal{S},\otimes)$, *and for any couple of weights* ω_1 *and* ω_2 *on* $(X,\mathcal{T})$ *and* $(Y,\mathcal{S})$ *there is a unique weight* $\omega_1\otimes\omega_2$ *on* $(X\otimes Y,\mathcal{T}\otimes\mathcal{S},\otimes)$ *such that*

$$\omega_1\otimes\omega_2(x\otimes y)=\omega_1(x)\cdot\omega_2(y),\ \ x\in X, y\in Y. \tag{4.21}$$

Proof. Let $L=[0,\ 1]$, and endow L with the natural ordering and the difference $t\ominus s=t-s$ iff $0\le s\le t\le 1$. Then $(L;\le,\ominus,0,1)$ is a complete D-poset. Let $\mathcal{U}$ be the set of all finite or countable sequences $\{t_i\}_i$ in $[0,\ 1]$ such that $t_i\ne 0$ and $\sum_i t_i=1$. Then $(L,\mathcal{U})$ is a D-algebraic D-test space.

Let ω_1 and ω_2 be weights on $(X,\mathcal{T})$ and $(Y,\mathcal{S})$, respectively, and define a mapping $\beta_{\omega_1\omega_2}(x,y)=\omega_1(x)\cdot\omega_2(y)$, $x\in X, y\in Y$. Then easy calculations prove that $\beta_{\omega_1\omega_2}$ is a bimorphism from $X\times Y$ into $[0,\ 1]$. According to Theorem 4.1.26, there is a tensor product $(X\otimes Y,\mathcal{T}\otimes\mathcal{S},\otimes)$ of $(X,\mathcal{T})$ and $(Y,\mathcal{S})$.

Due to the basic property of the tensor product, there is a morphism ϕ from $X\otimes Y$ into $[0,\ 1]$ such that $\beta_{\omega_1\omega_2}=\phi\circ\otimes$. We assert that ϕ is a weight on $X\otimes Y$. Indeed, since if $T=\{T(i)\}_i$ is a D-test in $X\otimes Y$, so $\phi\circ T$ is a D-test in $[0,\ 1]$, which but means that $\sum_i\phi(T(i))=1$. Let $x\in X, y\in Y$ be given, then $\phi(x\otimes y)=\beta_{\omega_1\omega_2}(x,y)=\omega_1(x)\cdot\omega_2(y)$ and ϕ is a unique weight on $X\otimes Y$ with the property (4.21). $\square$

We note that the former statement generalizes that from Corollary 4.1.27, where it has supposed that systems of weights are sufficient.

Let $(X,\mathcal{T})$ be a D-test space and let Δ be a set of weights on it. We will say that Δ is **full** if (i) for any finite index set I and $F\in X^I$, we have $F\in\mathcal{T}$ iff $\omega(F)=1$ for all $\omega\in\Delta$, and (ii) $\omega(x)=\omega(y)$ for all $\omega\in\Delta$ implies $x=y$.

Remark 4.2.22 1. A full system of weights is sufficient. Indeed, if, for $x\in X$, $\omega(x)=0$ for all $\omega\in\Delta$, then for any test T, $\{x\}\dot\cup T$ is a test, whence $\{x\}\sim\emptyset$.

2. A D-test space with a full system of states is D-algebraic; the proof is similar to that of Theorem 4.1.21.

3. Let $(X,\mathcal{T})$ be a D-algebraic D-test space. We will say that $(X,\mathcal{T})$ is **reduced** if, for $x,y\in X$, $\{x\}\approx\{y\}$ implies $x=y$. It is easy to see that for every D-algebraic D-test space there is a reduced D-algebraic D-test space with the same D-poset $\Pi(X)$. Clearly, a D-test space with a full set of weights is reduced.

Let $(X,\mathcal{T})$ and $(Y,\mathcal{S})$ be D-algebraic D-test spaces. Let $\mathcal{K}=(\mathcal{X},\mathcal{B})$ be a couple, where $\mathcal{X}$ is a nonvoid family of D-algebraic D-test spaces and $\mathcal{B}$ is a nonvoid family of bimorphisms on $X\times Y$ such that (i) for any $\beta\in\mathcal{B}$ there is a D-algebraic D-test space $(U,\mathcal{U})$ in $\mathcal{X}$ such that $\beta: X\times Y\to U$, and (ii) for any $(U,\mathcal{U})\in\mathcal{X}$ there is a bimorphism $\beta\in\mathcal{B}$ with $\beta: X\times Y\to U$; $\mathcal{K}$ is said to be a **consistent class** for $(X,\mathcal{T})$ and $(Y,\mathcal{S})$.

Definition 4.2.23 *Let $\mathcal{K} = (\mathcal{X}, \mathcal{B})$ be a consistent class for D-algebraic D-test spaces $(X,\mathcal{T})$ and $(Y,\mathcal{S})$. We say that the triple $(Z,\mathcal{U},\tau)$ is a* **tensor product** *of $(X,\mathcal{T})$ and $(Y,\mathcal{S})$* **in the class** *$\mathcal{K}$ if the following conditions are satisfied:*

(i) *$(Z,\mathcal{U})$ belongs to $\mathcal{X}$ and $\tau \in \mathcal{B}$.*

(ii) *For every D-algebraic D-test space $(V,\mathcal{W})$ in $\mathcal{X}$ and every bimorphism $\beta : X \times Y \to V$ from $\mathcal{B}$, there is a morphism $\phi : Z \to V$ such that $\beta = \phi \circ \tau$.*

(iii) *For any D-test $T \in \mathcal{U}, T \in Z^I$, there is a function $F \in (X \times Y)^I$ such that $T = \tau(F)$.*

We note that, similarly as for a general tensor product of D-test spaces, we can show that if a tensor product of $(X,\mathcal{T})$ and $(Y,\mathcal{S})$ exists in the class $\mathcal{K}$, it is unique up to isomorphism.

Theorem 4.2.24 *Let $(X,\mathcal{T})$ and $(Y,\mathcal{S})$ be D-algebraic D-test spaces having only finite D-tests with nonempty sets of weights Δ_X and Δ_Y, respectively. Let $\mathcal{X}$ be the class of all D-algebraic D-test spaces $(V,\mathcal{W})$ such that there is a bimorphism $\kappa : X \times Y \to V$, $V = \kappa(X \times Y)$, and the set*

$$\Delta_\kappa := \{\omega_1 \otimes_\kappa \omega_2 : \ \omega_1 \in \Delta_X, \, \omega_2 \in \Delta_Y\},$$

where $\omega_1 \otimes_\kappa \omega_2(\kappa(x,y)) = \omega_1(x) \cdot \omega_2(y)$, $(x,y) \in X \times Y$, is a full set of weights on $(V,\mathcal{W})$, where

$$\mathcal{W} = \{\kappa(F) : \ F \in (X \times Y)^I, \ \sum_{i \in I} \omega_1 \otimes_\kappa \omega_2(\kappa(F(i))) = 1, \ \forall\, (\omega_1, \omega_2) \in \Delta_X \times \Delta_Y\},$$

(I is any finite index set) and $\mathcal{B}$ be the set of all these bimorphisms κ's. Then $\mathcal{K} = (\mathcal{X}, \mathcal{B})$ is a consistent class for $(X,\mathcal{T})$ and $(Y,\mathcal{S})$, and there is a tensor product of $(X,\mathcal{T})$ and $(Y,\mathcal{S})$ in the class $\mathcal{K}$.

Proof. For $(x,y), (u,v) \in X \times Y$, put $(x,y) \equiv (u,v)$ iff $\omega_1(x)\omega_2(y) = \omega_1(u)\omega_2(v)$ for all $\omega_1 \in \Delta_X$ and $\omega_2 \in \Delta_Y$. Let $Z = (X \times Y)_\equiv$ be the quotient space and let $\mu : X \times Y \to Z$ be the canonical mapping. Then, for any $\omega_1 \in \Delta_X$ and any $\omega_2 \in \Delta_Y$ we have a well-defined mapping $\omega_1 \otimes \omega_2 : Z \to [0, 1]$ such that

$$\omega_1 \otimes \omega_2(z) = \omega_1(x) \cdot \omega_2(y)$$

whenever $z = \mu(x,y)$ for some $(x,y) \in X \times Y$.

Put

$$\mathcal{U} = \{F \in Z^I : \sum_{i \in I} \omega_1 \otimes \omega_2(F(i)) = 1, \ \forall\, (\omega_1, \omega_2) \in \Delta_X \times \Delta_Y, \ \forall \, \text{finite } I\}.$$

Then $(Z,\mathcal{U})$ is a D-algebraic D-test space with a full system of weights $\Delta := \{\omega_1 \otimes \omega_2 : \omega_1 \in \Delta_X,\ \omega_2 \in \Delta_Y\}$.

We claim that the triple $(Z,\mathcal{U},\mu)$ is the tensor product of $(X,\mathcal{T})$ and $(Y,\mathcal{S})$ in the class $\mathcal{K}$. Let $T_1, T_2 \in \mathcal{T}$, $F \in \mathcal{E}(Y)$. Let F' be any local complement of F. Then, for any $\omega_1 \otimes \omega_2 \in \Delta$,

$$\begin{aligned}\omega_1 \otimes \omega_2(\mu(T_1 \dot{\times} F)) &= 1 - \omega_1 \otimes \omega_2(\mu(T_1 \dot{\times} F')) = 1 - \omega_2(F') \\ &= \omega_1 \otimes \omega_2(\mu(T_2 \dot{\times} F)),\end{aligned}$$

hence, $\mu(T_1 \dot{\times} F')$ is a common local complement of $\mu(T_1 \dot{\times} F)$ and $\mu(T_2 \dot{\times} F)$. Similarly we prove that $\mu(E \dot{\times} S_1) \approx \mu(E \dot{\times} S_2)$ for any $E \in \mathcal{E}(X,\mathcal{T})$ and $S_1, S_2 \in \mathcal{S}$. Therefore, μ is a bimorphism, and since $\omega_1 \otimes_\mu \omega_2 = \omega_1 \otimes \omega_2$, and $\Delta_\mu = \Delta$ is a full set of weights for $(Z,\mathcal{U})$, this means that $(Z,\mathcal{U}) \in \mathcal{X}$ and $\mu \in \mathcal{B}$.

Let $(V,\mathcal{W})$ be a D-test space in $\mathcal{X}$ and let $\kappa : X \times Y \to V$ be any bimorphism from $\mathcal{B}$. Define a mapping $\phi : Z \to V$ via $\phi(\mu(x,y)) = \kappa(x,y)$. Since Δ_κ is full for $(V,\mathcal{W})$, it follows that $\mu(x,y) = \mu(u,v)$ implies $\kappa(x,y) = \kappa(u,v)$, hence ϕ is defined well. Moreover, if $F \in \mathcal{U}$, $F = Z^I$, then $F(i) = \mu(x_i, y_i)$ for some $(x_i, y_i) \in X \times Y$ and

$$\begin{aligned}\omega_1 \otimes_\kappa \omega_2(\phi(F)) &= \sum_{i \in I} \omega_1 \otimes_\kappa \omega_2(\kappa(x_i,y_i)) = \sum_{i\in I} \omega_1(x_i)\omega_2(y_i) \\ &= \sum_{i \in I} \omega_1 \otimes \omega_2(\mu(x_i,y_i)) = 1,\end{aligned}$$

hence, $\phi(F) \in \mathcal{W}$. That is, ϕ is a morphism, and $\kappa = \phi \circ \mu$.

Finally, let $T \in \mathcal{U}$, $T \in Z^I$. Then $T(i) = \mu(x_i, y_i)$ for some $(x_i, y_i) \in X \times Y$. Put $F(i) = (x_i, y_i)$ for chosen $(x_i, y_i) \in X \times Y$ such that $\mu(x_i, y_i) = F(i)$. Then $T = \mu(F)$. This proves that $(Z,\mathcal{U},\mu)$ is a tensor product of $(X,\mathcal{T})$ and $(Y,\mathcal{S})$ in the class $\mathcal{K}$. □

The tensor product $(Z,\mathcal{U},\tau)$ from Theorem 4.2.24 is called the **weight tensor product** of $(X,\mathcal{T})$ and $(Y,\mathcal{S})$ with respect to the nonempty sets of weights Δ_X and Δ_Y, and it will be denoted by $(X \otimes_\Delta Y, \mathcal{T} \otimes_\Delta \mathcal{S}, \otimes_\Delta)$, where $\Delta = \Delta_X \times \Delta_Y$.

Theorem 4.2.24 can be reformulated also for D-algebraic D-test spaces with not necessarily finite D-tests. We say that a nonempty set of weights Δ on $(X,\mathcal{T})$ is **completely full** (or **σ-completely full**) iff (i) for any index set (any countable index set) I and $F \in X^I$, we have $F \in \mathcal{T}$ iff $\omega(F) = 1$ for all $\omega \in \Delta$, and (ii) $\omega(x) = \omega(y)$ for all $\omega \in \Delta$ implies $x = y$.

For example, if H is a Hilbert space, $\dim H = \infty$, then $(\mathcal{S}(H), \mathcal{B}(H))$, where $\mathcal{S}(H)$ is the unit sphere in H and $\mathcal{B}(H)$ is the set of all orthonormal basis in H, is an algebraic test space (Theorem 4.1.18) with a completely full system of weights $\{\omega_y : y \in \mathcal{S}(H)\}$, where $\omega_y(x) = |(x,y)|^2$, $x \in \mathcal{S}(H)$, which is not a full system of weights.

For these D-test spaces we have the following result on tensor products.

Theorem 4.2.25 *Let $(X,\mathcal{T})$ and $(Y,\mathcal{S})$ be D-algebraic D-test spaces with nonempty sets of weights Δ_X and Δ_Y, respectively. Let $\mathcal{X}$ be the class of all D-algebraic D-test spaces $(V,\mathcal{W})$ such that there is a bimorphism $\kappa : X \times Y \to V$, $V = \kappa(X \times Y)$, and the set*

$$\Delta_\kappa := \{\omega_1 \otimes_\kappa \omega_2 : \ \omega_1 \in \Delta_X, \omega_2 \in \Delta_Y\},$$

where $\omega_1 \otimes_\kappa \omega_2(\kappa(x,y)) = \omega_1(x) \cdot \omega_2(y)$, $(x,y) \in X \times Y$, is a completely full set of weights on $(V,\mathcal{W})$, where

$$\mathcal{W} = \{\kappa(F) : \ F \in (X \times Y)^I, \ \sum_{i \in I} \omega_1 \otimes_\kappa \omega_2(\kappa(F(i))) = 1, \ \forall\, (\omega_1,\omega_2) \in \Delta_X \times \Delta_Y, \ \forall\, I\},$$

and $\mathcal{B}$ be the set of all these bimorphisms κ's. Then $\mathcal{K} = (\mathcal{X},\mathcal{B})$ is a consistent class for $(X,\mathcal{T})$ and $(Y,\mathcal{S})$, and there is a tensor product of $(X,\mathcal{T})$ and $(Y,\mathcal{S})$ in the class $\mathcal{K}$.

Proof. This follows the same ideas as the proof of Theorem 4.2.24. We change $\mathcal{U}$ form the proof of Theorem 4.2.24 to

$$\mathcal{U} = \{F \in Z^I : \sum_{i \in I} \omega_1 \otimes \omega_2(F(i)) = 1, \ \forall\, (\omega_1,\omega_2) \in \Delta_X \times \Delta_Y, \ \forall\, I\}.$$

□

The tensor product from Theorem 4.2.25 is said to be a **complete weight tensor product** of $(X,\mathcal{T})$ and $(Y,\mathcal{S})$ with respect to Δ_X and Δ_Y.

Proposition 4.2.26 *Let $(X,\mathcal{T})$ be a D-algebraic D-test space and let $L = \Pi(X)$ be the corresponding D-poset. For any weight ω on $(X,\mathcal{T})$ there is a unique state $\tilde{\omega}$ on L such that*

$$\tilde{\omega}(\pi(E)) = \omega(E)$$

for any $E \in \mathcal{E}(X)$. Conversely, if any test T in $\mathcal{T}$ is finite, any state μ on L defines a weight $\tilde{\mu}$ on $(X,\mathcal{T})$ such that

$$\tilde{\mu}(x) = \mu(\pi(x)), \ x \in X.$$

Proof. $\tilde{\omega}$ and $\tilde{\mu}$ are correctly defined, and the proofs are straightforward. □

The following result shows that state tensor products and weight tensor products are very close things.

Theorem 4.2.27 *Let P, Q and R be D-posets and let $(X,\mathcal{T}),(Y,\mathcal{S})$ and $(Z,\mathcal{U})$ be the corresponding standard D-test spaces. Let $\mathcal{P}_1$ and $\mathcal{P}_2$ be two nonempty sets of weights on $(X,\mathcal{T})$ and $(Y,\mathcal{S})$, respectively, and let $\tilde{\mathcal{P}}_i = \{\tilde{\omega} : \omega \in \mathcal{P}_i\}$, $i = 1,2$. Let (R,γ) be the state tensor product of P and Q in the category of D-posets with respect to $\tilde{\mathcal{P}} = \tilde{\mathcal{P}}_1 \times \tilde{\mathcal{P}}_2$ and let $(V,\mathcal{W},\delta)$ be the weight tensor product of D-algebraic D-test spaces $(X,\mathcal{T})$, and $(Y,\mathcal{S})$ with respect to $\mathcal{P}_1 \times \mathcal{P}_2$, and let D be the corresponding D-poset. Then D and R are isomorphic.*

Proof. Let π_X be the corresponding mapping from $\mathcal{E}(X)$ into $\Pi(X)$ defined by $\pi_X(F) := \{G \in \mathcal{E}(X) : G \approx F\}$; similarly we define π_Y, π_Z and π_V. Since $(V,\mathcal{W},\delta)$ is a weight tensor product, by Proposition 4.2.26 and the definitions of tensor products, there is a morphism $\phi : V \to Z$ such that $\phi \circ \delta = \kappa$, where $\kappa(x,y) = \gamma(\pi_X(\{x\}),\pi_Y(\{y\}))$, $(x,y) \in X \times Y$. Let $\tilde{\phi} : D \to R$ be the morphism defined by $\tilde{\phi}(\pi_V(E)) = \pi_Z(\phi(E))$. Let $\tilde{\delta} : P \times Q \to D$ be the bimorphism defined by $\tilde{\delta}(\pi_X(E),\pi_Y(F)) = \pi_V(\delta(E\dot{\times}F))$. Since (R,γ) is a state tensor product, there is a morphism $\psi : R \to D$ such that $\tilde{\delta} = \psi \circ \gamma$. Further, $\tilde{\kappa}(\pi_X(E),\pi_Y(F)) := \pi_Z(\kappa(E\dot{\times}F)) = \gamma(\pi_X(E),\pi_Y(F))$ implies $\tilde{\phi} \circ \tilde{\delta} = \tilde{\kappa} = \gamma$, whence $\gamma = \tilde{\phi} \circ \psi \circ \gamma$. Since every element in R is a finite orthogonal sum of elements of the form $\gamma(p,q)$, $p \in P$, $q \in Q$, we conclude $\tilde{\phi} \circ \psi = id_R$.

On the other hand, every test in $\mathcal{W}$ is finite. Let $a \in D$ and let $E \in \mathcal{E}(V)$ be such that $a = \pi_V(E)$. Now $E \preccurlyeq T$ for some $T \in \mathcal{W}$. By (ii) of the definition of a weight tensor product, there are finite E_0 and F, $E_0 \preccurlyeq F$, $F \in (X \times Y)^I$ such that $E = \delta(E_0)$ and $T = \delta(F)$. Then

$$\begin{aligned} a &= \pi_V(\delta(E_0)) = \bigoplus_i \pi_V(\{\delta(x_i,y_i)\}) \\ &= \bigoplus_i \tilde{\delta}(\pi_X(\{x_i\}),\pi_Y(\{y_i\})), \end{aligned}$$

when $E_0 = \{(x_i,y_i)\}_i$, which entails $\psi \circ \tilde{\phi}(a) = a$ (observe that $\tilde{\delta} = \psi \circ \tilde{\phi} \circ \tilde{\delta}$). This shows that ψ is surjective, and hence it is an isomorphism. □

4.2.6 Examples of State Tensor Products

In the present Section, we illustrate the above developed theory of tensor products on concrete examples.

Example 4.2.28 *Let $L = [0,1]$, $X = (0,1]$, and let any D-test in $\mathcal{T}$ consist of all finite sequences $\{x_i\}$ in X such that $\sum_i x_i = 1$. Then on $(X,\mathcal{T})$ there is a unique weight $\omega = id_X$. Therefore, the weight tensor product of $(X,\mathcal{T})$ with itself is again $(X,\mathcal{T})$. To prove that it is necessary to repeat the proof of Theorem* 4.2.15.

In addition, let f, g, h be generators on $[0, 1]$. Then $L_f \otimes L_g$ and $L_h \otimes L_h$ are isomorphic, and $L_f \otimes_{\mathcal{P}} L_g$ and L_h are isomorphic. This is true because all L_g's are isomorphic with L_{id}; the mapping $\phi_g : L_{id} \to L_g$, $t \mapsto g^{-1}(t)$, $t \in [0, 1]$, is an isomorphism.

Example 4.2.29 *Let S be a real or complex inner product space. Put $X = \{x \in S : ||x|| = 1\}$ and let $\mathcal{T}$ consist of the set of all maximal orthonormal systems in S. Then $(X, \mathcal{T})$ is a test space, and the following statements are equivalent ([Dvu], and Theorem 4.1.18):*

(i) *$(X, \mathcal{T})$ is an algebraic test space.*

(ii) *$(X, \mathcal{T})$ possesses at least one weight.*

(iii) *S is complete.*

If one statement is satisfied, then $\Pi(X)$ is isomorphic to the set of all closed subspaces in the Hilbert space S, which is a complete OMP.

Example 4.2.30 *Let $X = \mathcal{E}(H_1) \setminus \{O\}$ and $Y = \mathcal{E}(H_2) \setminus \{O\}$ and let D-tests on X and Y be the systems of all finite sequences of effects $\{A_i\}$ on Hilbert spaces H_1 and H_2 such that $\sum_i A_i = I$. Denote by $\mathcal{T}$ and $\mathcal{S}$ the systems of all D-tests on X and Y, respectively. Let*

$$\Delta_X = \{m_x : m_x(A) = (Ax, x),\ A \in X,\ x \in H_1, ||x|| = 1\}$$

and

$$\Delta_Y = \{m_y : m_y(A) = (Ay, y),\ A \in Y,\ x \in H_2, ||y|| = 1\}.$$

Then Δ_X and Δ_Y are full systems of weights on X and Y, respectively. The weight tensor product of $(X, \mathcal{T})$ and $(Y, \mathcal{S})$ with respect to Δ_X and Δ_Y exists and is equal to $(Z, \mathcal{U}, \otimes)$, where

$$Z = \{A \otimes B : A \in X,\ B \in Y\},$$

i.e., $A \otimes B$ is an operator in the tensor product $H_1 \otimes H_2$ of the Hilbert spaces H_1 and H_2,

$$\mathcal{U} = \left\{\{A_i \otimes B_i\}_{i \in I} :\ A_i \in X,\ B_i \in Y,\ \sum_{i \in I} A_i \otimes B_i = I_1 \otimes I_2, \forall \text{ finite } I\right\}$$

and $\otimes(A, B) = A \otimes B$. This follows from the fact that $(Ax, x)(By, y) = (Cx, x)(Dy, y)$, $x \in H_1, y \in H_2$, $||x|| = ||y|| = 1$ iff $(A \otimes Bx \otimes y, x \otimes y) = (C \otimes Dx \otimes y, x \otimes y)$ iff $(A \otimes B \sum_i x_i \otimes y_i, \sum_i x_i \otimes y_i) = (C \otimes D \sum_i x_i \otimes y_i, \sum_i x_i \otimes y_i)$ iff $A \otimes B = C \otimes D$ as operators in $H_1 \otimes H_2$. The last equivalence follows from the polar formulas for Hermitian bilinear forms generated by A, B, C, and D. Hence, $\{m_x \odot m_y : x \in$

$\mathcal{S}(H_1)$, $y \in \mathcal{S}(H_2)\}$, *where* $m_x \odot m_y(A \otimes B) := m_x(A) \cdot m_y(B) = (Ax, x)(By, y)$, *is a full system of weights on* $(Z, \mathcal{U})$.

For $\Pi(Z)$, *which by Theorem* 4.2.25 *is a state tensor product of* $\mathcal{E}(H_1)$ *and* $\mathcal{E}(H_2)$, *we have that it consists of all elements of the form* $\sum_i A_i \otimes B_i$, *where* $A_i \in \mathcal{E}(H_1), B_i \in \mathcal{E}(H_2)$ *for which there are* $A_j \otimes B_j \in \mathcal{E}(H_1) \otimes \mathcal{E}(H_2)$ *such that* $\sum_i A_i \otimes B_i + \sum_j A_j \otimes B_j = I_1 \otimes I_2$ *(all summations are over finite index sets).*

We note that the ordering $\le$ *on* $\Pi(Z)$ *is defined as follows* $\sum_i A_i \otimes B_i \le \sum_j C_j \otimes D_j$ *iff there is* $\{A'_k \otimes B'_k\}$ *such that* $\sum_i A_i \otimes B_i + \sum_k A'_k \otimes B'_k = \sum_j C_j \otimes D_j$, *and that can be different from the natural ordering on* $\mathcal{E}(H_1 \otimes H_2)$.

Example 4.2.31 *Let* $X = \mathcal{E}(H_1) \setminus \{O\}$ *and* $Y = \mathcal{E}(H_2) \setminus \{O\}$ *and let D-tests on X and Y be the systems of all summable effects* $\{A_i\}$ *on Hilbert spaces* H_1 *and* H_2 *such that* $\sum_i A_i = I$ *(the convergence is assumed to be, for example, weak). Denote by* $\mathcal{T}$ *and* $\mathcal{S}$ *the systems of all D-tests on X and Y, respectively. Let*

$$\Delta_X = \{m_x : m_x(A) = (Ax, x),\ A \in X,\ x \in H_1, ||x|| = 1\}$$

and

$$\Delta_Y = \{m_y : m_y(A) = (Ay, y),\ A \in Y,\ x \in H_2, ||y|| = 1\}.$$

Then Δ_X *and* Δ_Y *are completely full systems of weights on X and Y, respectively. The complete weight tensor product of* $(X, \mathcal{T})$ *and* $(Y, \mathcal{S})$ *with respect to* Δ_X *and* Δ_Y *exists and is equal to* $(Z, \mathcal{U}, \otimes)$, *where*

$$Z = \{A \otimes B : A \in X,\ B \in Y\},$$

i.e., $A \otimes B$ *is the operator in the tensor product* $H_1 \otimes H_2$ *of the Hilbert spaces* H_1 *and* H_2,

$$\mathcal{U} = \left\{ \{A_i \otimes B_i\}_i :\ A_i \in X,\ B_i \in Y,\ \sum_i A_i \otimes B_i = I_1 \otimes I_2 \right\}$$

and $\otimes(A, B) = A \otimes B$. *(Use the same arguments as Example* 4.2.30; $\{m_x \odot m_y : x \in \mathcal{S}(H_1),\ y \in \mathcal{S}(H_2)\}$ *is a completely full system of weights on* $(Z, \mathcal{U})$.*) For* $\Pi(Z)$ *we have that it consists of all elements of the form* $\sum_i A_i \otimes B_i$, *where* $A_i \in \mathcal{E}(H_1), B_i \in \mathcal{E}(H_2)$ *for which there are* $A_j \otimes B_j \in \mathcal{E}(H_1) \otimes \mathcal{E}(H_2)$ *such that* $\sum_i A_i \otimes B_i + \sum_j A_j \otimes B_j = I_1 \otimes I_2$.

We note that the ordering $\le$ *on* $\Pi(Z)$ *is defined as follows* $\sum_i A_i \otimes B_i \le \sum_j C_j \otimes D_j$ *iff there is* $\{A'_k \otimes B'_k\}$ *such that* $\sum_i A_i \otimes B_i + \sum_k A'_k \otimes B'_k = \sum_j C_j \otimes D_j$, *and it can be different from the natural ordering on* $\mathcal{E}(H_1 \otimes H_2)$.

Example 4.2.32 *Let* $L(H_i)$ *be the set of all closed subspaces of a Hilbert space* H_i, $i = 1, 2$. *Then* $L(H_i)$ *is a complete OMP. Let* $X = L(H_1) \setminus \{O\}$ *and* $Y = L(H_2) \setminus \{O\}$

and let D-tests on X and Y be all finite orthogonal decompositions of H_1 and H_2, respectively, which are in fact tests. Denote by $\mathcal{T}$ and $\mathcal{S}$ the systems of all tests on X and Y.

Let Δ_X and Δ_Y have the same meaning as those in Example 4.2.30. *Then the weight tensor product of $(X, \mathcal{T})$ and $(Y, \mathcal{S})$ is $(Z, \mathcal{U})$, where Z and $\mathcal{U}$ have the same form as those in Example* 4.2.30. *In addition, $\Pi(Z)$ is an orthoalgebra, we note that the authors do not know whether it is an OMP.*

4.3 Partition Logics, Orthoalgebras and Automata

The investigation of classical models for non-Boolean algebraic structures has brought up several interesting examples. Among them are Cohen's "firefly-in-a-box" model [Coh], Wright's urn model [WriR], as well as Aerts' vessel model [Aer 4] featuring stronger-than quantum correlations. Another type of classical objects are automata models, one of which has been intruduced by Moore [MooE] in an attempt to model quantum complementarity in the context of effective computation. D. Finkelstein and S.R. Finkelstein [FiFi], and subsequently Grib and Zapatrin [GrZa, GrZa 1] investigated the propositional structure of certain automaton models by lattice theoretical methods. Svozil [Svo] and Schaller and Svozil [ScSv] introduced partition logics, which appear to be a natural framework for the study of the propositional structure of Moore and Mealy type automata. Thereby, the set of automaton states is partitioned with respect to identifiability in input/output experiments; and the single partitions corresponding to Boolean algebras are pasted together to form more general structures.

We describe here how non-classical propositional structures, in particular partition logics of automata, fit into the scheme of orthoalgebras.

4.3.1 Boolean Atlases

According to Lock and Hardegree [LH1, LH2], we consider a family of Boolean algebras, a Boolean atlas, which will be equivalent to quasi orthoalgebras. Many considerations about co-measurable quantum propositional structures deal with Boolean subalgebras. In addition, these latter are intuitively better understandable than general quantum propositional logics.

A family $\mathcal{B} = \{B_i : i \in I\}$ of Boolean algebras is called a **Boolean atlas** if it satisfies the following conditions (here the operations in B_i are denoted by an index i):

(i) if $B_i \subseteq B_j$, then $B_i = B_j$;

(ii) if $a, b \in B_i \cap B_j$, then $a \leq_i b$ iff $a \leq_j b$;

(iii) $1_i = 1_j = 1$ and $0_i = 0_j = 0$ for all $i, j \in I$;

(iv) if $a \in B_i \cap B_j$, then $a^{\perp_i} = a^{\perp_j}$ for all $i, j \in I$;

(v) if $a, b \in B_i \cap B_j$ and if $a \wedge_i b = 0_i$, then $a \vee_i b = a \vee_j b$.

Note that $a, b \in B_i \cap B_j$ and yet $a \vee_i b \neq a \vee_j b$ and $a \wedge_i b \neq a \wedge_j b$. We define a **Boolean manifold** to be a Boolean atlas which satisfies the condition

if $a, b \in B_i \cap B_j$, then $a \vee_i b = a \vee_j b$ and $a \wedge_i b = a \wedge_j b$.

Let $\mathcal{B} = \{B_i : i \in I\}$ be a Boolean atlas, $a, b \in \bigcup_{i \in I} b_i$ and $S \subseteq \bigcup_{i \in I} B_i$. Then we say that

(i) a, b are **compatible** if there is $i \in I$ and $a, b \in B_i$;

(ii) a, b are **orthogonal** if there is $i \in I$ such that $a, b \in B_i$ and $a \wedge_i b = 0_i$. A subset S is called **pairwise orthogonal** if a, b are orthogonal for any $a, b \in S$;

(iii) S is **jointly compatible** if there is $i \in I$ with $S \subseteq B_i$; S is **pairwise compatible** if a, b are compatible for any $a, b \in S$;

(iv) S is **jointly orthogonal** if there is $i \in I$ with $S \subseteq B_i$ and S is pairwise orthogonal.

4.3.2 Quasi Orthoalgebras and Orthoalgebras

The notions of an orthoalgebra (or of a quasi orthoalgebra) goes back to axiomatic models of quantum mechanics introduced by Foulis and Randall [FoRa, RaFo 1] as special algebraic structures describing propositional logics.

A **quasi orthoalgebra** is a set L endowed with two special distinct elements $0, 1 \in L$ $(0 \neq 1)$ and equipped with a partially defined binary operation $\oplus$ satisfying the following conditions for all $a, b \in L$:

(oai) if $a \oplus b$ is defined, then $b \oplus a$ is defined and $a \oplus b = b \oplus a$ (commutativity law);

(oaii) $a \oplus 0$ is defined for any $a \in L$ and $a \oplus 0 = a$;

(oaiii) for any $a \in L$, there is a unique element $a' \in L$ such that $a \oplus a'$ is defined and $a \oplus a' = 1$ (orthocomplementation law);

(oaiv) if $a \oplus (a' \oplus b)$ is defined, then $b = 0$;

(oav) if $a \oplus (a \oplus b)$ is defined, then $a = 0$;

(oavi) if $a \oplus b$ is defined, then $a \oplus (a \oplus b)'$ is defined and $b' = a \oplus (a \oplus b)'$.

The following facts are true:

Proposition 4.3.1 *Let L be a quasi orthoalgebra, $a, b \in L$. Then*

(a) $0' = 1$, $1' = 0$;

(b) $(a')' = a$;

(c) *if* $a \oplus b = a \oplus c$, *then* $b = c$;

(d) *if* $a \oplus b = 1$, *then* $b = a'$.

The unique element a' is called the **orthocomplement** of $a \in L$, and the unary operation $' : L \to L$ defined by $a \mapsto a'$, $a \in L$, is said to be an **orthocomplementation**. We shall say that two elements $a, b \in L$ (i) are **orthogonal**, and write $a \perp b$, iff $a \oplus b$ is defined in L (it is clear that $a \perp b$ iff $b \perp a$), and (ii) $a \leq b$ iff there is an element $c \in L$ with $a \oplus c = b$.

It is easy to show that the relation $\leq$ is reflexive and antisymmetric, but not necessarily transitive. We recall that an associative quasi orthoalgebra, i.e., a quasi orthoalgebra, for which the associative law

(oavii) if $a \oplus b$, $(a \oplus b) \oplus c$ are defined in L, so are $b \oplus c$ and $a \oplus (b \oplus c)$, and $(a \oplus b) \oplus c = a \oplus (b \oplus c)$

holds, is said to be an **orthoalgebra** (OA in abbreviation), see Definition 1.5.1. In any orthoalgebra, $\leq$ is transitive. On the other hand it is possible to give an example of a quasi orthoalgebra with transitive $\leq$ which does not correspond to any orthoalgebra.

A subset A of a quasi OA (OA) L is a **quasi suborthoalgebra** (**suborthoalgebra**) of L if (i) $0, 1 \in A$; (ii) if $a \in A$, then $a' \in A$; (iii) $a, b \in A$ with $a \perp b$ implies $a \oplus b \in L$.

If a (quasi) suborthoalgebra A of L is, in addition, a Boolean algebra with respect to $\leq$, A is called a **Boolean suborthoalgebra** of L. Denote by $\vee_A$ and $\wedge_A$ the join and the meet taken only in A, respectively. Then, $a \oplus b = a \vee_A b$ whenever $a, b \in A$ and L is an OA. A maximal Boolean suborthoalgebra of L is called a **block**.

4.3.3 Examples of Orthoalgebras

We shall give a few examples of orthoalgebras having classical physical interpretations.

Firefly in a box. According to Cohen [Coh], consider a system consisting of a firefly in a box with a clear plastic window at the front and another one on the side pictured in Figure 4.1.

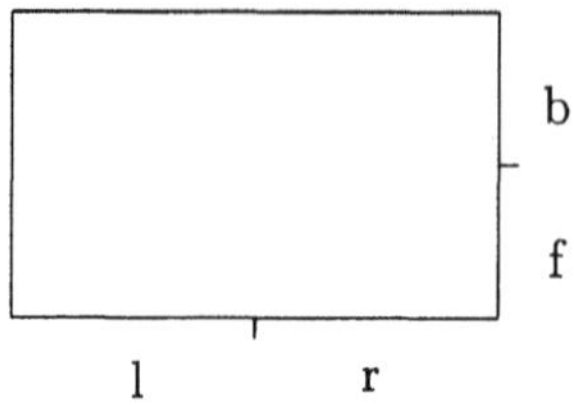

Fig. 4.1

Suppose each window has a thin vertical line drawn down the center to divide the window in half. We shall consider two experiments on the system: The experiment A: Look at the front window. The experiment B: Look at the side window. The outcomes of A and B are: See a light in the left half (l_A, l_B), right half (r_A, r_B) of the window or see no light (n_A, n_B). It is clear that $n_A = n_B =: n$ and we put $l_A =: l$, $r_A =: r$, $l_B =: f$, $r_B =: b$ (f for the front, b for the back).

The Greechie diagram of the corresponding propositional logic is given by Figure 4.2. (Recall that here the small circles on one smooth line denote mutually orthogonal atoms lying in the same block; for more details on Greechie diagrams, see [PtPu].) The associated Hasse diagram is given by Figure 4.3.

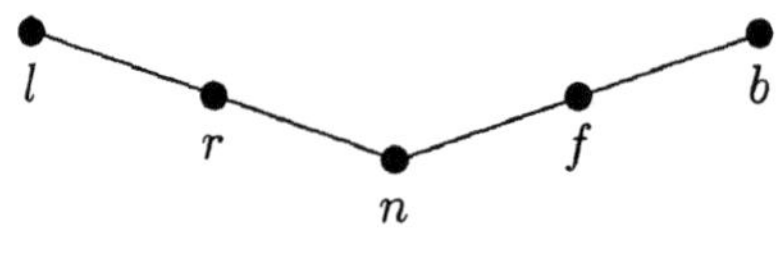

Fig. 4.2

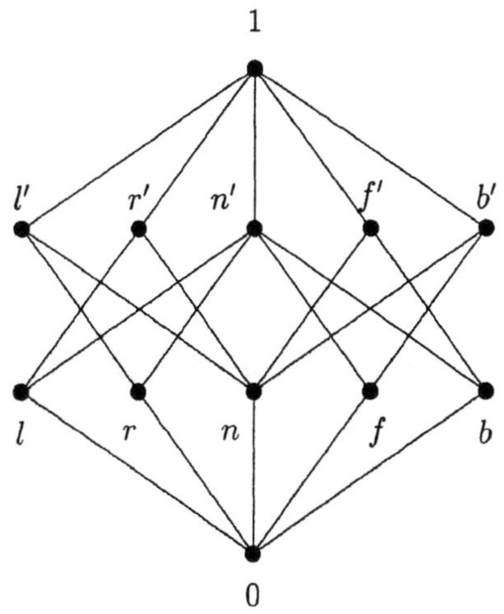

Fig. 4.3

A quantum mechanical realization of the above experiment has been given by Foulis and Randall [FoRa, Exam III]: Consider a device which, from time to time, emits a particle and projects it along a linear scale. We perform two experiments. Experiment A: We look to see if there is a particle present. If there is none, we record the outcome of A as the symbol n. If there is one, we measure its position coordinate x. If $x \geq 1$, we record the outcome of A as the symbol r, otherwise we record the symbol l. Similarly for experiment B: If there is no particle, we record the outcome of B as the symbol n. If there is, we measure the x–component p_x of its momentum. If $p_x \geq 1$, we write b as for the outcome, otherwise we write f. The propositional logic is the same as for the firefly box system.

Another interesting model equivalent to the firefly box system has been given by Wright [WriR]. It uses s generalized urn model. Consider an urn having balls which are all black except for one letter in red paint and one letter in green paint, limited to one of the five combinations of letters r, l, n, f, b as listed in Table 4.4.

Ball Type	Red	Green
1	l	b
2	l	f
3	r	b
4	r	f
5	n	n

Tab. 4.4

There are the two experiments Red and Green. To execute the Red experiment, draw a ball from the urn and examine it under a red filter and record the letter you see. Note that under the red filter, the green letter will appear black and will thus be invisible. There are three outcomes l, r, n. The Green experiment executes using a green filter (all red letters will appear invisible). The outcomes will be restricted to the letters b, f, n, which gives the propositional logic described by Figures 4.2 and 4.3.

Firefly in a three-chamber box. Consider again a firefly, but now in a three-chamber box pictured in Figure 4.5.

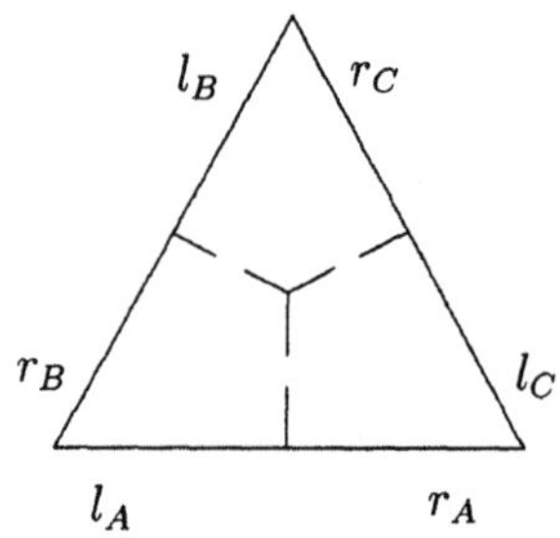

Fig. 4.5

The firefly is free to roam among the three chambers and to light up to will. The sides of the box are windows with vertical lines down their centers. We make three experiments, corresponding to the three windows A, B and C. For each experiment E, we record l_E, r_E, n_E if we see, respectively, a light to the left, right, of the center line or no light. It is clear that we can identify $r_A = l_C =: e$, $r_C = l_B =: c$, $r_B = l_A =: a$, but now we do not identify $f := n_A$, $b := n_B$, $d := n_C$.

The propositional logic of this model has the Greechie diagram given by Figure 4.6 and the corresponding Hasse diagram by Figure 4.7,

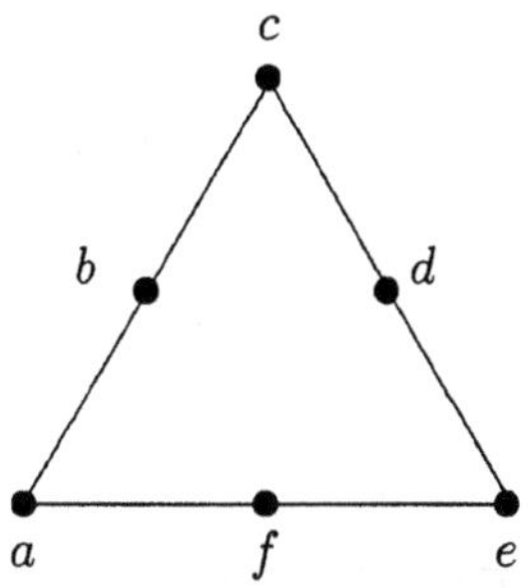

Fig. 4.6 Wright triangle

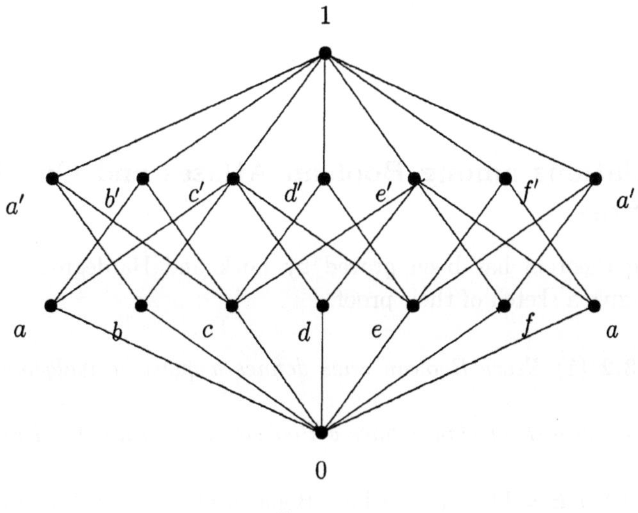

Fig. 4.7

which is an orthoalgebra, called the Wright triangle, being no OMP. It is the most simple case of an OA which is not an OMP. (Due to [HNP], an OA L is not an OML iff it contains the Wright triangle as a suborthoalgebra of L in such a way that, for atoms a, c, e on the corners of the triangle, $a \oplus (c \oplus e)$ is not defined in L.)

In analogy with the generalized urn models, Wright [WriR], we can describe the firefly three–chamber box system equivalently as follows. Consider an urn containing balls which are all black except for one letter in red paint, one letter in green paint and one letter in blue paint, limited to one of the following four combinations of letters

a, b, c, d, e, f according to Table 4.8. There are three experiments Red, Green and Blue using a red, green or blue filter. Assume now (somewhat unphysically) that each one of these three filters lets light through only in its own colour, and that different colours are invisible; i.e., they appear black. The corresponding propositional logic is again given by the Wright triangle.

Ball Type	Red	Green	Blue
1	a	a	d
2	c	f	c
3	b	e	e
4	b	f	d

Tab. 4.8

4.3.4 Relations among Boolean Atlases and Quasi Orthoalgebras

The following theorem has been proved by Lock and Hardegree and to be self-contained we give a sketch of their proof.

Theorem 4.3.2 (1) *Every Boolean atlas defines a quasi orthoalgebra in a natural way.*

(2) *Every quasi orthoalgebra defines a Boolean atlas in a natural way.*

Proof. (1) Let $\mathcal{B} = \{B_i : i \in I\}$ be a Boolean atlas. We define a quasi orthoalgebra L as follows: $L := \bigcup_{i \in I} B_i$, $0 = 0_i$, $1 = 1_i$, $a' = a^{\perp_i}$ for any $i \in I$ such that $a \in B_i$. We say $a \perp b$ iff there is an $i \in I$ such that $a, b \in B_i$ and $a \wedge_i b = 0$, and then $a \oplus b := a \vee_i b$. The operations are well-defined, and it can be shown that the properties of quasi orthoalgebras are satisfied.

We note that the relation $\leq$ on L is defined as follows: $a \leq b$ iff there is $x \in L$ with $a \oplus x = b$. This means the following: there is an $i \in I$ with $a, x \in B_i$, $a \wedge_i x = 0$, and $b = a \oplus x = a \vee_i x$. This implies $a \leq_i b$. On the other hand, if $a \leq_i b$ for some $i \in I$, then $a \wedge_i b^{\perp} = 0$, whence $a \perp_i b^{\perp}$. Therefore, $a \oplus b^{\perp} = a \vee_i b^{\perp}$ is defined, and, moreover, $a \perp_i (a \oplus b^{\perp})^{\perp}$, so that $a \oplus (a \oplus b^{\perp})^{\perp} = a \vee_i (a \vee_i b^{\perp})^{\perp} = a \vee_i (a^{\perp} \wedge_i b) = b$, whence $a \leq b$.

(2) Let L be a quasi orthoalgebra. Let $\{B_i : i \in I\}$ be the set of all blocks of L. Then $\mathcal{B} = \{B_i : i \in I\}$ is a Boolean atlas. □

Example 4.3.3 *Let* $\Omega = \{1,2,3,4,5,6\}$ *and let* B_1 *and* B_2 *be the Boolean algebras generated by* $\{1\},\{2\},\{3\},\{4\},\{5,6\}$ *and* $\{1\},\{2\},\{3,4\},\{5\},\{6\}$, *respectively (with respect the set-theoretic inclusion and* $1_1 = 1_2 = \Omega$*). Then* $\mathcal{B} = \{B_1, B_2\}$ *is a Boolean atlas, and* $L = B_1 \cup B_2$ *is, according to Theorem* 4.3.2, *a quasi orthoalgebra. An easy calculation shows that the order* $\leq$ *induced by* $\oplus$ *in* L *is not transitive. Indeed, we have* $\{3\} \leq \{3,4\}$, $\{3,4\} \leq \{3,4,5\}$ *but* $\{3\} \not\leq \{3,4,5\}$ *although* $\{3\} \subseteq \{3,4,5\}$, *consequently,* L *is not an OA.*

4.3.5 Partition Logics

In this Section, we present a notion of partition logics which will have an intimate connection with special types of automata, see also Svozil [Svo] and [ScSv].

Let L be a quasi orthoalgebra with $\leq$. A non-void subset I of L is said to be an **ideal** of L if

(i) if $a \in I$, $b \in L$, $b \leq a$, then $b \in I$;

(ii) $a, b \in I$ with $a \perp b$ imply $a \oplus b \in I$.

It is clear that $0 \in I$. An ideal I of L is said to be (i) **proper** if $I \neq L$ or, equivalently, $1 \notin I$; (ii) **prime** if, for any $a \in L$, either $a \in I$ or $a' \in I$. We denote by $\mathcal{P}(L)$ the set of all prime ideals in L.

A **probability measure** (or also a **state**) on L is a mapping $s : L \to [0, 1]$ such that (i) $s(1) = 1$, and (ii) $s(a \oplus b) = s(a) + s(b)$ whenever $a \perp b$. A probability measure s is **two-valued** if $s(a) \in \{0,1\}$ for any $a \in L$.

We note that there is a one-to-one correspondence between two-valued probability measures and prime ideals: If s is a two-valued probability measure, then $I_s = \{a \in L : s(a) = 0\}$ is a prime ideal; and if I is a prime ideal, then $s_I : L \to [0,1]$ defined via $s_I(a) = 0$ iff $a \in I$, otherwise $s_I(a) = 1$, is a two-valued probability measure on L.

A set $\mathcal{S}$ of probability measures on L is called **separating** if for all $a, b \in L$, $a \neq b$, there is a probability measure $s \in \mathcal{S}$ such that $s(a) \neq s(b)$. L is called **prime** iff it has a separating set of two-valued probability measures or, equivalently, for any different elements $a, b \in L$ there is a prime ideal I of L such that $a \in I$ and $b \notin I$.

Let $\mathcal{L}$ be a family of quasi orthoalgebras (or OAs, OMP, Boolean algebras, etc.) satisfying the following condition: For all $P, Q \in \mathcal{L}$, $P \cap Q$ is a quasi suborthoalgebra (subOA, sub OMP, Boolean subalgebra, etc.) of both P and Q, and the partial orderings and orthocomplementations coincide on $P \cap Q$. Define the set $L := \bigcup\{P : P \in \mathcal{L}\}$, a relation $\oplus$ and the unary operation $'$ as follows:

(i) $a \oplus b$ is defined iff there is a $P \in \mathcal{L}$ such that $a, b \in P$ and $a \perp_P b$, then $a \oplus b = a \oplus_P b$;

(ii) $a' = b$ iff there is a $P \in \mathcal{P}$ such that $a, b \in P$ and $a'^{P} = b$.

The set L with the above defined $\oplus$ is called the **pasting** of the family $\mathcal{L}$.

Let $\mathcal{R}$ be a family of partitions of a fixed set M. The pasting of the family of Boolean algebras $\{B_R : R \in \mathcal{R}\}$ is called a **partition logic**, and we denote it as a couple $(M, \mathcal{R})$.

Remark 4.3.4 If $\mathcal{B} = \{B_i : i \in I\}$ is a Boolean atlas, then $L = \bigcup_{i\in I} B_i$ with $\oplus$ and $'$ defined by the last above (i) and (ii) is a pasting of a family of Boolean algebras $\{B_i : i \in I\}$. Moreover, $a \oplus b$ is defined iff $a, b \in B_i$ for some $i \in I$ with $a \wedge_i b = 0$, and then $a \oplus b = a \vee_i b$.

We recall that two quasi orthoalgebras L_1 and L_2 are **isomorphic** iff there is a one-to-one mapping $\phi : L_1 \to L_2$ such that $a \oplus b$ is defined in L_1 iff $\phi(a) \oplus \phi(b)$ is defined in L_2 and $\phi(a \oplus b) = \phi(a) \oplus \phi(b)$.

Theorem 4.3.5 *A quasi orthoalgebra L is isomorphic to a partition logic if and only if L is prime.*

Proof. (i) Suppose that L is isomorphic to a partition logic $R = (M, \mathcal{R})$. Without loss of generality, we may assume that $L = R$. Take $A, B \in R$ such that $A \neq B$. Then there is a point $q \in (A \setminus B) \cup (B \setminus A)$. Put $P := \{C \in R : q \notin C\}$. Then P is a prime ideal in L. Indeed, let $C \in P$, and $D \leq C$. Then there is a partition $U \in \mathcal{R}$ such that the Boolean algebra $B(U)$ generated by U contains D, C, and $D \leq_{B(U)} C$ implies $D \subseteq C$. This follows $q \notin D$, hence $D \in P$.

If $E, F \in R$ and $E \perp F$, there is a Boolean algebra $B(V)$ generated by a partition V such that $E \cap_{B(V)} F = \emptyset$. Moreover, $E \oplus F = E \vee_{B(V)} F = E \cup F$ in M. Therefore, $q \notin E \cup F$, which gives $E \oplus F \in P$.

Finally, for every $C \in R$, either $q \in C$ or $q \in M \setminus C$, hence P is a prime ideal.

(ii) Conversely, suppose that L is prime. Let M be the set of all prime ideals in L, i.e., $M = \mathcal{P}(L)$. For $x \in L$, we set $p(x) := \{P \in \mathcal{P}(L) : x \notin P\}$. Since L is prime, the mapping $p : L \to 2^M$ is injective. Moreover, $x \perp y$ gives $p(x) \cap p(y) = \emptyset$ and $p(x \oplus y) = p(x) \cup p(y)$. Indeed, for any $P \in \mathcal{P}(L)$, $x, y \in P$ iff $x \oplus y \in P$, consequently, $x \oplus y \notin P$ iff either $x \notin P$ or $y \notin P$; since either $x \in P$ or $y \in P$ for any $P \in \mathcal{P}(L)$ and all orthogonal elements x and y.

In other words, we have proved that $x \perp y$ implies that the system $R(x, y) := \{p(x), p(y), p((x \oplus y)')\}$ is a partition of M. Let $\mathcal{R} = \{R(x, y) : x, y \in L, x \perp y\}$ and let R be the partition logic $(M, \mathcal{R})$. For every $x \in L$, $p(x) \in R(x, x')$, so that $p : L \to R$ is an injection, and by the definition, also a surjection.

Let $A, B \in R$ with $A \perp_R B$. That is, there is a partition $P \in \mathcal{R}$ with $A, B \in B(P)$, and $A \wedge_{B(\mathcal{P})} B = \emptyset$. By the definition of the partitions in $\mathcal{R}$, there are elements $x, y \in L$ such that $A = p(x), B = p(y)$ for some orthogonal elements $x, y \in L$. This proves that p is an isomorphism in question. $\square$

We say that two elements a and b of an OA L have a **Mackey decomposition** if there are three jointly orthogonal elements a_1, b_1, c in L such that $a = a_1 \oplus c$, $b = b_1 \oplus c$. In OMPs any Mackey decomposition is unique, for OAs this is not true, in general, however for prime orthoalgebras we have the following result.

Proposition 4.3.6 *Any Mackey decomposition in a prime orthoalgebra is uniquely defined.*

Proof. Assume that a and b have two Mackey decompositions, i.e., there are two jointly orthogonal systems $\{a_1, b_1, c_1\}$ and $\{a_2, b_2, c_2\}$ such that $a = a_1 \oplus c_1 = a_2 \oplus c_2$, $b = b_1 \oplus c_1 = b_2 \oplus c_2$. Put $d_1 := (a_1 \oplus b_1 \oplus c_1)'$ and $d_2 := (a_2 \oplus b_2 \oplus c_2)'$.

We assert that $d_1 = d_2$. Assume the converse. Then there is a two–valued probability measure s on L such that $s(d_1) = 1$ and $s(d_2) = 0$. Hence, $s(a_1) = s(b_1) = s(c_1) = 0$, but one of $s(a_2), s(b_2), s(c_2)$ is 1. This leads to a contradiction, since $a_1 \oplus c_1 = a = a_2 \oplus c_2$ and $b_1 \oplus c_1 = b = b_2 \oplus c_2$.

Therefore, $d_1 = d_2$, and hence $a_1 \oplus b_1 \oplus c_1 = a_2 \oplus b_2 \oplus c_2$. This entails $a \oplus b_1 = a \oplus b_2$, so that $b_1 = b_2$, and similarly $c_1 = c_2$, consequently, $a_1 = a_2$. □

4.3.6 Partition Logics and Automata Logics

Let an **alphabet** be a finite nonvoid set. The elements of the alphabet are called **symbols.** A **word** (or a **string**) is a finite (possibly empty) sequence of symbols. The **length** of a word w, denoted by $|w|$, is the number of symbols composing the string. The **empty word** is denoted by ϵ. Σ^* denotes the set of all words over an alphabet Σ. The **concatenation** of two words is the word formed by writing the first, followed by the second, with no intervening space. Let Σ be an alphabet. Σ^* with the concatenation as operation forms a monoid, where the empty word ϵ is the identity. A (**formal**) **language** over an alphabet Σ is a subset of Σ^*.

Definition 4.3.7 *A* **Moore automaton** *M is a five-tuple $M = (Q, \Sigma, \Delta, \delta, \lambda)$, where*

(i) *Q is a finite set, called the set of states;*

(ii) *Σ is an alphabet, called the input alphabet;*

(iii) *Δ is an alphabet, called the output alphabet;*

(iv) *δ is a mapping $Q \times \Sigma$ to Q, called the transition function;*

(v) *λ is a mapping Q to Δ, called the output function.*

Definition 4.3.8 *A* **Mealy automaton** *is a five-tuple $M = (Q, \Sigma, \Delta, \delta, \lambda)$, where Q, Σ, Δ, δ are as in the Moore automaton and λ is a mapping from $Q \times \Sigma$ to Δ.*

Informally, a Moore automaton is in a state $q \in Q$, emitting the output $\lambda(q) \in \Delta$ at any time. If an input $a \in \Sigma$ is applied to the machine, in the next discrete time step the machine instantly assumes the state $p = \delta(q, a)$ and emits the output $\lambda(p)$. A Mealy machine emits the output at the instant of the transition from one state to another, the output depending both on the previous state and the input.

Suppose now an observer is performing experiments with a Moore or Mealy automaton which is contained in a black box with input-output interface. Thus we are only allowed to observe the input and output sequences associated with the box. To conduct an experiment, the observer applies an input sequence and notes the resulting output sequence. Using this output sequence, the observer tries to interpret the information contained in the sequence to determine the values of the unknown parameters.

Suppose the observer conducts experiments on an automaton with a known transition table (i.e., the five-tuple $(Q, \Sigma, \Delta, \delta, \lambda)$) but unknown initial state. This will be called the **initial state identification problem.** Suppose further that only a single copy of the machine is available.

The logical structure of the initial-state identification problem can be defined as follows. Let us call a proposition concerning the initial state of the machine **experimentally decidable** if there is an experiment E which determines the truth value of that proposition. This can be done by performing E, i.e., by the input of a sequence of input symbols $a_1, a_2, a_3, \ldots, a_n$ associated with E, and by observing the output sequence $\lambda_E(q) = \lambda(a_1, q), \ldots, \lambda(\underbrace{\delta(\ldots\delta(q, a_1)\ldots, a_n)}_{n \text{ times}}, a_n)$. The most general form of a prediction concerning the initial state q of the machine is that the initial state q is contained in a subset P of the state set Q. Therefore, we may identify propositions concerning the initial state with subsets of Q. A subset P of Q is then identified with the proposition that the initial state is contained in P.

Definition 4.3.9 *Let E be an experiment (a preset or adaptive one), and let $\lambda_E(q)$ denote the obtained output of an initial state q. λ_E defines a mapping of Q to the set of output sequences Δ^*. We define an equivalence relation on the state set Q by*

$$q \stackrel{E}{\equiv} p \quad \textit{iff} \quad \lambda_E(q) = \lambda_E(p)$$

for any $q, p \in Q$. We denote the partition of Q corresponding to $\stackrel{E}{\equiv}$ by $Q/\stackrel{E}{\equiv}$. Obviously, the propositions decidable by the experiment E are the elements of the Boolean algebra generated by $Q/\stackrel{E}{\equiv}$, denoted by B_E. There is also another way to construct the experimentally decidable propositions of an experiment E. Let $\lambda_E(P) = \bigcup_{q \in P} \lambda_E(q)$ be the direct image of P under λ_E for any $P \subseteq Q$. We denote the direct image of Q by O_E, $O_E = \lambda_E(Q)$.

It follows that the most general form of a prediction concerning the outcome W of the experiment E is that W lays in a subset of O_E. Therefore, the experimentally decidable propositions consist of all inverse images $\lambda_E^{-1}(S)$ of subsets S of O_E, a procedure which can be constructively formulated (e.g., as an effectively computable algorithm), and which also leads to the Boolean algebra B_E. Let $\mathcal{B}$ be the set of all Boolean algebras B_E. We call the partition logic $R = (Q, \mathcal{B})$ an **automaton propositional calculus.**

Proposition 4.3.10 *To every partition logic R there exists an automaton M such that $R = R(M)$.*

Proof. Let $R = (Q, \mathcal{R})$ be a partition logic. Every $P \in \mathcal{R}$ can be rewritten as an indexed family $P = (P_i)_{i \in I_n}$, where the index set I_n denotes the set $\{1, \ldots, n\}$ of natural numbers. We assume that $P_i \neq P_j$ for $i \neq j$. N denotes the greatest number of elements in any partition $P \in \mathcal{R}$. Let $M = (Q, \mathcal{R}, I_N, \delta, \lambda)$ denote the automaton corresponding to the partition logic $R = (Q, \mathcal{R})$. What remains to be defined are the transition function δ and the output function λ. Let p be an arbitrary element of Q. Then, for all $q \in Q$ and for all $P \in \mathcal{R}$, let (i) $\delta(q, P) = p$ and (ii) $\lambda(q, P) = i$ iff $q \in P_i$.

4.3.7 Partition Logics in Examples

Example 4.3.11 *A "Fano plane" pictured at Fig. 4.9 is not a partition logic (it is not prime), it has only unique probability measure s, namely, $s(x) = 1/3$ for any atom $x \in L$.*

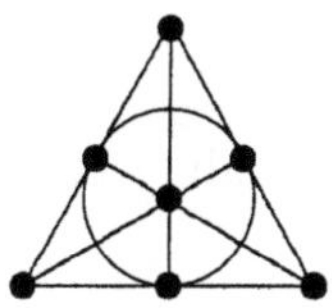

Fig. 4.9

Example 4.3.12 *The Wright triangle, pictured at Figure 4.6, is a partition logic. It has a separating set of two-valued probability measures given by Table 4.10.*

measure	a	b	c	d	e	f
1	1	0	0	1	0	0
2	0	0	1	0	0	1
3	0	1	0	0	1	0
4	0	1	0	1	0	1

Tab. 4.10

It is isomorphic to the following partition logic given by $\Omega = \{1, 2, 3, 4\}$ *and three decompositions of* Ω*:* $\{\{1\}, \{2\}, \{3, 4\}\}$, $\{\{2\}, \{3\}, \{1, 4\}\}$ *and* $\{\{1\}, \{3\}, \{2, 4\}\}$. *The transition and output table of a Mealy automaton realizing the Wright triangle is given by Table* 4.11.

δ	1	2	3	4
$\{\{1\}, \{2\}, \{3, 4\}\}$	1	1	1	1
$\{\{2\}, \{3\}, \{1, 4\}\}$	1	1	1	1
$\{\{1\}, \{3\}, \{2, 4\}\}$	1	1	1	1
λ	1	2	3	4
$\{\{1\}, \{2\}, \{3, 4\}\}$	1	2	3	3
$\{\{2\}, \{3\}, \{1, 4\}\}$	3	1	2	3
$\{\{1\}, \{3\}, \{2, 4\}\}$	1	3	2	3

Tab. 4.11

We recall that according to [WriR], *it cannot be modeled in a Hilbert space.*

Example 4.3.13 *An orthoalgebra given by Figure* 4.12 *is a partition logic. Its system of all two valued probability measures is given in Table* 4.13. *A possible Mealy automaton realization is given in Table* 4.14.

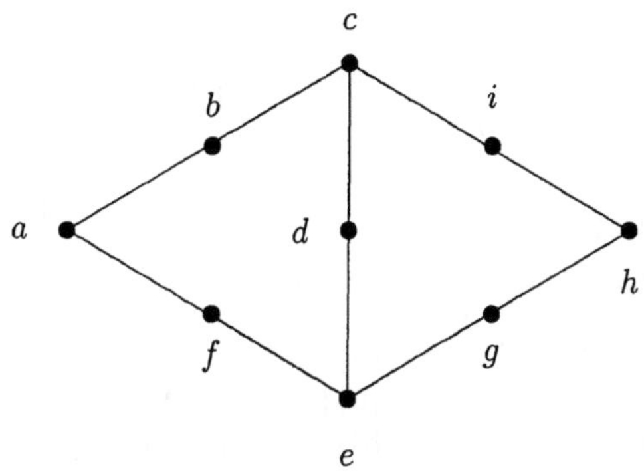

Fig. 4.12

The corresponding decompositions of $\Omega = \{1,2,3,4,5,6\}$ *are* $\{\{1,2\},\{3,4,6\},\{5\}\}$ *for the block* a,b,c, $\{\{5\},\{1,2,3,4\},\{6\}\}$ *for* c,d,e, $\{\{1,2\},\{3,4,5\},\{6\}\}$ *for* a,e,f, $\{\{6\},\{1,3,5\},\{2,4\}\}$ *for* e,g,h, $\{\{2,4\},\{1,3,6\},\{5\}\}$ *for* h,i,c.

measure	a	b	c	d	e	f	g	h	i
1	1	0	0	1	0	0	1	0	1
2	1	0	0	1	0	0	0	1	0
3	0	1	0	1	0	1	1	0	1
4	0	1	0	1	0	1	0	1	0
5	0	0	1	0	0	1	1	0	0
6	0	1	0	0	1	0	0	0	1

Tab. 4.13

δ	1	2	3	4	5	6
$\{\{1,2\},\{3,4,6\},\{5\}\}$	1	1	1	1	1	1
$\{\{5\},\{1,2,3,4\},\{6\}\}$	1	1	1	1	1	1
$\{\{1,2\},\{3,4,5\},\{6\}\}$	1	1	1	1	1	1
$\{\{6\},\{1,3,5\},\{2,4\}\}$	1	1	1	1	1	1
$\{\{2,4\},\{1,3,6\},\{5\}\}$	1	1	1	1	1	1
λ	1	2	3	4	5	6
$\{\{1,2\},\{3,4,6\},\{5\}\}$	1	1	2	2	3	2
$\{\{5\},\{1,2,3,4\},\{6\}\}$	2	2	2	2	1	3
$\{\{1,2\},\{3,4,5\},\{6\}\}$	1	1	2	2	2	3
$\{\{6\},\{1,3,5\},\{2,4\}\}$	2	3	2	3	2	1
$\{\{2,4\},\{1,3,6\},\{5\}\}$	2	1	2	1	3	2

Tab. 4.14

Example 4.3.14 *Orthoalgebras given by Figure* 4.15 *and Figure* 4.16 *are partition logics.*

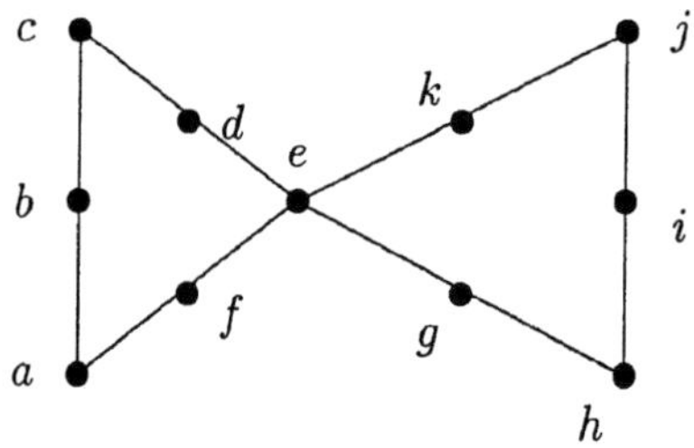

Fig. 4.15

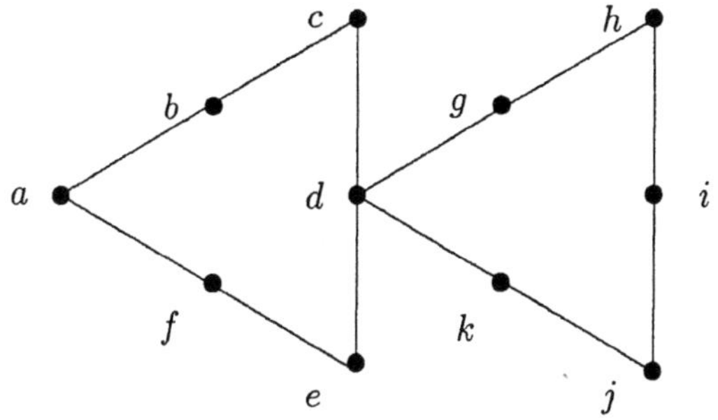

Fig. 4.16

We note that combining the Wright triangles we can obtain plenty of orthoalgebras which are partition logics.

4.3.8 Partition Test Spaces

We concentrate now on the relationship between partition logics with a special type of test spaces, i.e., with D-test spaces which describe orthoalgebras, see Remark 4.1.16.

Let X be a non-void set and Y a non-void family of subsets of a set X. A couple $(Y, \mathcal{T})$, where $\mathcal{T} \subseteq 2^Y$, is said to be a **partition test space** of X if

(i) every $T \in \mathcal{T}$ is a partition of X;

(ii) for every $y \in Y$, there is a $T \in \mathcal{T}$ such that $y \in T$.

Proposition 4.3.15 *A partition test space is a test space.*

Proof. We have to show that if $T_1 \subseteq T_2$, for $T_1, T_2 \in \mathcal{T}$, then $T_1 = T_2$. It follows from the fact that T_1 and T_2 are partitions of X. □

For an event E, put $\bigcup E := \{x \in X : x \in y, \ y \in E\}$.

Proposition 4.3.16 *Let $(Y, \mathcal{T})$ be a partition test space for X. If $E, F \in \mathcal{E}(Y, \mathcal{T})$ and $E \approx F$, then $\bigcup E = \bigcup F$.*

Proof. Let G be a common complement of E and F. Then $x \in \bigcup E$ iff $x \notin \bigcup G$ iff $x \in \bigcup F$. □

Proposition 4.3.17 *A partition test space $(Y, \mathcal{T})$ of X is algebraic if every partition of X consisting of elements of Y belongs to $\mathcal{T}$.*

Proof. Let E, F, G, H be events such that $E \approx_G F$ and $F \, loc \, H$.

From $E \approx_G F$ we obtain, for $x \in X$, $x \in \bigcup E$ iff $x \notin \bigcup G$ iff $x \in \bigcup F$, and from $F \, loc \, H$ we obtain $x \in \bigcup F$ implies $x \notin \bigcup H$.

From this it follows that $F \cup H$ is a partition of X and so $F \cup H \in \mathcal{T}$. □

Proposition 4.3.16 implies that every partition test space $(Y, \mathcal{T})$ of X can be enlarged to an algebraic partition test space $(Y, \mathcal{U})$, where $\mathcal{T} \subseteq \mathcal{U}$, and $\mathcal{U}$ contains all partitions of X which consist of elements of Y. The partition test space $(Y, \mathcal{U})$ with the latter property will be called a **completion** of $(Y, \mathcal{T})$. If $\mathcal{T}$ and $\mathcal{U}$ coincide, we say that $(Y, \mathcal{T})$ is **complete.**

If $(Y, \mathcal{T})$ is a complete partition test space, then for any events E, F with $\bigcup E = \bigcup F$ we have $E \approx F$. Indeed, let $\bigcup E = \bigcup F$, and let G be any local complement of E. Then $\bigcup G = (\bigcup E)^c = X \setminus \bigcup E = X \setminus \bigcup F$, hence G is also a local complement of F.

We note that $\Pi(Y)$ below is defined by (4.1).

Proposition 4.3.18 *Let $(Y, \mathcal{T})$ be a partition test space of the set X. Then*

(i) $\Pi(Y)$ *is an OMP if $E, F, G \in \mathcal{E}(Y)$ with $E \perp F$, $F \perp G$, $G \perp E$ imply $(E \cup F) \perp G$.*

(ii) $\Pi(Y)$ *is a concrete OMP*[4] *if the condition $(\bigcup E_1) \cap (\bigcup E_2) = \emptyset$ if and only if $E_1 \perp E_2$ holds.*

Proof. (i) This is evident.

(ii) According to Proposition 4.3.16, $\pi(E)$ can be identified with $\bigcup E \subseteq X$. □

Remark 4.3.19 The same set L can be the logic of several partition test spaces. A concrete logic L can have a test space not satisfying the condition (ii). Indeed, let $X = \{1, 2, 3, 4\}$ and take $(Y, \mathcal{T})$, where $Y = \{\{1\}, \{3, 4\}, \{2\}, \{2, 4\}, \{3\}\}$, $\mathcal{T} = \{T_1, T_2\}$ and $T_1 = \{\{1\}, \{3, 4\}, \{2\}\}$, $T_2 = \{\{1\}, \{2, 4\}, \{3\}\}$. Then $\Pi(Y)$ is a concrete OMP (it is isomorphic to Figure 4.2) with $\{2\} \cap \{3\} = \emptyset$, but $\{2\} \not\perp \{3\}$.

Theorem 4.3.20 *A test space $(X, \mathcal{T})$ is isomorphic to a partition test space if and only if it possesses a separating family of two–valued weights.*

Proof. Let $(Y, \mathcal{T})$ be a partition test space of X. If $y_1, y_2 \in Y$, $y_1 \neq y_2$, then $(y_1 \setminus y_2) \cup (y_2 \setminus y_1)$ possesses at least one point, say x, of X. Define a function $\omega : Y \to \{0, 1\}$ by putting $\omega(y) = 1$ iff $x \in y$, otherwise we put $\omega(y) = 0$. Then ω is a two–valued weight on $(Y, \mathcal{T})$, and $\omega(y_1) \neq \omega(y_2)$.

---

[4]An OMP L is a **concrete logic** if it is isomorphic to a family $\mathcal{L}$ of subsets of a set Ω such that: (i) $\Omega \in \mathcal{L}$. (ii) If $A, B \in \mathcal{L}$ and $A \cap B = \emptyset$, then $A \cup B \in \mathcal{L}$.

Conversely, let $(X, \mathcal{T})$ be a test space with a separating family Δ of two–valued weights. Define $\phi(x) := \{\omega \in \Delta : \omega(x) = 1\}, x \in X$, and $\phi(T) := \{\phi(x) : x \in T\}$, $T \in \mathcal{T}$.

Consider $(\phi(X), \phi(\mathcal{T}))$, where $\phi(X) := \{\phi(x) : x \in X\}$ and $\phi(\mathcal{T}) := \{\phi(T) : T \in \mathcal{T}\}$. We claim that $(\phi(X), \phi(\mathcal{T}))$ is a partition test space of X, where $\phi(X) \subseteq 2^{\Delta}$, $\phi(T)$ is a partition of Δ for any $T \in \mathcal{T}$. Observe that, for any $\omega \in \Delta$, $\omega(T) = 1 = \sum_{x\in T} \omega(x)$, so that there is a point $x_0 \in T$ such that $\omega(x_0) = 1$ and $\omega(x) = 0$ for any $x \neq x_0$. That is, for any $\omega \in \Delta$ and for any $T \in \mathcal{T}$, there is a unique $x \in T$ such that $\omega \in \phi(x)$. This implies that every $\phi(T)$ is a partition of Δ. □

Theorem 4.3.21 *There is a one-to-one correspondence (up to isomorphism) between partition logics and partition test spaces.*

Proof. Let $(Y, \mathcal{T})$ be a partition test space for a set X. For any event $E \subseteq T$, $T \in \mathcal{T}$, define $u(E) := \bigcup E$. We have if $E \approx F$, then $\bigcup E = \bigcup F$. Define $L := \{\bigcup E : E \in \mathcal{E}(Y)\}$. For every $T \in \mathcal{T}$, $u(T) := \{u(E) : E \subseteq T\}$ is a Boolean algebra. Indeed, every $u(E)$ is a union of some sets from the partition T of X. For $a, b \in L$, define $a \perp b$ iff there are disjoint $E, F \in \mathcal{E}(Y)$ with $E \cup F \subseteq T$ for some $T \in \mathcal{T}$, and $a = u(E)$, $b = u(F)$, and define $a \oplus b = u(E \cup F)$, $a' = u(T \setminus E)$ when $a = u(E)$, $E \subseteq T \in \mathcal{T}$. Clearly, $u(T) = X$ for every $T \in \mathcal{T}$ is the greatest elements in L (by the ordering $a \leq b$ iff $a \perp b'$). Clearly, L is a pasting of Boolean algebras $\{u(T) : T \in \mathcal{T}\}$. This L will be called the **logic** of $(Y, \mathcal{T})$ in X.

Conversely, if L is a partition logic, that is, L is a pasting of Boolean algebras $B(T_i)$, $i \in I$, where T_i is a partition of a set $X \neq \emptyset$ for any $i \in I$, then put

$$Y = \bigcup_{i\in I} \{y : y \in T_i\}.$$

The couple $(Y, \{T_i : i \in I\})$ is a partition test space of X, and its logic is isomorphic with L, and the proof is complete. □

We note that all examples in the previous section are arising by the way described in Theorem 4.3.20 and Theorem 4.3.21.

4.3.9 Product of Partition Logics

Assume that $(X, \mathcal{R})$ and $(Y, \mathcal{S})$ are two partition logics, where $\mathcal{R}$ and $\mathcal{S}$ are two systems of finite decompositions of X and Y, respectively. Choose $R \in \mathcal{R}$ and $S \in \mathcal{S}$. Then $R \times S := \{A \times B : A \in R,\ B \in S\}$ is a decomposition of $X \times Y$. Define

$$\mathcal{R} \times \mathcal{S} := \{R \times S : R \in \mathcal{R}, S \in \mathcal{S}\}.$$

Then the partition logic $\{X \times Y, \mathcal{R} \times \mathcal{S}\}$ is called the **product partition logic** of $(X, \mathcal{R})$ and $(Y, \mathcal{S})$.

Example 4.3.22 *Let $\Omega = \{1,2,3,4,5,6\}$ and let B_1 and B_2 be the Boolean algebras generated by $R_1 := \{\{1\},\{2\},\{3\},\{4\},\{5,6\}\}$ and $R_2 : \{\{1\},\{2\},\{3,4\},\{5\},\{6\}\}$, respectively. The system $(\Omega, \{R_1, R_2\})$ is a partition logic which is not a Boolean algebra, it is only a quasi orthoalgebra. Then $(\Omega\times\Omega, \{R_1\times R_1, R_1\times R_2, R_2\times R_1, R_2\times R_2\})$ is the product partition logic of $(\Omega, \{R_1, R_2\})$ with itself.*

Proposition 4.3.23 *Let μ and ν be probability measures on the partition logics $(X, \mathcal{R})$ and $(Y, \mathcal{S})$, respectively, where card $(Y) = n$ and $\mathcal{S} = \{\{y_1\},\ldots,\{y_n\}\}$, $y_i \in Y$ for $i = 1,\ldots,n$, $y_i \neq y_j$ for $i \neq j$, and $n \geq 1$. Then there is a unique probability measure $\mu\times\nu$ on $(X\times Y, \mathcal{R}\times\mathcal{S})$ such that*

$$(\mu\times\nu)(A\times B) = \mu(A)\,\nu(B),\ A\in(X,\mathcal{R}),\ B\in(Y,\mathcal{S}). \tag{4.22}$$

If, in addition, μ and ν are two-valued measures, so is $\mu\times\nu$.

Proof. Let U be any element of the product partition logic $L = (X\times Y, \mathcal{R}\times\mathcal{S})$. Without loss of generality we can suppose that $Y = \{1,\ldots,n\}$. It is easy to show that U can be represented uniquely in the form $U = \bigcup_{i=1}^n A_i\times\{i\}$, where all A_i's belong to the same Boolean algebra B_R generated by a decomposition $R\in\mathcal{R}$; the case $A_i = \emptyset$ is not excluded.

In addition, $U := \bigcup_{i=1}^n A_i\times\{i\} \perp \bigcup_{i=1}^n B_i\times\{i\} =: V$ iff $\{A_i\}_{i=1}^n$ and $\{B_i\}_{i=1}^n$ belong to the same Boolean algebra B_R and $A_i\cap B_i = \emptyset$ for $i = 1,\ldots,n$, and in this case $\bigcup_{i=1}^n A_i\times\{i\}\oplus\bigcup_{i=1}^n B_i\times\{i\} = \bigcup_{i=1}^n(A_i\cup B_i)\times\{i\} = \bigcup_{i=1}^n(A_i\oplus B_i)\times\{i\}$.

If now we define $(\mu\times\nu)(U) := \sum_{i=1}^n \mu(A_i)\nu(\{i\})$, then $\mu\times\nu$ is a well-defined mapping on L. It is simple to verify that $X\times Y = \bigcup_{i=1}^n X\times\{i\}$. Hence $(\mu\times\nu)(1) = (\mu\times\nu)(X\times Y) = \sum_{i=1}^n \mu(X)\nu(\{i\}) = \mu(X)\nu(Y) = 1$.

Similarly, $(\mu\times\nu)(U\oplus V) = (\mu\times\nu)(\bigcup_{i=1}^n(A_i\oplus B_i)\times\{i\}) = \sum_{i=1}^n\mu(A_i\oplus B_i)\nu(\{i\}) = \sum_{i=1}^n(\mu(A_i)+\mu(B_i))\nu(\{i\}) = (\mu\times\nu)(U)+(\mu\times\nu)(V)$, so that $\mu\times\nu$ is a probability measure on L satisfying (4.22). The uniqueness of $\mu\times\nu$ is now evident. □

The probability measure $\mu\times\nu$, if it exists, is said to be a **product probability measure** of μ and ν. We note that the extension of $\mu\times\nu$ from (4.22) to a probability measure on the product of general partition logics seems to be open.

The product of finite automata logics has an intuitive and rather simple realization: If the sets of states of two automata $M_1 = (Q_1,\Sigma_1,\Delta_1,\delta_1,\lambda_1)$ and $M_2 = (Q_2,\Sigma_2,\Delta_2,\delta_2,\lambda_2)$ are mutually disjunctive, then the automaton partition logic of the **parallel decomposition** $M_1\|M_2$ of the two automata is the product $\{Q_1\times Q_2, \mathcal{R}_1\times\mathcal{R}_2\}$ of the automaton partition logics $\{Q_1,\mathcal{R}_1\}$ and $\{Q_2,\mathcal{R}_2\}$ associated with M_1 and M_2, respectively.

Example 4.3.24 *To illustrate this, assume a Mealy machine* $M_s = (Q, \Sigma, \Delta, \delta, \lambda)$ *which has n states, n inputs and 2 outputs. That is*

$$\begin{aligned} Q &= \{1,2,\ldots,n\}, \\ \Sigma &= \{1,2,\ldots,n\}, \\ \Delta &= \{0,1\}. \end{aligned}$$

Its transition and output functions are ($\delta_{s,x}$ *stands for the Kronecker delta function)*

$$\begin{aligned} \delta(s,i) &= i, \\ \lambda(s,i) &= \delta_{s,i} = \begin{cases} 1, & \text{if } s = i, \\ 0, & \text{if } s \neq i. \end{cases} \end{aligned}$$

The flow table and state graph (for $n = 3$*) of this Mealy automaton is given in Figure* 4.17.

$M_s =$

s/i	1	2	⋯	n	1	2	⋯	n
1	1	2	⋯	n	1	0	⋯	0
2	1	2	⋯	n	0	1	⋯	0
⋮	1	2	⋯	n	0	0	⋯	0
n	1	2	⋯	n	0	0	⋯	1

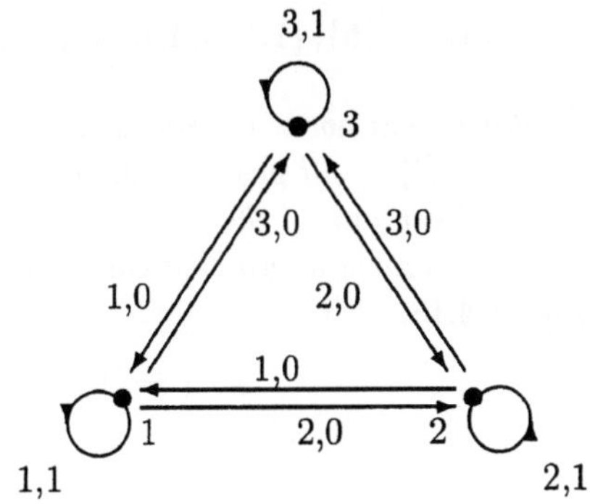

Fig. 4.17

Two automata $M_1 = (Q_1, \Sigma_1, \Delta_1, \delta_1, \lambda_1)$ and $M_2 = (Q_2, \Sigma_2, \Delta_2, \delta_2, \lambda_2)$ of the same type are **isomorphic** if and only if there exist three one-to-one mappings $f : Q_1 \longrightarrow Q_2$, $g : \Sigma_1 \longrightarrow \Sigma_2$, $h : \Delta_1 \longrightarrow \Delta_2$ such that $f[\delta_1(s_1, i_1)] = \delta_2[f(s_1), g(i_1)]$ and $f[\lambda_1(s_1, i_1)] = \lambda_2[f(s_1), g(i_1)]$, where $s_j \in Q_j$ and $i_j \in I_j$, $j \in \{1, 2\}$. The triple (f, g, h) is an **isomorphism** between M_1 and M_2. An isomorphism just renames the states, the inputs and the outputs. From a purely input/output point of view, g as well as h (or h^{-1}) are combinatory circuits and M_1 performs similar as the serial decomposition (see below) $h^{-1}M_2g$ of the machines g, M_2 and h^1.

The **serial connection** of the two machines $M_1 = (Q_1, \Sigma_1, \Delta_1, \delta_1, \lambda_1)$ and $M_2 = (Q_2, \Sigma_2, \Delta_2, \delta_2, \lambda_2)$ for which $\Delta_1 = \Sigma_2$ is the machine

$$M = M_1 \to M_2 = (Q_1 \times Q_2, \Sigma_1, \Delta_2, \delta, \lambda),$$

where $\delta[(s_1, s_2), i] = (\delta_1(s_1, i), \delta_2[s_2, \lambda(s_1, i)])$ and $\lambda[(s_1, s_2), i] = \lambda_2[s_2, \lambda_1(s, i)]$.

The **parallel connection** of the two machines $M_1 = (Q_1, \Sigma_1, \Delta_1, \delta_1, \lambda_1)$ and $M_2 = (Q_2, \Sigma_2, \Delta_2, \delta_2, \lambda_2)$ is the machine

$$M = M_1 \| M_2 = (Q_1 \times Q_2, \Sigma_1 \times \Sigma_2, \Delta_1 \times \Delta_2, \delta, \lambda),$$

where $\delta[(s_1, s_2), (i_1, i_2)] = (\delta_1(s_1, i_1), \delta_2(s_2, i_2))$ and $\lambda[(s_1, s_2), (i_1, i_2)] = (\lambda_1(s_1, i_1), \lambda_2(s_2, i_2))$.

Example 4.3.25 *In what follows, we explicitly construct the Mealy automaton M_s introduced before. Input/output experiments can be performed by the input of one symbol i (in this example, more inputs yield no finer partitions). Let us assume that one input $i = 5$. This experiment is able to distinguish between state $s = 5$ and all the other states; hence it induces a partition (suppose $n > 5$)*

$$v(5) = \{\{5\}, \{1, 2, 3, 4, 6, \ldots, n\}\}.$$

After this experiment, information about the initial state is lost (irreversible model). Consider now the partitions $v(i)$ of all possible experiments with one input x (all of them non-co-measurable). Every one of them generates a Boolean algebra of events with two atoms; e.g., $v(5)$ generates a two-element Boolean algebra 2^2 whose Hasse diagram is drawn in Figure 4.18.

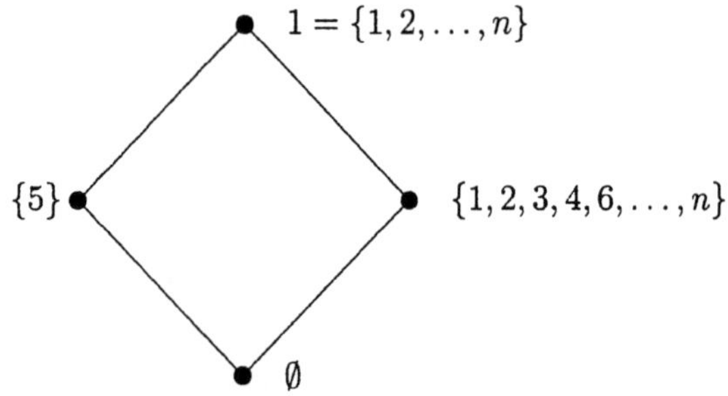

Fig. 4.18

The automaton propositional calculus and the associated partition logic *is the set of all partitions*

$$P = \{v(i) : i \in I\}.$$

Lattice theoretically, this amounts to a **pasting** *of all the $v(i)$'s. In the specific example, the pasting is just the horizontal sum– only the least and greatest elements $\emptyset$ and of each 2^2 is identified– and one obtains a Chinese lantern lattice MO_n.*

4.3.10 Exercises

1. Show that if $A = \{a_1, \ldots, a_n\}$ and $= \{b_1, \ldots, b_m\}$ are $\bigoplus$-orthogonal systems of elements in a difference poset L, then $\{a_1, \ldots, a_n, b_1, \ldots, b_m\}$ is $\bigoplus$-orthogonal iff $a := \bigoplus_{i=1}^n a_i \perp b := \bigoplus_{j=1}^m b_j$. In such a case,

$$\bigoplus\{a_1, \ldots, a_n, b_1, \ldots, b_m\} = a \oplus b.$$

2. Open problem: Find the tensor product $[0,1] \otimes [0,1]$. Is it equal to $[0,1]$?

3. Let L be an effect algebra. If, for any finite subset F of a subset G of L, $\bigoplus F$ exists, does $\bigoplus G$ exist ? (Hint: Take Chang's MV-algebra.)

4. Let S be an inner product space and let $D(S)$ be the set of all **Foulis-Randall subspaces** of S, i.e., of all subspaces M for which there exists an orthonormal system (ONS, for short) $\{u_i\}$ such that $M = \{u_i\}^{\perp\perp}$, which is a complete orthoposet. Any M of $D(S)$ possesses at least one local complement M', i.e., such an element $M' \in D(S)$, for which $M' \perp M$ and $M \vee M' = S$.

   Show that S is complete iff $D(S)$ is an orthomodular poset with respect to $\perp$: $M \to M^\perp$. ([DvPu, Thm 3.1])

5. Let $E(S)$ be the set of all **splitting subspaces** of S for which the condition $M + M^\perp = S$ holds which is an orthomodular poset, and which is not necessarily a σ-poset. Show that S is complete iff $E(S)$ is a σ-complete OMP. ([Dvu, Thm 4.1.6])

6. Show that there exists an incomplete S for which $E(S)$ is an OML and which consists of all finite- or co-finite-dimensional subspaces of S. ([PtWe])

7. Let E be a **quadratic space**, i.e., E is a vector space over a division ring K with a Hermitian form $(\cdot,\cdot)$. For any subset $M \subseteq E$, we put $M^\perp = \{x \in E : (x, y) = 0 \text{ for any } y \in M\}$. Let $L(E)$ denote the family of all orthogonally closed subspaces of E, i.e.,

$$L(E) = \{M \subseteq E : M^{\perp\perp} = M\},$$

and let $E(E)$ denote the set of all splitting subspaces of E, i.e.,

$$E(E) = \{M \subseteq E : M^\perp + M = E\}.$$

Then $E(E) \subseteq L(E)$, and E is said to be orthomodular iff $L(E) = E(E)$.

A non-zero vector $x \in E$ such that $(x, x) = 0$ is said to be isotropic. E is said to be anisotropic if it has no isotropic vectors.

We say that a quadratic space E is **Dacey** if, for any ONS $\{x_i\} \cup \{y_j\}$ in E with $\{x_i\} \cap \{y_j\} = \emptyset$, we have $\{x_i\}^{\perp\perp} = \{y_j\}^\perp$.

Let E be a quadratic space and define $E_0 := E \setminus \{0\}$ and let $\mathcal{T}(E_0)$ be the system of all MOSs in E. Then the pair $(E_0, \mathcal{T}(E_0))$ is a test space, and denote by $\mathcal{E}(E_0)$ the system of all events in E_0.

Show: The test space $(E_0, \mathcal{T}(E_0))$ is algebraic if and only if E is Dacey. ([Dvu 26, Thm 2.1])

8. **Solèr's theorem** ([Sol]): Any infinite-dimensional orthomodular space containing a sequence of orthonormal vectors[5] is either a real, complex or quaternionic Hilbert space.

   An anisotropic quadratic space E is **half-normal** if there is a sequence $\{e_i\}_{i=1}^\infty$ of mutually orthogonal vectors such that $(e_i, e_i) = 1$ for any i ($\{e_i\}_i$ is called an orthonormal sequence).

   Using Solèr's theorem show that if E is an infinite-dimensional half-normal anisotropic quadratic space such that any MOS in E is at most countable, then E is a real, complex or quaternionic separable Hilbert space, $\dim E = \aleph_0$, iff $(E_0, \mathcal{T}(E_0))$ is an algebraic test space.

9. Let X be a nonempty set of outcomes corresponding to a physical system and let $\mathcal{T} \subseteq \{0, 1, 2, \ldots\}^X$. We call $(X, \mathcal{T})$ an **E-test space** if the following two conditions hold:

   (1) For any $x \in X$ there exists a $t \in \mathcal{T}$ such that $t(x) \neq 0$.
   (2) If $s, t \in \mathcal{T}$ with $s \leq t$ (i.e., $s(x) \leq t(x)$ for all $x \in X$), then $s = t$.

   We call $f \in \{0, 1, 2, \ldots\}^X$ an **effect** if $f \leq t$ for some $t \in \mathcal{T}$, and let $\mathcal{E} = \mathcal{E}(X, \mathcal{T})$ be the set of all effects. We say that $f, g \in \mathcal{E}$ are (i) orthogonal ($f \perp g$) if $f + g \in \mathcal{E}$; (ii) local complements of each other (f *loc* g) if $f + g \in \mathcal{T}$, (iii) perspective ($f \cong g$) if they share a local complement. We say that $\mathcal{E}$ is algebraic if for $f, g, h \in \mathcal{E}$, $f \cong g$ and $h \perp f$ imply $h \perp g$.

[5] It is sufficient to suppose that there is an infinite sequence of mutually orthogonal vectors of the same length.

Show that if $\mathcal{E}$ is an algebraic E-test space, then $\cong$ defines an equivalence such that $\Pi(X) = \mathcal{E}/\cong$ can be organized into an effect algebra. Conversely, any effect algebra is isomorphic to the logic of an algebraic E-test space. ([Gud 11], [Gud 12])

10. Show that a real or complex inner product space S is complete iff there is a unit vector $x \in S$ and a positive number $W > 0$ such that, for any MONS $\{x_i\}$ in S,

$$0 < W \leq \sum_i |(x, x_i)|^2$$

holds. ([Dvu 3, Thm 2.1])

11. Find an example of a quasi orthoalgebra with the transitive order $\leq$ which does not correspond to any orthoalgebra.

Chapter 5

BCK-algebras

In 1966, Imai and Iséki [ImIs], [Ise] introduced the notion of a *BCK-algebra*. This notion originated from two different ways: (1) set theory, and (2) classical and non-classical propositional calculi. The BCK-operation $*$ is an analogue of the set-theoretical difference. Today BCK-algebras have been studied by many authors and they have been applied to many branches of mathematics, such as group theory, functional analysis, probability theory, topology, fuzzy set theory, and so on.

For a comprehensive source about BCK-algebras we can recommend the monograph [MeJu]. There are two important classes of BCK-algebras: commutative BCK-algebras and bounded commutative BCK-algebras. As noted in [CiTo], the latter are equivalent to MV-algebras, introduced by Chang [Cha], Wajsberg-algebras [FRT], and bricks [Bos], as well as to S-algebras [Buf]. Nowadays MV-algebras are important objects of investigation: [Mun, Mun 1], [Lac], [BDL], etc.

Lacava [Lac] presented a representation of MV-algebras via intervals $[0, u]$, where u is a strictly positive element in an Abelian lattice ordered group, and Mundici [Mun] strengthened this result showing that u can be chosen as a strong unit. In this case the lattice ordered group has the universal property, and it is unique up to isomorphism.

An analogous role is played by commutative BCK-algebras with the relative cancellation property [DvGr 2] as will be shown. In addition, commutative BCK-algebras have a close connection with compatible generalized difference posets.

5.1 Elements of BCK-algebras

5.1.1 Definitions and Elementary Properties of BCK-algebras

A **BCK-algebra** is a non-empty set X with a binary operation $*$ and with a constant element 0 such that the following axioms are satisfied: For all x, y, $z \in X$,

(BCK-1) $((x * y) * (x * z)) * (z * y) = 0$;

(BCK-2) $(x * (x * y)) * y = 0$;

(BCK-3) $x * x = 0$;

(BCK-4) $x * y = 0$ and $y * x = 0$ imply $x = y$;

(BCK-5) $0 * x = 0$.

We shall write $X = (X; *, 0)$. [1]

There is a binary relation $\leq$ in X defined by

$$x \leq y \quad \text{iff} \quad x * y = 0. \tag{5.1}$$

Moreover, $\leq$ is a partial order on X (called sometimes the **BCK-order**) with the least element 0, as follows from the following.

Proposition 5.1.1 *In a BCK-algebra $(X; *, 0)$, we have the following properties:*

(a) *$x \leq y$ implies $z * y \leq z * x$.*

(b) *$\leq$ is a partial order on X.*

Proof. (a) Let $x \leq y$. Then by (BCK-1), we have

$$(z * y) * (z * x) \leq x * y.$$

Hence, $(z*y)*(z*x) \leq 0$. Combining (BCK-5) and (BCK-4), we have $(z*y)*(z*x) = 0$, i.e., (a) holds.

(b) To show that $\leq$ is a partial order, it is necessary to show that $x \leq y$ and $y \leq z$ imply $x \leq z$. By (a), $y \leq z$ implies $x * z \leq x * y$. If $x \leq y$ then $x * y = 0$. Hence $x * z \leq 0$, and so $x \leq z$. Therefore, $\leq$ is transitive, and $\leq$ is a partial order with the least element 0. □

Hence, a BCK-algebra is a poset $(X; \leq)$ with the least element 0 and with a total binary operation $*$ such that

(BCK'-1) $(x * y) * (x * z) \leq (z * y)$;

(BCK'-2) $x * (x * y) \leq y$;

(BCK'-6) $x \leq y$ iff $x * y = 0$.

[1] In combinatory logic, there are various combinators. Some of their names are B, C, and K, and their corresponding formulas are: B $((y * z) * (x * z)) * (y * x) = 0$, C $((z * x) * y) * ((z * y) * x) = 0$, and K $(x * y) * x = 0$. An algebra $(X; *, 0)$ of type (2,0) is a BCK-algebra iff it satisfies axioms B,C, and K, and the rule (BCK-4); see Exercise 16.5.1.9.

If there exists a greatest element 1 in a BCK-algebra $(X; *, 0)$, then X is said to be a **bounded BCK-algebra**.

Proposition 5.1.2 *In a BCK-algebra* $(X; *, 0)$, *we have*

$$(x * y) * z = (x * z) * y, \tag{5.2}$$

for all elements $x, y, z \in X$.

Proof. By (BCK-2) we have $x * (x * z) \leq z$. Making use of (a) of Proposition 5.1.1 and (BCK-1), we have

$$(x * y) * z \leq (x * y) * (x * (x * z)) \leq (x * z) * y.$$

Interchanging y and z in the above inequality, we obtain $(x * z) * y \leq (x * y) * z$. By (BCK-4) we have $(x * y) * z = (x * z) * y$. □

Proposition 5.1.3 *For all elements* x, y, z *of a BCK-algebra* $(X; *, 0)$, *we have*

(a) $x * y \leq z$ *implies* $x * z \leq y$.

(b) $(x * z) * (y * z) \leq x * y$.

(c) $x \leq y$ *implies* $x * z \leq y * z$.

(d) $x * y \leq x$.

(e) $x * 0 = x$.

Proof. (a) This is a straightforward consequence of Proposition 5.1.2.

(b) This follows from (BCK-1) and (a).

(c) Let $x \leq y$. Then $x * y = 0$ and by (b),

$$(x * z) * (y * z) \leq x * y = 0.$$

Hence, $x * z \leq y * z$ which proves (c).

(d) By (a), Proposition 5.1.2, (BCK-3) and (BCK-5), we have

$$(x * y) * x = (x * x) * y = 0 * y = 0.$$

Consequently, $x * y \leq x$.

(e) By (BCK-2), we have $x * (x * 0) \leq 0$, that is, $x \leq x * 0$. In addition, by (d), we get $x * 0 \leq x$. Hence by (BCK-4), (e) holds. □

For any $x, y \in X$ denote

$$x \sqcap y := y * (y * x).$$

Obviously, $x \sqcap y$ is a lower bound of x and y, and $x \sqcap x = x$, $x \sqcap 0 = 0 \sqcap x = 0$.

But in general, $x \sqcap y \neq y \sqcap x$. X is said to be **commutative** if $x \sqcap y = y \sqcap x$ for any $x, y \in X$. In this case, X is a lower semilattice with respect to the partial order $\leq$, i.e., for all $x, y \in X$, the greatest lower bound of x and y, $x \wedge y$, exists in X and $x \wedge y = x \sqcap y$, as we will see in Section 5.1.3.

Proposition 5.1.4 *In any BCK-algebra, we have*

$$x * (y \sqcap x) = x * y.$$

Proof. Since $y \sqcap x \leq y$, by Proposition 5.1.1(a), we have

$$x * y \leq x * (y \sqcap x).$$

On the other hand, by (BCK-2) we have

$$x * (y \sqcap x) = x * (x * (x * y)) \leq x * y.$$

Therefore, $x * y = x * (y \sqcap x)$. □

Let X_0 be a non-void subset of a BCK-algebra $(X; *, 0)$. Then X_0 is said to be a **BCK-subalgebra** of X if, for any $x, y \in X_0$, $x * y \in X_0$.

5.1.2 Examples of BCK-algebras

Example 5.1.5 $([0, \infty); *_{\mathbb{R}}, 0)$, *where*

$$s *_{\mathbb{R}} t = \max\{0, s - t\}, \tag{5.3}$$

$s, t \in [0, \infty)$, *is an example of a commutative BCK-algebra.*

Example 5.1.6 *Let Ω be a non-void subset and let X be the set of all positive real valued functions on Ω. For $f, g \in X$, we define $f * g$ in the natural way*

$$(f * g)(\omega) = \begin{cases} 0, & \text{if } f(\omega) \leq g(\omega), \\ f(\omega) - g(\omega), & \text{if } f(\omega) \geq g(\omega), \end{cases}$$

for any $\omega \in \Omega$. Then $(X; *, 0_\Omega)$ is a commutative BCK-algebra, where $0_\Omega(\omega) = 0$ for any $\omega \in \Omega$.

Example 5.1.7 *Let $X = \{0, 1, 2, \ldots\}$ be ordered with the natural order. We define, for any $x, y \in X$,*

$$x * y = \begin{cases} 0, & \text{if } x \leq y, \\ 1, & \text{if } y < x,\ y \neq 0, \\ x, & \text{if } y < x,\ y = 0. \end{cases}$$

Then $(X; *, 0)$ is a non-bounded non-commutative BCK-algebra.

Example 5.1.8 *Let X be a usual ring of subsets of a non-void set Ω. Then $(X; \setminus, \emptyset)$ is a commutative BCK-algebra.*

Example 5.1.9 *Let $(X; *, 0)$ be a BCK-algebra and $1 \notin X$. We define the operation $*'$ on $X' = X \cup \{1\}$ as follows*

$$x *' y = \begin{cases} x * y, & \text{if } x, y \in X, \\ 0, & \text{if } x \in X,\ y = 1, \\ 1, & \text{if } x = 1,\ y \in X, \\ 0, & \text{if } x = 1 = y. \end{cases}$$

Then $(X; *, 0)$ is bounded BCK-algebra with the greatest element 1. We note that if X is commutative, then X' cannot be commutative.

Example 5.1.10 *Let $(X; \leq)$ be a poset with the least element 0. Define an operation $*$ as follows*

$$x * y = \begin{cases} 0, & \text{if } x \leq y, \\ x, & \text{otherwise.} \end{cases}$$

Then $(X; *, 0)$ is a BCK-algebra which is not necessary commutative but whose BCK-order coincides with the original one.

Example 5.1.11 *Let $X = \{0, 1, 2, 3, 4\}$ and $*$ operation be given by table 5.1. Then $(X; *, 0)$ is a bounded non-commutative BCK-algebra.*

*	0	1	2	3	4
0	0	0	0	0	0
1	1	0	0	0	0
2	2	1	0	1	0
3	3	3	3	0	0
4	4	4	4	4	0

Tab. 5.1

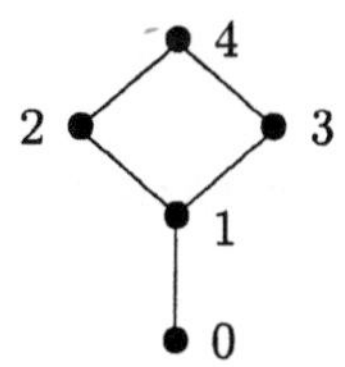

Fig. 5.1

Example 5.1.12 *Let $X = \{0, 1, 2, 3, 4\}$ and let the operation $*$ be given by table 5.2. Then $(X; *, 0)$ is a bounded non-commutative BCK-algebra with linear BCK-order.*

*	0	1	2	3	4
0	0	0	0	0	0
1	1	0	0	0	0
2	2	1	0	0	0
3	3	1	1	0	0
4	4	3	2	1	0

Tab. 5.2

Fig. 5.2

Example 5.1.13 *The example* B_{4-3-3} *from* [MeJu], *where* $*$ *is given by table* 5.3 *(next to it is its Hasse diagram), gives a commutative BCK-algebra which is not directed upwards,*[2] *consequently it is not a lattice.*

*	0	1	2	3
0	0	0	0	0
1	1	0	0	0
2	2	1	0	1
3	3	1	1	0

Tab. 5.3

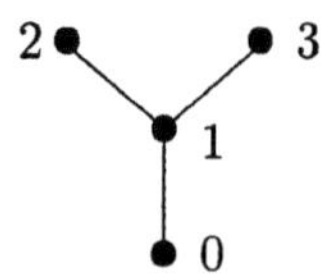

Fig. 5.3

Example 5.1.14 *Let* $(X;\oplus,\odot,^*,0,1)$ *be an MV-algebra. Define a binary operation* $*$ *by*

$$x * y := x \odot y^*, \ \ x, y \in X. \tag{5.4}$$

Then $(X;*,0)$ *is a bounded commutative BCK-algebra with greatest element* 1.

Conversely, let $(X;*,0)$ *be a bounded commutative BCK-algebra with greatest element* 1. *If we define the binary operation* $\odot$ *and the unary operation* * *as follows*

$$x \oplus y := 1 * ((1 * x) * y), \tag{5.5}$$

$$x^* := 1 * x, \tag{5.6}$$

[2]We recall that a poset $(X;\leq)$ is said to be **directed upwards** if given $a, b \in X$, there is an element $c \in X$ such that $a \leq c$ and $b \leq c$.

for all $x, y \in X$, *then* $(X; \oplus, \odot, ^*, 0, 1)$ *is an MV-algebra.*

In addition, starting from a bounded BCK-algebra and using successively (5.5), (5.6), *and* (5.4), *both the original and final* $*$ *operations coincide. Similarly, starting from an MV-algebra and makining the successive definitions* (5.4)–(5.6), *we obtain the original MV-algebra.*

Example 5.1.15 *Let* $(G; +, 0, \leq)$ *be an Abelian* ℓ*-group*[3] *with positive cone* $G^+ = \{g \in G : g \geq 0\}$. *Then* $(G^+; *_G, 0)$, *where*

$$g_1 *_G g_2 := 0 \vee \{g_1 - g_2\}, \; g_1, g_2 \in G^+,$$

is a commutative BCK-algebra.

5.1.3 Commutative BCK-algebras

We note that a BCK-algebra $(X; *, 0)$ is commutative iff, for all $x, y \in X$,

$$x * (x * y) = y * (y * x).$$

Proposition 5.1.16 *If a BCK-algebra* $(X; *, 0)$ *is commutative, then the infimum* $\inf\{x, y\}$ *is equal to* $x * (x * y)$.

Proof. According to (BCK-2) and Proposition 5.1.3(d), we have $x * (x * y) \leq x, y$. Suppose z is any element of X such that $z \leq x, y$. Then $z * x = z * y = 0$, so

$$z = z * 0 = z * (z * x) = x * (x * z).$$

By the same reason we have $z = y * (y * z)$, hence

$$z = x * (x * z) = x * (x * (y * (y * z))) \leq x * (x * y) = x \sqcap y,$$

which means that for all $x, y \in X$, $x \sqcap y$ is the greatest lower bound of x and y. □

Without danger of misunderstanding, for commutative BCK-algebras $x \wedge y$ will mean the infimum of x, y. We recall that if X is a BCK-algebra which is in addition a lattice, then it does not follow that X is commutative; see Example 5.1.12. In addition, if $(X; *, 0)$ is a commutative BCK-algebra, then it does not imply that X is a lattice, see Example 5.1.13.

Theorem 5.1.17 *Let* $(X; *, 0)$ *be a BCK-algebra. The following statements are equivalent: For all* $x, y, z \in X$,

[3] We recall that Abelian **lattice ordered group** (or ℓ**-group**) is an Abelian group G with an order $\leq$ such that G is a lattice and if $g \leq h$, then $v + g \leq v + h$ for $v, g, h \in G$.

(i) $x \leq z$ *and* $z * y \leq z * x$ *imply* $x \leq y$.

(ii) $x, y \leq z$ *and* $z * y \leq z * x$ *imply* $x \leq y$.

(iii) $x \leq y$ *implies* $x = y * (y * x)$.

(iv) X *is commutative.*

(v) $x * y = 0$ *implies* $x * (y * (y * x)) = 0$.

Proof. (i) $\Rightarrow$ (ii). This is trivial.

(ii) $\Rightarrow$ (iii). Suppose that $x \leq y$. By Proposition 5.1.3(d), $y*(y*x) \leq y$. Moreover, $y * (y * (y * x)) \leq y * x$. Hence by (ii) we have $x \leq y * (y * x)$. The converse inequality holds in any BCK-algebra, and so $x = y * (y * x)$.

(iii) $\Rightarrow$ (iv). In view of $y * (y * x) \leq x$, we have from (iii) that

$$y * (y * x) = x * (x * (y * (y * x))).$$

As

$$\begin{aligned}(y * (y * x)) * (x * (x * y)) &= (x * (x * (y * (y * x)))) * (x * (x * y)) \\ &= (x * (x * (x * y))) * (x * (y * (y * x))) \\ &= (x * y) * (x * (y * (y * x))) \\ &= (y * (y * x)) * y = 0,\end{aligned}$$

where we have applied Proposition 5.1.4, we have $y*(y*x) \leq x*(x*y)$. By symmetry of the last inequality, we have that X is commutative.

(iv) $\Rightarrow$ (i). If $x \leq z$ and $z * y \leq z * x$, then $x * z = 0$ and $(z * y) * (z * x) = 0$. From (iv) we have

$$x * y = (x * (x * z)) * y = (z * (z * x)) * y = (z * y) * (z * x) = 0,$$

giving $x \leq y$.

The equivalence (iii) $\Leftrightarrow$ (v) is obvious. □

The following result, due to Yutani [Yut], says that the class of commutative BCK-algebras forms a variety whose equational base is given below.

Theorem 5.1.18 *An algebra* $(X; *, 0)$ *of type* $(2, 0)$ *is a commutative BCK-algebra if and only if*

(i) $x * (x * y) = y * (y * x)$.

(ii) $(x * y) * z = (x * z) * y$.

(iii) $x * x = 0$.

(iv) $x * 0 = x$.

Proof. If X is a commutative BCK-algebra, then it satisfies (i)–(iv). Suppose now that an algebra $(X;*,0)$ of type (2,0) satisfies (i)–(iv). Using the first three conditions, we have

$$(x*(x*y))*y=(x*y)*(x*y)=0,$$

which is (BCK-2). If $x*y=y*x=0$, then by (i) and (iv),

$$x=x*0=x*(x*y)=y*(y*x)=y*0=y,$$

which is (BCK-4). (i) and (ii) give

$$(x*y)*(x*z)=(x*(x*z))*y=(z*(z*x))*y=(z*y)*(z*x),$$

so that

$$(x*y)*(x*z)=(z*y)*(z*x). \qquad (A)$$

Put $x=y$ and $z=0$ in the above equality, then by (iii), we have

$$0*x=(x*x)*(x*0)=(0*x)*(0*x)=0,$$

and (BCK-5) follows. By (A), (iv) and (BCK-5) we have

$$\begin{aligned}((x*y)*(x*z))*(z*y) &= ((z*y)*(z*x))*((z*y)*0)\\ &= (0*(z*x))*(0*(z*y))\\ &= 0*0=0.\end{aligned}$$

which is (BCK-1), and this completes the proof. □

It is worth recalling that the following minimal equational base for commutative BCK-algebras was given by [Cor 2], [MeJu, Thm VI.2.1]: (i) $x*(0*y)=y$, (ii) $(x*z)*(x*y)=(y*z)*(y*x)$.

A BCK-algebra $(X;*,0)$ is said to be (i) **positive implicative** if, for all $x,y\in X$,

$$x*y=(x*y)*y,$$

(ii) **implicative** if it is both commutative and positive implicative.

Example 5.1.8 gives an example of implicative BCK-algebras.

Some of the following results can be found in [CST] and [Pal].

Proposition 5.1.19 *Let $(X;*,0)$ be a commutative BCK-algebra, and let $x,y,z,z_1\in X$.*

(i) *Let* $x, y \leq z$. *Define*

$$x \vee_z y := z * ((z * x) \wedge (z * y)).$$

Then $x \leq x \vee_z y$, $y \leq x \vee_z y$ *and, for arbitrary* $z_1 \in X$ *with* $z_1 \geq z$, *we have*

$$x \vee_z y = x \vee_{z_1} y.$$

(ii) $x \vee y$ *exists in* X *if and only if there exists an element* $z \in X$ *such that* $x, y \leq z$, *in which case*

$$x \vee y = x \vee_z y. \tag{5.7}$$

(iii)

$$x * (y \wedge z) = (x * y) \vee (x * z). \tag{5.8}$$

(iv) *If* $x \vee y$ *exists in* X, *then, for arbitrary* $z \in X$, $(x * z) \vee (y * z)$ *is defined in* X, *and*

$$(x \vee y) * z = (x * z) \vee (y * z). \tag{5.9}$$

(v) *If* $x \vee y$ *exists in* X, *then*

$$(x * y) \wedge (y * x) = 0. \tag{5.10}$$

(vi) *If* $y \vee z$ *exists in* X, *then, for any* $x \in X$,

$$x * (y \vee z) = (x * y) \wedge (x * z). \tag{5.11}$$

(vii) *If* $x \vee y$ *exists in* X, *then* $(z \wedge x) \vee (z \wedge y)$ *exists in* X *for any* $z \in X$, *and*

$$z \wedge (x \vee y) = (z \wedge x) \vee (z \wedge y).$$

(viii) *If* $(X; *, 0)$ *is directed upwards, then* X *is a distributive lattice with respect to the BCK-order.*

Proof. (i)–(ii). Since X is commutative, it is a lower semilattice, and $x \wedge y = x * (x * y) = y * (y * x)$, Proposition 5.1.16. Let us define

$$x \vee_z y := z * ((z * x) \wedge (z * y))$$

if x, $y \leq z$. Since $x \vee_z y = z * ((z * x) \wedge (z * y)) \geq z * (z * x) = x$ and similarly $x \vee_z y \geq y$, $x \vee_z y$ is an upper bound of x and y. If $x, y \leq z_1$ and $z \leq z_1$, we have:

$z_1 * x \geq z_1 * z$ and $z_1 * y \geq z_1 * z$ so that $z_1 * z \leq (z_1 * x) \wedge (z_1 * y)$, and hence $z = z_1 * (z_1 * z) \geq z_1 * ((z_1 * x) \wedge (z_1 * y)) = x \vee_{z_1} y$.

Since $x \vee_z y \leq z \leq z_1$ and $x, y \leq x \vee_z y$, in the same way we see that $x \vee_z y \geq x \vee_{z_1} y$. From $x, y \leq x \vee_{z_1} y \leq z$ using similar arguments, it follows that $x \vee_{z_1} y \geq x \vee_z y$ and therefore $x \vee_z y = x \vee_{z_1} y$.

If x, $y \leq z, z_1$, there exists z_2 such that $z, z_1 \leq z_2$. Since $x \vee_z y = x \vee_{z_2} y$ and $x \vee_{z_1} y = x \vee_{z_2} y$, we have that $x \vee_z y = x \vee_{z_1} y$. Hence, $x \vee_z y$ is independent of the upper bound z.

We prove that $x \vee_z y$ is the supremum of x and y. Let us consider $u \in X$ such that $x, y \leq u$. Then $x \vee_z y = x \vee_u y = u * ((u * x) \wedge (u * y)) \leq u$ and, consequently, $x \vee y := x \vee_z y$ for all $z \in X$ with $x, y \leq z$.

(iii) Since $y \wedge z \leq y, z$, we have $x * y \leq x * (y \wedge z)$ and $x * z \leq (y \wedge z)$, so that $x * (y \wedge z)$ is an upper bound of $x * y$ and $x * z$. By (ii), $(x * y) \vee (x * z)$ exists in X. Let u be an element in X such that $x * y \leq u$ and $x * z \leq u$. Then $x * u \leq x * (x * y) = x \wedge y$ and $x * u \leq x * (x * z) = x \wedge z$, so that $x * u \leq (x \wedge y) \wedge (x \wedge z) = x \wedge y \wedge z$. Therefore,

$$x * (y \wedge z) \leq x * (x \wedge y \wedge z) \leq x * (x * u) \leq u,$$

which means that $x * (y \wedge z)$ is the least upper bound of $x * y$ and $x * z$.

(iv) Put $u = x \vee y$. Then $(x \vee y) * z$ is an upper bound of $x * z$ and $y * z$, and by (ii), $(x * z) \vee (y * z)$ is defined in X.

$$\begin{aligned}(x \vee y) * z &= (u * ((u * x) \wedge (u * y))) * z \\ &= (u * z) * (((u * x) \wedge (u * y))) \quad \text{by (5.2)} \\ &= ((u * z) * (u * x)) \vee ((u * z) * (u * y)) \quad \text{by (iii)} \\ &= (x * z) \vee (y * z).\end{aligned}$$

(v) From (iv) we have

$$(x \vee y) * y = x * y, \tag{5.12}$$

which implies that

$$\begin{aligned}(x * y) \wedge (y * x) &= (x * y) * ((x * y) * (y * x)) \\ &= (x * y) * (((x \vee y) * y) * ((x \vee y) * x)) \quad \text{by (5.12)} \\ &= (x * y) * (((x \vee y) * ((x \vee y) * x)) * y) \\ &= (x * y) * (x * y) = 0.\end{aligned}$$

(vi) It is clear that $y, z \leq y \vee z$, so that $x * (y \vee z) \leq x * y,\ x * z$, which says that $x * (y \vee z)$ is a lower bound of $x * y$ and $x * z$. Suppose now that $u \leq x * y$ and $u \leq x * z$. Then

$$u * (x * (y \vee z)) \leq (x * y) * (x * (y \vee z)) \leq (y \vee z) * y = z * y.$$

Similarly we have

$$u * (x * (y \vee z)) \leq y * z$$

which by (5.10) entails $u * (x * (y \vee z)) \leq (z * y) \wedge (y * z) = 0$. Hence, $u \leq x * (y \vee z)$ which proves (5.11).

(vii)–(viii).

$$\begin{aligned} x \wedge (y \vee z) &= x * (x * (y \vee z)) \\ &= x * ((x * y) \wedge (x * z)) \quad \text{by (5.11)} \\ &= (x * (x * y)) \vee (x * (x * z)) \\ &= (x \wedge y) \vee (x \wedge z). \end{aligned}$$

□

It is worth recalling that in (v), the existence of $x \vee y$ is essential for the validity of (5.10). Indeed, in Example 5.1.13, we have $(2 * 3) \wedge (3 * 2) = 1 \neq 0$. On the other hand, there exist commutative BCK-algebras with (5.10) which are not lattices; see Example 5.1.25 below. Such an equality will be crucial for commutative BCK-algebras with the relative cancellation property, which will be studied in Section 5.2.

5.1.4 BCK-algebras with the Condition (S)

A BCK-algebra $(X; *, 0)$ is said to satisfy **condition (S)** if, given any two elements $x, y \in X$, there exists an element $x \circ y$ in X such that (i) $(x \circ y) * x \leq y$, and (ii) if z is any element in X such that $z * x \leq y$, then $z \leq x \circ y$.

It is clear that (i) $x \circ y$ is a well-defined unique element for arbitrary $x, y \in X$, (ii) $x, y \leq x \circ y$, and (iii) $x \circ 0 = x = 0 \circ x$.

Proposition 5.1.20 *If $(X; *, 0)$ is a BCK-algebra with the condition (S), then $(X; \circ, 0)$ is a commutative ordered semigroup with a neutral element* 0.[4]

Proof. $z * x \leq y$ is equivalent to $z * y \leq x$ according to (5.2), which proves that $\circ$ is commutative.

Associativity. By the definition of $\circ$, we have

$$((x \circ y) \circ z) * z \leq x \circ y,$$

and so

$$(((x \circ y) \circ z) * y) * z = (((x \circ y) \circ z) * z) * y \leq (x \circ y) * y \leq x.$$

By symmetry we have the inverse inequality. Thus, $\circ$ is associative.

Finally, let $x \leq y$, then $(x \circ z) * y \leq (x \circ z) * x \leq z$, so that $x \circ z \leq y \circ z$. □

[4]A set X with a binary operation $\cdot$ and a partial order $\leq$ is said to be an **ordered semigroup** if $\cdot$ is associative, and $x \leq y$ entails $x \cdot z \leq y \cdot z$ for any $z \in X$.

Proposition 5.1.21 *Let* $(X; *, 0, 1)$ *be a bounded BCK-algebra such that* $1*(1*x) = x$ *for each* $x \in X$. *Then* $(X; *, 0)$ *satisfies condition (S). In this case* $x \circ y = x \oplus y$, *where* $\oplus$ *is defined by* (5.5).

Proof. Define $x \circ y = 1 * ((1 * x) * y)$ for all $x, y \in X$. Then $(x \circ y) * x = (1 * ((1 * x) * y)) * x = (1 * x) * ((1 * x) * y) \leq y$.

If, for $z \in X$, we have $z * x \leq y$, then $(z * x) * y = 0$. Hence $((1 * x) * y) * (1 * z) = ((1 * (1 * z)) * x) * y = (z * x) * y = 0$, which gives $(1 * x) * y \leq 1 * z$, and so

$$z = 1 * (1 * z) \leq 1 * ((1 * x) * y) = x \circ y.$$

Hence X satisfies condition (S). □

Remark 5.1.22 (1) Example 5.1.5 satisfies condition (S); we have $s \circ t = s + t$, $s, t \in [0, \infty)$.

(2) Example 5.1.6 satisfies condition (S); we have $f \circ g = f + g$, $f, g \in X$.

(3) Example 5.1.15 satisfies condition (S); we have $f \circ g = f + g$, $f, g \in G^+$.

(4) Each bounded commutative BCK-algebra satisfies condition (S), see Proposition 5.1.21.

Example 5.1.23 *Any bounded implicative BCK-algebra satisfies condition (S), and*

$$x \circ y = x \vee y.$$

Indeed, it is possible to show that a l.u.b of x *and* y, $x \vee y$, *exists in* X, *and* $x \vee y = 1 * ((1 * x) * y) = x \circ y$.

The commutative BCK-algebras satisfying condition (S) form a variety, [MeJu, Thm I.7.15], which is a proper subvariety of the variety of commutative BCK-algebras, as the following example shows.

Example 5.1.24 $([0, 1); *_{\mathbb{R}}, 0)$, *where* $*_{\mathbb{R}}$ *is defined by* (5.3), *is a commutative BCK-algebra for which condition (S) does not hold. Indeed, the element* $1/2 \circ 1/2$ *is not defined in* $[0, 1)$.

Similarly, Example 5.1.13 gives a commutative BCK-algebra not satisfying condition (S) since the element $1 \circ 1$ is not defined in X. On the other hand, Example 5.1.12 satisfies condition (S) (see Exercise 6.6.3.1).

5.1.5 Union BCK-algebras

Let $\{(X_i; *_i, 0_i)\}_{i\in I}$ be a system of BCK-algebras such that (i) $0_i = 0$ for any $i \in I$; and (ii) $X_i \cap X_j = \{0\}$ for $i \neq j$, $i, j \in I$. Put $X = \bigcup_{i\in I} X_i$ and define the total binary operation $*$ on X via

$$x * y = \begin{cases} x *_i y, & \text{if } x, y \in X_i,\ i \in I, \\ x, & \text{if } x \in X_i,\ y \in X_j,\ i \neq j,\ i, j \in I. \end{cases}$$

Then $(X; *, 0)$ is a BCK-algebra called a **union BCK-algebra**, and we write $X = \bigoplus_{i\in I} X_i$. A union BCK-algebra is commutative iff each X_i is commutative. It is clear that an arbitrary BCK-algebra $X = \bigoplus_{i\in I} X_i$, where $X_i = X$ and $|I| = 1$, is a union BCK-algebra. Moreover, each X_i is a BCK-subalgebra of $X = \bigoplus_{i\in I} X_i$, and if $u \in X_i$, then

$$X_u := \{x \in X : x \leq u\} \subseteq X_i.$$

Conversely, let $\{X_i\}_{i\in I}$ be a family of subsets of a BCK-algebra $(X; *, 0)$ such that

(a) $X = \bigcup_{i\in I} X_i$,
(b) $X_i \cap X_j = \{0\}$,
(c) $u \in X_i$ implies $X_u \subseteq X_i$,

then each X_i is a BCK-subalgebra of X, and X is the union BCK-algebra of $\{X_i\}_{i\in I}$.

More about union BCK-algebras can be found in [MeJu].

We note that if $X = \bigoplus_{i\in I} X_i$ and there are two different subalgebras X_i and X_j having at least two elements, then X does not satisfy the condition (S). Indeed, if $a \in X_i$ and $b \in X_j$ are non-zero elements, then $a * a \leq b$ and $b * a \leq b$ but there is no element $c \in X$ such that $a \leq c$ and $b \leq c$.

Example 5.1.25 *The example B_{5-2-7} from* [MeJu], *where $*$ is given by table* 5.4 *below, gives a commutative BCK-algebra which is a union BCK-algebra of $X = \{0, 1, 3\}$ and $X_2 = \{0, 2, 4\}$.*

*	0	1	2	3	4
0	0	0	0	0	0
1	1	0	1	0	1
2	2	2	0	2	0
3	3	1	3	0	3
4	4	4	2	4	0

Tab. 5.4

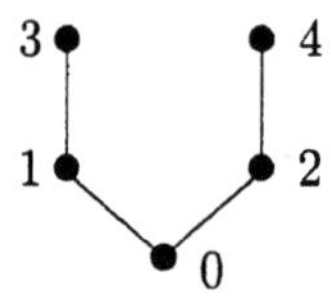

Fig. 5.4

We say that a BCK-algebra $(X; *, 0)$ has the **u.d. union property** (u.d. stands for upwards directed) if X can be expressed as $X = \bigoplus_{i \in I} X_i$, where each X_i is directed upwards.

Since $\{0\}$ is trivially an upwards directed BCK-subalgebra of X, using Zorn's lemma we conclude that in any BCK-algebra X there is a maximal upwards directed BCK-subalgebra of X.

Any BCK-algebra with the u.d. union property has a unique decomposition $X = \bigoplus_{i \in I} X_i$, where all X_i are directed upwards:

Lemma 5.1.26 *Let $X = \bigoplus_{i \in I} X_i$, where each X_i is directed upwards. If also $X = \bigoplus_{j \in J} Y_j$, where each Y_j is directed upwards, then for any $i \in I$, there is a unique $j \in J$ such that $X_i = Y_j$.*

Proof. Let $u \in X_i$ be non-zero. Then there is a unique Y_j such that $u \in Y_j$. Hence, $X_u \subseteq X_i$ and $X_u \subseteq Y_j$. Now let v be an arbitrary element of X_i. There is an element $w \in X_i$ which as an upper bound of u and v, i.e., $u, v \leq w$. Then $v \in X_w \subseteq X_i$ and since $u \in Y_j$, we conclude that $X_w \subseteq Y_j$, so that $X_i \subseteq Y_j$. By symmetrical reasoning we can show also that $Y_j \subseteq X_i$. □

5.1.6 Direct Product of BCK-algebras

We note that a **BCK-homomorphism** of two BCK-algebras $(X_1; *_1, 0_1)$ and $(X_2; *, 0_2)$ is a mapping $h: X_1 \to X_2$ such that, for all $x, y \in X_1$,

$$h(x *_1 y) = h(x) *_2 h(y).$$

If h is injective, h is said to be a **BCK-embedding**; if h is injective and surjective, h is said to be a **BCK-isomorphism.**

Let $\{(X_i; *_i, 0_i)\}_{i \in S}$ be a family of BCK-algebras and let $X = \prod_{i \in S} X_i$ be the cartesian product of $\{X_i\}_{i \in S}$. For $f, g \in X$, we define $(f * g)(i) := f(i) *_i g(i)$ for every $i \in S$, and 0 is defined via $0(i) = 0_i$. Then $(X; *, 0)$ is a BCK-algebra which is called the **direct product** of $\{(X_i; *_i, 0_i)\}_{i \in S}$.

It is easy to show that $\prod_{i \in S} X_i$ is a commutative BCK-algebra iff every X_i is commutative. Similarly, $\prod_{i \in S} X_i$ has the relative cancellation property iff every X_i has it.

A BCK-algebra $(X; *, 0)$ is said to a **subdirect product** of an indexed family $\{(X_i; *_i, 0_i)\}_{i \in S}$ of BCK-algebras if (i) X is a subalgebra of $\prod_{i \in S} X_i$, and (ii) for every $i \in S$, $\pi_i(X) = X_i$, where π_i is the i-th projection mapping from $\prod_{i \in S} X_i$ onto X_i. A BCK-embedding $\alpha: X \to \prod_{i \in S} X_i$ is **subdirect** if $\alpha(X)$ is a subdirect product of $\{(X_i; *_i, 0_i)\}_{i \in S}$; in this case X is called also a subdirect product of $\{(X_i; *_i, 0_i)\}_{i \in S}$.

Example 5.1.27 *The example B_{3-1-2} from* [MeJu] *with $X = \{0,1,2\}$ with table* 5.5 *below is a subdirect product of the direct product $\{0,1\} \times \{0,1\}$ with table* 5.5'.

*	0	1	2
0	0	0	0
1	1	0	1
2	2	2	0

Tab 5.5

*	0	1
0	0	0
1	1	0

Tab. 5.5'

Proposition 5.1.28 *Let $\{(X_i; *_i, 0_i)\}_{i \in S}$ be an indexed family of BCK-algebras. Then*

(a) $\prod_{i\in S} X_i$ *is bounded if and only if every X_i is bounded.*

(b) $\prod_{i\in S} X_i$ *is commutative if and only if every X_i is commutative.*

(c) $\prod_{i\in S} X_i$ *is positive implicative if and only if every X_i is positive implicative.*

(d) $\prod_{i\in S} X_i$ *is implicative if and only if every X_i is implicative.*

(e) $\prod_{i\in S} X_i$ *satisfies condition (S) if and only if every X_i satisfies condition (S).*

Proof. This is straightforward. □

5.1.7 Ideals of a BCK-algebra

Definition 5.1.29 *An* **ideal** *of a BCK-algebra $(X; *, 0)$ is a non-empty set I of X such that, for all $x, y \in X$,*

(i) $0 \in I$;

(ii) $x * y \in I$ *and* $y \in I$ *imply* $x \in I$.

It is clear that (i) for all $x, y \in X$ such that $x \le y$ and $y \in I$, then $x \in I$, (ii) if X satisfies condition (S), and $x, y \in I$, then $x \circ y \in I$, (iii) if X is commutative, $x, y \in I$ with $x \vee y$ defined in X, then $x \vee y \in I$, (iv) if h is a BCK-homomorphism, then the kernel of h, $Ker(h) := \{x \in X : h(x) = 0\}$, is an ideal of X.

We denote by $\mathcal{I}(X)$ the set of all ideals of I. It is clear that $\{0\}$, $X \in \mathcal{I}(X)$.

An ideal I of X is said to be **maximal** if it is a proper ideal of X and if it is not contained in any other proper ideal of X. A proper ideal of X is said to be a **prime** ideal of X if $x \wedge y \in I$ entails $x \in I$ or $y \in I$. We denote by $\mathcal{M}(X)$ and $\mathcal{P}(X)$ the sets of all maximal and prime ideals of X, respectively. It will be shown in Lemma 5.2.18 that if X is a commutative BCK-algebra, then $\mathcal{P}(X) \neq \emptyset$ but $\mathcal{M}(X)$ can be void; see Example 5.2.42 below.

Suppose that I is an ideal of X. For any x, y in X, we define $x \sim_I y$ if and only if

$$x * y \in I, \quad y * x \in I.$$

Lemma 5.1.30 *The relation $\sim_I$ is a congruence relation on X.*

Proof. Since $0 \in I$, we have $x \sim_I x$ for any $x \in X$. The symmetry of $\sim_I$ is trivial. Suppose now $x \sim_I y$ and $y \sim_I z$. By (BCK-1),

$$(x * z) * (x * y) \leq y * z,$$

Hence by definition of ideals, $x * z \in I$. In the same manner we can prove $z * x \in I$ which proves that $\sim_I$ is transitive.

Assume $x \sim_I y$ and $u \sim_I v$. We claim that $(x * u) \sim_I (y * v)$.

By (BCK-1) we have $(x * u) * (x * v) \leq v * u \in I$ and $(x * v) * (x * u) \leq u * v \in I$. Consequently, $(x * u) * (x * v) \in I$ and $(x * v) * (x * u) \in I$, i.e., $x * u \sim_I x * v$. By Proposition 5.1.3(b),

$$(x * v) * (y * v) \leq x * y \in I, \quad \text{and} \quad (y * v) * (x * v) \leq y * x \in I.$$

Consequently, $x * v \sim_I y * v$. □

Proposition 5.1.31 *Let I be an ideal of a BCK-algebra $(X; *, 0)$. Let $C_x := \{y \in X : y \sim_I x\}$, and let $X/I := \{C_x : x \in X\}$. Define*

$$C_x * C_y := C_{x*y}.$$

Then $(X/I; *, I)$ is a BCK-algebra with $I = C_0$, which is called a **quotient BCK-algebra**.

Proof. It is clear that $I = C_0$. Since $\sim_I$ is a congruence, the operation $*$ in X/I is well-defined. We define $C_x \leq C_y$ iff $C_x * C_y = C_0$ which shows that $C_0 \leq C_x$ for all $x \in X$. Let $C_x * C_y = C_y * C_x = C_0$. Then $x \sim_I y$. Hence, if $u \in C_x$ and $v \in C_y$, we have $u \in C_y$ and $v \in C_x$, i.e., $C_x = C_y$. It is evident that $C_x * C_x = C_0$. Therefore (BCK-3)–(BCK-5) hold. In addition, $x \leq y$ implies that $C_x \leq C_y$. Therefore, for all $x, y, z \in X$,

$$(C_x * C_y) * (C_x * C_z) = C_{x*y} * C_{x*z} = C_{(x*y)*(x*z)} \leq C_{z*y} = C_z * C_y,$$

which is (BCK-1). Since $C_x * (C_x * C_y) = C_{x*(x*y)} \leq C_y$, (BCK-2) holds. □

It is evident that, given a non-void subset A of a BCK-algebra $(X; *, 0)$, there exists an ideal $I(A)$ of X generated by A. The following result characterizes $I(A)$.

Theorem 5.1.32 *Let A be a non-void subset of a BCK-algebra $(X; *, 0)$. Then*

$$I(A) = \{x \in X : \exists\, a_1, \ldots, a_n \in A \text{ such that } (\cdots(x * a_1) * \cdots) * a_n = 0\} \quad (5.13)$$

Proof. Denote the right-hand side of (5.13) by B. Then $0 \in B$ and obviously $A \subseteq B$. Let $x * y \in B$ and $y \in B$. Then there exist elements $a_1, \ldots, a_n, b_1, \ldots b_m \in A$ such that

$$(\cdots(y * a_1) * \cdots) * a_m = 0, \quad (\cdots((x * y) * b_1) * \cdots) * b_n = 0,$$

hence

$$((\cdots(x * b_1) * \cdots) * b_n) * y = 0, \quad (\cdots(x * b_1) * \cdots) * b_n \leq y.$$

$*$ multiplying both sides of the above inequality by a_1 on the right, we have

$$((\cdots(x * b_1) * \cdots) * b_n) * a_1 \leq y * a_1.$$

Repeating successively the above argument with $a_2, \ldots, a_m$, we obtain $(\cdots(((\cdots(x * b_1) * \cdots) * b_n) * a_1) * \cdots) * a_m \leq (\cdots(y * a_1) * \cdots) * a_m = 0$, consequently, $(\cdots(((\cdots(x * b_1) * \cdots) * b_n) * a_1) * \cdots) * a_m = 0$. Hence $x \in B$ which proves that B is an ideal of X containing A.

Now let I be an ideal of X containing A and assume $x \in B$. Then there exist $c_1, \ldots, c_n \in A$ such that $(\cdots(x * c_1) * \cdots) * c_n = 0$. Hence, $(\cdots(x * c_1) * \cdots) * c_n \in I$. Since I is an ideal and $c_n \in I$, it follows that $(\cdots(x * c_1) * \cdots) * c_{n-1} \in I$. Repeating this argument n times, we obtain $x \in I$ which proves $B \subseteq I$, and consequently we have proved $B = I(A)$. □

For all $x, y \in X$, we make the recursive definitions

$$x *^0 y = x, \ x *^1 y = x * y, \ldots, \ x *^{n+1} y = (x *^n y) * y, \ n \geq 1.$$

Lemma 5.1.33 *Let I be an ideal of a BCK-algebra $(X; *, 0)$ and $a \in X$. Denote by*

$$I_a = \{x \in X : \exists\, n \geq 1 \ x *^n a \in I\}. \tag{5.14}$$

Then I_a is the ideal of X generated by I and a.

Proof. Since $a * a = 0 \in I$, and $x * a \leq x \in I$ for any $x \in I$, we conclude that I_a contains I and the element a. Suppose now that $x * y \in I_a$ and $y \in I_a$. Then there exist two integers m and n such that $y *^m a \in I$ and $(x * y) *^n a \in I$. There exists an element $e \in I$ such that $(x * y) *^n a = e$. Hence $((x * y) *^n a) * e = 0$, so that $(x *^n a) * e \leq y$. Therefore, $((x *^n a) * e) *^m a \leq y *^m a \in I$ and $(x *^{n+m} a) * e \in I$, which means that $x *^{n+m} a \in I$. This proves that I_a is an ideal of X. Of course, if J is another ideal containing I and a, we have $x *^n a \in I$ so that $x *^n a \in J$, and, consequently, $x \in J$. □

Lemma 5.1.34 *Let a be an element of a BCK-algebra $(X; *, 0)$. Then the ideal $I(a)$ of X generated by the element a is given by*

$$I(a) = \{x \in X : \exists\, n \geq 1 \ x *^n a = 0\}. \tag{5.15}$$

Proof. This follows from Lemma 5.1.33 when we put $I = \{0\}$. □

Theorem 5.1.35 (1) *For each $t \in T$, let I_t be an ideal of the BCK-algebra $(X_t; *_t, 0_t)$. Then $I := \prod_{t\in T} I_t$ is an ideal of $X := \prod_{t\in T} X_t$. Conversely, if I is an ideal of X, then $I_t := \pi_t(I)$, where π_t is the t-th projection of X onto X_t, is an ideal of X_t.*

(2) *Let I be an ideal of $X := \prod_{t\in T} X_t$ and $I_t = \pi_t(I)$. Then $I \subseteq \prod_{t\in T} I_t$. If T is a finite set, then $I = \prod_{t\in T} I_t$.*

(3) *An ideal $I = \prod_{t\in T} I_t$ is maximal in $X := \prod_{t\in T} X_t$ if and only if there is a unique index $s \in T$ such that I_s is a maximal ideal of X_s and $I_t = X_t$ for any $t \neq s$.*

Proof. (1) The first part of the assertion is clear. Suppose now I is an ideal of X, and let $I_t = \pi_t(I)$. Then $0_t \in I_t$. If $x_t *_t y_t \in I_t$ and $y_t \in I_t$, we define $x := (x(s))_s$, $y := (y(s))_s \in X$ by $x(s) := \delta_{st}x_t$ and $y(s) := \delta_{st}y_t$, for any $s \in T$. Then $x(t) *_t y(t) \in I_t$ and $y(t) \in I_t$, and so there exists an element $z \in I$ such that $\pi_t(z) = x_t *_t y_t \in I_t$. Then $x * y \leq z$ which implies that $x * y \in I$. Similarly there exists an element $u \in I$ such that $\pi_t(u) = y_t \in I_t$, which gives $y \in I$ while $y \leq u$. This means that I_t is an ideal of X_t.

(2) The first part follows from (1). If T is finite, then the statement follows from [MeJu; Thm II.10.4].

(3) Suppose that $I = \prod_t I_t$, where I_s is a maximal ideal of X_s and $I_t = X_t$ for all $t \neq s$. Then due to (1) and (2), I is an ideal of X, and it is clear that I is maximal.

Suppose now that $I = \prod_{t\in T} I_t$ is a maximal ideal of X. It is clear that there is at least one index t such that I_t is a maximal ideal of X_t. Suppose now that there are two indices s_0 and t_0 such that I_{s_0} and I_{t_0} are proper ideals of X_{s_0} and X_{t_0}, respectively. Then $J_{s_0} := \prod_{t\in T} I'_t$ and $J_{t_0} := \prod_{t\in T} I''_t$, where $I'_t = I_t$ if $t \neq s_0$ and $I'_{s_0} = X_{s_0}$, and $I''_t = I_t$ if $t \neq t_0$ and $I''_{t_0} = X_{t_0}$, are proper ideals of X containing I, which contradicts the maximality of I. □

Remark 5.1.36 It is possible to show that if $X = \prod_{t\in T} X_t$ and I is an ideal of X, and if T is an infinite index subset, then it can happen that I is a proper subset of $\prod_{t\in T} I_t$, where $I_t = \pi_t(I)$. Indeed, let $X = 2^{\{1,2,\ldots\}} \cong \prod_{n=1}^{\infty}\{0, 1_n\}$, where $1_n = 1$ for all n. Then $(X; \setminus, \emptyset)$ is a bounded commutative BCK-algebra, where $\setminus$ is the set-theoretic difference. Let I_0 be the set of all subsets of $\{1, 2, \ldots\}$ of finite cardinality. Then I_0 is a proper ideal of X, and X/I_0 is the a non-trivial bounded BCK-algebra (in fact, it is a Boolean algebra). Therefore there exists a two-valued measure μ on X/I_0. Define the measure m on X via $m(A) := \mu(A/I_0)$, $A \in X$. Then m is a two-valued measure on X, and $I := \{A \in X : m(A) = 0\}$ is a maximal ideal of X.

It is clear that $\pi_n : X \to \{0, 1_n\}$ is defined via $\pi_n(A) = 1_n\delta_n(A)$, $A \in X$. Therefore, $\pi_n(I) = \{0, 1_n\}$ since $\pi_n(\{n\}) = 1_n$, and $\{n\} \in I$, which means that $I \neq \prod_{n=1}^{\infty} \pi_n(I) = X$.

5.1.8 Measures on BCK-algebras

A mapping $m : X \to [0, \infty)$ such that, for all $x, y \in X$,

(i) $m(x * y) = m(x) - m(y)$ whenever $y \leq x$ is said to be a **measure**;

(ii) if $1 \in X$ and m is a measure with $m(1) = 1$, then m is said to be a **state**;

(iii) $m(x * y) = \max\{0, m(x) - m(y)\}$ is said to be a **measure-morphism**;

(iv) if $1 \in X$ and m is a measure-morphism with $m(1) = 1$, m is said to be a **state-morphism**.

Proposition 5.1.37 *Let $(X; *, 0)$ be a BCK-algebra.*

(i) *Any measure-morphism on X is a measure on X.*

(ii) *$m(0) = 0$ whenever m is a measure on X.*

(iii) *$m(x) \leq m(y)$ whenever $x \leq y$ and m is a measure on X.*

(iv) *$x \leq y$ implies $m(y * (y * x)) = m(x)$ whenever m is a measure on X.*

(v) *Let m be a measure on X, then $I_m := \{x \in X : m(x) = 0\}$ is an ideal of X.*

Proof. (i) This follows from the evident fact that a measure-morphism is in fact a BCK-homomorphism from X into $[0, \infty)$.

(ii) Since $0 \leq 0$, we have $m(0) = m(0 * 0) = m(0) - m(0) = 0$.

(iii) This is trivial.

(iv) We have

$$m(y * (y * x)) = m(y) - m(y * x) = m(y) - m(y) + m(x) = m(x).$$

(v) By (ii), $0 \in I_m$. Let $m(x * y) = 0$ and $m(y) = 0$. Due to Proposition 5.1.4, $0 = m(x * y) = m(x * (y \sqcap x)) = m(x) - m(y \sqcap x) = m(x)$, when we have used (iii) and $y \sqcap x \leq y$. Hence, $x \in I_m$. □

A measure (a state) on a BCK-algebra X is said to be a **σ-measure** (a **σ-state**), if for a sequence $\{x_n\}$ of elements from X with $\{x_n\} \nearrow x$, $x \in X$, (i.e., $x_1 \leq x_2 \leq \cdots \leq x$ and $x = \bigvee_n x_n$), we have $\lim_n m(x_n) = m(x)$.

Proposition 5.1.38 *Let m be a σ-measure on X. Then if $\{x_n\} \searrow x$ (i.e., $x_1 \geq x_2 \geq \cdots \geq x$ and $\bigwedge_n x_n = x$), we have $m(x_n) \searrow m(x)$.*

Proof. Let $\{x_n\} \searrow x$. Put $y_n = x_1 * x_n$, $y = x_1 * x$. We claim that $\bigvee_n y_n = y$. It is clear that $y_n \leq y$ for any $n \geq 1$. Let $y_n \leq z$ for some $z \in X$ and any $n \geq 1$. Then $x_1 * x_n \leq z$ so that $x_1 * z \leq x_n$ for any n. Hence $x_1 * z \leq x$, and, consequently, $x_1 * x \leq z$ so that $\bigvee_n y_n = x_1 * x$.

Calculate $\lim_n m(x_1 * x_n) = m(x_1 * x)$ so that $m(x_1) - \lim_n m(x_n) = m(x_1) - m(x)$ and $m(x_n) \searrow m(x)$. $\square$

We note that a measure on a commutative BCK-algebra which is continuous from above, is σ-additive. For general BCK-algebras this is not true (see (ii) of Example 5.1.44 below, where the measure is continuous from above, but it is not σ-additive).

A system $\mathcal{S}$ of measures (states, measure-morphisms, or state-morphisms, respectively) is an **order-determining system** (or **ordering**) on a BCK-algebra X if $m(x) \leq m(y)$ for all $m \in \mathcal{S}$ implies $x \leq y$.

Example 5.1.39 *Example* (5.1.6) *has an order-determining system of measure-morphisms* $\mathcal{S} = \{m_\omega : \omega \in \Omega\}$, *where* $m_\omega(f) := f(\omega),\ f \in X$.

Example 5.1.40 *Let* $X = ([0,1]; *_{\mathbb{R}}, 0, 1)$ *be the BCK-algebra, in which* $*_{\mathbb{R}}$ *is defined via* (5.3). *Then* $X = [0, 1]$ *has a unique state which is also a state-morphism* m, *namely,* $m(t) = t,\ t \in [0, 1]$, *and* $\mathcal{S} := \{m\}$ *is an order-determining system of state-morphisms on* X.

Proposition 5.1.41 *Let* $(X; *, 0)$ *be a BCK-algebra* ($(X; *, 0)$ *be a bounded BCK-algebra with the greatest element* 1). *If* X *possesses an order-determining system of measures (an order-determining system of states), then* X *is commutative.*

Proof. Let $\mathcal{S}$ be an order-determining system of measures on X. Suppose that, for $x, y \in X$ we have $x \leq y$. According to (iv) of Proposition 5.1.37, we have $m(y * (y * x)) = m(x)$ for all $m \in \mathcal{S}$. Since $\mathcal{S}$ is order-determining, $y * (y * x) = x$. Applying Theorem 5.1.17(iii), we see that X is commutative. $\square$

Theorem 5.1.42 *Let* m *be a measure on a BCK-algebra* $(X; *, 0)$. *For* $x, y \in X$, *we write* $x \sim_m y$ *if and only if* $m(x * y) = 0 = m(y * x)$. *Then* $\sim_m$ *is a congruence on* X *and* $(X_m; *_m, 0_m)$ *is a commutative BCK-algebra, where* $X_m := X/\sim_m := \{\tilde{x} := \{y \in X :\ y \sim_m x\} :\ x \in X\}$, $0_m := \tilde{0}$, *and* $*_m$ *is a binary operation on* X_m *induced by* $*$ *and* $\sim_m$. *In addition, the mapping* $\tilde{m} : X_m \to [0, \infty)$ *defined by* $\tilde{m}(\tilde{x}) := m(x)$, $\tilde{x} \in X_m$, *is a well-defined measure on* X_m.

Proof. Due to (v) of Proposition 5.1.37, $I_m := \{x \in X :\ m(x) = 0\}$ is an ideal of X. The property $x \sim_m y$ holds iff $x * y \in I_m$ and $y * x \in I_m$, so that $\sim_m$ is a congruence on X, and $X_m := X/\sim_m$ is a BCK-algebra, Proposition 5.1.31.

We now show that

$$x \sim_m y \quad \text{iff} \quad m(x) = m(y) = m(x \sqcap y) = m(y \sqcap x).$$

Indeed, let $\tilde{x} = \tilde{y}$. By (BCK-2), $x * (x * y) \leq y$, and since $x * y \leq x$, we have

$$m(y) \geq m(x * (x * y)) = m(x) - m(x * y) = m(x), \tag{5.16}$$

and by symmetry, $m(x) \geq m(y)$, so that $m(y) = m(y \sqcap x) = m(x \sqcap y) = m(x)$. The converse statement is now clear. Hence, $\tilde{m}(\tilde{x}) := m(x)$ is a well-defined mapping.

It is easy to see that $\tilde{x} \leq_m \tilde{y}$ iff $m(x * y) = 0$, where $\leq_m$ is the BCK-order in $X/\sim_m$. Define $x_0 = y \sqcap x = x * (x * y)$. We claim that $\widetilde{x_0} = \tilde{x}$. Indeed, it is sufficient to verify (5.16) for x and x_0.

$$\begin{aligned}
m(x_0) &= m(x * (x * y)) = m(x) - m(x * y) = m(x). \\
m(x_0 \sqcap x) &= m(x * (x * x_0)) = m\Big(x * \Big(x * (x * (x * y))\Big)\Big) \\
&= m(x) - m(x) + m((x * (x * y)) = m(x) - m(x * y) = m(x). \\
m(x \sqcap x_0) &= m(x_0 * (x_0 * x)) = m\Big((x * (x * y)) * \Big((x * (x * y)) * x\Big)\Big) \\
&= m(x * (x * y)) = m(x).
\end{aligned}$$

Since $\widetilde{x_0} = \tilde{x}$, we have $\widetilde{x_0} \leq_m \tilde{y}$ and then

$$\begin{aligned}
\tilde{m}(\tilde{y} *_m \tilde{x}) &= \tilde{m}(\tilde{y} *_m \widetilde{x_0}) = \tilde{m}(\widetilde{y * x_0}) \\
&= m(y * x_0) = m(y) - m(x_0) \\
&= m(y) - m(x) = \tilde{m}(\tilde{y}) - \tilde{m}(\tilde{x}),
\end{aligned}$$

so that $\tilde{m}$ is a measure on X_m.

We now show that $X/\sim_m$ is a commutative BCK-algebra. Let $\tilde{x} \leq_m \tilde{y}$. By (BCK-2), we have $\tilde{y} *_m (\tilde{y} *_m \tilde{x}) \leq_m \tilde{x}$. We prove that $\tilde{x} *_m (\tilde{y} *_m (\tilde{y} *_m \tilde{x})) = \tilde{0}$. Calculate

$$\begin{aligned}
m\Big(x * (y * (y * x))\Big) &= m(x * (x \sqcap y)) = m(x) - m(x \sqcap y) \\
&= m(x) - m(y) + m(y * x) = m(x) - m(y) + \tilde{m}(\tilde{y} *_m \tilde{x}) \\
&= m(x) - m(y) + \tilde{m}(\tilde{y}) - \tilde{m}(\tilde{x}) = 0,
\end{aligned}$$

which entails that $\tilde{x} *_m (\tilde{y} *_m (\tilde{y} *_m \tilde{x})) = \tilde{0}$, and $\tilde{x} \leq_m \tilde{y} *_m (\tilde{y} *_m \tilde{x})$ so that $\tilde{x} = \tilde{y} *_m (\tilde{y} *_m \tilde{x})$. This property is by Theorem 5.1.17(iii) the necessary and sufficient condition for X_m to be a commutative BCK-algebra. □

Example 5.1.43 *Let $X = \{0, 1, \ldots, K\}$ be integers ordered in the natural way, where K is either an integer or $\aleph_0$.*

(i) *X is a commutative BCK-algebra with the natural order if and only if*

$$n * m = \begin{cases} 0, & \text{if } n \leq m, \\ n - m, & \text{if } m < n. \end{cases}$$

(ii) *X is a positively implicative BCK-algebra with the natural order if and only if*

$$n * m = \begin{cases} 0, & \text{if } n \leq m, \\ n, & \text{if } m < n. \end{cases}$$

If K is an integer, then there is a unique state (a unique state-morphism) m, namely for the case (i) $m(i) = i/K$, *for* $i = 0, \ldots, K$, *and for the case* (ii) $m(i) = 0$ *for* $0 \leq i \leq K-1$ *and* $m(K) = 1$.

If K is not an integer, then there is no state-morphism, there is only a zero-valued measure-morphism.

(iii) *X cannot be an implicative BCK-algebra with the natural order.*

Example 5.1.44 *Let the interval of real numbers $X = [0,\ 1]$ be ordered in the natural way.*

(i) *X is a bounded commutative BCK-algebra with the BCK-order equal to the natural one if and only if there is a continuous, increasing function $g : [0,\ 1] \to [0,\ 1]$ with $g(0) = 0$ and $g(1) = 1$ (called a generator) such that*

$$x * y = x *_g y := g^{-1}(\max\{0, g(x) - g(y)\}), \tag{5.17}$$

for any $x, y \in [0,\ 1]$. If such a generator exists, it is unique. In addition, all commutative BCK-algebras on $[0,\ 1]$ with the natural order are isomorphic. Any such a BCK-algebra possesses a unique state-morphism, namely $m_g(x) := g(x)$, $x \in [0,\ 1]$, and $\mathcal{S}_g := \{m_g\}$ is an order-determining system of state-morphisms on it.

(ii) *X is a bounded positively implicative BCK-algebra with the BCK-order equal to the natural one if and only if*

$$x * y := \begin{cases} 0, & \text{if } x \leq y, \\ x, & \text{if } x > y, \end{cases} \tag{5.18}$$

for any $x, y \in [0,\ 1]$.

In this case, X possesses a unique state m, namely $m(1) = 1$ and $m(x) = 0$ for any $x \in [0,\ 1)$.

(iii) *X cannot be an implicative BCK-algebra with the natural order.*

Proof. (i) It is easy to verify that (5.17) entails that $[0,\ 1]$ is a commutative BCK-algebra. Suppose the converse, i.e., let $X = [0,\ 1]$ be a commutative BCK-algebra with the natural order. If $x \leq y$, then $x * y = 0$.

Define a partial binary operation $\ominus$ on $X = [0,\ 1]$ such that $b \ominus a$ is defined iff $a \leq b$ and then $b \ominus a := b * a$. It is not difficult to verify that, for all $a, b, c \in [0,\ 1]$, we have (i) $b \ominus a \leq b$ and $b \ominus (b \ominus a) = a$; (ii) if $a \leq b \leq c$ then $c \ominus b \leq c \ominus a$ and $(c \ominus a) \ominus (c \ominus b) = b \ominus a$ (see also [DvKi]), compare with the definition of D-posets, p. 10. The structure $([0,\ 1]; \ominus, 0, 1)$ is a D-poset [KoCh], and by Mesiar [Mes 2, Thm

1], there is a generator g such that, for $x \geq y$, (5.17) holds. Hence (5.17) holds for all $x, y \in [0, 1]$.

Define $X_g := ([0, 1]; *_g, 0)$ to be the BCK-algebra with $*_g$ defined via (5.17) for a fixed generator g, and let X_0 be the space corresponding to $g = id_{[0,1]}$. Then the mapping $h_g : X_g \to X_0$ defined as $h_g(x) := g(x)$, $x \in X_g$, is an isomorphism as claimed.

Using Example 5.1.40 and the established isomorphism, $\{m_g\}$ satisfies our assertion.

(ii) Since X is linearly ordered, it is clear that $x * y = 0$ whenever $x \leq y$. If $y = 0$ then $x * y = x$, so suppose that $1 \geq x > y > 0$. Since X is positively implicative, we have $y < x * y \leq x$. We claim that $x * y = x$.

Let $\epsilon > 0$ be an arbitrary real number less than $x - y$. Then from (5.2) we conclude that

$$(x * (x - \epsilon)) * y = (x * y) * (x - \epsilon).$$

Since the left-hand side of this equation is strictly positive, the right-hand side implies $x \geq x * y > x - \epsilon$, and hence $x * y = x$.

The converse is evident.

(iii) Now let m be a state on X and choose $x \in [0, 1)$. Then $1 * x = (1 * x) * x$ and therefore $m(1 * x) = m((1 * x) * x)$, so that $1 - m(x) = m(1 * x) - m(x) = 1 - 2\,m(x)$ which implies $m(x) = 0$. □

Example 5.1.45 *Let $X = [0, \infty)$ be ordered in the natural way.*

(i) If $g : [0, \infty) \to [0, \infty)$ is an increasing and continuous function with $g(0) = 0$, then $([0, \infty); *_g, 0)$ is a commutative BCK-algebra with BCK-order equal to the natural order, where

$$x *_g y := g^{-1}(\max\{0, g(x) - g(y)\}), \ x, y \in [0, \infty).$$

If $g_0 := id_{[0,\infty)}$, then any $([0, \infty); *_g, 0)$ is isomorphic to $([0, \infty); *_{g_0}, 0)$ if and only if g is unbounded. If g is bounded, then $([0, \infty); *_g, 0)$ is isomorphic to $([0, 1); *_{\mathbb{R}}, 0)$, where $*_{\mathbb{R}}$ is defined via (5.3).

Conversely, if $([0, \infty); *, 0)$ is a commutative BCK-algebra with the natural order, then there is an increasing and continuous function $g : [0, \infty) \to [0, \infty)$ with $g(0) = 0$ such that $* = *_g$, where $*_g$ is defined via (5.17). In addition, any measure m on $([0, \infty); *_g, 0)$ is of the form $m(x) = k\,g(x), x \in [0, \infty)$, where k is a non-negative real constant.

(ii) $([0, \infty); *, 0)$ is a positively implicative BCK-algebra with the natural order if and only if $*$ is defined via (5.18). In addition, the zero function on $[0, \infty)$ is the unique measure on it.

(iii) $([0, \infty); *, 0)$ cannot be any implicative BCK-algebra with the natural order.

Proof. (i) The first implication is straightforward. Suppose now that $([0, \infty); *, 0)$ is a commutative BCK-algebra with the natural order. Put $X_k := [0, k]$, where

$k \geq 1$ is an integer. The restriction of $*$ to $X_k \times X_k$ entails that $(X_k; *, 0, k)$ is a commutative and bounded BCK-subalgebra of $[0, \infty)$. According to (i) of Example 5.1.44, X_k is isomorphic to X_1, for any $k \geq 1$; an isomorphism is given for example by $h_k(x) := x/k$, $x \in X_k$. Due to the last example, there exists a unique increasing and continuous function $g_k : [0, k] \to [0, k]$ with $g_k(0) = 0$ and $g_k(k) = k$, such that, for all $x, y \in X_k$,

$$x * y = g_k^{-1}(\max\{0, g_k(x) - g_k(y)\}).$$

For any $k \geq 1$, we define $\hat{g}_k := g_k/g_k(1)$. We assert for any $k \geq 1$ that the restriction of $\hat{g}_{k+1}$ to X_k, $\hat{g}_{k+1}|X_k$, is equal to $\hat{g}_k$. Indeed, due to the uniqueness of g_k's on X_k with the condition $g_k(k) = k$, $k \geq 1$, we have $k\, g_{k+1}/g_{k+1}(1)|X_k = g_k$ for any $k \geq 1$. Let $t \in X_k$. Then

$$\begin{aligned} \hat{g}_{k+1}(t) &= \frac{g_{k+1}(t)}{g_{k+1}(1)} = \frac{k\, g_{k+1}(t)}{g_{k+1}(k)} \frac{g_{k+1}(k)}{k\, g_{k+1}(1)} \\ &= g_k(t) \frac{g_{k+1}(k)}{k\, g_{k+1}(1)} = \frac{g_k(t)}{g_k(1)} g_k(1) \left(\frac{k\, g_{k+1}(1)}{g_{k+1}(k)} \right)^{-1} \\ &= \hat{g}_k(t)\, g_k(1)\, (g_k(1))^{-1} = \hat{g}_k(t). \end{aligned}$$

Define a function $g : [0, \infty) \to [0, \infty)$ as follows: $g(t) = \hat{g}_k(t)$ if $t \in X_k$ for some $k \geq 1$. Then g is a well-defined function with $g(0) = 0, g(+\infty) = \infty$, g is increasing and continuous, and, in addition,

$$x * y = g^{-1}(\max\{0, g(x) - g(y)\}), \ x, y \in [0, \infty).$$

This equality follows from the observation that if $x, y \in X_k$, then

$$\begin{aligned} x * y &= g_k^{-1}(\max\{0, g_k(x) - g_k(y)\}) \\ &= \hat{g}_k^{-1}(\max\{0, \hat{g}_k(x) - \hat{g}_k(y)\}) \\ &= g^{-1}(\max\{0, g(x) - g(y)\}), \end{aligned}$$

which finishes the first part of the proof.

If g is unbounded, then $h_g : [0, \infty) \to [0, \infty)$ defined via $h_g(x) := g(x)$, $x \in [0, \infty)$, is an isomorphism between $([0, \infty); *_g, 0)$ and $([0, \infty); *_{id_{[0,\infty)}}, 0)$. In addition, for any $x, y \in [0, \infty)$, there exists a $z \in [0, \infty)$ with $z *_g x = y$. Indeed, we put $z := g^{-1}(g(x) + g(y))$.

Suppose now that g is bounded and let $k_0 := \sup\{g(x) : x \in [0, \infty)\}$. From the continuity of g we obtain that there is a $u > 0$ such that $g(u) > k_0/2$. But for $x = u$ and $y = u$ there is no z such that $z *_g u = u$ which entails that $([0, \infty); *_g, 0)$ is not isomorphic to $([0, \infty); *_{id_{[0,\infty)}}, 0)$. We claim that $([0, \infty); *_g, 0)$ is isomorphic to $([0, 1); *_{\mathbb{R}}, 0)$. Indeed, the mapping $h_g(x) := g(x)/k_0$ is an isomorphism in question.

The assertion on measures follows from the simple observation that any measure m on $X = ([0, \infty); *_{id_{[0,\infty)}}, 0)$ or on $X = ([0, 1); *_{\mathbb{R}}, 0)$ is of the form $m(x) = k\,x,\ x \in X$, where $k \geq 0$ is a constant.

(ii) This follows the same ideas as those in the proof of (ii) in Example 5.1.44.

(iii) This is the same as that in Example 5.1.45. □

Example 5.1.46 *Let the interval $X = [0, 1)$ be ordered in the natural way.*

(i) $([0, 1); *, 0)$ *is a commutative BCK-algebra with the natural order if and only if there is an increasing and continuous function $g:\ [0, 1) \to [0, \infty)$ with $g(0) = 0$ such that* (5.17) *holds for any $x, y \in [0, 1)$. Let $*_g$ be defined via* (5.17). *If g is bounded on $[0, 1)$, then $([0, 1); *_g, 0)$ is isomorphic to $([0, 1); *_{g_0}, 0)$, where $g_0 := id_{[0,1)}$, and if g is unbounded, then $([0, 1); *_g, 0)$ is isomorphic to $([0, \infty); *_{id_{[0,\infty)}}, 0)$.*

In addition, any measure on $([0, 1); *_g, 0)$ is of the form $m(x) = k\,g(x),\ x \in [0, 1)$, where k is a non-negative real constant.

(ii) $([0, 1); *, 0)$, *is a positively implicative BCK-algebra with the natural order if and only if $*$ is defined via* (5.18) *for all $x, y \in [0, 1)$. In addition, the zero function on $[0, 1)$ is a unique measure on it, and any measure m on $([0, 1); *_g, 0)$ is of the form $m(x) := k\,g(x),\ x \in [0, 1)$, where $k \geq 1$ is a constant.*

(iii) $([0, 1); *, 0)$ *cannot be any implicative BCK-algebra with the natural order.*

Proof. (i) This follows from [Mes 2, Thm 2, Cor 1] and from (ii) of Example 5.1.45, or it can be proved directly in a way different from that used in [Mes 2], who used a method of [Lin]. To do this let $\{d_n\}$ be a strictly increasing sequence of positive numbers such that $d_n \nearrow 1$. Since any $X_n = [0, d_n]$ is isomorphic to $[0, 1]$, there is a unique increasing and continuous function $g_n:\ X_n \to X_n$ with $g_n(0) = 0$ and $g_n(d_n) = d_n$ such that $x * y = \max\{0, g_n(x) - g_n(y)\}$ if $x, y \in X_n$. If we define $\hat{g}_n := g_n/g_n(d_1)$, then similarly as in the proof of Example 5.1.45, we have that $\hat{g}_{n+1}|X_n = \hat{g}_n$. The function $g(x) = \hat{g}_n(x)$ if $x \in X_n$, satisfies (5.17). Comparing this approach with that used in the proof of Example 5.1.45, we obtain the statements in question.

(ii) This follows the same ideas as those in the proof of (ii) in Example 5.1.44.

(iii) This is the same as that in Example 5.1.45. □

Example 5.1.47 *Let X be a linearly ordered set with least element 0. Then $(X; *, 0)$ is a positively implicative BCK-algebra with the original linear order if and only if $*$ is defined by* (5.18) *for all $x, y \in X$.*

If the greatest element 1 belongs to X, then m with $m(x) = 0$ if $x \in X \setminus \{1\}$ and $m(1) = 1$ is the only state on X, and any measure on X is a constant multiple of that state. If X is not bounded, then the zero function is the unique measure on X.

Proof. We claim that if $x > y$, then $x * y = x$. Suppose the converse. Then $x > x * y > y$. Let now y_1 be an arbitrary element of X such that $x > y_1 > y$. Then

$x > x * y \geq x * y_1 > y_1$. In particular, if $y_1 = x * y$, then $x > x * y \geq x * (x * y) > x * y$ which is a contradiction.

The statement about states and measures follows easily from the proof of Example 5.1.44. □

5.1.9 Exercises

1. Prove that for a commutative BCK-algebra $(X; *, 0)$, we have $I(x) \cap I(y) = I(x \wedge y)$ for all $x, y \in X$. ([MeJu, Cor II.8.6])
2. Prove that an algebra $(X; *, 0)$ of type (2,0) is a commutative BCK-algebra iff (i) $x * (0 * y) = x$, (ii) $(x * z) * (x * y) = (y * z) * (y * x)$. ([Cor 2], [MeJu, Thm VI.2.1])
3. Let $(X; *, 0)$ be a BCK-algebra. If there exists a binary operation $\sqcup$ such that for all $x, y, z \in X$, $(x * y) * z = x * (y \sqcup z)$, then X satisfies condition (S), and $\circ = \sqcup$. ([MeJu, Thm I.7.9])
4. Prove that if h is a BCK-homomorphism, then (i) $h(0) = 0$; (ii) $x \leq y$ entails $h(x) \leq h(y)$; (iii) $Ker(h) = \{x \in X : h(x) = 0\}$ is an ideal of X.
5. Prove that if $a_1, \ldots, a_n$ are elements of a BCK-algebra $(X; *, 0)$ satisfying condition (S), then for the ideal $I(a_1, \ldots, a_n)$ of X generated by $a_1, \ldots, a_n$ we have
$$I(a_1, \ldots, a_n) = \{x \in X : x \leq (a_1 \underbrace{\circ \cdots \circ}_{m_1 \text{ times}} a_1) \circ \cdots \circ (a_n \underbrace{\circ \cdots \circ}_{m_n \text{ times}} a_n)\}.$$
([MeJu, Thm II.7.11])
6. Let X be a positive implicative BCK-algebra satisfying condition (S). Then, for any $a_1, \ldots, a_n$ in X,
$$I(a_1, \ldots, a_n) = \{x \in X;\ x \leq a_1 \circ \cdots \circ a_n\}.$$
([MeJu, Thm II.7.12])
7. Let $(X; *, 0)$ be a bounded implicative BCK-algebra and $A \subseteq X$. Then
$$I(A) = \{x \in X : x \leq a_1 \vee \cdots \vee a_n \text{ for some } a_1, \ldots, a_n \in A\}.$$
([MeJu, Thm II.7.14])
8. Let I be an ideal of a commutative BCK-algebra $(X; *, 0)$, then X/I is a commutative BCK-algebra.
9. Let I be an ideal of a BCK-algebra $(X; *, 0)$ satisfying condition (S), then X/I satisfies condition (S). ([MeJu, Thm II.6.8])
10. Prove Theorem 5.1.28. ([MeJu, Thm I.12.2])

11. A commutative BCK-algebra $(X;*,0)$ is said to be **unitarily extendible** [Hoo 1] if, for $e \notin X$, we can extend the commutative BCK-algebra structure on X to the commutative BCK-algebra structure on $X \cup \{e\}$ with e as the greatest element in $X \cup \{e\}$. Prove: A non-zero commutative BCK-algebra $(X;*,0)$ is unitarily extendible if and only if there is a function $f : X \setminus \{0\} \to X \setminus \{0\}$ such that $f(f(x)) = x$ and $x * f(y) = y * f(x)$ for all $x, y \in X \setminus \{0\}$.[5] ([Hoo 1])

12. Prove: A bounded commutative BCK-algebra $(X;*,0)$ is unitarily extendible if and only if there exists an element $x_0 \in X \setminus \{0\}$ such that $x_0 \leq x$ for all $x \in X \setminus \{0\}$. ([Hoo 1])

13. Prove that $[0,1)$ in Example 5.1.46 is unitarily extendible iff g is bounded.

14. Prove that $[0,\infty)$ in Example 5.1.46 is unitarily extendible iff g is bounded.

15. Prove that $[0,1]$ in Example 5.1.46 is unitarily extendible iff g is bounded.

16. Prove: An algebra $(X;*,0)$ of type (2,0) is a BCK-algebra iff it satisfies axioms B,C, and K, from the footnote on p. 294 and the rule (BCK-4). ([MeJu, Thm 1.7])

5.2 Commutative BCK-algebras with the Relative Cancellation Property

In the present Section, we shall study a special family of commutative BCK-algebras, commutative BCK-algebras with the relative cancellation property, introduced by Dvurečenskij and Graziano [DvGr]: for $a, x, y \in X$, $a \leq x, y$ and $x * a = y * a$ imply $x = y$.

Such BCK-algebras form a proper subvariety of the variety of commutative BCK-algebras, and only they can be represented by Abelian ℓ-groups with the universal property, [DvGr 2]. We show that they are categorically equivalent to a special category of ℓ-groups [Dvu 11] generalizing a famous result of Mundici [Mun] for MV-algebras (= bounded commutative BCK-algebras). In addition, we will study their basic properties, like functional representations, ideals, measures, product BCK-algebras, etc.

We stress that given an MV-algebra M, its radical, *Rad*(M), carrying all important information on the MV-algebra, always forms a commutative BCK-algebra with the cancellation property, which is not necessarily bounded.

[5] We recall that in this case $e * x = f(x)$ for each $x \in X \setminus \{0\}$.

5.2.1 Basic Properties of BCK-algebras with the Relative Cancellation Property

Definition 5.2.1 *We say that a commutative BCK-algebra* $(X;*,0)$ *has the* **relative cancellation property** *if, for* $a, x, y \in X$, $a \le x, y$ *with* $x * a = y * a$ *we have* $x = y$.

Lemma 5.2.2 *Let* $(X;*,0)$ *be any upwards directed commutative BCK-algebra. Then it has the relative cancellation property.*

Proof. Suppose that there exist $a, c, c' \in X$ such that $a \le c, c'$ and $c * a = c' * a$. We have to prove that $c = c'$. Let $u \in X$ be such that $c, c' \le u$.

Then $(u * a) * (u * c) = (u * (u * c)) * a = (u \wedge c) * a = c * a$. Since $a \le c$, we have $u * c \le u * a$ and then $[u * (c * a)] * a = (u * a) * (c * a) = (u * a) * [(u * a) * (u * c)] = (u * a) \wedge (u * c) = u * c$.

In the same way we show that $[u * (c' * a)] * a = u * c'$. From $c * a = c' * a$, it follows that $u * c' = u * c$. Hence $c = u * (u * c) = u * (u * c') = c.'$ □

Example 5.2.3 *The example* B_{5-2-8} *from* [MeJu], *where* $*$ *is given by table* 5.6, *gives a commutative BCK-algebra which is not directed upwards but which has the relative cancellation property.*

$*$	0	1	2	3	4
0	0	0	0	0	0
1	1	0	1	0	0
2	2	2	0	0	2
3	3	2	1	0	2
4	4	1	4	1	0

Tab. 5.6

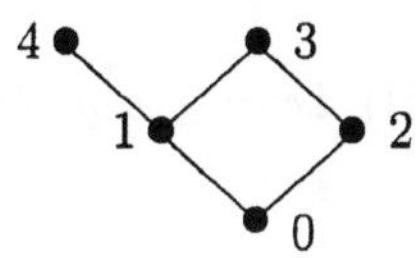

Fig. 5.6

Example 5.2.4 *Suppose that* $(G;+,\le,0)$ *is an Abelian* ℓ*-group with positive cone* $G^+ = \{g \in G : g \ge 0\}$. *Then* $(G^+;*_G,0)$ *is a commutative BCK-algebra with the relative cancellation property, where* $*_G$ *is defined via*

$$u *_G v := (u - v) \vee 0, \quad u, v \in G^+. \tag{5.19}$$

More generally, if G_0 *is a non-void subset of* G^+ *such that* $u, v \in G_0$ *imply* $u *_G v \in G_0$, *then* $(G_0;*_G,0)$ *is a commutative BCK-subalgebra of* $(G^+;*_G,0)$ *having the relative cancellation property.*

Remark 5.2.5 We note that Example 5.1.13 gives a commutative BCK-algebra which does not have the relative cancellation property. Indeed, we have $1 \leq 2, 3$ and $2 * 1 = 1 = 3 * 1$ but $2 \neq 3$. Consequently, it cannot be embedded into any Abelian ℓ-group.

Theorem 5.2.6 *Let $(X; *, 0)$ be a commutative BCK-algebra having the relative cancellation property. Define the partial binary operation $+$ with the domain $S = \{(a, b) \in X \times X :$ there exists a $c \in X$ with $c \geq b$ and $a = c * b\}$, whenever define $c = a + b$ if $c \geq b$, and $a = c * b$. Then, for all $a, b, c \in X$, we have*

(I) *$a + 0$, $0 + a$ exist for arbitrary $a \in X$, and $a + 0 = a = 0 + a$.*

(II) *$a + b$ exists if and only if $b + a$ exists, and then $a + b = b + a$ (commutativity).*

(III) *$a + b$ and $(a + b) + c$ exist in X if and only if $b + c$ and $a + (b + c)$ exist in X, and in this case $(a + b) + c = a + (b + c)$ (associativity).*

(IV) *If $a + b$ exists in X and if $a_1 \leq a$, $b_1 \leq b$, then $a_1 + b_1$ exists in X, and $a_1 + b_1 \leq a + b$.*

(V) *$a = b$ holds whenever $u + a = u + b$ for some $u \in X$ (cancellation property).*

(VI) *If $a + b \leq a + b'$ then $b \leq b'$. Conversely, if $b \leq b'$ and $a + b'$ exists in X, then $a + b \leq a + b'$.*

(VII) *If $a + c$ and $b + c$ exist in X, then $(a \wedge b) + c$ exists in X, and*

$$(a + c) \wedge (b + c) = (a \wedge b) + c. \tag{5.20}$$

(VIII) *Let $u \geq x, y$. Then $x \vee y$ exists in X, and*

$$(x \vee y) * y = x * (x \wedge y) = x * y, \tag{5.21}$$

and $(x * (x \wedge y)) + y$, $(x * y) + y$ exist in X, and

$$x \vee y = (x * (x \wedge y)) + y = (x * y) + y. \tag{5.22}$$

Proof. The relative cancellation property entails that the partial binary operation $+$ is defined correctly.

(I) and (II) are evident. To prove prove (III) suppose that $a + b$ and $(a + b) + c$ exist in X. Then there exist $x, y \in X$ such that $a \leq x$ and $x * a = b$, $a + b = x \leq y$ and $y * x = y * (a + b) = c$.

Since $c = y * x \leq y * a$ and $(y * a) * c = (y * a) * (y * x) = [y * (y * x)] * a = x * a = b$, then $b + c \in X$ and $b + c = y * a$.

From $a \leq y$ and $y * a = b + c$, it follows that $(a + (b + c)) \in X$ and $a + (b + c) = y = (a + b) + c$.

(IV) There are x, y in X such that $a_1+x = a$ and $b_1+y = b$. Using commutativity and associativity, we have $a + b = (a_1 + x) + (b_1 + y) = (a_1 + b_1) + (x + y)$ which entails the assertion.

(V) This follows easily from the definition of the relative cancellation property.

(VI) There exists $c \in X$ such that $(a + b) + c = a + b'$. Using associativity and commutativity of $+$, we have $a + (b + c) = a + b'$ and $(b + c) + a = b' + a$, so that $b + c = b'$ and, consequently, $b \leq b'$. The second part follows from (IV).

(VII) Since $a + c \geq c$ and $b + c \geq c$, we have $(a + c) \wedge (b + c) \geq c$. Then using property (5.2), we have

$$\begin{aligned}((a+c) \wedge (b+c)) * c &= ((a+c) * ((a+c) * (b+c))) * c \\ &= a * ((a+c) * (b+c)).\end{aligned}$$

We have

$$a * b = ((a+c) * c) * ((b+c) * c) \leq (a+c) * (b+c),$$

where we have used property Proposition 5.1.3(b) for BCK-algebras, so that

$$a * ((a+c) * (b+c)) \leq a * (a * b) = a \wedge b.$$

Hence,

$$((a+c) \wedge (b+c)) * c \leq a \wedge b. \tag{5.23}$$

Since $a \geq a \wedge b$ and $a + c$ exists in X, we have by (IV) that $(a \wedge b) + c$ exists in X. It is clear that $a + c \geq (a \wedge b) + c$ and $b + c \geq (a \wedge b) + c$, so that

$$\begin{aligned}(a+c) \wedge (b+c) &\geq (a \wedge b) + c, \\ ((a+c) \wedge (b+c)) * c &\geq ((a \wedge b) + c) * c, \\ ((a+c) \wedge (b+c)) * c &\geq a \wedge b,\end{aligned}$$

which in view of (5.23) entails $((a + c) \wedge (b + c)) * c = a \wedge b$, and, consequently, we have (5.20).

(VIII) Due to Proposition 5.1.4, we have $x * (x \wedge y) = x * y$. Using the basic properties of commutative BCK-algebras and (5.2), we have

$$\begin{aligned}(x \vee y) * y &= (u * ((u * x) \wedge (u * y)) * y \\ &= (u * y) * ((u * x) \wedge (u * y)) \\ &= (u * y) * (u * x) \\ &= (u * (u * x)) * y = x * y,\end{aligned}$$

which proves (5.21) and, consequently, (5.22).

□

5.2.2 BCK-algebras and the Riesz Decomposition Property

A poset $(X; \leq)$ has the **interpolation property** (or simply has **interpolation**), iff, for $x, y, p, q \in X$, $x, y \leq p, q$ implies that there exists $z \in X$ with $x, y \leq z \leq p, q$. A commutative BCK-algebra $(X; *, 0)$ with the relative cancellation property has the **Riesz decomposition property** iff, for $a, b, x \in X$, $x \leq a + b$ implies that there exist $a_1, b_1 \in X$ with $a_1 \leq a$, $b_1 \leq b$ and $x = a_1 + b_1$. (Compare with Lemma 1.7.5.)

It is evident that any commutative BCK-algebra $(X; *, 0)$ has interpolation. Indeed, let $x, y \leq p, q$ then $x, y \leq p \wedge q \leq p, q$.

Example 5.2.7 *Let $\Omega \neq \emptyset$ and let A, B be two non-void disjoint subsets of Ω. Define $E := \{X \subseteq \Omega : X \subseteq A \text{ or } X \subseteq B\}$. Then $(E; *, \emptyset)$, where $*$ is the set-theoretical difference, is a commutative BCK-algebra with the relative cancellation property which is not directed upwards, but it has the Riesz decomposition property.*

This example is not a special one because any commutative BCK-algebra with the relative cancellation property has the Riesz decomposition property, [DvGr 1]:

Theorem 5.2.8 *Any commutative BCK-algebra $(X; *, 0)$ with the relative cancellation property has the Riesz decomposition property.*

Proof. Suppose that $c \leq a+b$, and define $u := a+b$ and $X_u := \{x \in X : x \leq u\}$. Put now $v = c * (b \wedge c)$ and $a_1 = a \wedge v$, so that $a_1 \leq a$. On the other hand, $c * v = b \wedge c$ and if we put $b_1 = c * a_1$, then a_1, b_1 are defined in X_u and, moreover, $a_1 + b_1 = c$.

We must show that $b_1 \leq b$. Applying (5.22) and (5.20), we obtain

$$\begin{aligned} a_1 + b_1 = c &\leq (a+b) \wedge (b \vee c) \\ &= (a+b) \wedge [(c * (b \wedge c)) + b] \\ &= (a+b) \wedge (v+b) \\ &= (a \wedge v) + b = a_1 + b, \end{aligned}$$

and thus by 5.2.6(VI), $b_1 \leq b$. □

Lemma 5.2.9 *For $a_1, a_2, b_1, b_2 \in X$, let $a_1 + a_2 = b_1 + b_2$, where $(X; *, 0)$ is a commutative BCK-algebra with the relative cancellation property. Then there exist four elements $c_{11}, c_{12}, c_{21}, c_{22} \in X$ such that*

$$\begin{aligned} a_1 &= c_{11} + c_{12}, & b_1 &= c_{11} + c_{21}, \\ a_2 &= c_{21} + c_{22}, & b_2 &= c_{12} + c_{22}. \end{aligned}$$

Proof. Since $a_1 \le b_1 + b_2$, due to Theorem 5.2.8, we have that $a_1 = c_{11} + c_{12}$, where $c_{11} \le b_1$ and $c_{12} \le b_2$. Then there are $c_{21}, c_{22} \in X$ such that $c_{11} + c_{21} = b_1$ and $c_{12} + c_{22} = b_2$.

Calculate

$$(c_{11} + c_{12}) + a_2 = (c_{11} + c_{21}) + (c_{12} + c_{22}).$$

Using associativity and cancellation 5.2.6(III), we have $a_2 = c_{21} + c_{22}$. □

Lemma 5.2.10 *Let $(X; *, 0)$ be a commutative BCK-algebra with the relative cancellation property. If $a_1 + \cdots + a_n$ is defined in X, $c \le a_1 + \cdots + a_n$ for $c \in X$, then there exist $a'_1, \ldots, a'_n \in X$ such that $c = a'_1 + \cdots + a'_n$ and $a'_i \le a_i$ for any $i = 1, \ldots, n$.*

Proof. The case $n = 2$ was proved in Theorem 5.2.8. Now let the statement hold for any integer $\le n$, $n \ge 2$. Then $c \le (a_1 + \cdots + a_n) + a_{n+1}$ implies $c = b_1 + a'_{n+1}$, where $b_1 \le a_1 + \cdots + a_n$ and $a'_{n+1} \le a_{n+1}$. By the induction hypothesis, $b_1 = a'_1 + \cdots + a'_n$ where $a'_i \le a_i$ for any $i = 1, \ldots, n$, and by associativity, we have that $c = a'_1 + \cdots + a'_{n+1}$. □

Lemma 5.2.11 *Let $(X; *, 0)$ be a commutative BCK-algebra with the relative cancellation property. Let, for $y_1, \ldots, y_n$ and $z_1, \ldots, z_n$ in X with $z_i \le y_i$ for $i = 1, \ldots, n$, $y_1 + \cdots + y_n$ and $z_1 + \cdots + z_n$ be defined in X. Then $(y_1 * z_1) + \cdots + (y_n * z_n)$ exists in X, and*

$$(y_1 + \cdots + y_n) * (z_1 + \cdots + z_n) = (y_1 * z_1) + \cdots + (y_n * z_n).$$

Proof. Since $z_i \le y_i$ for any $i = 1, \ldots, n$, there exists $u_i \in X$ such that $z_i + u_i = y_i$ and $y_i * z_i = u_i$, $i = 1, \ldots, n$. Using associativity and commutativity of $+$, we have

$$y_1 + \cdots + y_n = (z_1 + u_1) + \cdots + (z_n + u_n) = (z_1 + \cdots + z_n) + (u_1 + \cdots + u_n).$$

According to Theorem 5.2.6(VI), $z_i \le y_i$ for $i = 1, \ldots, n$ implies $z_1 + \cdots + z_n \le y_1 + \cdots + y_n$. Therefore there exists an element $y \in X$ such that $(z_1 + \cdots + z_n) + y = y_1 + \cdots + y_n$ and $(y_1 + \cdots + y_n) * (z_1 + \cdots + z_n) = y$. Due to these equalities and due to the cancellation law Theorem 5.2.6(V), we have $y = (y_1 + \cdots + y_n) * (z_1 + \cdots + z_n) = u_1 + \cdots + u_n = (y_1 * z_1) + \cdots + (y_n * z_n)$ which proves the assertion. □

Lemma 5.2.12 *Let $(X; *, 0)$ be a commutative BCK-algebra with the relative cancellation property. Let $x_1 + \cdots + x_n = y_1 + \cdots + y_m$. Then there exist $z_{ij} \in X$ for $i = 1, \ldots, n$, and $j = 1, \ldots, m$ such that*

$$x_i = \sum_{j=1}^{m} z_{ij} \quad \text{and} \quad y_j = \sum_{i=1}^{n} z_{ij}$$

for any $i = 1, \ldots, n$ and any $j = 1, \ldots, m$.

Proof. The assertion was proved for $n = m = 2$ in Lemma 5.2.9. Now let $x_1 + \cdots + x_n = y_1 + \cdots + y_m$ with $n > 1$ and assume that the assertion holds when the number of x_i's is less than n. From $x_1 \leq x_1 + \cdots + x_n = y_1 + \cdots + y_m$ and Lemma 5.2.10 we have that $x_1 = z_{11} + \cdots + z_{1m}$ where $z_{1j} \leq y_j$ for each $j = 1, \ldots, m$. Then by Lemma 5.2.11, $x_2 + \cdots + x_n = (x_1 + \cdots + x_n) * x_1 = (y_1 + \cdots + y_m) * (z_{11} + \cdots + z_{1m}) = (y_1 * z_{11}) + \cdots + (y_m * z_{1m})$.

By the induction hypothesis, there exist z_{ij} such that $x_i = z_{i1} + \cdots + z_{im}$ for $i = 2, \ldots, n$, and $y_j * z_{1j} = z_{2j} + \cdots + z_{nj}$ for $j = 1, \ldots, m$. Then $y_j = z_{1j} + \cdots + z_{nj}$ and $x_i = z_{i1} + \cdots + z_{im}$ for $i = 1, \ldots, n$ and $j = 1, \ldots, m$. □

5.2.3 Ideals and Prime Ideals

Lemma 5.2.13 *Let a commutative BCK-algebra $(X; *, 0)$ have the relative cancellation property. Then, for all $x, y \in X$, we have*

$$(x * y) \wedge (y * x) = 0. \tag{5.24}$$

Proof. Let $z \leq x * y$ and $z \leq y * x$. Then $z \leq x * (x \wedge y)$ and $z \leq y * (x \wedge y)$, and by Theorem 5.2.6(I)–(VIII), we have $z + (x \wedge y) \leq x$, $z + (x \wedge y) \leq y$ which imply $z + (x \wedge y) \leq x \wedge y$ so that $z = 0$. □

We note that a proper ideal I of a commutative BCK-algebra $(X; *, 0)$ is said to be **prime** if $x \wedge y \in I$ implies $x \in I$ or $y \in I$.

Lemma 5.2.14 *A proper ideal I of a commutative BCK-algebra $(X; *, 0)$ with the relative cancellation property is prime if and only if, for each $x, y \in X$, either $x * y \in I$ or $y * x \in I$.*

Proof. Let I be a prime ideal. Then for all $x, y \in X$, (5.24) holds, so that $(x * y) \wedge (y * x) \in I$ and either $x * y \in I$ or $y * x \in I$.

Conversely, let $x \wedge y \in I$. Then either $x * y \in I$ or $y * x \in I$. Suppose that $x * y \in I$. Then $x * y = x * (y \wedge x) = x * (x \wedge y) \in I$ which entails $x \in I$. Similarly $y * x \in I$ yields $y \in I$. □

Lemma 5.2.15 *Let $(X; *, 0)$ be a commutative BCK-algebra such that (5.24) holds for all $x, y \in X$. Then a proper ideal I of X is prime if and only if, for all $x, y \in X$, $x * y \in I$ or $y * x \in I$.*

Proof. This follows from the same ideas as the proof of Lemma 5.2.14. □

Denote by $\mathcal{P}(X)$ the set of all prime ideals of X. A BCK-algebra $(X; *, 0)$ is said to be a **linear BCK-algebra** if its BCK-order is linear, i.e., for all $x, y \in X$, either $x * y = 0$ or $y * x = 0$.

Lemma 5.2.16 *Let I be a proper ideal of a commutative BCK-algebra $(X;*,0)$ with the relative cancellation property. Then I is a prime ideal if and only if $(X/I;*,C_0)$ is a linear commutative BCK-algebra.*

Proof. Let I be a prime ideal. Due to (5.24), $(x*y)\wedge(y*x)=0\in I$ so that either $x*y\in I$ or $y*x\in I$. Hence

$$C_x * C_y = C_{x*y} = C_0 \quad \text{or} \quad C_y * C_x = C_{y*x} = C_0,$$

and $(X/I;*,C_0)$ is a linear commutative BCK-algebra.

Now let $(X/I;*,C_0)$ be a linear commutative BCK-algebra. Then the above formula holds, which, by Lemma 5.2.14, means that I is a prime ideal. □

Lemma 5.2.17 *Let $(X;*,0)$ be a commutative BCK-algebra. Let $a*^n x=0$ and $a*^m y=0$ for natural numbers n and m. Then there exists a natural number p such that $a*^p(x\wedge y)=0$.*

Proof. Since, for $m\le n$, $a*^m x=0$ implies $a*^n x=0$, it suffices to verify when $a*^n x=0=a*^n y$, there exists a natural number p such that $a*^p(x\wedge y)=0$.

We proceed by mathematical induction on n.

Let $n=1$. Then $a*x=a*y=0$ and so $a\le x,y$, hence $a\le x\wedge y$, i.e., $a*(x\wedge y)=0$.

Assume now that the assertion in question holds for a natural number $n>1$. That is $a*^n x=0=a*^n y$ implies there exists a natural number p such that $a*^p(x\wedge y)=0$.

If $a*^{n+1}x=a*^{n+1}y=0$, then

$$0=a*^{n+1}x=((a*^n x)*^n y)*x.$$

In the same way we have

$$0=((a*^n x)*^n y)*y.$$

Due to the first induction step, we have

$$((a*^n x)*^n y)*(x\wedge y)=0,$$
$$((a*(x\wedge y))*^n x)*^n y=0,$$
$$(((a*(x\wedge y))*^n y)*^{n-1}x)*x=0.$$

By the equality $a*^{n+1}y=0$ we have

$$(((a*(x\wedge y))*^n y)*^{n-1}x)*y=0.$$

Hence

$$((a*^2(x\wedge y)*^n y))*^{n-1}x=0.$$

Repeating this procedure n times, we obtain

$$(a *^{n+1} (x \wedge y)) *^n y = 0, \tag{A}$$

as well as

$$(a *^{n+1} (x \wedge y)) *^n x = 0, \tag{B}$$

in the same way.

By the induction hypothesis and (A) and (B), we can find a natural number p such that $(a *^{n+1} (x \wedge y)) *^p (x \wedge y) = 0$ and $a *^{n+p+1} (x \wedge y) = 0$, which proves the assertion in question. □

Lemma 5.2.18 *Let $(X; *, 0)$ be a commutative BCK-algebra with the relative cancellation property. If $a \in X$, $a \neq 0$, then there exists a prime ideal I of X such that $a \notin I$.*

Proof. Consider an ideal I of X which is maximal with respect to the property that $a \notin I$. Due to Zorn's lemma, such an ideal always exists. We show that I is a prime ideal. Let $x, y \in X$ and assume $x * y \notin I$ and $y * x \notin I$. Then the ideal generated by I and the element $x * y$ contains the element a. By Lemma 5.1.33, there exists an integer n such that

$$a *^n (x * y) \in I.$$

Similarly, the ideal generated by I and $y * x$ contains the element a, and, therefore, there exists an integer m such that

$$a *^m (y * x) \in I.$$

Take the commutative quotient BCK-algebra $(X/I; *, C_0)$. From the last two equalities above we have that $C_a *^n C_{x*y} = C_0$ and $C_a *^m C_{y*x} = C_0$. Lemma 5.2.17 entails the existence of an integer p such that $C_a *^p (C_{x*y} \wedge C_{y*x}) = C_0$. Hence,

$$\begin{gathered} C_a *^p C_{(x*y)\wedge(y*x)} = C_0, \\ C_a *^p C_0 = C_0, \\ C_a = C_0, \end{gathered}$$

which implies the contradiction that $a \in I$. □

5.2.4 Subdirect Product of Linear Commutative BCK-algebras

Theorem 5.2.19 *Every commutative BCK-algebra $(X; *, 0)$ with the relative cancellation property is a subdirect product of linear commutative BCK-algebras.*

Proof. If $(X;*,0)$ is a linear commutative BCK-algebra, the statement of Theorem holds.

Let now X be not a linear BCK-algebra. Therefore the set $\{0\}$ is not a prime ideal of X, and due to Lemma 5.2.18,

$$\bigcap\{I : I \in \mathcal{P}(X)\} = \{0\}.$$

According to Lemma 5.2.16, the quotient BCK-algebra X/I ($I \in \mathcal{P}(X)$) is a linear commutative BCK-algebra. Let C_x^I denote the equivalence class of x in the quotient X/I.

Define the direct product $X_0 := \prod\{X/I : I \in \mathcal{P}(X)\}$ and define the mapping $\alpha : X \to X_0$ via

$$\alpha(x) := (C_x^I)_{I\in\mathcal{P}(X)}.$$

α is a homomorphism, and we claim that α is injective. Indeed, let $x_1 \neq x_2$ be two elements in X. Then $x_1 * x_2 \neq 0$ or $x_2 * x_1 \neq 0$. Suppose, e.g., $x_1 * x_2 \neq 0$. Then $x_1 * x_2 \notin \bigcap\{I : I \in \mathcal{P}(X)\}$, and there exists a prime ideal I of X such that $x_1 * x_2 \notin I$, so that $C_{x_1}^I \neq C_{x_2}^I$, and $\alpha(x_1) \neq \alpha(x_2)$.

If π_I is the projection from X_0 onto X/I then, for any $x \in X$, $\pi_I(\alpha(x)) = C_x^I$, so that $\pi_I(\alpha(X)) = X/I$. This proves that $\alpha(X)$ is a subdirect product of $\{X/I : I \in \mathcal{P}(X)\}$. □

We say that a BCK-algebra $(X;*,0)$ is **subdirectly irreducible** if $\bigcap\{I \in \mathcal{I}(X) : I \neq \{0\}\} \neq \{0\}$, where $\mathcal{I}(X)$ is the set of all ideals of X.

Lemma 5.2.20 *Let $(X;*,0)$ be a commutative, subdirectly irreducible BCK-algebra. Then X is linear if and only if X has the relative cancellation property.*

Proof. If X is linear, then obviously X has the relative cancellation property.

Assume now the converse. Given an element $a \in X$, let $I(a)$ denote the ideal of X generated by a. Let x and y be two non-zero elements of X. Due to Exercise 1. Section 5.1.9,

$$I(x*y) \cap I(y*x) = I((x*y) \wedge (y*x)) = I(0) = \{0\}.$$

Therefore, $I(x*y) = \{0\}$ or $I(y*x) = \{0\}$, so that $x*y = 0$ or $y*x = 0$ which gives $x \leq y$ or $y \leq x$. □

Example 5.1.13 gives a subdirectly irreducible BCK-algebra which is not linear. Famous Birkhoff's theorem [Bir, Thm VIII.8.15] asserts that any BCK-algebra having at least one element is a subdirect product of subdirectly irreducible BCK-algebras. Due to Lemma 5.2.20 and Theorem 5.2.19, we see that for commutative BCK-algebras, which have not the relative cancellation property, we do not obtain a subdirect product of linearly ordered commutative BCK-algebras.

5.2.5 Embedding of Commutative BCK-algebras into ℓ-groups

In Examples 5.1.15 and 5.2.4 we have seen that the positive cone of an Abelian ℓ-group forms a commutative BCK-algebra with the relative cancellation property. Example 5.2.4 is in some sense an archetype of commutative BCK-algebras with the relative cancellation property because the basic representation theorem with universal property for commutative BCK-algebras with the relative cancellation was proved in [DvGr 2]. In the present Section, we give such a representation.

We shall follow ideas of word techniques of Wyler [Wyl] who used Baer's approach [Bae], for more details see Section 1.6 in Chapter 1. We shall apply these methods for commutative BCK-algebras with the relative cancellation property from which we can derive a partial addition + according to Theorem 5.2.6. We note that in such a case $(X; \ominus, 0)$ is a generalized effect algebra in the sense of Section 1.1, where $x \ominus y := x * y$ if $y \leq x$. Then the partial addition $\oplus = +$.

Suppose that the alphabet is now the set X with the set of all words $\mathcal{W}(X) := \bigcup_{n=1}^{\infty} X^n$, forming a semigroup, and define the equivalence $\sim$ on $\mathcal{W}(X)$ by Section 1.6, in addition $\mathcal{W}(X)/\sim$ can be organized as a semigroup with $[(0)]$ as a neutral element; we denote by $[(a_1, \ldots, a_n)]$ the equivalence class determined by the word $(a_1, \ldots, a_n)$. It was proved, Corollary 1.6.9, that

$$(a_1, \ldots, a_n) \sim (a) \quad \text{iff} \quad a = a_1 + \cdots + a_n, \tag{5.25}$$

and $[(a)] = [(b)]$ iff $a = b$.

For upwards directed BCK-algebras we have the following result, compare with Lemma 1.6.10:

Lemma 5.2.21 *Let a and b from X have an upper bound in X. Then*

$$[(a)] + [(b)] = [(b)] + [(a)]. \tag{5.26}$$

In particular, if X is directed upwards, then $\mathcal{W}(X)/\sim$ is a commutative semigroup.

Proof. Let $u \in X$ be an upper bound of a and b. Then due to Proposition 5.1.19, $a \vee b$ exists in X. Using (5.21) and (5.22), we have

$$\begin{aligned}(a,b) &\sim (a \wedge b, a * (a \wedge b), b) = (a \wedge b, (a \vee b) * b, b) \\ &\sim (a \wedge b, a \vee b) \sim (a \wedge b, (a \vee b) * a, a) \\ &= (a \wedge b, b * (a \wedge b), a) \sim (b, a).\end{aligned}$$

The second assertion follows from (5.21) and (5.22). □

Lemma 5.2.22 *Let, for $A, B \in [X]$, we have $[A]+[B] = [(0)]$. Then $[A] = [(0)] = [B]$.*

Proof. This follows easily from (5.25). □

We note that if X is not directed upwards, then (5.26) can fail. Indeed, take X from Example 5.1.25. Then the partial binary operation $+$ is given by the table below, and here we have $[(1)] + [(2)] = [(1,2)] \neq [(2,1)] = [(2)] + [(1)]$.

+	0	1	2	3	4
0	0	1	2	3	4
1	1	3			
2	2		4		
3	3				
4	4				

Tab 5.7

We say that a partially ordered Abelian group $(G;+,\leq,0)$ with a mapping $h : X \to G^+$ is a **universal group** for a commutative BCK-algebra $(X;*,0)$ having the relative cancellation property if

(i) the positive cone G^+ is generating for G;

(ii) $h(X)$ generates G^+;

(iii) $h(x+y) = h(x) + h(y)$ whenever $x+y$ exists in X, $x, y \in X$;

(iv) for any partially ordered Abelian group G_1 and any order and $+$ preserving mapping $g : X \to G_1$ there is a group homomorphism $g' : G \to G_1$ such that $g = g' \circ h$.

The universal group, if it exists, is unique up to isomorphism, and g' from (iv) is a unique group homomorphism with that property. We denote the universal group $G = (G, h)$.

Using Wyler's technique and remarks in Section 1.6, we can prove the following important result [DvGr]:

Theorem 5.2.23 *Let $(X;*,0)$ be an upwards directed commutative BCK-algebra. Then, for all $A, B, C \in \mathcal{W}(X)$, we have*

(i) $[A] + [B] = [B] + [A]$.

(ii) *If* $[A] + [B] = [A] + [C]$, *then* $[B] = [C]$.

(iii) $\leq_X$, *where* $[A] \leq_X [B]$ *if and only if there is a* $C \in [X]$ *such that* $[A]+[C] = [B]$, *is a partial order on* $\mathcal{W}(X)/\sim$.

(iv) $[(a)] \wedge [(b)] = [(a \wedge b)]$, $a, b \in X$.

(v) $(\mathcal{W}(X)/\sim; +, \leq_X, [(0)])$ *is an Abelian partially ordered semigroup which is a lower semilattice, and it is the generating positive cone of a lattice ordered group* $G(X)$.

(vi) *A mapping* $h : X \to G(X)$ *defined via*

$$h(a) := [(a)], \ a \in X,$$

is injective and preserving $\wedge$, *and* $+$.

(vii) *The couple* $(G(X), h)$ *has the universal property.*

(viii) *If* X *is bounded with the greatest element* $1 \in X$, *then* $h(1)$ *is a strong unit*[6] *in* $G(X)$ *and* h *maps* X *onto the order interval* $[[(0)], [(1)]] := \{g \in G(X) : [(0)] \leq_X g \leq_X [(1)]\}$.[7]

(ix) *If* $g \in G(X)$, $0 \leq g \leq h(x)$ *for some* $x \in X$, *then there exists a unique* $y \in X$ *such that* $y \leq x$ *and* $h(y) = g$.

Proof. These results have been proved in [DvGr, Thm 3.5] using Wyler's ideas [Wyl], or it can follow ideas of Section 1.6. □

We are now able to prove the embedding theorem for commutative BCK-algebras with the relative cancellation property, [DvGr 2].

Theorem 5.2.24 *Let* $(X; *, 0)$ *be a commutative BCK-algebra with the relative cancellation property. Then there exists a lattice ordered group* $(G; +, \leq, 0)$ *with the (generating) positive cone* G^+ *and there is a (lower semilattice)* $G_0 \subseteq G$ *with the property that* $u, v \in G_0$ *implies* $u *_G v \in G_0$ *such that* X *can be injectively embedded onto* G_0 *preserving* $*$.

Proof. Let X be a commutative BCK-algebra with the relative cancellation property. According to Theorem 5.2.19, X is a subdirect product of linear commutative BCK-algebras $\{(X_i; *_i, 0_i)\}_{i \in S}$. Let $\alpha : X \to \prod_{i \in S} X_i$ be a subdirect embedding.

Every $(X_i; *_i, 0_i)$ is a linear commutative BCK-algebra. It is directed upwards, and by Theorem 5.2.23, there is a lattice ordered group $(G_i; +_i, \leq_i, 0_i)$ with the positive cone G_i^+ and with a (lower semilattice) $G_i^0 \subseteq G_i^+$ satisfying that $u, v \in G_i^0$ implies $(u - v) \vee_i 0_i \in G_0^i$ such that $(X_i; *_i, 0_i)$ is isomorphic to $(G_i^0; *_{G_i}, 0_i)$ by an isomorphism

[6] We recall that an element $u > 0$ of a lattice ordered group G is said to be a **strong unit** or an **order unit** if, given an element $g \in G$, there is an integer n such that $-nu \leq g \leq nu$.

[7] In this case, a commutative BCK-algebra X can be converted by a natural way into an MV-algebra and conversely, see Example 5.1.14, and (viii) gives Mundici's famous representation theorem of MV-algebras [Mun]. For more details see Section 5.3.2.

h_i. In addition, (G_i, h_i) is a universal group for $(X_i; *_i, 0_i)$. Moreover, any G_i is a lattice ordered group.

Define the product group $G := \prod_{i\in S} G_i$ with the ordering $\leq$ and the addition $+$ defined by coordinates. Then $(G; +, \leq, 0)$ is a lattice ordered group with the positive cone $G^+ = \prod_{i\in S} G_i^+$. The mapping $h : \prod_{i\in S} X_i \to G^+$ defined via

$$h((x_i)_{i\in S}) = (h_i(x_i))_{i\in S}$$

is an injective homomorphism from $\prod_{i\in S} X_i$ into G^+ preserving $*$ and the order in $\prod_{i\in S} X_i$. If we put $G_0 := h(\alpha(X))$, then G_0 is a lower semilattice of G^+ with an isomorphism $h \circ \alpha$ in question. □

Remark 5.2.25 Let $(X; *, 0)$ be a commutative BCK-algebra with the relative cancellation property. A partially ordered Abelian group $(G; +, \leq, 0)$ with an injective mapping $h : X \to G^+$ is said to be an **ambient group** for X if

(i) h preserves $+$ and the order on X;

(ii) $h(X)$ generates G^+;

(iii) G^+ generates G.

Theorem 5.2.23 asserts that any commutative BCK-algebra with the relative cancellation property has an ambient lattice ordered group with a mapping h satisfying (i)–(iii).

We note that an ambient group is not determined uniquely by X. Using a variation of an example from [BeFo 1], we have as follows. Let G be the set of all integers, $\mathbb{Z}$, with the non-standard positive cone $G^+ = \{0, 3, 4, 6, 7, 8, \ldots\}$. Take $X = \{0, 3, 4, 7\}$ and define $7 * 4 = 3,\ 7 * 3 = 4,\ 3 * 4 = 3,\ 4 * 3 = 4$ and $0 \leq x \leq 7$ for any $x \in X$. Then $(X; *, 0)$ is a commutative BCK-algebra with the relative cancellation property (in fact it is equivalent to the Boolean algebra 2^2), and G is an ambient group with the natural embedding, which is not a lattice ordered group. Simultaneously, $\mathbb{Z}^2$ with a mapping h defined via $h(0) = (0, 0)$, $h(2) = (1, 0)$, $h(4) = (0, 1)$, $h(7) = (1, 1)$ is another ambient group which is in fact a lattice ordered group and a universal group for X, and G and $\mathbb{Z}^2$ are not isomorphic.

It is worthy of recalling that Theorem 5.2.23 can be obtained also by another technique using semiclans introduced by Bosbach [Bos 1]. We say that $(C; \wedge, +)$ is a **semiclan** if it is a $\wedge$-semilattice and a partial groupoid with respect to $+$ if the following axioms are satisfied:

(C1) If $a \leq b$, then there exist $x, y \in C$ such that $b = a + x$ and $b = y + a$.

(C2) If $a+x,\ a+y \in C$, $a+x=a+y$, then $x=y$, and if $x+a,\ y+a \in C$, $x+a=y+a$, then $x=y$.

(C3) If $a+x,\ a+y \in C$, then $(a+x) \wedge (a+y) = a+(x \wedge y)$, and if $x+a,\ y+a \in C$, then $(x+a) \wedge (y+a) = (x \wedge y)+a$.

(C4) $a+b,\ (a+b)+c \in C$ iff $b+c,\ a+(b+c) \in C$, and in this case we have $(a+b)+c = a+(b+c)$.

(C5) If $(a \wedge b)+c = c$ and $a \vee b$ exists, then $a+b = a \vee b = b+a$.

For example if G is a (not necessarily Abelian) ℓ-group, then the positive cone G^+ is a semiclan. The basic result of [Bos 1], says that a semiclan C can be embedded into the positive cone G^+ of an ℓ-group G preserving partial addition $+$, $\wedge$ and existing $\vee$.

It is easy to verify that in view of Theorem 5.2.6 any commutative BCK-algebra with the relative cancellation property can be easily transformed into a semiclan in the sense of Bosbach which gives us another proof of Theorem 5.2.23, see Exercise 2.5.2.12.

5.2.6 Characterizations of Commutative BCK-algebras with the Relative Cancellation Property

Theorem 5.2.26 *A commutative BCK-algebra $(X;*,0)$ is a subdirect product of linear commutative BCK-algebras if and only if X has the relative cancellation property.*

Proof. If X has the relative cancellation property, the statement of Theorem follows from Theorem 5.2.19.

Conversely, let X be a subdirect product of linear commutative BCK-algebras. This product is obviously directed upwards, and therefore, it has the relative cancellation property. Consequently, X has the relative cancellation property. □

It is worth to recall that Example 5.1.13 gives a commutative BCK-algebra which is not a subdirect product of linear commutative BCK-algebras. Indeed, in the opposite case, it would have a lattice ordered group representation, which is impossible in view of $2*1=1=3*1$ while $2 \neq 3$.

Theorem 5.2.27 *Let $(X;*,0)$ be a commutative BCK-algebra. Then X has the relative cancellation property if and only if, for all $x,y \in X$,* (5.24) *holds.*

Proof. One direction is known from Lemma 5.2.13. Suppose now that (5.24) holds, and let I be a prime ideal of X. Then X/I is a commutative BCK-algebra which is linear. Indeed, since $(x*y) \wedge (y*x) = 0$, then by Lemma 5.2.14, either $x*y \in I$ or $y*x \in I$. This means that $C_x * C_y = C_{x*y} = C_0$ or $C_y * C_x = C_{y*x} = C_0$.

Repeating all arguments of the proof of Theorem 5.2.19, we can show that X is a subdirect product of linear commutative BCK-algebras, $\prod\{X/I : I \in \mathcal{P}(X)\}$, with an embedding α. According to Theorem 5.2.26, X has the relative cancellation property. □

The previous partial results can be joined in one.

Theorem 5.2.28 *Let $(X;*,0)$ be a commutative BCK-algebra. The following statements are equivalent.*

(i) *X has the relative cancellation property.*

(ii) *$(x*y)\wedge(y*x)=0$[8] for all $x,y\in X$.*

(iii) *X is a subdirect product of linear commutative BCK-algebras.*

(iv) *X can be embedded onto a commutative BCK-subalgebra of the positive cone G^+ of a lattice ordered group $(G;+,\leq,0)$.*

(v) *$(x*y)*(y*x)=x*y$ for all $x,y\in X$.*

(vi) *$(X;*,0)$ can be embedded into a commutative BCK-algebra $(\hat{X};*_\wedge,0)$, which is a lattice[9], and any element of $\hat{X}$ is a supremum of a finite subset of X, and $*_\wedge$ is an extension of $*$.*

Proof. The equivalence of (i)–(iii) has been proved in Theorems 5.2.19, 5.2.26 and 5.2.27. (i) implies (iv) by Theorem 5.2.23. (iv) yields evidently (i).

(ii) ⇒ (v). We have $0=(x*y)*((x*y)*(y*x))$ so that $x*y\leq(x*y)*(y*x)\leq x*y$.

(v) ⇒ (ii). $(x*y)\wedge(y*x)=(x*y)*((x*y)*(y*x))=(x*y)*(x*y)=0$.

It is clear that (vi) implies (iv).

(iv) ⇒ (vi). Let $(G_0;*_G,0)$ be the commutative BCK-subalgebra of $(G^+;*_G,0)$ which is isomorphic to $(X;*,0)$. Denote by $G_0(X)$ the least commutative BCK-subalgebra of $(G^+;*_G,0)$ which is a lattice under the BCK-order; it always exists. Let $\hat{X}$ be the set of all suprema of all finite subsets of G_0. We claim that $\hat{X}$ is under $*_G$ a commutative BCK-subalgebra of G^+ equal to $G_0(X)$. It is evident that $G_0(X)\subseteq\hat{X}$.

According to Proposition 5.1.19, two elements of a commutative BCK-algebra have a supremum in it iff they have an upper bound in it. Therefore, for all elements $x,y,z\in G^+$, we have $x*_G(y\wedge z)=(x*_Gy)\vee(x*_Gy)$ and $(x\vee y)*_Gz=(x*_Gz)\vee(y*_Gz)$.

[8] Such BCK-algebras are sometimes called Łukasiewicz BCK-algebras, [Pal].

[9] In fact, it is a distributive lattice by Proposition 5.1.19(viii), [Pal]; another proof of the equivalence of (ii) and (vi) was presented in [Stur] and [CST].

Let now $x = \bigvee_i x_i$ and $y = \bigvee_j y_j \in \hat{X}$. Then

$$\begin{aligned} x *_G y &= (\bigvee_i x_i) *_G y = \bigvee_i (x_i *_G y) \\ &= \bigvee_i (x_i *_G \bigvee_j y_j) = \bigvee_i \bigwedge_j (x_i *_G y_j) \in \hat{X} \end{aligned}$$

because, for any i, $\bigwedge_j (x_i *_G y_j) \in G_0$.

This means that $(\hat{X}; *_G, 0) = (G_0(X); *_G, 0)$ and putting $*_\wedge = *_G$, we have proved (vi). □

As a corollary we obtain the following result.

Theorem 5.2.29 *Commutative BCK-algebras with the relative cancellation property form a variety which is a proper subvariety of the variety of commutative BCK-algebras.*

Proof. From Theorem 5.2.28 we can show that an algebra $(X; *, 0)$ of type (2,0) is a commutative BCK-algebra with the relative cancellation property iff it satisfies (i) $x * x = 0$; (ii) $0 * x = 0$; (iii) $(x * y) * z = (x * z) * y$; (iv) $x * (x * y) = y * (y * x)$, and (v) $(x * y) * (y * x) = x * y$.

Example 5.1.13 shows that the variety of commutative BCK-algebras with the relative cancellation property is a proper subvariety of the variety of commutative BCK-algebras. □

We note that an algebra $(X; *, 0)$ of type (2,0) satisfying the conditions (i)–(v) from the last proof is said to be a T-*algebra*, [Buf], and Buff proved that every T-algebra is a subdirect product of linearly ordered T-algebras. In view of Theorem 5.2.29, it is our result from Theorem 5.2.19.

We note that the assertion that an MV-algebra is a subdirect product of linear MV-algebras has been proved by Chang [Cha 1]. An analogical statement for upwards directed commutative BCK-algebras was proved by Palasiński [Pal], Cornish [Cor], and later Fleischer [Fle] proved that (ii) implies (iii) from Theorem 5.2.28. An analogical assertion was presented in [CST] and by Raftery [Raf].

5.2.7 Universal Groups for Commutative BCK-algebras

In this Section, we strengthen the result of Theorem 5.2.23 showing that for any commutative BCK-algebra with the relative cancellation property we can find an ambient group which is an ℓ-group and, in addition, a universal group.

Theorem 5.2.30 *Any commutative BCK-algebra $(X; *, 0)$ with the relative cancellation property has a universal group (G, h) which is a lattice ordered ambient group for X. Moreover, h preserves $*$, $\wedge$ in X, and $\vee$ existing in X.*

Proof. According to Theorem 5.2.19, X is a subdirect product of linear commutative BCK-algebras $\prod_i X_i$. Similarly as in the proof of the implication (iv) $\Rightarrow$ (vi) of Theorem 5.2.28 (or using some results of [Stur] and [CST] under equivalent conditions), we can find in $\prod_i X_i$ the minimal commutative BCK-subalgebra of $\prod_i X_i$ containing the injective image od X which is a lattice. We denote it by $\hat{X}$; $\hat{X}$ consists of all suprema of all finite subsets of the image of X in $\prod_i X_i$. Without loss of generalization we can assume that $X \subseteq \prod_i X_i$. It is evident that $\prod_i X_i$ is a subdirect product of $\hat{X}$, and $\wedge$ and $\vee$ taken in X (if they exist) and in $\hat{X}$ are the same. It is clear that $\hat{X}$ has the relative cancellation property, and we can define a partial operation $\hat{+}$ which is an extension of $+$; without loss of clarity of exposition, we shall write $\hat{+} = +$.

The proof will proceed in several steps.

Claim 1. If $x \in \hat{X}$, $y \in X$ and $x \le y$, then $x \in X$.

Let $x = x_1 \vee \cdots \vee x_n$, where $x_1, \ldots, x_n \in X$. We know, Proposition 5.1.19 that $x_1 \vee x_2$ exists in X iff x_1 and x_2 have an upper bound, say z, in X. Then $x_1 \vee x_2 = z * ((z * x_1) \wedge (z * x_2))$. Using the induction argument, we obtain the claim in question.

Claim 2. Any element in $\hat{X}$ is a sum of finitely many elements of X.

Let $x \in \hat{X}$. Then $x = x_1 \vee \cdots \vee x_n$ with $x_i \in X$ for $i = 1, \ldots, n$. If $n = 1$, the claim is obvious. Let now the claim hold for any integer less than n. Using the identical arguments as those from the proof the implication (iv) $\Rightarrow$ (vi) in Theorem 5.2.28, we have

$$\begin{aligned}
&(a_1 \vee \cdots \vee a_n) * (a_1 \vee \cdots \vee a_{n-1}) = \\
&[(a_1 \vee \cdots \vee a_{n-1}) \vee a_n)] * (a_1 \vee \cdots \vee a_{n-1}) = \\
&[(a_1 \vee \cdots \vee a_{n-1}) * (a_1 \vee \cdots \vee a_{n-1})] \vee [a_n * (a_1 \vee \cdots \vee a_{n-1})] = \\
&a_n * (a_1 \vee \cdots \vee a_{n-1}) = \\
&a_n * [a_n \wedge (a_1 \vee \cdots \vee a_{n-1})] = \\
&a_n * [(a_n \wedge a_1) \vee \cdots \vee (a_n \wedge a_{n-1})] = \\
&(a_n * (a_n \wedge a_1)) \wedge \cdots \wedge (a_n * (a_n \wedge a_{n-1})).
\end{aligned}$$

The element $a_n' := (a_n * (a_n \wedge a_1)) \wedge \cdots \wedge (a_n * (a_n \wedge a_{n-1}))$ belongs to X, and by hypothesis, $a_1 \vee \cdots \vee a_{n-1} = a_1' + \cdots + a_{n-1}'$ with $a_i' \in X$ for $i = 1, \ldots, n-1$.

Using associativity of $+$, we obtain $a_1 \vee \cdots \vee a_n = (a_1 \vee \cdots \vee a_{n-1}) \vee a_n = (a_1 \vee \cdots \vee a_{n-1}) + a_n' = a_1' + \cdots + a_n'$ where $a_i' \in X$ for any $i = 1, \ldots, n$.

Step 3. Since $\hat{X}$ is an upwards directed commutative BCK-algebra, by Theorem 5.2.23, there exist a lattice ordered group G with the positive generating cone G^+ and an embedding mapping $h : \hat{X} \to G^+$ which preserves the order, the sum, and $*$ in $\hat{X}$; in fact, also it preserves $\wedge$ and $\vee$. Moreover, $h(\hat{X})$ generates G^+, and the couple (G, h) is a universal group for $\hat{X}$. It is evident that h preserves also the BCK-order and the sum in X. We claim that $h(X)$ generates G^+. Indeed, if $g \in G^+$, then $g = \sum_{i=1}^n n_i h(x_i)$, where $x_i \in \hat{X}$ for $i = 1, \ldots, n$. According to Claim 2, $x_i = \sum_{j=1}^m x'_{ij}$ where $x'_{ij} \in X$ (we can assume that for any i the number of x'_{ij} in every sum is the same). Then $g = \sum_{i=1}^n n_i h(\sum_{j=1}^m x'_{ij}) = \sum_{i=1}^n \sum_{j=1}^m n_i h(x'_{ij})$.

Suppose now that $f : X \to K$ is a mapping preserving $+$ and the order in X, where K is a partially ordered Abelian group. Define an extension $\hat{f} : \hat{X} \to K$ in the following way: If $x \in \hat{X}$ and $x = x_1 + \cdots + x_n$ with $x_1, \ldots, x_n \in X$, then

$$\hat{f}(x) := f(x_1) + \cdots + f(x_n).$$

The mapping $\hat{f}$ is defined well. Indeed, if x has another expression $x = y_1 + \cdots + y_m$, where $y_1, \ldots, y_m \in X$, according to Lemma 5.2.12 there exist elements $z_{ij} \in \hat{X}$ such that $x_i = \sum_{j=1}^m z_{ij}$ and $y_j = \sum_{i=1}^n z_{ij}$ for all $i = 1, \ldots, n$, and $j = 1, \ldots, m$. Since $z_{ij} \leq x_i \in X$, from Claim 1, we have $z_{ij} \in X$ for all i and j.

Since f is additive on X, we have

$$\hat{f}(x) = \sum_{i=1}^n f(x_i) = \sum_{i=1}^n \sum_{j=1}^m f(z_{ij}) = \sum_{j=1}^m f(y_j).$$

According to Claim 2 and Lemma 5.2.11, $\hat{f}$ preserves the sum in $\hat{X}$.

Let now $a, b \in \hat{X}$ be such that $a \leq b$. Then $b = b_1 + \cdots + b_m$, where $b_1, \ldots, b_m \in X$. Since $a \leq b_1 + \cdots + b_m$, by Lemma 5.2.10, $a = a_1 + \cdots + a_m$, where $a_i \in \hat{X}$ and $a_i \leq b_i$ for any $i = 1, \ldots, m$. In view of Claim 1, any a_i is from X. Therefore, $f(a_i) \leq f(b_i)$ and $\hat{f}(a) = f(a_1) + \cdots + f(a_m) \leq f(b_1) + \cdots + f(b_m) = \hat{f}(b)$.

Since (G, h) is a universal group for $\hat{X}$, there exists a unique group homomorphism $f' : G \to K$ such that $\hat{f} = f' \circ h$. Then $f = f' \circ h$ and f' is a unique group homomorphism with this property.

Finally, since h preserves $*$, $\wedge$ and $\vee$ in $\hat{X}$, Theorem 5.2.23, h preserves $*$ and $\wedge$ in X, as well existing $\vee$ in X. □

Corollary 5.2.31 *Let $(X; *, 0)$ be a commutative BCK-algebra with the relative cancellation property. Then $(X; *, 0)$ and $(\hat{X}; *, 0)$ have the same underlying ℓ-group G in their universal groups (G, h), and $(G, \hat{h})$, where $\hat{h}$ is a unique extension of h from $\hat{X}$ into G preserving $+$ and $\wedge$.*

Proof. This follows from the proof of Theorem 5.2.30. □

Remark 5.2.32 Since all universal groups for a commutative BCK-algebra $(X; *, 0)$ with the relative cancellation property are unique up to isomorphism, any universal group (G, h) for X consists of an injective embedding h and of a lattice ordered group G.

Example 5.2.33 *Let $(X;*,0)$ be the BCK-algebra from Example* 5.1.13. *Then X has two prime ideals $I_1 = \{0,1,4\}$ and $I_2 = \{0,2\}$. Then $X/I_1 = \{\{0,1,4\}, \{2,3\}\}$ and $X/I_2 = \{\{0,2\}, \{1,3\}, \{4\}\}$, and X has a universal group $(\mathbb{Z}^2, h)$, where $h(0) = (0,0)$, $h(1) = (0,1)$, $h(2) = (1,0)$, $h(3) = (1,1)$, $h(4) = (0,2)$.*

5.2.8 BCK-hull of Commutative BCK-algebras

In the previous Section, within the proof of Theorem 5.2.28, we have introduced a commutative BCK-algebra $(\hat{X};*,0)$ which is a lattice, "containing" the commutative BCK-algebra $(X;*,0)$ with the relative cancellation property as a BCK-subalgebra such that $\hat{X}$ consists of joins of all finite subsets from X. We call it the **BCK-hull** of the commutative BCK-algebra $(X;*,0)$ with the relative cancellation property. The BCK-hull for commutative BCK-algebras was introduced in [CST] and [Stur]. In [DvGr 2] we introduced it in another and simpler way. Any element $x \in \hat{X}$ is then expressible as the sum $x = x_1 + \cdots + x_n$ of elements $x_1, \ldots, x_n$ from X and the sum $+$ is taken in $\hat{X}$.

We note that the following result establishes the correctness of the definition of a BCK-hull using a different way of proof from that of [Stur].

Proposition 5.2.34 *Let $(X;*,0)$ be a commutative BCK-algebra with the relative cancellation property. Let $\iota_j : X \to X_j$ be an embedding preserving $*$, where $(X_j;*_j,0_j)$ are commutative BCK-algebras such that any element of X_j is the join of finitely many elements from $\iota_j(X)$ for $j = 1,2$. Then $(X_1;*_1,0_1)$ and $(X_1;*_2,0_2)$ are isomorphic commutative BCK-algebras.*

Proof. We show that if $\bigvee_{i=1}^n \iota_1(x_i) = \bigvee_{j=1}^m \iota_1(y_j)$, then

$$\bigvee_{i=1}^{n} \iota_2(x_i) = \bigvee_{j=1}^{m} \iota_2(y_j).$$

Due to Claim 2 of the proof of Theorem 5.2.30, there exist elements $x_1', \ldots, x_n' \in X$ and $y_1', \ldots, y_m' \in X$ such that $\iota_1(x_1') + \cdots + \iota_1(x_n') = \bigvee_{i=1}^n \iota_1(x_i) = \bigvee_{j=1}^m \iota_1(y_j) = \iota_1(y_1') + \cdots + \iota_1(y_m')$.

Since X_1 has the Riesz decomposition property, Lemma 5.2.12, there are elements $c_{ij} \in X_1$ for $i = 1, \ldots, n$, and $j = 1, \ldots, m$, such that $\iota_1(x_i) = \sum_{j=1}^m c_{ij}$ and $\iota_1(y_j) = \sum_{i=1}^n c_{ij}$. In addition, there are elements $c_{ij}' \in X$ such that $\iota_1(c_{ij}') = c_{ij}$ for every i and j. Then $x_i = \sum_{j=1}^m c_{ij}'$ and $y_j = \sum_{i=1}^n c_{ij}'$. This yields $\iota_2(x_i) = \sum_{j=1}^m \iota_2(c_{ij}')$ and $\iota_2(y_j) = \sum_{i=1}^n \iota_2(c_{ij}')$, so that $\bigvee_i \iota_2(x_i) \geq \iota_2(y_j)$ and $\bigvee_i \iota_2(x_i) \geq \bigvee_j \iota_2(y_j)$. By symmetry we have $\bigvee_j \iota_2(y_j) \geq \bigvee_i \iota_2(x_i)$.

Therefore the mapping $h : X_1 \to X_2$ defined by $h(\bigvee_i \iota_1(x_i)) = \bigvee_i \iota_2(x_i)$ is a well defined injective mapping of X_1 onto X_2.

We show now that h preserves $*_1$ in X_1.

(a) $h(\iota_1(a) *_1 \iota_1(b)) = h(\iota_1(a * b)) = \iota_2(a * b) = \iota_2(a) *_2 \iota_2(b) = h(\iota_1(a)) *_2 h(\iota_1(b))$.

(b) $h((\bigvee_i \iota_1(a_i)) *_1 \iota_1(b)) = h(\bigvee_i(\iota_1(a_i * b))) = \bigvee_i h(\iota_1(a_i * b)) = (\bigvee_i h(\iota_1(a_i))) *_2 h(\iota_1(b)) = h(\bigvee_i \iota_1(a_i)) *_2 h(\iota_1(b))$.

(c) $h(\iota_1(a) *_1 (\bigvee_j \iota_1(b_j))) = h(\bigwedge_j(\iota_1(a * b_j))) = \bigwedge_j \iota_2(a * b_j) = \bigwedge_j(h(\iota_1(a)) *_2 h(\iota_1(b_j))) = h(\iota_1(a)) *_2 h(\bigvee_j \iota_1(b_j))$.

(d) $h((\bigvee_i \iota_1(a_i)) *_1 (\bigvee_j \iota_1(b_j))) = \bigvee_i(h(\iota_1(a_i) *_1 (\bigvee_j \iota_1(b_j)))) = \bigvee_i \bigwedge_j(h(\iota_1(a_i)) *_2 h(\iota_1(b_j))) = h(\bigvee_i \iota_1(a_i)) *_2 h(\bigvee_j \iota_1(b_j))$.

□

Given a subset X_0 of X, we define

$$\widehat{X_0} := \{\bigvee_{i=1}^{n} x_i : \; x_i \in X_0, \; i = 1, \ldots, n, \; n \geq 1\}. \tag{5.27}$$

Let $\mathcal{M}(X)$, $\mathcal{P}(X)$, $\mathcal{M}(\hat{X})$, and $\mathcal{P}(\hat{X})$, be the sets of all maximal and prime ideals of X and $\hat{X}$, respectively. As it was shown in Lemma 5.2.18, $\mathcal{P}(X) \neq \emptyset$, but $\mathcal{M}(X)$ can be void, see Example 5.2.42.

Theorem 5.2.35 *Let $(X; *, 0)$ be a commutative BCK-algebra with the relative cancellation property.*

(i) *If I is an ideal of X, then $\hat{I}$ is an ideal of $\hat{X}$.*

(ii) *$\hat{I}_1 = \hat{I}_2$ for two ideals I_1 and I_2 of X if and only if $I_1 = I_2$.*

(iii) *If J is an ideal of $\hat{X}$, then $J_0 := J \cap X$ is an ideal of X, and $\widehat{J_0} = J$.*

(iv) *I is a prime ideal of X if and only if $\hat{I}$ is a prime ideal of $\hat{X}$.*

(v) *I is a maximal ideal of X if and only if $\hat{I}$ is a maximal ideal of $\hat{X}$.*

(vi) *The BCK-hull of X/I is isomorphic to $\hat{X}/\hat{I}$ for any ideal I of X.*

(vii) *X/I is isomorphic to $\hat{X}/\hat{I}$ for any prime ideal (any maximal ideal) I of X. In particular $\prod\{X/I : I \in \mathcal{P}(X)\}$ and $\prod\{\hat{X}/J : J \in \mathcal{P}(\hat{X})\}$ are isomorphic.*

Proof. (i) The proof consists of several steps. First of all we recall that any ideal I of X contains all existing joins of elements from I. In what follows, we show that $\hat{I}$ is an ideal of $\hat{X}$.

(1) Let $x \leq y$ and $y \in \hat{I}$. Then $x \in \hat{I}$. Indeed, let $y = \bigvee_i e_i$. Then $x = x \wedge y = x \wedge (\bigvee_i e_i) = \bigvee_i(x \wedge e_i)$. Since $x \wedge e_i \leq e_i$ and $e_i \in X$, then $x \wedge e_i \in X$ and from $x \wedge e_i \in I$ it follows that $x \in \hat{I}$.

(2) Let $a * b \in \hat{I}$ for $a \in X$ and $b \in I$. Then $a * b = \bigvee_k g_k$, where $g_k \in I$. Since $a * b \in X$, we conclude that $\bigvee_k g_k \in I$, which implies that $a \in I \subseteq \hat{I}$.

(3) Let $a * b \in \hat{I}$ for $a \in \hat{X}$ and $b \in I$. Then $a = \bigvee_i a_i$, where $a_i \in X$. Therefore $(\bigvee_i a_i) * b = \bigvee_i (a_i * b) \in \hat{I}$. Since $a_i * b \leq a * b \in \hat{I}$, by (1), $a_i * b \in \hat{I}$ for any i which entails by (2) $a_i \in I$ for each i, consequently, $a = \bigvee_i a_i \in \hat{I}$.

(4) Let $a * b \in \hat{I}$ for $a \in X$ and $b \in \hat{I}$. Then $a * b = a * (a \wedge b)$. On the other hand $a \wedge b \leq b \in \hat{I}$ which by (1) gives $a \wedge b \in \hat{I}$. Simultaneously, $a \wedge b \leq a \in X$ entails $a \wedge b \in X$, therefore $a \wedge b \in I$. From $a * (a \wedge b) \in I$ we obtain that $a \in I$.

(5) Let $a * b \in \hat{I}$ for $a \in \hat{X}$ and $b \in \hat{I}$. Then $a = \bigvee_i a_i$, where $a_i \in X$ for any i and $a * b = (\bigvee_i a_i) * b = \bigvee_i (a_i * b) \in \hat{I}$. Hence $a_i * b \leq a * b \in \hat{I}$ gives by (1) $a_i * b \in \hat{I}$ and by (4) we conclude that $a_i \in I$ for any i. Finally $a \in \hat{I}$ which proves (i).

(ii) Let $\hat{I}_1 = \hat{I}_2$ for two ideals I_1 and I_2 of X. Given $x \in I_1$, $x = \bigvee_j y_j$, where $y_j \in I_2$. Since $x \in X$, the join $\bigvee_j y_j$ is defined in X, consequently $x = \bigvee_j y_j \in I_2$, i.e., $I_1 \subseteq I_2$. By symmetry we obtain $I_2 \subseteq I_1$.

(iii) It is evident that J_0 is an ideal of X, and $\widehat{J_0} \subseteq J$. Take $y \in J$, then $y = \bigvee_j y_j$, where $y_j \in X$. Therefore $y_j \in J$ and $y_j \in J_0$ for any j. Hence $y \in \widehat{J_0}$.

(iv) To prove that I prime implies $\hat{I}$ prime: Let $a, b \in \hat{X}$. We claim that $a * b \in \hat{I}$ or $b * a \in \hat{I}$. Let $a = \bigvee_i a_i$ and $b = \bigvee_j b_j$, where $a_i, b_j \in X$. Then $a * b = \bigvee_i a_i * b$. If $a_i * b \in \hat{I}$ for all i then $a * b \in \hat{I}$. Let i be such that $a_i * b$ does not belong to $\hat{I}$. Since $a_i * b = a_i * \bigvee_j b_j = \bigwedge_j (a_i * b_j) \leq a_i * b_j$, by (1) we have that $a_i * b_j$ does not belong to $\hat{I}$ for all j. Since I is prime, we have that $a_i * b_j$ does not belong to I for all j, and then $b_j * a_i \in I$ for any j and for this particular i. We have $\bigvee_j (b_j * a_i) = (\bigvee_j b_j) * a_i = b * a_i \in \hat{I}$. Since $b * a \leq b * a_i$, by (1) we have that $b * a \in \hat{I}$.

Let now $\hat{I}$ be a prime ideal of $\hat{X}$. Then, for any $a, b \in X$, $a * b \in \hat{I}$, or $b * a \in \hat{I}$, which easily proves I is a prime ideal of X.

(v) To prove that I maximal implies $\hat{I}$ is maximal: Suppose that $\hat{I} \subseteq J$, where J is an ideal of $\hat{X}$. Due to (ii), J_0 is an ideal of X, and it contains I. The maximality of I entails either $J_0 = I$, or $J_0 = X$, and again (iii) gives $J = \widehat{J_0}$ is either $\hat{I}$ or $\hat{X}$.

Let now $\hat{I}$ be a maximal ideal of $\hat{X}$, and let $I \subseteq I'$, where I' is an ideal of X. Then $\widehat{I'}$ contains $\hat{I}$, consequently $\widehat{I'}$ is either $\hat{I}$ or $\hat{X}$. Applying (ii), we obtain that I is maximal.

(vi) Define a mapping $\Psi_I : X/I \to \hat{X}/\hat{I}$ by $x/I \mapsto x/\hat{I}$, where $x \in X$, and x/I and $x/\hat{I}$ are cosets. Then Ψ_I is a homomorphism, and $\Psi_I(x/I) = \Psi_I(y/I)$ for $x, y \in X$ iff $x/I = y/I$. Indeed, $x/\hat{I} = y/\hat{I}$ iff $x * y \in \hat{I}$ and $y * x \in \hat{I}$ which by (i) gives $x * y \in I$ and $y * x \in I$, so that $x/I = y/I$.

Take now $a \in \hat{X}$, then $a = \bigvee_i a_i$, where $a_i \in X$, and we have $a/\hat{I} = \bigvee_i (a_i/\hat{I})$ which proves that the BCK-hull of X/M is isomorphic to $\hat{X}/\hat{I}$.

(vii) Suppose that I is a prime ideal of X. Then X/I is linearly ordered, so that X/I is equal to the BCK-hull of X/I, which by (vi) gives X/I is isomorphic to $\hat{X}/\hat{I}$.
□

5.2.9 Relations among BCK-algebras and Universal Groups

The following two results characterize elements of a BCK-algebra via its universal group.

Proposition 5.2.36 *Let $(X;*,0)$ be a commutative BCK-algebra with the relative cancellation property and let $(G(X),h)$ be its universal group. If $0 \leq g \leq h(x)$, where $g \in G(X)$ and $x \in X$, then there is a unique element $x_0 \in X$ such that $h(x_0) = g$.*

Proof. Let us assume that $0 \leq g \leq h(x)$, where $x \in X$. Then $(G(X), \hat{h})$, where $\hat{h}$ is a unique extension of h from the BCK-hull $\hat{X}$ of X into $G(X)$, is a universal group for $(\hat{X};*,0)$, Corollary 5.2.31, and $\hat{X}$ is a lattice. According to Theorem 5.2.23(ix), there exists a unique element x_0 in $\hat{X}$ such that $h(x_0) = g$. Since h is injective and preserves the order in $\hat{X}$, we have $x_0 \leq x$. Since $x_0 = \bigvee_{i=1}^{k} x_i$, where $x_1, \ldots, x_k \in X$, we conclude that x_0 is defined in X. □

Proposition 5.2.37 *Let $x, x_1, \ldots, x_n$ be elements of a commutative BCK-algebra $(X;*,0)$ with the relative cancellation property. Let $(G(X),h)$ be a universal group for X.*

(1) *If $h(x) = h(x_1) + \cdots + h(x_n)$, then $x_1 + \cdots + x_n$ is defined in X, and $x = x_1 + \cdots + x_n$.*

(2) *If $h(x) \geq h(x_1) + \cdots + h(x_n)$, then $x_1 + \cdots + x_n$ is defined in X, and $x \geq x_1 + \cdots + x_n$.*

Proof. (1) If $n = 1$, the statement is evident. Now let the statement hold for any integer $\leq n$, $n \geq 1$. Then $h(x) = h(x_1)+\cdots+h(x_n)+h(x_{n+1})$ which gives $h(x) \geq h(x_i)$ for any i, so that $x \geq x_i$. Then $h(x) - h(x_{n+1}) = h(x_1) + \cdots + h(x_n)$, so that $h(x*x_{n+1}) = h(x_1)+\cdots+h(x_n)$. The induction argument gives $x*x_{n+1} = x_1+\cdots+x_n$, so that $x = x_1 + \cdots + x_n + x_{n+1}$.

(2) There exists $g \in G^+(X)$ such that $h(x) = h(x_1) + \cdots + h(x_n) + g$. Since $G(X)$ is a universal group for X as well as for the BCK-hull $\hat{X}$ of X, $g \in G^+(X)$ iff $g = \sum_{j=1}^{k} h(y_j)$ by 5.2.23(ix) for $y_1, \ldots, y_k \in X$. Therefore by (1), $x = x_1 + \cdots + x_n + y_1 + \cdots + y_k$, so that $x_1 + \cdots + x_n$ exists in X and is less than x. □

Theorem 5.2.38 *Let $(X;*,0)$ be a BCK-algebra with the relative cancellation property and with a universal group $(G(X),h)$. The following statements are equivalent.*

(i) *X is linear.*

(ii) *$G(X)$ is linear.*

(iii) *$\hat{X}$ is linear.*

Proof. (ii) $\Rightarrow$ (i). If $G(X)$ is a linear, then clearly X is linear.

(i) $\Rightarrow$ (ii). Suppose now that X is linear and let the lattice ordered group $G(X)$ be not linear. Then there exist two elements $f_1, f_2 \in G(X)$ such that $f_1 \neq f_1 \wedge f_2 \neq f_2$. Let $f_1 = f_1^+ - f_1^-$, $f_2 = f_2^+ - f_2^-$, where $f_i^+, f_i^- \geq 0$ for $i = 1,2$. Then $f_1^- = \sum_{i=1}^n f_i^1$, $f_2^- = \sum_{i=1}^m f_i^2$, where $f_i^1, f_i^2 \in h(X)$. Without loss of generality we can assume that $n = m$ and $f_1^1 \leq f_2^1 \leq \cdots \leq f_n^1$, $f_1^2 \leq f_2^2 \leq \cdots \leq f_n^2$, with $f_n^1 \leq f_n^2$. Then $f_1^-, f_2^- \leq nf_n^2 =: f_0$. Therefore, $f_i = f_i^+ - f_i^- \geq f_i^+ - f_0$ and $f_i + f_0 \geq f_i^+$ for $i = 1,2$. Then $f_1 + f_0 \neq (f_1 + f_0) \wedge (f_2 + f_0) \neq f_2 + f_0$.

Since $f_1 + f_0$ and $f_2 + f_0$ both are positive, there exist $x_1, \ldots, x_k \in X$ such that $f_i + f_0 \leq v := \sum_{j=1}^k h(x_j)$ for $i = 1,2$. Then $v_0 := (v - f_1 - f_0) \wedge (v - f_2 - f_0) \geq 0$, and $0 = (v - f_1 - f_0 - v_0) \wedge (v - f_1 - f_0 - v_0)$. Put $g_i = v - f_i - f_0 - v_0$ for $i = 1,2$. Then $g_1 > 0$, $g_2 > 0$ and $g_1 \wedge g_2 = 0$.

Due to the Riesz decomposition property [Goo 1, Prop 2.2(b)], there exist positive elements g_j^i in G with $g_j^i \leq h(x_j)$ for $i = 1, \ldots, k$ and $i = 1,2$ such that

$$g_i = \sum_{j=1}^k g_j^i.$$

Since $g_1 \wedge g_2 = 0$, there are two elements $g_{j_1}^1 > 0$ and $g_{j_2}^2 > 0$ such that $g_{j_1}^1 \wedge g_{j_2}^2 = 0$. In view of Proposition 5.2.36, there exist two non-zero elements $x, y \in X$ such that $g_{j_1}^1 = h(x)$ and $g_{j_2}^2 = h(y)$. Then $0 = g_{j_1}^1 \wedge g_{j_2}^2 = h(x) \wedge h(y) = h(x \wedge y)$ which gives $x \wedge y = 0$ and this contradicts the linearity of X.

(i) $\Leftrightarrow$ (iii). Since $(G(X), \hat{h})$ is also a universal group for $(\hat{X}; *, 0)$, where $\hat{h}$ is the extension of h from $\hat{X}$ into $G(X)$ by Corollary 5.2.31, due to the above, we conclude that X is linear iff $\hat{X}$ is linear. $\square$

We say that a commutative BCK-algebra $(X; *, 0)$ with the relative cancellation property is **Archimedean** if, whenever for two elements a and b in X such that $na = \sum_{i=1}^n a_i$, where $a_i = a$, exists in X $na \leq b$ and for any integer $n \geq 1$, it follows that $a = 0$. It is evident that any BCK-subalgebra of an Archimedean BCK-algebra is also Archimedean.

The following example gives a bounded commutative BCK-algebra (= MV-algebra) with linear order which is not Archimedean.

Example 5.2.39 *Let $X = \{n, \dot{n} : n \geq 0\}$. Define the total binary operation $*$ on X by $n * m = \max\{0, n - m\}$, $n * \dot{m} = 0$, $\dot{n} * m = (\dot{n+m})$, $\dot{n} * \dot{m} = 0$ if $n \geq m$, and $\dot{n} * \dot{m} = m - n$ if $n < m$. Then $(X; *, 0)$ is a bounded commutative BCK-algebra, where $0 < 1 < 2 < \cdots < n < n+1 < \cdots < (\dot{n+1}) < \dot{n} < \cdots < \dot{2} < \dot{1} < \dot{0}$. On the other hand, X is not Archimedean since $nm \leq \dot{0}$ for any $n \geq 1$ and any $m \geq 1$.*

Converting it into an MV-algebra via Example 5.1.14, *we obtain a so-called* **Chang** *MV-algebra,* [Cha]. *Here $n \oplus m = n + m$, $\dot{n} \oplus m = \dot{\max}\{0, m - n\}$, and $\dot{n} \oplus \dot{m} = \dot{0}$.*

Theorem 5.2.40 *Let $(X; *, 0)$ be a commutative BCK-algebra with the relative cancellation property and with a universal group $(G(X), h)$. The following statements are equivalent.*

(i) *X is Archimedean.*

(ii) *$G(X)$ is an Archimedean lattice ordered group.*

(iii) *$\hat{X}$ is Archimedean.*

Proof. (ii) $\Rightarrow$ (i). Assume that $G(X)$ is an Archimedean lattice ordered group. If na is defined in X for some $a \in X$ and for any integer $n \geq 1$, and if $na \leq b$ for $b \in X$, then we have $nh(a) \leq h(b)$, which yields $h(a) = 0$, and consequently $a = 0$.

(i) $\Rightarrow$ (ii). Suppose now that X is Archimedean. To prove that $G(X)$ is Archimedean, it is sufficient to verify that if, for any $n \geq 1$, $nu \leq v$ for $u \geq 0$, $u, v \in G(X)$, then $u = 0$. Indeed, the general case of $u \in G(X)$ follows from the simple considerations: $nu^+ = (nu)^+ = (nu) \vee 0 \leq v \vee 0 = v^+$ which implies $u^+ = 0$ and, therefore, $u = u^+ + u^- = u^- \leq 0$.

Assume that $nu \leq v$ for $u, v \in G^+(X)$ and for each $n \geq 1$. Since $h(X)$ generates the positive cone $G^+(X)$, there exist finitely many elements $y_1, \ldots, y_k$ in X such that

$$v = \sum_{i=1}^{k} h(y_i). \tag{A}$$

The proof will use induction on the integer k.

Case 1. Assume that $v = h(y)$ for some $y \in X$. Since $u \leq h(y)$, according to Proposition 5.2.36, there is a unique $x \in X$ such that $u = h(x)$. Since $nh(x) \leq h(y)$ for any integer $n \geq 1$ and $nh(x) \in G^+(X)$, applying again Proposition 5.2.36, we conclude that nx is defined in X for any integer $n \geq 1$. Hence $nh(x) = h(nx) \leq h(y)$ which entails $nx \leq y$ for any $n \geq 1$. The hypothesis gives $x = 0$, and consequently $u = 0$.

Case 2. Suppose we have proved the problem in question for any element $u \in G^+(X)$ and for any element v of the form (A), where the summation in (A) is over integers $\leq k$. Assume now $nu \leq v = \sum_{i=1}^{k} h(y_i) + h(y)$ for $n \geq 1$, where $y = y_{k+1} \in X$. Then $nu - h(y) \leq \sum_{i=1}^{k} h(y_i)$ for any integer $n \geq 1$. In lattice ordered groups we have $j(g \vee 0) = (jg) \vee (j-1)g \vee \cdots \vee g \vee 0$ for any integer $j \geq 1$ (see [Bir, p. 293]). Therefore, for all $n, m \geq 1$, we have

$$\begin{aligned} &m[(nu - h(y)) \vee 0] \\ &= m(nu - h(y)) \vee (m-1)(nu - h(y)) \vee \cdots \vee (nu - h(y)) \vee 0 \\ &= [mnu - mh(y)] \vee [(m-1)nu - (m-1)h(y)] \vee \cdots \vee (nu - h(y)) \vee 0 \end{aligned}$$

$$\leq [mnu - h(y)] \vee [(m-1)nu - h(y)] \vee \cdots \vee (nu - h(y)) \vee 0$$
$$\leq \sum_{i=1}^{k} h(y_i) \vee \sum_{i=1}^{k} h(y_i) \vee \cdots \vee \sum_{i=1}^{k} h(y_i) \vee 0$$
$$= \sum_{i=1}^{k} h(y_i).$$

Since $(nu - h(y)) \vee 0 \in G^+(X)$ applying the induction hypothesis, we conclude that $(nu - h(y)) \vee 0 = 0$ for any integer $n \geq 1$. Therefore $nu \leq h(y)$ for all $n \geq 1$. We can now apply Case 1 to conclude $u = 0$.

(ii) $\Leftrightarrow$ (iii). Finally, $(G(X), \hat{h})$ is a universal group for $\hat{X}$, then due to the equivalence of (i) and (ii), we conclude the equivalence of (ii) and (iii). □

We say that a commutative BCK-algebra $(X; *, 0)$ is **Dedekind monotone σ-complete (Dedekind σ-complete)** if, for every non-decreasing sequence (every sequence) $\{a_n\}_n$ of elements in X bounded from above, $\bigvee_{n=1}^{\infty} a_n$ exists in X. It is worth recalling that X is Dedekind monotone σ-complete iff X is Dedekind σ-complete.

Proposition 5.2.41 *Let a commutative BCK-algebra $(X; *, 0)$ with the relative cancellation property be Dedekind σ-complete. Then $X, \hat{X}$, and $G(X)$ are Archimedean.*

Proof. Assume that na is defined in X for some $a \in X$ and for any $n \geq 1$, and let $na \leq b$. Since $\{a_n\}_n$, where $a_n := na$, is non-decreasing, $\bigvee_{n=1}^{\infty} na$ exists in X. For any integer $k \geq 1$, we have $(k+1)a \leq \bigvee_{n=1}^{\infty} na$ so that $ka \leq (\bigvee_{n=1}^{\infty} na) * a$ which implies $\bigvee_{k=1}^{\infty} ka \leq (\bigvee_{n=1}^{\infty} na) * a$, and whence $a = 0$.

The rest of the proof follows from Theorem 5.2.40. □

In what follows, we show that there is an upwards directed commutative BCK-algebra $(X; *, 0)$ which is Dedekind σ-complete (and consequently Archimedean) such that $\mathcal{M}(X) = \emptyset$.

Example 5.2.42 *Let $\mathbb{Z}^+ = \{0, 1, 2, \ldots\}$ and let $H = (\mathbb{Z}^+)^{\mathbb{R}}$. Set $K = \{f \in H :\ f$ has a countable support $\}$. Then K is an ideal of the upwards directed commutative BCK-algebra H. Put $X := H/K$, then X is also an upwards directed commutative BCK-algebra. Using ideas of* [Goo 1, Ex 9.6], *it is possible to show that X is Dedekind σ-complete with $\mathcal{M}(X) = \emptyset$.*

5.2.10 Ideals of BCK-algebras and ℓ-ideals

In the present section, we compare ideals in commutative BCK-algebras with ℓ-ideals[10] of underlying lattice ordered groups.

[10] A non-void subset K of an Abelian lattice ordered group G is said to be an **ℓ-ideal** of G if it is a subgroup and if $y_1 \leq x \leq y_2$, $y_1, y_2 \in K$, then $x \in K$.

A prime ideal P of X which does not contain any prime ideal of X different from P is said to be a **minimal prime ideal.**

Given a subset I of X, let $h_0(I)$ be the ℓ-ideal of $G(X)$ generated by the image $h(I)$ in $G(X)$.

Proposition 5.2.43 *Let $(X;*,0)$ be a commutative BCK-algebra with the relative cancellation property. Let $(G(X),h)$ be a universal group for X. Let K be an ℓ-ideal of $G(X)$. Then*

$$I_K := h^{-1}(h(X)\cap K) \tag{5.28}$$

is an ideal of X such that $h_0(I_K)=K$.

Proof. It is evident that $0\in I_K$. Suppose that $y\in I_K$, and $z\le y$, where $z\in X$. Then $0\le h(z)\le h(y)$, which entails $h(z)\in K$, so that $z\in I_K$. Let now $x*y\in I_K$ and $y\in I_K$. Then $x*y=x*(x\wedge y)=: x_0$, so that $x\wedge y\in I_K$, and $x=x_0+(x\wedge y)$. Then $h(x)=h(x_0)+h(x\wedge y)\in K$ which gives $x\in I_K$.

It is clear that $h(I_K)\subseteq K$, so that $h_0(I_K)\subseteq K$. Let now $g\in K$, $g\ge 0$. Then there exist finitely many elements $x_1,\dots,x_n\in X$ such that $g=\sum_{i=1}^k h(x_i)$. Therefore, $x_1,\dots,x_k\in I_K$ which gives $g\in h_0(I_K)$. If g is an arbitrary element of K, then $g=g^+-g^-$, where $g^+,g^-\in K$, and $g^+,g^-\ge 0$. In any rate, $g^+,g^-\in h_0(I_K)$, and, consequently, $g\in h_0(I_K)$. □

Proposition 5.2.44 *Let I be an ideal of a commutative BCK-algebra $(X;*,0)$ with the relative cancellation property. Let $K_I := h_0(I)$. Then*

$$I=h^{-1}(h(X)\cap K_I). \tag{5.29}$$

Proof. It is clear that $I\subseteq h^{-1}(h(X)\cap K_I)$. Choose now $x\in h^{-1}(h(X)\cap K_I)$. Then $h(x)\in K_I$, so that there are $x_1,\dots,x_k\in I$ such that $h(x)\le h(x_1)+\cdots+h(x_k)$. Applying the Riesz interpolation property [Goo 1, Prop 2.2(b)], we find $g_1,\dots,g_k\in G^+(X)$ such that $h(x)=g_1+\cdots+g_k$, where $g_i\le h(x_i)$. Due to Proposition 5.2.36, there exist unique elements $x_1',\dots,x_k'\in X$ such that $h(x_i')=g_i$. Since $h(x_i')\le h(x_i)$, we have $x_1',\dots,x_k'\in I$, and by Proposition 5.2.37, $x_1'+\cdots+x_k'$ is defined in X and it is equal to x, so that $x=x_1'+\cdots+x_k'\in I$, which proves (5.29). □

We note that by a **maximal ℓ-ideal** of an Abelian lattice ordered group $(G;+,\le,0)$ we mean any proper ℓ-ideal of G which is not contained properly in any proper ℓ-ideal of G. We denote by $\mathcal{M}(G)$ and by $\mathcal{I}(G)$ the sets of all maximal ℓ-ideals and of all ℓ-ideals of G, respectively.

Theorem 5.2.45 *The mapping $\phi_h : I\mapsto h_0(I)$, $I\in\mathcal{I}(X)$, is a one-to-one correspondence from the set of all ideals of X, $\mathcal{I}(X)$, onto the set of all ℓ-ideals of $G(X)$,*

$\mathcal{I}(G(X))$, *preserving the set-theoretical inclusion of ideals: If* $I_1 \subseteq I_2$, $I_1, I_2 \in \mathcal{I}(X)$, *then* $h_0(I_1) \subseteq h_0(I_2)$.

In addition, I is a maximal ideal of X if and only if $h_0(X)$ is a maximal ℓ-ideal of $G(X)$.

Proof. This follows directly from Propositions 5.2.43 and 5.2.44; the inverse mapping to ϕ_h is given by (5.28) $K \mapsto h^{-1}(h(X) \cap K)$. □

Combining Theorem 5.2.45 with Proposition 5.2.44 we have:

Theorem 5.2.46 *Let $(G(X), h)$ be a universal group of a commutative BCK-algebra $(X; *, 0)$ with the relative cancellation property, and let $\hat{X}$ be the BCK-hull of X. There exists a one-to-one correspondence between (maximal) ideals of X, (maximal) ideals of $\hat{X}$ and (maximal) ℓ-ideals of $G(X)$.*

Proof. Since $(G(X), \hat{h})$, where $\hat{h}$ is a unique extension of h to $\hat{X} \to G(X)$, Corollary 5.2.31, is a universal group for $\hat{X}$, the correspondence in question is $I \mapsto \hat{I} \mapsto \hat{h}_0(\hat{I})$. From (5.28), we have $h_0(I) = \hat{h}_0(\hat{I})$. □

It is noteworthy that there exist commutative BCK-algebras with the relative cancellation property that have no maximal ideal as we have seen in Example 5.2.42.

Theorem 5.2.47 *Let I be an ideal of a commutative BCK-algebra $(X; *, 0)$ with the relative cancellation property. Let $(G(X), h)$ be a universal group for X. Then $(G(X)/h_0(I), h_I)$ is a universal group for X/I, where*

$$h_I(x/I) := h(x)/h_0(I) \quad (x \in X). \tag{5.30}$$

Proof. *Step 1.* According to Theorem 5.2.28, to show that X/I is with the relative cancellation property, it is necessary and sufficient to verify that

$$((x/I) * (y/I)) * ((y/I) * (x/I)) = (x/I) * (y/I).$$

But this property holds from the equality $(x * y) * (y * x) = x * y$ holding for all x, y in any commutative BCK-algebra X with the relative cancellation property.

Step 2. It is well-known that $G(X)/h_0(I)$ is an Abelian lattice ordered group.

We show that h_I from (5.30) is a well-defined mapping on X/I. Let $x/I = y/I$ for some $x, y \in X$. Then $x * y \in I$ and $y * x \in I$, so that $h(x * y) \in h_0(I)$ and $h(y * x) \in h_0(I)$. Hence $h(x) - h(x \wedge y) \in h_0(I)$, and $h(y) - h(x \wedge y) \in h_0(I)$, which give $h(x) - h(y) \in h_0(I)$, i.e., $h(x)/h_0(I) = h(y)/h_0(I)$. In addition, $h_I(x/I)$ is a positive element of $G(X)/h_0(I)$.

Step 3. We show that h_I is injective. Suppose that $h_I(x/I) = h_I(y/I)$. Then $h(x) - h(y) \in h_0(I)$, so that $(h(x) - h(x \wedge y)) - (h(y) - h(x \wedge y)) = h(x * y) - h(y * x) \in$

$h_0(I)$. In any rate, there exist $x_1, \ldots, x_n \in I$ such that $h(x*y) - h(y*x) \leq \sum_{i=1}^n h(x_i)$. Then

$$h(x * y) \leq \sum_{i=1}^{n} h(x_i) + h(y * x).$$

Applying the Riesz interpolation property [Goo 1, Prop 2.2(b)], we can find finitely many elements $g_1, \ldots, g_n, g' \in G^+(X)$ such that $h(x * y) = \sum_{i=1}^n g_i + g'$, where $g_i \leq h(x_i)$ and $g' \leq h(y * x)$. Using Proposition 5.2.36 there are unique elements $x_1', \ldots, x_n', x' \in X$ such that $g_i = h(x_i')$ for $i = 1, \ldots, n$, and $g' = h(x')$. Therefore, $x_1', \ldots, x_n' \in I$ and $x' \leq y * x$, and by Proposition 5.2.37, $x * y = x_1' + \cdots + x_n' + x'$. Since $x' \leq x * y$ and simultaneously $x' \leq y * x$, by Theorem 5.2.28(ii), $x' = 0$, i.e., $x * y = x_1' + \cdots + x_n' \in I$. By symmetry, we conclude $y * x \in I$, which gives $x/I = y/I$.

Step 4. We now prove that h_I is monotone. Let $x/I \leq y/I$. Then $x * y \in I$ which yields $h(x*y) \in h_0(I)$, i.e., $h(x) - h(x \wedge y) \in h_0(I)$, and $h(x)/h_0(I) = h(x \wedge y)/h_0(I) \leq h(y)/h_0(I)$.

In addition, let $h_I(x/I) \leq h_I(y/I)$. Then $h(x)/h_0(I) \leq h(y)/h_0(I)$, so that $h_I((x \wedge y)/I) = h(x \wedge y)/h_0(I) = (h(x) \wedge h(y))/h_0(I) = (h(x)/h_0(I)) \wedge (h(y)/h_0(I)) = h(x)/h_0(I)$.

By Step 3, we have $(x \wedge y)/I = x/I$, i.e., $x/I \leq y/I$.

Step 5. Let now $z/I = (x/I) + (y/I)$, i.e., $z/I \geq x/I$ and $(z/I) * (x/I) = y/I$. Therefore, $x/I = (x \wedge z)/I$. Calculate

$$\begin{aligned} h_I(y/I) &= h_I((z * x)/I) = h(z * x)/h_0(I) \\ &= (h(z) - h(z \wedge x))/h_0(I) = (h(z) - h(x))/h_0(I) \\ &= h_I(z/I) - h_I(x/I), \end{aligned}$$

which gives $h_I((x/I) + (y/I)) = h_I(x/I) + h_I(y/I)$.

Step 6. Suppose now that ϕ is a mapping from X/I into a partially ordered Abelian group G_1 which preserves the BCK-order and existing $+$ in X/I.

Since $(G(X), h)$ is a universal group for X, there is a group-homomorphism $\psi : G(X) \to G_1$ such that $\phi(x/I) = \psi(h(x))$, $x \in X$. Define the mapping $\psi_I : G(X)/h_0(I) \to G_1$ by

$$\psi_I(g/h_0(I)) = \psi(g) \quad (g \in G(X)).$$

We show that ψ_I is defined correctly. So let $g_1 - g_2 \in h_0(I)$, where $g_1, g_2 \in G(X)$. Then there are $x_1, \ldots, x_n \in I$ such that

$$-\sum_{i=1}^{n} h(x_i) \leq g_1 - g_2 \leq \sum_{i=1}^{n} h(x_i).$$

Applying the group-homomorphism ψ to this inequality, we obtain the inequality $-\sum_{i=1}^n \psi(h(x_i)) \leq \psi(g_1) - \psi(g_2) \leq \sum_{i=1}^n \psi(h(x_i))$. Therefore, $0 \leq \psi(g_1) - \psi(g_2) \leq 0$, so that $\psi(g_1) = \psi(g_2)$. It is clear that ψ_I is a group-homomorphism, and $\psi_I \circ h_I = \psi$.

Summarizing Steps 1–6, we see that $(G(X)/h_0(I), h_I)$ is a universal group for X/I. □

We say that an ℓ-ideal K of an Abelian lattice ordered group $(G; +, \leq, 0)$ is a **prime ℓ-ideal** of G iff K is a proper ℓ-ideal of G and G/K is linear. We denote by $\mathcal{P}(G)$ the set of all prime ℓ-ideals of G. A **minimal prime ℓ-ideal** of G is a prime ℓ-ideal of G which does not contain any prime ℓ-ideal different from itself.

Theorem 5.2.48 *Let $(X; *, 0)$ be a commutative BCK-algebra with the relative cancellation property. Let $(G(X), h)$ be a universal group for X. The mapping $P \mapsto h_0(P)$, $P \in \mathcal{P}(X)$, is a one-to-one correspondence from the set of all prime ideals of X onto the set of all prime ℓ-ideals of $G(X)$ preserving the set-theoretical inclusion. The inverse mapping is given by (5.28). In addition, there is a one-to-one correspondence between the set of all minimal prime ideals of X and the set of all minimal prime ℓ-ideals of $G(X)$.*

Proof. Let P be a prime ideal of X. Then X/P is a linear commutative BCK-algebra. By Theorem 5.2.38, $G(X/P)$ is linear, and by Theorem 5.2.47 $G(X/P) \cong G(X)/h_0(P)$ which yields that $G(X)/h_0(P)$ is linear.

Since P is a proper ideal of X, there is a non-zero element $x_0 \notin P$. Then $h(x_0) \notin h_0(P)$. If not, then there exist $x_1, \ldots, x_n \in P$ such that $h(x_0) \leq h(x_1) + \cdots + h(x_n)$. As in the previous proofs, we can find $x'_1, \ldots, x'_n \in I$ such that $x'_i \leq x_i$ and $h(x_0) = h(x'_1) + \cdots + h(x'_n)$, which by Proposition 5.2.37 entails $x_0 = x'_1 + \cdots + x'_n \in I$. So $h_0(P)$ is a proper ℓ-ideal of $G(X)$.

Conversely, suppose now that K is a prime ℓ-ideal of $G(X)$. Define by (5.28) the ideal I_K of X. Then I_K is a proper subset of X, and $h_0(I_K) = K$. Therefore in view of Theorem 5.2.47 $G(X/I_K) \cong G(X)/h_0(I_K) = G(X)/K$ and $G(X/I_K)$ is linear, which by Theorem 5.2.38 proves that X/I_K is linear, so that I_K is a prime ideal of X. □

The following result gives an interesting application of Theorem 5.2.48.

Theorem 5.2.49 *Let $(X; *, 0)$ be a commutative BCK-algebra with the relative cancellation property. Then there is a bounded commutative BCK-algebra $(Y; *, 0)$ such that the space of all prime ideals of X is inclusion-isomorphic to the space of all prime ideals of Y.*

Proof. Let $(G(X), h)$ be a universal group for X. Theorem 5.2.48 says that $\mathcal{P}(X)$ is inclusion-isomorphic to the set of all prime ℓ-ideals of $G(X)$. Due to [CiTo, Cor 2.7], there is an Abelian lattice ordered group G with a strong unit u in G such that $\mathcal{P}(G(X)) \cong \mathcal{P}(G)$ preserving the inclusion. Define the bounded BCK-algebra $Y := \{y \in G^+ : 0 \leq g \leq u\}$. By Exercise 1.5.2.1, the universal group of Y is again G, so that $\mathcal{P}(G) \cong \mathcal{P}(Y)$. Therefore we have $\mathcal{P}(X) \cong \mathcal{P}(G(X)) \cong \mathcal{P}(G) \cong \mathcal{P}(Y)$. □

It is worth recalling that the assertion of Theorem 5.2.49 does not hold for the case of maximal ideals of X, in general, since there is a non-trivial commutative BCK-algebra X with $\mathcal{M}(X) = \emptyset$, while any bounded non-trivial commutative BCK-algebra has at least one maximal ideal.

5.2.11 Quasi Strong Units of BCK-algebras

In the present Section, we introduce quasi strong elements for BCK-algebras. These elements in some sense characterize the whole information involved in X.

An element u of a commutative BCK-algebra $(X; *, 0)$ with the relative cancellation property is said to be a **strong unit** for X if, for any $a \in X$, there is an integer $n \geq 1$ such that $n\,u$ is defined in X and $a \leq n\,u$.

An element u of X is said to be a **quasi strong unit** for X iff $h(u)$ is a strong unit for $G(X)$. It is evident that any strong unit for X is a quasi strong unit; the converse does not hold, in general.

Remark 5.2.50 (1) For $([0, 1); *_{\mathbb{R}}, 0)$, any non-zero element is a quasi strong unit, but there is no strong unit.

(2) If $X = \{0, 1, \ldots, n\}$, then for $(X; *_{\mathbb{R}}, 0)$ any non-zero element is a quasi strong unit. The element i is a strong unit for X iff i divides n.

(3) The BCK-algebra $(\{0, 1, 2\}; *, 0)$, where $1 * 2 = 1 * 0 = 1$, $2 * 1 = 2 * 0 = 2$, $1 * 1 = 2 * 2 = 0 * 0 = 0$ and $0 * 1 = 0 * 2 = 0$, has no quasi strong unit; its underlying universal group is $\mathbb{Z}^2$. But this BCK-algebra can be embedded into a bounded commutative BCK-algebra (= its BCK-hull) with four elements whose underlying universal group is again $\mathbb{Z}^2$.

(4) Example 5.1.13 gives a commutative BCK-algebra, where the element 1 is a quasi strong unit, but X is without the relative cancellation property.

(5) There exists a commutative BCK-algebra with the relative cancellation property which has no quasi strong unit (take, e.g., the positive cone of an Abelian ℓ-group which has no strong unit).

(6) $X = \{(n, m) : 0 \leq n, m \leq 2\} \setminus \{(2, 2)\}$ with $(n_1, m_1) * (n_2, m_2) := (\max(0, n_1 - n_2), \max(0, m_1 - m_2))$ gives an example of a BCK-algebra, where any non-zero element is a quasi strong unit, but X is not a lattice.

Proposition 5.2.51 *An element $u > 0$ of a commutative BCK-algebra $(X; *, 0)$ with the relative cancellation property is a quasi strong unit for X if and only if given $x \in X$ there is an integer $n \geq 0$ such that $x *^n u \leq u$.*

Proof. Suppose that u is a quasi strong unit for X. Given an $x \in X$, there is an integer $n \geq 1$ such that $h(x) \leq nh(u)$. Therefore, $h(x) *_G^{n-1} h(u) \leq (nh(u)) *_G^{n-1} h(u)$ which gives $h(x *^{n-1} u) \leq h(u)$, so that $x *^{n-1} u \leq u$.

Suppose the converse, i.e., given $x \in X$, there is an integer $n \geq 0$ such that $x*^n \leq u$. Using arguments of mathematical induction, we can show that $h(x *^n u) \leq h(u)$ implies $h(x) \leq (n+1)h(u)$, i.e., u is a quasi strong unit for X. □

Lemma 5.2.52 *Let u be a quasi strong unit u for a commutative BCK-algebra $(X; *, 0)$ with the relative cancellation property.*

(i) *An ideal I of X is proper if and only if $u \notin I$.*

(ii) *Let I be a proper ideal of X. Then there is a maximal ideal M of X containing I. In particular, the set of all maximal ideals of X $\mathcal{M}(X)$ is non-void.*

(iii) *$u \notin M$ for any maximal ideal M of X.*

Proof. (i) If $u \notin I$, then I is trivially proper. Suppose now that $u \notin I$. Since $h(u)$ is a strong unit in $G(X)$, then, in view of the Riesz interpolation property, $h(u) \notin h_0(I)$. Therefore, $h_0(I)$ is a proper ℓ-ideal of $G(X)$. Using (5.28), we have that I is also a proper ideal of X.

(ii) According to (i), $u \notin I$. Define $h_0(I)$. Using Zorn's lemma, we can show that there exists an ℓ-ideal K of $G(X)$ containing $h_0(I)$ and which is maximal under the condition $h(u) \notin K$. Let K' be an ℓ-ideal of $G(X)$ containing K. Then either $h(u) \notin K'$, or $h(u) \in K'$. In the first case we obtain that $K' = K$, and in the second one $K' = G(X)$, respectively, which proves that K is a maximal ℓ-ideal of $G(X)$. Theorem 5.2.45 yields that $M := I_K$, where I_K is defined by (5.28), is a maximal ideal of X containing I.

(iii) This follows from (i). □

5.2.12 Exercises

1. Prove: Let $(G; +, \leq, 0)$ be an Abelian lattice ordered group with the positive cone G^+. Let G_0 be a non-empty subset of G^+ generating G^+ such that $u, v \in G_0$ imply $u *_G v := (u - v) \vee 0 \in G_0$. Then $(G_0; *, 0)$ is a commutative BCK-algebra with the relative cancellation property, and a universal group of G_0 is (G, id_{G_0}), where id_{G_0} is the embedding of G_0 into G.

   In particular, if u is a strong unit in G, then, for the bounded commutative BCK-algebra $(G(u); *_G, 0)$, where $G(u) := \{g \in G : 0 \leq g \leq u\}$, $(G, id_{G(u)})$ is its universal group. ([Dvu 15])

2. Using Bosbach's notion of semiclans [Bos 1] show in details that any commutative BCK-algebra with the relative cancellation property can be embedded into the positive cone of an Abelian ℓ-group as its BCK-subalgebra.

3. Show that if X is an algebra of subsets of a set $\Omega \neq \emptyset$, then $(X; \setminus, \emptyset)$ has a universal group $(G(X), h)$, where $G(X)$ consists of all step-functions measurable with respect to X and with integer-valued coefficients, and $h(A) := \chi_A$, $A \in X$.

4. Prove: Let nx and ny exist in a commutative BCK-algebra $(X;*,0)$ with the relative cancellation property. Then $n(x \wedge y)$ is defined in X, and $(nx) \wedge (ny) = n(x \wedge y)$. (Hint: Use the ℓ-group representation.)
5. Show that an element u of X is a quasi strong unit iff $I(u) = X$.
6. Let $X = \bigoplus_i X_i$ be a union algebra with the u.d. union property, where each X_i is a commutative upwards directed BCK-algebra with a universal group (G_i, h_i). Define the direct sum G of the system of groups $\{G(X_i)\}_i$, $G = G(X) := \bigoplus_i G(X_i)$. That is, $G(X)$ is the subset of the product $\prod_i G(X_i)$ consisting of all elements $(w_i)_i$, where $w_i \in X_i$, with finitely many non-zero w_i's, and the addition and the ordering are by coordinates. In addition, $G(X)$ is a lattice ordered group and let $h : X \to G$ is defined via $h(a) := (b_i)_i$, where $b_i = [(a)]$ if $a \in X_i$ and $b_i = [(0)]$ if $a \notin X_i$. Prove that (G, h) is a universal group for X. ([DvGr 1, Thm 5.5])
7. Let $X = \bigoplus_i X_i$ be a union algebra, where each X_i is either directed upwards or a non-zero and unitarily extendible commutative BCK-subalgebra of X (see Example 11 in Section 5.1.9). Let $\tilde{X}_i := X_i$ if X_i is directed upwards and $\tilde{X}_i := X_i \cup \{e_i\}$ when X_i is only unitarily extendible, where e_i is the greatest element in $X_i \cup \{e_i\}$. Embed X into $\hat{X} := \bigoplus_{i\in I} \tilde{X}_i$ and let (G, h) be a universal group for $\tilde{X}$. Show that (G, h_0) is a universal group for X, where h_0 is the restriction of h to X. ([DvGr 1, Thm 6.7])
8. Show: Let $(X;*,0)$ be a bounded positively implicative BCK-algebra with the greatest element 1. Then there are a non-void set Ω, a Boolean algebra $\mathcal{B} \subseteq 2^\Omega$ and a homomorphism h from X onto $\mathcal{B}$.

   Hint: Put $\Omega = \mathcal{M}(X)$ and $\widehat{x} : \mathcal{M}(X) \to \{0, 1\}$ via

   $$\widehat{x}(M) = \begin{cases} 0, & \text{if } x \in M, \\ 1, & \text{if } x \notin M, \end{cases} \quad M \in \mathcal{M}(X).$$

   ([Dvu 21])
9. Show that in the previous example h is injective iff $\bigcap \mathcal{M}(X) = \{0\}$, iff X possesses an order-determining system of states.

5.3 Representation of Commutative BCK-algebras

In Sections 5.2.6 and 5.2.7, we have shown that any commutative BCK-algebra with the relative cancellation property can be embedded into the positive cone of an Abelian lattice ordered group as its BCK-subalgebra. Moreover, we have presented that any our BCK-algebra admits a universal group. In the present section, we give its categorical representation via a special category of Abelian ℓ-groups. Our methods

[Dvu 11] are based on Wyler's approach [Wyl] and give as a particular case the categorical representation of the conical BCK-algebras of Cornish [Cor 1] via ℓ-groups and as well as the famous Mundici [Mun] categorical representation of MV-algebras (= bounded commutative BCK-algebras) via unital ℓ-groups. We note that our method is different from Mundici's approach who used the method of good sequences.

In addition, we introduce also an additional total binary operation called the product on BCK-algebras for bounded as well as unbounded ones, and we give its representation via ℓ-rings as well as its categorical characterization.

5.3.1 Categorical Equivalences of Commutative BCK-algebras

The first kind of categorical representation of a special family of commutative BCK-algebras was done by Cornish [Cor 1] for conical BCK-algebras. We note that a **conical BCK-algebra** is a commutative BCK-algebra X satisfying the condition (S), and, in addition, for all $x, y \in X$,

(1) $(x \circ y) * x = y$;

(2) $(y * (x \wedge y)) \circ (x \wedge y) = y$.

Remark 5.3.1 (i) Condition (1) entails that $x + y$ is defined in X for all $x, y \in X$, and $x + y = x \circ y$.

(ii) Condition (2) is superfluous.

(iii) A commutative BCK-algebra $(X; *, 0)$ is conical if and only if $(X; *, 0)$ is with the relative cancellation property and $x + y$ is defined for all $x, y \in X$.

It is worth recalling that the variety of conical BCK-algebras forms a category which is categorically equivalent to the category of all Abelian ℓ-groups, [Cor 1, Thm 2.1]. In addition, this variety is a proper subvariety of the variety of commutative BCK-algebras, as it follows from Example 5.1.24.

We denote by $\mathcal{BCK}$ the category whose objects are commutative BCK-algebras and morphisms are BCK-homomorphisms.

Let G_1 and G_2 be two Abelian ℓ-groups. A mapping $h : G_1 \to G_2$ is said to be an **ℓ-group homomorphism** iff h is both a group-homomorphism and a lattice-homomorphism; in other words, for each $a, b \in G$, $h(a+b) = h(a) + h(b)$, $h(a \wedge b) = h(a) \wedge h(b)$ (as well as $h(a \vee b) = h(a) \vee h(b)$).

We denote by $\mathcal{LG}$ the category whose objects are pairs (G, G_0), where G is an Abelian ℓ-group and G_0 is a non-void subset of the positive cone G^+ of G such that G_0 generates G^+ and $(G_0; *_G, 0)$ is a BCK-algebra (in fact a BCK-subalgebra of $(G^+; *_G, 0)$, see Examples 5.1.15 and 5.2.4). A morphism from (G, G_0) into (G', G'_0) is an ℓ-group homomorphism $h : G \to G'$ such that $h(G_0) \subseteq G'_0$.

Let now (G, G_0) be an object of $\mathcal{LG}$ and define a morphism $\mathcal{X}$ from the category $\mathcal{LG}$ into the category $\mathcal{BCK}$ as follows

$$\mathcal{X}(G, G_0) = (G_0; *_G, 0), \tag{5.31}$$

where $*_G$ is defined via (5.19), i.e.,

$$u *_G v := 0 \vee (u - v).$$

Let h be a morphism from (G, G_0) into (G', G'_0). We define $\mathcal{X}(h)$ as a mapping from $\mathcal{X}(G, G_0)$ into $\mathcal{X}(G', G'_0)$ via

$$\mathcal{X}(h)(a) := h(a), \; a \in G_0. \tag{5.32}$$

Theorem 5.3.2 *$\mathcal{X}$ is a faithful and full functor from the category $\mathcal{LG}$ of Abelian ℓ-groups into the category $\mathcal{BCK}$ of commutative BCK-algebras with the relative cancellation property.*

Proof. Let h be a morphism from (G, G_0) into (G', G'_0), that is, h is an ℓ-group homomorphism from G into G' such that $h(G_0) \subseteq G'_0$. Then, for all $u, v \in G_0$, we have

$$\begin{aligned}\mathcal{X}(h)(u *_G v) &= h(u *_G v) = h(u - (u \wedge v)) = h(u) - h(u \wedge v) \\ &= h(u) - (h(u) \wedge h(v)) = h(u) *_{G'} h(v),\end{aligned}$$

which proves that $\mathcal{X}(h)$ is a morphism from $hom(\mathcal{X}(G, G_0), \mathcal{X}(G', G'_0))$. It is easy to verify that if h' is a morphism from (G', G'_0) into (G'', G''_0), then $\mathcal{X}(h' \circ h) = \mathcal{X}(h') \circ \mathcal{X}(h)$, and $\mathcal{X}(1_{(G,G'_0)}) = 1_{\mathcal{X}(G,G'_0)}$, which proves that $\mathcal{X}$ is a functor.

Let now h_1 and h_2 be two morphisms from (G, G_0) into (G', G'_0) such that $\mathcal{X}(h_1) = \mathcal{X}(h_2)$. Then $h_1(u) = h_2(u)$ for any $u \in G_0$. Since G_0 generates G^+, it is clear that $h_1(g) = h_2(g)$ for any $g \in G$, which proves that $\mathcal{X}$ is faithful.

To prove that $\mathcal{X}$ is a full functor, suppose that f is a morphism from $\mathcal{X}(G, G_0)$ into $\mathcal{X}(G', G'_0)$, i.e., f is a BCK-homomorphism from $(G_0; *_G, 0)$ into $(G'_0; *_{G'}, 0)$. Then f preserves $+$ and meets $\wedge$ in G_0, moreover, the "BCK"-plus $+$ coincides with the original group $+$. Since G_0 generates G^+, using the Riesz decomposition property, Lemma 5.2.12, f can be uniquely extended to a group-homomorphism $\hat{f} : G \to G'$.

We claim that $\hat{f}$ is a lattice-homomorphism. The proof will proceed in several steps.

Step 1. Let $a, b, u \in G^+$. If $\hat{f}(a \wedge b) = \hat{f}(a) \wedge \hat{f}(b)$ and if $\hat{f}(u \wedge (b - (a \wedge b))) = \hat{f}(u) \wedge \hat{f}(b - (a \wedge b))$, then

$$\hat{f}((a + u) \wedge b) = \hat{f}(a + u) \wedge \hat{f}(b).$$

Indeed, we have $(a \wedge b) + [u \wedge (b - (a \wedge b))] = ((a \wedge b) + u) \wedge b = (a + u) \wedge (b + u) \wedge b = (a + u) \wedge b$.

Therefore

$$\begin{aligned} \hat{f}((a+u) \wedge b) &= \hat{f}(a \wedge b) + \hat{f}(u \wedge (b - (a \wedge b))) \\ &= (\hat{f}(a) \wedge \hat{f}(b)) + [\hat{f}(u) \wedge (\hat{f}(b) - (\hat{f}(a) \wedge \hat{f}(b)))] \\ &= [((\hat{f}(a) \wedge \hat{f}(b)) + \hat{f}(u)] \wedge \hat{f}(b) \\ &= (\hat{f}(a) + \hat{f}(u)) \wedge (\hat{f}(b) + \hat{f}(u)) \wedge \hat{f}(b) = \hat{f}(a+u) \wedge \hat{f}(b). \end{aligned}$$

Step 2. Let $v_1, \ldots, v_k, a \in G_0$ satisfy $v_1 + \cdots + v_k \leq a$. Then $v := v_1 + \cdots + v_k$ is defined in G_0. Indeed, there exist $c_1, \ldots, c_j \in G_0$ such that $v_1 + \cdots + v_k + c_1 + \cdots + c_j = a$. Applying j times the operation $*_G$ successively to this equality and to the elements $c_j, \ldots, c_1$ and using the associativity of the "BCK"-plus in G_0, we see that v is defined in G_0.

Step 3. $\hat{f}(a \wedge b) = \hat{f}(a) \wedge \hat{f}(b)$ whenever $a \in G^+$ and $b \in G_0$.

Since G_0 is generating for G^+, a is of the form $a = a_1 + \cdots + a_n$ for some $a_1, \ldots, a_n \in G_0$. The proof will follow mathematical induction on n.

If $n = 1$, the statement is trivial. Suppose now that the statement holds for any $a' = a_1 + \cdots + a_i$ with $1 \leq i \leq n$. Put $a = a_1 + \cdots + a_n$, $u = a_{n+1}$. Then there exist $v_1, \ldots, v_k \in G_0$ such that $b = (v_1 + \cdots + v_k) + (a \wedge b)$. Since $v := v_1 + \cdots + v_k \leq b \in G_0$, by Step 2, $v \in G_0$. Hence $v = b - (a \wedge b)$. Since $\hat{f}$ preserves meets in G_0, we have $\hat{f}(u \wedge v) = \hat{f}(u) \wedge \hat{f}(v)$, so that $\hat{f}(u \wedge (b - (a \wedge b))) = \hat{f}(u) \wedge \hat{f}(b - (a \wedge b)) = \hat{f}(u) \wedge (\hat{f}(b) - (\hat{f}(a) \wedge \hat{f}(b)))$, where we have used induction hypothesis for v. By Step 1, $\hat{f}((a+u) \wedge b) = \hat{f}(a+u) \wedge \hat{f}(b)$, i.e., $\hat{f}((a_1 + \cdots + a_{n+1}) \wedge b) = \hat{f}(a_1 + \cdots + a_{n+1}) \wedge \hat{f}(b)$ for any n.

Step 4. $\hat{f}(a \wedge b) = \hat{f}(a) \wedge \hat{f}(b)$ whenever $a, b \in G^+$.

Let $a = a_1 + \cdots + a_n$, $b = b_1 + \cdots + b_k$. The proof will follow complete induction on k.

If $k = 1$, we apply Step 3. Suppose now that the assertion holds for any j with $1 \leq j \leq k$. Put $B = a$, $A = b_1 + \cdots + b_k$, $u = b_{k+1}$. By Step 3, $\hat{f}(u \wedge (B - (A \wedge B))) = \hat{f}(u) \wedge \hat{f}(B - (A \wedge B))$ and $\hat{f}(A \wedge B) = \hat{f}(A) \wedge \hat{f}(B)$. Therefore the conditions of Step 1 are satisfied, so that $\hat{f}((A + u) \wedge B) = \hat{f}(A + u) \wedge \hat{f}(B)$ which proves $\hat{f}((a_1 + \cdots + a_n) \wedge (b_1 + \cdots + b_{k+1})) = \hat{f}(a_1 + \cdots + a_n) \wedge \hat{f}(b_1 + \cdots + b_{k+1})$ for each n and each k.

Step 5. $\hat{f}(a \wedge b) = \hat{f}(a) \wedge \hat{f}(b)$ whenever $a, b \in G$. There exist $a', a'', b', b'' \in G^+$ such that $a = a' - a''$ and $b = b' - b''$. By Step 4, $\hat{f}((a' + b'') \wedge (b' + a'')) = \hat{f}(a' + b'') \wedge \hat{f}(b' + a'')$. Subtracting $\hat{f}(b'')$ and $\hat{f}(a'')$ we obtain the assertion in question.

Consequently the extension $\hat{f}$ is an ℓ-group homomorphism, and $\hat{f}$ is a morphism from (G, G_0) into (G', G'_0) such that $\mathcal{X}(\hat{f}) = f$, which proves that the functor $\mathcal{X}$ is full. ☐

Theorem 5.3.3 *The functor $\mathcal{X}$ from the category $\mathcal{LG}$ into the category $\mathcal{BCK}$ is a right-adjoint.*

Proof. We show that given a BCK-algebra $(X;*,0)$ with the relative cancellation property there is a universal arrow $((G,G_0),f)$, i.e., f is a BCK-homomorphism from X into $\mathcal{X}(G,G_0)$ such that if (G',G'_0) is an object from $\mathcal{LG}$ and f' is a BCK-homomorphism from X into $\mathcal{X}(G',G'_0)$, then there is a unique morphism $f^*:(G,G_0)\to(G',G'_0)$ such that $\mathcal{X}(f^*)\circ f=f'$. We claim that $((G(X),h(X),h))$ is a universal arrow for X.

Let $(G(X),h)$ be a universal group for $(X;*,0)$ guaranteed by Theorem 5.2.30. Then $(G(X),h(X))$ is an object of $\mathcal{LG}$. Since $h:X\to h(X)$ and h is injective and preserves meets in X, h is a BCK-homomorphism from X into $(h(X);*_{G(X)},0)$.

Let now (G',G'_0) be an object in $\mathcal{LG}$ and let f' be a morphism from X into $\mathcal{X}(G',G'_0)$. Then f' maps X into $G'_0\subseteq G'$, so that according to Theorem 5.2.30, there is a unique group-homomorphism $f^*:G(X)\to G'$ such that $f^*\circ h=f'$.

We claim that f^* preserves the meets in $G(X)$.

Step 1. Let $x,y\in X$, then $f^*(h(x)\wedge h(y))=f^*(h(x\wedge y))=f'(x\wedge y)=f'(x)\wedge f'(y)=f^*(h(x))\wedge f^*(h(y))$.

Step 2. Put for a moment $G=G(X)$, $G_0=h(X)$, $\hat{f}=f^*$, and use all Steps 1–5 from the proof of Theorem 5.3.2. This proves that f^* is a lattice-homomorphism.

Therefore f^* is a unique morphism from $(G(X),h(X))$ into (G',G'_0) satisfying $\mathcal{X}(f^*)\circ h=f'$ which proves that $\mathcal{X}$ is a (right) adjoint. □

Define a morphism $\mathcal{G}:\mathcal{BCK}\to\mathcal{LG}$ via $\mathcal{G}(X):=(G(X),h(X))$, where $(G(X),h)$ is a universal group. From the construction of $\mathcal{G}$ and from the proof of Theorem 5.3.3 we conclude the following result.

Theorem 5.3.4 *$\mathcal{G}$ is a functor from the category $\mathcal{BCK}$ into $\mathcal{LG}$ which is left-adjoint to $\mathcal{X}$.*

Theorem 5.3.5 *$\mathcal{X}$ is a categorical equivalence of the category $\mathcal{LG}$ of Abelian ℓ-groups and of the category $\mathcal{BCK}$ of commutative BCK-algebras with the relative cancellation property.*

Proof. According to [MaL, Thm IV.4.1], it is necessary to show that for any BCK-algebra $(X;*,0)$ with the relative cancellation property there is an object (G,G_0) in $\mathcal{LG}$ such that $\mathcal{X}(G,G_0)$ is isomorphic to X. To show that it is sufficient to put $(G,G_0)=(G(X),h(X))$, where $(G(X),h)$ is a universal group for X. By Theorem 5.2.30, h is an isomorphism in question. □

As a corollary we have that a genuine addition can be uniquely recovered from BCK-structure of commutative BCK-algebras with the relative cancellation property.

The category of conical BCK-algebras can be in view of Remark 5.3.1 and Theorem 5.3.5 characterized as follows. Let $\mathcal{LG}_0$ consist of all objects (G, G^+) of $\mathcal{LG}$ and of all morphisms from (G, G^+) into (H, H^+), $(H, H^+) \in \mathcal{LG}$ and let $\mathcal{CBCK}$ consist of all conical commutative BCK-algebras. Because any BCK-homomorphism of conical BCK-algebras preserves $+$, we see that $\mathcal{LG}_0$ is a subcategory of $\mathcal{LG}$ and $\mathcal{CBCK}$ is a subcategory of $\mathcal{BCK}$. As a direct consequence of Theorem 5.2.30 we have the Cornish result [Cor 1, Thm 2.1], because $\mathcal{LG}_0$ is isomorphic to the category of Abelian ℓ-groups, where objects are Abelian ℓ-groups G and morphisms are ℓ-group homomorphisms of ℓ-groups.

Corollary 5.3.6 *The restriction of $\mathcal{X}$ on $\mathcal{LG}_0$ defines a categorical equivalence of Abelian ℓ-groups $\mathcal{LG}_0$ and the category of conical BCK-algebras.*

Proof. Take into account that if $(G(X), h)$ is a universal group for a conical BCK-algebra X, then $h(X) = G^+(X)$. □

5.3.2 Categorical Representation of Bounded Commutative BCK-algebras

Using methods developed in the previous Section, we can present another proof of a celebrated result of Mundici [Mun] on the categorical equivalence of the category of bounded commutative BCK-algebras (= MV-algebras) and unital ℓ-groups.

For sake of brevity, for a bounded BCK-algebra $(X; *, 0)$ with the greatest element 1 we shall write $(X; *, 0, 1)$.

So let $\mathcal{BBCK}$ be the category of bounded commutative BCK-algebras whose objects are bounded commutative BCK-algebras $(X; *, 0, 1)$ and a morphism from $(X; *, 0, 1)$ into $(X_1; *_1, 0_1, 1_1)$ is any morphisms f from $(X; *, 0)$ into $(X_1; *_1, 0_1)$ such that $f(1) = 1_1$. The $\mathcal{BBCK}$ is a subcategory of $\mathcal{BCK}$.

We note that a **strong unit** for an Abelian ℓ-group G is an element $u \in G^+$ such that, for any $g \in G$, there exists an integer $n \geq 1$ with $-nu \leq g \leq nu$. We define

$$[0, u]_G := \{g \in G : 0 \leq g \leq u\}.$$

Then $([0, u]_G; *_G, 0, u)$ is a bounded commutative BCK-algebra. By $\mathcal{UG}$ we denote the category of unital Abelian ℓ-groups, whose objects are pairs $(G; [0, u]_G)$, where u is a strong unit for G, and a morphism $h : (G, [0, u]_G) \to (H; [0, v]_H)$ is an ℓ-group homomorphism $h : G \to H$ such that $h(u) = v$. Then $\mathcal{UG}$ is a subcategory of $\mathcal{LG}$.

We define the restriction $\mathcal{X}_b$ of $\mathcal{X}$ as a morphism from $\mathcal{UG}$ into $\mathcal{BBCK}$ as follows

$$\mathcal{X}_b(G, [0, u]_G) := ([0, u]_G; *_G, 0)$$

and

$$\mathcal{X}_b(h)(a) = h(a), \ a \in [0, u]_G.$$

Theorem 5.3.7 *The morphism $\mathcal{X}_b$ defines a categorical equivalence of the category $\mathcal{UG}$ and the category $\mathcal{BBCK}$.*

Proof. It is easy to verify that $\mathcal{X}_b$ is a faithful and full functor.

Let now $(X; *, 0, 1)$ be a bounded commutative BCK-algebra and let $(G(X), h)$ be its universal group. Then $h(X) = [0, h(1)]_{G(X)}$ and $h(1)$ is a strong unit for $G(X)$. Modifying arguments of the proofs from the previous sections, we can prove the assertion in question. □

We note that the category $\mathcal{MV}$ of MV-algebras has MV-algebras as objects and morphisms are mappings preserving $\oplus, \odot, ^*, 0$, and 1. The category of Abelian unital ℓ-groups consists of objects which are pairs (G, u), where u is a strong unit for the ℓ-group G, and of morphisms which are all ℓ-group homomorphisms preserving fixed strong units.[11] As a corollary of Theorem 5.3.7 we have the following fundamental result of Mundici.

Corollary 5.3.8 (Mundici [Mun]) *The category of MV-algebras is categorically equivalent with the category of unital Abelian ℓ-groups. The equivalence is defined by* **Mundici's functor** Γ *which is defined via*

$$\Gamma(G, u) := ([0, u]_G; \oplus_G, \odot_G, ^{*_G}, 0, u), \tag{5.33}$$

where, for $a, b \in [0, u]_G = \{g \in G : 0 \leq g \leq u\}$, we have

$$\begin{aligned} a \oplus_G b &:= u \wedge (a + b), \\ a \odot_G b &:= 0 \vee (a + b - u), \\ a^{*_G} &:= u - a, \end{aligned}$$

and Γ defines a categorical equivalence in question.

We note that Mundici's main tool were good sequences, and this notion was not possible to be applied for BCK-algebras.

5.3.3 Categorical Representation of Commutative BCK-algebras with Quasi Strong Unit

We note that an element u of a commutative BCK-algebra $(X; *, 0)$ is said to be a quasi strong unit for X if, for each element $x \in X$, there exists an integer $n \geq 1$ such that $x *^n u = 0$.

[11] We note that two definitions of unital Abelian ℓ-groups, one as $\mathcal{UG}$ and the second one having objects pairs (G, u), are only formally different.

Proposition 5.3.9 *Let $(G(X), h)$ be a universal group for a commutative BCK-algebra $(X; *, 0)$ with the relative cancellation property. An element $u \in X$ is a quasi strong unit for X if and only if $h(u)$ is a strong unit for $G(X)$.*

Proof. Suppose that u is a quasi strong unit for X. Then given $x \in X$, there is an integer $n \geq 1$ such that $x *^n u = 0$. Using arguments of mathematical induction, we can prove that $h(x) \leq nh(u)$. Since $h(X)$ generates $G(X)$, $h(u)$ is a strong unit for $G(X)$.

Conversely, let $h(u)$ be a strong unit in $G(X)$. Given $x \in X$, there is an integer $n \geq 1$ such that $h(x) \leq nh(u)$. Then $0 \leq h(x) *_G^n h(u) = h(x *^n u) = 0$, which gives $x *^n u = 0$ because h is injective. □

Define a category $\mathcal{ULG}$ whose objects are triplets (G, G_0, u) such that (G, G_0) is an object of $\mathcal{LG}$ and u is a strong unit for G belonging to G_0. A morphism from (G, G_0, u) into (H, H_0, v) is any ℓ-group homomorphism $h : G \to H$ such that $h(G_0) \subseteq H_0$ and $h(u) = v$.

The category $\mathcal{BCKQ}$ consists of objects of the form $((X; *, 0), u)$, where $(X; *, 0)$ is a commutative BCK-algebra with the relative cancellation property with a fixed quasi strong unit $u \in X$, and a morphism from $((X; *, 0), u)$ into $((Y; *, 0), v)$ is any BCK-homomorphism $f : X \to Y$ such that $f(u) = v$.

It is clear that $\mathcal{BCKQ}$ can be understood as a subcategory of $\mathcal{BCK}$, and $\mathcal{BBCK}$ is a subcategory of $\mathcal{BCKQ}$.

Define a morphism $\mathcal{X}_q : \mathcal{ULG} \to \mathcal{BCKQ}$ as follows

$$\mathcal{X}_q((G, G_0, u)) = ((G_0; *_G, 0), u)$$

and

$$\mathcal{X}_q(f) = f|G_0,$$

if f is a morphism from (G, G_0, u) into (H, H_0, v).

Theorem 5.3.10 *$\mathcal{X}_q$ defines a categorical equivalence of $\mathcal{ULG}$ and $\mathcal{BCKQ}$.*

5.3.4 Product on MV-algebras

Let $M = [0, 1]^\Omega$ be the system of all functions on Ω with values in the interval $[0, 1]$. Then M can be converted into an MV-algebra with $0 = 0_\Omega$, $1 = 1_\Omega$ (constant functions equal to 0 and 1, respectively), and $(f \oplus g)(\omega) = \min\{f(\omega) + g(\omega), 1\}$, $\omega \in \Omega$, and $f^* := 1 - f$. Then M, viewed as a bounded BCK-algebra, has the derived partial addition $+$ defined $f + g = f \oplus g$ iff $f + g \leq 1$. Then M is closed under pointwise multiplication $f \cdot g \in M$.

Motivated by this we will introduce product MV-algebras, which were studied in [DvRi 2], [Rie 5], [Rie 6], [Jak 11]. Here we will follows ideas from [DiDv].

We note that a non-void subset I of an MV-algebra M is said to be an **ideal** of M if

(i) $x, y \in I$ imply $x \oplus y \in I$;

(ii) $x \in I$, $y \leq x$ imply $y \in I$.

We note that the notions of ideals of the MV-algebra M and of ideals for the bounded BCK-algebra $(M; *, 0, 1)$ (see Example 5.1.14) coincide (see Exercise 15.3.7). In the usual way we define maximal ideals of M. Since 1 is a quasi strong unit, by Lemma 5.2.52(ii) the set of all maximal ideals $\mathcal{M}(M)$ on M is non-void.

Denote by

$$Rad(M) := \bigcap \{A : \ A \in \mathcal{M}(M)\},$$

and we call $Rad(M)$ the **radical** of M. A non-zero element a of M is said to be **infinitesimal** if na exists in M for any integer $n \geq 1$. The set of all infinitesimals in M is denoted by $Infinit(M)$. Then according to [CDM, Prop 3.6.4], we have

$$Rad(M) = Infinit(M) \cup \{0\}.$$

An MV-algebra M is said to be (i) **semisimple** if $Rad(M) = \{0\}$, (ii) **perfect** if $M = Rad(M) \cup Rad(M)^*$, where $Rad(M)^* := \{a^* : \ a \in Rad(M)\}$. In [DiLe], it is shown that the category of perfect MV-algebras is categorically equivalent to the category of all Abelian (not necessarily unital) ℓ-groups.

We note that the partial addition $+$ in M which we can derived as from a bounded commutative BCK-algebra is equivalently defined: $a + b$ is defined in M iff $a \leq b^*$, and in such the case

$$a + b := a \oplus b. \tag{5.34}$$

Proposition 5.3.11 *Let $(M; \oplus, ^*, 0, 1)$ be an MV-algebra. If $a, b \in Rad(M)$, then $a + b$ exists in M, and $a + b \in Rad(M)$.*

Proof Let $a, b \in Rad(M)$. Since any maximal ideal is closed under $\oplus$, we have that $a \oplus b \in A$ for any maximal ideal A of M. Then $a \oplus b \in Rad(M)$, so that by the above, $a \oplus b$ is infinitesimal. Hence $a \leq a \oplus b \leq (a \oplus b)^* = a^* \odot b^* \leq b^*$, which proves that $a + b$ is defined in M. □

Definition 5.3.12 *We say that an MV-algebra $(M; \oplus, ^*, 0, 1)$ admits a* **product** *if there is a binary operation $\cdot$ on M satisfying for all $a, b, c \in M$*

(i) *if $a + b$ is defined in M, then $a \cdot c + b \cdot c$ and $c \cdot a + c \cdot b$ exist and*

$$\begin{aligned} (a + b) \cdot c &= a \cdot c + b \cdot c, \\ c \cdot (a + b) &= c \cdot a + c \cdot b; \end{aligned}$$

(ii) $(a \cdot b) \cdot c = a \cdot (b \cdot c)$,

and we say that M is a **product MV-algebra**, *and we write* $(M; \oplus,^*, \cdot, 0, 1)$.

It is worth saying that if $\cdot$ is a product on M, then

(iii) $a \cdot 0 = 0 = 0 \cdot a$;

(iv) if $a \leq b$, then for any $c \in M$, $a \cdot c \leq b \cdot c$ and $c \cdot a \leq c \cdot b$.

Property (iii) follows easily from the following. $a \cdot 0 = a \cdot (0+0) = a \cdot 0 + a \cdot 0$, and the cancellation property gives $a \cdot 0 = 0$. Similarly, $0 \cdot a = 0$.

Theorem 5.3.13 *The class of product MV-algebras is equationally definable.*

Proof. The class of MV-algebras is equationally definable. The condition (i) is equivalent to the identities, for all $c, x, y \in M$,

$$\begin{aligned} c \cdot (x \odot (x \wedge y)^*) &= (c \cdot x) \odot (c \cdot (x \wedge y))^*, \\ (x \odot (x \wedge y)^*) \cdot c &= (x \cdot c) \odot ((x \wedge y) \cdot c)^*. \end{aligned}$$

Hence the class of product MV-algebras is equationally definable. □

We note that an **ℓ-ring** is a usual ring $(R; +, \cdot, 0)$ with a partial order $\leq$ such that $(R; +, 0, \leq)$ is an ℓ-group and, if $0 \leq a$ and $0 \leq b$, then $0 \leq a \cdot b$. A **function ring** or an **f-ring** is an ℓ-ring such that $a \wedge b = 0$ and $0 \leq c$ imply $(c \cdot a) \wedge b = (a \cdot c) \wedge b = 0$. An ℓ-ring R is Archimedean if $(R; +, 0, \leq)$ is Archimedean. For an ℓ-ring R we shall write $R = (R; +, \cdot, 0, \leq)$.

We note that if unity 1 of an ℓ-ring R is a strong unit, then R is an f-ring, [Bir, Lem XVII.5.2], and any Archimedean f-ring is commutative and associative, [Bir, Thm XVII.10]. More about ℓ-rings see, e.g., [Bir] or [Fuc].

Theorem 5.3.14 *Let $(R; +, \cdot, 0, \leq)$ be an associative ℓ-ring with a strong unit u such that $u \cdot u \leq u$. Then $M = \Gamma(R, u)$ is an MV-algebra with a product $\cdot$, which is the restriction of $\cdot$ to $M \times M$.*

Conversely, let $(M; \oplus,^*, 0, 1)$ be an MV-algebra with a product $\cdot$. Then there exists (up to isomorphism) a unique associative ℓ-ring $(R; +, \cdot, 0, \leq)$ with a strong unit u such that $u \cdot u \leq u$, $M \cong \Gamma(R, u)$, and $\phi(a \cdot b) = \phi(a) \cdot \phi(b)$, $a, b \in M$, where ϕ is an MV-isomorphism of M onto $\Gamma(R, u)$.

Proof. (1) Let $M = \Gamma(R,u)$. For $a, b \in M$ we have $a \le u$, $b \le u$, so that $a \cdot b \le u \cdot u \le u$, which says that the restriction of $\cdot$ to $M \times M$ defines a product on the MV-algebra.

(2) Let M be an MV-algebra with a product $\cdot$. According to Mundici [Mun] (see also Corollary 5.3.8), there is an ℓ-group $(G; +, 0, \le)$ with a strong unit u and an MV-isomorphism ϕ from M onto $\Gamma(G, u)$. We can define the product $\cdot$ on $\Gamma(G,u)$ as follows

$$\phi(a) \cdot \phi(b) := \phi(a \cdot b), \ a, b \in M.$$

Because $[0, u]_G$ generates the positive cone G^+ of G, ϕ preserves all existing $+$ in M, we see that $\cdot$ is a product on $\Gamma(G, u)$.

Given $g \in G^+$ there exist $a_1, \ldots, a_n \in M$ such that $g = \sum_{i=1}^n \phi(a_i)$. For any $\phi(c)$, where $c \in M$, we define

$$g \cdot \phi(c) = \phi(a_1 \cdot c) + \cdots + \phi(a_n \cdot c). \tag{A}$$

We claim that (A) is defined unambiguously. Indeed, if $g = \sum_{j=1}^m \phi(b_j)$, for some $b_1, \ldots, b_m \in M$, due to the Riesz decomposition property holding on ℓ-groups [Goo 1, Prop 2.2], there exist elements $c_{ij} \in M$ such that $a_i = \sum_{j=1}^m c_{ij}$ and $b_j = \sum_{i=1}^n c_{ij}$ for all i, $1 \le i \le n$, and all j, $1 \le j \le m$. Then

$$\begin{aligned}
\sum_{i=1}^n \phi(a_i \cdot c) &= \sum_{i=1}^n \phi((\sum_{j=1}^m c_{ij}) \cdot c) = \sum_{i=1}^n \phi(\sum_{j=1}^m (c_{ij} \cdot c)) \\
&= \sum_{i=1}^n \sum_{j=1}^m \phi(c_{ij} \cdot c) = \sum_{j=1}^m \sum_{i=1}^n \phi(c_{ij} \cdot c) \\
&= \sum_{j=1}^m \phi(\sum_{i=1}^n c_{ij} \cdot c) = \sum_{j=1}^m \phi((\sum_{i=1}^n c_{ij}) \cdot c) \\
&= \sum_{j=1}^m \phi(b_j \cdot c),
\end{aligned}$$

which proves that $\cdot$ on $G^+ \times \Gamma(G, u)$ is correct. We now extend $\cdot$ to $G \times \Gamma(G, u)$ as follows: If $g = g_1 - g_2$, $g_1, g_2 \in G^+$, then

$$g \cdot \phi(c) := g_1 \cdot \phi(c) - g_2 \cdot \phi(c).$$

Since if $g_1 - g_2 = h_1 - h_2$ for $g_i, h_i \in G^+$, $i = 1, 2$, then $g_1 + h_2 = h_1 + g_2$, by (A) we have

$$\begin{aligned}
(g_1 + h_2) \cdot \phi(c) &= (h_1 + g_2) \cdot \phi(c), \\
g_1 \cdot \phi(c) + h_2 \cdot \phi(c) &= h_1 \cdot \phi(c) + g_2 \cdot \phi(c), \\
g_1 \cdot \phi(c) - g_2 \cdot \phi(c) &= h_1 \cdot \phi(c) - h_2 \cdot \phi(c).
\end{aligned}$$

Let now $c \in G^+$. Then $c = \phi(c_1) + \cdots + \phi(c_s)$, where $c_t \in M$, $t = 1, \ldots, s$. We extend $\cdot$ to $G \times G^+$ as follows

$$g \cdot c := \sum_{t=1}^{s} g \cdot \phi(c_t),\ g \in G.$$

If $c = \sum_{w=1}^{v} \phi(d_w)$, using the Riesz decomposition property, we have as above

$$\sum_{t=1}^{s} g \cdot \phi(c_t) = \sum_{w=1}^{v} g \cdot \phi(d_w).$$

It is clear that the "multiplication" $\cdot$ can be extended to the whole $G \times G$: If $c = c_1 - c_2$, where $c_1, c_2 \in G^+$, let

$$g \cdot c := g \cdot c_1 - g \cdot c_2.$$

It is evident that if $g, h \in G^+$, then $g \cdot h \in G^+$, and $u \cdot u \leq u$, and due to (ii), $\cdot$ is associative on G, so that $(G; +, \cdot, 0, \leq)$ is an ℓ-ring with a strong unit u, which proves Theorem. □

An MV-algebra M is said to be **Archimedean** if existence of $na = a_1 + \cdots + a_n$, where $a_1 = \cdots = a_n = a$ for any $n \geq 1$ implies $a = 0$. Due to [Dvu 10] (or Exercise 2.5.3.7), an MV-algebra M is Archimedean iff its representation group (G, u) is an Archimedean ℓ-group. In addition, M is Archimedean iff M is semisimple [Dvu 10].

Theorem 5.3.15 *Let $(M; \oplus, ^*, \cdot, 0, 1)$ be an MV-algebra with a product $\cdot$, and let ϕ and $(R; +, \cdot, 0, \leq)$ be as in Theorem* 5.3.14.

(1) *If $a \cdot b = b \cdot a$, for all $a, b \in M$, then $(R; +, \cdot, 0, \leq)$ with strong unit* 1 *is a commutative ℓ-ring.*

(2) *If, for any $a \in M$, we have $a \cdot 1 = 1 \cdot a = a$, then $(R; +, \cdot, 0, \leq)$ is an f-ring with strong unit $\phi(1)$ which is unity in R.*

(3) *If M is semisimple and $a \cdot 1 = 1 \cdot a = a$ for any $a \in M$, then $(R; +, \cdot, 0, \leq)$ is a commutative f-ring.*

(4) *If there is an element $e \in M$ such that $e \cdot a = a \cdot e = a$ for any $a \in M$, then $\phi(e)$ is unity in the ℓ-ring $(R; +, \cdot, 0, \leq)$.*

Proof. (1) This follows from the construction of $(R; +, \cdot, 0)$ and from the proof of Theorem 5.3.14.

(2) Since $a \cdot 1 = 1 \cdot a = a$ for any $a \in M$, it is possible to show that $u = \phi(1)$ from the proof of Theorem 5.3.14 is a unity in R. From [Bir, Lem XVII.5.2] we conclude that R is an f-ring.

(3) According to (2), R is an f-ring. Due to the semisimplicity of M we infer that $(R;+,0)$ is an Archimedean ℓ-group, and by [Bir, Thm XVII.10], the multiplication $\cdot$ on R is commutative.

(4) This is evident. □

To illustrate our notions, we investigate the possibility of the existence of a product on finite MV-algebras. It is evident that every MV-algebra under the trivial product, i.e., $a \cdot b = 0$ for all $a, b \in M$, is a product MV-algebra, called a **zero-product MV-algebra**.

Let a be an element of M. If there exists a greatest integer n such that $na := a + \cdots + a$ is defined in M,

$$ord(a) := n;$$

if na is defined in M for any integer $n \geq 1$, we put

$$ord(a) = \infty.$$

A non-zero element a of M is said to be an **atom** of M if $b \leq a$, $b \in M$, imply $b = 0$ or $b = a$. An MV-algebra M is said to be **atomic** if given a non-zero element b of M there is an atom a in M such that $a \leq b$.

An element e of a product MV-algebra M is said to be a **unity** if $a \cdot e = e \cdot a = a$ for any $a \in M$.

Proposition 5.3.16 *If e is unity of a product MV-algebra M, then $e = 1$. Moreover, 1 is unity for $\cdot$ if and only if, for any $a \in M$, $a \cdot 1 \geq a$ and $1 \cdot a \geq a$.*

Proof. Let e be unity for a product $\cdot$. Then $1 \cdot 1 \leq 1 = 1 \cdot e \leq 1 \cdot 1$ which proves $1 \cdot 1 = 1$. Let $a \in M$. Then $a \cdot 1 \geq a \cdot e = a$ and $1 \cdot a \geq e \cdot a = a$. In addition, $a + a^* = 1 = 1 \cdot 1 = (a + a^*) \cdot 1 = a \cdot 1 + a^* \cdot 1 \geq a + a^*$, which entails $a \cdot 1 = a$. By symmetry, we have $1 \cdot a = a$ for any $a \in M$, and in addition $1 = 1 \cdot e = e$.

Assume now that for any $a \in M$, $a \cdot 1 \geq a$ and $1 \cdot a \geq a$. Then $1 = a + a^* \leq a \cdot 1 + a^* \cdot 1 = (a + a^*) \cdot 1 = 1 \cdot 1 \leq 1$ which proves that 1 is unity for $\cdot$ on M. □

Theorem 5.3.17 *A finite MV-algebra $(M; \oplus, ^*, 0, 1)$ admits a product $\cdot$ such that $a \cdot 1 = a = 1 \cdot a$ for any $a \in M$ if and only if $(M; \oplus, ^*, 0, 1)$ is a Boolean algebra, i.e., $a \oplus a = a$ for any $a \in M$. If this is the case, then $a \cdot b = a \wedge b \in M$.*

Proof. Suppose that M is Boolean and define $a \cdot b := a \odot b$ for all $a, b \in M$. Then is a product in question.

Conversely, assume that the product $\cdot$ satisfies $a \cdot 1 = a = 1 \cdot a$ for any $a \in M$. The finiteness of M yields that M is atomic. If a and b are two different atoms, then $(na) \wedge (mb) = 0$ whenever na and mb are defined in M. Therefore $((na) \vee$

$(mb)) - (na) = (mb) - ((na) \wedge (mb)) = (mb)$ so that $na + mb$ is defined in M and $na + mb = (na) \vee (mb)$.

Given an atom a, let $j_a(x)$ denote the greatest integer n such that na is defined in M and $na \leq x$. Since M is finite, $j_a(x)$ is finite for any atom a and for any element x, and the element x of M can be uniquely expressed in the form

$$x = j_{a_1}(x)a_1 + \cdots + j_{a_k}(x)a_k,$$

where $a_1, \ldots, a_k$ denote the set of all atoms in M.

Assume again that a and b are two different atoms. Hence $a \cdot b \leq a \cdot 1 = a$ and $a \cdot b \leq 1 \cdot b = b$ which gives $a \cdot b \leq a \wedge b = 0$. Similarly $a \cdot a \leq a$ which entails $a \cdot a \in \{0, a\}$.

If M is not Boolean, then there is an atom a of M such that $n := ord(a) \geq 2$. If $a \cdot a = a$, then $(na) \cdot (na) = n^2(a \cdot a) = n^2 a$ which yields $n = ord(a) \geq n^2$. If $a \cdot a = 0$, then $0 \neq a = a \cdot 1 = a \cdot (ord(a_1)a_1 + \cdots + ord(a_k)a_k) = 0$ which is again absurd. Consequently, on M, which is not Boolean, there is no product in question.

Assume now that M is Boolean, and let $a \cdot a = 0$ for some atom a of M. Then $a = a \cdot 1 = a \cdot (a + a^*) = a \cdot a^* = a \cdot (\sum_{a_i \neq a} a_i) = 0$, which is absurd. Hence $a \cdot a = a$ for any atom a. Therefore $x \cdot y = (j_{a_1}(x)a_1 + \cdots + j_{a_k}(x)a_k) \cdot (j_{a_1}(y)a_1 + \cdots + j_{a_k}(y)a_k) = j_{a_1}(x \wedge y)a_1 + \cdots + j_{a_k}(x \wedge y)a_k = x \wedge y$ for any $x, y \in M$. □

As a corollary we have that a product $\cdot$ such that $a \cdot 1 = a = 1 \cdot a$, for any $a \in M$, on a non-Boolean MV-algebra M can exists only if M is an infinite MV-algebra.

Let G be an Abelian ℓ-group and define the lexicographical product

$$G(\mathbb{Z}) := \mathbb{Z} \times_{lex} G,$$

where $\mathbb{Z}$ is the group of all integers. Then the element $(1, 0)$ is a strong unit in $G(\mathbb{Z})$ and $\Gamma(G(\mathbb{Z}), (1, 0))$ is a perfect MV-algebra. Due to [DiLe], for any perfect MV-algebra M there is a (unique up to ℓ-group isomorphism) ℓ-group G such that $M \cong \Gamma(G(\mathbb{Z}), (1, 0))$.

Since each element of $\Gamma(G(\mathbb{Z}), (1, 0))$ is of the form either $(0, g)$ or $(1, -g)$, where $g \in G^+$, we have that $(0, g_1) + (0, g_2)$ is always defined in $\Gamma(G(\mathbb{Z}), (1, 0))$, and $(0, g_1) + (0, g_2) = (0, g_1 + g_2)$. The sum $(0, g_1) + (1, -g_2)$ is defined iff $g_1 \leq g_2$, and then $(0, g_1) + (1, -g_2) = (1, -(g_2 - g_1))$, and $(1, -g_1) + (1, -g_2)$ is never defined in $\Gamma(G(\mathbb{Z}), (1, 0))$.

Example 5.3.18 *Let $(M; \oplus, ^*, 0, 1)$ be a finite MV-algebra.*

(1) *If a is a unique atom of M of order n, then M admits a non-zero product if and only if $n = 1$.*

(2) *Let a and b be the only two atoms of M and of order $n > 1$ and n^2, respectively. If we define $\cdot$ via $a \cdot a = b$ and $a \cdot b = b \cdot a = b \cdot b = 0$, then $\cdot$ is a non-zero, commutative and associative product on M without unity for $\cdot$.*

(3) *Let a and b be the only two atoms of M and of order 1 and $n \geq 1$, respectively. Define $a \cdot a = 0 = b \cdot a = b \cdot b$ and $a \cdot b = b$. Then $\cdot$ is a non-zero product which is associative, but not commutative, and without unity.*

Theorem 5.3.19 *Let M be a perfect MV-algebra and define a binary operation $\cdot$ as follows. For $a, b \in Rad(M)$, let*

$$\begin{aligned} a \cdot b &= 0, \\ a \cdot b^* &= a, \\ a^* \cdot b &= b, \\ a^* \cdot b^* &= (a+b)^*. \end{aligned} \tag{5.35}$$

Then $\cdot$ is a commutative product on M such that $a \cdot 1 = a = 1 \cdot a$ for any $a \in M$.

Let $M = \Gamma(G(\mathbb{Z}), (1,0))$ and let G be an Archimedean ℓ-group. Then on M there is a unique product $\cdot$ such that $a \cdot 1 = a = 1 \cdot a$ for any $a \in M$; this product is defined by (5.35).

Proof. Since $Rad(M) \cap Rad(M)^* = \emptyset$, we see that $\cdot$ is correctly defined by (5.35), and it is easy to verify that it is a product in question on M.

Let a and b be two element of $Rad(M)$. Then nb is defined in M for any integer $n \geq 1$, Proposition 5.3.11, so that

$$a = a \cdot 1 = a \cdot ((nb) + (nb)^*) = n(a \cdot b) + a \cdot (nb)^* \geq n(a \cdot b).$$

The Archimedeanicity of G entails that $a \cdot b = 0$. Hence $a = a \cdot b + a \cdot b^* = a \cdot b^*$. Similarly $a = (b^* \cdot a) + (b \cdot a) = b^* \cdot a$, and $a^* = a^* \cdot b^* + a^* \cdot b = a^* \cdot b^* + b$ which gives $a^* \cdot b^* = (a+b)^*$. □

Proposition 5.3.20 *Let M be an MV-algebra such that $Rad(M)$ is Archimedean, i.e., if $na \leq b$ for any $n \geq 1$ and for some $b \in Rad(M)$, then $a = 0$. If $\cdot$ is a product on M such that $a \cdot 1 = a = 1 \cdot a$ for any $a \in M$, then, for any $a, b \in Rad(M)$,* (5.35) *holds.*

Proof. The first three identities follow the same ideas as the proof of Theorem 5.3.19. For the last one we have

$$1 = (a + a^*) \cdot (b + b^*) = a \cdot b + a \cdot b^* + a^* \cdot b + a^* \cdot b^* = (a+b) + (a^* \cdot b^*),$$

which entails that $a^* \cdot b^* = (a+b)^*$. □

Remark 5.3.21 It is worth recalling that if an MV-algebra M has an Archimedean radical, then there is a unique binary operation $\cdot : (Rad(M) \cup Rad(M)^*) \times (Rad(M) \cup Rad(M)^*) \to Rad(M) \cup Rad(M)^*$ which is commutative, associative, and for all $a, b, c \in Rad(M) \cup Rad(M)^*$ we have $(a+b) \cdot c = a \cdot c + b \cdot c$, $a \cdot 1 = a$; it is defined via (5.35). On the other hand it does not mean that $\cdot$ can be extended to a product on whole M satisfying $a \cdot 1 = a = 1 \cdot a$ for any $a \in M$. For example, if M is finite, then $Rad(M) = \{0\}$, and $\cdot$ can be extended iff M is a Boolean MV-algebra, Theorem 5.3.17. An extension of $\cdot$ is possible, e.g., if M is a perfect MV-algebra with an Archimedean radical, Theorem 5.3.19.

Remark 5.3.22 We note that in [BDG] an MV*-algebra $(M; \oplus, ^*, 0, 1)$ has been introduced, i.e., M is a perfect MV-algebra with a binary operation $* : Rad(M) \times Rad(M) \to Rad(M)$ such that

(i) $*$ is associative;

(ii) $a * (b \oplus c) = (a * b) \oplus (a * c)$, $(b \oplus c) * a = (b * a) \oplus (c * a)$;

(iii) $a * 0 = 0 = 0 * a$

for all $a, b, c \in Rad(M)$. In view of Proposition 5.3.11, $b \oplus c = b + c$ for all $b, c \in Rad(M)$, therefore, (iii) is superfluous, and $*$ coincides with our product restricted to $Rad(M) \times Rad(M)$, because values of this restricted product are from $Rad(M)$.

Remark 5.3.23 It is worth recalling that a product $\cdot$ on an MV-algebra M can be defined only with condition (i) of the product in Definition 5.3.12. Then Theorem 5.3.14 holds also without the assumption of associativity of the ℓ-ring R. In addition, the product $\cdot$ on M is associative, i.e., $(a \cdot b) \cdot c = a \cdot (b \cdot c)$, $a, b, c \in M$, iff R is associative.

Denote by $\mathcal{PMV}$ the category of MV-algebras with a product, i.e., its objects are MV-algebras with a product, and morphisms are MV-homomorphisms of product MV-algebras preserving also $\cdot$. We denote by $\mathcal{R}$ the category of associative ℓ-rings with a distinguished strong unit u such that $u \cdot u \leq u$, and its morphisms are group-homomorphisms which preserve $\cdot, \wedge, \vee$ and distinguished strong units. We denote by $\Gamma_{\mathcal{R}}$ a map from $\mathcal{R}$ into $\mathcal{PMV}$ defined by $\Gamma_{\mathcal{R}}((R; +, \cdot, 0, \leq, u)) := (\Gamma(R, u); \oplus, ^*, \cdot, 0, u)$, and $\Gamma_{\mathcal{R}}(f) := f|\Gamma(R, u)$.

Theorem 5.3.24 *$\Gamma_{\mathcal{R}}$ is a faithful and full functor from $\mathcal{R}$ to $\mathcal{PMV}$.*

Proof. Let h_1 and h_2 be two morphism from $(R; +, \cdot, 0, \leq, u)$ into $(R'; +, \cdot, 0, \leq, u')$ such that $\Gamma_{\mathcal{R}}(h_1) = \Gamma_{\mathcal{R}}(h_2)$. Then $h_1(a) = h_2(a)$ for any $a \in \Gamma(R, u)$. Since $\Gamma(R, u)$ generates R^+ and R, we have that $h_1(g) = h_2(g)$ for any $g \in R$ which proves that $\Gamma_{\mathcal{R}}$ is faithful.

To prove that $\Gamma_{\mathcal{R}}$ is a full functor, suppose that f is a morphism from $\Gamma(R, u)$ into $\Gamma(R', u')$. Since $\Gamma(R, u)$ generates R, due to the Riesz decomposition property, f can be uniquely extended to a group-homomorphism $\hat{f}$ from R into R'.

Claim 1. $\hat{f}$ is a lattice-homomorphism. The proof will proceed in several steps.

Step 1. f preserves meets in $\Gamma(R,u)$. This follows from the observation that $a \wedge b = a \odot (a^* \oplus b)$ which entails f preserves meets in $\Gamma(R,u)$.

Step 2. Let $a, b, u_0 \in R^+$. If $\hat{f}(a \wedge b) = \hat{f}(a) \wedge \hat{f}(b)$ and if $\hat{f}(u_0 \wedge (b - (a \wedge b))) = \hat{f}(u_0) \wedge \hat{f}(b - (a \wedge b))$, then

$$\hat{f}((a + u_0) \wedge b) = \hat{f}(a + u_0) \wedge \hat{f}(b).$$

Indeed, we have $(a \wedge b) + [u_0 \wedge (b - (a \wedge b))] = ((a \wedge b) + u_0) \wedge b = (a + u_0) \wedge (b + u_0) \wedge b = (a + u_0) \wedge b$.

Therefore

$$\begin{aligned}
\hat{f}((a + u_0) \wedge b) &= \hat{f}(a \wedge b) + \hat{f}(u_0 \wedge (b - (a \wedge b))) \\
&= (\hat{f}(a) \wedge \hat{f}(b)) + [\hat{f}(u_0) \wedge (\hat{f}(b) - (\hat{f}(a) \wedge \hat{f}(b)))] \\
&= [((\hat{f}(a) \wedge \hat{f}(b)) + \hat{f}(u_0))] \wedge \hat{f}(b) \\
&= (\hat{f}(a) + \hat{f}(u_0)) \wedge (\hat{f}(b) + \hat{f}(u_0)) \wedge \hat{f}(b) = \hat{f}(a + u_0) \wedge \hat{f}(b).
\end{aligned}$$

Step 3. $\hat{f}(a \wedge b) = \hat{f}(a) \wedge \hat{f}(b)$ whenever $a \in R^+$ and $b \in \Gamma(R,u)$.

Since $\Gamma(R,u)$ is generating for R^+, a is of the form $a = a_1 + \cdots + a_n$ for some $a_1, \ldots, a_n \in \Gamma(R,u)$. The proof will follow mathematical induction on n.

If $n = 1$, the statement is trivial. Suppose now that the statement holds for any $a' = a_1 + \cdots + a_i$ with $1 \le i \le n$. Put $a = a_1 + \cdots + a_n$, $u_0 = a_{n+1}$. Then there exist $v_1, \ldots, v_k \in \Gamma(R,u)$ such that $b = (v_1 + \cdots + v_k) + (a \wedge b)$. Since $v := v_1 + \cdots + v_k \le b \in \Gamma(R,u)$, $v \in \Gamma(R,u)$. Hence $v = b - (a \wedge b)$. Since $\hat{f}$ preserves meets in $\Gamma(R,u)$, we have $\hat{f}(u_0 \wedge v) = \hat{f}(u_0) \wedge \hat{f}(v)$, so that $\hat{f}(u_0 \wedge (b - (a \wedge b))) = \hat{f}(u_0) \wedge \hat{f}(b - (a \wedge b)) = \hat{f}(u_0) \wedge (\hat{f}(b) - (\hat{f}(a) \wedge \hat{f}(b)))$ when we have used induction hypothesis. By Step 1, $\hat{f}((a + u_0) \wedge b) = \hat{f}(a + u_0) \wedge \hat{f}(b)$, i.e., $\hat{f}((a_1 + \cdots + a_{n+1}) \wedge b) = \hat{f}(a_1 + \cdots + a_{n+1}) \wedge \hat{f}(b)$ for any n.

Step 4. $\hat{f}(a \wedge b) = \hat{f}(a) \wedge \hat{f}(b)$ whenever $a, b \in R^+$.

Let $a = a_1 + \cdots + a_n$, $b = b_1 + \cdots + b_k$. The proof will follow complete induction on k.

If $k = 1$, we apply Step 3. Suppose now that the assertion holds for any j with $1 \le j \le k$. Put $B = a$, $A = b_1 + \cdots + b_k$, $u_0 = b_{k+1}$. By Step 3, $\hat{f}(u_0 \wedge (B - (A \wedge B))) = \hat{f}(u_0) \wedge \hat{f}(B - (A \wedge B))$ and $\hat{f}(A \wedge B) = \hat{f}(A) \wedge \hat{f}(B)$. Therefore the conditions of Step 1 are satisfied, so that $\hat{f}((A + u_0) \wedge B) = \hat{f}(A + u_0) \wedge \hat{f}(B)$ which proves $\hat{f}((a_1 + \cdots + a_n) \wedge (b_1 + \cdots + b_{k+1})) = \hat{f}(a_1 + \cdots + a_n) \wedge \hat{f}(b_1 + \cdots + b_{k+1})$ for each n and each k.

Step 5. $\hat{f}(a \wedge b) = \hat{f}(a) \wedge \hat{f}(b)$ whenever $a, b \in R$. There exist $a', a'', b', b'' \in R^+$ such that $a = a' - a''$ and $b = b' - b''$. By Step 4, $\hat{f}((a' + b'') \wedge (b' + a'')) = \hat{f}(a' + b') \wedge \hat{f}(b' \wedge a'')$. Subtracting $\hat{f}(b'')$ and $\hat{f}(a'')$ we obtain the assertion in question.

Claim 2. $\hat{f}$ preserves the product $\cdot$ in R.

Let $a, b \in R^+$. There exist $a_1, \ldots, a_n, b_1, \ldots, b_m \in \Gamma(R, u)$ such that $a = a_1 + \cdots + a_n$ and $b = b_1 + \cdots + b_m$. Then $a \cdot b = \sum_{i=1}^n \sum_{j=1}^m a_i \cdot b_j$ and $a_i \cdot b_j \in \Gamma(R, u)$. Calculate $\hat{f}(a \cdot b) = \sum_{i=1}^n \sum_{j=1}^m \hat{f}(a_i \cdot b_j) = \sum_{i=1}^n \sum_{j=1}^m \hat{f}(a_i) \cdot \hat{f}(b_j) = (\sum_{i=1}^n \hat{f}(a_i)) \cdot (\sum_{j=1}^m \hat{f}(b_j)) = \hat{f}(a) \cdot \hat{f}(b)$.

If now $a, b \in R$, then $a = a_1 - a_2$ and $b = b_1 - b_2$, where $a_1, a_2, b_1, b_2 \in R^+$. Then $\hat{f}(a \cdot b) = \hat{f}(a_1 \cdot b_1 - a_1 \cdot b_2 - a_2 \cdot b_1 + a_2 \cdot b_2) = \hat{f}(a_1 \cdot b_1) - \hat{f}(a_1 \cdot b_2) - \hat{f}(a_2 \cdot b_1) + \hat{f}(a_2 \cdot b_2) = \hat{f}(a_1) \cdot \hat{f}(b_1) - \hat{f}(a_1) \cdot \hat{f}(b_2) - \hat{f}(a_2) \cdot \hat{f}(b_1) + \hat{f}(a_2) \cdot \hat{f}(b_2) = \hat{f}(a) \cdot \hat{f}(b)$.

Consequently, we have proved that $\hat{f}$ is a morphism from $(R; +, \cdot, 0, \leq, u)$ into $(R'; +, \cdot, 0, \leq, u')$ such that $\Gamma_{\mathcal{R}}(\hat{f}) = f$. □

Theorem 5.3.25 *The functor $\Gamma_{\mathcal{R}}$ defines a categorical equivalence of the category $\mathcal{R}$ of unital associative ℓ-rings with a strong unit u such that $u \cdot u \leq u$ and the category $\mathcal{PMV}$ of product MV-algebras.*

Proof. According to [MaL, Thm IV.4.1], to prove that $\Gamma_{\mathcal{R}}$ is an equivalence of the categories in question, it is necessary and sufficient to show that $\Gamma_{\mathcal{R}}$ is faithful and full, and each object M from $\mathcal{PMV}$ is isomorphic to $\Gamma_{\mathcal{R}}((R; +, \cdot, 0, \leq, u))$ for some object $(R; +, \cdot, 0, \leq, u)$ in $\mathcal{R}$.

Due to Theorem 5.3.24, $\Gamma_{\mathcal{R}}$ is faithful and full, and by Theorem 5.3.14, there exists an object $(R; +, \cdot, 0, \leq, u)$ in $\mathcal{R}$ such that $\Gamma_{\mathcal{R}}((R; +, \cdot, 0, \leq, u))$ is isomorphic to M. That proves Theorem. □

We note that using methods of the proofs of Theorems 5.3.14 and 5.3.24, it is possible to show that if $(G(M), h)$ is a universal group for the product MV-algebra M, then G can be converted into an ℓ-ring such that h preserves the product in M. In addition, the pair $(G(M), h)$ is a universal arrow for M, so that $\Gamma_{\mathcal{R}}$ is right-adjoint, and the morphism $\mathcal{P} : \mathcal{PMV} \to \mathcal{R}$ defined by $\mathcal{P}(M) = G(M)$ is an opposite functor from $\mathcal{PMV}$ into $\mathcal{R}$ such that $\mathcal{P}\mathcal{R} \cong 1_{\mathcal{R}}$ and $\mathcal{R}\mathcal{P} \cong 1_{\mathcal{P}}$.

Theorem 5.3.25 holds also for the category of product MV-algebras which are not necessarily associative (see Remark 5.3.23):

Theorem 5.3.26 *Let $\mathcal{R}$ be the category of (not necessarily associative) ℓ-rings with a fixed strong unit u such that $u \cdot u \leq u$, and let $\mathcal{PMV}$ be the category of MV-algebras having a (not necessarily associative) product. Then $\Gamma_{\mathcal{R}} : \mathcal{R} \to \mathcal{PMV}$ defined by $\Gamma_{\mathcal{R}}((R; +, \cdot, 0, \leq, u)) := (\Gamma(R, u); \oplus, ^*, \cdot, 0, u)$ defines a categorical equivalence between $\mathcal{R}$ and $\mathcal{PMV}$.*

We finish this section with a comparison of product MV-algebras with MV*-algebras from the categorical point of view.

According to [BeDi], we say that a perfect MV-algebra M has a **principal radical** if there is an element $a \in Rad(M)$ such that the ideal of M generated by a coincides with $Rad(M)$. The category $\mathcal{MV}_{pr}$ of perfect MV-algebras with principal radical is a category whose objects are pairs (M, a), where M is a perfect MV-algebra with a fixed element $a \in Rad(M)$ generating $Rad(M)$ as an ideal, and morphisms are MV-homomorphisms preserving the fixed elements a in radicals. [12]

Due to [BeDi, Prop 27], this category is equivalent with the category $\mathcal{MV}$ of all MV-algebras, or equivalently with the category of all unital Abelian ℓ-groups (Mundici's representation, Corollary 5.3.8).

Let now $\mathcal{MV}_p^*$ be the category of MV*-algebras whose morphisms are MV-homomorphisms preserving a binary operation $*$ on the radical as given in Remark 5.3.22. According to [DiLe, Thm 3.5], the category $\mathcal{MV}_p$ of perfect MV-algebras is categorically equivalent with the category of all (not necessarily unital) ℓ-groups, and $\mathcal{MV}_p^*$ is categorically equivalent to the category of all associative (not necessarily unital) ℓ-rings [BDG, Thm 2.6]. Let $\mathcal{MV}_{pr}^*$ denote the category having as objects pairs (A, g), where A is a perfect MV*-algebra with principal radical with the distinguished generator g such that $g * g \le g$, and morphisms are $\mathcal{MV}_{pr}^*$-homomorphisms preserving the distinguished generator of the radical.

Using ideas developed above and in [BDG], we have the following result.

Theorem 5.3.27 (1) *The category $\mathcal{MV}_p^*$ of perfect MV*-algebras is categorically equivalent to the category of all associative ℓ-rings.*

(2) *The category $\mathcal{MV}_{pr}^*$ is categorically equivalent with the category of associative unital ℓ-rings with a strong unit u such that $u \cdot u \le u$, or equivalently, $\mathcal{MV}_{pr}^*$ is categorically equivalent with the category $\mathcal{PMV}$ of product MV-algebras.*

5.3.5 Product on Commutative BCK-algebras

The system $X = [0,1]^\Omega$ of all fuzzy sets on $\Omega \neq \emptyset$ has a natural BCK-operation $*$ defined by $(f * g)(\omega) := \max\{0, f(\omega) - g(\omega)\}$, $\omega \in \Omega$. The partial sum $f + g$ is defined in X iff $f \le 1 - g$, and X admits a natural multiplication $f \cdot g$, which is a total operation and is left and right distributive with respect to the addition $+$.

Motivated by that as well as by the product on MV-algebras, we shall study in this Section products on commutative BCK-algebras ([Dvu 14]).

Definition 5.3.28 *We say that a commutative BCK-algebra $(X; *, 0)$ with the relative cancellation property admits a* **product** *if there is a binary operation $\cdot$ on X satisfying for all $a, b, c \in X$ the following:*

(i) *If $a + b$ is defined in X, then $a \cdot c + b \cdot c$ and $c \cdot a + c \cdot b$ exist and*

$$\begin{aligned}(a+b)\cdot c &= a\cdot c + b\cdot c,\\ c\cdot(a+b) &= c\cdot a + c\cdot b,\end{aligned}$$

[12] We note that any MV-homomorphism maps radicals into radicals (Proposition 5.3.11).

where + is the derived partial operation on the BCK-algebra X, and we say that X is a **product BCK-algebra**. *Sometimes we write $X = (X; *, \cdot, 0)$. An element u of a product BCK-algebra X is said to be a* **unity**, *if $a \cdot u = u \cdot a = a$ for any $a \in X$.*

A product $\cdot$ on X is

(ii) **associative** if $(a \cdot b) \cdot c = a \cdot (b \cdot c)$, $a, b, c \in X$;

(iii) **commutative** if $a \cdot b = b \cdot a$, $a, b \in X$.

It is worth saying that if $\cdot$ is a product on X, then

(iv) $a \cdot 0 = 0 = 0 \cdot a$;

(v) if $a \leq b$, then for any $c \in X$, $a \cdot c \leq b \cdot c$ and $c \cdot a \leq c \cdot b$.

Property (iv) follows easily from the following: $a \cdot 0 = a \cdot (0 + 0) = a \cdot 0 + a \cdot 0$, and the cancellation property gives $a \cdot 0 = 0$. Similarly, $0 \cdot a = 0$.

Theorem 5.3.29 *The class of product BCK-algebras is equationally definable.*

Proof. The class of commutative BCK-algebras with the relative cancellation property is equationally definable, Theorem 5.2.29. The condition (i) is equivalent to the identities

$$\begin{aligned} c \cdot (x * (x \wedge y)) &= (c \cdot x) * (c \cdot (x \wedge y)), \\ (x * (x \wedge y)) \cdot c &= (x \cdot c) * ((x \wedge y) \cdot c) \end{aligned}$$

for all $c, x, y \in X$. Hence the class of product BCK-algebras is equationally definable. □

Denote by $\mathcal{PBCK}$ the category of product BCK-algebras, i.e., its objects are BCK-algebras with product, and morphisms are BCK-homomorphisms of product BCK-algebras preserving also $\cdot$.

We denote by $\mathcal{PR}$ the category whose objects are pairs (R, R_0), where R is an Abelian ℓ-ring and R_0 is a non-void subset of the positive cone R^+ of R such that R_0 generates R^+ and $(R_0; *_R, \cdot, 0)$ is a product BCK-algebra under $\cdot$. A morphism from (R, R_0) into (R', R'_0) is an ℓ-group homomorphism $h : R \to R'$, preserving $\cdot, \wedge, \vee$, such that $h(R_0) \subseteq R'_0$.

We denote by $\mathcal{X}_{\mathcal{R}}$ a morphism from $\mathcal{PR}$ into $\mathcal{PBCK}$ defined by

$$\mathcal{X}_{\mathcal{R}}(R, R_0) = (R_0; *_R, \cdot, 0)^{13},$$

and $\mathcal{X}_{\mathcal{R}}(f) := f|R_0$.

[13]We note that $*_R$ is understood as that defined by (5.19).

Theorem 5.3.30 *Let* $(R;+,\cdot,0,\leq)$ *be an (associative)* ℓ*-ring with* $\emptyset \neq R_0 \subseteq R^+$ *such that* $a, b \in R_0$ *entails* $a *_R b \in R_0$ *and* $a \cdot b \in R_0$*. Then* $(X;*,0) := \mathcal{X}(R, R_0)$ *is a product BCK-algebra with an (associative) product* $\cdot$*, which is the restriction of* $\cdot$ *to* $R_0 \times R_0$*.*

Conversely, let $(X;*,0)$ *be a product BCK-algebra with an (associative) product* $\cdot$*. Then there exists (up to isomorphism) a unique (associative)* ℓ*-ring* $(R;+,\cdot,0,\leq)$ *with a non-empty set* $R_0 \subseteq R^+$ *generating* R^+ *as a group-cone such that* $a, b \in R_0$ *entails* $a *_R b \in R_0$*,* $a \cdot b \in R_0$*, and* $X \cong \mathcal{X}(R, R_0)$*, and* $\phi(a \cdot b) = \phi(a) \cdot \phi(b)$*,* $a, b \in X$*, where* ϕ *is a BCK-isomorphism of* X *onto* $\mathcal{X}(R, R_0)$*.*

Proof. (1) Let $X = \mathcal{X}(R, R_0)$. For $a, b \in X$ we have $a \cdot b \in X$ which says that the restriction of $\cdot$ to $X \times X$ defines an (associative) product on the BCK-algebra X.

(2) For the rest we follows ideas of the proof of Theorem 5.3.14. □

Theorem 5.3.31 *The functor* $\mathcal{X}_{\mathcal{R}}$ *defines a categorical equivalence of the category* $\mathcal{PR}$ *of* ℓ*-rings and the category* $\mathcal{PBCK}$ *of product BCK-algebras.*

Proof. Using ideas of Theorem 5.3.24, we can prove that $\mathcal{X}_{\mathcal{R}}$ is a faithful and full functor from $\mathcal{PR}$ to $\mathcal{PBCK}$.

According to [MaL, Thm IV.4.1], to prove that $\mathcal{X}_{\mathcal{R}}$ is an equivalence of the categories in question, it is necessary and sufficient to show that $\mathcal{X}_{\mathcal{R}}$ is faithful and full, and each object X from $\mathcal{PBCK}$ is isomorphic to $\mathcal{X}_{\mathcal{R}}(R, R_0')$ for some object (R, R_0) in $\mathcal{PR}$.

By Theorem 5.3.30, there exists an object (R, R_0) in $\mathcal{PR}$ such that $\mathcal{X}_{\mathcal{R}}(R, R_0)$ is isomorphic to X which proves Theorem. □

Remark 5.3.32 Theorem 5.3.31 holds also for the category of product BCK-algebras which are also associative. In this case $\mathcal{APR}$ stands for the category of associative elements from $\mathcal{PR}$.

We now compare product BCK-algebras and MV*-algebras introduced in [BDG].

Denote by $\mathcal{PCBCK}$ the category of all product conical BCK-algebras, and by $\mathcal{PCBCK}_u$ the category of all product conical BCK-algebras whose objects are pairs (X, u), where u is a fixed quasi strong unit such that $u \cdot u \leq u$, and morphisms are the BCK-homomorphisms preserving the product and fixed quasi strong units.

Finally, let $\mathcal{R}$ denote the category of associative ℓ-rings, where the objects are associative ℓ-rings, and let $\mathcal{R}_u$ be the category of associative ℓ-rings with a fixed strong unit u such that $u \cdot u \leq u$, i.e., objects are pairs (R, u), where u is a fixed strong unit in R, and morphisms are ℓ-ring morphisms preserving fixed strong units.

Theorem 5.3.33 (1) *The categories* $\mathcal{MV}^*_p$ *of perfect MV*-algebras,* $\mathcal{PCBCK}$ *of product conical BCK-algebras, and* $\mathcal{R}$ *of associative* ℓ*-rings are categorically equivalent.*

(2) *The categories* $\mathcal{MV}^*_{pr}$*,* $\mathcal{PCBCK}_u$*, and* $\mathcal{R}_u$ *are categorical equivalent.*

Proof. This follows ideas developed in above and in [BDG], and from the observation that an element u of X is a quasi strong unit iff the ideal of X generated by u is equal to X. □

5.3.6 Ideals in Product BCK-algebras and BCKf-algebras

Let X be a product BCK-algebra. A non-empty subset I of X is said to be a **·-ideal** of X if

(i) $0 \in I$;

(ii) $a * b \in I$, $a \in X$, and $b \in I$, entail $a \in I$;

(iii) $a \in I$ and $b \in X$ entail $a \cdot b \in I$ and $b \cdot a \in I$.

We denote by $\mathcal{I}_p(X)$ the set of all ·-ideals of X.

Let $(R; +, \cdot, 0, \leq)$ be an ℓ-ring. An L-**ideal** of R is a non-void subset J of R such that

(i) $x, y \in J$ entails $x \pm y \in J$;

(ii) $x \in R$, $y \in J$, $|x| \leq |y|$ entail $x \in J$;

(iii) $x \in J$ and $y \in R$ entail $x \cdot y \in J$ and $y \cdot x \in J$.

We denote by $\mathcal{I}_L(R)$ the set of all L-ideals of R.

Let $(X; *, 0)$ be a commutative BCK-algebra with the relative cancellation property and let $(G(X), h)$ be its universal group. Given a subset I of X, let $h_0(I)$ be the ℓ-ideal of $G(X)$ generated by the image $h(I)$ of I in $G(X)$.

Theorem 5.3.34 *Let $(R; +, \cdot, 0, \leq)$ be an ℓ-ring with $\emptyset \neq R_0 \subseteq R^+$ generating R^+ such that $a, b \in R_0$ entails $a *_R b \in R_0$ and $a \cdot b \in R_0$, and let $X = \mathcal{X}_{\mathcal{R}}(R, R_0)$. Then the correspondence Φ defined by*

$$\Phi(I) := h(I), \; I \in \mathcal{I}_p(X), \tag{5.36}$$

where h is the embedding of X into R, is an order-isomorphism from the set $\mathcal{I}_p(X)$ of ·-ideals of the product BCK-algebra X, ordered by inclusion, onto $\mathcal{I}_L(R)$. The inverse isomorphism Ψ is given by

$$\Psi(H) := X \cap H, \; H \in \mathcal{I}_L(R). \tag{5.37}$$

Proof. In Theorem 5.2.45 we have proved that the mappings (5.36) and (5.37) define inclusion isomorphisms between the set of ideals of X and ℓ-ideals of R. We show that they preserve also "·-ideal property".

We assert that $x \in \Phi(I)$ iff $x = x_1 + \cdots + x_j - y_1 - \cdots - y_k$, where $x_i, y_s \in I$. It is clear that if x has a given form, then $x \in \Phi(I)$. Conversely, let $x \in \Phi(I)$. Assume that $x \geq 0$. Then there exist $x_1, \ldots, x_n \in X$ such that $x = x_1 + \cdots + x_n$. We show that $x_1, \ldots, x_n \in I$. The proof will follow mathematical induction on n. If $n = 1$, the statement is evident. Assume thus that whenever $x \in \Phi(I)$ and $x = x_1 + \cdots + x_i$ for $i \leq n$, then $x_1, \ldots, x_i \in I$. Let now $x = x_1 + \cdots + x_n + x_{n+1}$. Then $0 \leq x_{n+1} \leq x$, so that $x_{n+1} \in \Phi(I)$ and $x - x_{n+1} \in \Phi(I)$ and by induction $x_1, \ldots, x_n \in I$. Hence $x = x_1 + x_2 + \cdots + x_{n+1}$ which proves also $x_{n+1} \in I$.

If x is an arbitrary element of $\Phi(I)$, then $x = x^+ - x^-$ and $x^+, x^- \in \Phi(I)$, so that $x^+ = x_1 + \cdots + x_j$, $x^- = y_1 + \cdots + y_k$.

Let now $x \in \Phi(I)$ and $y \in R$, then $x = x_1 + \cdots + x_j - y_1 - \cdots - y_k$ and $y = u_1 + \cdots + u_s - v_1 - \cdots - v_k$, where $x_1, \ldots, x_j, y_1, \ldots, y_k \in I$ and $u_1, \ldots, u_s, v_1, \ldots, v_k \in X$. Then evidently $x \cdot y$, $y \cdot x \in \Phi(I)$, which proves that Φ preserves $\cdot$-ideals. □

A product BCK-algebra X is said to be a BCKf-**algebra** if, for all $a, b, c \in X$,

$$a \wedge b = 0 \text{ implies } (a \cdot c) \wedge b = 0 = (c \cdot a) \wedge b. \tag{5.38}$$

For example, any linearly ordered product BCK-algebra is a BCKf-algebra.

Theorem 5.3.35 *Let $(R; +, \cdot, 0, \leq)$ be an ℓ-ring with $\emptyset \neq R_0 \subseteq R^+$ such that $a, b \in R_0$ entails $a *_R b \in R_0$ and $a \cdot b \in R_0$. Let $X = \mathcal{X}_R(R, R_0)$. Then X is a BCKf-algebra if and only if R is an f-ring.*

Proof. One direction is trivial.

Suppose now that X is a BCKf-algebra, and let for $a, b, c \in R$ we have $a \wedge b = 0$ and $c \geq 0$. Let $x \leq a \cdot c$ and $x \leq b$ and express $a = \sum_i a_i$, $b = \sum_j b_j$, and $c = \sum_k c_k$, where $a_i, b_j, c_k \in X$.

Since $x \leq \sum_{i,k} a_i \cdot c_k$, the Riesz decomposition property, [Goo 1, Prop. 2.2], entails $x = \sum_{i,k} c_{ik}$, where $c_{ik} \leq a_i \cdot c_k$. Then $c_{ik} \leq \sum_j b_j$ and applying again the Riesz decomposition property, we have $c_{ik} = \sum_j c_{ijk}$, where $c_{ijk} \leq b_j$, for any i and any k. Since $a_i \wedge b_j = 0$, we conclude from $c_{ijk} \leq a_i \cdot c_k$ and $c_{ijk} \leq b_j$ that $c_{ijk} = 0$ for all i, j, k. Hence, $c_{ik} = 0$ for all i, k which yields $x = 0$.

In a similar way we can prove that $(c \cdot a) \wedge b = 0$. □

Corollary 5.3.36 *In any BCKf-algebra X we have for all $a, b, c \in X$*

$$\begin{aligned}
a \cdot (b \vee c) &= (a \cdot b) \vee (a \cdot c), \\
(b \vee c) \cdot a &= (b \cdot a) \vee (c \cdot a), \\
a \cdot (b \wedge c) &= (a \cdot b) \wedge (a \cdot c), \\
(b \wedge c) \cdot a &= (b \cdot a) \wedge (c \cdot a), \\
a \wedge b = 0 &\Rightarrow a \cdot b = 0.
\end{aligned}$$

Proof. Use Theorem 5.3.35 and [Bir, Cor XVII.5.1, Lem XVII.5.1]. □

Given a non-void subset A of X, we define

$$A^\perp := \{a \in X : a \wedge x = 0 \text{ for any } x \in A\}.$$

Theorem 5.3.37 *Let $(R; +, \cdot, 0, \leq)$ be an ℓ-ring with $\emptyset \neq R_0 \subseteq R^+$ such that $a, b \in R_0$ entails $a *_R b \in R_0$ and $a \cdot b \in R_0$. Let $X = \mathcal{X}_\mathcal{R}(R, R_0)$. Then X is a BCKf-algebra if and only if, for every subset A of X, the set $A^\perp$ is a $\cdot$-ideal of X.*

Proof. Let X be a BCKf-algebra. Let $a * b \in A^\perp$, $b \in A^\perp$ and $x \in A$. Then $a \wedge b \in A^\perp$. We show that $a \in A^\perp$. Let $z \leq a$ and $z \leq x$. Then $z \leq a = a \wedge b + a * b$. The Riesz decomposition property entails that $z = z_1 + z_2$, where $z_1 \leq a \wedge b$ and $z_2 \leq a * b$. Since we also have $z_2 \leq x$, we conclude $z_2 = 0$, and $z_1 \leq a \wedge b \in A^\perp$, consequently, $z_1 = z_2 = 0 = z$ which yields $a \in A^\perp$.

Take now $a \in A^\perp$, $b \in X$, and $x \in A$. Then $a \wedge x = 0$, which gives $(a \cdot b) \wedge x = (b \cdot a) \wedge x = 0$.

Conversely, assume that for any non-void $A \subseteq X$, $A^\perp$ is a $\cdot$-ideal of X. Let $a \wedge b = 0$, $c \in X$. Then $a \in \{b\}^\perp$. Since $\{b\}^\perp$ is a $\cdot$-ideal of X, $a \cdot c$, $c \cdot a \in \{b\}^\perp$, i.e., $(a \cdot c) \wedge b = (c \cdot a) \wedge b = 0$. □

Theorem 5.3.38 *The class of BCKf-algebras is equationally definable.*

Proof. Any product BCK-algebra is equationally definable by Theorem 5.3.29. The implication (5.38) is equivalent to the identities

$$\begin{aligned} (c \cdot (x * (x \wedge y))) \wedge (y * (x \wedge y)) &= 0, \\ ((x * (x \wedge y)) \cdot c) \wedge (y * (x \wedge y)) &= 0. \end{aligned}$$

Hence the class of BCKf-algebras is equationally definable. □

Corollary 5.3.39 *Any BCKf-algebra is a subdirect product of subdirectly irreducible BCKf-algebras.*

Proof. This follows from Theorem 5.3.38 and [Bir, Thm VIII.15]. □

Proposition 5.3.40 *Any subdirectly irreducible BCKf-algebra is linearly ordered.*

Proof. Let X be an irreducible BCKf-algebra and suppose that X is not linearly ordered. Then there exist two non-zero elements $a, b \in X$ such that $a \wedge b = 0$. Put $I = \{b\}^\perp$ and $J = I^\perp$. Then $I \cap J = \{0\}$, and $a \in I$ and $b \in J$. By Theorem 5.3.37, I and J are both $\cdot$-ideals of X. The irreducibility of X entails that $I = \{0\}$ or $J = \{0\}$ which is a contradiction. Therefore, X is linearly ordered. □

Theorem 5.3.41 *A product BCK-algebra is a BCKf-algebra if and only if it is a subdirect product of linearly ordered product BCK-algebras.*

Proof. By Corollary 5.3.39 and Proposition 5.3.40, any BCKf-algebra is a subdirect product of linearly ordered product BCK-algebras.

Conversely, any linearly ordered product BCK-algebra is an BCKf-algebra. By Theorem 5.3.38, (5.38) holds in every subdirect product of linearly ordered BCKf-algebras, i.e., X is a BCKf-algebra whenever X is a subdirect product of linearly ordered product BCK-algebras. $\square$

A weaker form of BCKf-algebras are **almost BCKf-algebras**, i.e., such product BCK-algebras X that $a \wedge b = 0$, $a, b \in X$, imply $a \cdot b = 0$. Similarly, an associative ℓ-ring R is said to be an **almost f-ring** if $x \wedge y = 0$ implies $x \cdot y = 0$. We recall that not every almost f-ring has to be an f-ring [Bir, § XVII.6].

A non-zero element u of a BCK-algebra X is a **weak unit** if $u \wedge a \neq 0$. Similarly a non-zero positive element u of an ℓ-algebra R is a weak unit if $x \wedge u \neq 0$ for any positive element $x \in R$.

Theorem 5.3.42 *Let $(R; +, \cdot, 0, \leq)$ be an associative ℓ-ring with $\emptyset \neq R_0 \subsetneq R^+$ generating R^+ such that $a, b \in R_0$ entails $a *_R b \in R_0$ and $a \cdot b \in R_0$. Let $X = \mathcal{X}_{\mathcal{R}}(R, R_0)$. Then X is an associative almost BCKf-algebra if and only if R is an almost f-ring. If X has a unity e, then X is an almost BCKf-algebra if and only if e is a weak unit for R.*

Proof. One direction is evident.

Suppose now that X is an almost BCKf-algebra, and let $a \wedge b = 0$ for $a, b \in R$. Then $a = \sum_i a_i$ and $b = \sum_j b_j$, where $a_i, b_j \in X$. Since $a_i \wedge b_j = 0$ for all i, j, we have $a \cdot b = \sum_{i,j} a_i \cdot b_j = 0$.

Let now e be a unity for X. Then e is a unity for R, and e is a weak unit for X iff e is a weak unit for R. Applying [Bir, Thm XVII.12], R is an almost f-ring iff e is a weak unit for R. $\square$

5.3.7 Exercises

1. Show that for an MV-algebra M, ideals defined for MV-algebras and ideals defined for bounded commutative BCK-algebras coincide by Example 5.1.14.
2. Prove that an MV-algebra M is semisimple iff M is Archimedean. ([Dvu 10])
3. Prove directly that the category of conical BCK-algebras is categorically equivalent with the category of all Abelian ℓ-groups. (Hint: Use Remark 5.3.1(ii) and $\circ = +$). ([Cor 1, Thm 2.1])
4. Prove that the category of perfect MV-algebras with principal radical is categorically equivalent with the category of all MV-algebras. ([BeDi, Prop 27])

5. Prove that the category of perfect MV-algebras is categorically equivalent with the category of Abelian ℓ-groups. (Hint: Use Proposition 5.3.11.) ([DiLe, Thm 3.5])

6. Prove that if M is an MV-algebra, then $Rad(M) = Infinit(M) \cup \{0\}$. ([CDM, Prop 3.6.4])

7. Let $u > 0$ be a quasi strong unit for a commutative BCK-algebra $(X; *, 0)$ with the relative cancellation property and $X_u := \{x \in X : \ x \leq u\}$. Prove that $x \in Rad(X)$ if $x \wedge u \in Infinit(X_u) \cup \{0\}$.

8. Let $u > 0$ be a quasi strong unit for a commutative BCK-algebra $(X; *, 0)$ with the relative cancellation property and $X_u := \{x \in X : x \leq u\}$. Prove that X is semisimple iff X_u is semisimple.

9. Let $u > 0$ be a quasi strong unit for a commutative BCK-algebra $(X; *, 0)$ with the relative cancellation property. Then $Rad(X) = Rad(X_u) = Infinit(X_u) \cup \{0\}$, and if $x, y \in Rad(X)$, then $x + y$ is defined in X and $x + y \in Rad(X)$. ([Dvu 13, Thm 4.7])

Chapter 6

BCK-algebras in Applications

We apply general methods of BCK-algebras to obtain a finer analysis of BCK-algebras to describe semisimple BCK-algebras, simple BCK-algebras, bounded commutative BCK-algebras, to make a comparison with difference posets and to describe pseudo MV-algebras.

6.1 Algebraic Properties of Commutative BCK-algebras

In the present Section, we give other characterizations of commutative BCK-algebras showing their functional representations, semisimple BCK-algebras, simple BCK-algebras which follow from their ℓ-group representation.

6.1.1 Semisimple BCK-algebras

Semisimple MV-algebras are those which have a fuzzy set representation. In the present section, we study semisimple BCK-algebras and their representations by positive-valued functions, and their connection with Archimedeanicity and the existence of ordering systems of measure-morphisms and measures, respectively.

If $\mathcal{M}(X) \neq \emptyset$, the set

$$Rad(X) := \bigcap\{M : M \in \mathcal{M}(X)\}$$

is called the **radical** of X. If $Rad(X) = \{0\}$, X is said to be **semisimple**.

We note that a BCK-algebra $(X; *, 0)$ is **simple** if there is no non-trivial proper ideal of X. It is clear that any simple BCK-algebra is semisimple.

Proposition 6.1.1 *Let $X \subseteq [0, \infty)^{\Omega}$, $\Omega \neq \emptyset$, be a non-void system of functions such that $f, g \in X$ implies that $f * g \in X$, where $(f * g)(\omega) := \max\{f(\omega) - g(\omega), 0\}$, $\omega \in$*

Ω. Then $(X;*,0_\Omega)$, where $0_\Omega(\omega)=0$, $\omega\in\Omega$, is a commutative BCK-algebra with the relative cancellation property which is Archimedean. If X possesses at least two elements, then X is semisimple.

Proof. It is simple to verify that $(X;*,0_\Omega)$ is a commutative BCK-algebra which has the relative cancellation property and which is Archimedean.

Assume $X\neq\{0_\Omega\}$. For each point $\omega\in\Omega$, the set $I_\omega:=\{f\in X:\ f(\omega)=0\}$ is an ideal of X. We claim that any I_ω is either equal to X, or it is a maximal ideal of X. Indeed, assume that I_ω is a proper subset of X. Take $f\in X\setminus I_\omega$ and let $I_{f,\omega}$ be the ideal generated by I_ω and by the function f. According to Lemma 5.1.33 (5.14), $I_{f,\omega}=\{g\in X:\ \exists\, n\geq 1\ g(\omega)*^n_{\mathbb{R}} f(\omega)=0\}$. It is evident that $I_{f,\omega}=X$: given $g\in X$, there is an integer $n\geq 1$ such that $(n-1)f(\omega)\leq g(\omega)<nf(\omega)$. Then $g*^n f\in I_\omega$ which proves that $I_{f,\omega}=X$, so that I_ω is a maximal ideal of X.

Therefore, $\bigcap\{M:\ M\in\mathcal{M}(X)\}\subseteq\bigcap\{I_\omega:\ \omega\in\Omega\}=\{0_\Omega\}$, which proves that X is semisimple. □

Theorem 6.1.2 *Let M be a maximal ideal of a commutative BCK-algebra $(X;*,0)$ with the relative cancellation property. Then X/M is isomorphic to a BCK-subalgebra of $([0,\infty);*_{\mathbb{R}},0)$.*

Proof. Since any maximal ideal M of a commutative BCK-algebra X is prime, Exercise 4.6.1.8, the relative cancellation property of X entails that X/M is linear, Lemma 5.2.16. In addition, an ideal M of X is maximal iff X/M is simple, Exercise 3.6.1.8. It is simple to verify that X/M also satisfies the relative cancellation property, and by Exercise 1.6.1.8, X/M is Archimedean. By Theorems 5.2.38 and 5.2.40, X/M has a universal group G which is linear and Archimedean. Due to Hölder's theorem, [Bir, Thm XIII.12], G is isomorphic to a subgroup of the additive group of all real numbers. This implies that X/M is isomorphic to some BCK-subalgebra of $([0,\infty);*_{\mathbb{R}},0)$. □

Let $\mathcal{M}(X)\neq\emptyset$. By Theorem 6.1.2, for each maximal ideal M of X, we have established a homomorphism

$$\phi_M: X\to X/M\to[0,\infty). \tag{6.1}$$

Fix a family $\{\phi_M:\ M\in\mathcal{M}(X)\}$ of homomorphims from X into $[0,\infty)$ given by (6.1).[1] For each $x\in X$, we associate a function $\hat{x}\in[0,\infty)^{\mathcal{M}(X)}$ by

$$\hat{x}:\mathcal{M}(X)\to[0,\infty),\quad \text{where}\quad \hat{x}(M)=\phi_M(x),\ M\in\mathcal{M}(X). \tag{6.2}$$

Let $\mathcal{F}(X)=\{\hat{x}:\ x\in X\}$. Then $\mathcal{F}(X)$ is a commutative BCK-algebra of positive functions on $\mathcal{M}(X)$ because $\hat{x}*\hat{y}=\widehat{x*y}$, $x,y\in X$.

[1]For additional information on such representations see Lemma 6.1.21 and Section 2.2.4.

Theorem 6.1.3 *For any commutative BCK-algebra* $(X; *, 0)$ *with the relative cancellation property and with* $\mathcal{M}(X) \neq \emptyset$ *there is a commutative BCK-algebra* $\mathcal{F}(X) \subseteq [0, \infty)^{\Omega}$ *for some* $\Omega \neq \emptyset$ *and a homomorphism of* X *onto* $\mathcal{F}(X)$*. The homomorphism* $\Phi : x \mapsto \hat{x}$ *is one-to-one if and only if* X *is semisimple.*

Proof. It is easy to see that the mapping $\Phi : x \mapsto \hat{x}$ is a homomorphism of BCK-algebras. The equality $\hat{x} = \hat{y}$ holds iff, for each maximal ideal M of X, $x * y \in M$ and $y * x \in M$. Hence, if $Rad(X) = \{0\}$, we obtain $x * y = 0 = y * x$ which gives $x = y$.

If $Rad(X) \neq \{0\}$, choose a non-zero element $y \in Rad(X)$. Then $\hat{y}(M) = \phi_M(y) = 0 = \hat{0}(M)$, for some $M \in \mathcal{M}(X)$, which contradicts that Φ is one-to-one. □

Theorem 6.1.4 *Let* $(X; *, 0)$ *be a non-trivial commutative BCK-algebra with the relative cancellation property. The following statements are equivalent.*

(i) X *is isomorphic to some commutative BCK-algebra of functions on some* $\Omega \neq \emptyset$.

(ii) X *is semisimple.*

(iii) X *has an order-determining system of measure-morphisms.*

(iv) $\hat{X}$ *is semisimple.*

(v) $\hat{X}$ *is isomorphic to some commutative BCK-algebra of functions on some* $\Omega \neq \emptyset$.

(vi) $\hat{X}$ *has an order-determining system of measure-morphisms.*

Proof. (i) $\Leftrightarrow$ (ii). By Proposition 6.1.1, any function-representable commutative BCK-algebra is semisimple. On the other hand, by Theorem 6.1.3, any semisimple BCK-algebra is isomorphic to some commutative BCK-algebra of functions.

(i) $\Leftrightarrow$ (iii). Suppose now that X is a system of functions from $[0, \infty)^{\Omega}$. Given $\omega \in \Omega$, the mapping $m_{\omega}(f) := f(\omega)$, $f \in X$, is a measure-morphism on X, and the system $\mathcal{S} = \{m_{\omega} : \omega \in \Omega\}$ is order-determining.

Conversely, let $\mathcal{S}$ be an order-determining system of measure-morphisms on X. Given $x \in X$, define the mapping $\bar{x} : \mathcal{S} \to [0, \infty)$ by $\bar{x}(m) := m(x)$, $m \in \mathcal{S}$. Then $\bar{X} := \{\bar{x} : x \in X\}$ is a commutative BCK-algebra of positive functions defined on $\mathcal{S}$. Since $\mathcal{S}$ is order determining, the mapping $x \mapsto \bar{x}$ is an isomorphism of X onto $\bar{X}$.

(ii) $\Leftrightarrow$ (iv). Let X be semisimple, then by Theorem 5.2.35(v), $\mathcal{M}(\hat{X}) = \{\hat{M} : M \in \mathcal{M}(X)\}$. Let $x \in Rad(\hat{X})$, then $x = \bigvee_i x_i$, where $x_i \in X$. Since $x_i \leq x \in \hat{M}$ for any $M \in \mathcal{M}$, we have $x_i \in M$ for any M and any i, so that $x_i = 0$ and $x = 0$.

Suppose now that $\hat{X}$ is semisimple. Then $Rad(X) \subseteq Rad(\hat{X}) = \{0\}$.

The remaining equivalences are now evident. □

6.1.2 Topologies and Semisimplicity

Following Belluce [Bel], who worked with MV-algebras, we introduce the following topologies (called also hull-kernel or spectral topologies) on the sets of ideals also for commutative BCK-algebras and we show their connection to semisimplicity.

For each $x \in X$, let

$$M(x) = \{M : M \in \mathcal{M}(X), x \notin M\}.$$

Then $M(0) = \emptyset$, $M(a) \subseteq M(b)$ whenever $a \leq b$, $M(a \wedge b) = M(a) \cap M(b)$, $a, b \in X$, $M(a \vee b) = M(a) \cup M(b)$ whenever $a \vee b$ is defined in X, and $\{M(x) : x \in X\}$ is the base of a topology, say $\mathcal{T}_{\mathcal{M}}$, on the set $\mathcal{M}(X)$ of all maximal ideals on X. It is possible to show that a subset C of $\mathcal{M}(X)$ is closed iff

$$C = C(I) := \{M \in \mathcal{M}(X) : M \supseteq I\}, \tag{6.3}$$

where I is an arbitrary ideal of X. Similarly, O is open in $\mathcal{M}(X)$ if

$$O = O(I) := \{M \in \mathcal{M}(X) : M \not\supseteq I\}. \tag{6.4}$$

If $\mathcal{P}(X)$ is the set of all prime ideals on X, then $\mathcal{P}(X) \neq \emptyset$ (see Lemma 5.2.16), and define, for each $x \in X$,

$$P(x) = \{P : P \in \mathcal{P}(X), x \notin P\}.$$

Then, similarly to above, $P(0) = \emptyset$, $P(a) \subseteq P(b)$ whenever $a \leq b$, $P(a \wedge b) = P(a) \cap P(b)$, $a, b \in X$, $P(a \vee b) = P(a) \cup P(b)$ whenever $a \vee b$ is defined in X, and $\{P(x) : x \in X\}$ is the base of a topology, say $\mathcal{T}_{\mathcal{P}}$, on the set $\mathcal{P}(X)$. In addition, $P(x) = \emptyset$ iff $x = 0$.

Similarly as for $\mathcal{T}_{\mathcal{M}}$, it is possible to show that a subset C of $\mathcal{P}(X)$ is closed iff

$$C = C(I) := \{M \in \mathcal{P}(X) : M \supseteq I\}, \tag{6.5}$$

where I is an arbitrary ideal of X, and O is open in $\mathcal{P}(X)$ iff

$$O = O(I) := \{M \in \mathcal{P}(X) : M \not\supseteq I\}. \tag{6.6}$$

Proposition 6.1.5 *Let us define the mappings $\Phi_{\mathcal{M}} : \mathcal{M}(X) \to \mathcal{M}(\hat{X})$ and $\Phi_{\mathcal{P}} : \mathcal{P}(X) \to \mathcal{P}(\hat{X})$ by (5.27). Then $\Phi_{\mathcal{M}}$ and $\Phi_{\mathcal{P}}$ are homeomorphisms.*

Proof. If $\mathcal{M}(X) = \emptyset$, the statement is trivial. Suppose thus $\mathcal{M}(X) \neq \emptyset$. By (iv) and (v) of Theorem 5.2.35, the mappings $\Phi_{\mathcal{M}}$ and $\Phi_{\mathcal{P}}$ are one-to-one. Take a basic neighbourhood $\tilde{M}(x)$, where $x \in \hat{X}$. Then $x = \bigvee_{i=1}^{n} x_i$ with $x_i \in X$. Then $\Phi_{\mathcal{M}}^{-1}(\tilde{M}(x)) = \Phi_{\mathcal{M}}^{-1}(\bigcup_{i=1}^{n} \tilde{M}(x_i)) = \bigcup_{i=1}^{n} \Phi_{\mathcal{M}}^{-1}(\tilde{M}(x_i)) = \bigcup_{i=1}^{n} M(x_i)$, so that $\Phi_{\mathcal{M}}$ is continuous.

Similarly, $\Phi_{\mathcal{M}}(M(x)) = \tilde{M}(x)$, where $x \in X$.

In an analogous way we prove that $\Phi_{\mathcal{P}}$ is a homeomorphism. □

Theorem 6.1.6 *Let $(X;*,0)$ be a commutative BCK-algebra with the relative cancellation property. The topology $\mathcal{T}_{\mathcal{P}}$ is a T_0-space, and if $\mathcal{M}(X) \neq \emptyset$, then the topology $\mathcal{T}_{\mathcal{M}}$ is a T_2-space.*

Proof. (i) Let $M_1 \neq M_2$, $M_1, M_2 \in \mathcal{P}(X)$. Then either $M_1 \setminus M_2 \neq \emptyset$ or $M_2 \setminus M_1 \neq \emptyset$. In the first case take $0 \neq x \in M_1 \setminus M_2$. Then $M_2 \in P(x)$ and $M_1 \notin P(x)$.

Similarly we proceed in the second case $M_2 \setminus M_1 \neq \emptyset$.

(ii) Take two different maximal ideals M_1 and M_2 of X. We can find two different points $x \in M_1 \setminus M_2$ and $y \in M_2 \setminus M_1$. Therefore, $x \in M_1$, $M_2 \in M(x)$ and $y \in M_2$, $M_1 \in M(y)$. We have

$$\emptyset = M(0) = M((x * y) \wedge (y * x)) = M(x * y) \cap M(y * x).$$

Since $x * y \in M_1$, we conclude that $x * y \notin M_2$. Similarly, $y * x \notin M_1$. This proves $M_2 \in M(x * y)$ and $M_1 \in M(y * x)$. □

Theorem 6.1.7 *Let $(X;*,0)$ be a commutative BCK-algebra with the relative cancellation property. The following statements are equivalent.*

(i) $\mathcal{M}(X) = \mathcal{P}(X)$.

(ii) $\mathcal{P}(X)$ *is a T_2-space.*

(iii) $\mathcal{M}(\hat{X}) = \mathcal{P}(\hat{X})$.

(iv) $\mathcal{P}(\hat{X})$ *is a T_2-space.*

Proof. The implication (i) $\Rightarrow$ (ii) follows from Theorem 6.1.6.

(ii) $\Rightarrow$ (i). (a) First we claim that if $M_1 \subseteq M_2$ for two prime ideals M_1 and M_2 of X, then $M_1 = M_2$. Indeed, let $M_1 \neq M_2$. Then there are two mutually disjoint basic neighbourhoods $P(x)$ and $P(y)$ of M_1 and M_2, respectively. Therefore $\emptyset = P(x) \cap P(y) = P(x \wedge y)$ which implies $x \wedge y = 0$. Since M_1 and M_2 are prime ideals, then $x \wedge y = 0 \in M_1$ entails $x \in M_1$ or $y \in M_1$, and because $x \notin M_1$, we have $y \in M_1 \subseteq M_2$ which contradicts $M_1 \in P(y)$.

(b) Let now M be a prime ideal of X and let I be a proper ideal of X containing M. Take $y \in X \setminus I$. Due to Lemma 5.2.18 and Exercise 26.1.8, there is a prime ideal P of X containing I such that $y \notin P$. Therefore by step (a), $P = I = M$ which gives that M is a maximal ideal of X.

We obtain the remaining implications using the homeomorphisms $\Phi_{\mathcal{P}}$ and $\Phi_{\mathcal{M}}$ of Proposition 6.1.5. □

We now present a topological characterization of semisimple BCK-algebras.

Theorem 6.1.8 *Let $(X;*,0)$ be a commutative BCK-algebra with the relative cancellation property. The following statements are equivalent:*

(i) *X is semisimple.*

(ii) *$\mathcal{M}(X)$ is dense in $\mathcal{P}(X)$.*

(iii) *$\mathcal{M}(\hat{X})$ is dense in $\mathcal{P}(\hat{X})$.*

(iv) *$\hat{X}$ is semisimple.*

Proof. (i) $\Rightarrow$ (ii). Suppose that X is semisimple. According to Theorem 6.1.4, X is isomorphic to a commutative BCK-algebra of functions of $[0,\infty)^{\Omega}$ for some $\Omega \neq \emptyset$. Let M be a prime ideal of X and consider a basic neighbourhood $P(x)$ of M in $\mathcal{P}(X)$, that is, $x \notin M$. Clearly $x \neq 0$, and since x is in fact a non-zero function of $[0,\infty)^{\Omega}$, there exists $\omega \in \Omega$ such that $x(\omega) \neq 0$. Then $x \notin I_{\omega} := \{y \in X : y(\omega) = 0\}$, and in addition, I_{ω} is a maximal ideal of X. Then $I_{\omega} \in P(x)$ and, hence, $\mathcal{M}(X)$ is dense in $\mathcal{P}(X)$.

(ii) $\Rightarrow$ (i). Conversely, suppose that $\mathcal{M}(X)$ is dense in $\mathcal{P}(X)$ and take $x \in Rad(X)$. Suppose that $x \neq 0$. Then there is an ideal I which is maximal with respect to the property that $x \notin I$. As was shown in the proof of Lemma 5.2.18, I is a prime ideal of X. This means that $I \in P(x)$. Since $\mathcal{M}(X)$ is dense in $\mathcal{P}(X)$, it follows that there exists $M \in \mathcal{M}(X)$ such that $M \in P(x)$, that is $x \notin M$, which contradicts $x \in Rad(X)$.

The remaining equivalences are proved using the homeomorphism $\Phi_{\mathcal{P}}$ from Proposition 6.1.5. □

We now show that the existence of a quasi strong unit is closely connected with the compactness of $\mathcal{P}(X)$ and $\mathcal{M}(X)$. We note that $\hat{X}$ is the BCK-hull of X, see Section 5.2.8.

Theorem 6.1.9 *If $\hat{X}$ possesses a quasi strong unit, then $\mathcal{M}(X) \neq \emptyset$ and the topologies $\mathcal{T}_{\mathcal{P}}$ and $\mathcal{T}_{\mathcal{M}}$ are compact. If $\mathcal{T}_{\mathcal{P}}$ is compact, then $\hat{X}$ possesses a quasi strong unit. If any proper ideal of X can be embedded into maximal one, then the compactness of $\mathcal{T}_{\mathcal{M}}$ implies that $\hat{X}$ possesses a quasi strong unit.*

Proof. In view of Proposition 6.1.5, the spaces of prime ideals (maximal ideals) on X and $\hat{X}$ are homeomorphic.

Suppose that u is a quasi-strong unit for $\hat{X}$. An ideal I of $\hat{X}$ is proper iff $u \notin I$, and by that, using Lemma 5.2.18, we can show that given a proper ideal I there is a maximal ideal M of $\hat{X}$ containing I.

Assume now that $\{O(I_i)\}$ is an open covering of $\mathcal{M}(\hat{X})$. Then $\mathcal{M}(\hat{X}) = O(I)$, where $I = \bigvee_i I_i$. Because $\hat{X}$ has a quasi strong unit, we have $I = \hat{X}$. Since the ideal I is generated by the family $\{I_i\}$ by Theorem 5.1.32, it is of the form

$$I = \{x \in \hat{X} : \exists\, x_1 \in I_{i_1}, \ldots, x_n \in I_{i_n} \text{ such that } (\cdots(x * x_1) * \cdots) * x_n = 0\},$$

and we have that $u \in I$. Hence, there are indices $i_1, \ldots, i_n$ and $x_1 \in I_{i_1}, \ldots, x_n \in I_{i_n}$ such that

$$(\cdots(u * x_1) * \cdots) * x_n = 0,$$

which means that $u \in \bigvee_{j=1}^{n} I_{i_j}$. Hence

$$\bigcup_{j=1}^{n} O(I_{i_j}) = \mathcal{M}(\hat{X}),$$

and this proves the compactness of $\mathcal{M}(\hat{X})$ and of $\mathcal{M}(X)$.

Conversely, let $\mathcal{M}(\hat{X})$ be a non-empty compact set. Since $\bigcup_{x\in\hat{X}} M(x) = \mathcal{M}(\hat{X})$, the compactness of $\mathcal{M}(\hat{X})$ entails that there is finitely many elements $x_1, \ldots, x_n$ in $\hat{X}$ such that $\bigcup_{i=1}^{n} M(x_i) = \mathcal{M}(\hat{X})$. Then $\mathcal{M}(\hat{X}) = M(u)$, where $u = \bigvee_{i=1}^{n} x_i$. Since $M(u) = O(I_u)$, where I_u is the ideal of $\hat{X}$ generated by u, we see that u is a quasi strong unit for $\hat{X}$.

Similarly we proceed for the set of prime ideals. □

6.1.3 Fuzzy set representation of BCK-algebras

We say that X has a **fuzzy set representation** if X is isomorphic to a commutative BCK-algebra of fuzzy sets on some set $\Omega \neq \emptyset$, i.e., X can be embedded into $[0,1]^{\Omega}$.

We say that a commutative BCK-algebra $(X;*,0)$ with the relative cancellation property is **compact** if $\mathcal{M}(X) \neq \emptyset$ and, given a maximal ideal M of X, the quotient X/M can be embedded into the closed interval $[0,c]$ for some $c > 0$. Without loss of generality we can assume that $c = 1$.

For example, bounded commutative BCK-algebras, the interval $[0,1)$, and finite commutative BCK-algebras with the relative cancellation property are compact BCK-algebras, while the set of all non-negative integers is not compact.

Theorem 6.1.10 *For any compact commutative BCK-algebra $(X;*,0)$ with the relative cancellation property there is a commutative BCK-algebra $\mathcal{F}(X)$ of fuzzy sets and a homomorphism of X onto $\mathcal{F}(X)$.*

Proof. This is identical with the proof of Theorem 6.1.3 □

Theorem 6.1.11 *A non-trivial commutative BCK-algebra $(X;*,0)$ with the relative cancellation property possesses at least one non-zero measure-morphism if and only if $\mathcal{M}(X) \neq \emptyset$. If, in addition, X is compact, then there is at least one non-trivial bounded measure-morphism on X.*

Proof. Suppose $\mathcal{M}(X) \neq \emptyset$. According to Theorem 6.1.3 and equation (6.2), the mapping $x \mapsto \hat{x} \in [0,\infty)^{\mathcal{M}(X)}$, $x \in X$, where $\hat{x}(M) = \phi_M(x)$, $M \in \mathcal{M}(X)$, is a homomorphism of X onto the commutative BCK-algebra $\mathcal{F}(X) = \{\hat{x} : x \in X\}$.

Given $M \in \mathcal{M}(X)$, the mapping $m_M(x) := \phi_M(x)$, $x \in X$, is a measure-morphism. Since X is non-trivial, there exist a non-zero $x \in X$ and a maximal ideal M of X such that $x \notin M$. Then $\hat{x}(M) \neq 0$ which proves that there is a non-trivial measure-morphism on X.

Conversely, let m be a non-trivial measure-morphism on X. Denote by $I_m = \{x \in X : m(x) = 0\}$. Then I_m is a proper ideal of X. We assert that I_m is a maximal ideal. Indeed, take $x \in X$ such that $m(x) \neq 0$, and denote by $I_{m,x}$ the ideal of X generated by I_m and x. Given $y \in X$, there is an integer $n \geq 1$ such that $(n-1)m(x) \leq m(y) < nm(x)$. Then $m(y *^n x) = 0$, so that $I_{m,x} = X$, consequently, I_m is a maximal ideal of X.

If X is compact, we proceed in an analogous way. □

Theorem 6.1.12 *Any semisimple commutative BCK-algebra with the relative cancellation property is Archimedean.*

Proof. Let na be defined in X for any integer $n \geq 1$, and let $na \leq b$ for some element $b \in X$. We claim $a = 0$. If not, then $a \notin Rad(X) = \{0\}$, which yields that there is a maximal ideal M_a of X such that $a \notin M_a$. The ideal $M_{a,a}$ of X generated by M_a and a is equal to X. Hence $b \in M_{a,a}$, and consequently, there is an integer n_0 such that $b *^{n_0} a \in M_a$.

Since $na \leq b$ for any $n \geq 1$, we have $b *^n a = b * (na)$. On the other hand $n_0a \leq (n_0+1)a$ which gives $b *^{n_0+1} a \leq b *^{n_0} a \in M_a$. Therefore $(b *^{n_0} a) * (b *^{n_0+1} a) \in M_a$. On the other hand, $(b *^{n_0} a) * (b *^{n_0+1} a) = (b * (n_0 a)) * (b * (n_0+1)a) = [b * (b * (n_0 + 1)a)] * (n_0 a) = ((n_0+1)a) * (n_0 a) = a \in M_a$ which is a contradiction with $a \notin M_a$. Therefore, $a = 0$ and X is Archimedean. □

Remark 6.1.13 Not every Archimedean (even Dedekind σ-complete) commutative BCK-algebra with the relative cancellation property is semisimple. Example 5.2.42 gives an X with $\mathcal{M}(X) \neq \emptyset$, so that this X is not semisimple.

6.1.4 Atomic and Semisimple BCK-algebras

We note that a non-zero element a of X is said to be an **atom** of X if $x \leq a$ implies either $x = 0$ or $x = a$. Let $\mathcal{A}(X)$ denote the set of all atoms of X. X is said to be **atomic** if given a non-zero element $x \in X$ there is an atom a of X such that $a \leq x$. It is clear that $\mathcal{A}(X) = \mathcal{A}(\hat{X})$.

The **BCK-order** of an element $x \in X$, denoted by $ord_X(x)$, is the greatest integer $n \geq 1$ such that nx is defined in X. If nx is defined for any integer n, we define $ord(x) = \infty$, i.e.

$$ord_X(x) := \sup\{n \geq 1 : \; nx \in X\}.$$

We recall that our definition of order is different from that for MV-algebras.

A BCK-algebra X is **BCK-locally finite** if every element of X different from 0 has a finite BCK-order.

Proposition 6.1.14 *Let $(X; *, 0)$ be a commutative BCK-algebra with the relative cancellation property. Then X is BCK-locally finite if and only if $\hat{X}$ has the same property.*

Proof. Let X be locally finite and let $x \in \hat{X}$, $x \neq 0$. Then $x = x_1 + \cdots + x_k$, where $x_i \in X$ for any $1 \leq i \leq k$, and at least one of the x_i is non-zero. Assume that nx exists for an integer n. Then $nx = (nx_1) + \cdots + (nx_k)$ which implies that $ord_{\hat{X}}(x) = \min\{ord_X(x_1), \ldots, ord_X(x_k)\} < \infty$.

The converse statement is evident. □

The following result generalizes in the first part the similar assertion of Hoo [Hoo, Cor 4.6] for MV-algebras.

Theorem 6.1.15 *Let $(X; *, 0)$ be a commutative BCK-algebra with the relative cancellation property and let I_a be the ideal of X generated by an atom a of X. Then there exists either an integer $n \geq 1$ such that na is defined in X, but $(n+1)a$ does not exist in X, and then $I_a = \{0, a, 2a, \ldots, na\}$, or na is defined for any integer $n \geq 1$, and then $I_a = \{0, a, 2a, \ldots\}$.*

In addition, $ord_X(a) = ord_{\hat{X}}(a)$ for any atom a of X, and if a and b are two mutually different atoms of X, then

$$(na) \wedge (mb) = 0 \tag{6.7}$$

whenever na and mb are defined in X.

Proof. It is known that $I_a = \{x \in X : \exists \, n \geq 1 \; x *^n a = 0\}$. Take a non-zero element $x \in I_a$ and let $n \geq 1$ be the least integer such that $x *^n a = 0$. We claim that $x = na$.

Indeed, suppose first that $n = 1$. Then $x * a = 0$ so that $x \leq a$, which gives $x = a$. Suppose by induction that for an arbitrary non-zero element $y \in I_a$ and for any integer i with $1 \leq i \leq n$ which is minimal under the condition $y *^i a = 0$ we have proved that $y = ia$. Then $x *^{n+1} a = (x * a) *^n a = 0$, which gives, by hypothesis, $x * a = na$ if $x * a \neq 0$. Therefore, $na \leq x$, and $x = na + a = (n+1)a$.

Hence, either there exists an integer $n \geq 1$ such that na is defined in X but $(n+1)a$ is not, or na is defined for any integer n. According to the above we obtain $I_a = \{0, a, 2a, \ldots, na\}$ or $I_a = \{0, a, 2a, \ldots\}$.

Let now a be an atom of X. It is clear that $ord_X(a) \leq ord_{\hat{X}}(a)$. The ideal of X generated by a is $I_a = \{0, a, 2a, \ldots, ord_X(a)a\}$. Its BCK-hull $\widehat{I_a}$, given by (5.27), is equal to I_a. Due to Theorem 5.2.35 (i), $\widehat{I_a}$ is an ideal of $\hat{X}$ containing a. Therefore it also contains the ideal J_a of $\hat{X}$ generated by a. We know already that $J_a = \{0, a, 2a, \ldots, ord_{\hat{X}}(a)a\}$, hence, $J_a = I_a$ which proves $ord_X(a) = ord_{\hat{X}}(a)$.

Assume that $a, b \in \mathcal{A}(X)$ and let $a \neq b$. First we show that $(na) \wedge b = 0$. Indeed, if not, then $b \leq na$ and b belongs to the ideal I_a generated by a. By the first part of Theorem, $b = ia$ for some integer i which entails $a \leq b$, and this contradicts $b \neq a$.

Since $(na) \wedge (mb) \leq mb$, by the Riesz decomposition property, $(na) \wedge (mb) = b_1 + \cdots + b_m$, where $b_i \in \{0, b\}$. Then $(na) \wedge (mb) = kb$, where $0 \leq k \leq m$. If $k > 0$, then $0 = (na) \wedge b \wedge (mb) = (na) \wedge (mb) \wedge b = (kb) \wedge b = b$ which yields $b = 0$, a contradiction. Hence, $k = 0$. □

Given an integer $n \geq 1$, we denote by L_n the finite commutative BCK-subalgebra

$$L_n := \{0, \tfrac{1}{n}, \tfrac{2}{n}, \ldots, \tfrac{n}{n}\}$$

of $([0, \infty); *_{\mathbb{R}}, 0)$. Similarly we define

$$L_\infty = \{0, 1, 2, \ldots\}.$$

All L_n's and L_∞ are commutative BCK-algebras with linear order, and we call them **basic BCK-chains**.

The following results generalize results of Belluce [Bel, Thm 34] and Cignoli [Cig, Thm 2.3] for MV-algebras to commutative BCK-algebras with the relative cancellation property.

Proposition 6.1.16 *Let a be an atom of a semisimple commutative BCK-algebra $(X; *, 0)$ with the relative cancellation property. Then there is a maximal ideal M_a of X such that $a \notin M_a$. Moreover, $\mathcal{A}(X) \setminus \{a\} \subseteq M_a$, a/M_a is an atom of X/M_a,*

$$ord_X(a) = ord_{X/M_a}(a/M_a),$$

and X/M_a is isomorphic to L_n, where $n = ord_X(a)$.

Proof. Let X be semisimple and let a be an atom of X. Since $a \neq 0$, $a \notin Rad(X)$, there is a maximal ideal M_a of X such that $a \notin M_a$. Let b be an arbitrary atom of X different from a. Then $a \wedge b = 0 \in M_a$ which gives $b \in M_a$ since M_a is also a prime ideal of X. Hence, $\mathcal{A}(X) \setminus \{a\} \subseteq M_a$.

We assert that a/M_a is an atom of X/M_a. Indeed, suppose $x/M_a \leq a/M_a$ for some $x \in X$. Then $x/M_a = (x/M_a) \wedge (a/M_a) = (x \wedge a)/M_a$. If $x \wedge a = 0$, then $x \in M_a$ because $a \notin M_a$, and so $x/M_a = 0/M_a$; if $x \wedge a = a$, then $a \leq x$, and $a/M_a \leq x/M_a \leq a/M_a$ so that $a/M_a = x/M_a$.

Next we show by induction that if $na \in X$ and $(na)/M_a + a/M_a = b/M_a$ for some $c \in X$, then $(n+1)a$ is defined in X and $b \geq (n+1)a$.

Let $n = 1$, and let $a/M_a + a/M_a = b/M_a$. Then $a/M_a = (b*a)/M_a$. If $(b*a) \wedge a = 0$ then $b * a = 0 \in M_a$ which gives $b * a \in M_a$ and we obtain the contradiction $a/M_a = 0/M_a$. Hence $(b * a) \wedge a = a$ which entails $b * a \geq a$ and $b \geq b * a \geq a$, so that $b \geq 2a$.

Assume now that the assertion hold for any integer $i \leq n$. Then

$$\begin{aligned} ((n+1)a)/M_a + a/M_a &= c/M_a, \\ ((n+1)a)/M_a &= (c*a)/M_a, \\ (na)/M_a + a/M_a &= (c*a)/M_a. \end{aligned}$$

By the induction hypothesis, we have $c*a \geq (n+1)a$ which gives $c \geq c*a \geq (n+1)a \geq a$. Consequently, $(n+2)a$ exists in X and $c \geq (n+2)a$.

Since if na is defined in X, then so is $n(a/M_a)$ in X/M_a, and $n(a/M_a) = (na)/M_a$, which proves that $ord_X(a) \leq ord_{X/M_a}(a)$. From the above we have that $ord_X(a) = ord_{X/M_a}(a)$.

Let $I_{a/M}$ be the ideal of X/M generated by the atom a/M_a. According to Theorem 6.1.15, $I_{a/M} = \{0, a/M_a, 2(a/M_a), \ldots, ord(a/M_a)(a/M_a)\}$. On the other hand, x/M_a is simple because M_a is maximal, i.e., X/M_a does not contain any non-zero proper ideal of X/M_a. This gives $I_{a/M_a} = X/M_a$ and proves the problem in question. □

Theorem 6.1.17 *Every semisimple atomic commutative BCK-algebra $(X;*,0)$ with the relative cancellation property is a subdirect product of basic BCK-chains. Moreover, the following statements are equivalent:*

(i) *$(X;*,0)$ is a subdirect product of finite basic BCK-chains.*

(ii) *X is compact.*

(iii) *$ord(a) < \infty$ for any $a \in \mathcal{A}(X)$.*

Proof. Let M_a be a maximal ideal as in Proposition 6.1.16. Since $\mathcal{A}(X) \setminus \{a\} \subseteq M_a$, we have $\bigcap\{M_a : a \in \mathcal{A}(X)\} = \{0\}$. It is straightforward to verify that X is a subdirect product of $\{X/M_a : a \in \mathcal{A}(X)\}$. Any factor X/M_a is isomorphic to $L_{ord(a)}$ which proves the first part of Theorem.

(ii) $\Rightarrow$ (i). Suppose now X is compact. Then X/M_a is a bounded subset of $[0, \infty)$, i.e., X/M_a is isomorphic to a finite L_n.

(i) $\Rightarrow$ (ii). Conversely, let X be a subdirect product of finite BCK-chains, i.e., X can be subdirectly embedded into $\prod_{i \in I} L_{n_i}$, where $n_i < \infty$ for any $i \in I$. Suppose that $x = (x_i)_{i \in I} \in \prod_{i \in I} L_{n_i}$. Then given $x \in X$, we can define the function $\hat{x} \in [0,1]^I$ by $\hat{x}(i) = x_i$. Then X has a fuzzy set representation $\{\hat{x} : x \in X\}$, which by Exercise 5.6.1.8 means that X is compact.

(iii) $\Rightarrow$ (ii). Suppose now $ord(a) < \infty$ for any $a \in \mathcal{A}(X)$. Then $X/M_a \cong L_{ord(a)}$ which means by Exercise 6.6.1.8 that X is compact.

(i) $\Rightarrow$ (iii). This is evident. □

Theorem 6.1.18 *The radical $Rad(X)$ of an Archimedean commutative BCK-algebra $(X; *, 0)$ with the relative cancellation property does not contain any atom of X.*

If, in addition, X is atomic, then X is semisimple, and it is a subdirect product of basic BCK-algebras. Moreover, for atomic commutative BCK-algebras with the relative cancellation property the notions of Archimedeanicity and semisimplicity coincide.

Proof. Suppose that $a \in Rad(X)$ for some atom a of X. There exists a prime ideal M_a of X such that $a \notin M_a$. We claim that M_a is a maximal ideal of X.

Let x be an arbitrary element of X. Denote by $j_a(x)$ the greatest integer $n \geq 0$ such that $na \leq x$. Since X is Archimedean, $j_a(x)$ is always finite. Consequently, $x = x_1 + x_2$, where $x_2 = j_a(x)a$ and $x_1 = x * x_2$. Moreover, $x_1 \wedge a = 0$, because if $x_1 \wedge a = a$, then $j_a(x) = j_a(x) + 1$. Therefore, $x_1 \in M_a$.

Suppose now $x \in I$, where I is a proper ideal of X containing M_a. Hence, $j_a(x) = 0$, if not then $a \in I$ which proves that $I = X$. Since $x = x_1$ and $x_1 \in M_a$, we have $I = M_a$.

Therefore, $a \in Rad(X) \subseteq M_a$ which proves $a \in M_a$. Consequently, we have $Rad(X) = \{0\}$.

The rest of the assertions follows from Theorem 6.1.17. □

6.1.5 Measure-Morphisms and Maximal Ideals

We note that according to 5.1.37 (v), if m is a measure on a BCK-algebra X, then

$$I_m := \{x \in X : m(x) = 0\} \tag{6.8}$$

is an ideal of X.

Proposition 6.1.19 *A measure m on a commutative BCK-algebra $(X; *, 0)$ is a measure-morphism on X if and only if*

$$m(x \wedge y) = \min\{m(x), m(y)\}, \; x, y \in X. \tag{6.9}$$

Proof. Suppose that m is a measure-morphism on X. Then $m(y * x) = m(y * (x \wedge y)) = m(y) - m(x \wedge y)$. On the other hand, $m(y * x) = m(y) *_{\mathbb{R}} m(x) = m(y) - \min\{m(x), m(y)\}$, which forces $m(x \wedge y) = \min\{m(x), m(y)\}$.

Suppose now (6.9) holds. Then $m(x * y) = m(x * (y \wedge x)) = m(x) - m(y \wedge x) = m(x) - \min\{m(x), m(y)\} = m(x) *_{\mathbb{R}} m(y)$. □

Proposition 6.1.20 *Let m be a non-trivial measure on a commutative BCK-algebra $(X;*,0)$ with the relative cancellation property. Then m is a measure-morphism if and only if I_m is a maximal ideal of X.*

Proof. Due to the hypothesis, I_m is always a proper ideal of X. Let m be a measure-morphism on X, and let $x \in X$ be such an element that $m(x) \neq 0$. Denote by $I_{m,x}$ the ideal of X generated by I_m and x, i.e., $I_{m,x} = \{y \in X : \exists\, n \geq 1,\ y *^n x \in I_m\}$. Let z be an arbitrary element in X. There exists an integer $n \geq 1$ such that $(n-1)m(x) \leq m(z) < nm(x)$. Then $m(z *^n x) = m(z) *_{\mathbb{R}}^n m(x) = 0$, i.e., $z \in I_{m,x}$, which says that $X = I_{m,x}$. Consequently, I_m is a maximal ideal of X.

Suppose now I_m is a maximal ideal of X. Then X/I_m is linear, i.e., for all $x, y \in X$, $m(x*y) = 0$ or $m(y*x) = 0$. Let $m(x*y) = 0$. Then $m(x*y) = m(x*(x \wedge y)) = m(x) - m(x \wedge y) = 0$. Hence, $m(x) = m(x \wedge y) \leq \min\{m(x), m(y)\}$, so that $m(x \wedge y) = \min\{m(x), m(y)\}$, and m is a measure-morphism. □

Lemma 6.1.21 *Let G_1 and G_2 be two Abelian lattice ordered subgroups of $(\mathbb{R};+)$ each containing a common non-zero element g_0. If there is an injective group homomorphism ϕ of G_1 into G_2 preserving the order such that $\phi(g_0) = g_0$, then $G_1 \subseteq G_2$ and ϕ is the identity on G_1. If, in addition, ϕ is surjective, then $G_1 = G_2$.*

Proof. Suppose the converse. Then there is a non-zero element $g \in G_1$ such that $\phi(g) = h \neq g$. In this case, there are two integers p and q, $q > 1$, such that either $h < \frac{p}{q} g_0 < g$ or $g < \frac{p}{q} g_0 < h$. In the first case we obtain $qh < pg_0 < qg$, so that $\phi(pg_0) < \phi(qg)$ which gives $p\phi(g_0) = pg_0 < q\phi(g) = qh$, and this is a contradiction. The second case is similar. □

Lemma 6.1.22 (1) *Let $(X;*,0)$ be a commutative BCK-subalgebra of $([0,\infty);*_{\mathbb{R}},0)$. Then there is an Abelian lattice ordered subgroup G of $(\mathbb{R};+)$ containing X such that (G, id_X) is a universal group for $(X;*_{\mathbb{R}},0)$.*

(2) *Let $(X_1;*_{\mathbb{R}},0)$ and $(X_2;*_{\mathbb{R}},0)$ be two commutative BCK-subalgebras of $([0,\infty);*_{\mathbb{R}},0)$ containing a non-zero element x_0. If there is an injective BCK-homomorphism ψ from X_1 into X_2 such that $\psi(x_0) = x_0$, then $X_1 \subseteq X_2$ and ψ is the identity on X_1. If, in addition, ψ is surjective, then $X_1 = X_2$.*

Proof. (1) Let $(G(X), h)$ be a universal group for $(X;*_{\mathbb{R}},0)$. Since id_X maps X into $\mathbb{R}$, there is a group homomorphism $g : G(X) \to \mathbb{R}$ such that $x = g(h(x))$, $x \in X$. We claim that g is injective. Indeed, $g(h(x)) = g(h(y))$ iff $x = y$. Thus $g(\sum_{i=1}^n h(x_i)) = 0$ iff $\sum_{i=1}^n h(x_i) = 0$ iff $x_1 = \cdots = x_n = 0$. Therefore, $g(\sum_{i=1}^n h(x_i)) = g(\sum_{j=1}^m h(y_j))$ iff $\sum_{i=1}^n h(x_i) = \sum_{j=1}^m h(y_j)$ which entails that g is injective.

Put $G := g(G(X))$, then (G, id_X) is a universal group for X containing X.

(2) Due to (1), there are two lattice ordered subgroups of $(\mathbb{R};+)$, G_1 and G_2, containing X_1 and X_2, respectively, such that (G_1, id_{X_1}) and (G_2, id_{X_2}) are universal

groups for X_1 and X_2. The universal property entails that there is a group homomorphism $\hat{\psi} : G_1 \to G_2$ such that $\hat{\psi}(x) = \psi(x)$ for any $x \in X_1$. Similarly as in part (1), we can show that $\hat{\psi}$ is injective. Moreover, $\hat{\psi}$ preserves the order and $\hat{\psi}(x_0) = \psi(x_0) = x_0$, which by Lemma 6.1.21 entails that $G_1 \subseteq G_2$, and $\hat{\psi}$ is the identity on G_1, consequently, ψ is the identity on X_1, and $X_1 \subseteq X_2$.

The rest of the proof is now evident. □

Proposition 6.1.23 *Let m_1 and m_2 be two non-zero measure-morphisms on a commutative BCK-algebra $(X; *, 0)$ with the relative cancellation property. Suppose $I_{m_1} = I_{m_2}$. If there is an element $x_0 \in X$ such that $m_1(x_0) = m_2(x_0) > 0$, then $m_1 = m_2$.*

Proof. First of all we shall assume that $I_{m_1} = I_{m_2} = \{0\}$. Hence $(m_1(X); *_{\mathbb{R}}, 0)$ and $(m_2(X); *_{\mathbb{R}}, 0)$ are two commutative BCK-subalgebras of $([0,\infty); *_{\mathbb{R}}, 0)$. Define the mapping $\phi_0 : m_1(X) \to m_2(X)$ via $\phi_0(m_1(x)) = m_2(x)$ $(x \in X)$. Then ϕ_0 is a BCK-isomorphism of $m_1(X)$ onto $m_2(X)$. Since $\phi_0(m_1(x_0)) = m_2(x_0) = m_1(x_0) > 0$, using Lemma 6.1.22, ϕ_0 is the identity mapping on $m_1(X)$ and $m_1(X) = m_2(X)$.

Suppose now that m_1 and m_2 satisfy the conditions of Lemma. Then, according to Proposition 6.1.20, I_{m_1} is a maximal ideal of X, and X/I_{m_1} is linear and Archimedean. In addition, we can define a measure-morphism $\overline{m_1}$ on X/I_{m_1} via $\overline{m_1}(x/I_{m_1}) = m_1(x)$, $x \in X$. Due to Theorem 5.1.42, $\overline{m_1}$ is a measure-morphism on X/I_{m_1} such that $I_{\overline{m_1}} = \{0\}$. Similarly we define $\overline{m_2}$ on $X/I_{m_2} = X/I_{m_1}$. Then $I_{\overline{m_1}} = \{0\} = I_{\overline{m_2}}$, and applying the first part of the proof, we obtain $\overline{m_1} = \overline{m_2}$, and consequently,

$$m_1(x) = \overline{m_1}(x/I_{m_1}) = \overline{m_2}(x/I_{m_2}) = m_2(x),$$

which finishes the proof. □

Corollary 6.1.24 *Let m_1 and m_2 be two non-zero measure-morphisms on a commutative BCK-algebra $(X; *, 0)$ with the relative cancellation property. If $I_{m_1} = I_{m_2}$, then there is a constant $c > 0$ such that $m_1(x) = c\,m_2(x)$, $x \in X$.*

Proof. Let x_0 be an element of X such that $m_1(x_0) > 0$. Then $m_2(x_0) > 0$. Let $c := m_1(x_0)/m_2(x_0)$ and put $\tilde{m}_2 = c\,m_2$. Thus $m_1(x_0) = \tilde{m}_2(x_0)$, and m_1 and $\tilde{m}_2$ satisfy the conditions of Proposition 6.1.23. Then $m_1 = \tilde{m}_2 = c\,m_2$. □

Let m_1 and m_2 be two measure-morphisms on $(X; *, 0)$. We shall write $m_1 \approx m_2$ iff there is a constant $c > 0$ such that $m_1 = c\,m_2$. It is evident that $\approx$ is an equivalence on the set $\mathcal{M}_m(X)$ of all measure-morphisms on X. We define $\mathcal{M}_{\tilde{m}}(X) := \mathcal{M}_m(X)/\approx$.

Theorem 6.1.25 *Let $(X; *, 0)$ be a non-trivial commutative BCK-algebra with the relative cancellation property. Then there is a one-to-one correspondence between the set of all maximal ideals $\mathcal{M}(X)$ and $\mathcal{M}_{\tilde{m}}(X)$.*

In particular, if $(X; *, 0)$ is a bounded commutative BCK-algebra, then $\mathcal{M}(X) \neq \emptyset$, and there is a one-to-one correspondence between $\mathcal{M}(X)$ and the set of all state-morphisms on X.

Proof. It is evident that $\mathcal{M}_m(X)$ is non-void iff $\mathcal{M}(X) \neq \emptyset$.

Suppose that m is a non-trivial measure-morphism on X. According to Proposition 6.1.20, Lemma 6.1.21, and Corollary 6.1.24, the mapping $m/\approx \mapsto I_m := \{x \in X : m(x) = 0\}$ is one-to-one and onto.

Suppose now that $1 \in X$ is the greatest element in X. The second assertion follows directly from Lemma 6.1.21 and Corollary 6.1.24. □

6.1.6 State Space of a BCK-algebra

Proposition 6.1.26 *Let u be a non-zero element of a commutative BCK-algebra $(X; *, 0)$ with the relative cancellation property. Denote by $\mathcal{S}_u(X)$ the set of all measures on X such that $m(u) = 1$. Then $\mathcal{S}_u(X)$ is a convex set, and any measure-morphism $m \in \mathcal{S}_u(X)$ is an extremal point of $\mathcal{S}_u(X)$.*

Proof. Suppose that $m = \lambda m_1 + (1 - \lambda) m_2$ for some $0 < \lambda < 1$ and for $m_1, m_2 \in \mathcal{S}_u(X)$. Then $m \in \mathcal{S}_u(X)$, and $I_m = I_{m_1} \cap I_{m_2}$, and the maximality of I_m entails $I_m = I_{m_1} = I_{m_2}$. By Proposition 6.1.20, m_1 and m_2 are measure-morphisms on X, and in view of Proposition 6.1.23, $m_1 = m_2$. □

Proposition 6.1.27 *Any measure (any measure-morphism) on a commutative BCK-algebra $(X; *, 0)$ with the relative cancellation property can be uniquely extended to a measure (a measure-morphism) on the BCK-hull $\hat{X}$. Conversely, the restriction of a measure (a measure-morphism) on $\hat{X}$ to X is so on X.*

Proof. Let m be a measure-morphism on X. Then $m(x + y) = m(x) + m(y)$ whenever $x + y$ is defined in X. Since any element $z \in \hat{X}$ can be expressed in the form $z = z_1 + \cdots + z_n$, where $z_i \in X$, using the Riesz decomposition property, we can show that the mapping $\hat{m} : \hat{X} \to [0, \infty)$, defined by $m(z) = m(z_1) + \cdots + m(z_n)$, is a well-defined mapping which is a measure on $\hat{X}$. It is clear that $\hat{m}$ is the unique extension of m to $\hat{X}$.

If $X = \{0\}$, then $\hat{X} = X$. Assume thus $X \neq \{0\}$, and suppose now that m is, in addition, a measure-morphism on X. According to Proposition 6.1.20, I_m is a maximal ideal of X, consequently, by (v) of Theorem 5.2.35, $\widehat{I_m}$ is a maximal ideal in $\hat{X}$. It is clear that $\widehat{I_m} = I_{\hat{m}}$, so that $\hat{m}$ is a measure-morphism on $\hat{X}$.

The second part of the assertion is evident. □

A measure (a measure-morphism) m on X is said to be a **quasi state** (a **quasi state-morphism**) if $\sup\{m(x) : x \in X\} = 1$. If X is bounded, quasi states and states on X coincide. Denote by $\mathcal{S}_q(X)$ the set of all quasi states on X. In general, $\mathcal{S}_q(X)$ can be empty.

We note that $\mathbb{Z}^+ := \{0, 1, 2, \ldots\}$ with the natural operation $*$ defined by $n * m := \max\{0, n - m\}$, $n, m \in \mathbb{Z}^+$, gives an example of a linear commutative BCK-algebra such that any measure on $\mathbb{Z}^+$ is of the form $m_t(n) := t\,n,\ n \in \mathbb{Z}^+$, where t is any non-negative real number. Hence m_t is a measure-morphism, and $\mathcal{S}_q(\mathbb{Z}^+) = \emptyset$.

Proposition 6.1.28 *Let $(X; *, 0)$ be an upwards directed commutative BCK-algebra and let $\mathcal{S}_q(X) \neq \emptyset$. Then $\mathcal{S}_q(X)$ is a convex set, and any quasi state-morphism on X is an extremal point of $\mathcal{S}_q(X)$.*

Proof. Suppose that $m_1, m_2 \in \mathcal{S}_q(X)$, and let $\lambda \in [0,1]$. We claim $m := \lambda\, m_1 + (1-\lambda)\, m_2 \in \mathcal{S}_q(X)$. Indeed, given ϵ with $0 < \epsilon < 1$, there are x_1 and x_2 in X such that $m_i(x_i) > 1-\epsilon$ for $i = 1, 2$. Then, for $x = x_1 \vee x_2$, we have $m_i(x) > 1-\epsilon$, $i = 1, 2$, and hence, $m(x) > \lambda\,(1-\epsilon) + (1-\lambda)\,(1-\epsilon) = 1-\epsilon$, so that $\sup\{m(x) : x \in X\} = 1$.

Assume now that s is a quasi state-morphism, and let $s = \mu\, s_1 + (1-\mu)\, s_2$ for some $\mu \in (0,1)$ and for $s_1, s_2 \in \mathcal{S}_q(X)$. Then $I_s = I_{s_1} \cap I_{s_2}$, and similarly as in the proof of Proposition 6.1.26, we have $I_s = I_{s_1} = I_{s_2}$, and s_1 and s_2 are quasi state-morphisms. We assert that $s_1 = s_2$. Indeed, by Corollary 6.1.24, $s_1 = c\, s_2$ for some $c > 0$. Then $1 = \sup\{s_1(x) : x \in X\} = c\, \sup\{s_2(x) : x \in X\} = c$, which gives $s_1 = s_2$. □

A **measure** on an Abelian lattice ordered group G is any positive (group) homomorphism $s : G \to \mathbb{R}$, i.e., (i) $s(g_1 + g_2) = s(g_1) + s(g_2)$, $g_1, g_2 \in G$, and (ii) $s(g) \geq 0$ whenever $g \in G^+$. If u is a strong unit on G, i.e., $u > 0$, and given $g \in G$, there is an integer n such that $g \leq n\, u$, a measure s with the property $s(u) = 1$ is said to be a **state** on (G, u).

A net of measures $\{m_\alpha\}$ on X (on G) **converges weakly** to a measure m iff $m_\alpha(a) \to m(a)$ for every $a \in X$ $(a \in G)$.

Theorem 6.1.29 *Let $(X; *, 0)$ be a commutative BCK-algebra with the relative cancellation property and let $(G(X), h)$ be its universal group. Then, for any measure s on $G(X)$, the mapping s' on X defined by $s'(x) := s(h(x))$, $x \in X$, is a measure on X. The mapping $s \mapsto s'$ is a one-to-one correspondence between measures on $G(X)$ and measures on X preserving convex linear combinations.*

If, in addition, $u > 0$, $u \in X$, and if $\mathcal{S}_u(G(X))$ is the set of all measures s on $G(X)$ with the property $s(h(u)) = 1$, then $\mathcal{S}_u(G(X))$ is a convex set, and under the above correspondence, the extremal measures on $\mathcal{S}_u(X)$ are mapped onto extremal measures of $\mathcal{S}_u(G(X))$.

Proof. The assertion follows directly from the fact that $(G(X), h)$ is a universal group for X. □

If u is a quasi strong unit for X, it is easy to see that $\mathcal{S}_u(X)$ in the weak topology is a compact space: Let $(G(X), h)$ be a universal group for X. Since any measure m on X preserves $+$ and the order on X, then the universal property of $(G(X), h)$

entails that there is a unique measure [2] $\hat{m} : G(X) \to R$ such that $\hat{m}(h(x)) = m(x)$ for any $x \in X$.

We now characterize the state space of MV-algebras (= bounded commutative BCK-algebras) (for a more general case, see Theorem 6.1.32).

Theorem 6.1.30 *Let $(X; *, 0)$ be a non-trivial bounded commutative BCK-algebra with the greatest element 1. Then $\mathcal{S}_1(X)$ is a non-empty compact convex Hausdorff space with respect to the weak topology, and the space of all state-morphisms on X is a non-void compact Hausdorff space. Any state is a weak limit of convex linear combinations of the set of extremal points of $\mathcal{S}_1(X)$. In addition, the following statements are equivalent:*

(i) *m is an extremal state on X.*

(ii) *m is a state-morphism on X.*

(iii) $m(x \wedge y) = \min\{m(x), m(y)\},\ x, y \in X.$

Proof. Let I be an ideal of X which is maximal under the condition $1 \notin I$. Due to Lemma 5.2.18, I is a maximal ideal of X, which by Theorem 6.1.25 forces that there is at least one state-morphism on X.

Let $(G(X), h)$ be a universal group for $(X; *, 0)$. Then by Theorem 5.2.23 (or by Mundici [Mun] when we use the language of MV-algebras) $u := h(1)$ is a strong unit in $G(X)$, and X is isomorphic (as a BCK-algebra) with the interval $[0, h(1)] := \{g \in G(X) : 0 \leq g \leq h(1)\}$.

Using the Krein–Mil'man theorem, [ReSi], we obtain the fact that any state on X is a weak limit point of the convex hull of the extremal points of $\mathcal{S}_1(X)$. In addition, it is now easy to verify that state-morphisms form a non-void compact Hausdorff space.

Suppose now that m is a state on X and let s_m be a unique state on $G(X)$ such that $s_m(h(x)) = m(x)$ $(x \in X)$. Due to [Goo 1, Thm 12.18] and Theorem 6.1.29, we obtain the equivalence of (i)–(iii). □

Remark 6.1.31 If $(X; *, 0)$ is a non-trivial bounded commutative BCK-algebra with the greatest element 1, then by Theorems 6.1.25 and 6.1.30, there is a one-to-one correspondence between the maximal ideals on X, state-morphisms on X, and extremal states on X, respectively.

We say that $(X; *, 0, 1)$ is **Archimedean in the sense of Belluce** if for each $x, y \in X$, if $n \odot x \leq y$ [3] for all $n \geq 0$, then $x \odot y = x$. Belluce [Bel, Thms 31, 33] proved that X is semisimple iff X is Archimedean in the sense of Belluce. The following result

[2] We note that a mapping s on a group G is a **measure** on G if $m(g_1 + g_2) = m(g_1) + m(g_2)$ for all $g_1, g_2 \in G$, and $m(g) \geq 0$ for any $g \geq 0$.

[3] We recall that $n \odot x := x_1 \oplus \cdots \oplus x_n$, where $x_1 = \cdots = x_n = x$, $n \geq 1$, $x \in M$.

compares these two notions of Archimedeanicity, which coincide in MV-algebras, and it gives another criteria of semisimplicity of X using an order determining system of measure-morphisms (compare also to [Dvu]).

Theorem 6.1.32 *Let $(X; *, 0, 1)$ be a bounded non-trivial commutative BCK-algebra. The following statements are equivalent:*

(i) *X is semisimple.*

(ii) *X is Archimedean.*

(iii) *X Archimedean in the sense of Belluce.*

(iv) *There exists an order-determining system of state-morphisms on X.*

(v) *There exists an order-determining system of states on X.*

(vi) *X is isomorphic to some commutative BCK-algebra of fuzzy sets on some $\Omega \neq \emptyset$.*

Proof. The equivalence of (i) and (iii) has been proved by Belluce [Bel, Thms 31, 33]. The statements (i), (iv), (vi) are equivalent according to Theorem 6.1.4, and (i) implies (ii) in view of Theorem 6.1.12.

(v) $\Rightarrow$ (iv). Let $\mathcal{S}$ be an order-determining system of states on X. Let $\mathcal{S}_{sm}$ be the space of all state-morphisms on X, and suppose $m(x) \leq m(y)$ for all $m \in \mathcal{S}_{sm}$. The inequality $m(x) \leq m(y)$ holds for any m from the convex hull of $\mathcal{S}_{sm}$ and, consequently, for any weak limit of such measures. Due to Theorem 6.1.30, $m(x) \leq m(y)$ holds for any state $m \in \mathcal{S}$, so that $x \leq y$, and $\mathcal{S}_{sm}$ is order-determining.

The implication (iv) $\Rightarrow$ (v) is evident.

(ii) $\Rightarrow$ (v). Let now X be Archimedean. According to Theorem 5.2.40, its universal group $(G(X), h)$ is an Archimedean lattice ordered group, and, in addition, $u := h(1)$ is a strong unit in $G(X)$.

Due to [Goo 1, Thm 4.14], $\mathcal{S}_u(G(X))$ is an order determining system of states on $G(X)$, and according to Theorem 6.1.29, $\mathcal{S}_u(X)$ is an order-determining system of states on X. □

As a matter of interest we note that Belluce [Bel, p. 1378] has posed an open problem, whether any complete MV-algebra (= bounded commutative BCK-algebra) is semisimple (his suspicion was probably not). Cignoli [Cig, Cor 2.2] gave a positive answer. Since due to Proposition 5.2.41, any Dedekind σ-complete MV-algebra is Archimedean, in view of Theorem 6.1.32, we have another confirmation of Cignoli's result.

We now characterize the state space of BCK-algebras having quasi strong units.

Proposition 6.1.33 *A linear commutative BCK-algebra $(X; *, 0)$ is Archimedean if and only if any non-zero element u of X is a quasi strong unit.*

Proof. Due to Theorem 5.2.38, X is linear iff $G(X)$ is linear.

Assume first that, for any positive element $u \in X$, $h(u)$ is a strong unit in $G(X)$. We show that any positive element g in $G(X)$ is a strong unit in $G(X)$. Indeed, let g' be any element of $G^+(X)$. Then $g = \sum_{i=1}^k n_i h(x_i)$, where $n_i > 0$ and $x_i \in X \setminus \{0\}$ for any $i = 1, \dots, k$. Since any $h(x_i)$ is a strong unit in $G(X)$, there exists an integer $n \geq 1$ such that $g' \leq nh(x_i)$ for any i. Then $kg' \leq ng \leq kng$ which gives $g' \leq n\, g$, so that g is a strong unit in $G(X)$.

It is now simple to verify that $G(X)$ is an Archimedean group, which by Theorem 5.2.40 forces X to be Archimedean.

Suppose now that X is Archimedean. In view of Theorem 5.2.40, $G(X)$ is Archimedean. Choose a non-zero element u in X. Then $h(u) > 0$. If for some $g \in G^+(X)$, $ng \leq h(u)$ for any $n \geq 1$, then the Archimedeanicity of $G(X)$ entails that $g = 0$, and, consequently, u is a quasi strong unit in X. □

Proposition 6.1.34 *Let $(X; *, 0)$ be a linear Archimedean commutative BCK-algebra with an atom a. If $ord_X(a) < \infty$, then na is a strong unit in X if and only if n divides $ord_X(a)$; otherwise na is a quasi strong unit in X. If $ord_X(a) = \infty$, then na is a strong unit in X for any integer $n \geq 1$.*

Proof. Let I_a be an ideal of X generated by a. According to Theorem 6.1.15, $I_a = \{0, a, 2a, \dots, ord_X(a)a\}$ if $ord_X(a) < \infty$, or $I_a = \{0, a, 2a, \dots\}$ if $ord_X(a) = \infty$. Let x be a non-zero element of X. The linearity of X entails that $a \leq x$. Since X is Archimedean, there is a greatest integer $j_a(x) \geq 1$ such that $j_a(x)a \leq x$. It is not hard to verify that, in fact, $j_a(x)a = x$, which proves that $I_a = X$. The assertions of Proposition are now evident because the universal group of X is $(\mathbb{Z}, j_a)$. □

The following result extends Theorem 6.1.30 to commutative BCK-algebras which need not be bounded.

Theorem 6.1.35 *Let u be a quasi strong unit of a non-trivial commutative BCK-algebra $(X; *, 0)$ with the relative cancellation property. Then $\mathcal{S}_u(X)$ is a non-empty compact convex Hausdorff space, and the space of all measure-morphisms from $\mathcal{S}_u(X)$ is a non-void compact Hausdorff space. Any measure from $\mathcal{S}_u(X)$ is a weak limit point of the convex hull of the set of extremal points of $\mathcal{S}_u(X)$. In addition, the following statements are equivalent:*

(i) *m is an extremal measure from $\mathcal{S}_u(X)$.*

(ii) *m is a measure-morphism from $\mathcal{S}_u(X)$.*

(iii) $m(x \wedge y) = \min\{m(x), m(y)\}$, $x, y \in X$, $m(u) = 1$.

Proof. If u is a strong unit of X, let I be an ideal of X which is maximal under the condition $u \notin I$. We claim that I is a maximal ideal of X. Indeed, suppose that

$I \subseteq J$, where is J an ideal of X. There are two cases: (a) $u \in J$, in this case given $x \in X$ there is an integer $n \geq 1$ such that $x \leq n\,u \in J$, which gives $x \in J$; (b) $u \notin J$, then the maximality of I entails $J = I$.

Therefore I is a maximal ideal of X, and by Theorem 6.1.25, there is at least one measure-morphism m such that $m(u) = 1$.

If u is quasi strong unit of X, then $h(u)$ is a strong unit of $G(X)$. By [Goo 1, Cor 4.4], there exists a state s on $G(X)$ such $s(h(u)) = 1$. Then $s'(x) := s(h(x))$, $x \in X$, gives a measure s' from $\mathcal{S}_u(X)$, such that $\mathcal{S}_u(X) \neq \emptyset$. It is now easily verified that $\mathcal{S}_u(X)$ is a non-empty, convex, compact Hausdorff space in the weak topology of measures.

Applying the Krein–Mil'man theorem to $\mathcal{S}_u(X)$, we see that any measure $m \in \mathcal{S}_u(X)$ is a weak limit point of measures from the convex hull of extremal points of $\mathcal{S}_u(X)$. In addition, all measure-morphisms from $\mathcal{S}_u(X)$ form a compact Hausdorff space.

Suppose now that m is a measure on X with $m(u) = 1$ and let s_m be the unique state on $G(X)$ such that $s_m(h(x)) = m(x)$ $(x \in X)$. Due to [Goo 1, Thm 12.18] and Theorem 6.1.29, we obtain the equivalence of (i)–(iii). □

Proposition 6.1.36 *Let $u > 0$ be a quasi strong unit in a commutative BCK-algebra with the relative cancellation property. Let $X_u := \{x \in X : x \leq u\}$. Then X is Archimedean if and only if X_u is Archimedean.*

Proof. Let $(G(X), h)$ be a universal group of X. Since u is a quasi strong unit in X, then $G_u := \{g \in G(X) : 0 \leq g \leq h(u)\}$ is a bounded commutative BCK-algebra with the greatest element $h(u)$ which is isomorphic with X_u. Since $h(u)$ is a strong unit in $G(X)$, it is possible to show that $(G(X), h_u)$ is a universal group for $(X_u; *_G, 0)$, where $h_u = h|X_u$. Therefore, in view of Theorem 5.2.40, X is Archimedean iff $G(X)$, so that $G(X)$ is Archimedean iff X_u is Archimedean. □

We extend the characterization of semisimple commutative BCK-algebras to commutative BCK-algebras with quasi strong unit. The following result generalizes Theorem 6.1.32, and it shows that the notion of Archimedeanicity of a BCK-algebra with quasi strong unit is crucial for semisimplicity.

Theorem 6.1.37 *Let u be a quasi strong unit of a non-trivial commutative BCK-algebra $(X; *, 0)$ with the relative cancellation property. The following statements are equivalent:*

(i) *X is semisimple.*

(ii) *X is Archimedean.*

(iii) *X has an order-determining system of measure-morphisms from $\mathcal{S}_u(X)$.*

(iv) *X has an order-determining system of measures from $\mathcal{S}_u(X)$.*

(v) *X is isomorphic to some commutative BCK-algebra of functions on some $\Omega \neq \emptyset$.*

Proof. This follows the same ideas as the proof of Theorem 6.1.32, and use Theorem 6.1.35. □

6.1.7 Simple Commutative BCK-algebras

In this section we shall study measures on simple BCK-algebras.

Theorem 6.1.38 *Let $(X; *, 0)$ be a simple commutative BCK-algebra with the relative cancellation property. (1) If a is an atom of X, then $X = \{0, a, 2a, \ldots, ord_X(a)a\}$ if $ord_X(a) < \infty$, or $X = \{0, a, 2a, \ldots\}$ if $ord_X(a) = \infty$, and for a universal group $(G(X), h)$, for X we have $G(X) = \mathbb{Z}$.*

(2) If X is atomless, then X is isomorphic to some BCK-subalgebra of $([0, \infty); *_{\mathbb{R}}, 0)$ which is dense in the interval $[0, t_0]$, or in $[0, \infty)$, where $t_0 := \sup\{\alpha(x) : x \in X\}$ and α is the embedding of X into $[0, \infty)$.

(3) Conversely, if a commutative BCK-algebra $(X; *, 0)$ is isomorphic to some BCK-subalgebra of $([0, \infty); *_{\mathbb{R}}, 0)$, then X is simple.

Proof. (1) If a is an atom of X, the assertion follows from the proof of Proposition 6.1.34.

(2) Suppose now that X is atomless. Then for its universal group $(G(X), h)$ we have that $G(X)$ is isomorphic to an additive subgroup of $(\mathbb{R}; +)$. Without loss of generality we can assume that $G(X)$ is a subgroup of $(\mathbb{R}; +)$.

We claim that $G(X)$ is dense in $\mathbb{R}$. If not, then $G(X)$ cannot contain arbitrarily small positive real numbers, so there exists a real number $\epsilon > 0$ such that the open interval $(0, \epsilon)$ does not contain any element of $G(X)$. We assert that there is a least positive real number in $G(X)$. Indeed, if not, there exist positive real numbers $g_1 > g_2 > \cdots$ that belong to $G(X)$. Therefore also $g_i - g_{i+1}$ belongs to $G(X)$, and all but finitely many of them must be less than $\epsilon > 0$, which is impossible.

Hence, there exists a least positive real numbers $g \in G(X)$. Any real number b may be expresses in the form $b = ng + c$, where $0 \leq c < g$. If $b \in G(X)$, then also $c \in G(X)$ which forces $c = 0$ by the minimality of g, and hence $b = ng$. Therefore the group $G(X)$ is isomorphic to $\mathbb{Z}$, and X can be embedded into $\mathbb{Z}^+$, which entails that X has an atom.

We now show that the image $\alpha(X)$ is dense in $[0, t_0]$. Since X is atomless, there is a sequence $x_1 > x_2 > \cdots \searrow 0$, therefore, $\alpha(x_n) \searrow 0$, which proves that given $\epsilon > 0$, $(0, \epsilon)$ contains at least one element $\alpha(x_n)$.

Due to the definition of t_0, given $\epsilon > 0$, the interval $(t_0 - \epsilon, t_0)$ contains at least one element $\alpha(x)$, $x \in X$.

Let now $0 < t < t_0$. Therefore, there exists x_0 such that $t < \alpha(x_0) \leq t_0$. Given $\epsilon < \alpha(x_0) - t$, in view of the first part of the present proof, there is an element

$g \in G(X)$ such that $|t - g| < \epsilon$. Then either $g \leq t < \alpha(x_0)$ or $t < g < \alpha(x_0)$. In both cases, since X is a lattice, due to Theorem 5.2.23(ix), there is a unique $x_1 \in X$ such that $\alpha(x_1) = g$.

Therefore the image of X into $[0, \infty)$ is dense in $[0, t_0]$.

(3) Without loss of generality we can assume that X is a BCK-subalgebra of $([0, \infty); *_{\mathbb{R}}, 0)$. Suppose that I is a non-empty ideal of X. Choose a non-zero element $x_0 \in I$ and let $x \in X$. If $x \leq x_0$, then $x \in I$. Otherwise, let $j_{x_0}(x)$ be the greatest integer $n \geq 1$ such that $j_{x_0}(x)x_0 \leq x$. Then $x *_{\mathbb{R}} (j_{x_0}(x)x_0) < x_0 \in I$ so that $x = j_{x_0}(x)x_0 + (x *_{\mathbb{R}} (j_{x_0}(x)x_0) \in I$, which proves that $I = X$. □

Corollary 6.1.39 (1) *Every simple commutative BCK-algebra $(X; *, 0)$ with the relative cancellation property has at most the cardinality of the continuum.*

(2) *A commutative BCK-algebra $(X; *, 0)$ with the relative cancellation property is finite and simple if and only if X is isomorphic to $(\{0, \ldots, n\}; *_{\mathbb{R}}, 0)$ for some integer $n \geq 0$.*

Proof. (1) If $X = \{0\}$, the statement is trivial. If $X \neq \{0\}$, it follows directly from Theorem 6.1.38 because the zero ideal is a maximal ideal of X and $X = X/\{0\}$.

(2) This follows from (1) of Theorem 6.1.38. □

A measure m on X is **faithful** if $m(x) = 0$ entails that $x = 0$.

Proposition 6.1.40 *Let $(X; *, 0)$ be a simple non-trivial commutative BCK-algebra with the relative cancellation property. Then there is a faithful measure-morphism m_0 such that, for any measure m on X, $m = c\,m_0$ for some constant $c \geq 0$.*

In addition, if u is a non-zero fixed element of X, then, for any measure m on X and for any BCK-automorphism ϕ of X with $\phi(u) = u$, we have $m(\phi(x)) = m(x)$, $x \in X$.

Proof. Since X is simple, the zero-ideal of X is the only maximal ideal of X. Then $X = X/\{0\}$, and X is linear and Archimedean. The group $G(X)$ is also linear and Archimedean. According to Hölder's theorem, [Bir, Thm XIII.12], $G(X)$ is isomorphic to a subgroup of the additive group of all real numbers. Therefore, X is isomorphic to some BCK-subalgebra of $([0, \infty); *_{\mathbb{R}}, 0)$, and if m_0 is this isomorphism, then m is a non-trivial measure-morphism on X.

Suppose now that m is any measure on X. Since the kernel I_m of m is an ideal of X, we obtain that either $I_m = \{0\}$ or $I_m = X$. In the first case we have according to Theorem 6.1.29, $m = c\,m_0$ for some $c > 0$, and in the second one, $m = 0\,m_0$.

It is clear that m_0 is faithful.

Suppose now that ϕ is an automorphism of X with $\phi(u) = u$. Using the first part of Proposition and Lemma 6.1.21, we have $m(\phi(x)) = m(x)$, $x \in X$. □

Proposition 6.1.41 *Let $(X_1; *_{\mathbb{R}}, 0, 1)$ and $(X_2; *_{\mathbb{R}}, 0, 1)$ be two bounded commutative BCK-algebras. Then X_1 and X_2 are isomorphic if and only if $X_1 = X_2$. If this is the case, then the identity is the only BCK-automorphism on X_1.*

Proof. One direction is clear.

Suppose thus that ϕ is a BCK-isomorphism from X_1 onto X_2. Then $\phi(1) = 1$. Let (G_1, h_1) and (G_2, h_2) be their universal groups. Since X_1 and X_2 are linear and Archimedean, so are G_1 and G_2, which means that G_1 and G_2 are subgroups of $(\mathbb{R}; +)$. The mapping ϕ can be extended uniquely to a group isomorphism $\hat{\phi} : G_1 \to G_2$ which by Lemma 6.1.21 means that G_1 and G_2 are identical, and $\hat{\phi}$ is the identity. Therefore, $X_1 = X_2$, and ϕ is the identity on X_1. □

Let $(X; *, 0)$ be a BCK-algebra. For any non-void subset M of X, we put

$$M^* := \{x * y : x, y \in M\}.$$

Proposition 6.1.42 *Let $(X; *, 0)$ be a BCK-algebra. For any non-void subset M of X, we put $M_0 := M$, and, for any integer $n \geq 1$, $M_n := (M_{n-1})^*$. Then $M_n \subseteq M_{n+1}$, and $X_0(M) := \bigcup_{n=0}^{\infty} M_n$ is a BCK-subalgebra of X generated by M.*

Proof. Straightforward. □

Theorem 6.1.43 *Let α be a real number, $0 \leq \alpha \leq 1$, and let $X(\alpha)$ be the bounded commutative BCK-subalgebra of $([0, 1]; *_{\mathbb{R}}, 0)$ generated by α with the greatest element 1. If $\alpha = p/q$, where p and q are relatively prime, then $X(\alpha) = \{0, 1/q, 2/q, \ldots, q/q\}$. If α is an irrational number, then $X(\alpha)$ is countable and dense in $[0, 1]$. In addition $X(\alpha) = X(\beta)$ if and only if either $\alpha = \beta$, or $\alpha = 1-\beta$. $X(\alpha)$ and $X(\beta)$ are isomorphic if and only if $X(\alpha) = X(\beta)$.*

Moreover, in both cases, $X(\alpha)$ is simple, and there is a unique state on $X(\alpha)$, namely $m(t) := t$, $t \in X(\alpha)$.

Proof. Without loss of generality we can assume that $\alpha \leq 1/2$; otherwise we change it into $1 - \alpha$.

Suppose that $\alpha = p/q$. If $p = 1$, we have finished. Let now $1 < p < q$. Since p does not divide q, p does not divide $q - p$. Hence $q - p = n_1 p + p_1$, where $0 < p_1 < p$. Hence $p_1/q \in X(p/q)$. If $p_1 = 1$, we have finished. Let $0 < p_1 < p$. Since p and q are relatively prime, then either p_1 does not divide q, or p_1 does not divide $q - p$. Then either $p = n_2 p_1 + p_2$, or $q - p = n_2 p_1 + p_2$, where $0 < p_2 < p_1$. Hence $p_2/q \in X(p/q)$.

If $p_2 = 1$, we have finished. Otherwise, we repeat the last step with p_2, p and $q - p$, respectively.

Applying this process sufficiently many times, we obtain a finite sequence of integers $p_k := 1 < p_{k-1} < \cdots < p_2 < p_1$ such that $p_i/q \in X(p/q)$ for any $i = 1, \ldots, k$, which gives that $1/q \in X(p/q)$.

Let now α be an irrational number. It follows from Proposition 6.1.42 that $X(\alpha)$ is at most countable. If $X(\alpha)$ were finite, then there would be a unique atom a in $X(\alpha)$. If $x \in X(\alpha)$ is non-zero, then there is a greatest integer $j_a(x)$ such that $j_a(x)a \in X(\alpha)$ and $j_a(x)a \leq x$. In fact, $j_a(x)a = x$. Then $a = 1/j_a(1)$, which proves that α is rational. Hence, $|X(\alpha)| = \aleph_0$, and $X(\alpha)$ is atomless.

Define

$$S(\alpha) := \{m + n\alpha : \ m, n \in \mathbb{Z}, \ 0 \leq m + n\alpha \leq 1\}.$$

Then $0, 1, \alpha, 1-\alpha \in S(\alpha)$, and $S(\alpha)$ is a commutative BCK-subalgebra of $([0,1]; *_{\mathbb{R}}, 0)$. Therefore, $S(\alpha) \supseteq X(\alpha)$.

Now put

$$G(\alpha) := \{m + n\alpha : \ m, n \in \mathbb{Z}\}.$$

Then $G(\alpha)$ is an Abelian lattice ordered subgroup of $(\mathbb{R}; +)$ with the strong unit 1. Due to Mundici's representation theorem [Mun] or Corollary 5.3.8, $(G(\alpha), id_{S(\alpha)})$ is a universal group for $S(\alpha)$.

According to Lemma 6.1.22 there is an Abelian lattice ordered subgroup G of $(\mathbb{R}; +)$ containing $X(\alpha)$ such that $(G, id_{X(\alpha)})$ is a universal group for $X(\alpha)$ and, in addition, 1 is also a strong unit for G.

Since $G(\alpha)$ is generated by α and 1, we conclude that $G(\alpha) \subseteq G$. On the other hand, there is a group homomorphism $\psi : G \to G(\alpha)$ such that $\psi(x) = x$ for any $x \in X(\alpha)$. It is not hard to show that ψ is injective, so that by Lemma 6.1.22 $G \subseteq G(\alpha)$. Applying again Mundici's representation theorem, we conclude that $G = G(\alpha)$, consequently $X(\alpha) = S(\alpha)$.

If α is rational, then $S(\alpha) = \{0, 1/q, \ldots, q/q\}$ is finite, where $\alpha = p/q$ and p, q are relatively primes. So $X(\alpha) = X(\beta)$ iff $\alpha = \beta$ or $\alpha = 1 - \beta$.

Suppose that α is irrational, $S(\alpha) = S(\beta)$ and let $0 < \alpha < \beta < 1$. Then $\alpha = m+n\beta$ and $\beta = p + q\alpha$, which yields that $\alpha = m + np + nq\alpha$. Hence $m + np = 0$ and $nq = 1$. In the case $n = q = 1$, we get the contradiction $0 < \beta - \alpha = p < 1$. In the second case when $n = q = -1$, we get $\alpha = m - \beta < 1$, which gives $m = 1$, whence $\alpha = 1 - \beta$ as required.

Applying Proposition 6.1.41 we obtain that $X(\alpha) = X(\beta)$ iff $X(\alpha)$ and $X(\beta)$ are isomorphic.

We show that $X(\alpha)$ is simple. Choose a non-zero ideal I of $X(\alpha)$, and let $x_0 \in I$ be a non-zero element. If $1/2 \leq x_0 \leq 1$, then $1 *_{\mathbb{R}} x_0 \leq x_0 \in I$ which implies $1 \in I$. If $0 < x_0 < 1/2$, there is an integer n_0 such that $1/2 \leq n_0 x_0 \leq 1$. Then $1 *_{\mathbb{R}} (n_0 x_0) \leq n_0 x_0 \in I$ which again implies $1 \in I$, so that $I = X(\alpha)$.

Applying Theorem 6.1.38, we see that $X(\alpha)$ is dense in $[0, 1]$.

It is evident that m, where $m(t) := t, \ t \in X(\alpha)$, is the only state on $X(\alpha)$; in addition, m is a state-morphism. □

We note that according to Proposition 6.1.41 and Theorem 6.1.43, there are uncountably many non-isomorphic, countable, bounded commutative BCK-subalgebras of $([0,1]; *_{\mathbb{R}}, 0, 1)$ with one generator.

A measure m on a BCK-algebra $(X; *, 0)$ is said to be **discrete** if there is a constant $c \geq 0$ such that $m(x) \in \{0, c, 2c, \ldots\}$ for any $x \in X$.

Theorem 6.1.44 *Let $(X; *, 0)$ be a non-trivial simple commutative BCK-algebra with the relative cancellation property. The following statements are equivalent:*

(i) *There is an atom in X.*

(ii) *There exists a non-zero discrete measure on X.*

(iii) *$(\mathbb{Z}, h)$ is the universal group for X.*

Proof. (i) $\Rightarrow$ (ii). Suppose that a is an atom of X. Then by Theorem 6.1.38, $X = \{0, a, \ldots, ord_X(a)a\}$ if $ord_X(a) < \infty$, otherwise, $X = \{0, a, 2a, \ldots\}$. Then the mapping $m : X \to [0, \infty)$ defined by $m(ka) := k$, $k \in 0, 1, \ldots, ord_X(a)$, is a non-zero discrete measure on X.

(ii) $\Rightarrow$ (i). Let now m be a non-zero discrete measure on X. Since X is simple, $I_m = \{x \in X : m(x) = 0\} = \{0\}$, and so m is faithful. The discreteness of m implies that there is the least positive integer $m(a) > 0$, so that a is an atom of X.

(i) $\Rightarrow$ (iii). This follows from Theorem 6.1.38.

(iii) $\Rightarrow$ (ii). For any $x \in X$, we define $m(x) := k$ iff $h(x) = k$, where $(\mathbb{Z}, h)$ is the universal group of X. Then m is a non-zero discrete measure on X. □

6.1.8 Exercises

1. Prove that a simple commutative BCK-algebra with the relative cancellation property is Archimedean.
2. Prove that every proper ideal M of a commutative BCK-algebra $(X; *, 0)$ with the relative cancellation property is the intersection of all prime ideals containing M.
3. Prove that an ideal M of X is maximal iff X/M is simple.
4. Prove that any maximal ideal M of a commutative BCK-algebra X is prime.
5. Prove: Let $(X; *, 0)$ be a commutative BCK-algebra with the relative cancellation property. Then X is compact if and only if $\hat{X}$ is compact. The following statements are equivalent.

   (i) X has a fuzzy set representation.
   (ii) X is semisimple and compact.
   (iii) There exists an order-determining system of measure-morphisms on X having values in the interval $[0, 1]$.
   (iv) X is compact and $\mathcal{M}(X)$ is dense in $\mathcal{P}(X)$.
   (v) $\hat{X}$ has a fuzzy set representation.

(vi) $\hat{X}$ is semisimple and compact.

(vii) There exists an order-determining system of measure-morphisms on $\hat{X}$ having values in the interval $[0,1]$.

(viii) $\hat{X}$ is compact and $\mathcal{M}(\hat{X})$ is dense in $\mathcal{P}(\hat{X})$.

6. Prove: Let $(X;*,0)$ be a commutative BCK-algebra with the relative cancellation property. Let there exist a non-void subset $\mathcal{M} \subseteq \mathcal{M}(X)$ such that $\bigcap\{M : M \in \mathcal{M}\} = \{0\}$ and let X/M be a bounded subset of $[0,\infty)$ for arbitrary $M \in \mathcal{M}$. Then X is compact and semisimple.
7. Prove: A commutative BCK-algebra $(X;*,0)$ with the relative cancellation property such that any non-zero element is the join of finitely many sums of atoms is semisimple and is a subdirect product of basic BCK-chains. Furthermore, the universal group for X is $(\mathbb{Z}^{|\mathcal{A}(X)|}, h)$ with the restricted direct product $\mathbb{Z}^{|\mathcal{A}(X)|}$ and $h(x) = (j_a(x))_{\mathcal{A}(X)}$, where $\mathbb{Z}$ is the additive group of all integers. If, in addition, any atom is of finite BCK-order, X is a subdirect product of finite BCK-chains.
8. Prove: Each finite commutative BCK-algebra $(X;*,0)$ with the relative cancellation property is semisimple and is a subdirect product of finitely many finite BCK-chains, the universal group of X is $\mathbb{Z}^n$, where n the the number of all atoms in X, and the BCK-hull $\hat{X}$ of X is the direct product of finitely many finite BCK-chains.
9. Prove: Let u be a non-zero quasi strong unit of a commutative BCK-algebra $(X;*,0)$ with the relative cancellation property. Then $u \notin Rad(X)$.
10. Prove: Any commutative BCK-algebra with the relative cancellation property and with strong unit is upwards directed.
11. Prove: Let $u > 0$ be a quasi strong unit in a commutative BCK-algebra $(X;*,0)$ with the relative cancellation property. If I is a prime ideal of X, then u/I is a non-zero strong unit in X/I, moreover, X is a subdirect product of linear commutative BCK-algebras with strong unit.
12. A measure m on a BCK-algebra X is faithful if $m(x) = 0$ entails that $x = 0$. Prove: Let $u > 0$ be a quasi strong unit in a commutative BCK-algebra $(X;*,0)$ with the relative cancellation property. If X is not semisimple, then X has no faithful state.
13. A BCK-algebra $(X;*,0)$ is said to be **local** iff X has exactly one maximal ideal. If X is simple, then X is local. Example 5.2.39 gives a commutative BCK-algebra which is local, but neither simple nor semisimple.

    Prove: Let a commutative BCK-algebra $(X;*,0)$ with the relative cancellation property have a non-zero quasi strong unit u. Then X is local if and only if X possesses only one measure m such that $m(u) = 1$. If this is the case, then m is a measure-morphism.

6.2 Dedekind Complete BCK-algebras

We will study Dedekind complete commutative BCK-algebras with the relative cancellation property and their connection with corresponding universal groups. We shall characterize Dedekind orthogonally complete atomic, Archimedean BCK-algebras which generalizes results of Jakubík known for MV-algebras. Finally, we show when Dedekind complete and atomic commutative BCK-algebras are isomorphic to direct products of basic BCK-chains; this generalizes a result of Cignoli [Cig] known for MV-algebras.

6.2.1 Dedekind Complete Commutative BCK-algebras

We start with some lattice properties of the derived sum +.

Proposition 6.2.1 *Let $\{x_i\}$ be a non-empty family of elements of a commutative BCK-algebra $(X;*,0)$ with the relative cancellation property. The following identities hold in the sense that if either side exists, then the other exists, in which case the two sides agree.*

(i)

$$(\bigvee_i x_i) + x = \bigvee_i (x_i + x).$$

(ii) *If $\bigwedge_i (x_i + x)$ exists, then so does $(\bigwedge_i x_i) + x$, and*

$$(\bigwedge_i x_i) + x = \bigwedge_i (x_i + x).$$

If $\bigwedge_i x_i$ and $x_i + x$ exist for any i, so do $(\bigwedge_i x_i) + x$ and $\bigwedge_i (x_i + x)$, and the equality in (ii) *holds.*

(iii) *If $x \geq x_i$ for any i, then*

$$x * \bigwedge_i x_i = \bigvee_i (x * x_i).$$

(iv) *If $x \geq x_i$ for any i, then*

$$x * \bigvee_i x_i = \bigwedge_i (x * x_i).$$

Proof. (i) Suppose that $x_0 := (\bigvee_i x_i) + x$ exists in X. Then by Theorem 5.2.6, for any i, $x_i + x$ is defined in X, and $x_i + x \leq x_0$. Assume now that $x_i + x \leq y$ for any i and some $y \in X$. Then $x \leq y$ and hence $x_i \leq y * x$, so that $\bigvee_i x_i \leq y * x$, and consequently, $(\bigvee_i x_i) + x \leq y$.

Let now $\bigvee_i (x_i + x)$ be defined in X. Then $x \leq \bigvee_i (x_i + x)$, and put $b := (\bigvee_i (x_i + x)) * x$. Then $\bigvee_i (x_i + x) = b + x$ and $x_i + x \leq b + x$ for any i, i.e., $x_i \leq b$. Let now $x_i \leq u$ for some $u \in X$. Then $x_i \leq u \wedge b$ for any i. Since $u \wedge b \leq b$, we have that $(u \wedge b) + x$ is defined in X. Hence, $x_i + x \leq (u \wedge b) + x$ for any i, which gives $\bigvee_i (x_i + x) \leq (u \wedge b) + x$ and $(\bigvee_i (x_i + x)) * x \leq u \wedge b$. Therefore, $b = u \wedge b$, and $b \leq u$, which proves that $\bigvee_i x_i$ exists in X and, in addition, $b = \bigvee_i x_i$. Hence $(\bigvee_i (x_i + x)) * x = \bigvee_i x_i$, which entails $\bigvee_i (x_i + x) = (\bigvee_i x_i) + x$.

(ii) Let $\bigwedge_i (x_i + x)$ exist in X. Then $\bigwedge_i (x_i + x) \leq x_i + x$ for any i, and $(\bigwedge_i (x_i + x)) * x \leq x_i$. Let now $b \leq x_i$ for any i. Then $b + x \leq x_i + x$ and $b + x \leq \bigwedge_i (x_i + x)$ which yields $b \leq (\bigwedge_i (x_i + x)) * x$. Thus $\bigwedge_i x_i$ is defined in X, and $\bigwedge_i x_i = (\bigwedge_i (x_i + x)) * x$, consequently, $(\bigwedge_i x_i) + x = \bigwedge_i (x_i + x)$.

Assume now that $\bigwedge_i x_i$ and $x_i + x$ are defined in X for any i. Then $(\bigwedge_i x_i) + x$ is defined in X, and $(\bigwedge_i x_i) + x \leq x_i + x$. Let now $z \leq x_i + x$ for any i. Then $z * x \leq x_i$ so that $z * x \leq \bigwedge_i x_i$, i.e., $z * (z \wedge x) \leq \bigwedge_i x_i$. Hence, $z \leq (\bigwedge_i x_i) + (z \wedge x) \leq (\bigwedge_i x_i) + x$ which proves (ii).

(iii) We have $x * \bigwedge_i x_i \geq x * x_i$ for any i. Suppose that $u \geq x * x_i$ for any i. Then $x * u \leq x * (x * x_i) = x_i$, which gives $x * u \leq \bigwedge_i x_i$. Hence

$$0 = (x * u) * (\bigwedge_i x_i) = (x * (\bigwedge_i x_i)) * u,$$

which entails $x * (\bigwedge_i x_i) \leq u$, and it proves that $\bigvee_i (x * x_i)$ exists in X and equals $x * (\bigwedge_i x_i)$.

Conversely, assume that $\bigvee_i (x * x_i)$ exists in X. Then $\bigvee_i (x * x_i) \geq x * x_i$ for any i, and consequently, $x * \bigvee_i (x * x_i) \leq x * (x * x_i) = x_i$. Let now $u \leq x_i$ for any i. Then $x * u \geq x * x_i$ and $x * u \geq \bigvee_i (x * x_i)$. Therefore,

$$\begin{aligned} x * (x * u) &\leq x * \bigvee_i (x * x_i), \\ x \wedge u &\leq x * \bigvee_i (x * x_i), \\ u &\leq x * \bigvee_i (x * x_i), \end{aligned}$$

which proves that $x * \bigvee_i (x * x_i) = \bigwedge_i x_i$, consequently $\bigvee_i (x * x_i) = x * \bigwedge_i x_i$.

(iv) We have $\bigwedge_i (x * x_i) \leq x * x_i$ for any i. Then $x * \bigwedge_i (x * x_i) \geq x_i$. Suppose now that $b \geq x_i$ for any i. Then $x * b \leq x * x_i$ and $x * b \leq \bigwedge_i (x * x_i)$.

Hence $0 = (x*b)*\bigwedge_i(x*x_i) = (x*\bigwedge_i(x*x_i))*b$ which gives $x*\bigwedge_i(x*x_i) \leq b$, so that $\bigvee_i x_i$ exists in X and $\bigvee_i x_i = x*\bigwedge_i(x*x_i)$. Consequently, $x*\bigvee_i x_i = \bigwedge_i(x*x_i)$.

Let now $x*\bigvee_i x_i$ be defined in X. Then $x*\bigvee_i x_i \leq x*x_i$ for any i. Let $u \leq x*x_i$ for any i. Then $x*u \geq x*(x*x_i) = x_i$, so that $x*u \geq \bigvee_i x_i$. Hence

$$\begin{aligned} x*(x*u) &\leq x*\bigvee_i x_i, \\ x\wedge u &\leq x*\bigvee_i x_i, \\ u &\leq x*\bigvee_i x_i, \end{aligned}$$

which proves that $\bigwedge_i(x*x_i)$ exists in X and equals $x*\bigvee_i x_i$. □

We note that a BCK-algebra $(X;*,0)$ is **Dedekind upwards complete** provided that any non-empty subset of X which has an upper bound in X has a supremum in X. If any non-empty subset of X has an infimum in X, we say that X is **Dedekind downwards complete**. If X is Dedekind upwards and downwards complete, we say that X is **Dedekind complete**. In a similar way we say that X is **Dedekind upwards σ-complete, Dedekind downwards σ-complete**, or **Dedekind σ-complete** assuming that corresponding statements as above hold for any non-empty countable subset of elements in X with necessary assumptions.

We note that in any commutative BCK-algebra X, $x\wedge y$ is defined for all x and y in X, and $x\vee y$ exists iff x and y have an upper bound in X. A similar statement holds for Dedekind (σ-) completeness, see Proposition 5.1.19(ii).

Two elements $x,y\in X$ are said to be **disjoint** if $x\wedge y = 0$. X is **Dedekind orthogonally complete** (**Dedekind orthogonally σ-complete**) if any (countable) family of mutually disjoint elements having an upper bound in X has a supremum in X.

Proposition 6.2.2 *A commutative BCK-algebra $(X;*,0)$ with the relative cancellation property is Dedekind complete (Dedekind σ-complete) if and only if X is Dedekind upwards complete (Dedekind upwards σ-complete) if and only if X is Dedekind downwards complete (Dedekind downwards σ-complete).*

Proof. Suppose that X is Dedekind upwards complete. Let $\{x_i\}$ be a non-empty family of elements in X. Choose an element x_{i_0} from $\{x_i\}$ and define $x_i' := x_i\wedge x_{i_0}$. The element x_{i_0} is an upper bound for $\{x_i'\}$. Similarly, $\{x_{i_0}*x_i'\}$ has an upper bound, e.g., x_{i_0}. Therefore, $x = \bigvee_i(x_{i_0}*x_i')$ exists in X, and by (iii) of Proposition 6.2.1, we have $x = x_{i_0}*\bigwedge_i x_i'$ which says that $\bigwedge_i x_i'$ is defined in X. It is clear that $\bigwedge_i x_i' = \bigwedge_i x_i$.

Suppose now that X is Dedekind downwards complete and assume that $\{x_i\}$ has an upper bound x in X. Then $\bigwedge_i(x * x_i)$ is defined in X, and by (iv) of Proposition 6.2.1, we have that $\bigwedge_i(x * x_i) = x * \bigvee_i x_i$ and $\bigvee_i x_i$ is defined in X.

The rest of Proposition is evident. □

A lattice-ordered group $(G; +, \leq, 0)$ is **Dedekind complete** (**Dedekind σ-complete**) provided that any non-empty (countable) subset of G which is bounded above in G has a supremum in G. Equivalently, G is Dedekind complete (Dedekind σ-complete) iff every non-empty (countable) subset of G which is bounded below in G has an infimum in G. In a similar way we say that G is **Dedekind orthogonally complete** (**Dedekind orthogonally σ-complete** if any (countable) system of positive mutually disjoint elements in G^+ having an upper bound in G has a supremum in G.

Theorem 6.2.3 *Let $(X; *, 0)$ be a commutative BCK-algebra with the relative cancellation property and let $(G(X), h)$ be its universal group. The following statements are equivalent:*

(i) *X is Dedekind complete.*

(ii) *$G(X)$ is Dedekind complete.*

(iii) *$\hat{X}$ is Dedekind complete.*

Proof. (ii) $\Rightarrow$ (i). Let $\{x_i\}$ be a non-empty subset such that $x_i \leq x$ for any i, where $x \in X$. Then there exists a supremum g of $\{h(x_i)\}$ in $G(X)$. Since $h(x_i) \leq h(x)$ for any i, we have $g \leq h(x)$. According to Proposition 5.2.36, there is a unique $y \in X$ such that $h(y) = g$, which means that y is a supremum of $\{x_i\}$ in X.

(i) $\Rightarrow$ (ii). First of all we show that if a subset $\{x_i\}$ of X has a supremum x in X, then $h(x)$ is the supremum of $\{h(x_i)\}$ in $G(X)$. Indeed, let $g \in G(X)$ be an upper bound of $\{h(x_i)\}$. Then $h(x_i) \leq h(x) \wedge g \leq h(x)$ for any i. Applying Proposition 5.2.36, we have that $h(x) \wedge g = h(z)$ for a unique element $z \in X$. This implies $x_i \leq z$ for any i, so that $x \leq z$, consequently, $h(x) \leq h(z) \leq h(x)$. Therefore, $h(x) \wedge g = h(x)$ and $h(x) \leq g$.

Similarly we can show that if a subset $\{x_i\}$ of X has an infimum x in X, then $h(x)$ is an infimum of $\{h(x_i)\}$ in $G(X)$. Indeed, suppose that $h(x_i) \geq g$ for any i, where $g \in G(X)$. Then $h(x_i) = h(x_i) \vee 0 \geq g \vee 0 = g^+$ for any i. Due to Proposition 5.2.36, there is a unique $u \in X$ such that $g^+ = h(u)$, which implies $x_i \geq u$ for any i. So $x \geq u$ and $h(x) \geq h(u) = g^+ \geq g$.

We now claim that under our hypothesis, any subset $\{g_i\}$ of $G^+(X)$, which is bounded above in $G^+(X)$, has a supremum in $G^+(X)$. So assume that $g_i \leq v$ for any i, where $v \in G^+(X)$. Since $h(X)$ generates the positive cone $G^+(X)$, there exist finitely many elements $y_1, \ldots, y_k$ in X such that

$$v = \sum_{j=1}^{k} h(y_j).$$

The proof will be by induction on the integer k.

Case 1. Assume that $v = h(y)$ for some $y \in X$. Due to Proposition 5.2.36, given g_i there is a unique $x_i \in X$ such that $h(x_i) = g_i$, which gives $x_i \leq y$ for any i. Therefore, there is a supremum x of $\{x_i\}$ in X. Due to the above considerations, $h(x)$ is a supremum of $\{g_i\}$ in $G^+(X)$.

Case 2. Suppose that we have found a supremum of $\{g_i\}$ in $G^+(X)$ whenever its upper bound v consists of at most k summands. Assume now that $v = \sum_{j=1}^{k+1} h(y_j)$, where $y_{k+1} = y$. Then the interval $[0, \sum_{j=1}^{k} h(y_j)]$ in $G^+(X)$ is by assumptions complete. Since the interval $[\sum_{j=1}^{k} h(y_j), \sum_{j=1}^{k+1} h(y_j)]$ is isomorphic to $[0, h(y)]$, we obtain that $[\sum_{j=1}^{k} h(y_j), \sum_{j=1}^{k+1} h(y_j)]$ is also Dedekind complete.

Let $a = \sum_{j=1}^{k} h(y_j)$ and, for any i, we put

$$g_i' = g_i \wedge a, \quad g_i'' = g_i \vee a.$$

Then $g_i' \in [0, a]$ and $g_i'' \in [a, a + h(y)]$ for any i, and $g_i = g_i' + g_i'' - a$.

By the induction assumptions, the elements

$$g' = \bigvee_i g_i' \quad \text{and} \quad g_i'' = \bigvee_i g_i''$$

exist in $G^+(X)$. Then $g_i = g_i' + (g_i'' - a) \leq g$, where

$$g = g' + (g'' - a).$$

Suppose now that $g_i \leq b$ for any i, where $b \in G^+(X)$. If we put $b' = b \wedge a$ and $b'' = b \vee a$, then $b = b' + (b'' - a)$, and $b' \geq g_i'$ and $b'' \geq g_i''$ for any i. This implies that $g' \leq b'$, $g'' \leq b''$ and thus $g \leq b$.

We now show that any subset $\{g_i\}$ in $G^+(X)$ has an infimum in $G^+(X)$. Fix an element g_{i_0} from $\{g_i\}$ and define $\hat{g}_i := g_i \wedge g_{i_0}$ for any i. Since $\hat{g}_i \leq g_{i_0}$, then $g_{i_0} \geq g_{i_0} - \hat{g}_i \geq 0$ which means that $\{g_{i_0} - \hat{g}_i\}$ has a supremum g in $G^+(X)$. Then $g = \bigvee_i (g_{i_0} - \hat{g}_i) = g_{i_0} - \bigwedge_i \hat{g}_i$ which proves that $c := \bigwedge_i \hat{g}_i$ exists in $G^+(X)$. It is evident that c is also an infimum of $\{g_i\}$ in $G^+(X)$.

Finally, assume that $\{g_i\}$ is an arbitrary non-empty subset in $G(X)$ with an upper bound v. Then $g_i^+ = g_i \vee 0 \leq v \vee 0 = v^+$ and $g_i^- = -(g_i \wedge 0) \geq -(v \wedge 0) = v^- \geq 0$ for any i. Thus there is a supremum $g_1 \in G^+(X)$ of $\{g_i^+\}$ and an infimum $g_2 \in G^+(X)$ of $\{g_i^-\}$. Since $g_i = g_i^+ - g_i^- \leq g_1 - g_2$, we see that $g := g_1 - g_2$ is an upper bound for $\{g_i\}$.

We claim that g is a supremum of $\{g_i\}$. Suppose that $g_i \leq m$ for any i, where $m \in G(X)$. We have $g_i^+ \leq m^+$ and $g_i^- \geq m^-$ so that $g_1 \leq m^+$ and $g_2 \geq m^-$ which yield $g = g_1 - g_2 \leq m^+ - m^- = m$.

The equivalence of (ii) and (iii) follows from the equivalence (i) and (ii) because $G(X)$ is also the underlying group of the universal group for the BCK-hull $\hat{X}$ of X, Corollary 5.2.31. □

Theorem 6.2.4 *Let $(X;*,0)$ be a commutative BCK-algebra with the relative cancellation property and let $(G(X),h)$ be its universal group. The following statements are equivalent:*

(i) *X is Dedekind σ-complete.*

(ii) *$G(X)$ is Dedekind σ-complete.*

(iii) *$\hat{X}$ is Dedekind σ-complete.*

Proof. This follows the same ideas as the proof of Theorem 6.2.3 □

It is worth recalling that an Archimedean BCK-algebra need not be Dedekind σ-complete. For example, $([0,\infty) \cap \mathbb{Q}; *_{\mathbb{R}}, 0)$ is Archimedean but not Dedekind σ-complete.

Let a be an atom of X. Given an element $x \in X$, we define

$$j_a(x) = \sup\{n \geq 0 : n\,a \text{ is defined in } X,\ n\,a \leq x\}. \tag{6.10}$$

Proposition 6.2.5 *Let $(X;*,0)$ be an Archimedean atomic commutative BCK-algebra with the relative cancellation property. Then, given $x \in X$, we have*

$$\bigvee_{a \in \mathcal{A}(X)} j_a(x)\, a = x. \tag{6.11}$$

Proof. Due to the Archimedeanicity of X, we have that $j_a(x)$ is finite for any $x \in X$ and for any atom $a \in \mathcal{A}(X)$. It is clear that $j_a(x)a \leq x$ for any $a \in \mathcal{A}(X)$. Suppose that $j_a(x)a \leq y$ for all $a \in \mathcal{A}(X)$, where $y \in X$. Then $j_a(x)a \leq x \wedge y \leq x$ for any $a \in \mathcal{A}(X)$. In view of definition of $j_a(x)$, we see that $x \wedge y = x$, i.e., $x \leq y$. Indeed, if not, then there exists an atom b of X such that $b \leq x * (x \wedge y) = x * y$. Since $x = (x * y) + (x \wedge y)$, in view of the assumption $j_b(x)b \leq x \wedge y$, we conclude that

$$j_b(x) \geq j_b(x * y) + j_b(x \wedge y) \geq 1 + j_b(x \wedge y) \geq 1 + j_b(j_b(x)b) = 1 + j_b(x),$$

which is a contradiction. □

The following result has been proved by Jakubík [Jak 4, Thm 3] for bounded commutative BCK-algebras (= MV-algebras). Here we present its generalization for commutative BCK-algebras, and methods of our proof will be completely different from Jakubík's one.

Theorem 6.2.6 *Let $(X;*,0)$ be an atomic commutative BCK-algebra with the relative cancellation property. The following statements are equivalent:*

(i) *X is Dedekind orthogonally complete and Archimedean.*

(ii) *X is Dedekind complete.*

Proof. (i) $\Rightarrow$ (ii). Let $\{x_i\}$ be any bounded subset of X with an upper bound u. Since X is Archimedean, $j_a(x)$ is finite for any $x \in$X and any atom $a \in \mathcal{A}(X)$. Since $j_a(x_i) \le j_a(u) < \infty$ for any i, we see that

$$x_a := \bigvee_i j_a(x_i)a$$

is defined in X. Since $x_a = n_a a$, where n_a is a non-negative integer, we see by (6.7) that, for two different atoms a and b, $x_a \wedge x_b = 0$, consequently, the system $\{x_a : a \in \mathcal{A}(X)\}$ consists of mutually disjoint elements, so that the Dedekind orthogonally completeness of X entails that there is a supremum

$$x = \bigvee_{a \in \mathcal{A}(X)} x_a.$$

We claim that x is a supremum of $\{x_i\}$. Indeed, we have $x \ge x_a \ge j_a(x_i)a$ for any i and any atom a of X. Therefore by (6.11), $x \ge x_i$ for any i. Suppose now that $x_i \le y$ for any i, where $y \in X$. Then $j_a(x_i)a \le y$ for any i and any atom a, consequently, $x_a \le y$, so that $x \le y$.

(ii) $\Rightarrow$ (i). This is evident in view of Proposition 5.2.41. □

Theorem 6.2.7 *Let $(X;*,0)$ be a commutative BCK-algebra with the relative cancellation property and let $(G(X),h)$ be its universal group. The following statements are equivalent:*

(i) *X is Dedekind orthogonally complete (Dedekind orthogonally σ-complete).*

(ii) *$G(X)$ is Dedekind orthogonally complete (Dedekind orthogonally σ-complete).*

(iii) *$\hat{X}$ is Dedekind orthogonally complete (Dedekind orthogonally σ-complete).*

Proof. (ii) $\Rightarrow$ (i). Let $\{x_i\}$ be a family of mutually disjoint elements of X which has an upper bound $x \in X$. Then $\{h(x_i)\}$ is a family of mutually orthogonal elements of $G(X)$ with an upper bound $h(x)$. Therefore, there is a supremum $g \in G^+(X)$ of $\{h(x_i)\}$. Since $g \le h(x)$, we have according to Proposition 5.2.36 that there is a unique element $y \in X$ such that $h(y) = g$. Then y is a supremum of $\{x_i\}$ in X.

(i) $\Rightarrow$ (ii). Suppose now that $\{g_i\}$ is a non-empty subset of mutually disjoint elements in $G(X)$ which is bounded above by an element $v \in G^+(X)$. Then

$$v = \sum_{j=1}^{k} h(y_j),$$

where $y_1, \ldots, y_k \in X$. The proof will be by the induction argument on the integer k.

Case 1. $k = 1$. Then $v = h(y)$ for some $y \in X$. In view of Proposition 5.2.36, $g_i = h(x_i)$ for any i, and $\{x_i\}$ is a family of mutually disjoint elements of X which is bounded above by y. Then $x = \bigvee_i x_i$ exists in X, so that $h(x) = \bigvee_i h(x_i)$.

Case 2. Suppose now that we have found a supremum of $\{g_i\}$ in $G^+(X)$ whenever its upper bound v consists of at most k summands. Assume now that $v = \sum_{j=1}^{k+1} h(y_j)$, and $y_{k+1} = y$. Put $a = h(y)$, and define

$$a_i = g_i \wedge a \quad \text{and} \quad b_i = g_i - a_i.$$

Then $a_i, b_i \in [0, g_i]$ and the families $\{a_i\}$ and $\{b_i\}$ both consist of mutually disjoint elements.

Since $g_i \le \sum_{j=1}^{k+1} h(y_j) = \sum_{j=1}^{k} h(y_j) + a$, then $a \vee g_i \le \sum_{j=1}^{k} h(y_j) + a$, and

$$\sum_{j=1}^{k} h(y_j) = \sum_{j=1}^{k} h(y_j) + a - a \ge (a \vee g_i) - a = g_i - (a \wedge g_i) = b_i.$$

By the induction hypothesis, there exist

$$b = \bigvee_i b_i \quad \text{and} \quad a_0 = \bigvee_i a_i.$$

Let $a_i^* = \bigvee_{s \neq i} a_s$. Then $a_i^* \in [0, a]$. Since $a_i \wedge a_i^* = 0$, we have

$$a_i + a_i^* = a_i \vee a_i^* = a_0.$$

Moreover, in view of the fact that $b_i \wedge a_i^* = 0$, we have

$$b_i + a_i^* = b_i \vee a_i^*.$$

Therefore,

$$\begin{aligned} b_i + a_0 &= b_i + (a_i \vee a_i^*) = (b_i + a_i) \vee (b_i + a_i^*) \\ &= g_i \vee (b_i \vee a_i^*) = (g_i \vee b_i) \vee a_i^* \\ &= g_i \vee a_i^*, \end{aligned}$$

and $b + a_0 = (\bigvee_i b_i) + a_0 = \bigvee_i (b_i + a_0) = \bigvee_i (g_i \vee a_i^*)$.

If $d \in G^+(X)$ is any upper bound of $\{g_i\}$, then $d \geq a_i^*$ and $d \geq g_i \vee a_i^*$. Hence $d \geq b + a_0$, and in view of $b + a_0 \geq b_i + a_i = g_i$, we have

$$b + a_0 = \bigvee_i g_i.$$

The equivalence of (ii) and (iii) follows from the equivalence of (i) and (ii) and from the fact that $G(X)$ is also an underlying group of the universal group for $\hat{X}$, Corollary 5.2.31. □

We now present another generalization of Jakubík's result concerning characterization of Dedekind complete and Dedekind orthogonally σ-complete MV-algebras [Jak 6, Thm 3.6] to BCK-algebras.

Theorem 6.2.8 *Let $(X; *, 0)$ be a commutative BCK-algebra with the relative cancellation property. The following statements are equivalent:*

(i) *X is Dedekind complete.*

(ii) *X is Dedekind σ-complete and Dedekind orthogonally complete.*

Proof. Due to Theorems 6.2.4 and 6.2.7 we see that X is Dedekind σ-complete and Dedekind orthogonally complete iff $G(X)$ is Dedekind σ-complete and Dedekind orthogonally complete. In view of [Jak, Thm 4], this is necessary and sufficient for $G(X)$ to be Dedekind complete. Consequently by Theorem 6.2.3, X is Dedekind complete. □

6.2.2 Decomposition of Dedekind Complete Atomic BCK-algebras

The main aim of the present section is to exhibit conditions when a Dedekind complete and atomic BCK-algebra to be representable as a direct product of basic BCK-chains. This will generalize a corresponding result of Cignoli [Cig] for bounded BCK-algebras (= MV-algebras).

We start with by presenting two infinite distributive laws for commutative BCK-algebras.

Theorem 6.2.9 *Let $\{x_i\}$ be a non-empty family of elements of a commutative BCK-algebra $(X; *, 0)$ with the relative cancellation property. Then*

(i) *If $\bigvee_i x_i$ exists in X, then so does $\bigvee_i (x \wedge x_i)$, and*

$$x \wedge (\bigvee_i x_i) = \bigvee_i (x \wedge x_i). \tag{6.12}$$

(ii) *If* $\bigwedge_i x_i \in X$ *and if there is a* $z \in X$ *such that* $z \geq x, x_i$ *for any* i, *then* $x \vee (\bigwedge_i x_i)$ *and* $\bigwedge_i (x \vee x_i)$ *exist in* X, *and*

$$x \vee (\bigwedge_i x_i) = \bigwedge_i (x \vee x_i). \tag{6.13}$$

Proof. (i) Let $v = \bigvee_i x_i$ be defined in X. Then $x \wedge v = x \wedge (\bigvee_i x_i) \geq x \wedge x_i$ for any i. Assume now that $x \wedge v \geq m \geq x \wedge x_i$ for any i. Then by property (i) of BCK-algebras

$$(x \wedge v) * (x \wedge x_i) = [x * (x * v)] * [x * (x * x_i)] \leq (x * x_i) * (x * v) \leq v * x_i,$$

and $(x \wedge v) * m \leq (x \wedge v) * (x \wedge x_i) \leq v * x_i$ for any i. Using (iv) of Proposition 6.2.1, we have

$$\bigwedge_i (v * x_i) = v * \bigvee_i x_i = v * v = 0.$$

Therefore, $(x \wedge v) * m = 0$, which entails $x \wedge v = m$. Then $\bigvee_i (x \wedge x_i)$ exists in X, and $x \wedge v = x \wedge (\bigvee_i x_i) = \bigvee_i (x \wedge x_i)$.

(ii) By Proposition 5.1.19, $x \vee (\bigwedge_i x_i)$ and $x \vee x_i$ are defined in X for each i. Then by Proposition 6.2.1, $z * (x \vee (\bigwedge_i x_i)) = (z * x) \wedge (z * (\bigwedge_i x_i)) = (z * x) \wedge \bigvee_i (z * x_i) = \bigvee_i ((z * x) \wedge (z * x_i)) = \bigvee_i (z * (x \vee x_i))$. Therefore, $x \vee (\bigwedge_i x_i) = z * (z * (x \vee (\bigwedge_i x_i))) = z * \bigvee_i (z * (x \vee x_i)) = \bigwedge_i (z * (z * (x \vee x_i)) = \bigwedge_i (x \vee x_i)$. □

It is worth recalling that the existence of the expressions on the right-hand side of (6.12) is not sufficient to the existence of the left-hand side of (6.12). Indeed, take $([0, \infty) \cap \mathbb{Q}; *_{\mathbb{R}}, 0)$ and choose $x_n \nearrow \sqrt{2}/2$ and $x = 0.5$. Then $\bigvee_n (x \wedge x_n) = 0.5$ but $\bigvee_n x_n$ does not exist in X.

We recall that a finite indexed system $F = \{a_1, \ldots, a_n\}$ of a commutative BCK-algebra $(X; *, 0)$ with the relative cancellation property is **$\bigoplus$-orthogonal** if $a_1 + \cdots + a_n$ exists in X, and we write $\bigoplus_{i=1}^n a_i := a_1 + \cdots + a_n$.

We note that an arbitrary indexed system $G = \{a_i\}_{i \in I}$ of not necessarily different elements of X is **$\bigoplus$-orthogonal** iff, for every finite subset F of I, the system $\{a_i\}_{i \in F}$ is $\bigoplus$-orthogonal. A $\bigoplus$-orthogonal system $G = \{a_i\}_{i \in I}$ of X has a **$\bigoplus$-sum** in X, written as $\bigoplus_{i \in I} a_i$, iff there exists in X a supremum

$$\bigoplus_{i \in I} a_i := \bigvee_F \bigoplus_{i \in F} a_i, \tag{6.14}$$

where F runs over all finite subsets in I. In this case we also write $\bigoplus G := \bigoplus_{i \in I} a_i$.

Example 6.2.10 *Let us consider* $([0, 1]; *_{\mathbb{R}}, 0)$, *where* $s *_{\mathbb{R}} t := \max\{0, s - t\}$. *Then an indexed system* $\{x_i\}$ *of* $[0, 1]$ *is* $\bigoplus$*-orthogonal iff* $\sum_i x_i$ *exists and is less than* 1. *In this case,* $\bigoplus_i x_i = \sum_i x_i$. *Similarly, a system* $\{x_i\}$ *in* $([0, \infty); *_{\mathbb{R}}, 0)$ *has a* $\bigoplus$*-sum iff* $\sum_i x_i < \infty$; *then* $\bigoplus_i x_i = \sum_i x_i$. *Analogous properties hold for* $([0, \infty); *_{\mathbb{R}}, 0)$, *i.e.,* $\{x_i\}$ *has a* $\bigoplus$*-sum iff* $\sum_i x_i < \infty$.

Proposition 6.2.11 *Let $(X;*,0)$ be an Archimedean atomic commutative BCK-algebra with the relative cancellation property. Let j_a be defined for any atom a in X by* (6.10). *Then, for $x, y \in X$, we have:*

(i) $x = y$ *iff* $j_a(x) = j_a(y)$ *for any* $a \in \mathcal{A}(X)$.

(ii) $x \leq y$ *iff* $j_a(x) \leq j_a(y)$ *for any* $a \in \mathcal{A}(X)$.

(iii) $j_a(x+y) = j_a(x) + j_a(y)$ *if* $x + y \in X$.

(iv) $j_a(x \wedge y) = \min\{j_a(x), j_a(y)\}$.

(v) $j_a(x * y) = j_a(x) - j_a(y)$ *if* $y \leq x$.

(vi) $j_a(x * y) = j_a(x) *_{\mathbb{R}} j_a(y)$.

(vii) $\{j_a(x)a : a \in \mathcal{A}(X)\}$ *is a* $\bigoplus$*-orthogonal system of disjoint elements of* X *such that*

$$\bigoplus_{a \in \mathcal{A}(X)} j_a(x)a = x. \tag{6.15}$$

Proof. Due to the Archimedeanicity of X, $j_a(x)$ is a finite function for any atom a in X.

In view of (6.11), (i) and (ii) are evident.

(iii) It is clear that $j_a(x)a + j_a(y)a \leq x + y$. If $j_a(x)a + j_a(y)a + a \leq x + y$, then $a \leq x *^{j_a(x)} a + y *^{j_a(y)} a$. Using the Riesz decomposition property 5.2.8, there are two elements a_1 and a_2 in X with $a_1 \leq x *^{j_a(x)} a$ and $a_2 \leq y *^{j_a(y)} a$ such that $a = a_1 + a_2$. Since a is an atom, we have either $a_1 = 0$ so that $j_a(y) = j_a(y) + 1$, or $a_2 = 0$ which gives $j_a(x) = j_a(x) + 1$. Therefore, $j_a(x+y) = j_a(x) + j_a(y)$ for any atom a.

(iv) Using (6.11) and (i) of Theorem 6.2.9, and (6.7), we have

$$\begin{aligned}
x \wedge y &= x \wedge (\bigvee_a j_a(y)a) = \bigvee_a (x \wedge j_a(y)a) \\
&= \bigvee_a [\bigvee_b (j_b(x)b \wedge j_a(y)a] = \bigvee_a \bigvee_b (j_b(x)b \wedge j_a(y)a) \\
&= \bigvee_a [j_a(x)a \wedge j_a(y)a] = \bigvee_a [\min\{j_a(x), j_a(y)\}a] \\
&= \min\{j_a(x), j_a(y)\}a \vee \bigvee_{b \neq a} [\min\{j_b(x), j_b(y)\}b] \\
&= \min\{j_a(x), j_a(y)\}a + \bigvee_{b \neq a} [\min\{j_b(x), j_b(y)\}b],
\end{aligned}$$

since $a \wedge \bigvee_{b \neq a} [\min\{j_b(x), j_b(y)\}b] = 0$. Hence, $j_a(x \wedge y) = \min\{j_a(x), j_a(y)\}$.

(v) This follows from the fact that $(x * y) + y = x$ and from (iii).

(vi) Use the equality $x * y = x * (x \wedge y)$ and apply (v) and (iv).

(vii) In view of (6.7), we have that, for two different atoms a and b, $j_a(x)a \wedge j_b(y)b = 0$, and due to (6.11), x is a supremum for $\{j_a(x)a : a \in \mathcal{A}(X)\}$.

Take now an arbitrary non-empty finite subset $F = \{a_1, \ldots, a_n\}$ of atoms of X. We claim that

$$\bigvee_{i=1}^{n} j_{a_i}(x)a_i = \bigoplus_{i=1}^{n} j_{a_i}(x)a_i.$$

We prove it using the induction argument on n. Let $n = 2$. Put $b_i = j_{a_i}a_i$ for $i = 1, 2$. Then $b_1 \vee b_2$ is defined in X, and by Theorem 5.2.6(VIII), we have $(b_1 \vee b_2) * b_2 = b_1 * (b_1 \wedge b_2) = b_1$ so that $b_1 \vee b_2 = b_1 + b_2$.

The case for $n + 1$ follows from the fact

$$((b_1 \vee \cdots \vee b_n) \vee b_{n+1}) * (b_1 \vee \cdots \vee b_n) = b_{n+1} * (b_{n+1} \wedge (b_1 \vee \cdots \vee b_n)) = b_{n+1}.$$

Thus $b_1 \vee \cdots \vee b_{n+1} = (b_1 \vee \cdots \vee b_n) + b_{n+1} = b_1 + \cdots + b_n + b_{n+1}$.

Applying (6.11), we see that (6.15) holds. □

We say that an atomic commutative BCK-algebra is **atomically bounded** if any system $\{i_a a : a \in \mathcal{A}(X)\}$, where i_a is a non-negative integer such that $i_a a$ is defined in X, has an upper bound in X. For example, any bounded atomic commutative BCK-algebra is atomically bounded. On the other hand, $L_\infty \times L_\infty$ is an unbounded commutative BCK-algebra which is atomically bounded.

Proposition 6.2.12 *A Dedekind complete atomic BCK-algebra $(X; *, 0)$ with the relative cancellation property is atomically bounded if and only if, any system $\{i_a a : a \in \mathcal{A}(X)\}$, where i_a is any non-negative integer such that $i_a a$ is defined in X, has a $\bigoplus$-sum x in X. If this is the case, then $j_a(x) = i_a$ and $x = \bigoplus\{i_a a : a \in \mathcal{A}(X)\} = \bigvee\{i_a a : a \in \mathcal{A}(X)\}$.*

Proof. Let X be Dedekind complete and atomically bounded. Then there is a supremum $x = \bigvee\{i_a a : a \in \mathcal{A}(X)\}$. Using the ideas used in the proof of (vii) of Proposition 6.2.11, we can show that $x = \bigoplus\{i_a a : a \in \mathcal{A}(X)\}$, and therefore, $j_a(x) = i_a$ for any atom a.

Conversely, let now $\{i_a a : a \in \mathcal{A}(X)\}$ have a $\bigoplus$-sum x. Then x is by definition of a $\bigoplus$-sum an upper bound for $\{i_a a : a \in \mathcal{A}(X)\}$. Moreover, $x = \bigvee\{i_a a : a \in \mathcal{A}(X)\}$, which proves the assertion in question. □

In what follows, we generalize for (unbounded) BCK-algebras the result of Cignoli [Cig], who proved that a bounded commutative BCK-algebra (= MV-algebra) is complete and atomic iff it is a direct product of finite basic BCK-chains.

Theorem 6.2.13 *Let $(X; *, 0)$ be a commutative BCK-algebra with the relative cancellation property. Then X is Dedekind complete and atomically bounded if and only if X is isomorphic to a direct product of basic BCK-chains.*

Proof. Suppose that $X \cong \prod_i L_{n_i}$, where L_{n_i} is either a finite or infinite basic BCK-chain. Then X is atomic, Dedekind complete and atomically bounded.

Conversely, assume that X is Dedekind complete and atomically bounded. Due to Proposition 5.2.41, X is Archimedean, and for any atom a in X, $j_a(x)$, defined by (6.10), is finite. Due to (vi) of Proposition 6.2.11, for any atom a, the mapping j_a is a BCK-homomorphism from X onto $L_a := L_{ord(a)}$.

Define now the mapping $\Phi : X \to \prod_a L_a$ via

$$\Phi(x) := \{j_a(x) : a \in \mathcal{A}(X)\}.$$

In view of Proposition 6.2.11(vii), Φ is an injective BCK-homomorphism of X into $\prod_a L_a$.

We claim that Φ is, in addition, surjective. Indeed, take an element $\{i_a\} \in \prod_a L_a$, where i_a is an integer from L_a. Our assumptions entail that $\{i_a a : a \in \mathcal{A}(X)\}$ has a supremum $x = \bigvee_a i_a a$. In addition, there exists an element

$$x_a := \bigvee_{b \neq a} i_b b.$$

Applying the infinite distributive law (6.12), we have $a \wedge x_a = 0$ for any $a \in \mathcal{A}(X)$, so that $x = (i_a a) \vee x_a = (i_a a) + x_a$. Hence, $j_a(x) = i_a$ for any a, which means that $\Phi(x) = \{i_a\}$. □

Theorem 6.2.13 yields the above mentioned result of Cignoli [Cig] for bounded BCK-algebras:

Corollary 6.2.14 *A bounded commutative BCK-algebra $(X; *, 0)$ is complete and atomic if and only if it is isomorphic to a direct product of finite basic BCK-chains,*

Proof. The boundedness of X entails that $j_a(x) \leq j_a(1)$, where 1 is the greatest element of X. In this case $j_a(1) = ord(a)$ for any atom a of X. The statements in question follow now from Theorem 6.2.13. □

If X is atomic, but not atomically bounded, Theorem 6.2.13 can fail: Indeed, see Example 5.1.27.

Corollary 6.2.15 *Any finite commutative BCK-algebra $(X; *, 0)$ is a direct product of finitely many finite BCK-chains if and only if X is bounded.*

6.2.3 Exercises

1. Prove: Let a be an atom of an Archimedean commutative BCK-algebra $(X; *, 0)$ with the relative cancellation property. Then j_a defined via (6.10) is a non-trivial measure-morphism on X, and there exists a unique maximal ideal M_a of X such that $a \notin M_a$. Moreover,

(i) $\mathcal{A}(X) \setminus \{a\} \subseteq M_a$.

(ii) $x \in M_a$ if and only if $j_a(x) = 0$.

(iii) $X/M_a \cong L_{ord(a)}$.

In addition, X is atomic if and only if $\{j_a : a \in \mathcal{A}(X)\}$ is order-determining. In this case, X is semisimple, and X is a subdirect product of $\prod_a X/M_a$. ([DvGr 4])

2. An atomic commutative BCK-algebra $(X; *, 0)$ with the relative cancellation property is semisimple if and only if X is Archimedean.

3. Let $X = \prod_a X_a$. Then

   (i) X is commutative if and only if each X_a is commutative.

   (ii) X is commutative with the relative cancellation property if and only if each X_a is so. In this case $\{x_a\} + \{y_a\} = \{x_a + y_a\}$.

   (iii) Let $x_i = \{x_a^i\}$ and $x = \{x_a\}$. Then $\bigvee_i x_i = x$ if and only if $\bigvee_i x_a^i = x_a$ for each a, and in this case $\bigvee_i \{x_a^i\} = \{\bigvee_i x_a^i\}$.

   (iv) X is Dedekind complete (σ-complete) if and only if each X_a is so.

   (v) Let $x_i = \{x_a^i\}$ and $x = \{x_a\}$. Then $x = \bigoplus_i x_i$ if and only if $x_a = \bigoplus_i x_a^i$ for each a, and in this case $\bigoplus_i \{x_a^i\} = \{\bigoplus_i x_a^i\}$.

   (vi) X is Archimedean if and only if each X_a is so.

4. Prove: Let $(X; *, 0)$ be an Archimedean atomic commutative BCK-algebra with the relative cancellation property and let $\mathcal{A}(X)$ be finite. Then any finitely additive measure on X is a convex combination of j_a, where j_a is defined by (6.10).

5. Prove: Let u be a quasi strong unit for a Dedekind complete atomically bounded commutative BCK-algebra $(X; *, 0)$ with the relative cancellation property and let $\mathcal{A}(X)$ be finite. Then $Ext(\mathcal{S}_u(X)) = \{j_a^u : a \in \mathcal{A}(X)\}$, and $\mathcal{S}_u(X) = Con(\{j_a^u : a \in \mathcal{A}(X)\})$. In particular, any measure m on X is completely additive.

6. Let u be a non-zero element of X. An element x of X, $x \le u$, is said to be a **characteristic element with respect to** u if $x \wedge (u * x) = 0$. We denote by $\mathcal{B}_u(X)$ the set of all characteristic elements of X with respect to u.

   Prove: For any $u > 0$, the set $\mathcal{B}_u(X)$ is a non-degenerated Boolean algebra, where the complement of x is defined via $x^{*_u} = u * x$. If, for elements $\{x_i\}$ of $\mathcal{B}_u(X)$, $\bigvee_i x_i$ is defined in X, then $\bigvee_i x_i \in \mathcal{B}_u(X)$. Moreover, if X is Dedekind σ-complete (Dedekind complete), then so is $\mathcal{B}_u(X)$.

7. Prove: Let $(X;*,0)$ be a commutative BCK-algebra with the relative cancellation property.

   (1) Let x and y be two elements of X having an upper bound. Then $x+y$ exists and equals $x \vee y$ iff $x \wedge y = 0$.

   (2) Let $x,\ y \in \mathcal{B}_u(X)$, where $u > 0$. Then $x \wedge y = 0$ iff $x+y$ is defined in X and $x+y \leq u$. If it is the case, then $x \vee y = x+y$.

   (3) Let $\{x_i\}_{i\in I}$ be a system of mutually disjoint elements of X such that $\bigvee_{i\in I} x_i$ is defined in X. Then $\bigoplus_{i\in I} x_i$ is defined in X and $\bigoplus_{i\in I} x_i = \bigvee_{i\in I} x_i$.

8. Prove: Let $(X;*,0)$ be a Dedekind complete commutative BCK-algebra with the relative cancellation property and let u be a quasi strong unit for X. Then any two-valued measure m on $\mathcal{B}_u(X)$ with $m(u) = 1$ can be uniquely extended to a measure-morphism from $\mathcal{S}_u(X)$. (Hint: Use [Goo 1, Thm 9.9, Lem 8.12, Thm 8.14].) ([DvGr 4])

9. Let $(X;*,0)$ be a Dedekind σ-complete commutative BCK-algebra with the relative cancellation property and let u be a quasi strong unit for X. Then any probability measure from $\mathcal{B}_u(X)$ can be extended to a state m from $\mathcal{S}_u(X)$. ([DvGr 4])

10. Show that in Proposition 6.2.1: (a) the right-hand side of (ii) implies always the left-hand and the equality, while the converse is not always true. (b) (iii) holds without any restriction, and in (iv) the left-hand side entails the second one and the equality without any restriction. (Hint: Use Theorem 6.2.9.)

6.3 Connections between BCK-algebras and Difference Posets

In the present Section, we discuss the interrelations between BCK-algebras studied in Chapter 5 and posets with difference studied in Chapter 1, and interrelations among bounded commutative BCK-algebras, difference posets, MV-algebras, quantum MV-algebras and orthoalgebras, respectively. We show that only commutative BCK-algebras coincide either with D-posets, or MV-algebras, or QMV-algebras, respectively.

Theorem 6.3.1 *Let $(X;*,0)$ be a BCK-algebra. We define a partial binary operation $\div$ on X such that, for $x,y \in X$, $y \div x$ is defined if and only if $x \leq y$, and in this case*

$$y \div x := y * x. \tag{6.16}$$

Then

(i) $y \div x \leq y$ *if* $x \leq y$.

(ii) $y \div (y \div x) \leq x$ *if* $x \leq y$.

(iii) *if* $x \leq y \leq z$, *then* $z \div y \leq z \div x$ *and* $(z \div x) \div (z \div y) \leq y \div x$.

The partial binary operation $\div$ *is a difference on* $(X; \leq)$ *if and only if* $(X; *, 0)$ *is a commutative BCK-algebra.*

Proof. (i). According to (5.2), $(y * x) * y = (y * y) * x = 0 * x = 0$, so that $y * x \leq y$.
(ii). By (BCK-2), $y \div (y \div x) = y * (y * x) \leq x$.
(iii) By Proposition 5.1.3(c), $x \leq y$ implies $z \div y \leq z \div x$ and (BCK-1) entails that $(z \div x) \div (z \div y) \leq y \div x$.

Suppose now that $\div$ is a difference on X. Let $x \leq y$. (D2) entails that $y * (y * x) = y \div (y \div x) = x$. Hence, by (iii)–(iv) of Theorem 5.1.17, X is commutative.

Conversely, let X be a commutative BCK-algebra. Since $y \div (y \div x) \leq x$ if $x \leq y$, it is enough to show that $x \leq y \div (y \div x)$.

$$\begin{aligned} x * (y \div (y \div x)) &= x * (y * (y * x)) \\ &= x * (x \sqcap y) \\ &= x * (y \sqcap x) \\ &= x * y \\ &= 0. \end{aligned}$$

This means $x \leq y \div (y \div x)$. Hence the partial binary operation $\div$ is a difference on $(X; \leq)$. □

Corollary 6.3.2 *Let* $(X; *, 0)$ *be a BCK-algebra and let* $\div$ *be defined via* (6.16). *The following statements are equivalent:*

(i) $\div$ *is a difference on* $(X; \leq)$.

(ii) $x = y \div (y \div x)$ *whenever* $x \leq y$, $x, y \in X$.

(iii) $(X; *, 0)$ *is a commutative BCK-algebra.*

Proof. (i) implies (ii) by definition of a difference. The implication (ii) $\Rightarrow$ (iii) has been established in the proof of Theorem 6.3.1, and the implication (iii) $\Rightarrow$ (i) follows from Theorem 6.3.1. □

Theorem 6.3.3 *Let* $(X; *, 0, 1)$ *be a bounded BCK-algebra with the induced order* $\leq$. *We define*

(i) *a partial binary operation* $\div$ *defined via* (6.16);

(ii) *a binary operation* $\oslash$ *on* X *defined via*

$$y \oslash x := y \div (x \sqcap y), \quad x, y \in X; \tag{6.17}$$

(iii) *a unary operation* $^* : X \to X$ *defined via*

$$x^* := 1 \div x, \quad x \in X; \tag{6.18}$$

(iv) *binary operations* $\oplus$ *and* $\odot$ *on* X *defined via*

$$x \oplus y := (x^* \oslash y)^*, \quad x, y \in X, \tag{6.19}$$
$$x \odot y := x \oslash y^*, \quad x, y \in X. \tag{6.20}$$

Then, for all $x, y \in X$,

$$y \oslash x = y * x, \quad x \oplus y = (x^* * y)^*, \quad x \odot y = (x^* \oplus y^*)^*,$$

and the following statements are equivalent:

(a) $(X; \leq, \div, 0, 1)$ *is a D-poset.*

(b) $(X; *, 0, 1)$ *is a bounded commutative BCK-algebra.*

(c) $(X; \oplus, ^*, 0, 1)$ *is an MV-algebra.*

If it is the case, then the orders determined by the BCK-algebra X*, the D-poset* X *and the MV-algebra* X *coincide.*

Proof. (a) $\Leftrightarrow$ (b). Note that the greatest element of the D-poset $(X; \leq, \div, 0, 1)$ is the greatest element of the BCK-algebras $(X; *, 0, 1)$, and vice versa. By applying Corollary 6.3.2, we can see (a) $\Leftrightarrow$ (b).

(b) $\Rightarrow$ (c). We check all MV-axioms. Let $x, y, z \in X$. Then

$$\begin{aligned}
(x^*)^* &= (1 \div x)^* = 1 \div (1 \div x) = x, \\
0^* &= 1 \div 0 = 1 * 0 = 1, \\
x \oplus y &= (x^* \oslash y)^* = ((1 \div x) \oslash y)^* = ((1 * x) * y)^* \\
&= ((1 * y) * x)^* = y \oplus x, \\
(x \oplus y) \oplus z &= (y \oplus x) \oplus z = [(y \oplus x)^* \oslash z]^* \\
&= [(y^* \oslash x) \oslash z]^* = [(y^* * x) * z]^* = [(y^* * z) * x]^* \\
&= [(y \oplus z)^* \oslash x]^* = (y \oplus z) \oplus x = x \oplus (y \oplus z), \\
x \oplus 0 &= (x^* \oslash 0)^* = (x^* * 0)^* = (x^*)^* = x, \\
x \oplus 1 &= (x^* \oslash 1)^* = (x^* * 1)^* = 0^* = 1, \\
x \oplus x^* &= (x^* \oslash x^*)^* = (x^* * x^*) = 0^* = 1.
\end{aligned}$$

To prove $(x^* \oplus y)^* \oplus y = (x \oplus y^*)^* \oplus x$, we use the equality

$$u \oslash (u \oslash v) = v \oslash (v \oslash u),$$

holding for all $u, v \in X$.

$$\begin{aligned}
y \oplus (x^* \oplus y)^* &= [y^* \oslash (y \oplus x^*)^*]^* = [y^* \oslash (y^* \oslash x^*)]^* \\
&= [x^* \oslash (x^* \oslash y^*)]^* \\
&= [x^* \oslash (x \oplus y^*)^*]^* = x \oplus (x \oplus y^*)^* \\
&= (x \oplus y^*)^* \oplus x.
\end{aligned}$$

It is easy to verify that $x \odot y = (x^* \oplus y^*)^*$ for all $x, y \in X$. We note that the order $\leq_M$ determined in the MV-algebra $(X; \oplus, ^*, 0, 1)$ is given by $x \leq_M y$ iff $y \oplus x^* = 1$. Suppose that $x \leq_M y$, then $(y^* \oslash x^*)^* = 1$ so that $y^* * x^* = 0$ and $y^* \leq x^*$. Hence $1 * x^* \leq 1 * y^*$, i.e., $x \leq y$, and vice versa.

(c) $\Rightarrow$ (b). Since $(X; \oplus, ^*, 0, 1)$ is an MV-algebra, using the axiom (MVviii), we have $(x \oplus y^*)^* \oplus y^* = (x^* \oplus y)^* \oplus x^*$. Hence

$$\begin{aligned}
y^* \oplus (y^* \oplus x)^* &= x^* \oplus (x^* \oplus y)^*, \\
y^* \oplus (y * x) &= x^* \oplus (x * y), \\
(y * (y * x))^* &= (x * (x * y))^*, \\
y * (y * x) &= x * (x * y),
\end{aligned}$$

so that $(X; *, 0, 1)$ is a bounded commutative BCK-algebra. □

Theorem 6.3.4 *Let a poset with difference $(X; \leq, \ominus)$ be a lower semilattice with respect to $\leq$. Define a binary operation $*$ on X by*

$$x * y := x \ominus (x \wedge y), \quad x, y \in X. \tag{6.21}$$

Then

$$x * y = x \ominus y \quad \text{if} \quad y \leq x, \quad x, y \in X,$$

and X possesses a least element 0. Moreover, $(X; *, 0)$ is a commutative BCK-algebra if and only if, for all $x, y, z \in X$, we have

$$(x * y) * z = (x * z) * y. \tag{6.22}$$

Proof. It is possible to show that under our conditions, for any $b \in X$, $b \ominus b$ is the least element of X, i.e., $(X; \leq, \ominus)$ possesses a least element $0 = b \ominus b$.

Suppose now (6.22). We now show that then X satisfies the Yutani conditions (i)–(iii) of Theorem 5.1.18 on commutative BCK-algebras. Since X is a poset with difference, $x * x = x \ominus (x \wedge x) = x \ominus x = 0$ and $x * 0 = x \ominus (x \wedge 0) = x \ominus 0 = x$, which proves (iii) and (iv). Since X is a lower semilattice and by applying (D2) we have $x*(x*y) = x\ominus(x*y) = x\ominus(x\ominus(x\wedge y)) = x\wedge y = y\wedge x = y\ominus(y*x) = y*(y*x)$, which proves (i). Hence if $(X; *, 0)$ satisfies (6.22), then it is a commutative BCK-algebra. The converse easily follows from Proposition 5.1.2. □

Corollary 6.3.5 *Let a D-poset $(X; \leq, \ominus, 0, 1)$ be a bounded lower semilattice. Let $*$ be a binary operation on X defined by (6.21), and let $\oplus$, $\odot$, * be defined by (6.17)–(6.20), respectively. Then the following statements are equivalent:*

(i) $(x * y) * z = (x * z) * y, \quad x, y, z \in X.$

(ii) *$(X; *, 0, 1)$ is a bounded commutative BCK-algebra.*

(iii) *$(X; \oplus, ^*, 0, 1)$ is an MV-algebra.*

In addition, if this is the case, X is a lattice with respect to $\leq$.

Proof. This follows from Theorem 6.3.3 and Theorem 6.1.8. □

We remark that Chovanec and Kôpka [ChKo] proved that a D-poset being simultaneously a lattice can be organized equivalently into an MV-algebra iff (6.22) holds. We note that the D-poset $\mathcal{E}(H)$ from Example 1.3.16 cannot be equivalently reformulated as a BCK-algebra (respectively, as an MV-algebra), because $\mathcal{E}(H)$ is not a lower semilattice.

In what follows, we shall study relations between BCK-algebras and QMV-algebras.

Proposition 6.3.6 *Let $(X; \oplus, ^*, 0, 1)$ be a QMV-algebra. Define a binary operation $*$ on X via*

$$x * y := x \odot y^*, \quad x, y \in X. \tag{6.23}$$

Then, for all $x, y, z \in X$, we have

(i) $x * 0 = 0.$

(ii) $x * x = 0.$

(iii) $(x * y) * z = (x * z) * y.$

(iv) $(x * (x * y)) * y = 0.$

(v) $x \Cap y = y * (y * x).$

Proof. (i) (ii), and (v) are evident.
(iii) $(x*y)*z = (x\odot y^*)\odot z^* = x\odot(y^*\odot z^*) = x\odot(z^*\odot y^*) = (x\odot z^*)\odot y^* = (x*z)*y.$
(iv) According to (iii), $(x * (x * y)) * y = (x * y) * (x * y) = 0.$ □

Theorem 6.3.7 *A QMV-algebra* $(X;\oplus,^*,0,1)$ *is an MV-algebra if and only if, for* $x,y\in X$,

$$y \Cap x = x \quad \textit{implies} \quad x = x \Cap y. \tag{6.24}$$

Proof. First we show that $y \Cap x = x$ iff $x * y = 0$, where $*$ is as in 6.23. Indeed, let $x*y=0$. Then $y\Cap x = x*(x*y) = x*0 = x$. Conversely, let $y \Cap x = x$. Then

$$x*y = (x*(x*y))*y = (x*y)*(x*y) = 0,$$

where we have used (iii) and (ii) of Proposition 6.3.6.

If now $(X;\oplus,^*,0,1)$ is an MV-algebra, then $(X;\leq,\wedge,\vee,0,1)$ is a distributive lattice in which

$$u \Cap v = u\wedge v = v \Cap u, \;\; u,v\in X,$$

so that condition (6.24) is satisfied.

Conversely, let (6.24) hold. We show that $u\Cap v = v\Cap u$ for all $u,v\in X$. For that it is sufficient to establish that $(u\Cap v)*(v\Cap u)=0$. Calculate

$$\begin{aligned}(v\Cap u)*(u\Cap v) &= ((v\Cap u)^*\oplus(u\Cap v))^*\\ &= ((v\oplus u^*)^*\oplus u^*\oplus(u\Cap v))^*\\ &= ((v\oplus u^*)^*\oplus(u^*\oplus v))^* = 1^* = 0.\end{aligned}$$

By hypothesis, $v\Cap u \leq_Q u\Cap v$.[4]. Similarly we can prove that $u\Cap v\leq_Q v\Cap u$, which gives $u\Cap v = v\Cap u$. □

Remark 6.3.8 Condition (6.24) says that, for x,y from a QMV-algebra $(X;\oplus,^*,0,1)$

$$x\leq_Q y \quad \Leftrightarrow \quad y\Cap x = x, \tag{6.25}$$

and Theorem 6.3.7 says that (6.25) is a necessary and sufficient condition for $(X;\oplus,^*,0,1)$ to be an MV-algebra. (See also Chapter 2.)

Theorem 6.3.9 *Let* $(X;\oplus,^*,0,1)$ *be a QMV-algebra and define a binary operation* $*$ *on* X *via* (6.23). *Then* $(X;*,0)$ *is a BCK-algebra if and only if* $(X;\oplus,^*,0,1)$ *is an MV-algebra. If it is the case, then* $(X;*,0,1)$ *is a bounded commutative BCK-algebra.*

---

[4]We note that for all $a,b\in X$ we define

$$a\leq_Q b \;\text{ iff }\; a = a\Cap b.$$

Then $\leq_Q$ is a partial order on X, see also Chapter 2.

Proof. We note that in view of Proposition 6.3.6, (BCK-2), (BCK-3), (BCK-5) and (5.2) hold in any QMV-algebra. If $(X;\oplus,^*,0,1)$ is an MV-algebra, then $(X;*,0,1)$ is evidently a bounded commutative BCK-algebra.

Let now $(X;*,0,1)$ be a bounded BCK-algebra. We prove that $x*y=0$ implies $x=x\Cap y$. Calculate

$$\begin{aligned}(x\Cap y)*x &= (y*(y*x))*x=(y*x)*(y*x)=0.\\ x*(x\Cap y) &= x\odot(x\Cap y)^*=(x^*\oplus(x\Cap y))^*\\ &= (x^*\oplus y)^*=x\odot y^*=x*y=0.\end{aligned}$$

We show that the condition $y\Cap x=x$ entails $x*y=0$. Check

$$y\Cap x=(y\oplus x^*)\odot x=x=1\odot x.$$

Since $0=1^*\leq y\oplus x^*$ and $0=1^*\leq x$, by cancellation law, (ii) of Proposition 2.2.3, we have $y\oplus x^*=1$, i.e., $x*y=0$.

By the last two steps we have $x*y=0$ iff $y\Cap x=x$, and by Theorem 6.3.7, we have that $(X;\oplus,^*,0,1)$ is an MV-algebra, and consequently, $(X;*,0,1)$ is a bounded commutative BCK-algebra. □

Theorem 6.3.10 *Let $(X;*,0,1)$ be a bounded BCK-algebra with the induced BCK-order $\leq$. Define operations $^*,\oplus,\odot$ on X via (6.17)–(6.20). The following statements are equivalent:*

(i) $(X;\oplus,^*,0,1)$ *is a QMV-algebra.*

(ii) $(X;\oplus,^*,0,1)$ *is an MV-algebra.*

Proof. It is clear that (ii) entails (i).

Suppose thus (i). As (6.23) holds by definition of $\odot$, we can apply Theorem 6.3.7; so we conclude that $(X;\oplus,^*,0,1)$ is an MV-algebra. □

We now present an interrelation between BCK-algebras and orthoalgebras, for definitions see Section 1.5. We note that if L is an orthoalgebra, and $a\leq b$ are two elements in L, then there exists a unique element $b\ominus a$ defined by

$$b\ominus a=(a\oplus b^{\perp})^{\perp}. \tag{6.26}$$

In addition, $(L;\leq,\ominus,0,1)$ is D-poset.

Theorem 6.3.11 *Let $(X;*,0,1)$ be a bounded BCK-algebra with the induced order $\leq$. Define a partial binary operation $\boxplus$ on X as follows: $x\boxplus y$ is defined in X if and only if $x\leq 1*y$, and then we put*

$$x\boxplus y:=1*((1*y)*x). \tag{6.27}$$

Define operations $^*,\oplus,\odot$ via (6.17)–(6.20). Then the following statements are equivalent:

(a) $(X;\boxplus,0,1)$ *is an orthoalgebra.*

(b) $(X;*,0,1)$ *is an implicative BCK-algebra.*

(c) $(X;*,0,1)$ *is a positively implicative BCK-algebra.*

(d) $(X;\leq,\vee,\wedge,^*,0,1)$ *is a Boolean algebra.*

(e) $(X;\oplus,^*,0,1)$ *is an MV-algebra such that* $x\oplus y=x\vee y$ *for all* $x,y\in X$.

Proof. (a) $\Rightarrow$ (b). Since $(X;\leq,\ominus,0,1)$ is a D-poset, where $\ominus$ is defined by (6.26) for $\oplus=\boxplus$, by (b) of Theorem 6.3.3, $(X;*,0,1)$ is a bounded commutative BCK-algebra, with $x^\perp:=1\ominus x=x^*$ for any $x\in X$. We assert that, for any $x\in X$, $x\wedge x^*=0$. Indeed, suppose $z\leq x,x^*$. Then $z\leq x\leq z^*$ and therefore, $z\boxplus z$ exists in X, which by axioms of ortholgebras is possible iff $z=0$. Applying (i) and (iv) of Theorem 5.1.17, this implies that $(X;*,0,1)$ is an implicative BCK-algebra.

By the definition of an implicative BCK-algebra, a BCK-algebra is implicative iff it is both commutative and positive implicative, so that (b) and (c) are equivalent.

(b) $\Rightarrow$ (d). This follows from Example 5.1.23, Exercise 8.5.2.12,

(d) $\Rightarrow$ (e). If $(X;\leq,\vee,\wedge,^*,0,1)$ is a Boolean algebra, then $(X;*,0,1)$ is a bounded commutative BCK-algebra. Calculate by (6.19)

$$x\oplus y=(x^* * y)^*=(x^* *(x^* * y^*))^*=(x^*\wedge y^*)^*=x\vee y.$$

We note that the equivalence of (d) and (e) is well known due to Chang [Cha].

(e) $\Rightarrow$ (a). It is simple to verify that $(X;\boxplus,0,1)$ satisfies the axioms of orthoalgebras (OA1)–(OA4) (see, Definition 1.5.1). Let now $x\boxplus x$ exists in X. Then by (e),

$$1=x\oplus x^*=x\vee x^*=x^*,$$

which implies $x=0$, and $(X;\boxplus,0,1)$ is an orthoalgebra. □

Theorem 6.3.12 *Let an orthoalgebra* $(X;\oplus,0,1)$ *be a lower semilattice. Define a binary operation* $*$ *on* X *via*

$$x*y=(x^\perp\oplus(x\wedge y))^\perp,\ x,y\in X.$$

Then the following statements are equivalent:

(i) $(x*y)*z=(x*z)*y,\ x,y,z\in X$.

(ii) $(X;*,0,1)$ *is a bounded implicative BCK-algebra.*

(iii) $(X;\oplus,^*,0,1)$ *is an MV-algebra with* $x\oplus y=x\vee y$ *for all* $x,y\in X$.

(iv) $(X;\leq,\vee,\wedge,^*,0,1)$ *is a Boolean algebra.*

In addition, if this is the case, X *is a lattice with respect to* $\leq$.

Proof. Since $(X; \leq, \ominus, 0, 1)$ is a D-poset, where $\ominus$ is defined via (6.26), we have from Corollary 6.3.5 that (ii) implies (i).

Suppose now (i). Then $(X; *, 0, 1)$ is a bounded commutative BCK-algebra. It is evident that $x^\perp = x^*$, $x \in X$. We assert that $x \wedge x^* = 0$ for any $x \in X$, which is necessary and sufficient for $(X; *, 0, 1)$ to be an implicative BCK-algebra. Indeed, if $z \leq x, x^*$, then $z \leq x \leq z^\perp$ which implies by (OA4) $z = 0$,

The rest of the proof follows similar ideas as those in the proof of Theorem 6.3.11. □

Corollary 6.3.13 *Let $(X; \leq, ^\perp, 0, 1)$ be an orthomodular lattice. Then it determines a BCK-algebra via Theorem* 6.3.12 *if and only if X is a Boolean algebra.*

Proof. This follows from Theorem 6.3.12. □

6.3.1 Exercises

1. Show that a GDL P can be converted into a commutative BCK-algebra with the relative cancellation property such that $x * y = x \ominus (x \wedge y)$, $x, y \in L$, iff P satisfies the Riesz property iff any two elements in P are compatible. (Hint: Use Theorem 1.10.14.)

2. Prove: Let $(E; +, S, \leq, 0)$ be a positive commutative minimal clan (called also a Vitali space, see [Con]). Then $(E; *, 0)$, where $\ominus$ is defined via $x*y = x-(x \wedge y)$, $x, y \in E$, is an upwards directed commutative BCK-algebra.

3. Prove: Let $(X; *, 0)$ be an upwards directed commutative BCK-algebra. Then $(X; +, S, \leq, 0)$, where $+$ is a partial addition derived from X and $S \subseteq X \times X$, where $+$ is defined, and $\leq$ is the BCK-order, is a positive commutative minimal clan.

4. Show that any positive commutative minimal clan E can be embedded into the positive cone G^+ of an Abelian ℓ-group G such that the image of E generates G^+ as an ℓ-ideal.

5. Show that the D-poset $(\mathcal{E}(H); \ominus, O, I)$ is not a lattice ([GuGr]) and not a commutative BCK-algebra, in which the BCK-order coincides with the original partial order of effect operators. (See Exercise 17, Section 1.1.)

6. Show that the BCK-algebra $(X; *, 0)$ from Example 5.1.12 is non-commutative, with the condition (S), and $(X; \oplus, ^*, 0, 4)$ with $\oplus := \circ$, $a^* := 4 * a$, where $\circ$ is derived from the condition (S), is an S-algebra which is neither an MV-algebra nor a QMV-algebra. In addition, $\leq_{BCK} \neq \leq_Q$.

7. Show that if a bounded BCK-algebra $(X; *, 0, 1)$ satisfies $1 * (1 * a) = a$ for any $a \in X$, then $(X; *, 0)$ is with the condition (S), and $(X; \oplus, ^*, 0, 1)$, where

$\oplus = \circ$, $a^* := 1 * a$, $a \in X$, is an S-algebra. In addition, $(X; \oplus, ^*, 0, 1)$ is an MV-algebra iff it is a QMV-algebra iff $(X; *, 0, 1)$ is a commutative BCK-algebra iff $\leq_{BCK} = \leq_{QMV}$. (Hint: Use Proposition 5.1.21, and Proposition 6.3.6.)

6.4 Pseudo MV-algebras

Recently, G. Georgescu and A. Iorgulescu [GeIo] introduced pseudo MV-algebras which are a generalization of MV-algebras introduced in 1958 by Chang [Cha] and which are a non-commutative generalization of a two-valued reasoning. We recall that recently J. Rachůnek [Rac] introduced non-commutative MV-algebras which are equivalent with pseudo MV-algebras (see Exercise 12.6.4.5).

6.4.1 On Partial Addition in Pseudo MV-algebras

We introduce a partial binary operation which models the addition in pseudo MV-algebras. Such an addition in classical MV-algebras entails an ℓ-group representation of MV-algebras (see Mundici's representation, Corollary 5.3.8).

Definition 6.4.1 *A* **pseudo MV-algebra** *is an algebra* $(M; \oplus, ^-, ^\sim, 0, 1)$ *of type* $(2, 1, 1, 0, 0)$ *such that the following axioms hold for all* $x, y, z \in M$ *with an additional binary operation* $\odot$ *defined via*

$$y \odot x := (x^- \oplus y^-)^\sim$$

(A1) $x \oplus (y \oplus z) = (x \oplus y) \oplus z$;

(A2) $x \oplus 0 = 0 \oplus x = x$;

(A3) $x \oplus 1 = 1 \oplus x = 1$;

(A4) $1^\sim = 0$; $1^- = 0$;

(A5) $(x^- \oplus y^-)^\sim = (x^\sim \oplus y^\sim)^-$;

(A6) $x \oplus x^\sim \odot y = y \oplus y^\sim \odot x = x \odot y^- \oplus y = y \odot x^- \oplus x$;

(A7) $x \odot (x^- \oplus y) = (x \oplus y^\sim) \odot y$;

(A8) $(x^-)^\sim = x$.

If we define $x \leq y$ iff $x^- \oplus y = 1$, then $\leq$ is a partial order such that M is a distributive lattice with $x \vee y = x \oplus x^\sim \odot y$. For basic properties of pseudo MV-algebras see [GeIo] or Exercises 3.6.4.5.

For example, if u is a strong unit of a (not necessarily Abelian) ℓ-group G,

$$\Gamma(G, u) = [0, u]$$

and

$$\begin{aligned} x \oplus y &:= (x+y) \wedge u, \\ x^- &:= u - x, \\ x^\sim &:= -x + u, \\ x \odot y &:= (x - u + y) \vee 0, \end{aligned}$$

then $(\Gamma(G,u); \oplus, ^-, ^\sim, 0, u)$ is a pseudo MV-algebra [GeIo], see also Exercise 13.6.4.5.

Let $(M; \oplus, ^-, ^\sim, 0, 1)$ be a pseudo MV-algebra. Define a partial binary operation $+$ on M via: $x + y$ is defined iff $x \le y^-$, and in this case

$$x + y := x \oplus y.$$

It is clear that $x + y$ is defined iff $x \le y^-$ iff $y \le x^\sim$, and the partial addition $+$ coincides with that derived for MV-algebras or BCK-algebras, Section 5.2.

Proposition 6.4.2 *The following properties hold in any pseudo MV-algebra M.*

(i) $x + 0 = x = 0 + x$ *for any* $x \in M$.

(ii) $x + x^\sim = 1 = x^- + x$ *for any* $x \in M$.

(iii) *If* $x^- + y = 1$, *then* $y = x$.

(iv) *If* $x + y = 1$, *then* $y = x^\sim$ *and* $x = y^-$.

Proof. (i) Since $x \le 1 = 0^-$, then $x + 0 = x \oplus 0 = x$. Similarly, $0 \le x^-$ implies $0 + x = 0 \oplus x = x$.

(ii) $(x^\sim)^- = x \ge x$ so that $x + x^\sim = x \oplus x^\sim = 1$, and $x^- \ge x^-$ implies $x^- + x = x^- \oplus x = 1$.

(iii) Let $x^- + y = 1$, then $x \le y$. On the other hand $x^- \le y^-$ so that $y^{-\sim} \le x^{-\sim}$, and $y \le x$. The last two inequalities give $x = y$.

(iv) Let $x + y = 1$. Then $x \le y^-$ entails $x^= \le (y^-)^= = (y^=)^-$,[5] which means that $x^= + y^=$ is defined in M, and $x^= + y^= = 1$. By (iii), we conclude that $(x^-)^- + y^= = 1$ yields $x^- = y^=$ so that $(x^-)^\sim = (y^-)^{-\sim}$ and $x = y^-$, $y = x^\sim$. □

Proposition 6.4.3 *Let $x + y$ and $(x + y) + z$ be defined in M. Then $y + z$ and $x + (y + z)$ are defined in M, and*

$$(x + y) + z = x + (y + z).$$

Similarly, if $y + z$ and $x + (y + z)$ are defined in M, so are $x + y$ and $(x + y) + z$.

[5] We note that $x^= := (x^-)^-$ and $x^\approx := (x^\sim)^\sim$.

Proof. From the existence of $(x+y)+z$ we conclude that $z \leq (x+y)^{\sim} = y^{\sim} \odot x^{\sim} \leq y^{\sim}$ and $y+z$ is defined in M. In addition, from $z \leq y^{\sim} \odot x^{\sim}$ we have $y+z = y \oplus z \leq y \oplus (y^{\sim} \odot x^{\sim}) = y \vee x^{\sim} = x^{\sim}$. Hence $x+(y+z) \in M$ and $x+(y+z) = x \oplus (y \oplus z) = (x \oplus y) \oplus z = (x+y)+z$. □

Proposition 6.4.4 *The following properties hold in any pseudo MV-algebra M.*

(i) *If $x+b=y$, then $b = x^{\sim} \odot y$.*

(ii) *If $x+y_1 = x+y_2$, then $y_1 = y_2$.*

(iii) *If $x_1+y = x_2+y$, then $x_1 = x_2$.*

(iv) *If $y = a+x$, then $a = y \odot x^{-}$.*

Proof. (i) If $x+b=y$, then $y^{-}+(x+b) = y^{-}+y = 1$, $(y^{-}+x)+b = 1$, and by (iv) of Proposition 6.4.2. $b = (y^{-}+x)^{\sim} = x^{\sim} \odot y^{-\sim} = x^{\sim} \odot y$.

(ii) There exists an element $z \in M$ such that $z+(x+y_1) = z+(x+y_2) = 1$. Hence $(z+x)+y_1 = (z+x)+y_2 = 1$, and by (iv) of Proposition 6.4.2 $y_1 = y_2$.

(iii) There exists an element $u \in M$ such that $(x_1+y)+u = (x_2+y)+u = 1$, and hence $x_1 = x_2$.

(iv) Calculate $1 = y+y^{\sim} = (a+x)+y^{\sim} = a+(x+y^{\sim})$, so that $a = (x+y^{\sim})^{-} = y \odot x^{-}$. □

Remark 6.4.5 $x \leq y$ if and only if there is an element b with $x+b=y$. In that case, b is unique. (See also Exercise 11.6.4.5.)

Indeed, it follows from the following observation: If $x \leq y$ then $y = x \oplus x^{\sim} \odot y = x + x^{\sim} \odot y$, so put $b = x^{\sim} \odot y$. The converse statement is evident and the uniqueness follows from (i) of Proposition 6.4.4.

Proposition 6.4.6 (i) *If $a+b_1$ and $a+b_2$ exist in M, then $a+(b_1 \vee b_2)$ is defined in M, and*

$$a+(b_1 \vee b_2) = (a+b_1) \vee (a+b_2). \tag{6.28}$$

(ii) *If b_1+a and b_2+a exist in M, then $(b_1 \vee b_2)+a$ is defined in M, and*

$$(b_1 \vee b_2)+a = (b_1+a) \vee (b_2+a). \tag{6.29}$$

Proof. (i) Since $a+b_i$ is defined in M for $i=1,2$, $a \leq b_i^{-}$. Hence $a \leq b_1^{-} \wedge b_2^{-} = (b_1 \vee b_2)^{-}$, which yields $a+(b_1 \vee b_2) \in M$. By [GeIo, (a) Prop 1.13], or Exercise 5.6.4.5,

$$\begin{aligned} a+(b_1 \vee b_2) &= a \oplus (b_1 \vee b_2) = (a \oplus b_1) \vee (a \oplus b_2) \\ &= (a+b_1) \vee (a+b_2). \end{aligned}$$

(ii) Similarly, if b_i+a is defined in M for $i=1,2$, then $b_i \leq a^{-}$, so that $b_1 \vee b_2 \leq a^{-}$ and $(b_1 \vee b_2)+a$ is defined in M. Using again [GeIo, (a) Prop 1.13], we have (6.29). □

Proposition 6.4.7 (i) *If $a+b_1$ and $a+b_2$ exist in M, then $a+(b_1 \wedge b_2)$ is defined in M, and*

$$a+(b_1 \wedge b_2) = (a+b_1) \wedge (a+b_2). \tag{6.30}$$

(ii) *If b_1+a and b_2+a exist in M, then $(b_1 \wedge b_2)+a$ is defined in M, and*

$$(b_1 \wedge b_2)+a = (b_1+a) \wedge (b_2+a). \tag{6.31}$$

Proof. (i) Let $a+b_i$ be defined in M for $i=1,2$. Then $a \leq b_i^-$ and $a \leq b_1^- \vee b_2^- = (b_1 \wedge b_2)^-$, so that $a+(b_1 \wedge b_2)$ exists in M. Applying distributivity of $\oplus$ with respect to $\wedge$, [GeIo, Prop 1.15], we have (6.30).

(ii) $b_1 \wedge b_2 \leq b_1 \leq a^-$ so that $(b_1 \wedge b_2)+a \in M$. □

Proposition 6.4.8 *For all x, y of a pseudo MV-algebra $(M; \oplus, ^-, ^\sim, 0, 1)$ we have:*

(a) $x \odot (x \wedge y)^- = x \odot y^-$.

(b) $x \odot (x \wedge y)^- + y = x \odot y^- + y = x \vee y$.

(c) $(x \wedge y)^\sim \odot y = x^\sim \odot y$, $x \vee y = x + (x \wedge y)^\sim \odot y$.

(d) $(x \vee y) \odot y^- = x \odot (x \wedge y)^- = x \odot y^-$.

(e) $y^\sim \odot (x \vee y) = y^\sim \odot x = (x \wedge y)^\sim \odot x$.

Proof. (a) $x \odot (x \wedge y)^- = x \odot (x^- \vee y^-) = x \odot x^- \vee x \odot y^- = x \odot y^-$.

(b) Evident.

(c) $(x \wedge y)^\sim \odot y = (x^\sim \vee y^\sim) \odot y = x^\sim \odot y \vee y^\sim \odot y = x^\sim \odot y$.

(d) From (b) we conclude that $x \vee y = x \odot (x \wedge y)^- + y$. By (iv) of Proposition 6.4.4, we conclude that $x \odot (x \wedge y)^- = (x \vee y) \odot y^-$.

(e) Since by (c) $x \vee y = x + (x \wedge y)^\sim \odot y$, from (i) of Proposition 6.4.4 we have $(x \wedge y)^\sim \odot y = x^\sim \odot (x \vee y)$. □

Corollary 6.4.9 *If $a \wedge b = 0$, then $a+b, b+a$ exist in $(M; \oplus, ^-, ^\sim, 0, 1)$, and*

$$a+b = a \vee b = b+a. \tag{6.32}$$

Proof. We have $a \vee b = a + a^\sim \odot b = a + (b \wedge a)^\sim \odot b = a + 0^\sim \odot b = a+b$. By symmetry we have $a \vee b = b+a$. □

Proposition 6.4.10 *In any pseudo MV-algebra $(M; \oplus, ^-, ^\sim, 0, 1)$ we have the following cancellation laws.*

(i) *If $a+b_1 \leq a+b_2$, then $b_1 \leq b_2$.*

(ii) *If $a_1+b \leq a_2+b$, then $a_1 \leq a_2$.*

(iii) *If $a+b$ is defined in M, and if $a_1 \leq a$ and $b_1 \leq b$, then a_1+b_1 is defined in M, and $a_1+b_1 \leq a+b$.*

Proof. (i) There exists an element $c \in M$ such that $a+b+c = a+b_2$. By cancellation, Proposition 6.4.4, $b_1 + c = b_2$, which yields $b_1 \leq b_2$. In a similar way we prove (ii).

(iii) We have $a_1 \leq a \leq b^- \leq b_1^-$, so that $a_1 + b_1$ is defined in M. There exist two elements $a_2, b_2 \in M$ such that $a+b = (a_2+a_1)+(b_1+b_2) \geq a_1+(b_1+b_2) \geq a_1+b_1$. □

We say that a pseudo MV-algebra $(M; \oplus, ^-, ^\sim, 0, 1)$ has the **Riesz decomposition property** iff $c, a, b \in M$ with $c \leq a+b$ imply that there exist $a_1, b_1 \in M$ with $a_1 \leq a$ and $b_1 \leq b$ such that $c = a_1 + b_1$.

Theorem 6.4.11 *Any pseudo MV-algebra has the Riesz decomposition property.*

Proof. Suppose that $c \leq a+b$, and define $v := c \odot (b \wedge c)^-$ and $a_1 = a \wedge v$. Then $a_1 \leq a$ and $a_1 \leq c$. Put $b_1 := a_1^\sim \odot c$. Then $c = a_1 + b_1$.

To finish the proof, we have to show that $b_1 \leq b$. Applying distributivity with respect to $+$, we obtain

$$\begin{aligned} a_1 + b_1 = c &\leq (a+b) \wedge (b \vee c) \\ &= (a+b) \wedge (c \odot b^- + b) \quad \text{by (A6)} \\ &= (a+b) \wedge (c \odot (b \wedge c)^- + b) \quad \text{Proposition 6.4.8}(a) \\ &= (a+b) \wedge (v+b) \\ &= (a \wedge v) + b = a_1 + b, \end{aligned}$$

and by Proposition 6.4.10 we have $b_1 \leq b$. □

Theorem 6.4.12 *Let $a_1+a_2 = b_1+b_2$. Then there exist four elements $c_{11}, c_{12}, c_{21}, c_{22} \in M$ such that*

$$a_1 = c_{11} + c_{12}, \qquad b_1 = c_{11} + c_{21},$$
$$a_2 = c_{21} + c_{22}, \qquad b_2 = c_{12} + c_{22}.$$

Moreover, we may assume that $c_{12} \wedge c_{21} = 0$, and under this condition the c_{ij} are determined uniquely.

Proof. We define $c_{11} := a_1 \wedge b_1$, $c_{12} := b_1^\sim \odot a_1 = (b_1 \wedge a_1)^\sim \odot a_1 = c_{11}^\sim \odot a_1$, $c_{22} := a_2 \wedge b_2$, and $c_{21} := a_2 \odot b_2^- = a_2 \odot (a_2 \wedge b_2)^- = a_2 \odot c_{22}^-$.

Then $a_1 = c_{11} + c_{12}$ and $a_2 = c_{21} + c_{22}$. We now show that $c_2 \odot b_2^- = a_1^- \odot b_1$. Put $y = a_1 + a_2 = b_1 + b_2$. Then $y = b_1 + b_1^\sim \odot y = y \odot b_2^- + b_2 = a_1 + a_1^\sim \odot y = y \odot a_2^- + a_2$. Hence by the cancellation property, Proposition 6.4.10, we conclude that $b_2 = b_1^\sim \odot y$, $b_1 = y \odot b_2^-$, $a_1 = y \odot a_2^-$, and $a_2 = a_1^\sim \odot y$.

Calculate and use (12) of Exercise 3.6.4.5.

$$\begin{aligned} c_{21} = a_2 \odot b_2^- &= a_1^\sim \odot y \odot (b_1^\sim \odot y)^- = a_1^\sim \odot y \odot (y^- \oplus b_1^{\sim -}) \\ &= a_1^\sim \odot y \odot (y^- + b_1) = a_1^\sim \odot b_1 \odot (b_1^- \oplus y) \\ &= a_1^\sim \odot b_1 \odot (b_1^- \oplus b_1 \oplus b_2) = a_1^\sim \odot b_1 \odot 1 \\ &= a_1^\sim \odot b_1 = (a_1 \wedge b_1)^\sim \odot b_1 = c_{11}^\sim \odot b_1. \end{aligned}$$

By symmetry we have

$$b_2 \odot a_2^- = b_1^\sim \odot a_1 = c_{12} = (b_1 \wedge a_1)^\sim \odot a_1 = c_{11}^\sim \odot a_1.$$

Then $c_{12} \wedge c_{21} = c_{11}^\sim \odot b_1 \wedge c_{11}^\sim \odot a_1 = c_{11}^\sim \odot (b_1 \wedge a_1) = 0$. Hence by (6.32), $c_{12} + c_{21} = c_{21} + c_{12}$, so that $a_1 + a_2 = c_{11} + c_{12} + c_{21} + c_{22} = c_{11} + c_{21} + c_{12} + c_{22} = b_1 + b_2$.

Uniqueness. Adding the elements c_{11} and c_{21}, respectively, to the equality $c_{12} \wedge c_{21} = 0$, we obtain $(c_{11} + c_{12}) \wedge (c_{11} + c_{21}) = c_{11}$, so that $c_{11} = a_1 \wedge b_1$, and similarly $c_{22} = a_2 \wedge b_2$. Using the cancellation property, we see that c_{12} and c_{21} are defined consequently in the same way as at the beginning of the present proof. □

Theorem 6.4.12 can be extended in the following form.

Theorem 6.4.13 *Let $(M; \oplus, ^-, ^\sim, 0, 1)$ be a pseudo MV-algebra and let $a_1, \ldots, a_m$ and $b_1, \ldots, b_n$ be elements of M such that*

$$a_1 + \cdots + a_m = b_1 + \cdots + b_n. \tag{6.33}$$

Then there exist elements c_{ij} in M for $1 \leq i \leq m$ and $1 \leq j \leq n$ such that for all i and j

$$\begin{aligned} a_i &= c_{i1} + \cdots + c_{in}, \\ b_j &= c_{1j} + \cdots + c_{mj}. \end{aligned}$$

Moreover, we may assume that

$$(c_{i+1,j} + \cdots + c_{mj}) \wedge (c_{i,j+1} + \cdots + c_{in}) = 0 \tag{6.34}$$

for all $i < m$ and all $j < n$, and under this condition the c_{ij} are uniquely determined.

Proof. We will follow ideas of Fuchs [Fuc, p. 68] (see also the proof of Theorem 1.10.14). The case $m = 2 = n$ was proved in Theorem 6.4.12. Suppose thus that the theorem is true for integers m' and n' with $m' \leq m$, $n' < n$ ($n \geq 3$), and write the equality (6.33) in the form

$$a_1 + \cdots + a_m = b_1 + \cdots + b_{n-2} + (b_{n-1} + b_n).$$

By the induction hypothesis, there exist elements c_{ij} ($1 \leq i \leq m$, $1 \leq j \leq n-2$) and elements d_i ($1 \leq i \leq m$) in M such that

$$\begin{aligned} a_i &= c_{i1} + \cdots + c_{i,n-2} + d_i, \\ b_j &= c_{1j} + \cdots + c_{mj} \end{aligned}$$

for $1 \leq j \leq n-2$, and $b_{n-1} + b_n = d_1 + \cdots + d_m$.

We again apply the induction hypothesis, now to the last equation, to conclude the existence of elements c_{ij} in M, $1 \leq i \leq m$, $j = n-1, n$, with the property

$$d_i = c_{i,n-1} + c_{in}, \quad b_j = c_{1j} + \cdots + c_{mj}.$$

We now prove (6.34). The induction hypothesis yields

$$(c_{i+1,j} + \cdots + c_{mj}) \wedge (c_{i,j+1} + \cdots + c_{i,n-2} + d_i) = 0$$

for $i < m$ and $j \leq n-2$ and

$$(c_{i+1,n-1} + \cdots + c_{m,n-1}) \wedge c_{in} = 0,$$

respectively, which completes the proof of (6.34).

The uniqueness can be obtained as follows. Adding c_{ij} to (6.34), we obtain

$$(c_{ij} + c_{i+1,j} + \cdots + c_{mj}) \wedge (c_{ij} + c_{i,j+1} + \cdots + c_{in}) = c_{ij},$$

and

$$c_{ij} + c_{i+1,j} + \cdots + c_{mj} = (c_{1,j} + \cdots + c_{i-1,j})^{\sim} \odot b_j$$

and

$$c_{ij} + c_{1,j+1} + \cdots + c_{in} = (c_{i1} + \cdots + c_{i,j-1})^{\sim} \odot a_i.$$

This shows that the elements c_{ij} can be computed recursively from (6.34), which finishes the proof. □

A pseudo MV-algebra M is said to be (i) **Archimedean** if $n\,a := a + \cdots + a$, where addition is n-fold over a, exists in M for any integer $n \geq 1$ only for $a = 0$; [6] (ii) **σ-complete** if $\bigvee_{n=1}^{\infty} a_n \in M$ for any sequence $\{a_n\}_n$ of elements in M, and (iii) complete if $\bigvee_t a_t \in M$ for any system $\{a_t\}_t$ of elements in M. It is clear that any complete pseudo MV-algebra is σ-complete.

[6] We define $0a := 0$.

Proposition 6.4.14 *Any σ-complete pseudo MV-algebra M is Archimedean.*

Proof. Let $n\,a$ exist in M for any integer $n \geq 1$. Then $a_0 := \bigvee_{n=1}^{\infty} n\,a$ is defined in M, and by Exercise 5.6.4.5, we have

$$a_0 \odot a^- = (\bigvee_{n=1}^{\infty} n\,a) \odot a^- = \bigvee_{n=1}^{\infty} (n\,a \odot a^-) = \bigvee_{n=2}^{\infty} (n-1)\,a = a_0.$$

Since $a_0 \geq a$, we have $a = 0$ by Proposition 6.4.4. $\square$

6.4.2 Commutativity of Atomic Pseudo MV-algebras

A non-zero element a of M is said to be an **atom** of M if $b \leq a$ for $b \in M$ entails either $b = 0$ or $b = a$. A pseudo MV-algebra M is said to be **atomic** if for each non-zero element b of M there exists an atom a of M such that $a \leq b$. We denote by $\mathcal{A}(M)$ the set of all atoms in M.

We say that a pseudo MV-algebra M is **commutative** if $a \oplus b = b \oplus a$ for all $a, b \in M$. By [GeIo, Prop 1.29], every commutative pseudo MV-algebra is an MV-algebra, see Exercise 6.6.4.5.

Given an integer $n \geq 1$ and $x \in M$, we put

$$n \odot x := x \oplus \cdots \oplus x.$$

We note that $n \odot x$ is due to associativity of $\oplus$ defined well.

Proposition 6.4.15 *Let a and b be two different atoms of a pseudo MV-algebra M. Then*

$$(n \odot a) \wedge (m \odot b) = 0 \tag{6.35}$$

for any integer $n, m \geq 1$. In particular, if $n\,a$ and $m\,b$ are defined in M, then

$$(n\,a) \wedge (m\,b) = 0. \tag{6.36}$$

In addition, $a + b$ and $b + a$ are defined in M, and

$$a \oplus b = b \oplus a = a \vee b = a + b = b + a; \tag{6.37}$$

moreover, $(n \odot a) + (m \odot b)$ and $(m \odot b) + (n \odot a)$ are defined in M, and

$$(n \odot a) + (m \odot b) = (n \odot a) \vee (m \odot b) = (m \odot b) + (n \odot a). \tag{6.38}$$

Proof. Suppose that $x \leq n \odot a$ and $x \leq m \odot b$. In view of Theorem 6.4.11, $x = a_1 \oplus \cdots \oplus a_n$ where $a_i \leq a$ for $i = 1, \ldots, n$. Similarly $a_i = b_1^i \oplus \cdots \oplus b_m^i$, where $b_j^i \leq b$ for $j = 1, \ldots, m$, $i = 1, \ldots, n$, so that $b_j^i \leq a \wedge b$ which gives $b_j^i = 0$ for all i and j. Consequently, $x = 0$, and this proves (6.35). (6.36) follows directly from (6.35).

We have $a \oplus b = a + a^{\sim} \odot (a \oplus b)$. On the other hand, $a^{\sim} \odot (a \oplus b) = a^{\sim} \wedge b$, and either $a^{\sim} \wedge b = 0$ or $a^{\sim} \wedge b = b$. In the first case we would have $b \leq a \oplus b = a$ which is impossible, hence $b \leq a^{\sim}$. Applying Corollary 6.4.9, $a+b = b+a$ which proves (6.37).

(6.38) follows from similar reasoning as (6.37). □

Let a be an atom of M and let $x \in M$. We define

$$j_a(x) := \sup\{n \geq 0 : \; n\,a \leq x\}.$$

If M is Archimedean, then $j_a(x) < \infty$ for any $x \in M$. Similarly, we define

$$ord(x) := \sup\{n \geq 1 : \; n\,x \in M\}, \; x \in M.$$

Then $ord(0) = \infty$, and M is Archimedean iff $ord(x) < \infty$ for any $0 \neq x \in M$.

Theorem 6.4.16 *Every finite pseudo MV-algebra M is commutative.*

Proof. It is clear that M is atomic, and let $a_1, \ldots, a_n$ be the set of all distinct atoms of M. Then by Proposition 6.4.15, $j_a(x)a \wedge j_b(x)b = 0$ for any two distinct atoms a and b. Using the distributivity of M and (6.38), we have

$$\bigvee_{i=1}^{n} j_{a_i}(x)a_i = j_{a_1}(x)a_1 \oplus \cdots \oplus j_{a_n}(x)a_n = j_{a_1}(x)a_1 + \cdots + j_{a_n}(x)a_n.$$

In addition,

$$x = \bigvee_{i=1}^{n} j_{a_i}(x)a_i = j_{a_1}(x)a_1 \oplus \cdots \oplus j_{a_n}(x)a_n,$$

which entails $x \oplus y = j_{a_1}(x)a_1 \oplus \cdots \oplus j_{a_n}(x)a_n \oplus j_{a_1}(y)a_1 \oplus \cdots \oplus j_{a_n}(y)a_n = (j_{a_1}(x)a_1 \oplus j_{a_1}(y)a_1) \oplus \cdots \oplus (j_{a_n}(x)a_n \oplus j_{a_n}(y)a_n) = y \oplus x$, where we have used (6.38). □

Theorem 6.4.16 can be extended as follows.

Theorem 6.4.17 *Every atomic and complete pseudo MV-algebra M is commutative.*

Proof. As in Theorem 6.4.16, using Proposition 6.4.15, we can show that

$$x = \bigvee_{a \in \mathcal{A}(M)} j_a(x)a. \tag{6.39}$$

Indeed, let $x_0 := \bigvee_{a \in \mathcal{A}(M)} j_a(x)a$. Then $x_0 \leq x$. If $x_0 \neq x$, then due to the fact that $x = x_0 + x_0^{\sim} \odot x$, we conclude that $x_0^{\sim} \odot x > 0$ and there exists an atom a of M with $a \leq x_0^{\sim} \odot x \leq x$ which entails that $j_a(x) = j_a(x) + 1$ giving a contradiction.

Suppose now that A and B are two mutually disjoint non-void subsets of atoms of M, and let $\{n_a : n_a \geq 0,\ a \in A\}$ and $\{n_b : n_b \geq 0,\ b \in B\}$ be sets of integers such that $n_a a$ and $n_b b$ are defined in M for all $a \in A$ and all $b \in B$. Then

$$\bigvee_{a \in A} n_a a + \bigvee_{b \in B} n_b b = \bigvee_{a \in A} n_a a \vee \bigvee_{b \in B} n_b b = \bigvee_{b \in B} n_b b + \bigvee_{a \in A} n_a a.$$

Indeed, $\bigvee_{a \in A} n_a a \wedge \bigvee_{b \in B} n_b b = \bigvee_{a \in A} \bigvee_{b \in B} (n_a a \wedge n_b b) = 0$, and calling Corollary 6.4.9, we have the assertion in question.

Hence by Proposition 6.4.15

$$\begin{aligned} x \oplus y &= (\bigvee_a j_a(x)a) \oplus y = \bigvee_a (j_a(x)a \oplus y) \\ &= \bigvee_a (j_a(x)a \oplus \bigvee_b j_b(y)b) = \bigvee_a \bigvee_b j_a(x)a \oplus j_b(y)b \\ &= \bigvee_a \bigvee_b j_b(y)b \oplus j_a(x)a = \cdots = y \oplus x. \end{aligned}$$

□

Theorem 6.4.18 *Every atomic and σ-complete pseudo MV-algebra M with the countable set of atoms is commutative.*

Proof. This follows the same ideas as the proof of Theorem 6.4.17. □

Given an integer $n \geq 2$, we define $L_n := \{0, 1/n, 2/n, \ldots, (n-1)/n, 1\}$ as a finite chain equipped with the usual MV-structure.

Corollary 6.4.19 *The pseudo MV-algebras in Theorems* 6.4.16–6.4.18 *are representable via Mundici's representation of MV-algebras by ℓ-groups. In addition, in view of Cignoli* [Cig, Thm 2.6], *or Corollary* 6.2.14, *they are isomorphic to the direct product of finite chains $\prod_{a \in \mathcal{A}(M)} L_{ord(a)}$.*

It is well-known that any poset M has the **Mac Neille completion** by cuts, say $\hat{M}$. $\hat{M}$ is a complete lattice in which M can be embedded preserving all existing joins and meets in M. By Schmidt [SchJ], the Mac Neille completion $\hat{M}$ is (up to isomorphism) a lattice in which M can be supremum-densely and infimum-densely embedded (i.e., for any element $x \in \hat{M}$ there exist two subsets U, Q of M such that $x = \bigvee_{\hat{M}} \phi(U) = \bigwedge_{\hat{M}} \phi(Q)$ (where ϕ is the embedding of M into $\hat{M}$ preserving all existing joins and meets in M).

We note that if M is an MV-algebra, then the Mac Neille completion $\hat{M}$ of M is an MV-algebra iff M is semisimple [Höl], and then $ord_M(a) = ord_{\hat{M}}(a)$. Indeed, $x+y$ exists in M iff $x+y$ is defined in $\hat{M}$: $\phi(x) \leq \phi(y^*) = \phi(y)^*$.

Theorem 6.4.17 can be extended as follows:

Theorem 6.4.20 *Every atomic Archimedean pseudo MV-algebra M is commutative. In addition, the Mac Neille completion $\hat{M}$ of M is isomorphic as an MV-algebra to the direct product $\prod_{a\in\mathcal{A}(M)} L_{ord(a)}$.*

Proof. We show that given an element $x \in M$, equality (6.39) holds. Indeed, $j_a(x)a \leq x$ for any atom $a \in \mathcal{A}(M)$. If now $j_a(x)a \leq y$ for any atom a, then $j_a(x)a \leq x \wedge y$ for any atom $a \in \mathcal{A}(M)$. We assert that $x \wedge y = x$. If not, then there would exist an atom $a \leq x \odot y^- = x \odot (x \wedge y)^-$ which is a contradiction with the definition of $j_a(x)$, so that (6.39) holds.

Similarly as in the proof of Theorem 6.4.17, we can now prove that $x \oplus y = y \oplus x$.

According to [Höh, Thm 6.3], the semisimplicity of M is a necessary and sufficient condition for $\hat{M}$ to be an MV-algebra. In addition, the semisimplicity of MV-algebras is equivalent to the Archimedeanicity of M. Using Schmidt's [SchJ] characterization of the Mac Neille completion, we see that each atom of M is an atom of $\hat{M}$, and conversely, each atom in $\hat{M}$ is an atom of M. Moreover, $\hat{M}$ is atomic, and $ord_M(a) = ord_{\hat{M}}(a)$, $a \in \mathcal{A}(M)$. Applying Corollary 6.2.14, the complete atomic MV-algebra $\hat{M}$ is isomorphic to the direct product of finite chains $\{L_{ord(a)} : a \in \mathcal{A}(M)\}$, which finishes the proof.

□

6.4.3 Commutativity of Linear Pseudo MV-algebras

We say that a pseudo MV-algebra M is **linear** if the partial ordering on M is linear.

Proposition 6.4.21 *Let $(M; \oplus, ^-, ^\sim, 0, 1)$ be a linear pseudo MV-algebra. If $x \oplus y < 1$, then $x + y$ is defined in M.*

Proof. Suppose the converse. Then $y^- < x$. Hence $x = y^- + (y^-)^\sim \odot x = y^- + y \odot x$. But for linear pseudo MV-algebras $x \oplus y < 1$ entails $y \odot x = 0$ [GeIo, Prop 1.30]. Hence, $x = y^-$, which is a contradiction. □

Proposition 6.4.22 *A linear pseudo MV-algebra M is Archimedean if and only if, for each $x, y \in M$, if $n \odot x \leq y$ for all $n \geq 1$, then $x \odot y = x$.*

Proof. Suppose that M is Archimedean and let $n \odot x \leq y$ for all $n \geq 1$. If $y = 1$, then $x \odot 1 = x$. If $y < 1$, then by Proposition 6.4.21 $n \odot x = n\,x$ for any $n \geq 1$, and the Archimedeanicity of M entails $x = 0$. So $x \odot y = x$.

Conversely, let now $n\,x$ be defined in M for each $n \geq 1$. Hence by definition of the addition $+$, we have that $nx \leq x^-$ for any $n \geq 1$. So $n \odot x \leq x^-$ for any $n \geq 1$, which entails $x \odot x^- = x$ so that $x = 0$ due to the fact that $x \odot x^- = 0$. □

Theorem 6.4.23 *Any Archimedean linear pseudo MV-algebra M is commutative.*

Proof. (1) Suppose that a is an atom of an Archimedean linear pseudo MV-algebra M. The Archimedeanicity of M yields that $n := ord(a) < \infty$. We assert that $n\,a = 1$. Indeed, if not, then $n\,a < 1$. Due to definition of $ord(a)$, we have $a^- < n\,a$. Then $(n\,a)^\sim < (a^-)^\sim = a$ so that $(n\,a)^\sim = 0$, and hence $n\,a = 1$.

Take an element $x \in M$. We assert that $x = i\,a$ for some integer $i \in \{0, 1, \ldots, n\}$. There exists an integer $i \in \{0, 1, \ldots, n-1\}$ such that $x \in [i\,a, (i+1)a]$, i.e., $i\,a \leq x \leq (i+1)a$. Hence $0 \leq x \odot (i\,a)^- \leq (i+1)a \odot (i\,a)^- = a$ which yields that either $x \odot (i\,a)^- = 0$ or $x \odot (i\,a)^- = a$, and consequently, $x = i\,a$ or $x = (i+1)a$.

(2) Suppose that M has no atom. Then, for every $x \in M$, $x > 0$, we can find an element $y \in M$ such that $0 < y < x$. We have $x = x \odot y^- + y$, so that $x \odot y^+ > 0$, and there are two elements $y_1, y_2 \in M$ such that $0 < y_1 \leq x \odot y^-$ and $y_2 \leq y$. If we put $z = y_1 \wedge y_2$, then $2z$ is defined in M and $0 < 2z \leq x$. Thus, for each $x > 0$, there exists a non-zero z such that $2\,z < x$.

Assume that a and b are elements of M with $a \oplus b \neq b \oplus a$, say $b \oplus a < a \oplus b$. Then $a \oplus b = x + (b \oplus a)$, where $x = (a \oplus b) \odot (b \oplus a)^- = (a \oplus b) \odot a^- \odot b^-$, and choose a z to this x such that $2z \leq x$. The Archimedeanicity of M and Proposition 6.4.21 entail $ord(z) < \infty$, and therefore there exist two integers $m, n \geq 1$ such that $m \odot z \leq a < (m+1) \odot z$ and $n \odot z \leq b < (n+1) \odot z$. Hence, $0 < x \leq (m+n+2) \odot z \odot [(m+n) \odot z]^-$.

Case 1. $(m+n+2) \odot z < 1$. Then $x \leq 2\,z$, which is a contradiction.

Case 2. $(m+n+2) \odot z = 1$. There are three subcases.

Case 2a. $(m+n) \odot z = 1$. Then $x = 0$.

Case 2b. $(m+n) \odot z < 1$ and $(m+n+1) \odot z < 1$. Then $(m+n+2) \odot z \odot [(m+n)z]^- = 1 \odot [(m+n)z]^- = [z_1 + z + (m+n)z] \odot [(m+n)z]^- = z_1 + z \leq 2\,z$, where $z_1 \leq z$, which is a contradiction.

Case 2c. $(m+n) \odot z < 1$ and $(m+n+1) \odot z = 1$. Then $(m+n+2) \odot z \odot [(m+n)z]^- = 1 \odot [(m+n)z]^- = [z_1 + (m+n)z] \odot [(m+n)z]^- = z_1 \leq z < 2\,z$, which is a contradiction.

Summarizing, M is commutative. □

6.4.4 Ideals of Pseudo MV-algebras

According to [GeIo], we say that an **ideal** of a pseudo MV-algebra M is a subset I of M such that (i) $0 \in I$, (ii) if $x, y \in I$, then $x \oplus y \in I$ and $y \oplus x \in I$, and (iii) if $x \in I$, $y \in M$, and $y \leq x$, then $y \in I$.

An ideal I of M is said to be (i) **normal** if $x \oplus I = I \oplus x$ [7] for any $x \in M$, (ii) **prime** if it is proper and if $x \wedge y \in I$ entails $x \in I$ or $y \in I$, (iii) **maximal** if I is a proper ideal of M and if it is not included in any proper ideal of M different from I, (iv) **minimal prime** if it is non-trivial and prime not containing as a proper subset any non-trivial prime ideal.

For example, the zero ideal and M are normal ideals of M.

[7] We define $x \oplus I := \{x \oplus y : \ y \in I\}$; similarly we define $I \oplus x$.

We note that the set of all ideals of M, $\mathcal{I}(M)$, is linear under the set-theoretical inclusion whenever M is linear. Indeed, if I_1 and I_2 are two ideals of M which are not comparable, then there exists $x \in I_1 \setminus I_2$ and $y \in I_2 \setminus I_1$ which would entail a contradiction $x \not\leq y$ and $y \not\leq x$.

Proposition 6.4.24 (i) *An ideal I of a pseudo MV-algebra M is normal if and only if*

(*) *for each $x, y \in M$, $y \odot x^- \in I$ iff $x^\sim \odot y \in I$.*

(ii) *In addition, if I is normal, then for each $h \in I$, $h^= \in I$ and $h^\approx \in I$;*

Proof. Let I be a normal ideal. Assume that $h_1 := y \odot x^- \in I$. Then $x \vee y = y \odot x^- + x = h_1 + x = x + x^\sim \odot y$, and there exists $h_2 \in I$ such that $x \vee y = x \oplus h_2$. Then $x^\sim \odot y = x^\sim \odot (x \oplus h_2) = x^\sim \wedge h_2 \leq h_2 \in I$.

Similarly we can prove that $x^\sim \odot y \in I$ entails $y \odot x^- \in I$.

Choose now $h \in I$. Then $1 = h + h^\sim$ so that there exists $h' \in I$ such that $1 = h^\sim \oplus h'$, but $1 = h^\sim + h^\approx$, which yields $h^\approx \leq h' \in I$. Similarly we have $h^= \in I$.

The opposite direction: Let $x \in M$ and $h \in I$ and let $y = h \oplus x$. Then $x \leq y$ and $y \odot x^- + x = x \vee y = y = x + x^\sim \odot y$. Hence if $y \odot x^- \in I$, then by the assumption, $h' = x^\sim \odot y \in I$, which gives $h \oplus x = x \oplus h'$. In a similar way we have $x \oplus I \subseteq I \oplus x$. □

As it was proved in [GeIo, Thm 3.8], normal ideals and congruences are in a one-to-one correspondence; the quotient pseudo MV-algebra over a normal ideal I, M/I, is defined as the set of all elements of the form $x/I := \{y \in M : x \odot y^- \oplus y \odot x^- \in I\}$, or equivalently, $x/I := \{y \in M : x^\sim \odot y \oplus y^\sim \odot x \in I\}$, see Exercise 8.6.4.5.

Proposition 6.4.25 *For any proper normal ideal J of a pseudo MV-algebra M, the following conditions are equivalent:*

(i) *J is a maximal ideal of M.*

(ii) *For each $z \in M$, $z \notin J$ iff $(n \odot z)^- \in J$ for some integer $n \geq 1$.*

(iii) *For each $z \in M$, $z \notin J$ iff $(n \odot z)^\sim \in J$ for some integer $n \geq 1$.*

Proof. (i) $\Rightarrow$ (ii). Suppose that J is a maximal ideal of M. If $z \notin J$, then by Exercise 9.6.4.5(iii), $1 \in I(J, z)$ so that $1 = h \oplus n \odot z$ for some integer $n \geq 1$ and $h \in J$. Hence, $1 = h \oplus ((n \odot z)^-)^\sim$, i.e., $(n \odot z)^- \leq h$, and $(n \odot z)^- \in J$.

If $z \in J$, then $n \odot z \in J$ for any $n \geq 1$. Since J is proper, it follows that $(n \odot z)^- \notin J$.

(ii) $\Rightarrow$ (i). Let H be an ideal of M containing J and let $J \neq H$. For every $z \in H \setminus J$, we have by the assumption that $(n \odot z)^- \in J$ for some integer $n \geq 1$. Since $n \odot z \in H$ and H contains J, we conclude that $1 \in H$, and $H = M$.

The equivalence of (ii) and (iii) follows from the fact that J is a normal ideal. □

Let X be a non-void subset of M. The set

$$X^\perp := \{a \in M : \ a \wedge x = 0 \text{ for any } x \in X\}$$

is said to be the **polar** of X. As it was shown in [GeIo, Prop 2.13], $X^\perp$ is always an ideal of M.

We note some easy properties of polars: (i) $X \subseteq X^{\perp\perp}$; (ii) if $X \subseteq Y$, then $Y^\perp \subseteq X^\perp$, and (iii) $X^\perp = X^{\perp\perp\perp}$.

For $g \in M$, we define $g^\perp := \{g\}^\perp$ and $g^{\perp\perp} := \{g\}^{\perp\perp}$, and $g^{\perp\perp}$ is said to be a **principal polar**. It is clear that if $a \le b$, then $b^\perp \subseteq a^\perp$.

An ideal I of M is said to **closed** if $\bigvee_i x_i \in I$ whenever $\{x_i\}$ is a family of elements of I having $\bigvee_i x_i \in M$. For example, each polar is a closed ideal of M.

In accordance with group theory, we say that M is a **direct product** or **cardinal sum** of ideals $I_1, \ldots, I_n$ if $M = I_1 \times \cdots \times I_n$, and the operations are taken componentwise. The set I_i is identified with the set $\{(0, \ldots, 0, a, 0, \ldots, 0) \in M : \ a \in I_i\}$. Since $(a_1, \ldots, a_n) = (b_1, \ldots, b_n)$ iff $a_i = b_i$ for $i = 1, \ldots, n$, a_i is said to be a **component** of a in I_i.

If J is an ideal of M, then

$$J = J_1 \times \cdots \times J_n$$

with $J_i = I_i \cap J$ for $i = 1, \ldots, n$.

Proposition 6.4.26 *Let $I_1, \ldots, I_n$ be ideals of a pseudo MV-algebra M such that $M = I_1 \times \cdots \times I_n$. Then each I_i is a normal ideal of M. If $n = 2$, then $I_2 = I_1^\perp$.*

Proof. We identify $I_i := \{(0, \ldots, 0, a, 0, \ldots, 0) \in M : \ a \in I_i\}$. Then $I_i \cap I_j = \{0\}$ for $i \neq j$.

Suppose that $n = 2$, and take $x \in M$ and $h \in I_1$. Then

$$\begin{aligned} x \oplus h &= x_1 \oplus x_2 \oplus h = x_1 \oplus (x_2 \oplus h) \\ &= x_1 \oplus (h \oplus x_2) = (x_1 \oplus h) \oplus x_2 \\ &= ((x_1 \oplus h) \odot x_1^- \oplus x_1) \oplus x_2 = h_1 \oplus (x_1 \oplus x_2) = h_1 \oplus x, \end{aligned}$$

where $h_1 = (x_1 \oplus h) \odot x_1^- \in I_1$ because $x_1 \oplus h \in I_1$.

Similarly

$$\begin{aligned} h \oplus x &= h \oplus (x_1 \oplus x_2) = (x_1 \oplus x_1^\sim \odot (h \oplus x_1)) \oplus x_2 \\ &= x_1 \oplus h_2 \oplus x_2 = x_1 \oplus x_2 \oplus h_2 = x \oplus h_2, \end{aligned}$$

where $h_2 = x_1^\sim \odot (h \oplus x_1) \in I_1$, i.e., I_1 is a normal ideal of M.

Since if $x = x_1 \oplus x_2$, where $x_1 \in I_1$ and $x_2 \in I_2$, then $x_1 \wedge x_2 = 0$ and $x = x_2 \oplus x_1$, so that I_2 is a normal ideal of M, too.

Let now $n \geq 2$. Putting $J_1 = I_1 \times \cdots \times I_{n-1}$, $J_2 = I_n$, by induction we can show that each of I_i is a normal ideal of M.

Finally, let $M = I_1 \times I_2$. Then $I_1 \subseteq I_2^{\perp}$. Let now $x \in I_2^{\perp}$, then $x = (x_1, x_2)$, where $x_1 \in I_1$ and $x_2 \in I_2$. Then $(x_1, x_2) \wedge (0, b) = 0$ for any $b \in I_2$, in particular for $b = x_2$ which proves $x_2 = 0$ and $x \in I_1$. So $I_1 = I_2^{\perp}$. □

Proposition 6.4.27 *If I is a closed ideal of a complete pseudo MV-algebra M, then $I \cap I^{\perp} = \{0\}$, and $M \cong I \times I^{\perp}$.*

In addition, each polar of M is a closed ideal of M, consequently a normal ideal.

Proof. (1) It is clear that $I \cap I^{\perp} = \{0\}$. For any element $a \in M$, let $a_1 = \bigvee_{\alpha}(a \wedge t_{\alpha})$, where t_{α} runs over all elements of I. Since $a \wedge t_{\alpha} \in I$, by hypothesis, $a_1 \in I$. Define $a_2 = a_1^{\sim} \odot a$. For arbitrary $t \in I$, $a_1 \oplus t \in I$, therefore $a_1 \geq a \wedge (a_1 \oplus t) = (a_1 \oplus a_2) \wedge (a_1 \oplus t) = a_1 \oplus (a_2 \wedge t)$. Hence $a_2 \wedge t = 0$ for any $t \in I$, i.e., $a_2 \in I^{\perp}$.

The normality of I and $I^{\perp}$ follows from Proposition 6.4.26.

(2) Let J be a polar. Then $J = J^{\perp\perp}$. Let $a = \bigvee_i a_i$, where $a_i \in J$ for any i. For each $x \in J^{\perp}$, we have $x \wedge a = \bigvee_i (x \wedge a_i) = 0$, so that $a \in J^{\perp\perp} = J$. □

Proposition 6.4.28 *An ideal I of a complete pseudo MV-algebra M is closed if and only if $I = I^{\perp\perp}$.*

Proof. If I is a closed ideal, then by Proposition 6.4.27, $I^{\perp}$ is a closed ideal of M so that $M \cong I \times I^{\perp} = I^{\perp\perp} \times I^{\perp\perp\perp} = I^{\perp\perp} \times I^{\perp}$. Since $I \subseteq I^{\perp\perp}$, we have by Proposition 6.4.26, $I = I^{\perp\perp}$.

Let now $I = I^{\perp\perp}$, then I is a polar, and by Proposition 6.4.27, I is a closed ideal. □

Let $\{(M_t; \oplus_t, ^{-_t}, ^{\sim_t}, 0_t, 1_t)\}_{t \in T}$ be a family of pseudo MV-algebras. The Cartesian product $M := \prod_{t \in T} M_t$, where $\oplus, ^{-}, ^{\sim}, 0, 1$ are defined in a usual way by coordinates, is said to be a **direct product** of $\{(M_t; \oplus_t, ^{-_t}, ^{\sim_t}, 0_t, 1_t)\}_{t \in T}$. A pseudo MV-algebra M is a **subdirect product** of a family $\{(M_t; \oplus_t, ^{-_t}, ^{\sim_t}, 0_t, 1_t)\}_{t \in T}$ of pseudo MV-algebras iff there exists a one-to-one homomorphism $h : M \to \prod_{t \in T} M_t$ of pseudo MV-algebras [8] such that, for each $t \in T$, $\pi_t \circ h$ is a homomorphism from M onto M_t, where π_t is the t-th projection from $\prod_{t \in T} M_t$ onto M_t.

According to [GeIo], we say that a pseudo MV-algebra M is **representable** if it can be represented as a subdirect product of linear pseudo MV-algebras.

Theorem 6.4.29 *For a pseudo MV-algebra M, the following statements are equivalent.*

[8] We note that a mapping $h : M_1 \to M_2$ of two pseudo MV-algebras M_1 and M_2 is said to be a **homomorphism** if h preserves $\oplus$, $^{-}$, $^{\sim}$ and 0 and 1.

(i) *M is representable.*

(ii) *There exists a family* $\{H_t : t \in T\}$ *of normal prime ideals M with* $\bigcap_{t\in T} H_t = \{0\}$.

(iii) *Polars are normal ideals of M.*

(iv) *Minimal prime ideals are normal.*

Proof. (i) $\Rightarrow$ (ii). Let $h : M \to \prod_{t\in T} M_t$ be a pseudo MV-embedding of M into a subdirect product of linearly ordered MV-algebras $\{M_t\}_{t\in T}$. Define $H_t := Ker(\pi_t \circ h)$. Then H_t is a normal ideal, and since $\pi_t \circ h$ is surjective, we conclude that M/H_t [Gelo, Lem 3.13] is isomorphic to M_t so that each H_t is prime. If $x \in \bigcap_{t\in T} H_t$, then $\pi_t \circ h(x) = 0$ for all $t \in T$ so that $h(x) = 0$ implying $x = 0$.

(ii) $\Rightarrow$ (i). Let $\{H_t\}_{t\in T}$ be a family of normal prime ideals of M with $\bigcap_{t\in T} H_t = \{0\}$. Then $h : M \to \prod_{t\in T} M/H_t$ is a subdirect product in question.

(i) $\Rightarrow$ (iii). Suppose that M is a subdirect product of linear pseudo MV-algebra $\{M_t\}_{t\in T}$. It si easy to verify that $\{x_t\}_t \wedge \{y_t\}_t = \{0_t\}_t$ iff $\{t : x_t > 0\} \cap \{t : y_t > 0\} = \emptyset$.

Let $J^\perp$ be a polar of M. We show that $J^\perp$ is normal. Let $x \in M$, $h \in J^\perp$, and $y \in J$. Then $x \oplus h = (x \oplus h) \odot x^- + x$. But $\{t : (x_t \oplus_t h_t) \odot_t x_t^{-t} > 0\} \subseteq \{t : h_t > 0\}$. Hence $\{t : (x_t \oplus_t h_t) \odot_t x_t^{-t} > 0\} \cap \{t : y_t > 0\} \subseteq \{t : h_t > 0\} \cap \{t : y_t > 0\} = \emptyset$ which proves $(x \oplus h) \odot x^- \in J^\perp$.

In a similar way we can prove that $h \oplus x = x \oplus x^\sim \odot (h \oplus x)$ and $x^\sim \odot (h \oplus x) \in J^\perp$.

(iii) $\Rightarrow$ (iv). By [Gelo, Thm 2.20], an ideal P of M is minimal prime iff $P = \bigcup\{g^\perp : g \notin P\}$. Hence, if $x \in M$ and $h \in P$, then there exists $g \notin P$ such that $h \in g^\perp$, so that $x \oplus h = h_1 \oplus x$ for some $h_1 \in g^\perp$. In the same way we prove $h \oplus x = x \oplus h_2$ for some $h_2 \in g^\perp$.

(iv) $\Rightarrow$ (i). This is evident. (See also [Gelo, Prop 3.13].) □

Corollary 6.4.30 *Each complete pseudo MV-algebra M is representable.*

Proof. This follows from Proposition 6.4.27 and Theorem 6.4.29. □

A pseudo MV-algebra M is **projectable** if, for every $g \in M$, $M = g^\perp \times g^{\perp\perp}$.

Proposition 6.4.31 *Every projectable pseudo MV-algebra M is representable.*

Proof. We show that (iv) of Theorem 6.4.29 is satisfied, i.e., each minimal prime ideal P of M is normal. P is of the form $P = \{g^\perp : g \notin P\}$. Hence, if $x \in M$ and $h \in P$, then $h \in g^\perp$ for some $g \notin P$, and $x = x_1 + x_2$, where $x_1 \in g^\perp$, $x_2 \in g^{\perp\perp}$. Calculate $x \oplus h = x_1 \oplus x_2 \oplus h = x_1 \oplus h \oplus x_2 = h_1 \oplus x_1 \oplus x_2 = h_1 \oplus x$, for some $h_1 \in g^\perp$. In a similar way, $h \oplus x = x \oplus h_2$ for some $h_2 \in g^\perp$. □

Finally we present some open problems on pseudo MV-algebras.[9]

[9]Note added in proofs. All above problems were positively solved in the paper A. Dvurečenskij, Pseudo MV-algebras are intervals in ℓ-groups.

(i) Is any pseudo MV-algebra M isomorphic to some interval $\Gamma(G,u)$, where u is a strong unit of a (not necessarily Abelian) ℓ-group G ?

(ii) Is any Archimedean pseudo MV-algebra M commutative, i.e. $a \oplus b = b \oplus a$ for all $a, b \in M$?

(iii) Is the Mac Neille completion of a pseudo MV-algebra M a pseudo MV-algebra iff M is Archimedean ?

6.4.5 Exercises

1. Prove: Let G be an ℓ-group, then $(G^+; *_G, 0)$, where $u *_G v := 0 \vee (u - v)$, $u, v \in G^+$, is a BCK-algebra iff G is an Abelian ℓ-group. ([Cor 1, Thm 1.1])

2. An algebra $(R; \Delta, \cdot, 0, 1)$ of type (2,2,0,0) is called a generalized Boolean quasiring, [DLM], [DLM 1], if, for all $x, y, z \in R$, we have (1) $x\Delta y = y\Delta x$; (2) $x\Delta 0 = x$; (3) $(xy)z = x(yz)$; (4) $xy = yx$; (5) $xx = x$; (6) $x0 = 0$; (7) $x1 = x$; (8) $1\Delta(1\Delta xy)(1\Delta y) = y$.

   Show:

   (a) Let g be a continuous and strictly increasing function from $[0,1]$ into $[0,\infty)$ such that $g(0) = 0$. Define two total binary operations $\oplus_g$ and Δ_g on $[0,1]$ by $a \oplus_g b := g^{-1}(\min\{g(a) + g(b), g(1)\})$ and $a\Delta_g b := g^{-1}(|g(a) - g(b)|)$. Then $([0,1]; \Delta_g, \min)$ is a generalized Boolean quasiring for which property (A) from [DLM]:

   $$(1\Delta(1\Delta x) \wedge (1\Delta y)) \wedge (1\Delta(x \wedge y)) = x\Delta y \tag{A}$$

   does not hold.

   (b) Let (G,u) be an Abelian unital ℓ-group with strong unit u. Define $\Gamma(G,u) := \{g \in G : 0 \le g \le u\}$, and define $a \oplus b := (a+b) \wedge u$ and

   $$a\Delta b := (a-b) \vee 0 + (b-a) \vee 0 = |a-b|.$$

   Then $(\Gamma(G,u); \Delta, \wedge)$ is a generalized Boolean quasiring. It satisfies (A) iff $\Gamma(G,u)$ is a Boolean algebra.

3. Let M be a pseudo MV-algebra. Show (1) $y \odot x = (x^\sim \oplus y^\sim)^-$; (2) $(x^\sim)^- = x$; (3) $0^\sim = 0^- = 1$; (4) $x \odot 1 = 1 \odot x = x$; (5) $x \odot 0 = 0 \odot x = 0$; (6) $x \oplus x^\sim = 1$, $x^- \oplus x = 1$; (7) $x \odot x^- = 0$, $x^\sim \odot x = 0$; (8) $(x \oplus y)^- = y^- \odot x^-$, $(x \oplus y)^\sim = y^\sim \odot x^\sim$; (9) $(x \odot y)^- = y^- \oplus x^-$, $(x \odot y)^\sim = y^\sim \oplus x^\sim$; (10) $x \oplus y = (y^- \odot x^-)^\sim = (y^\sim \odot x^\sim)^-$; (11) $x^\sim \odot y \oplus y^\sim = y^\sim \odot x \oplus x^\sim$; (12) $x \odot (x^- \oplus y) = y \odot (y^- \oplus x)$; (13) $x \odot (y \odot z) = (x \odot y) \odot z$. ([GeIo])

4. Show that $\leq$ defined by $x \leq y$ iff $x^- \oplus y = 1$ is a partial order on a pseudo MV-algebra M with the least and last elements 0 and 1, respectively, and M is a distributive lattice. Moreover, $x \leq y$ iff $y = x \oplus x^\sim \odot y$ iff $x = x \odot (x^- \oplus y)$ iff there exists $a \in M$ such that $y = x \oplus a$. $x \vee y = x \oplus x^\sim \odot y$ and $x \wedge y = x \odot (x^- \oplus y) = y \odot (y^- \oplus x)$.

5. Prove: Let M be a pseudo MV-algebra. Then $a \odot (\bigvee_i a_i) = \bigvee_i (a \odot a_i)$ and $(\bigvee_i a_i) \odot a = \bigvee_i (a_i \odot a)$.

6. Show that if $a \oplus b = b \oplus a$, for all $a, b \in M$, then $a^- = a^\sim$ and $(M; \oplus, \odot, ^-, 0, 1)$ is an MV-algebra.

7. Let M be a pseudo MV-algebra. Then there exists an ℓ-group G (not necessarily Abelian) and an element $u > 0$ such that M can be embedded as pseudo MV-algebras into $\Gamma(G, u)$. (Hint: Use $+$ and show that $(M; \wedge, +)$ is a semiclan, and apply Bosbach's result [Bos 1], see Exercise 2.5.2.12.)

8. Let H be a normal ideal. We define $x \sim_H y$ iff $x \odot y^- \oplus y \odot x^- \in H$, or equivalently, iff $x^\sim \odot y \oplus y^\sim \odot x$. Show that $\sim_I$ is a congruence, and M/H is a pseudo MV-algebra.

9. (i) The ideal $I(a)$ of M generated by an element $a \in M$ is the set

$$I(a) = \{x \in M : \ x \leq n \odot a, \ n \geq 1\}.$$

(ii) The ideal $I(J, a)$ of M generated by an ideal J of M and an element $a \in M$ is the set

$$I(J, a) = \{x \in M : \ x \leq \bigoplus_{i=1}^{n} (a_i \oplus n_i \odot a), \ a_i \in J, \ n_i \geq 0, i = 1, \ldots, n, \ n \geq 1\}.$$

(iii) If J is a normal ideal of M, then the ideal $I(J, a)$ generated by I and and element $a \in M$ is the set

$$I(J, a) = \{x \in M : \ x \leq h \oplus n \odot a, \ h \in J, \ n \geq 1\}.$$

10. Show that any polar in a pseudo MV-algebra M is an ideal of M.

11. Show that the following statements are equivalent in a pseudo MV-algebra M.

 (i) $x \leq y$.
 (ii) There exists a unique element $b \in M$ such that $x + b = y$.
 (iii) There exists an element $b \in M$ such that $x \oplus b = y$.
 (iv) There exists a unique element $a \in M$ such that $a + x = y$.
 (v) There exists an element $a \in M$ such that $a \oplus x = y$.

12. A non-commutative MV-algebra in the sense of Rachůnek is an algebra $(M; \oplus, \odot, ^{\neg}, ^{\sim}, 0, 1)$ of type (2,2,1,1,0,0) such that, for all $x, y, z \in M$, we have: (1) $x \oplus (y \oplus z) = (x \oplus y) \oplus z$, (2) $x \oplus x^{\sim} = 1 = x^{\neg} \oplus x$; (3) $x \oplus 1 = 1 = 1 \oplus x$, (4) $x \oplus 0 = x = 0 \oplus x$, (5) $(x \oplus y)^{\neg} = x^{\neg} \odot y^{\neg}$, (6) $(x \oplus y)^{\sim} = x^{\sim} \odot y^{\sim}$, (7) $(x^{\sim})^{\neg} = x = (x^{\neg})^{\sim}$, (8) $0^{\neg} = 1 = 0^{\sim}$. (9) $y \oplus (x \odot y^{\sim}) = (y^{\neg} \odot x) \oplus y$, (1') $x \odot (y \odot z) = (x \odot y) \odot z$, (2') $x \odot x^{\sim} = 0 = x^{\neg} \odot x$, (3') $x \odot 0 = 0 = 0 \odot x$, (4') $x \odot 1 = x = 1 \odot x$, (5') $(x \odot y)^{\neg} = x^{\neg} \oplus y^{\neg}$, (6') $(x \odot y)^{\sim} = x^{\sim} \oplus y^{\sim}$, (9') $y \odot (x \oplus y^{\sim}) = (y^{\neg} \oplus x) \odot y$.

    Show that if $(M; \oplus, \odot, ^{-}, ^{\sim}, 0, 1)$ is a pseudo MV-algebra, then $(M; \oplus, \odot', ^{\neg}, ^{\sim}, 0, 1)$, where $x \odot' y := y \odot x$, and $^{\neg} = ^{-}$, is a non-commutative MV-algebra. Conversely, if $(M; \oplus, \odot, ^{\neg}, ^{\sim}, 0, 1)$ is a non-commutative MV-algebra, then $(M; \oplus, \odot', ^{-}, ^{\sim}, 0, 1)$, where $x \odot' y = y \odot x$ and $^{-} = ^{\neg}$, is a pseudo MV-algebra.

13. Let $G = (\mathbb{Z} \times \mathbb{Z} \times \mathbb{Z}; +, (0,0,0), \leq)$ be the Scrimger 2-group, i.e.

$$(k_1, m_1, n_1) + (k_2, m_2, n_2) := \begin{cases} (m_1 + k_2, m_2 + k_1, n_1 + n_2), & \text{if } n_2 \text{ is odd} \\ (k_1 + k_2, m_1 + m_2, n_1 + n_2), & \text{if } n_2 \text{ is even.} \end{cases}$$

    Then $0 = (0,0,0)$ is the neutral element, and

$$-(k, m, n) = \begin{cases} (-m, -k, -n), & \text{if } n \text{ is odd} \\ (-k, -m, -n), & \text{if } n \text{ is even,} \end{cases}$$

    and G is a non-Abelian ℓ-group with the positive cone $G^+ = \mathbb{Z} \times \mathbb{Z} \times \mathbb{Z}^+_{>0} \cup \mathbb{Z}^+ \times \mathbb{Z}^+ \times \{0\}$, or equivalently, $(k_1, m_1, n_1) \leq (k_2, m_2, n_2)$ iff (i) $n_1 < n_2$, or (ii) $n_1 = n_2$, $k_1 \leq k_2$, $m_1 \leq m_2$.

    Show that $u = (1,1,1)$ is a strong unit in G, and $\Gamma(G, u)$ is a non-Archimedean pseudo MV-algebra which is not an MV-algebra.

Chapter 7

Loomis-Sikorski Theorems for MV-algebras and BCK-algebras

We recall that σ-complete MV-algebras are MV-algebras which are σ-complete lattices. Such MV-algebras are always semisimple algebras, and they are exactly those for which there exists an MV-isomorphism with a Bold algebra, i.e., with an algebra of fuzzy sets on a crisp set Ω which contains 1_Ω, and which is closed under the fuzzy complementation and formation of $\min\{f+g,1\}$. Belluce [Bel] showed that every semisimple MV-algebra M can be always represented as a Bold algebra of continuous fuzzy sets on the compact Hausdorff space of all maximal ideals of M. And this is an analogue of Stone's representation theorem for Boolean algebras. Situation with σ-complete MV-algebras is more complicated as we will see below.

Tribes are Bold algebras of fuzzy sets which are roughly speaking closed under pointwise suprema, and they are a direct generalization of a σ-algebra of crisp subsets.

The famous Loomis–Sikorski theorem [Sik] plays a crucial role for analysis of Boolean σ-algebras, and it says that every Boolean σ-algebra is a σ-homomorphic image of a σ-algebra of subsets.

In the present Chapter, we show that every σ-complete MV-algebra is an MV-σ-homomorphic image of a tribe, which gives a generalization of the Loomis–Sikorski theorem for σ-complete MV-algebras. In addition, we extend this result also for Dedekind σ-complete ℓ-groups with strong unit, and commutative BCK-algebras with the relative cancellation property.

In addition, we describe weakly divisible MV-algebras and characterize MV-observables and present a calculus for them.

7.1 Loomis–Sikorski's Theorem for σ-complete MV-algebras

MV-algebras were introduced in Section 2.2, and in an equivalent formulation as bounded commutative BCK-algebras in Example 5.1.44, and studied in Section 5.3.2.

We note that if $(M; \oplus, \odot, ^*, 0, 1)$ is an MV-algebra, then it is a distributive lattice with respect to the partial order $\le$ defined by $a \le b$ iff $a \odot b^* = 0$, and with the least and greatest element 0 and 1, respectively. The partial operation $+$ is defined in M by (5.34), i.e., $a + b$ exists in M iff $a \le b^*$, and in such the case,

$$a + b = a \oplus b.$$

The basic result of the theory of MV-algebras is the famous Mundici categorical representation by unital Abelian ℓ-groups (G, u), where u is a strong unit of G, via

$$M \cong \Gamma(G, u) = \{g \in G : \ 0 \le g \le u\},$$

where Γ is Mundici's functor (5.33). Then $(\Gamma(G, u); \oplus_G, \odot_G, ^{*_G}, 0, u)$ is its representation, where

$$\begin{aligned} a \oplus_G b &:= u \wedge (a + b), \\ a \odot_G b &:= 0 \vee (a + b - u), \\ a^{*_G} &:= u - a. \end{aligned}$$

7.1.1 Bold Algebras of Fuzzy Sets

We recall that according to Subsection 2.2.4 and Subsection 6.1.1, an important family of MV-algebras are **semisimple** MV-algebras, i.e., those MV-algebras for which $Rad(M) = \{0\}$, where

$$Rad(M) := \bigcap \{A : \ A \in \mathcal{M}(M)\},$$

and $\mathcal{M}(M)$ is the set of all maximal ideals of M.

If A is an ideal of M, then $x \sim_A y$ means $(x \odot y^*) \oplus (y \odot x^*) \in A$. The quotient MV-algebra M/A is defined in the usual sense, see Subsection 2.2.2 or Subsection 5.1.7.

According to Belluce [Bel], we say that a subset $\mathcal{F} \subseteq [0,1]^\Omega$, where $\Omega \ne \emptyset$, is a **Bold algebra** if

(i) $0_\Omega \in \mathcal{F}$;

(ii) $f \in \mathcal{F}$ entails $1_\Omega - f \in \mathcal{F}$;

(iii) $f, g \in \mathcal{F}$ entails $f \oplus g \in \mathcal{F}$, where

$$(f \oplus g)(\omega) = \min\{f(\omega) + g(\omega), 1\}, \ \omega \in \Omega.$$

Then $\mathcal{F}$ is an MV-algebra in which $0 = 0_\Omega$, $1 = 1_\Omega$, and

$$\begin{aligned} f \oplus g &= \min\{f + g, 1\}, \\ f \odot g &= \max\{f + g - 1, 0\}, \\ f^* &= 1_\Omega - f, \end{aligned}$$

which is semisimple or, equivalently, Archimedean. [1]

In particular, if X is a topological space, by $C(X)$ we denote the set of all continuous fuzzy subsets on X, and the Bold algebra of all continuous fuzzy subsets is of special interest, as we will see below. They were studied by many authors, see e.g., [DiSe], [Jak 12].

A **state** on MV-algebra M is a mapping $m : M \to [0, 1]$ such that $m(1) = 1$, and $m(a + b) = m(a) + m(b)$ whenever $a + b$ is defined in M. Denote by $\mathcal{S}(M)$ the set of all states on M, then $\mathcal{S}(M) \neq \emptyset$. A state m on M is said to be **σ-additive** if $a_n \nearrow a$ entails $m(a_n) \nearrow m(a)$.

A **state-morphism** is a mapping $m : M \to [0, 1]$ such that $m(1) = 1$, and $m(a \oplus b) = \min\{1, m(a) + m(b)\}$, $a, b \in M$. Any state-morphism is a state, but the converse does not hold, in general. There is a one-to-one correspondence between state-morphisms and maximal ideals, see Remark 6.1.31:

Theorem 7.1.1 (1) *A state m on M is a state-morphism iff $Ker_m := \{a \in M : m(a) = 0\}$ is a maximal ideal.*

(1) *Given a maximal ideal A of M, the mapping $x \mapsto x/A$ is a state-morphism.*

(2) *The mapping $m \mapsto Ker_m$ is a one-to-one correspondence from the set of all state-morphisms on the set of all maximal ideals of M.*

(3) *A state m on M is an extremal point (= extremal state) of the set $\mathcal{S}(M)$ if and only if m is a state-morphism.*

Denote by $Ext(\mathcal{S}(M))$ the set of all extremal states (= state-morphisms) on M. Then by [Mun 3, Thm 2.5] or Theorem 6.1.29,

$$Ext(\mathcal{S}(M)) \neq \emptyset,$$

[1] We note that for MV-algebras (= bounded commutative BCK-algebras) we have two notions of Archimedeanicity: one in the sense of Belluce [Bel], see also Theorem 6.1.32, and the second one as that for BCK-algebras using the derived partial addition $+$. This kind of Archimedeanicity means that if $na = a + \cdots + a$ exists in M for any integer $n \geq 1$, then $a = 0$. By Theorem 6.1.32 all these notions coincide and, in addition, they are equivalent to semisimplicity, see also Theorems 2.2.39 and 7.1.3 below.

and it is a compact Hausdorff space with respect to the weak topology of states (i.e., $m_\alpha \to m$ iff $m_\alpha(a) \to m(a)$ for any $a \in M$), and any state m on M is in the closure of the convex hull of $Ext(\mathcal{S}(M))$.

We introduced in Section 6.1.2 a topology $\mathcal{T}_\mathcal{M}$ on the set $\mathcal{M}(M)$ of all maximal ideals of M. Given an ideal I of M, let

$$O(I) := \{A \in \mathcal{M}(M) : A \not\supseteq I\},$$

and let $\mathcal{T}_\mathcal{M}$ be the collection of all subsets of the above form. It is possible to show that $\mathcal{T}_\mathcal{M}$ gives a compact Hausdorff topological space. Moreover, any closed subset of $\mathcal{M}(M)$ is of the form

$$C(I) = \{A \in \mathcal{M}(M) : A \supseteq I\},$$

where I is any ideal of M. It is worthy of recalling the following assertion:

Proposition 7.1.2 *$\mathcal{M}(M)$ and $Ext(\mathcal{S}(M))$ are homeomorphic spaces; the homeomorphism is given by $m \mapsto Ker_m$, $m \in Ext(\mathcal{S}(M))$.*

Proof. This follows from [Goo 1, Thm 15.32] and Mundici's ℓ-group representation of MV-algebras, see Corollary 5.3.8. □

The following characterizations of semisimple MV-algebras can be found in [Bel], [Mun 3] and Theorem 5.1.32.

Theorem 7.1.3 *Let M be an MV-algebra. The following statements are equivalent:*

(i) *M is semisimple.*

(ii) *M is Archimedean.*

(iii) *M is Archimedean in the sense of Belluce.*

(iv) *There exists an order-determining system of state-morphisms on M.*

(v) *There exists an order-determining system of states on M.*

(vi) *M is isomorphic to some Bold algebra of fuzzy subsets of some set $\Omega \neq \emptyset$.*

(vii) *M is isomorphic to some Bold algebra of continuous functions defined on some compact Hausdorff spaces X.*

(viii) *M is isomorphic to some Bold algebra of $C(\mathcal{M}(M))$, the set of all continuous fuzzy subsets defined on $\mathcal{M}(M)$.*

Let M be semisimple. The property (vi) means the following: Let $a \in M$ and $A \in \mathcal{M}(M)$. Then $a \mapsto \bar{a}$, where $\bar{a} \in [0,1]^{\mathcal{M}(M)}$ is defined as follows

$$\bar{a}(A) := a/A, \; A \in \mathcal{M}(M), \tag{7.1}$$

is an MV-isomorphism between M and $\{\bar{a} : a \in M\}$. We note that if A is a maximal ideal of M, then using (6.1)–(6.2) and Hölder's theorem, [Bir, Thm XIII.12], a/A, $a \in M$, can be represented as a unique number in $[0,1]$ with the condition that $1/A$ maps to the number 1, see also Lemma 6.1.22.

Or, equivalently, since there is a one-to-one correspondence between $Ext(\mathcal{S}(M))$ and $\mathcal{M}(M)$ given by the homeomorphism $m \mapsto Ker_m$, we have the embedding $a \mapsto \hat{a}$, of M into $[0,1]^{Ext(\mathcal{S}(M))}$, where $\hat{a}$ is defined as follows

$$\hat{a}(m) := m(a), \ m \in Ext(\mathcal{S}(M)). \tag{7.2}$$

In addition, in view of the correspondence $m \mapsto Ker_m$, $\hat{a}(m) = \bar{a}(Ker_m)$, see Exercise 7.7.4.3.

7.1.2 σ-complete MV-algebras and Tribes

The following forms of distributive laws are known, see Exercise 1.7.4.3.

Proposition 7.1.4 *Let $\bigvee_i a_i$ be defined in an MV-algebra $(M;\oplus,^*,0,1)$. Then for any $b \in M$, we have*

$$\begin{aligned} b \wedge (\bigvee_i a_i) &= \bigvee_i (b \wedge a_i), \\ (\bigvee_i a_i) \odot b^* &= \bigvee_i (a_i \odot b^*), \\ b \odot (\bigvee_i a_i)^* &= \bigwedge_i (b \odot a_i^*). \end{aligned}$$

The equalities hold in that sense that the expressions on the right-hand side exist, and are equal to the left-hand ones.

We say that an MV-algebra M is **σ-complete (complete)** if M is a σ-complete (complete) lattice; M is σ-complete (complete) iff $G = (G,u)$ is Dedekind σ-complete (complete) [2] ℓ-group with strong unit u, Proposition 6.2.2, where $M \cong \Gamma(G,u) = \{g \in G : 0 \le g \le u\}$, and Γ is Mundici's functor (5.33).

An **MV-homomorphism** of two MV-algebras M_1 and M_2 is any mapping $h : M_1 \to M_2$ preserving $0, 1$, $\oplus$, and *. An **MV-σ-homomorphism** is any MV-homomorphism of σ-complete MV-algebras preserving also countable joins (and meets).

We give another proof of the following statement (see e.g., [Cig, Lem 2.1]).

[2] We note that an ℓ-group G is **Dedekind σ-complete** if, for any sequence $\{g_n\}_n$ of elements of G with an upper bound in G, $\bigvee_n g_n \in G$; similarly for Dedekind complete.

Proposition 7.1.5 *Any σ-complete MV-algebra is Archimedean.*

Proof. This follows directly from Theorem 6.1.12. □

The following notion is a direct generalization of a σ-algebra of crisp subsets. A **tribe** of fuzzy sets on a set $\Omega \neq \emptyset$ is a non-void system $\mathcal{T} \subseteq [0,1]^\Omega$ such that

(i) $1_\Omega \in \mathcal{T}$,

(ii) if $a \in \mathcal{T}$, then $1 - a \in \mathcal{T}$,

(iii) if $\{a_n\}_{n=1}^\infty$ is a sequence of elements of $\mathcal{T}$, then

$$\min\left\{\sum_{n=1}^\infty a_n, 1\right\} \in \mathcal{T}.$$

We note that all above operations with fuzzy sets are defined pointwisely on Ω. They were intensively studied in [KlNa] (with a nice characterization), [RiNe], [Nav], etc.

The basic properties of tribes are:

Proposition 7.1.6 *Let $\mathcal{T}$ be a tribe of fuzzy subsets of Ω. Then*

(i) $a \vee b = \max\{a, b\} \in \mathcal{T}$, $a \wedge b = \min\{a, b\} \in \mathcal{T}$.

(ii) $b - a \in \mathcal{T}$ *if* $a \leq b$, *i.e., if* $a(\omega) \leq b(\omega)$ *for all* $\omega \in \Omega$.

(iii) *If* $a_n \in \mathcal{T}$, $n \geq 1$, *and* $a_n \nearrow a$ *(pointwisely), then* $a = \lim_n a_n \in \mathcal{T}$.

(iv) $\mathcal{T}$ *is a Bold algebra, in addition a σ-complete MV-algebra closed under point suprema of sequences from it.*

Proof. (i), (ii), (iv). It is clear that, for the MV-algebraic operation $\oplus$, we have $f \oplus g = \min\{f + g, 1\} \in \mathcal{T}$, which proves that $\mathcal{T}$ is a Bold algebra, consequently, (i) holds, and in this case $f \odot g := \max\{0, f + g - 1\} \in \mathcal{T}$, and $f^* = 1 - f$.

(ii) $b - a = b \odot a^* \in \mathcal{T}$.

(iii) Put $a_0 = 0$. By (ii), $a_n - a_{n-1} \in \mathcal{T}$ for any $n \geq 1$. Hence

$$a = \lim_n a_n = \lim_n \sum_{i=1}^n (a_i - a_{i-1}) = \min\left\{\sum_{n=1}^\infty (a_n - a_{n-1}), 1\right\} \in \mathcal{T}.$$

□

Let $\mathcal{T}$ be a tribe and denote by

$$\mathcal{S}_0(\mathcal{T}) := \{A \subseteq \Omega : \chi_A \in \mathcal{T}\}. \tag{7.3}$$

Theorem 7.1.7 *Let $\mathcal{T}$ be a tribe of fuzzy subsets of a set Ω. Then*

(1) *$\mathcal{S}_0(\mathcal{T})$ is a σ-algebra of crisp subsets of Ω.*

(2) *If $f \in \mathcal{T}$, then f is $\mathcal{S}_0(\mathcal{T})$-measurable.*

(3) *$\mathcal{T}$ contains all $\mathcal{S}_0(\mathcal{T})$-measurable fuzzy functions on Ω if and only if $\mathcal{T}$ contains all constant functions with values in $[0,1]$.*

Proof. (1). Since $1_\Omega = \chi_\Omega \in \mathcal{T}$, we obtain $\Omega \in \mathcal{S}_0(\mathcal{T})$. Let $A \in \mathcal{S}_0(\mathcal{T})$. Then $\chi_{\Omega\setminus A} = 1 - \chi_A \in \mathcal{T}$, that is $\Omega \setminus A \in \mathcal{S}_0(\mathcal{T})$. If $\{A_n\}$ is a sequence of subsets in $\mathcal{S}_0(\mathcal{T})$, then

$$\chi_{\bigcup_n A_n} = \min\{\sum_n \chi_{A_n}, 1\} \in \mathcal{T},$$

that is $\bigcup_n A_n \in \mathcal{S}_0(\mathcal{T})$.

(2) Let $f \in \mathcal{S}_0(\mathcal{T})$. We have to prove that $A_\alpha = \{\omega : \ f(\omega) \geq \alpha\} \in \mathcal{S}_0(\mathcal{T})$ for any $\alpha \in [0,1]$. If $\alpha = 0$, then $A_0 = \Omega \in \mathcal{S}_0(\mathcal{T})$. If $\alpha = 1$, then

$$\chi_{\{\omega:\ f(\omega)=1\}} = 1 - \min\left\{\sum_{n=1}^{\infty}(1-f),\ 1\right\} \in \mathcal{T},$$

hence $A_1 = \{\omega : \ f(\omega) = 1\} \in \mathcal{S}_0(\mathcal{T})$.

Let $0 < \alpha < 1$. Then there is a sequence $\{r_n\}$ of rational numbers of the form $\sum_{i=1}^k a_i/2^i$, $a_i \in \{0,1\}$, such that $r_n \nearrow \alpha$. Due to $A_\alpha = \bigcap_n A_{r_n}$, it suffices to prove $A_{r_n} \in \mathcal{S}_0(\mathcal{T})$, i.e., to show that

$$\chi_{\{\omega:\ f(\omega)\geq\alpha\}} \in \mathcal{S}_0(\mathcal{T})$$

for any number $\alpha = \sum_{i=1}^k a_i/2^i$, $a_i \in \{0,1\}$.

We shall prove that by induction with respect to k. If $k = 1$, then $\alpha = a_1/2$ and we have two possibilities: $a_1 = 0$, or $a_1 = 1$. In the first case $A_0 = \Omega \in \mathcal{S}_0(\mathcal{T})$. In the second one

$$\begin{aligned} A_{\frac{1}{2}} &= \{\omega : \ f(\omega) \geq 1/2\} = \{\omega : \ (f(\omega) + f(\omega)) \wedge 1 = 1\} \\ &= \{\omega : \ (f \oplus f)(\omega) = 1\} \in \mathcal{S}_0(\mathcal{T}), \end{aligned}$$

where we used the above assertion for $\alpha = 1$.

Let $A_\alpha \in \mathcal{S}_0(\mathcal{T})$ whenever α has a dyadic development of length k and assume that

$$\alpha = \sum_{i=1}^{k+1} a_i/2^i.$$

If $a_1 = 0$, then $\alpha = \beta/2$, where $\beta = \sum_{i=2}^{k+1} a_i/2^{i-1}$ is a dyadic number of length k. Hence,

$$\{\omega : f(\omega) \geq \alpha\} = \{\omega : 2f(\omega) \geq \beta\} = \{\omega : (f \oplus f)(\omega) \geq \beta\} \in \mathcal{S}_0(\mathcal{T}).$$

If $a_1 = 1$, then $\alpha = (1+\beta)/2$, hence

$$\{\omega : f(\omega) \geq \alpha\} = \{\omega : (f \odot f)(\omega) \geq \beta\} \in \mathcal{S}_0(\mathcal{T}).$$

(3) Let $\mathcal{T}$ contain all $\mathcal{S}_0(\mathcal{T})$-measurable functions $f : \Omega \to [0,1]$. Since constant functions are $\mathcal{S}_0(\mathcal{T})$-measurable, they belong to $\mathcal{T}$. Conversely, let $\mathcal{T}$ contain all [0,1]-valued constant functions. We have to prove that $\mathcal{T}$ contains every $\mathcal{S}_0(\mathcal{T})$-measurable function $f : \Omega \to [0,1]$.

There exists a sequence $\{f_n\}_n$ of simple $\mathcal{S}_0(\mathcal{T})$-measurable functions $f_n : \Omega \to [0,1]$ such that $f_n \nearrow f$. It is sufficient to prove that any simple $\mathcal{S}_0(\mathcal{T})$-measurable functions $g : \Omega \to [0,1]$ belongs to $\mathcal{T}$. Let

$$g = \sum_{i=1}^{k} \alpha_i \chi_{A_i},$$

$A_i \in \mathcal{S}_0(\mathcal{T})$, $A_i \cap A_j = \emptyset$ for $i \neq j$, $\alpha_i \in [0,1]$. Put $g_i : \Omega \to [0,1]$ via $g_i(\omega) = \alpha_i$. Then

$$g = (g_1 \odot \chi_{A_1}) \oplus \cdots \oplus (g_k \odot \chi_{A_k}).$$

Therefore g belongs to $\mathcal{T}$, because $g_i \in \mathcal{T}$ is a constant function and in view that $\chi_{A_i} \in \mathcal{T}$ is a characteristic function of a set $A_i \in \mathcal{S}_0(\mathcal{T})$. □

We now introduce t-conorms and t-norms which play a crucial role in fuzzy set investigation. More on t-conorms and t-norms see [KMP].

A function $S : [0,1] \times [0,1] \to [0,1]$ is said to be a **t-conorm** if it is commutative, associative, 0 is a neutral element for S, i.e., $S(0,x) = x$ for any $x \in [0,1]$, and is nondecreasing in both components. A function $T : [0,1] \times [0,1] \to [0,1]$ is said to be a **t-norm** if it is commutative, associative, 1 is a neutral element for T, i.e., $T(1,x) = x$ for any $x \in [0,1]$, and is nondecreasing in both components. If S is a t-conorm, then

$$T(x,y) = 1 - S(1-x, 1-y), \; x, y \in [0,1],$$

is a t-norm, and vice versa. t-conorms and t-norms model logical connectives "or" and "and", respectively.

A t-conorm S is said to be **Archimedean** if S is continuous and $S(a,a) > a$ for any $a \in (0,1)$. An Archimedean t-conorm S is said to be **strict** if S is strictly increasing, i.e., $S(a,c) < S(b,c)$ whenever $a < b$, $c < 1$, otherwise, it is called **nilpotent**. The following representation of t-conorms holds [Lin]:

Theorem 7.1.8 *A function $S: [0,1]\times[0,1]\to[0,1]$ is an Archimedean t-conorm if and only if there exists an increasing and continuous function $g: [0,1]\to[0,\infty]$ with $g(0)=0$ such that*

$$S(a,b) := g^{-1}(\min\{g(a)+g(b), g(1)\}),\ a,b\in[0,1].$$

Moreover, S is strict if and only if $g(1)=\infty$.

Example 7.1.9 *Examples of t-conorms are the following*

(i) *L, where*

$$L(a,b)=\min\{a+b,1\},\quad a,b\in[0,1],$$

is a **Łukasiewicz** *t-conorm which is Archimedean with a generator $g=id_{[0,1]}$. The corresponding t-norm is T_L, where*

$$T_L(a,b)=\max\{a+b-1,0\},\quad a,b\in[0,1].$$

(ii) *Max, where*

$$Max(a,b)=\max\{a,b\},\quad a,b\in[0,1],$$

is a continuous t-conorm which is not Archimedean, and $S\ge Max$ for any t-conorm S. The corresponding t-norm is Min, $Min(a,b)=\min\{a,b\}$.

(iii) *S_P, where*

$$S_P(a,b)=a+b-ab,\quad a,b\in[0,1],$$

is a strict Archimedean t-conorm (in this case $g(x)=-\ln(1-x)$). The corresponding t-norm is the product $T_P(a,b)=ab$.

(iv) *The strongest t-conorm S_W is given by*

$$S_W(a,b)=\begin{cases}\max\{a,b\}, & \text{if } \min\{a,b\}=0,\\ 1, & \text{otherwise,}\end{cases}\qquad a,b\in[0,1],$$

and $S\le S_W$ for any t-conorm S.

(v) *Another important family of t-conorms is the family of* **Frank t-conorms**,

$$\{S^F_\lambda:\ \lambda\in[0,\infty]\},$$

where $S^F_0=Max$, $S^F_1=S_P$, $S^F_\infty(a,b)=\min\{a+b,1\}$, $S^F_\lambda(a,b)=1-\log_\lambda(1+(\lambda^{1-a}-1)(\lambda^{1-b}-1)/(\lambda-1))$ if $\lambda\neq 0,1,\infty$. They are continuous and they are the unique solution (in some sense) to the equation $T(a,b)+S(a,b)=a+b$, $a,b\in[0,1]$.

(vi) *A* **Yager** *t-conorm is a function* $S_p^Y(a,b) = \min\{(a^p + b^p)^{1/p}, 1\}$, $p > 0$. *Its generator is* $g(t) = t^p$, *and the corresponding t-norm* $T_p^Y(a,b) = \max^{1/p}\{0, a^p + b^p - 1\}$.

We note that some of the t-conorms S generate σ-complete MV-algebras on $[0,1]$ with $\oplus = S$ and $a^* = 1 - a$, see Exercise 4.7.4.3. On the other hand, e.g., Max does not produce an MV-algebra in this way.

t-conorms and t-norm are closely connected with so-called BL-algebras (basic logic algebras) recently invented by Hájek [Haj], [Haj 1] in order to provide an algebraic proof of the completeness theorems of a class of $[0,1]$-valued logics arising in fuzzy logic. We mention that a **BL-algebra** [3] is a lattice L with the least and largest elements 0 and 1, respectively, endowed with two binary operations $\odot$ (called product) and $\to$ (called residuum) such that

(i) $\odot$ is associative, commutative and isotone and, for any $x \in L$, $x \odot 1 = x$,

(ii) $x \odot y \leq z$ iff $x \leq y \to z$,

(iii) $x \wedge y = x \odot (x \to y)$,

(iv) $x \vee y = [(x \to y) \to y] \wedge [(y \to x) \to x]$,

(v) $(x \to y) \vee (y \to x) = 1$.

An MV-algebra generates a BL-algebra, if the residuum is defined by $x \to y = x^* \oplus y$. In addition, Höhle [Höl] showed that a BL-algebra L generates via $x^* := x \to 0$ and $x \oplus y := x^* \to y$, an MV-algebra $([0,1]; \oplus, ^*, 0, 1)$ iff $x = x^{**}$.

On the other hand, any continuous t-norm T on $[0,1]$ with no idempotents, except 0, 1 and at least one nilpotent, is equivalent with $\oplus = S$.

For example, for the Łukasiewicz t-norm $x \odot y = \max\{0, x + y - 1\}$, $x, y \in [0,1]$, the residuum is defined via

$$x \to y = \begin{cases} 1, & \text{if } x \leq y, \\ 1 - x + y, & \text{elsewhere.} \end{cases}$$

Similarly, if $\odot = Min$ and

$$x \to y = \begin{cases} 1, & \text{if } x \leq y, \\ y, & \text{elsewhere,} \end{cases}$$

we obtain a BL-algebra on $[0,1]$ generated by Min.

In the same way, if T_P is the product t-norm, then $\odot = T_P$, and

$$x \to y = \begin{cases} 1, & \text{if } x \leq y, \\ y/x, & \text{elsewhere.} \end{cases}$$

[3] Pseudo BL-algebras were introduced recently by Georgescu and Iorgulescu in [GeIo 1].

Any t-conorm S can be extended due to associativity and commutativity of S to an n-ary operation S_n $(n \geq 0)$ via

$$S_n(a_1, \ldots, a_n) := S(S_{n-1}(a_1, \ldots, a_{n-1}), a_n),\ a_1, \ldots, a_n \in [0,1],\ n \geq 1,$$

and $S_0 = 0$. Moreover, we can define $S_\infty :\ [0,1]^{\aleph_0} \to [0,1]$ via

$$S_\infty(a_1, a_2, \ldots) = \lim_n S_n(a_1, \ldots, a_n);$$

it exists due to boundedness and isotonicity of S.

We now present another type of tribes, which is connected with t-conorms. Let S be a t-conorm. An S-**tribe** on $\Omega \neq \emptyset$ is a collection $\mathcal{T}_S$ of fuzzy sets from $[0,1]^\Omega$ such that

(i) $1_\Omega \in \mathcal{T}_S$,

(ii) if $f \in \mathcal{T}_S$, then $1 - f \in \mathcal{T}_S$;

(iii) if $\{f_n\}_n$ is sequence from $\mathcal{T}_S$, then $S_\infty(f_1, f_2, \ldots) \in \mathcal{T}_S$.

For example, if L is the Łukasiewicz t-conorm, then the L-tribe (= Łukasiewicz tribe) is exactly a tribe. On the other side, any Łukasiewicz tribe is a Max-tribe, [BuKl, Prop 2]. If $\mathcal{T}_S$ is an S-tribe, then

$$\mathcal{S}(\mathcal{T}_S) := \{A \subseteq \Omega :\ \chi_A \in \mathcal{T}_S\}$$

is a σ-algebra of crisp subsets of Ω.

The following characterization of tribes is due to Klement and Navara [KlNa].

Theorem 7.1.10 *A collection $\mathcal{T}$ of fuzzy sets on Ω is a tribe if and only if there exist a σ-algebra $\mathcal{B}$ of subsets of Ω and a sequence $\{\nabla_n\}_n$ of filters in $\mathcal{B}$ with $\nabla_m \subseteq \nabla_n$ whenever n is a divisor of m, such that*

$$\mathcal{T} = \{f \in [0,1]^\Omega :\ f \text{ is } \mathcal{B}\text{-measurable},\ f^{-1}(\{0, 1/n, 2/n, \ldots, n/n\}) \in \nabla_n,\ n \geq 1\}.$$

We continue with tribes. Let $\mathcal{F}$ be a family of fuzzy subsets of Ω. We define $\mathcal{F}_0 := \mathcal{F} \cup \{0_\Omega, 1_\Omega\}$, and for any ordinal number $\alpha > 0$, we define

$$\mathcal{F}_\alpha = (\bigcup_{\beta<\alpha} \mathcal{F}_\beta)^*,$$

where $\mathcal{C}^*$ for $\mathcal{C} \subseteq \mathcal{F}$ denotes the set of all functions of the form $\min\{\sum_{n=1}^\infty f_n, 1\}$, where either f_n or $1 - f_n \in \mathcal{C}$ for any $n \geq 1$. Then

(i) $\mathcal{F} \subseteq \mathcal{F}_\beta \subseteq \mathcal{F}_\alpha \subseteq \mathcal{T}(\mathcal{F})$ for $0 < \beta < \alpha$, where $\mathcal{T}(\mathcal{F})$ is a tribe generated by $\mathcal{F}$.

(ii) $\mathcal{T}(\mathcal{F}) = \bigcup_{\beta<\omega_1} \mathcal{F}_\beta$, where ω_1 is the first uncountable ordinal number.

The following result characterizes the tribe generated by $C(X)$, the space of all continuous fuzzy functions on X, where X is a compact Hausdorff space. We note that by $\mathcal{B}(X)$ we will mean the **Baire σ-algebra**, i.e., the σ-algebra generated by compact G_δ sets on X, or equivalently, by $\{f^{-1}([a,\infty)) : f \in C(X),\ a \in \mathbb{R}\}$.

Proposition 7.1.11 *Let X be a compact Hausdorff space. Let $\mathcal{T}(X)$ be the tribe generated by $C(X)$, and let $\mathcal{M}(X)$ be the set of all Baire measurable fuzzy sets on X. Then $\mathcal{T}(X) = \mathcal{M}(X)$.*

Proof. It is obvious that $\mathcal{M}(X)$ is a tribe containing $C(X)$, hence, $\mathcal{T}(X) \subseteq \mathcal{M}(X)$. Let $\mathcal{S}_\mathcal{T}$ and $\mathcal{S}_\mathcal{M}$ be the systems of all crisp subsets $A \subseteq X$ such that $\chi_A \in \mathcal{T}(X)$ and $\chi_A \in \mathcal{M}(X)$, respectively. Then $\mathcal{S}_\mathcal{T}$ and $\mathcal{S}_\mathcal{M}$ are σ-algebras of subsets of X, and $\mathcal{S}_\mathcal{T} \subseteq \mathcal{S}_\mathcal{M} \subseteq \mathcal{B}(X)$.

Since each $f \in \mathcal{T}(X)$ is $\mathcal{S}_\mathcal{T}$-measurable, then each $f \in C(X)$ is $\mathcal{S}_\mathcal{T}$-measurable, so that $\mathcal{B}(X) \subseteq \mathcal{S}_\mathcal{T}$.

On the other hand, because $\mathcal{T}(X)$ contains $C(X)$, $\mathcal{T}(X)$ contains all constant functions taking values in the interval $[0,1]$. By Theorem 7.1.7 this is a necessary and sufficient condition in order $\mathcal{T}(X)$ consists of all $\mathcal{S}_\mathcal{T}$-measurable fuzzy subsets of X. Consequently, $\mathcal{T}(X) = \mathcal{M}(X)$. □

7.1.3 Loomis–Sikorski's Theorem

In the present Section, we give a generalization of the Loomis–Sikorski theorem for σ-complete MV-algebras. First, we present some partial results.

For any $a \in M$, we put

$$M(a) := \{A \in \mathcal{M}(M) : a \notin A\}. \tag{7.4}$$

Then $\{M(a) : a \in M\}$ is a base of $\mathcal{T}_\mathcal{M}$ of the compact Hausdorff space $\mathcal{M}(M)$, and for $a, b \in M$,

(i) $M(0) = \emptyset$,

(ii) $M(a) \subseteq M(b)$ whenever $a \leq b$,

(iii) $M(a \wedge b) = M(a) \cap M(b)$, $M(a \vee b) = M(a) \cup M(b)$.

An element $a \in M$ is **idempotent** iff $a \vee a^* = 1$. It is possible to show that a is idempotent iff $a \oplus a = a$ iff $a \odot a = a$ iff $a \wedge a^* = 0$. Denote by $B(M)$ the set of all idempotent elements of M.

Proposition 7.1.12 $(B(M); \wedge, \vee, ^*, 0, 1)$ *is a Boolean algebra with the least and greatest elements* 0 *and* 1 *and with unary operation* * *taken from* M. *In addition, if* $\bigvee_i a_i$ *of elements* $\{a_i\}$ *of* $B(M)$ *exists in* M, *then* $\bigvee_i a_i \in B(M)$. *Consequently, if* M *is* σ*-complete or complete, so is* $B(M)$.

Proof. Let $\{a_i\}$ be a family of elements of $B(M)$ such $a = \bigvee_i a_i \in M$. Then

$$a \wedge a^* = \bigvee_i (a_i \wedge a^*) \le \bigvee_i (a_i \wedge a_i^*) = 0,$$

so $a \in B(M)$. □

Proposition 7.1.13 *Let* $(M; \oplus, ^*, 0, 1)$ *be an MV-algebra. If* $a \in M$, *for the set theoretical complement* $M(a)^c$ *of* $M(a)$ *we have*

$$M(a)^c \subseteq M(a^*).$$

If a *is idempotent, then* $M(a)^c = M(a^*)$. *In general, if* $a \le b$, *then*

$$M(b) \setminus M(a) \subseteq M(b * a).$$

If a *and* b *are idempotent, and* $a \le b$, *then* $b * a^* = b \wedge a^*$ *and* $M(b) \setminus M(a) = M(b * a)$.

If M *is semisimple, then* a *is idempotent if and only if* $M(a^*) = M(a)^c$.

Proof. Let $a \le b$. Take $A \in M(b) \setminus M(a)$. Then $b \notin A$ and $a \in A$. We claim that $b * a \notin A$. If not, then $b * a \in A$, so that $b = a + (b * a) \in A$, which is a contradiction.

Let a and b be two idempotent elements of M and take $A \in M(b * a)$. Then $b * a \notin A$ and $b \notin A$. Since A is a prime ideal of M, then $0 = (b * a) \wedge a \in A$ entails $a \in A$ so that $A \in M(b) \setminus M(a)$.

The first condition follows from the above for $b = 1$.

Let M be semisimple and $M(a^*) = M(a)^c$. Then for any maximal ideal I of M either $a \in I$ or $a^* \in I$ which entails $a \wedge a^* \in I$, and the semisimplicity of M gives $a \wedge a^* = 0$. □

Remark 7.1.14 We note that if $M(a)$ is clopen, then not necessarily $M(a^*) = M(a)^c$. Indeed, take $M = [0, 1]$. Then $\mathcal{M}(M) = \{\{0\}\}$, and for any non-zero $a \in M$, $M(a) = \mathcal{M}(M)$ is clopen, and $M(a)^c = \emptyset$.

Similarly, if M is not semisimple, then the equality $M(a^*) = M(a)^c$ does not entail that a is idempotent. Indeed, take the Chang MV-algebra [Cha] (see Example 5.2.39) $M = \{0, 1, 2, \ldots, n, \ldots, \dot{n}, \ldots, \dot{2}, \dot{1}, \dot{0}\}$. Then $\mathcal{M}(M) = \{\{0, 1, \ldots\}\}$, and $M(1) = \emptyset = M(\dot{1})^c$ and 1 is not idempotent.

A topological space Ω is said to be

(i) **totally disconnected** if every two different points are separated by a clopen subset of Ω;

(ii) **basically disconnected** if the closure of every open F_σ subset of Ω is open;

(iii) **extremally disconnected** if the closure of every open set is open.

Proposition 7.1.15 *If M is a σ-complete or complete MV-algebra, then the space $\mathcal{M}(M)$ is basically disconnected or extremally disconnected, respectively.*

Proof. Since M is σ-complete (complete) if and only if its representation ℓ-group G is Dedekind σ-complete (complete), according to [Goo 1, Thm 8.14], the space $Ext(\mathcal{S}(M))$ is homeomorphic to the set $\mathcal{M}(B(M))$ of all maximal ideals of the Boolean algebra $B(M)$. Hence, by [Sik, Thm 22.4], the Boolean algebra $B(M)$ is σ-complete (complete) iff $\mathcal{M}(B(M))$ is basically disconnected (extremally disconnected). □

Proposition 7.1.16 *Let M be a semisimple MV-algebra. If $a = \bigvee_t a_t \in M$, then*

$$M(a) \setminus \bigcup_t M(a_t),$$

where $M(a)$ is defined by (7.4), is a nowhere dense subset of $\mathcal{M}(M)$.

Proof. Let $a = \bigvee_t a_t$ and suppose $M(a) \setminus \bigcup_t M(a_t)$ is not nowhere dense. Since $\{M(a) : a \in M\}$ is a base of the topological space $\mathcal{M}(M)$, there exists a non-zero element $b \in M$ such that $\emptyset \neq M(b) \subseteq M(a) \setminus \bigcup_t M(a_t)$. Due to $M(b) = M(b) \cap M(a) = M(b \wedge a)$, we take $b_0 := b \wedge a$, which is a non-zero element of M. Then $M(b_0) \cap M(a_t) = \emptyset$ for any t, so that $M(b_0 \wedge a_t) = \emptyset$ and the semisimplicity of M yields $b_0 \wedge a_t = 0$ for any t.

Then

$$b_0 = b_0 \wedge a = b_0 \wedge \bigvee_t a_t = \bigvee_t (b_0 \wedge a_t) = 0,$$

which gives $M(b) = \emptyset$, a contradiction, so that our assumption was false, and consequently, $M(a) \setminus \bigcup_t M(a_t)$ is a nowhere dense set. □

We note that the converse to Proposition 7.1.16 is in general not true for a semisimple MV-algebra. Take $M = [0,1]$, then $M(0.3) \setminus (M(0.1) \cup M(0.2)) = \emptyset$ but $0.3 \neq 0.1 \vee 0.2$.

Proposition 7.1.17 *Let M be a semisimple MV-algebra and let $a_t \leq a$ for any t. If $\bigcap_t M(a * a_t)$ is a nowhere dense subset of $\mathcal{M}(M)$, then $a = \bigvee_t a_t$.*

Proof. It is clear that in order to prove $a = \bigvee_t a_t$ it is sufficient to verify that $a_t \leq b \leq a$ for any t implies $b = a$.

So let $\bigcap_t M(a * a_t)$ be nowhere dense, and let $b < a$ for some $b \geq a_t$, $b \leq a$. Then $a * b \neq 0$ and $M(a * b)$ is a nonempty open subset of $\mathcal{M}(M)$. By assumptions, there exists a non-zero open subset $O \subseteq M(a * b)$ such that $O \cap \bigcap_t M(a * a_t) = \emptyset$. Consequently, there is a non-zero element $c \in M$ such that $M(c) \subseteq O$. Hence, for any $I \in M(c) \subseteq M(a * b)$, we have $c \notin I$, $a * b \notin I$ and $I \notin \bigcap_t M(a * a_t)$. This entails that there is an index t such that $a * a_t \in I$. Since $a_t \leq b$, we have $a * b \leq a * a_t \in I$ which implies $a * b \in I$, and this is a contradiction with $a * b \notin I$. Finally, our assumption $b < a$ was false, and whence $b = a$, and so $a = \bigvee_t a_t$. □

The converse statement holds, e.g., for the case that $\{a_t\}$ is a system of idempotents, see Propositions 7.1.13 and 7.1.16. But we note that the converse to Proposition 7.1.17 is not true in general for a semisimple MV-algebra. Take $M = [0, 1]$, then $\mathcal{M}(M) = \{\{0\}\}$. If $a_n = 0.5 - 1/n$, then $a := \bigvee_n a_n = 0.5$, and $M(a * a_n) = M(1/n) = \mathcal{M}(M)$ for any $n \geq 1$, so that $\bigcap_n M(a * a_n)$ is not nowhere dense.

It is worthy of recalling that if M is a Boolean algebra, then $a = \bigvee_t a_t$ iff $\bigcap_t M(a * a_t) = M(a) \setminus \bigcup_t M(a_t)$ is a nowhere dense set, and this observation is a corner-stone for the Loomis–Sikorski theorem. As we have seen, for semisimple MV-algebras, the analogical statement is not true, in general. Hence, for the validity of the Loomis–Sikorski theorem on σ-complete MV-algebras we have to develop below a more detailed analysis of σ-complete MV-algebras.

Let $a \in M$ and $k \geq 1$. We recall that

$$k \odot a := a_1 \oplus \cdots \oplus a_k,$$

where $a_1 = \cdots = a_k = a$.

Proposition 7.1.18 *For any $a \in M$,*

$$M(a) = \bigcup_{k=1}^{\infty} M((k \odot a)^*)^c.$$

If, in addition, M is σ-complete, then the closure of any $M(a)$ is open.

Proof. If $a = 0$, the statement is evident. Let now $a \neq 0$. Let $I \in M(a)$. Then $a \notin I$, and consequently, there is an integer $k \geq 1$ such that $(k \odot a)^* \in I$, and $I \in M((k \odot a)^*)^c$.

Conversely, let $I \in \bigcup_{k=1}^{\infty} M((k \odot a)^*)^c$. There exists an integer $k \geq 1$ such that $I \in M((k \odot a)^*)^c$. Hence, $(k \odot a)^* \in I$. Since I is a maximal ideal, we conclude that $a \notin I$.

The second part of the assertion follows from Proposition 7.1.15 and from the fact that $M(a)$ is an open F_σ set. □

Remark 7.1.19 We note that a Boolean algebra M is σ-complete (complete) iff $\mathcal{M}(M)$ is basically disconnected (extremally disconnected) [Sik, Thm 22.4]. For MV-algebras such an assertion is not true, in general. Indeed, take the Chang MV-algebra $M = \{0, 1, 2, \ldots, n, \ldots, \dot{n}, \ldots, \dot{2}, \dot{1}, \dot{0}\}$. Then $\mathcal{M}(M) = \{\{0, 1, \ldots\}\}$, and it is basically disconnected (extremally disconnected), but M is not σ-complete.

Proposition 7.1.20 *An element a of a semisimple MV-algebra M is idempotent if and only if $\bar{a}$ is a characteristic function.*

Proof. Let $a \in B(M)$. For any maximal ideal A of M, from $0 = a \wedge a^* \in A$ we conclude that either $a \in A$ or $a^* \in A$ which implies that $\bar{a}(A) \in \{0, 1\}$. Conversely, let $\bar{a}$ be a characteristic function. Then $\bar{a}(A) = a/A$ is either 0 or 1, or equivalently, either $a \in A$ or $a^* \in A$. Hence $a \wedge a^* \in A$ for any $A \in \mathcal{M}(M)$, and the semisimplicity of M entails $a \wedge a^* = 0$. □

For a bounded function $g : X \to \mathbb{R}$ on a topological space X we define

$$\tilde{g}(x) = \inf_{U \in \mathcal{N}(x)} \sup\{g(y) : y \in U\}, \tag{7.5}$$

where $\mathcal{N}(x)$ is the system of open neighbourhoods for $x \in X$. Then

(i) $g(x) \le \tilde{g}(x)$ for any $x \in X$.

(ii) $\tilde{g}(x) = g(x)$ if g is continuous at x.

If $D(g)$ is the set of discontinuity points for g, then

$$\{x \in X : g(x) \neq \tilde{g}(x)\} \subseteq D(g) = \bigcup_n \{g^{-1}(R_n) - Int(g^{-1}(R_n))\}, \tag{7.6}$$

where $\{R_n\}$ is an open basis in $\mathbb{R}$.

Let f be a real-valued function on $\Omega \neq \emptyset$. We define

$$N(f) := \{\omega \in \Omega : |f(\omega)| > 0\}.$$

Suppose that $\mathcal{F}$ is a Bold algebra of fuzzy sets on Ω. Then for all $f, g \in \mathcal{F}$ we have

(i) $N(f \oplus g) = N(f) \cup N(g)$.

(ii) $N(f * g) = \{\omega \in \Omega : f(\omega) > g(\omega)\}$.

(iii) $(f * g) \oplus (g * f) = (f * g) + (g * f)$.

(iv) $N((f * g) \oplus (g * f)) = N(f - g)$.

(v) $N(f) \subseteq N(g)$ if $f \le g$.

We note from Proposition 7.1.20 that if M is a semisimple algebra, then

$$M(a) = \{I \in \mathcal{M}(M) : a \notin I\} = N(\bar{a}).$$

A state s on (G, u) is **discrete** if $s(G) = \frac{1}{n}\mathbb{Z}$ for some integer $n \geq 1$, where $\mathbb{Z}$ is the set of all integers.

To present the main assertion of the chapter, we give the following two important statements of theory of unital ℓ-groups, [Goo 1, Lem 9.12, Cor 9.14].

Theorem 7.1.21 *Let (G, u) be a unital Abelian Dedekind σ-complete ℓ-group and set $H = \mathcal{C}(Ext(\mathcal{S}(G,u)))$ for the set of all continuous functions on $Ext(\mathcal{S}(G,u))$, where $Ext(\mathcal{S}(G,u))$ is the set of all extremal states on (G,u).*

(1) *Then H is a Dedekind σ-complete ℓ-group, and the mapping $\psi : G \to H$ defined via $\psi(g)(m) := m(g)$, for $m \in Ext(\mathcal{S}(G,u))$, $g \in G$, preserves infima and suprema.*

(2) *Set*

$$B = \{f \in H : f(s) \in s(G) \text{ for all discrete states } s \in Ext(\mathcal{S}(G,u))\}. \qquad (7.7)$$

Then the mapping $\psi : G \to H$ defined via $\psi(g)(m) := m(g)$, $m \in Ext(\mathcal{S}(G,u))$, $g \in G$, provides an isomorphism of (G,u) onto $(B,1)$ as ordered groups with strong units.

We are now ready to formulate the principal result, the Loomis–Sikorski representation theorem for σ-complete MV-algebras. We will follow ideas developed in [Dvu 16]. In an independent way the same result was proved by Mundici [Mun 5].

Theorem 7.1.22 (Loomis-Sikorski Theorem) *For every σ-complete MV-algebra M there exist a tribe $\mathcal{T}$ of fuzzy sets and an MV-σ-homomorphism h from $\mathcal{T}$ onto M.*

Proof. Let $\mathcal{T}$ be the tribe of fuzzy sets on $Ext(\mathcal{S}(M))$ generated by the Bold algebra $\{\hat{a} : a \in M\}$. Denote by $\mathcal{T}'$ the class of all functions $f \in \mathcal{T}$ with the property that for some $b \in M$, $N((f * \hat{b}) \oplus (\hat{b} * f))$ is a meager set.

If b_1 and b_2 are two elements of M such that $N((f * \hat{b}_i) \oplus (\hat{b}_i * f))$ is a meager set for $i = 1, 2$. Then

$$N((\hat{b}_1 * \hat{b}_2) \oplus (\hat{b}_2 * \hat{b}_1)) = N(\hat{b}_1 - \hat{b}_2) \subseteq N(\hat{b}_1 - f) \cup N(f - \hat{b}_2)$$

is a meager set. Due to the Baire theorem saying that any non-empty open set of a compact Hausdorff space cannot be a meager set, we conclude that $\hat{b}_1 = \hat{b}_2$, i.e., $b_1 = b_2$.

It is clear that $\mathcal{T}'$ is closed under the formation of complement $f \mapsto 1-f$ and it is a Bold algebra containing $\{\hat{a}:\ a \in M\}$.

To show that $\mathcal{T}'$ is a tribe, it is necessary to verify that $\mathcal{T}'$ is closed under limits of non-decreasing sequences from $\mathcal{T}'$.

Let $\{f_n\}_n$ be a sequence of non-decreasing elements from $\mathcal{T}'$. Choose $b_n \in M$ such that $N(f_n - \hat{b}_n)$ is a meager set. Without loss of generality we can assume that $b_n \le b_{n+1}$. Denote $f = \lim_n f_n$, $b = \bigvee_{n=1}^{\infty} b_n$, $b_0 = \lim_n \hat{b}_n$. Then $f, b_0 \in \mathcal{T}$ and $b \in M$. We have

$$N(f-\hat{b}) \subseteq N(f-b_0) \cup N(\hat{b}-b_0)$$

and $N(f-b_0) = \{m:\ f(m) < b_0(m)\} \cup \{m:\ b_0(m) < f(m)\}$.

If $m \in \{m:\ f(m) < b_0(m)\}$, then there is an integer $n \ge 1$ such that $f(m) < \hat{b}_n(m) \le b_0(m)$. Hence $f_n(m) \le f(m) < \hat{b}_n(m) \le b_0(m)$ so that $m \in \{m:\ f_n(m) < \hat{b}_n(m)\}$.

Similarly we can prove that if $m \in \{m:\ b_0(m) < f(m)\}$, then there is an integer $n \ge 1$ such that $m \in \{m:\ \hat{b}_n(m) < f_n(m)\}$.

The last two cases imply

$$N(f-b_0) \subseteq \bigcup_{n=1}^{\infty} N(\hat{b}_n - f_n),$$

which is a meager set.

Apply now (7.5) to the function b_0 to obtain $\tilde{b}_0$, i.e.

$$\tilde{b}_0(m) := \inf_{U \in \mathcal{N}(m)} \sup\{b_0(y):\ y \in U\}.$$

Since $Ext(\mathcal{S}(M))$ is basically disconnected, compact, Hausdorff, and $b_0^{-1}((\alpha,\infty)) = \bigcup_n \hat{b}_n^{-1}((\alpha,\infty))$ for any $\alpha \in \mathbb{R}$, $b_0^{-1}((\alpha,\infty))$ is an open F_σ set, by [Goo 1, Lem 9.1], (see Exercise 8.7.4.3) $\tilde{b}_0$ is continuous. Since b_0 is a point limit of a sequence of continuous functions, by [Kur, pp. 86, 405-6], $D(b_0) \supseteq N(\tilde{b}_0 - b_0)$ is a meager set,

$$b_0 \le \tilde{b}_0 \le \hat{b},$$

and $N(\hat{b} - b_0) \subseteq N(\hat{b} - \tilde{b}_0) \cup D(b_0)$.

Finally we show $\tilde{b}_0 = \hat{b}$. Define $\mathcal{C}(X)$, where $X = Ext(\mathcal{S}(M))$, as the set of all continuous real valued functions defined on X. Since X is a basically disconnected, compact, Hausdorff space, $\mathcal{C}(X)$ is a Dedekind σ-complete ℓ-group with a strong unit 1_X, [Goo 1, Cor 9.3] (see Exercise 9.7.4.3). Using Mundici's functor Γ, we see that the system of all continuous fuzzy functions on X, $C(X) = \Gamma(\mathcal{C}(X), 1_X)$, is a σ-complete MV-algebra. Applying [Goo 1, Lem 9.12] (see Theorem 7.1.21(1)), we conclude the mapping $a \mapsto \hat{a}$, $a \in M$, preserves countable suprema and infima. Consequently, $\tilde{b}_0 = \hat{b}$, which proves that $\mathcal{T}'$ is a tribe, and whence, $\mathcal{T}' = \mathcal{T}$.

Due to definition of $\mathcal{T}'$, for any $f \in \mathcal{T}$ there is a unique element $h(f) := b \in M$ such that $N(f - \hat{b})$ is meager, which proves that $h : \mathcal{T} \to M$ is a surjective MV-σ-homomorphism in question. □

We note that if M is an atomic σ-complete MV-algebra with the countable set of atoms, then M is σ-isomorphic with some tribe [DCR].

7.1.4 Loomis–Sikorski Theorem for Dedekind σ-complete ℓ-groups

In the present Section, we apply the Loomis–Sikorski theorem for MV-algebras to Dedekind σ-complete ℓ-groups.

A **g-tribe** is a non-void system $\mathcal{T}$ of bounded functions on a set $\Omega \neq \emptyset$ such that

(i) $0_\Omega, 1_\Omega \in \mathcal{T}$;

(ii) $f \pm g \in \mathcal{T}$ whenever $f, g \in \mathcal{T}$;

(iii) if $\{f_n\}_n$ is a sequence of elements from $\mathcal{T}$ for which there exists $f \in \mathcal{T}$ with $f_n \leq f$ (pointwisely), $n \geq 1$, then $\sup_n f_n \in \mathcal{T}$.

It is evident that $\mathcal{T}$ with respect to the pointwise ordering, i.e., $f \leq g$ iff $f(\omega) \leq g(\omega)$ for every $\omega \in \Omega$, is a Dedekind σ-complete ℓ-group with strong unit 1_Ω. In addition, $\Gamma(\mathcal{T}, 1_\Omega) := \{f \in \mathcal{T} : 0_\Omega \leq f \leq 1_\Omega\}$ is a tribe of fuzzy functions.

Let (G, u) be an ℓ-group with strong unit u. We note that a **state** on G is any mapping $s : G \to \mathbb{R}$ such that $s(g_1 + g_2) = s(g_1) + s(g_2)$, $s(u) = 1$, $s(g) \geq 0$ for $g \in G^+$. We define the weak topology on the set $\mathcal{S}(G, u)$ of all states on a unital ℓ-group (G, u): A net $\{m_\alpha\}$ of states on (G, u) **converges weakly** to a state m on (G, u) iff $m_\alpha(x) \to m(x)$ for any $x \in G$; and we write $m = \text{w-lim}_\alpha m_\alpha$.

If m is a state on $M = \Gamma(G, u)$, then it can be uniquely extended to a state $\hat{m}$ on G and conversely, any restriction of a state s on G to M gives a state on M.

Similarly as for Theorem 6.1.30, we can prove that $\mathcal{S}(G, u)$ is a compact space.

Theorem 7.1.23 *There exists a one-to-one correspondence $m \leftrightarrow \hat{m}$ between $\mathcal{S}(M)$ and $\mathcal{S}(G, u)$, and extremal points of $\mathcal{S}(M)$ are mapped onto the extremal points of $\mathcal{S}(G, u)$ and vice-versa. In addition, this mapping is a homeomorphism.*

Proof. This follows from Theorem 6.1.29 and Remark 6.1.31. In addition, since u is a strong unit for G, it is easy to see that a net $\{m_\alpha\}$ of states on G converges weakly to a state m iff their restrictions to the interval $[0, u]$ converge weakly to the restriction $m|[0, u]$. □

Since an ℓ-group is Dedekind σ-complete iff M is a σ-complete MV-algebra, we conclude that if G is Dedekind σ-complete, then the set $X = Ext(\mathcal{S}(G, u))$, the set of all extremal points of $\mathcal{S}(G,u)$, is a compact Hausdorff basically disconnected space, see also [Goo 1, Cor 9.3] or Exercise 9.7.4.3.

We now present an analogue of the Loomis–Sikorski theorem for Dedekind σ-complete ℓ-group.

Theorem 7.1.24 *For any Dedekind σ-complete ℓ-group G with strong unit u there exists a g-tribe $\mathcal{T}$ of bounded functions on a compact Hausdorff space X and an ℓ-group σ-homomorphism* [4] *$\hat{h}$ from $\mathcal{T}$ onto G with $h(1_X) = u$.*

Proof. Put $X = Ext(\mathcal{S}(G,u))$ and let $\mathcal{C}(X)$ be the set of all continuous functions on the compact space X. Set

$$B = \{f \in \mathcal{C}(X) : f(s) \in s(G) \text{ for all discrete states } \ s \in X\}.$$

Then B is an ℓ-group with strong unit 1_X. According to [Goo 1, Cor 9.14] (see Theorem 7.1.21(2)), the mapping $\psi : G \to \mathcal{C}(X)$, defined by $\psi(g)(s) := s(g)$, $g \in G$ ($s \in X$), defines an isomorphism of (G,u) onto $(B, 1_X)$ (as ordered groups with ordered units).

Let $\mathcal{T}(B)$ be the g-tribe generated by B. We assert that for the tribe $\mathcal{T}(M)$ of fuzzy sets generated by the set $\hat{M} := \{\psi(g) : g \in M\}$, where $M = \Gamma(G,u)$, we have

$$\mathcal{T}(M) = \Gamma(\mathcal{T}(B), 1_X). \qquad (A)$$

Step 1. Since for any $g \in G$, $\psi(g) \in \mathcal{T}(M)$ and $\psi(g) \in \Gamma(\mathcal{T}(B), 1_X)$, we conclude from the fact that $\Gamma(\mathcal{T}(B), 1_X)$ is a tribe containing $C(X)$ that

$$\mathcal{T}(M) \subseteq \Gamma(\mathcal{T}(B), 1_X). \qquad (B)$$

Let now $G(\mathcal{T}(M))$ be the ℓ-group generated by $\mathcal{T}(M)$. In view of (B), we have

$$G(\mathcal{T}(M)) \subseteq \mathcal{T}(B). \qquad (C)$$

Step 2. Since 1_X is a strong unit for $G(\mathcal{T}(M))$, to prove that $G(\mathcal{T}(M))$ is closed under $\sup_n f_n$ for every sequence $\{f_n\}_n$ of elements from $G(\mathcal{T}(M))$ with an upper bound in $G(\mathcal{T}(M))$, it is sufficient to verify that for each sequence $\{f_n\}_n$ of functions belonging to $[0,\, k1_X]$ for $k \geq 1$.

We proceed by induction on k. The case $k = 1$ is trivial. Suppose now that $k > 1$ and that the interval $[0, (k-1)1_X]$ is closed under $\sup_n$. Since $[(k-1)1_X,\, k1_X]$ is isomorphic to $[0, 1_X]$, we conclude that $[(k-1)1_X, k1_X]$ is closed under $\sup_n$.

Let $\{f_n\}_n$ be a sequence of functions of $[0, k1_X]$. For any $k \geq 1$, we put

$$f_n^1 = \min\{f_n, a\}, \quad f_n^2 = \max\{f_n, a\},$$

[4] In our case this means: (i) $\hat{h}(f \pm g) = \hat{h}(f) \pm \hat{h}(g)$; (ii) $\hat{h}(\sup_n f_n) = \bigvee_n \hat{h}(f_n)$; (iii) $\hat{h}(1_X) = u$.

where $a = (n-1)1_X$. Then $f_n = f_n^1 + (f_n^2 - a)$ for each $n \ge 1$. Define $f^1 = \sup_n f_n^1$ and $f^2 = \sup_n f_n^2$ and $f = f^1 + f^2$. Then $f^1 \in [0,(k-1)1_X]$, $f^2 \in [(k-1)1_X, k1_X]$ and $f \ge f_n$ for each $n \ge 1$. Let now $s \in X$ and $\epsilon > 0$. Then there exists an integer $n \ge 1$ such that $f^i(s) - f_n^i(s) \le \epsilon/2$ for $i = 1,2$. Hence $f(s) - f_n(s) = f^1(s) - f_n^1(s) + f^2(s) - f_n^2(s) < \epsilon$ which proves that $f = \sup_n f_n$. Similarly we prove that $\inf_n f_n$ exists in $[0, k1_X]$.

Step 3. In fact we have verified that $G(\mathcal{T}(M))$ is a g-tribe containing B which entails

$$\mathcal{T}(B) \subseteq G(\mathcal{T}(M)), \tag{D}$$

which by (C) yields $\mathcal{T}(B) = G(\mathcal{T}(M))$, consequently, $\mathcal{T}(M) = \Gamma(\mathcal{T}(B), 1_X)$.

Step 4. By the Loomis–Sikorski Theorem 7.1.22, there exists an MV-σ-homomorphism h from $\mathcal{T}(M)$ onto M. Using arguments of Mundici's categorical equivalence of ℓ-groups with strong unit with MV-algebras, given by $M \cong G \leftrightarrow \Gamma(G,u)$, see Corollary 5.3.8 and (5.33), we conclude that there exists an extension $\hat{h}$ of h from $\mathcal{T}(B)$ onto G such that $\hat{h}$ is an ℓ-group homomorphism, i.e., for $\hat{h}$ we have (i) $\hat{h}(f \pm g) = \hat{h}(f) \pm \hat{h}(g)$, (ii) $\hat{h}(\max\{f,g\}) = \hat{h}(f) \vee \hat{h}(g)$, and (iii) $\hat{1}_X = u$.

To prove that $\hat{h}(\sup_n f_n) = \sup_n \hat{h}(f_n)$ whenever $\sup_n f_n \in \mathcal{T}(B)$, we proceed in the same way as in the Step 2. □

Comparing this Theorem with Proposition 7.1.11, we have the following statement.

Proposition 7.1.25 *Let X be a compact Hausdorff space and let $\mathcal{T}_g(X)$ be the g-tribe generated by the set $C(X)$ of all continuous functions on X. Then $\mathcal{T}_g(X) = \mathcal{M}_b(X)$, where $\mathcal{M}_b(X)$ denotes the set of all bounded Baire measurable functions on X. In addition, $\mathcal{M}(X) = \Gamma(\mathcal{T}_g(X), 1_X)$.*

Proof. Since $\mathcal{M}_b(X)$ is a g-tribe containing $C(X)$, we have $\mathcal{T}_g(X) \subseteq \mathcal{M}_b(X)$. $\mathcal{T}_g(X)$ is an ℓ-group with strong unit 1_X, whence $\Gamma(\mathcal{T}_g(X), 1_X)$ is a tribe containing the set $C(X)$ of all continuous fuzzy sets on X, which by Proposition 7.1.11 entails

$$\mathcal{M}(X) \subseteq \Gamma(\mathcal{T}_g(X), 1_X) \subseteq \mathcal{M}_b(X).$$

If f is a non-negative function from $\mathcal{M}_b(X)$, then there is an integer $k \ge 1$ such that $f \le k1_X$, and by Proposition 7.1.11 this means that $f/k \in \mathcal{M}(X)$. Consequently, f belongs to the ℓ-group generated by $\mathcal{M}(X)$. This is also true for any function $f \in \mathcal{M}_b(X)$. Consequently, if $f \in \Gamma(\mathcal{T}_g(X), 1_X)$, then $f \in \mathcal{M}(X)$. Hence, $\mathcal{M}_b(X) = \mathcal{T}_g(X)$ and $\mathcal{M}(X) = \Gamma(\mathcal{T}_g(X), 1_X)$. □

7.2 Loomis–Sikorski's Theorem for MV-algebras and BCK-algebras

We generalize the Loomis–Sikorski theorem for perfect MV-algebras, Dedekind σ-complete BCK-algebras, and Dedekind σ-complete ℓ-rings showing that there exist special kinds of tribes, BCK-tribes or r-tribes consisting of fuzzy sets, or of bounded functions, defined on a compact Hausdorff space. Tribes are closed under point suprema of bounded sequences and they are a direct generalization of σ-algebras of crisp subsets.

7.2.1 Loomis–Sikorski's Theorem for Perfect MV-algebras

Perfect MV-algebras were introduced in Section 5.3.4. We note that an MV-algebra M is said to be **perfect** if

$$M = Rad(M) \cup Rad(M)^*,$$

where $Rad(M)^* := \{a^* : \ a \in Rad(M)\}$. If M is a perfect MV-algebra, then

$$Rad(M) \cap Rad(M)^* = \emptyset.$$

For example, let G be an Abelian ℓ-group and define the lexicographical product

$$G(\mathbb{Z}) := \mathbb{Z} \times_{lex} G,$$

where $\mathbb{Z}$ is the group of all integers. Then the element $(1,0)$ is a strong unit in $G(\mathbb{Z})$, and

$$\Gamma(G(\mathbb{Z}),(1,0))$$

is a perfect MV-algebra, in fact it is the Chang MV-algebra, and $Rad(M) = \{(0,g) : g \in G^+\}$ and $Rad(M)^* = \{(1,g) : \ g \in G^-\}$. The operation $\oplus$ and * are defined as follows: $(a,b)\oplus(c,d) = (1,0)\wedge(a+c,\ b+d)$ and $(a,b)^* = (1,0)-(a,b) = (1-a,-b)$.

The famous result of Di Nola and Lettieri [DiLe] says that, for any perfect MV-algebra M, there is a (unique up to ℓ-group isomorphism) ℓ-group G such that $M \cong \Gamma(G(\mathbb{Z}),(1,0))$.

The following result is due to [GeLe].

Proposition 7.2.1 *Let $\{x_i\}$ be a family of elements of $Rad(M)$ of a perfect MV-algebra M. If $x = \bigvee_i x_i$ exists in M, then $x \in Rad(M)$.*

Proof. If $x \in Rad(M)^*$, then due to $x_i \le y$ for any $y \in Rad(M)^*$, we have $x \le y$. In particular, $x \le x_i^*$ for any i, so that $x_i \le x^*$ which gives the contradiction $x \le x^*$. □

Proposition 7.2.2 *If h is an MV-homomorphism from an MV-algebra M_1 into a second one M_2, then h maps $Rad(M_1)$ into $Rad(M_2)$.*

Proof. Let $a \in Rad(M_1)$. Then, for any $n \geq 1$, $n\,a \in Rad(M_1)$ and $a + b \in Rad(M_1)$ for each $b \in Rad(M_1)$. Hence, for any integer $n \geq 1$, $n\,h(a) \leq h(b^*) = h(b)^*$, which implies $h(a) \in Rad(M_2)$. □

Proposition 7.2.3 *Let M_1 and M_2 be two perfect MV-algebras and let h be a mapping from $Rad(M_1)$ into $Rad(M_2)$ preserving $+$ and $\wedge$ and $\vee$. Then h can be uniquely extended to the MV-homomorphism $\hat{h}$ from M_1 into M_2.*

Proof. *Claim 1.* Let G_1^+ and G_2^+ be the positive cones of Abelian ℓ-groups G_1 and G_2, respectively, and let $f : G_1^+ \to G_2^+$ preserve $+$, $\wedge$ and $\vee$. Then f can be uniquely extended to an ℓ-group homomorphism $\hat{f} : G_1 \to G_2$.

Indeed, extend f to $\hat{f}$ via $\hat{f}(a_1 - a_2) = f(a_1) - f(a_2)$, whenever $a_1, a_2 \in G_1^+$; then $\hat{f}$ is a group-homomorphism. We show that it preserves meets. Let $a = a' - a''$ and $b = b' - b''$, where $a', a'', b', b'' \in G_1^+$. Then

$$\hat{f}((a' + b'') \wedge (b' + a'')) = f((a' + b'') \wedge (b' + a'')) = (f(a') + f(b'')) \wedge (f(b') + f(a'')).$$

Subtracting $\hat{f}(b'')$ and $\hat{f}(a'')$, we obtain $\hat{f}(a \wedge b) = \hat{f}(a) \wedge \hat{f}(b)$.

Claim 2. If we define $\hat{h}(a) = h(a^*)^*$, $a \in Rad(M_1)^*$, then $\hat{h}$ is the MV-homomorphism in question.

Step 1. Let $a, b \in Rad(M_1)$. Then $a \oplus b = a + b$, so that $h(a \oplus b) = h(a) + h(b) = h(a) \oplus h(b)$.

Step 2. Let $a, b \in Rad(M_1)^*$. Then $a \oplus b = 1$, and $\hat{h}(a), \hat{h}(b) \in Rad(M_1)^*$, so that $\hat{h}(a \oplus b) = 1 = \hat{h}(a) \oplus \hat{h}(b)$.

Step 3. Let $a \in Rad(M_1)$ and $b \in Rad(M_1)^*$. Without loss of generality, we can assume that $M_1 = \Gamma(\mathbb{Z} \times_{lex} G_1, (1,0))$ and $M_2 = \Gamma(\mathbb{Z} \times_{lex} G_2, (1,0))$. Then $a \cong (0, a)$ and $b \cong (1, b)$, where $a \in G_1^+$ and $b \in G_1^-$, and $a \oplus b \cong (1,0) \wedge (1, a + b) = (1, 0 \wedge (a + b)) \in Rad(M_1)^*$.

From the above we conclude there exists a unique mapping $f : G_1^+ \to G_2^+$ such that $h((0,a)) = (0, f(a))$ and $h((1,b)) = (1, -f(-b))$.

It is clear that f preserves $+$, $\wedge$ and $\vee$ in G_1^+, so that by Claim 1, it can be uniquely extended to an ℓ-group-homomorphism $\hat{f} : G_1 \to G_2$. Hence

$$\begin{aligned} \hat{h}(a \oplus b) &= \hat{h}(1, 0 \wedge (a + b)) = (1, \hat{f}(0 \wedge (a + b))) \\ &= (1, \hat{f}(0) \wedge (\hat{f}(a) + \hat{f}(b))) = (0, \hat{f}(a)) \oplus (1, \hat{f}(b)) = \hat{h}(a) \oplus \hat{h}(b). \end{aligned}$$

□

An MV-algebra M is said to be (i) **Rad-Dedekind σ-complete (complete)** if, for any sequence (any family) $\{a_n\}$ of elements of $Rad(M)$ such that $a_n \le b$ for any $n \ge 1$, where $b \in Rad(M)$, we have $\bigvee_n a_n \in Rad(M)$; (ii) **Rad-Archimedean** if $n\, a := a + \cdots + a \le b$ $(n \ge 1)$, for $a, b \in Rad(M)$ entails $a = 0$. It is evident that if $M = \Gamma(\mathbb{Z} \times_{lex} G, (1,0))$, then M is Rad-Dedekind σ-complete (complete) iff G is Dedekind σ-complete (complete). We note that if M is Rad-Dedekind σ-complete, then it is not necessarily σ-complete, see e.g., Chang's MV-algebra $\Gamma(\mathbb{Z}(\mathbb{Z}), (1,0))$.

According to [BeDi], we say that a perfect MV-algebra M is with **principal radical** if there is an element $a \in Rad(M)$ such that the ideal of M generated by a coincides with $Rad(M)$. The category of perfect MV-algebras with principal radical is a category whose objects are pairs (M, a), where M is a perfect MV-algebra with a fixed element $a \in Rad(M)$ generating $Rad(M)$, as an ideal, and morphisms are MV-homomorphisms preserving fixed elements a in radicals. This category is categorically equivalent to the category of unital ℓ-groups, see Exercises 5.3.6.

We now present the Loomis–Sikorski theorem for perfect MV-algebras.

Theorem 7.2.4 *Let M be a Rad-Dedekind σ-complete perfect MV-algebra with principal radical. Then there exists a g-tribe $\mathcal{T}$ of bounded functions on a compact Hausdorff space X and an MV-homomorphism h from the perfect MV-algebra $M(\mathcal{T}) := \Gamma(\mathbb{Z} \times_{lex} \mathcal{T}, (1, 0_X))$ with principal radical onto M such that h preserves all existing countable joins in $M(\mathcal{T})$. In addition, h maps $Rad(M(\mathcal{T}))$ onto $Rad(M)$, and $h(1_X) = u$, where u is a fixed element of $Rad(M)$ generating it as an ideal.*

Proof. Suppose that $M = \Gamma(\mathbb{Z} \times_{lex} G, (1,0))$ and let u be a fixed element in $Rad(M)$ generating $Rad(M)$ as an ideal. Hence u is a strong unit for G. Since M is Rad-Dedekind σ-complete, then G is Dedekind σ-complete, and by the Loomis–Sikorski theorem 7.1.24, for the Dedekind σ-complete unital ℓ-group (G, u), there exists a g-tribe $\mathcal{T}$ of bounded real-valued functions defined on the compact Hausdorff space X of all extremal states on (G, u) and there exists an ℓ-group-homomorphism h' from $(\mathcal{T}, 1_X)$ onto (G, u) such that if $\sup_n f_n \in \mathcal{T}$, then $h'(\sup_n f_n) = \bigvee_n h'(f_n)$, and $h(1_X) = u$.

In other words, the mapping $h'' :\ Rad(M(\mathcal{T})) \cong \mathcal{T}^+$ onto $G^+ \cong Rad(M)$ defined by $h''((0, f)) := h'(f)$, $f \in \mathcal{T}^+$, preserves $+$, $\wedge$, and $\vee$. Hence by Proposition 2.4, there exists a unique extension h of h'' from the perfect MV-algebra $M(\mathcal{T})$ into G. In any rate, h is surjective, Theorem 7.1.22.

Suppose now that $\bigvee_n f_n$ is defined in $M(\mathcal{T})$. Then either all $f_n \in Rad(M(\mathcal{T}))$ or there exists $f_n \in Rad(M(\mathcal{T}))^*$. In the first case by Proposition 7.2.1, $\bigvee_n f_n \in Rad(M(\mathcal{T}))$, so that $\bigvee_n f_n = \sup_n f_n$, and $h(\bigvee_n f_n) = \bigvee_n h(f_n)$.

In the second case, $\bigvee_n f_n = \bigvee_{n \in D} f_n$, where $D := \{n :\ f_n \in Rad(M(\mathcal{T}))^*\}$. For $n \in D$, $f_n^* \in Rad(M(\mathcal{T}))$, so that $\bigwedge_{n \in D} f_n^* = \inf_{n \in D} f_n^*$ and $\bigvee_{n \in D} f_n = (\bigwedge_{n \in D} f_n^*)^*$, which entails

$$h(\bigvee_n f_n) = h(\bigvee_{n\in D} f_n) = h((\bigwedge_{n\in D} f_n^*)^*) = (h(\bigwedge_{n\in D} f_n^*))^* = (\bigwedge_{n\in D} h(f_n^*))^* = \bigvee_{n\in D} h(f_n).$$

□

7.2.2 Loomis-Sikorski's Theorem for Product BCK-algebras

In the present Section, we give the Loomis–Sikorski theorem for product BCK-algebras which were introduced in [Dvu 14] and studied in Section 5.3.5. As a corollary, we obtain the Loomis–Sikorski theorem for product σ-complete MV-algebras proved by Mundici [Mun 5].

Let u be a quasi strong unit for X. A measure $m \in \mathcal{S}_u(X)$ is said to be **discrete** if there exists an integer $n \geq 1$ such that $m(X) \subseteq \frac{1}{n}\mathbb{Z}$. If m is a discrete measure-morphism, then there exists an integer n such that $m([0,u]) = \{0, 1/n, 2/n, \ldots, n/n\}$.

We now present the Loomis–Sikorski theorem for product BCK-algebras. We note that a **product BCK-tribe** is a BCK-tribe $\mathcal{X}$ such that if $f, g \in \mathcal{X}$, then the usual product fg is in $\mathcal{X}$.

Theorem 7.2.5 *Let $(X; *, 0, \cdot)$ be a Dedekind σ-complete product BCK-algebra with a quasi strong unit u which is a product unity for $\cdot$. Then there exists a product BCK-tribe $\mathcal{X}$ of bounded positive functions defined on a compact Hausdorff space $\Omega \neq \emptyset$ and a BCK-σ-homomorphism h from $\mathcal{X}$ onto X which preserves product in $\mathcal{X}$ and $h(1_\Omega) = u$.*

Proof. Let u be a quasi strong unit for X, and let $\Omega := Ext(\mathcal{S}_u(X))$. Then Ω is a compact Hausdorff space. Denote by $X_u := [0,u] = \{x \in X : 0 \leq x \leq u\}$. It is clear that X_u is a bounded product commutative BCK-algebra. Any restriction of $\mathcal{S}_u(X)$ onto X_u gives a measure on X_u, and conversely, any measure (measure-morphism) on X_u can be uniquely extended to a measure (measure-morphism) on X.

Let $(R; +, 0, \cdot, \leq)$ be an Archimedean commutative and associative ℓ-ring from Theorem 5.3.30 with $R_0 \subseteq R^+$ generating R^+ as a group-cone such that $(X; *, 0, \cdot) \cong (R_0, *_R, 0, \cdot)$. Without loss of generality, we can assume that $X = R_0$. Since u is a quasi strong unit for X, so is u for the ℓ-ring R.

Let now m be a measure-morphism from $\mathcal{S}_u(X)$. Then Ker_m is a maximal ideal of X. We show that any maximal ideal I of X is a $\cdot$-ideal. Indeed, let $a \in I$ and $b \in X$. Then by Lemma 5.2.10, there exists an integer $n \geq 1$ such that $b = b_1 + \cdots + b_n$, where $b_i \in X_u$, $i = 1, \ldots, n$. Hence, $a \cdot b = a \cdot (b_1 + \cdots + b_n) = a \cdot b_1 + \cdots + a \cdot b_n \leq a \cdot u + \cdots + a \cdot u = n\, a$, which proves that na is in the ℓ-ideal of R generated by I, consequently by Theorem 5.2.46 $a \cdot b \in I$. In a similar way we can prove that $b \cdot a \in I$.

Hence $m(x \cdot y) = m(x) \cdot m(y)$ for all $x, y \in X$. This can be seen due to property that the quotient BCK-algebra X/Ker_m is a linear Archimedean BCK-algebra, so that it is isomorphic to a BCK-subalgebra of $\mathbb{R}$ containing the number 1 as a quasi strong unit which is closed under the natural multiplication. (We can apply Theorem 5.3.30 to Pickert–Hion's theorem, [Fuc, p. 126].)

Consequently, a measure-morphism m is discrete iff $m(X_u) = \{0,1\}$.

Hence, the mapping $x \mapsto \hat{x} \in [0,\infty)^\Omega$, $x \in X$, where $\hat{x}(m) := x/Ker_m = m(x)$, $m \in Ext(\mathcal{S}_u(X))$, gives a one-to-one representation of X in $\hat{X} := \{\hat{x} : x \in X\}$, which is a subset of $C(\Omega)$, the space of all real-valued continuous functions on Ω. Moreover, $\widehat{x \cdot y} = \hat{x} \cdot \hat{y}$. According to [Goo 1, Lem 9.12] (see Theorem 7.1.21(1)) and Theorem 6.2.3, we can show that a continuous function $f : \Omega \to [0,\infty)$ belongs to $\hat{X}$ iff $f(m) \in \mathbb{Z}^+$ for any discrete measure m from $Ext(\mathcal{S}_u(X))$.

Let now $\mathcal{X}$ be a minimal BCK-tribe of bounded non-negative functions on Ω generated by $\hat{X}$, and consider by $\mathcal{X}'$ the class of all functions $f \in \mathcal{X}$ with the property that for some $b \in X$, $N(f - \hat{b}) := \{m \in \Omega : |f(m) - \hat{b}(m)| > 0\}$ is a meager set.

Let b_1 and b_2 be two elements of X such that $N(f - \hat{b}_i)$ is a meager set for $i = 1,2$. Then $N(\hat{b}_1 - \hat{b}_2) \subseteq N(\hat{b}_1 - f) \cup N(f - \hat{b}_2)$ is a meager set. Due to the Baire theorem saying that any non-empty open set of a compact Hausdorff space cannot be a meager set, we can conclude that $\hat{b}_1 = \hat{b}_2$, i.e., $b_1 = b_2$.

We show that if $f, g \in \mathcal{X}'$, then $f * g \in \mathcal{X}'$, where $(f * g)(\omega) := \max\{0, f(\omega) - g(\omega)\}$, $\omega \in \Omega$. In fact, let $\mathcal{I}$ be the set of all $f \in \mathcal{X}$ such that $N(f)$ is a meager set. Then (i) $0_\Omega \in \mathcal{I}$; (ii) if $f \in \mathcal{I}$ and $g \in \mathcal{X}$, then $\min\{f,g\} \in \mathcal{I}$; (iii) if $f * g \in \mathcal{I}$ and $g \in \mathcal{I}$, then $f * g = f - \min\{f,g\}$ which proves that $0 < f(m) = (f(m) - \min\{f,g\}(m)) + (\min\{f,g\})(m)$, so that $f \in \mathcal{I}$. Since $N(f - \hat{b}) = N(f * \hat{b}) \cup N(\hat{b} * f)$, $N(f - \hat{b})$ is a meager set iff $N(f * \hat{b})$ and $N(\hat{b} * f)$ are meager sets, which entails by Lemma 5.1.30 (on congruences) $\mathcal{X}'$ is a BCK-algebra.

To show that $\mathcal{X}'$ is a BCK-tribe, it is necessary to verify that $\mathcal{X}'$ is closed under limits of non-decreasing sequences from $\mathcal{X}'$.

Let $\{f_n\}_n$ be a sequence of non-decreasing elements from $\mathcal{X}'$ with $f_n \leq f$, $n \geq 1$, for some $f \in \mathcal{X}$. Choose $b_n \in X$, $n \geq 1$, such that $N(f_n - \hat{b}_n)$ is a meager set. Without loss of generality we can assume that $b_n \leq b_{n+1}$ for any n. Denote $f = \lim_n f_n$, $b = \bigvee_{n=1}^\infty b_n$, $b_0 = \lim_n \hat{b}_n$. Using the proof of Theorem 7.1.22, we can show that $N(f - \hat{b})$ is a meager set. So $\mathcal{X}' = \mathcal{X}$.

We now show that $\mathcal{X}$ is closed under formation of the natural product of functions. Indeed, let $f, g \in \mathcal{X}$ and let $a, b \in X$ be such elements that $N(f - \hat{a})$ and $N(g - \hat{b})$ are meager sets. Then in view of the inequalities $|f \cdot g - \widehat{a \cdot b}| = |f \cdot g - \hat{a} \cdot \hat{b}| \leq |f \cdot g - f \cdot \hat{b}| + |f \cdot \hat{b} - \hat{a} \cdot \hat{b}| = |f|\,|g - \hat{b}| + |f - \hat{a}|\,|\hat{b}|$ we have that $N(f \cdot g - \widehat{a \cdot b})$ is also a meager set.

Consequently, $\mathcal{X}$ is a product BCK-tribe, and the mapping $h : f \mapsto b$ whenever $N(f - \hat{b})$ is a meager set defines a BCK-σ-morphism from the product BCK-tribe $\mathcal{X}$

onto X with $h(1_\Omega) = u$. □

As any MV-algebra can be considered as a bounded commutative BCK-algebra, as a corollary of Theorem 7.2.5 we obtain the Loomis–Sikorski theorem for product MV-algebras obtained originally by Mundici [Mun 1, Cor 6.3].

Theorem 7.2.6 *Let M be a σ-complete product MV-algebra such that 1 is a product unity. Then there exists a tribe $\mathcal{T}$ of fuzzy sets on a compact Hausdorff space $\Omega \neq \emptyset$ which is closed under the natural product of functions, and there exists an MV-σ-homomorphism h from $\mathcal{T}$ onto M preserving product in $\mathcal{T}$.*

Proof. Let $\mathcal{T}$ be a tribe generated by $\hat{M} = \{\hat{a} : a \in M\}$, where $\hat{a}(m) := m(a)$, $m \in \mathcal{S}_1(M)$. Then $\mathcal{T}$ is a BCK-tribe generated by $\hat{M}$, consequently, we can apply Theorem 7.2.5. □

An **r-tribe** is a g-tribe $\mathcal{T}$ such that $fg \in \mathcal{T}$ whenever $f, g \in \mathcal{T}$. It is evident that an r-tribe is a Dedekind σ-complete ℓ-ring with the natural product of functions with unity 1_Ω, which is also a strong unit.

Theorem 7.2.7 *Let $(R; +, 0, \cdot, \leq)$ be a Dedekind σ-complete ℓ-ring with unity 1 which is a strong unit for R. Then there exists an r-tribe $\mathcal{T}$ of bounded functions on a compact Hausdorff space $\Omega \neq \emptyset$ and an ℓ-ring σ-homomorphism h from $\mathcal{T}$ onto R with $h(1_\Omega) = 1$.*

Proof. According to Theorem 7.2.5, there exists a g-tribe $\mathcal{T}$ of bounded functions on the compact Hausdorff space $\Omega \neq \emptyset$ of all measures s on the product BCK-algebra $(R^+; *_R, 0, \cdot)$ with $s(1) = 1$ and an ℓ-group σ-homomorphism h from $\mathcal{T}$ onto R with $h(1_\Omega) = 1$. Then by construction, h maps $\Gamma(\mathcal{T}, 1_\Omega)$ onto the interval $[0, 1]$ in R, and in addition, $\Gamma(\mathcal{T}, 1_\Omega)$ is a tribe generated by $\{\hat{x} : x \in [0, 1]\}$, where $[0, 1]$ is understood as a product MV-algebra. Consequently, by the proof of Theorem 7.2.6, h preserves products in $\Gamma(\mathcal{T}, 1_\Omega)$. Consequently, in view that 1_Ω is a strong unit for $\mathcal{T}$, h preserves product in the whole $\mathcal{T}$, which proves that $\mathcal{T}$ is an r-tribe. □

We note that according to Theorem 5.3.19, we can define a product on a perfect MV-algebra via formula (5.35). Then $\cdot$ is a commutative product on M such that $a \cdot 1 = a = 1 \cdot a$ for any $a \in M$. If M is Rad-Archimedean, then on M there is a unique product $\cdot$ such that $a \cdot 1 = a = 1 \cdot a$ for any $a \in M$; this product is defined as above.

Consequently, on non-trivial Rad-Archimedean perfect MV-algebras there is no commutative product with unity 1. Hence, a commutative and associative product with unit 1 on perfect algebras can be defined only on their radicals as it was done in [BDG]: A **product** on a perfect MV-algebra is a binary operation $\cdot : Rad(M) \times Rad(M) \to Rad(M)$ such that $a \cdot (b + c) = (a \cdot b) + (a \cdot b)$ and $(b + c) \cdot a = (b \cdot a) + (c \cdot a)$, $a, b, c \in Rad(M)$, and M is said to be a **product perfect MV-algebra**. The following form of the Loomis–Sikorski theorem holds for perfect MV-algebras.

Theorem 7.2.8 *Let M be a Rad-Dedekind σ-complete product perfect algebra with a fixed element u generating the radical as an ideal and let u be a unity for the product. Then there exists an r-tribe $\mathcal{T}$ of bounded functions on a compact Hausdorff space X and an MV-homomorphism h from the product perfect algebra $M(\mathcal{T}) := \Gamma(\mathbb{Z} \times_{lex} \mathcal{T}, (1, 0_X))$, with principal radical with 1_X generating the radical of $M(\mathcal{T})$ as an ideal, onto M such that h preserves all existing countable joins and products in $M(\mathcal{T})$. In addition, h maps $Rad(M(\mathcal{T}))$ onto $Rad(M)$, and $h(1_X) = u$.*

Proof. In view of Theorem 5.3.19 we can suppose that $M = \Gamma(\mathbb{Z} \times_{lex} R, (1,0))$, where R is an ℓ-group and let u be a fixed element in $Rad(M)$ generating $Rad(M)$ as an ideal. Hence u is a strong unit for R. Since M is Rad-Dedekind σ-complete, R is Dedekind σ-complete, and by the Loomis–Sikorski theorem 7.1.22, for the Dedekind σ-complete unital ℓ-group (R, u), there exists a g-tribe $\mathcal{T}$ of bounded real-valued functions defined on a compact Hausdorff space X equal to the set of all extremal states on (R, u) and there exists an ℓ-group-homomorphism h' from $(\mathcal{T}, 1_X)$ onto (R, u) such that if $\sup_n f_n \in \mathcal{T}$, then $h'(\sup_n f_n) = \bigvee_n h'(f_n)$, and $h(1_X) = u$.

In other words, the mapping $h'' : Rad(M(\mathcal{T})) \cong \mathcal{T}^+$ onto $R^+ \cong Rad(M)$ defined by $h''((0, f)) := h'(f)$, $f \in \mathcal{T}^+$, preserves $+$, $\wedge$, and $\vee$. Therefore on R^+ there is a product $\cdot$. Hence by Proposition 7.2.3 and Theorem 7.2.4, there exists a unique extension h of h'' from the perfect MV-algebra $M(\mathcal{T})$ onto R which preserves countable suprema in $Rad(M)$. Moreover, h preserves the product on $Rad(M) \cong R^+$, consequently R is an ℓ-ring with a unity u, i.e., 1_X is a unity for the natural product of functions in $\mathcal{T}$, i.e., $\mathcal{T}$ is an r-tribe. □

7.3 Weakly Divisible MV-algebras

According to [DiSe], we say that an MV-algebra M is **divisible** if, for each element $a \in M$ and for each integer $n \geq 1$, there exists $x \in M$ such that $n \odot x = a$ and $a^* \oplus ((n-1) \odot x) = x^*$.[5] Using arguments of mathematical induction it is possible to show that M is divisible iff for each element $a \in M$ and for each integer $n \geq 1$ there exists $x \in M$ such that nx is defined in M and $nx = a$. We note that using Mundici's representation of M, via (5.33), it is possible to show that if $nx = a = ny$ then $x = y$, and in this case we write $x = \frac{a}{n}$. It is worthy of recalling that if $\frac{1}{n}$ exists in M, then $m\frac{1}{n}$ is a constant element in M for any m with $0 \leq m \leq n$.

Similarly, we say that an MV-algebra M is **weakly divisible** if, for any integer $n \geq 1$, there is an element $v \in M$ such that nv is defined in M and $nv = 1$. It is clear that any divisible MV-algebra is weakly divisible. Under some conditions, as we show below, the converse is also true.

[5] We note that given an integer $n \geq 1$ and $a \in M$, we define $n \odot a := a_1 \oplus \cdots \oplus a_n$, where $a_i = a$, $i = 1, \ldots, n$.

We note that an ℓ-group G is **divisible** if given an element g and given an integer $n \geq 1$ there is an $h \in G$ such that $g = nh$.

For example, the MV-subalgebra $\mathbb{Q}_1$, the set of all rational numbers in $[0,1]$, is a divisible MV-algebra, while the MV-subalgebra $\mathbb{D}_1$, the set of all dyadic rational numbers in $[0,1]$, is not divisible.

Example 7.3.1 *Let α be an irrational number in $[0,1]$, and let $G(\alpha) = \{\frac{p}{q} + n\alpha : p, n, \in \mathbb{Z},\ q \in \mathbb{Z}^+ \setminus \{0\}\}$. Then $G(\alpha)$ is an ℓ-subgroup of $\mathbb{R}$, and 1 is a strong unit in $G(\alpha)$. Define $M = \Gamma(G(\alpha), 1)$. M is a countable semisimple MV-algebra which is weakly divisible but neither divisible, nor σ-complete.*

Indeed, e.g., the element $\frac{\alpha}{2}$ is not defined in M. Moreover, the set M is dense in $[0,1]$ within the natural topology.

Example 7.3.2 *Let $\alpha \in \mathbb{R}$. Define $G_1(\alpha) := \{m + n\alpha : m, n \in \mathbb{Z}\}$ and put $M(\alpha) = \Gamma(G_1(\alpha), 1)$. If $\alpha = p/q \in [0,1]$, where p and q are relatively primes, then $M(\frac{p}{q}) = \{0, 1/q, 2/q, \ldots, q/q\}$ and $M(\frac{p}{q})$ is not weakly divisible, and $M(\frac{p}{q})$ is not closed under the natural multiplication, whenever $\frac{p}{q} \notin \{0,1\}$. Moreover, $M(\frac{p}{q})$ is σ-complete.*

If $\alpha \in \mathbb{R}$ is irrational, then $M(\alpha)$ is countable and dense in $[0,1]$ and also not weakly divisible and not σ-complete. $M(\sqrt{2})$ is not weakly divisible but it is closed under the natural multiplication. Similarly, this is the case if $\alpha = \sqrt{k}$ for some k which is not a square.

Proposition 7.3.3 *Let (G, u) be an ℓ-unital group. Then $M = \Gamma(G, u)$ is divisible if and only if G is divisible.*

Proof. If G is divisible, then for any $a \in M$ and for any $n \geq 1$ there exists an element $b \in G$ such that $a = nb$. Hence $a \in M$.

Conversely, let M be divisible. Any element $g \in G^+$ is of the form $g = a_1 + \cdots + a_k$, where $a_1, \ldots, a_k \in M$. Given an integer $n \geq 1$, there are element $b_1, \ldots, b_k \in M$ such that $a_i = nb_i$ for $i = 1, \ldots, k$. Hence $g = nb_1 + \cdots + nb_k = nb$, where $b = b_1 + \cdots + b_k \in G^+$. The rest of the proof is now evident. $\square$

Following [DiSe], we say that a subalgebra A of an MV-algebra M is **dense** in M if, for each non-zero $b \in M$, there exists a non-zero $a \in A$ such that $a \leq b$. The following result has been proved in [DiSe, Thm 2.5]:

Theorem 7.3.4 *Let M be a divisible and semisimple MV-algebra. Then M is (up to isomorphism) a dense subalgebra of $C(\mathcal{M}(M))$. In addition, any element $f \in C(\mathcal{M}(M))$ is the supremum of all the elements of M majorized by f.*

Proposition 7.3.5 *Any weakly divisible and σ-complete MV-algebra contains all r-constant elements, $r \in [0,1]$.*

Proof. Let M be a weakly divisible and σ-complete MV-algebra. According to (vi) of Theorem 7.1.3, M is isomorphic to some Bold algebra $\mathcal{F}$ of fuzzy sets on some Ω, so let ϕ be an isomorphism from M onto $\mathcal{F}$. Since $\frac{1}{n} \in M$ for any $n \geq 1$, we have that $(1/n)_\Omega \in \mathcal{F}$ for any $n \geq 1$, so that, for any rational number $m/n \in [0,1]$, we have $(m/n)_\Omega \in \mathcal{F}$. The σ-completeness of M entails that all constant functions on Ω, r_Ω, where $r \in [0,1]$, belong to $\mathcal{F}$. Consequently $\phi^{-1}(r_\Omega) \in M$ for any $r \in [0,1]$, and $\phi^{-1}(r_\Omega)$ is an r-constant element in M. □

Proposition 7.3.6 *Let m be a state on an MV-algebra $M = \Gamma(G,u)$ and let $\hat{m}$ be its unique extension to G as a state on it. Then either $\hat{m}(G) \cap [0,1]$ is dense in $[0,1]$ or there is an integer $n \geq 1$ such that $\hat{m}(G) \cap [0,1] = \{0, \frac{1}{n}, \ldots, \frac{n}{n}\}$.*

Proof. Suppose that $M = \Gamma(G,u)$ and extend m to a state $\hat{m}$ on G. Then $\hat{m}(G)$ is an additive subgroup of $\mathbb{R}$. It is well-known (see e.g., [Goo 1, Lem 4.21]) that any additive subgroup of $\mathbb{R}$ is either a dense subset of $\mathbb{R}$ or is a cyclic group. In the first case we have that $\hat{m}(G) \cap [0,1]$ is a dense subset of $[0,1]$. In the second one $\hat{m}(G) \cap [0,1] = \{0, \frac{1}{n}, \ldots, \frac{n}{n}\}$. □

We note that a state m on M is said to be **discrete** if there exists an integer $n \geq 1$ such that $m(M) \subseteq \{0, \frac{1}{n}, \ldots, \frac{n}{n}\}$. A state s on (G,u) is said to be **discrete** if $s(G) = \frac{1}{n}\mathbb{Z}$ for some integer $n \geq 1$.

We note that if m is a discrete state and n is the least integer such that $m(M) \subseteq \{0, \frac{1}{n}, \ldots, \frac{n}{n}\}$, then this does not mean that $m(M) = \{0, \frac{1}{n}, \ldots, \frac{n}{n}\}$. Indeed, let $M = \{0, a, a^*, 1\}$ be the Boolean algebra 2^2, and define a state m such that $m(a) = 0.3$ and $m(a^*) = 0.7$. $M \cong \Gamma(\mathbb{Z}^2, (1,1))$, and if $\hat{m}$ is a unique extension of m to $\mathbb{Z}^2$, then $\hat{m}(\mathbb{Z}^2) = \frac{1}{10}\mathbb{Z}$ while $m(M) \subset \{0, \frac{1}{10}, \ldots, \frac{10}{10}\}$. In addition, a state m on this MV-algebra is discrete iff $m(a)$ is a rational number as it follows from the following result.

Proposition 7.3.7 *A state m on M is discrete if and only if m is a rational convex combination* [6] *of discrete extremal states.*

Proof. Suppose that $M = \Gamma(G,u)$. Then any discrete state m on M can be uniquely extended to a discrete state $\hat{m}$ on G. According to [Goo 1, Prop 6.22], $\hat{m}$ is a discrete state iff $\hat{m}$ is a rational combination of discrete extremal states on G. Therefore, m is a discrete state on M iff it is a rational convex combination of discrete extremal states on M. □

Proposition 7.3.8 *Let M be a weakly divisible MV-algebra. Then M admits no discrete state.*

---

[6] That is, $m = \sum_{i=1}^{k} \alpha_i m_i$, where all α_i's are nonnegative rational numbers such that $\sum_{i=1}^{k} \alpha_i = 1$.

Proof. Since for any integer $n \geq 1$, the element $\frac{1}{n}$ is defined in M, for any state m we have $m(\frac{1}{n}) = 1/n$; therefore, m cannot be a discrete state. □

Theorem 7.3.9 *Let M be a weakly divisible σ-complete MV-algebra. Then there exists a compact Hausdorff topological space X such that M is isomorphic as MV-algebras with $C(X)$.*

Proof. Denote by X the set of all extremal states (= state-morphisms) on M. If $M = \Gamma(G,u)$, then X is homeomorphic to the set of all extremal states on (G,u).

Let B be the set of all continuous functions on X such that $f(s) \in s(M)$ for any discrete state s on M. Then B is an ℓ-group with the strong unit 1_X and is σ-complete, and by [Goo 1, Cor 9.14], see also Theorem 7.1.21(1), (G,u) is isomorphic with $(B,1_X)$; the isomorphism $\psi : G \to B$ in question is given by $\psi(g) : s \mapsto \hat{s}(g)$, $g \in G$, where $\hat{s}$ is a unique extension of s onto G. It is clear that $\psi(u) = 1_X$.

Since by Proposition 7.3.8, M has no discrete states, we conclude that $M = \Gamma(G,u)$ is isomorphic as MV-algebras with $C(X) = \Gamma(B,1_X)$. □

Theorem 7.3.10 *Any weakly divisible and σ-complete MV-algebra $(M;\oplus,^*,0,1)$ is isomorphic as MV-algebras to $C(\mathcal{M}(M))$.*

Proof. Since the spaces $Ext(\mathcal{S}(M))$ and $\mathcal{M}(M)$ are homeomorphic, the assertion follows from the proof of Theorem 7.3.9. □

Theorem 7.3.11 *Any weakly divisible σ-complete MV-algebra is divisible.*

Proof. According to Theorem 7.3.10, $M \cong C(\mathcal{M}(M))$ as MV-algebras. Consequently, given $f \in C(\mathcal{M}(M))$, $\frac{f}{n}$ is defined in $C(\mathcal{M}(M))$ for any integer $n \geq 1$, whence $\frac{a}{n}$ is defined in M for any $a \in M$ and any integer $n \geq 1$. □

As another corollary we obtain the following result, which is a strengthening of [DiSe, Thm 3.2], where it was argued that any divisible σ-complete MV-algebra can be embedded into $C(\mathcal{M}(M))$ as a dense subalgebra, and if M is complete, then it is isomorphic to $C(\mathcal{M}(M))$.

Theorem 7.3.12 *Any divisible σ-complete MV-algebra M is isomorphic as MV-algebras with $C(\mathcal{M}(M))$.*

Proof. Any divisible MV-algebra is weakly divisible, so we can apply Theorem 7.3.10. □

Theorem 7.3.13 *An MV-algebra M is weakly divisible and σ-complete if and only if $\mathcal{M}(M)$ is basically disconnected and $M \cong C(\mathcal{M}(M))$.*

Proof. If M is weakly divisible and σ-complete, then by Theorem 7.3.10, M is isomorphic with $C(\mathcal{M}(M))$. According to [Goo 1, Cor 9.3], or Exercise 9.7.4.3, $C(\mathcal{M}(M))$ is σ-complete iff $\mathcal{M}(M)$ is basically disconnected.

On the other hand, it is clear that $C(\mathcal{M}(M))$ is always weakly divisible. □

7.4 MV-observables

In the present Section, we introduce MV-observables as σ-homomorphisms from the Borel tribe generated by the Borel sets of $\mathbb{R}$ and constant functions from $[0,1]$ into an MV-algebra M. We show that it is possible to define such observables only for weakly divisible MV-algebras. For them we present a representation as well as a so-called calculus of MV-observables (see [Dvu 28]).

7.4.1 MV-observables

It is evident that given a family T of fuzzy subsets of Ω there exists a tribe generated by T. Let $\mathcal{B}(\mathbb{R})$ be the σ-algebra of Borel subsets of the real line $\mathbb{R}$. We denote by $\mathcal{T}_{\mathcal{B}}(\mathbb{R})$ the tribe generated by $\{\chi_E : E \in \mathcal{B}(\mathbb{R})\}$ and by all constant fuzzy sets taking values in the interval $[0,1]$. We called it the **Borel tribe**, and the elements of $\mathcal{T}_{\mathcal{B}}(\mathbb{R})$ **Borel fuzzy subsets** of $\mathbb{R}$. In the same manner we define $\mathcal{T}_{\mathcal{B}}(\mathbb{R}^n)$ and Borel fuzzy subsets of $\mathbb{R}^n$.

Proposition 7.4.1 *The tribe $\mathcal{T}_{\mathcal{B}}(\mathbb{R}^n)$ of Borel fuzzy subsets of $\mathbb{R}^n$ consists of all Borel measurable fuzzy sets on $\mathbb{R}^n$.*

Proof. It is evident that $\mathcal{B}(\mathbb{R}^n) \subseteq \mathcal{S}_0(\mathcal{T}_{\mathcal{B}}(\mathbb{R}^n))$, and according to (3) of Theorem 7.1.7, $\mathcal{T}_{\mathcal{B}}(\mathbb{R}^n)$ consists of all $\mathcal{S}_0(\mathcal{T}_{\mathcal{B}}(\mathbb{R}^n))$-measurable fuzzy sets on $\mathbb{R}$. Denote by $\mathcal{T}$ the set of all elements from $\mathcal{T}_{\mathcal{B}}(\mathbb{R}^n)$ which are Borel measurable. It is clear that the characteristic functions of all Borel sets and all constant functions from $[0,1]$ belong to $\mathcal{T}$, and in addition, $\mathcal{T}$ is a tribe, consequently $\mathcal{T} = \mathcal{T}_{\mathcal{B}}(\mathbb{R}^n)$. □

An **MV-observable** on a σ-complete MV-algebra M is any MV-σ-homomorphism x from $\mathcal{T}_{\mathcal{B}}(\mathbb{R})$ into M. It is clear that the **range**, $R(x)$, of an MV-observable is an MV-subalgebra of M. In addition, if a is a crisp element in $\mathcal{T}_{\mathcal{B}}(\mathbb{R})$, then $x(a)$ is a Boolean element in M, i.e., $x(a) \oplus x(a) = x(a)$.

Theorem 7.4.2 *Let $\mathcal{T}$ be a tribe of fuzzy sets on a crisp set $\Omega \neq \emptyset$ containing all constant fuzzy sets on Ω, and let $f: \Omega \to \mathbb{R}$ be an $\mathcal{S}_0(\mathcal{T})$-measurable function. Then the mapping x_f defined by*

$$x_f(a) := a \circ f, \ a \in \mathcal{T}_{\mathcal{B}}(\mathbb{R}), \tag{7.8}$$

is an MV-observable on the tribe $\mathcal{T}$. Conversely, let x be an MV-observable on the tribe $\mathcal{T}$. Then there exists a unique real-valued function f on Ω which is $\mathcal{S}_0(\mathcal{T})$-measurable such that $x = x_f$.

Proof. Let $E \in \mathcal{B}(\mathbb{R})$, by assumptions we have $f^{-1}(E) \in \mathcal{S}_0(\mathcal{T})$, which entails $x_f(\chi_E) = \chi_E \circ f = \chi_{f^{-1}(E)} \in \mathcal{T}$. Similarly, $c \circ f \in \mathcal{T}$ for any constant $c \in [0,1]$. Since the set of all Borel fuzzy sets a on $\mathbb{R}$ such that $a \circ f \in \mathcal{T}$ is a tribe containing the generators of $\mathcal{T}_{\mathcal{B}}(\mathbb{R})$, x_f is a well-defined MV-observable on $\mathcal{T}$.

Conversely, let x be an MV-observable on $\mathcal{T}$. Then x maps all crisp sets from $\mathcal{T}_{\mathcal{B}}(\mathbb{R})$ onto crisp sets from $\mathcal{T}$. Let $\mathbb{Q}$ be the set of all rational numbers in $\mathbb{R}$, and let $r_1, r_2, \ldots$ be an enumeration of $\mathbb{Q}$. We set $\chi_{A_i} = x(\chi_{(-\infty, r_i)})$ for any $i \geq 1$. We define a mapping $f: \Omega \to \mathbb{R}$ as follows

$$f(\omega) = \inf\{r_j : \omega \in A_j\}.$$

Then f is a well-defined function such that $x(\chi_{(-\infty,r_j)}) = \chi_{f^{-1}(A_j)}$, and f is $\mathcal{S}_0(\mathcal{T})$-measurable. In addition, $x(\chi_E) = \chi_{f^{-1}(E)}$ for any $E \in \mathcal{B}(\mathbb{R})$.

For any integer $n \geq 1$, we have $1_\Omega = x(1_\mathbb{R}) = x(n\frac{1}{n}) = x(\frac{1}{n} \oplus \cdots \oplus \frac{1}{n}) = n\,x(\frac{1}{n})$ which proves that $x(\frac{1}{n}) = \frac{1}{n}$. Consequently, $x(\frac{m}{n}) = \frac{m}{n}$ for any $0 \leq m \leq n$. Using the density of rational numbers, we have that $x(c) = c$ for any constant $c \in [0,1]$. Hence, the set of all $a \in \mathcal{T}_{\mathcal{B}}(\mathbb{R})$ such that $x(a) = x_f(a)$ is a tribe containing the generators of $\mathcal{T}_{\mathcal{B}}(\mathbb{R})$, consequently, $x(a) = x_f(a)$ for any $a \in \mathcal{T}_{\mathcal{B}}(\mathbb{R})$.

The uniqueness of f can be proved as follows. Let g be a real-valued function on Ω such that $x_f = x_g$. Then $\{\omega : f(\omega) \neq g(\omega)\} = \bigcup_{r\in\mathbb{Q}}(\{\omega : f(\omega) < r < g(\omega)\} \cup \{\omega : g(\omega) < r < f(\omega)\}) = \emptyset$. □

Theorem 7.4.3 *Let x be an MV-observable on a σ-complete MV-algebra M. Then the range $R(x)$ is a σ-complete MV-subalgebra of M with a countable generating set. Conversely, if M_1 is a σ-complete MV-subalgebra of M with a countable generator, then there exists an MV-observable x on M such that $M_1 = R(x)$.*

Proof. It is clear that $R(x)$ is a σ-complete MV-subalgebra of M with the countable generator $\{x(s\chi_{(-\infty,r)}) : r \in \mathbb{Q},\ s \in [0,1] \cap \mathbb{Q}\}$.

Suppose conversely that M_1 is a σ-complete MV-subalgebra of M. By the Loomis–Sikorski theorem 7.1.22 for σ-complete MV-algebras, there exists a tribe $\mathcal{T}$ of fuzzy sets on a non-void subset Ω and there exists an MV-σ-homomorphism h from $\mathcal{T}$ onto M_1. Let $\{a_n : n \geq 1\}$ be a generator of M_1, and let $\{b_n : n \geq 1\}$ be a set of such elements from $\mathcal{T}$ that $h(b_n) = a_n$ for any $n \geq 1$. We denote by $\mathcal{T}_0$ the smallest tribe containing all b_n. The function $f: \Omega \to [0,1]^{\aleph_0}$ defined by

$$f(\omega) := (b_1(\omega), b_2(\omega), \ldots)$$

is a Borel-measurable mapping from Ω into the compact metric space $Y = [0,1]^{\aleph_0}$. We note that $\mathcal{S}_0(\mathcal{T}_0)$ is equal to the σ-algebra generated by $\{b_i^{-1}(E) : E \in \mathcal{B}([0,1]),\ i \geq 1\}$. In addition, for any $E \in \mathcal{B}([0,1])$ we have $f^{-1}(\pi_i^{-1}(E)) = b_i^{-1}(E)$, where π_i is the i-th projection from Y onto $[0,1]$. Consequently, $\mathcal{S}_0(\mathcal{T}_0) = \{f^{-1}(E) : E \in \mathcal{B}(Y)\}$. Now, by a classical theorem of Kuratowski [Kur], there exists a Borel isomorphism

d of Y onto $\mathbb{R}$, so that the function $f_1(\omega) := d(f(\omega))$ is $\mathcal{S}_0(\mathcal{T}_0)$-measurable, and $\mathcal{S}_0(\mathcal{T}_0) = \{f_1^{-1}(E) : E \in \mathcal{B}(\mathbb{R})\}$. If we now define a mapping x by

$$x(a) := h(a \circ f_1), \; a \in \mathcal{T}_{\mathcal{B}}(\mathbb{R}),$$

then x is an MV-observable on M whose range is M_1. $\square$

We now present a finer version of the Loomis–Sikorski theorem for σ-complete MV-algebras.

Theorem 7.4.4 *Let M be a σ-complete MV-algebra. Then there exists a tribe $\mathcal{T}$ of fuzzy sets on a compact Hausdorff space $\Omega \neq \emptyset$ and an MV-σ-homomorphism h from $\mathcal{T}$ onto M such that h maps $\mathcal{S}_0(\mathcal{T})$ onto $B(M)$.*

Proof. We define $\Omega := Ext(\mathcal{S}(M))$ and let $\hat{M} = \{\hat{a} \in [0,1]^\Omega : a \in M\}$, where $\hat{a}(m) := m(a)$, $m \in Ext(\mathcal{S}(M))$. According to the proof of the Loomis–Sikorski theorem 7.1.22, if we define $\mathcal{T}$ as the tribe of fuzzy subsets of Ω generated by $\hat{M}$, a mapping $h : \mathcal{T} \to M$ can be defined by $h(f) = a$ whenever $f \in \mathcal{T}$, $a \in M$, and $\{m \in \Omega : f(m) \neq m(a)\}$ is a meager set, h is a well-defined MV-σ-homomorphism from $\mathcal{T}$ onto M (see the proof of Theorem 7.1.22). It is evident that an element a of a semisimple MV-algebra M is Boolean iff $\hat{a}$ is a two-valued function on Ω, so that $\hat{a} = \chi_A$ for some $A \in \mathcal{S}_0(\mathcal{T})$, and in addition, A is a clopen subset of Ω. We define by $\mathcal{S}_0$ the set of all $A \in \mathcal{S}_0(\mathcal{T})$ such that $\chi_A = \hat{a}$ for some $a \in M$.

According to [Goo 1, Thm 8.14], the space $Ext(\mathcal{S}(M))$ is homeomorphic with the set $Ext(\mathcal{S}(B(M)))$ of all extremal states on the Boolean σ-algebra $B(M)$. In addition, any restriction of $m \in Ext(\mathcal{S}(M))$ to $B(M)$ gives an element of $Ext(\mathcal{S}(B(M)))$, and conversely, any element of $Ext(\mathcal{S}(B(M)))$ can be uniquely extended to an extremal state on M, and this correspondence defines the mentioned homeomorphism.

Consequently, by the proof of the classical Loomis–Sikorski theorem, $\mathcal{S}_0$ is a σ-algebra of crisp subsets of Ω, and due to the definition of h, its restriction onto $\mathcal{S}_0$ defines a σ-homomorphism from $\mathcal{S}_0$ onto $B(M)$.

By definition $\mathcal{S}_0 \subseteq \mathcal{S}_0(\mathcal{T})$. On the other hand, $A \in \mathcal{S}_0(\mathcal{T})$ iff $\chi_A \in \mathcal{T}$, i.e., $h(\chi_A) = a$ for some $a \in M$. Since χ_A is a Boolean element of $\mathcal{T}$, so is a in M. Consequently, $\hat{a} = \chi_A$ and $A \in \mathcal{S}_0$. $\square$

Let M be a σ-complete MV-algebra. The triplet $(\Omega, \mathcal{T}, h)$, where $\Omega = Ext(\mathcal{S}(M))$, $\mathcal{T}$ is the tribe, and h is the MV-σ-homomorphism from $\mathcal{T}$ onto M described in the proof of Theorem 7.4.4, is said to be a **canonical representation** of M.

Another version of the Loomis–Sikorski theorem is the following statement.

Theorem 7.4.5 *Let $\mathcal{T}$ be a tribe of fuzzy sets on $\Omega \neq \emptyset$ containing all constant fuzzy sets on Ω. Let h be an MV-σ-homomorphism from $\mathcal{T}$ onto a σ-complete MV-algebra M. Then h maps $\mathcal{S}_0(\mathcal{T})$ onto $B(M)$.*

Proof. Let $a \in B(M)$. Then there exists an element $g \in \mathcal{T}$ with $h(g) = a$. According to (3) of Theorem 7.1.7, there are two cases.

Case 1. Let $g = \sum_{i=1}^{k} \alpha_i \chi_{A_i}$, where $\alpha_i \in [0,1]$, $A_i \in \mathcal{S}_0(\mathcal{T})$ and $A_i \cap A_j = \emptyset$ for $i \neq j$, $i,j = 1,\dots,k$. There exists an integer s such that $s\,\alpha_j \geq 1$ for $j = 1,\dots,k$. Then $s \odot g = \chi_A$, where $A = \bigcup_{j=1}^{k} A_j$. Hence, $h(\chi_A) = h(s \odot g) = s \odot a = a$.

Case 2. There exists a non-decreasing sequence of simple functions $\{g_n\}_n$ of elements of $\mathcal{T}$ such that $\lim_n g_n = g$. Let $g_n = \sum_{i=1}^{k_n} \alpha_i^n \chi_{A_i^n}$, where $\alpha_i^n \in [0,1]$ and $A_i^n \cap A_j^n = \emptyset$ for $i \neq j$. We define $A_n = \bigcup_{i=1}^{k_n} A_i^n$ and $A = \bigcup_{n=1}^{\infty} A_n$. As in Case 1, there exists an integer $s \geq 1$ such that $s\,\alpha_i^1 \geq 1$ for $i = 1,\dots,k_1$. Hence, $s\alpha_i^n \geq 1$ for any $i = 1,\dots,k_n$ and any $n \geq 1$. Therefore, $s \odot g_n \leq s \odot g_{n+1} \leq s \odot g$, i.e., $\chi_{A_n} \leq \chi_{A_{n+1}} \leq s \odot g$, which entails $\chi_{A_n} \nearrow s \odot g$ and $h(\chi_A) = h(\lim_n \chi_{A_n}) = h(\lim_n s \odot g_n) = h(s \odot g) = s \odot a = a$. $\square$

Let $c \in [0,1]$. An element a in a semisimple MV-algebra M is said to be a *c*-**constant** if $m(a) = c$ for any state $m \in M$. If M is a tribe, then the *c*-constant is equal to $c1_\Omega$.

We note that $\mathcal{T}_\mathcal{B}(\mathbb{R}^n)$ is weakly divisible. In addition, if M is a σ-complete, weakly divisible MV-algebra, then any *c*-constant, where $c \in [0,1]$, belongs to M.

In the proof of Theorem 7.3.10, it was shown that a weakly divisible σ-complete MV-algebra M is isomorphic as MV-algebras to the space $C(Ext(\mathcal{S}(M)))$ of all continuous functions on the basically disconnected space $Ext(\mathcal{S}(M))$ with values in the interval $[0,1]$. In this case, $C(Ext(\mathcal{S}(M)))$ is the Bold algebra, and the tribe generated by $C(Ext(\mathcal{S}(M)))$ is equal to the set of all Baire measurable fuzzy sets on $Ext(\mathcal{S}(M))$, Proposition 7.1.11.

Proposition 7.4.6 *Let h be an MV-σ-homomorphism from a tribe $\mathcal{T}$ of fuzzy sets on $\Omega \neq \emptyset$ containing all constant fuzzy sets into a σ-complete MV-algebra M. Then M is weakly divisible, and $h(c1_\Omega)$ is a c-constant for any $c \in [0,1]$.*

Proof. Let $n \geq 1$ be given. Then $1 = h(1_\Omega) = h(n\frac{1}{n}1_\Omega)$, which proves that M is weakly divisible. Hence, $h(\frac{m}{n}1_\Omega)$ is an $\frac{m}{n}$-constant in M. Consequently, $h(c1_\Omega)$ is a *c*-constant for any $c \in [0,1]$. $\square$

The following statement is an easy consequence of the above proposition.

Corollary 7.4.7 *Let x be an MV-observable of a σ-complete MV-algebra M. Then M is weakly divisible, and $x(c)$ is a c-constant for any $c \in [0,1]$.*

For example, let $L_n := \{0, 1/n, 2/n, \dots, n/n\}$ be the so-called basic MV-algebra. If $n \geq 2$, then on L_n there exists no MV-observable. Similarly, since any finite MV-algebra M is a direct product of finitely many basic MV-algebras, [Cig] (see Corollary 6.2.15), if M is a finite non Boolean MV-algebra, then it does not admit any MV-observable.

Theorem 7.4.8 *Let $\mathcal{T}$ be a tribe of fuzzy sets on a non-void set Ω containing all constant fuzzy sets and let h be an MV-σ-homomorphism from $\mathcal{T}$ onto a σ-complete MV-algebra M. Then, for any MV-observable x on M, there exists an $\mathcal{S}_0(\mathcal{T})$-measurable real-valued function f defined on Ω such that*

$$x(a) = h(a \circ f), \ a \in \mathcal{T}_{\mathcal{B}}(\mathbb{R}). \tag{7.9}$$

If g is any $\mathcal{S}_0(\mathcal{T})$-measurable real-valued function defined on Ω such that $x(a) = h(a \circ g)$, $a \in \mathcal{T}_{\mathcal{B}}(\mathbb{R})$, then $h(\chi_E) = 0$, where $E = \{\omega : \ f(\omega) \neq g(\omega)\}$.

Conversely, if f is a real-valued $\mathcal{S}_0(\mathcal{T})$-measurable mapping on Ω, then the mapping x_f defined by

$$x_f(a) := h(a \circ f), \ a \in \mathcal{T}_{\mathcal{B}}(\mathbb{R}), \tag{7.10}$$

defines an MV-observable on M.

Proof. *Step 1.* Suppose that $A, B \in \mathcal{S}_0(\mathcal{T})$, $A \subseteq B$, and $c \in B(M)$ are such that $h(\chi_A) \leq c \leq h(\chi_B)$. Then there exists an element $C \in \mathcal{S}_0(\mathcal{T})$ such that $A \subseteq C \subseteq B$, and $h(\chi_C) = c$. In fact, since h maps $\mathcal{S}_0(\mathcal{T})$ onto $B(M)$, there exists $C_1 \in \mathcal{T}$ such that $h(C_1) = c$. If we define $C = (C_1 \cap B) \cup A$, then $A \subseteq C \subseteq B$ and $h(C) = c$.

Step 2. Similarly as in the proof of Theorem 7.4.2, let $r_1, r_2, \ldots$ be an enumeration of the rational numbers of $\mathbb{R}$. It is clear that $x(\chi_{(-\infty,r_i)}) \leq x(\chi_{(-\infty,r_j)})$ whenever $r_i < r_j$, and any $x(\chi_{(-\infty,r_i)})$ is a Boolean element of M. We shall now construct crisp sets $A_1, A_2, \ldots$ from $\mathcal{S}_0(\mathcal{T})$ such that (a) $h(\chi_{A_i}) = x(\chi_{(-\infty,r_i)})$ for any i; (b) $A_i \subseteq A_j$ whenever $r_i < r_j$. According to Theorem 7.4.5, we can find a set A_1 in $\mathcal{S}_0(\mathcal{T})$ such that $h(A_1) = x(\chi_{(-\infty,r_1)})$. Suppose that $A_1, \ldots, A_n$ in $\mathcal{S}_0(\mathcal{T})$ have been constructed such that (i) $h(\chi_{A_i}) = x(\chi_{(-\infty,r_i)})$ for $i = 1, \ldots, n$; (ii) $A_i \subseteq A_j$ whenever $r_i < r_j$, $1 \leq i, j, \leq n$. We shall construct A_{n+1} as follows. Let $(i_1, \ldots, i_n)$ be the permutation of $(1, \ldots, n)$ such that $r_{i_1} < r_{i_2} < \cdots < r_{i_n}$. Then, there exists a unique k such that $r_{i_k} < r_{n+1} < r_{i_{k+1}}$ (we define $r_{i_0} = -\infty$ and $r_{i_{n+1}} = \infty$), and by Step 1, we can select $A_{n+1} \in \mathcal{S}_0(\mathcal{T})$ such that $A_{i_k} \subseteq A_{n+1} \subseteq A_{i_{k+1}}$ (we define $A_{i_0} = \emptyset$ and $A_{i_{n+1}} = \Omega$). By induction, we have proved that there exists a sequence $A_1, A_2, \ldots$ in $\mathcal{S}_0(\mathcal{T})$ with the properties (a) and (b).

As

$$h(\chi_{\bigcap_j A_j}) = \bigwedge_j h(\chi_{A_j}) = \bigwedge_j x(\chi_{(-\infty,r_j)}) = 0,$$

we may, by replacing every A_k by $A_k \setminus \bigcap_j A_j$ if necessary, assume that $\bigcap_j A_j = \emptyset$. In addition, we have $h(\chi_{\bigcup_j A_j}) = \bigvee_j x(\chi_{(-\infty,r_i)}) = 1$.

We now define a function f on Ω as follows

$$f(\omega) = \begin{cases} \inf\{r_j : \ \omega \in A_j\}, & \text{if } \omega \in \bigcup_j A_j, \\ 0, & \text{otherwise.} \end{cases}$$

Then f is a real-valued function on Ω, and, for any k

$$f^{-1}((-\infty, r_k)) \cap \bigcup_j A_j = \bigcup_{j:\ r_j<r_k} A_j,$$

so that f is $\mathcal{S}_0(\mathcal{T})$-measurable. Further,

$$f(\chi_{f^{-1}((-\infty,r_k))}) = h(\chi_{\bigcup_{j:r_j<r_k} A_j}) = \bigvee_{j:r_j<r_k} x(\chi_{(-\infty,r_j)}) = x(\chi_{(-\infty,r_k)}).$$

Therefore, $h(\chi_E \circ f) = x(\chi_E)$ whenever $E = (-\infty, r_k)$ for some k. Similarly, by Proposition 7.4.6, $h(c \circ f) = h(c1_\Omega) = x(c)$ for any $c \in [0,1]$. Hence, the system of all $a \in \mathcal{T}_\mathcal{B}(\mathbb{R})$ such that $x(a) = h(a \circ f)$ is a tribe containing the generator of $\mathcal{T}_\mathcal{B}(\mathbb{R})$, consequently, it coincides with $\mathcal{T}_\mathcal{B}(\mathbb{R})$.

The uniqueness of a function f in the given sense is proved in a similar way as in the proof of Theorem 7.4.2.

The second statement is now evident. □

It is worthy of recalling that according to Theorem 7.4.8, a σ-complete MV-algebra M admits an MV-observable iff M is weakly divisible:

Theorem 7.4.9 *A σ-complete MV-algebra M admits an MV-observable if and only if M is weakly divisible.*

Proof. Let x be an MV-observable of M. Due to Corollary 7.4.7, M is weakly divisible.

Suppose now M is a weakly divisible σ-complete MV-algebra. By Theorem 7.3.10, M is isomorphic to the system $C(\Omega)$ of all continuous fuzzy sets on $\Omega := Ext(\mathcal{S}(M))$. The tribe $\mathcal{T}(M)$ generated by $C(\Omega)$ is equal to the system of all Baire measurable functions on Ω, Proposition 7.1.11. In addition, $\mathcal{T}(M)$ contains all constant functions taking values in the interval $[0,1]$. Take now a Baire functions $f : \Omega \to \mathbb{R}$, and define x_f via (7.10). According to Theorem 7.4.8, x_f is an MV-observable on M, where h is defined in the same way as that in the proof of Theorem 7.4.4. □

7.4.2 Calculus of MV-observables

Let f be a Borel measurable function from $\mathbb{R}$ into $\mathbb{R}$. Then $f \circ x$, where x is an MV-observable of M and

$$(f \circ x)(a) := x(a \circ f),\ a \in \mathcal{T}_\mathcal{B}(\mathbb{R}),$$

is an MV-observable of M. In this manner we can define x^2, x^3 or e^x, etc.

Theorem 7.4.10 *Let x and y be two MV-observables of a σ-complete MV-algebra M. Then $R(x) \subseteq R(y)$ if and only if there exists a Borel measurable function $f: \mathbb{R} \to \mathbb{R}$ such that $x = f \circ y$.*

Proof. One direction is evident. Suppose now that $R(x) \subseteq R(y)$. We define $\mathcal{T} = \mathcal{T}_{\mathcal{B}}(\mathbb{R})$, $h = y$, $M = R(y)$ and apply Theorem 7.4.8. □

Theorem 7.4.11 *Let $\{x_n\}_n$ be a sequence of MV-observables on a σ-complete MV-algebra M. Then there exists an MV-observable x and a sequence of Borel measurable functions $\{f_n\}_n$ from $\mathbb{R}$ into $\mathbb{R}$ such that $x_n = f_n \circ x$ for any n.*

Proof. Let M_0 be the MV-subalgebra of M generated by $\bigcup_n R(x_n)$. According to Theorem 7.4.3, there exists an MV-observable x such that $R(x) = M_0$. Then we apply Theorem 7.4.10 and we obtain the statement in question. □

The following result can be of a particular interest for many valued analysis and for a so-called calculus of MV-observables.

Theorem 7.4.12 *Let $x_1, \ldots, x_n$ be MV-observables of a σ-complete MV-algebra M. Then there exists a unique MV-σ-homomorphism x from $\mathcal{T}_{\mathcal{B}}(\mathbb{R}^n)$ into M such that, for any Borel fuzzy set $a \in \mathcal{T}_{\mathcal{B}}(\mathbb{R})$, $x_i(a) = x(a \circ \pi_i)$, where π_i is the i-th projection of $\mathbb{R}^n$ onto $\mathbb{R}$, $i = 1, \ldots, n$.*

Proof. Let M_0 be the MV-subalgebra of M generated by all ranges $R(x_1), \ldots, R(x_n)$. Since M_0 has a countable generator, according to Theorem 7.4.3, there exists an MV-observable y of M such that $R(y) = M_0$. Consequently, by Theorem 7.4.8, there exist Borel measurable functions $f_1, \ldots, f_n$ from $\mathbb{R}$ into $\mathbb{R}$ such that $x_i(a) = (f_i \circ y)(a)$, $a \in \mathcal{T}_{\mathcal{B}}(\mathbb{R})$. Define the Borel measurable mapping $f: \mathbb{R} \to \mathbb{R}^n$ by $f(t) := (f_1(t), \ldots, f_n(t))$, $t \in \mathbb{R}$, and set $x: \mathcal{T}_{\mathcal{B}}(\mathbb{R}^n) \to M$ via

$$x(b) := y(b \circ f), \; b \in \mathcal{T}_{\mathcal{B}}(\mathbb{R}^n). \tag{7.11}$$

Then x is an MV-σ-homomorphism from $\mathcal{T}_{\mathcal{B}}(\mathbb{R}^n)$ into M. Moreover, let $a \in \mathcal{T}_{\mathcal{B}}(\mathbb{R})$, then

$$x(a \circ \pi_i) = y(a \circ \pi_i \circ f) = y(a \circ f_i) = x_i(a).$$

To prove the uniqueness, let z be any MV-σ-homomorphism from $\mathcal{T}_{\mathcal{B}}(\mathbb{R})$ into M such that $z(a \circ \pi_i) = x_i(a)$, $a \in \mathcal{T}_{\mathcal{B}}(\mathbb{R})$, $i = 1, \ldots, n$. Then, for $E_i \in \mathcal{B}(\mathbb{R})$ $(i = 1, \ldots, n)$, we have

$$\begin{aligned} z(\chi_{E_1 \times \cdots \times E_n}) &= z(\min_{1 \le i \le n} (\chi_{E_i} \circ \pi_i)) = \bigwedge_{i=1}^{n} z(\chi_{E_i} \circ \pi_i) = \bigwedge_{i=1}^{n} x(\chi_{E_i} \circ \pi_i) \\ &= x(\min_{1 \le i \le n} (\chi_{E_i} \circ \pi_i)) = x(\chi_{E_1 \times \cdots \times E_n}). \end{aligned}$$

On the other hand, let $c \in [0,1]$. Then $x(c\chi_{\mathbb{R}^n}) = x(c\chi_{\mathbb{R}} \circ \pi_i) = x_i(c\mathbb{R}) = z(c\chi_{\mathbb{R}} \circ \pi_i) = z(c\chi_{\mathbb{R}^n})$.

Hence the set of all Borel fuzzy sets b from $\mathcal{T}_{\mathcal{B}}(\mathbb{R}^n)$ such that $x(b) = z(b)$ is a tribe containing the generator of $\mathcal{T}_{\mathcal{B}}(\mathbb{R}^n)$, so that $x = z$. □

The unique MV-σ-homomorphism x from Theorem 7.4.12 is said to be a **joint MV-observable** of $x_1, \ldots, x_n$.

Theorem 7.4.13 *Let x be a joint MV-observable of the MV-observables $x_1, \ldots, x_n$ of a σ-complete MV-algebra M. Let y be an MV-observable of M and $f_1, \ldots, f_n$ let be Borel measurable functions from $\mathbb{R}$ into $\mathbb{R}$ such that $x_i = f_i \circ y$ for $i = 1, \ldots, n$. Let $\phi : \mathbb{R}^n \to \mathbb{R}$ be a Borel function. Define $\phi \circ x : \mathcal{T}_{\mathcal{B}}(\mathbb{R}) \to M$ and $\phi(f_1, \ldots, f_n) \circ y : \mathcal{T}_{\mathcal{B}}(\mathbb{R}) \to M$, where $\phi(f_1, \ldots, f_n)(t) := \phi(f_1(t), \ldots, f_n(t))$, $t \in \mathbb{R}$, by*

$$(\phi \circ x)(a) := x(a \circ \phi), \ a \in \mathcal{T}_{\mathcal{B}}(\mathbb{R}), \tag{7.12}$$

$$(\phi(f_1, \ldots, f_n) \circ y)(a) := y(a \circ \phi(f_1, \ldots, f_n)), \ a \in \mathcal{T}_{\mathcal{B}}(\mathbb{R}). \tag{7.13}$$

Then $\phi \circ x$ and $\phi(f_1, \ldots, f_n) \circ y$ are identical MV-observables of M.

Proof. According to the proof of Theorem 7.4.8 and Theorems 7.4.10–7.4.11, it is evident that the construction of the joint MV-observable x of $x_1, \ldots, x_n$ does not depend on the used MV-observable y and Borel functions $f_1, \ldots, f_n$. It is clear that $\phi \circ x$ and $\phi(f_1, \ldots, f_n) \circ y$ are MV-observables of M. Consequently, by (7.11), for any $a \in \mathcal{T}_{\mathcal{B}}(\mathbb{R})$, we have

$$\begin{aligned}(\phi \circ x)(a) &= x(a \circ \phi) = y(a \circ \phi \circ f)\\ &= y(a \circ \phi(f_1, \ldots, f_n))\\ &= (\phi(f_1, \ldots, f_n) \circ y)(a),\end{aligned}$$

where $f(t) = (f_1(t), \ldots, f_n(t))$, $t \in \mathbb{R}$. □

The MV-observable $\phi \circ x$ from Theorem 7.4.13 we will denote by $\phi \circ (x_1, \ldots, x_n)$, i.e., as a function of $x_1, \ldots, x_n$. We have seen that it can be defined equivalently by (7.12) or by (7.13).

For example, if $\phi(t_1, t_2) = t_1 + t_2$, we can define addition of two MV-observables x_1 and x_2 by

$$x_1 + x_2 := \phi \circ (x_1, x_2). \tag{7.14}$$

In a similar way we can define product, difference, etc., of MV-observables.

Theorem 7.4.14 *Let $x_1, \ldots, x_n$ be MV-observables of a σ-complete MV-algebra M, $g_1, \ldots, g_k$ be real-valued Borel functions on $\mathbb{R}^n$, and h be a real-valued functions on $\mathbb{R}^k$. Then $y_i = g_i \circ (x_1, \ldots, x_n)$, $i = 1, \ldots, k$, are MV-observables of M, and*

$$h \circ (y_1, \ldots, y_k) = (h(g_1, \ldots, g_k)) \circ (x_1, \ldots, x_n), \tag{7.15}$$

where $h(g_1, \ldots, g_k)$ is the function from $\mathbb{R}^n$ into $\mathbb{R}$ defined by the correspondence $(t_1, \ldots, t_n) \mapsto h(g_1(t_1, \ldots, t_n), \ldots, g_k(t_1, \ldots, t_n))$.

Proof. Suppose $y_i = g_i \circ (x_1, \ldots, x_n)$ and let g be the mapping from $\mathbb{R}^n$ into $\mathbb{R}^k$ defined by $g(t_1, \ldots, t_n) = (g_1(t_1, \ldots, t_n), \ldots, g_k(t_1, \ldots, t_n))$. Then the mapping $y : \mathcal{T}_{\mathcal{B}}(\mathbb{R}^k) \to M$ given by

$$y(b) := x(b \circ g), \; b \in \mathcal{T}_{\mathcal{B}}(\mathbb{R}^k),$$

is an MV-σ-homomorphism such that, for $a \in \mathcal{T}_{\mathcal{B}}(\mathbb{R})$,

$$y(a \circ \pi_i^k) = x(a \circ \pi_i^k \circ g) = x(a \circ g_i) = y_i(a),$$

where π_i^k is the i-th projection from $\mathbb{R}^k$ on $\mathbb{R}$. Therefore, y is the unique joint MV-observable of $y_1, \ldots, y_k$. Consequently, for $h \circ (y_1, \ldots, y_k)$, we have

$$(h \circ (y_1, \ldots, y_k))(a) = y(a \circ h) = x(a \circ h \circ g),$$

which proves the theorem. □

If we apply Theorems 7.4.12–7.4.13, we see that, for example, the sum $x_1 + \cdots + x_n$ of MV-observables $x_1, \ldots, x_n$ is well-defined, and, in addition, $x_1 + \cdots + x_n = x_{i_1} + \cdots + x_{i_n}$, where $(i_1, \ldots, i_n)$ is any permutation of $(1, \ldots, n)$.

We now present a representation theorem for MV-observables. According to Theorem 7.4.9, we can limit ourselves to weakly divisible σ-complete MV-algebras.

Theorem 7.4.15 *Let M be a weakly divisible σ-complete MV-algebra with the canonical representation $(\Omega, \mathcal{T}, h)$. If f is a Baire measurable real-valued function on Ω, then*

$$x_f(a) := h(a \circ f), \; a \in \mathcal{T}_{\mathcal{B}}(\mathbb{R}),$$

defines an MV-observable of M. Conversely, given an MV-observable x on M, there exists a real-valued Baire measurable function f on Ω such that $x = x_f$. If g is another real-valued Baire measurable function on Ω with $x_f = x_g$, then $\{\omega : \; f(\omega) \neq g(\omega)\}$ is a meager set.

Proof. The statement follows directly from Theorems 7.4.8–7.4.9. □

Theorem 7.4.16 *Let the conditions of Theorem 7.4.15 be satisfied. Let $x_1, \ldots, x_n$ be MV-observables with the joint MV-observable x. Then there exist real-valued Baire measurable functions $f_1, \ldots, f_n$ on Ω such that $x_i = x_{f_i}$, $i = 1, \ldots, n$, and $x = x_f$, where $f(\omega) = (f_1(\omega), \ldots, f_n(\omega))$, and*

$$x_f(b) = h(b \circ f), \; b \in \mathcal{T}_{\mathcal{B}}(\mathbb{R}^n). \tag{7.16}$$

If $g : \Omega \to \mathbb{R}^n$ is Baire measurable such that $x_f = x_g$, then $\{\omega : \; f(\omega) \neq g(\omega)\}$ is a meager set.

Proof. According to Theorem 7.4.15, there exist Baire measurable functions $f_i : \Omega \to \mathbb{R}$, $i = 1, \ldots, n$, such that $x_i = x_{f_i}$, $(i = 1, \ldots, n)$. Due to Theorem 7.4.12, the joint MV-observable $x : \mathcal{T}_\mathcal{B}(\mathbb{R}^n) \to M$ is uniquely determined by $x(a \circ \pi_i) = x_i(a)$, $a \in \mathcal{T}_\mathcal{B}(\mathbb{R})$. Define x_f by (7.16). Then x_f is an MV-σ-homomorphism from $\mathcal{T}_\mathcal{B}(\mathbb{R}^n)$ into M with $x_f(a \circ \pi_i) = h(a \circ \pi_i \circ f) = h(a \circ f_i) = x_{f_i}(a) = x_i(a)$ for $a \in \mathcal{T}_\mathcal{B}(\mathbb{R})$. The uniqueness of a joint MV-observable entails $x = x_f$.

If $x = x_g$ for some Baire measurable function $g : \Omega \to \mathbb{R}^n$, then $\{\omega : f(\omega) \neq g(\omega)\} = \bigcup_{i=1}^n \{\omega : f_i(\omega) \neq g_i(\omega)\}$, where $g = (g_1, \ldots, g_n)$ with $g_i : \Omega \to \mathbb{R}$, so that the set under question is meager. □

Let x be an MV-observable of M. The **spectrum** of x is the set $\sigma(x) := \bigcap\{C : C$ is a closed crisp subset of $\mathbb{R}$ with $x(\chi_C) = 1\}$. Since the natural topology of $\mathbb{R}$ satisfies the second countability axiom, there exists a sequence of closed sets $C_1, C_2, \ldots$ such that $x(\chi_{C_n}) = 1$ for all n, and $\sigma(x) = \bigcap_n C_n$. x is said to be (i) **bounded** if $\sigma(x)$ is a bounded set of $\mathbb{R}$; (ii) **discrete** if $\sigma(x) = \{c_1, c_2, \ldots\}$; (iii) **simple** if $\sigma(x) = \{c_1, \ldots, c_n\}$; (iv) **question** if $\sigma(x) \subseteq \{0, 1\}$ and (v) **constant** if $\sigma(x) = \{c\}$.

Proposition 7.4.17 *Let the conditions of Theorem* 7.4.15 *be satisfied. Let $f : \Omega \to \mathbb{R}$ be a Baire measurable function. The MV-observable x_f is*

(i) *bounded if and only if f is a bounded function up to a meager set;*

(ii) *discrete if and only if $f = \sum_i c_i \chi_{A_i}$, $A_i \cap A_j = \emptyset$ for $i \neq j$, $A_i \in \mathcal{S}_0(\mathcal{T})$;*

(iii) *simple if and only if $f = \sum_{i=1}^n c_i \chi_{A_i}$, $A_i \cap A_j = \emptyset$ for $i \neq j$, $A_i \in \mathcal{S}_0(\mathcal{T})$;*

(iv) *question if and only if $f = \chi_A$ for some $A \in \mathcal{S}_0(\mathcal{T})$;*

(v) *constant if and only if $f = c\chi_\Omega$.*

(All statements (ii)–(v) *are valid up to a meager set.)*

Proof. (i) Let x be bounded. Then there exists a constant c such that $\sigma(x) \subseteq [-c, c]$. Hence

$$1 = x_f(\chi_{[-c,c]}) = h(\chi_{[-c,c]} \circ f) = h(\chi_{f^{-1}([-c,c])}),$$

which proves that $\{\omega : f(\omega) \notin [-c, c]\}$ is a meager set. The converse is clear.

From the rest we prove only (ii). Put $A_i = f^{-1}(\{c_i\})$, and calculate, $x_f(\chi_{\{c_i\}}) = h(\chi_{\{c_i\}} \circ f) = h(\chi_{f^{-1}(\{c_i\})}) = h(\chi_{A_i})$. Therefore, $f = \sum_i c_i \chi_{A_i}$. □

Let m be a σ-additive state on a σ-complete MV-algebra M. If x is an MV-observable on M, then the composition m_x of m with x, $m_x := m \circ x$, is a σ-additive state on $\mathcal{T}_\mathcal{B}(\mathbb{R})$. In addition, the restriction $\widehat{m_x}$ of m_x onto $\mathcal{B}(\mathbb{R})$, defined via $\widehat{m_x}(E) := m_x(\chi_E)$, $E \in \mathcal{B}(\mathbb{R})$, is a usual probability measure on the Borel σ-algebra $\mathcal{B}(\mathbb{R})$.

In a similar way, if h is an MV-σ-homomorphism of a tribe $\mathcal{T}$ onto M, then $m_h := m \circ h$ is a σ-additive state, and $\widehat{m_h} : \mathcal{S}_0(\mathcal{T}) \to [0,1]$ defined by $\widehat{m_h}(A) := m_h(\chi_A)$, $A \in \mathcal{S}_0(\mathcal{T})$, is a usual probability measure on $\mathcal{S}_0(\mathcal{T})$

Let x be an MV-observable; the **mean value** of x in a σ-additive state m we understand the expression

$$m(x) := \int_{\mathbb{R}} t \, d\widehat{m_x}(t) \tag{7.17}$$

whenever the expression on the right-hand side of (7.17) exists and is finite.

The following result gives a way how to prove many of known results from classical probability theory (convergence of MV-observables, limit theorems) for probability theory on MV-algebras.

Proposition 7.4.18 *Let h be an MV-σ-homomorphism of a tribe containing all constants onto a σ-complete MV-algebra M. Let x_f be defined by (7.8) for some $\mathcal{S}_0(\mathcal{T})$-measurable function $f : \Omega \to \mathbb{R}$, and let m be a σ-additive state on M. Then $m(x_f)$ exists and is finite if and only if $\int_\Omega f(\omega)\, d\widehat{m_h}(\omega)$ is so. In such a case*

$$m(xf) = \int_\Omega f(\omega)\, d\widehat{m_h}(\omega).$$

Proof. This follows from the equality (7.17) and from transforms of integrals. □

7.4.3 Exercises

1. Prove Proposition 7.1.4.
2. Show that an MV-algebra M is σ-isomorphic to a tribe $\mathcal{T}$ iff M has an order-determining system of σ-additive states. (Hint: Take into account the system $\{s_\omega : \omega \in \Omega\}$, where $s_\omega(f) := f(\omega)$, $f \in \mathcal{T}$.)
3. Show that the Mac Neille completion of an MV-algebra is an MV-algebra iff M is semisimple. (Hint: An ℓ-group has the Mac Neille completion as an ℓ-group iff G is Archimedean; [Fuc, Thm V.19].) ([Höl])
4. Let S be an Archimedean t-conorm on $[0,1]$ with a generator g for which $g(1) < \infty$. Then $[0,1]$ with $0, 1, \oplus = S$, and $^*g$ defined via $a^{*_g} = g^{-1}(g(1) - g(a))$ is an MV-algebra isomorphic with the MV-algebra $\Gamma(\mathbb{R}, 1)$.
5. Show that any BL-algebra is distributive. ([Tur 1])
6. Show that a BL-algebra generates an MV-algebra with the same $\odot$ and $x^* := x \to 0$ iff $x^{**} = x$.
7. Let M be a semisimple MV-algebra. Show that if $m \leftrightarrow Ker_m$, then $\hat{a}(m) = \bar{a}(Ker_m)$.

8. Let X be a compact, Hausdorff, basically disconnected, and let f be a bounded real-valued function on X. For each $x \in X$, let $\mathcal{N}(x)$ be the collection of open neighborhoods of x, and define

$$g(x) = \inf_{U \in \mathcal{N}(x)} \sup_{y \in U} f(y).$$

Assume for all $\alpha \in \mathbb{R}$ that $f^{-1}((\alpha, \infty))$ is an open F_σ subset of X. Then g is a continuous real-valued function on X. ([Goo 1, Lem 9.1])

9. Let X be a compact Hausdorff space. Then the space $C(X)$ of all real-valued continuous functions on X is a Dedekind σ-complete ℓ-group iff X is basically disconnected. ([Goo 1, Cor 9.3])

10. Show that any atomic σ-complete MV-algebra with the countable set of all atoms is σ-isomorphic with a tribe. ([DCR])

11. Show that for any σ-additive state m on an atomic σ-complete MV-algebra M with the countable set of all atoms $\{a_n\}$ there exists a sequence $\{t_n\}_{n=1}^{\infty}$ of non-negative real numbers with $\sum_{n=1}^{\infty} t_n = 1$ such that

$$s(a) = \sum_{n=1}^{\infty} t_n j_{a_n}(a)/ord(a_n), \; a \in M,$$

where j_{a_n} is defined by (6.10). Conversely, $\sum_{n=1}^{\infty} t_n j_{a_n}(a)/ord(a_n)$, $a \in M$, defines a σ-additive state on M. ([DCR])

12. A ring $\mathcal{S}$ of subsets of a non-void subset Ω is said to be a **Dedekind σ-complete ring** if $\bigcup_n A_n \in \mathcal{S}$ whenever there is an $A \in \mathcal{S}$ such that $A_n \subseteq A$ for any n. For example, any σ-ring of subsets is a Dedekind σ-complete ring.

   A **BCK-observable** is defined as a mapping h from a Dedekind σ-complete ring $\mathcal{S}$ of subsets of $\mathbb{R}$ into a BCK-algebra X such that

$$h(\bigcup_{n=1}^{\infty} E_n) = \bigoplus_{n=1}^{\infty} h(E_n)$$

   whenever $\{E_n\}$ is a sequence of mutually disjoint subsets of $\mathcal{S}$ such that $\bigcup_n E_n \in \mathcal{S}$. If a BCK-algebra X has a greatest element 1 we set $\mathcal{S} = \mathcal{B}(\mathbb{R})$ and $h(\mathbb{R}) = 1$.

   Prove: Any atomically bounded Dedekind σ-complete commutative BCK-algebra with the countable set of atoms is the range of some BCK-observable defined on a Dedekind σ-complete ring of subsets of $\mathbb{R}$. ([DCR 1])

13. Prove: Let $(X; *, 0)$ be an atomic Dedekind σ-complete BCK-algebra with the countable set of all atoms. Let $\{t_n\}_n$ be a sequence of non-negative numbers such that $\sum_n t_n < \infty$. Then m defined by

$$m(a) := \sum_n t_n j_{a_n}(a), \; a \in X,$$

is a measure on X. Conversely, suppose that m is a measure on X. Then m can be uniquely expressed in the above form. ([DCR 1])

14. Show that a BL-algebra L can be converted into a D-poset via $a \ominus b := a \odot b^*$, $b \leq a$ iff $x^{**} = x$ for any $x \in L$. (Hint: See Exercise 6.)
15. Show that a BL-algebra L can be converted into a BCK-algebra via $a*b := a\odot b^*$, $a, b \in L$, iff $x^{**} = x$ for any $x \in L$.

Bibliography

[Aar] J.F. AARNES, *Quasi-states on C^*-algebras*, Trans. Amer. Math. Soc. **149** (1970), 601–625.

[Abb] J.C. ABBOTT, *Semi-Boolean algebras*, Mat. Vesnik **4** (1967), 177–198.

[Abi] A. ABIAN, *Order in a special class of rings and a structure theorem*, Proc. Amer. Math. Soc. **52** (1975), 45–49.

[Ada] J. ADÁMEK, *"Theory of Mathematical Structures"*, D. Reidel Publ. Co., Dordrecht, Boston, Lancaster, 1983.

[Aer] D. AERTS, *Construction of the tensor product for lattices of properties of physical entities*, J. Math. Phys. **25** (1984), 1434–1441.

[Aer 1] D. AERTS, *A possible explanation for the probabilities of quantum mechanics,* J. Math. Phys. **27** (1986), 202–210.

[Aer 2] D. AERTS, *Quantum structures: An attempt to explain the origin of their appearance in nature*, Inter. J. Theor. Phys. **34** (1995), 1165–1186.

[Aer 3] D. AERTS, *The description of joint quantum entities and the formulation of a paradox*, Inter. J. Theor. Phys. **39** (2000), 485–496.

[Aer 4] D. AERTS, *Example of a macroscopic classical situation that violates Bell inequalities,* Lett. Nuovo Cimento **34** (1982), 107–111.

[ACVS] D. AERTS, E. COLEBUNDERS, A. Van der VOORDE, B. Van STEIRTEGHEM, *State property systems and closure spaces: A study of categorical equivalence, Inter. J. Theor. Phys.* **38** (1999), 359–385.

[AeDa] D. AERTS, I. DAUBECHIES, *Physical justification for using tensor product to describe quantum systems as one joint system*, Helv. Phys. Acta **51** (1978), 661–675.

[AevS] D. AERTS, B. Van STEIGRTEGHEM, *Quantum axiomatics and a theorem of M.P. Solèr*, Inter. J. Theor. Phys. **39** (2000), 497–502.

[Alf] E.M. ALFSEN, *"Compact Convex Sets and Boundary Integrals"*, Springer-Verlag, Berlin-Heidelberg-New York, 1971.

[And] T. ANDO, *Problem of infimum in the positive cone.* In: Analytic and Geometric Inequalities and Applications, (edited by) T. M. Rassias and H.M. Srivastava, Kluwer Academic Publisher, Dordrecht, 1999, pp. 1–12.

[Bae] R. BAER, *Free sums of groups and their generalizations. An analysis of the associative law,* Amer. J. Math. **41** (1949), 706–742.

[Bat] M. BATTAGLIA, *Order theoretic type decomposition of JBW*-triples*, Quart. J. Math. Oxford **42** (1991), 129–147.

[Bela] V. P. BELAVKIN, *Mathematical foundations of general system theory*, MIEM, Moscow, 1987, pp. 78 (in Russian).

[Bela 1] V. P. BELAVKIN, *The semilogics, quasilogics and other quantum structures*, Tatra Mt. Math. Publ. **10** (1997), 199–223.

[Bel] L.P. BELLUCE, *Semisimple algebras of infinite valued logic and Bold fuzzy set theory*, Canad. J. Math. **38** (1986), 1356–1379.

[BeDi] L.P. BELLUCE, A. Di NOLA, *Yosida type representation for perfect MV-algebras*, Math. Log. Quart. **42** (1996), 551–563.

[BDG] L.P. BELLUCE, A. Di NOLA, G. GEORGESCU, *Perfect MV-algebras and ℓ-rings*, J. Non-Classical Logic **9** (1999), 159–172.

[BDL] L.P. BELLUCE, A. di NOLA, A. LETTIERI, *Local MV-algebras*, Rend. Circ. Mat. Palermo, Ser. II. **42** (1993), 347–361.

[BeBu] E.G. BELTRAMETTI, S. BUGAJSKI, *Quantum observables in classical frameworks*, Inter. J. Theor. Phys. **34** (1995), 1221–1229.

[BeBu 1] E.G. BELTRAMETTI, S. BUGAJSKI, *A classical extension of quantum mechanics*, J. Phys. A: Math. Gen. **28** (1995), 3329–3343.

[BeBu 2] E.G. BELTRAMETTI, S. BUGAJSKI, *The Bell phenomena in classical frameworks,* J. Phys. A: Math. Gen. **29** (1996), 247–261.

[BeBu 3] E. BELTRAMETTI, S. BUGAJSKI, *Effect algebras and statistical physical theories,* **38** (1997), 3020–3030.

[BBGP] E. BELTRAMETTI, S. BUGAJSKI, S. GUDDER, S. PULMANNOVÁ, *Convex and linear effect algebras*, Rep. Math. Phys. **44** (1999), 359–379.

[BeCa] E.G. BELTRAMETTI, G. CASSINELLI, *"The Logic of Quantum Mechanics"*, Addison-Wesley, Reading, 1981.

[BeFo] M.K. BENNETT, D.J. FOULIS, *Phi-symmetric effect algebras*, Found. Phys. **25** (1995), 1699–1722.

[BeFo 1] M.K. BENNETT, D.J. FOULIS, *Interval and scale effect algebras*, Advan. Math. **19** (1997), 200–215.

[BeFo 2] M.K. BENNETT, D.J. FOULIS, *A generalized Sasaki projection for effect algebras*, Tatra Mt. Math. Publ. **16** (1998), 55–66.

[BeFo 3] M.K. BENNETT, D.J. FOULIS, R.J. GREECHIE, *Quotients of interval effect algebras*, Inter. J. Theor. Phys. **35** (1996), 2321–2338.

[Ber] L. BERAN, *"Orthomodular Lattices - Algebraic Approach"*, Academia, Czechoslovak Academy of Sciences & D. Reidel Publishing Company, Praha, Czechoslovakia & Dordrecht, Holland, 1984.

[Berb] S.K. BERBERIAN, *"Baer *-Rings"*, Springer-Verlag, Berlin, Heidelberg, New York, 1972.

[Bir] G. BIRKHOFF, *"Lattice Theory"*, Amer. Math. Soc. Coll. Publ., Vol. **25**, Providence, Rhode Island, 1967.

[BiNe] G. BIRKHOFF, J. von NEUMANN, *The logic of quantum mechanics*, Ann. Math. **37** (1936), 823–834.

[Bla] B. BLACKADAR, *"K-Theory for Operator Algebras"*, Springer-Verlag, New York, 1986.

[Boo] G. BOOLE, *"An Investigation of the Laws of Thought"*, Macmillan, 1854, reprinted by Dover Press, New York, 1967.

[Bos] B. BOSBACH, *Concerning bricks*, Acta Math. Acad. Sci. Hungar. **38** (1981), 89–104.

[Bos 1] B. BOSBACH, *Concerning semiclans*, Arch. Math. **37** (1981), 316–324.

[Bos 2] B. BOSBACH, *Concerning cone algebras*, Algebra Universalis **15** (1982), 58–66.

[Bos 3] B. BOSBACH, *Residuationsstrukturen*, Univ. Kassel, 1997.

[Bra] O. BRATTELI, *Inductive limits of finite-dimensional C^*-algebras*, Trans. Amer. Math. Soc. **171** (1972), 195–234.

[BrRo] O. BRATTELI, D.W. ROBINSON, *"Operator Algebras and Quantum Statistical Mechanics, I"*, Springer-Verlag, New York, Heidelberg, Berlin, 1979.

[BGHR] G. BRUNS, R. GREECHIE, J. HARDING, M. RODDY, *Completions of orthomodular lattices,* Order **7** (1990), 67–76.

[Buf] H.W. BUFF, *Decidable and undecidable MV-algebras*, Algebra Universalis **21** (1985), 234–249.

[Bug] S. BUGAJSKI, *Fundamentals of fuzzy probability theory,* Inter. J. Theor. Phys. **35** (1996), 2229–2244.

[BgHS] S. BUGAJSKI, K.-E. HELLWIG, W. STULPE, *On fuzzy random variables and statistical maps,* Rep. Math. Phys. **41** (1998), 1–11.

[Bur] P. BURMEISTER, *"A Model Theoretic Oriented Approach to Partial Algebras"*, Academic Verlag, Berlin, 1986.

[BuMa] P. BURMEISTER, M. MĄCZYŃSKI, *Orthomodular (partial) algebras and their representations*, Demonstratio Math. **27** (1994), 701–722.

[BGL] P. BUSCH, M. GRABOWSKI, P. J. LAHTI, *"Operational Quantum Physics"*, Springer-Verlag, Berlin, 1995.

[BGL 1] P. BUSCH, M. GRABOWSKI, P. J. LAHTI, *Who is afraid of POV measures? Unified approach to quantum phase observables,* Ann. Phys. **237** (1995), 1–11.

[BuHS] P. BUSCH, K.-E. HELLWIG, W. STULPE, *On classical representation of finite dimensional quantum mechanics*, Inter. J. Theor. Phys **32** (1993), 399–405.

[BLM] P. BUSCH, P.J. LAHTI, P. MITTELSTAEDT, *"The Quantum Theory of Measurement"* Lecture Notes in Physics, Springer–Verlag, Berlin, Heidelberg, New York, London, Budapest, 1991.

[BuKl] D. BUTNARIU, E.P. KLEMENT, *"Triangular Norm-Based Measures and Games with Fuzzy Coalitions"*, Kluwer Academic Publ., Dordrecht, 1993.

[CMG] J.C. CARREGA, R. MAYET, R.J. GREECHIE, *Minimal orthomodular lattices from quadratic spaces over finite fields,* Inter. J. Theor. Phys. **39** (2000), 517–524.

[CaMo] G. CASTAGNOLI, D. MONTI, *Quantum non-deterministic computation,* Inter. J. Theor. Phys. **39** (2000), 525–530.

[Cat] G. CATTANEO, *A unified framework for the algebra of unsharp quantum mechanics*, Inter. J. Theor. Phys. **36** (1997), 3085–3117.

[CDG] G. CATTANEO, M.L. DALLA CHIARA, R. GIUNTINI, *Some algebraic structures for many-valued logics*, Tatra Mt. Math. Publ. **15** (1998), 173–195.

[CDGP] G. CATTANEO, M.L. DALLA CHIARA, R. GIUNTINI, S. PULMANNOVÁ, *Effect algebras and paraboolean manifolds*, Inter. J. Theor. Phys. **39** (2000), 551–564.

[CFM] G. CATTANEO, G. FRANCO, G. MARINO, *Ordering on families of subspaces of pre-Hilbert spaces and Dacey pre-Hilbert spaces,* Boll. Unione Mat. Ital. **1-B** (1987), 167–183.

[CaLo] G. CATTANEO, F. LOMBARDO, *Independnt axiomatization of MV-algebras*, Tatra Mt. Math. Publ. **15** (1998), 227–232.

[CaNi] G. CATTANEO, G. NISTICÒ, *Brower–Zadeh posets and three-valued Łukasiewicz posets*, Fuzzy Sets and Systems **33** (1989), 165–190.

[Cha] C.C. CHANG, *Algebraic analysis of many-valued logics.* Trans. Amer. Math. Soc. **88** (1958), 467–490.

[Cha 1] C.C. CHANG, *A new proof of the completeness of the Łukasiewicz axioms*, Trans. Amer. Math. Soc. **93** (1959), 74–80.

[Che] G. CHEVALIER, *Order and orthogonality relations in rings and algebras*, Tatra Mt. Math. Publ. **3** (1993), 1–16.

[Che 1] G. CHEVALIER, *Orthosymmetric ortholattices and Rickart ∗-rings*, Inter. J. Theor. Phys. **34** (1995), 1289–1295.

[Che 2] G. CHEVALIER, *Congruence relations in orthomodular lattices*, Tatra Mt. Math. Publ. **15** (1998), 197–225.

[Che 3] G. CHEVALIER, *Different orthocomplementations on the subspace lattice of a finite dimensional complex vector space,* Inter. J. Theor. Phys. **39** (2000), 565–570.

[Che 4] G. CHEVALIER, *Different orthocomplementations on a lattice of subspaces,* submitted.

[CDS] G. CHEVALIER, A. DVUREČENSKIJ, K. SVOZIL, *Piron's and Bell's geometric lemmas and Gleason's theorem,* Found. Phys. (2000), to appear.

[ChPu] G. CHEVALIER, S. PULMANNOVÁ, *Some ideal lattices on partial abelian manifolds and effect algebras*, submitted.

[Cho] F. CHOVANEC, *States and observables on MV-algebras*, Tatra Mt. Math. Publ. **3** (1993), 55–65.

[ChJu] F. CHOVANEC, M. JUREČKOVÁ, *Law of large numbers on D-posets of fuzzy sets*, Tatra Mt. Math. Publ. 1 (1992), 15–18.

[ChKo] F. CHOVANEC, F. KÔPKA, *On a representation of observables in D-posets of fuzzy sets*, Tatra Mt. Math. Publ. **1** (1992), 19–23.

[ChKo 1] F. CHOVANEC, F. KÔPKA, *D-lattices*, Inter. J. Theor. Phys. **34** (1995), 1297–1302.

[ChKo 2] F. CHOVANEC, F. KÔPKA, *Boolean D-posets*, Tatra Mt. Math. Publ. **10** (1997), 183–197.

[ChKo 3] F. CHOVANEC, F. KÔPKA, *On representation of finite Boolean D-posets*, Tatra Mt. Math. Publ. **15** (1998), 107–121.

[ChKo 4] F. CHOVANEC, F. KÔPKA, *Difference posets in the quantum structures background*, Inter. J. Theor. Phys. **39** (2000), 571–583.

[ChKo 5] F. CHOVANEC, E. RYBÁRIKOVÁ, *On observables on finite Boolean D-posets.* In: Proc. 7th IFSA World Congress, Prague, June 26–29, 1997, Eds. M. Mareš et al., Academia, Prague, 1997, vol. I, pp. 141–143.

[ChRy] F. CHOVANEC, E. RYBÁRIKOVÁ, *Ideals and filters in D-posets*, Inter. J. Theor. Phys. **37** (1998), 17–22.

[Cig] R. CIGNOLI, *Complete and atomic algebras of the infinite valued Łukasiewicz logic*, Studia Logica **50** (1991), 375–384.

[CDM] R. CIGNOLI, I.M.L. D'OTTAVIANO, D. MUNDICI, *"Algebraic Foundations of Many-valued Reasoning"*, Kluwer Academic Publ., Dordrecht, 2000.

[CiTo] R. CIGNOLI, A. TORRENS, *Boolean products of MV-algebras*, J. Math. Anal. Appl. **199** (1996), 637–653.

[CiTon] R. CIGNOLI, A. TORRENS, *The poset of prime ℓ-ideals of an abelian ℓ-group with a strong unit*, J. Algebra **184** (1996), 604–612.

[Coe] B. COECKE, *Representation for pure and mixed states of quantum physics in Euclidean space,* Inter. J. Theor. Phys. **34** (1995), 1165.

[Coh] D.W. COHEN, *"An Introduction to Hilbert Space and Quantum Logic"*, Springer-Verlag, New York, Berlin, Heidelberg, London, Paris, Tokyo, 1989.

[Con] C. CONSTANTINESCU, *"Some Properties of Spaces of Measures"*, Suppl. Atti Sem. Mat. Fis. Univ. Modena **35** (1989), 1–289.

[Cor] W.H. CORNISH, *On Iséki's BCK-algebras*. In: *Algebraic Structures and Applications* (Nedlands (1980)), Lecture Notes in Pure and Appl. Math. 74, Dekker, New York, 1982, pp. 101–122.

[Cor 1] W.H. CORNISH, *Lattice ordered groups and BCK-algebras*, Math. Japonica **4** (1980), 471–476.

[Cor 2] W.H. CORNISH, *A minimal equational base for commutative BCK-algebras*, Math. Seminar Notes **10** (1982), 251–256.

[CST] W.H. CORNISH, T. STURM, T. TRACZYK, *Embedding of commutative BCK-algebras into distributive lattice BCK-algebras*, Math. Japonica **29** (1984), 309–320.

[DaCh] M.L. DALLA CHIARA, *Unsharp quantum logics*, Inter. J. Theor. Phys. **34** (1995), 1331–1336.

[DaGi] M.L. DALLA CHIARA, R. GIUNTINI, *Łukasiewicz' theory of truth, from the quantum logical point of view*. In: Alfred Tarski and the Vienna Circle, Eds. J. Woleński, E. Köhler, Kluwer Acad. Publ., 1999, pp. 127–134.

[DaGi 1] M.L. DALLA CHIARA, R. GIUNTINI, *Quantum logics*. In: Handbook of Philosophical Logic, Eds. D. Gabbay, F. Guenthner, Kluwer Acad. Publ., Dordrecht, to appear.

[dAPu] A.B. D'ANDREA, S. PULMANNOVÁ, *Boolean quotients of orthomodular lattices*, Algebra Universalis **37** (1995), 485–495.

[dAPu 1] A.B. D'ANDREA, S. PULMANNOVÁ, *Quasivarieties of orthomodular lattices*, Rep. Math. Phys. **27** (1994), 688–699.

[Dar] M.R. DARNEL, *"Theory of Lattice-Ordered Groups"*, Marcel Dekker, Inc., New York, Basel, Hong-Kong, 1995.

[dLdS] P. de LUCIA, A. de SIMONE, *Measures on weakly orthocomplete OMP with a filtering sub-OMP*, Tatra Mt. Math. Publ. **15** (1998), 233–244.

[dLDv] P. de LUCIA, A. DVUREČENSKIJ, *Decompositions of Riesz-space-valued measures on orthomodular posets*, Tatra Mt. Math. Publ. **2** (1993), 229–239.

[dLDv 1] P. de LUCIA, A. DVUREČENSKIJ, *Yosida-Hewitt decompositions of Riesz-space-valued measures on orthoalgebras,* Tatra Mt. Math. Publ. **3** (1993), 101–110.

[dLPa] P. de LUCIA, E. PAP, *Noncommutative version of Nikodym boundedness theorem for uniform space-valued functions*, Inter. J. Theor. Phys. **34** (1995), 981–993.

[dLPa 1] P. de LUCIA, E. PAP, *Nikodým convergence theorem for uniform space valued functions defined on D-poset*, Math. Slovaca **45** (1995), 367–376.

[dLPt] P. de LUCIA, P. PTÁK, *Quantum probability spaces that are nearly classical*, Bull. Polish Acad. Sciences–Math. **40** (1992), 163–173.

[Deu] D. DEUTSCH, *Quantum theory, the Church-Turing principle and the universal quantum computer*, Proc. Roy. Soc. London A **400** (1985), 69–117.

[Deu 1] D. DEUTSCH, *Quantum computational networks*, Proc. Roy. Soc. London A **425** (1989), 96–117.

[DiDv] A. Di NOLA, A. DVUREČENSKIJ, *Product MV-algebras,* Multi-Valued Logic, (2000), to appear.

[DiLe] A. Di NOLA, A. LETTIERI, *Perfect MV-algebras are categorical equivalent to abelian ℓ-groups,* Studia Logica **53** (1994), 417–432.

[DiGe] A. Di NOLA, G. GEORGESCU, *Ideals, ℓ-rings and MV-algebras,* preprint 1998.

[DGL] A. Di NOLA, G. GEORGESCU, A. LETTIERI, *Extending probabilities to states of MV-algebras*, Collegium Logicum, Ann. Kurt-Gödel Soc. **3** (1999), 31–50.

[Dir] P. DIRAC, *"The Principles of Quantum Mechanics"*, Oxford, Clarendon, 1930.

[DiSe] A. Di NOLA, S. SESSA, *On MV-algebras of continuos functions.* In: Non-Classical Logics and Their Appl. to Fuzzy Subsets, eds. U. Höhle, E.P. Klement Kluwer Academic Publ., Dordrecht, 1995, pp. 23–32.

[Dix] J. DIXMIER, *On some C^*-algebras considered by Glimm*, J. Func. Anal. **1** (1967), 182–203.

[DDL] G. DORFER, A. DVUREČENSKIJ, H. LÄNGER, *Symmetric difference in orthomodular lattices,* Math. Slovaca **46** (1996), 435–444.

[Dor] D. DORNINGER, *Sublogics of ring-like logics*, Tatra Mt. Math. Publ. **15** (1998), 75–83.

[DoMu] D. DORNINGER, W.B. MÜLLER, *"Allgemeine Algebra und Andwedungen"*, B.G. Teubner, Stuttgart, 1984.

[DLM] D. DORNINGER, H. LÄNGER, M. MĄCZYŃSKI, *On ring-like structures occurring in axiomatic quantum mechanics*, Österr. Akad. Wiss. Sitzungsber. Math-naturw. Kl. Abt. II **206** (1997), 279–289.

[DLM 1] D. DORNINGER, H. LÄNGER, M. MACZYŃSKI, *The logic induced by a system of homomorphisms and its various algebraic characterizations*, Demonstratio Math. **30** (1997), 215–232.

[DrSi] J. DRAVECKÝ, J. ŠIPOŠ, *On the additivity of Gudder integral*, Math. Slovaca **30** (1980), 299-303.

[Dra] M.P. DRAZIN, *Natural structures on semigroups with involution*, Bull. Amer. Math. Soc. **84** (1978), 139–141 .

[Dri] M. DRIESCHNER, *Symmetry and quantum composition - A key to the structure of physical logic*, Inter. J. Theor. Phys. **37** (1998), 427–433.

[DDdL] M. DUCHOŇ, A. DVUREČENSKIJ, P. de LUCIA, *Moment problem for effect algebras*, Inter. J. Theor. Phys. **36** (1997), 1941–1958.

[Dvu] A. DVUREČENSKIJ, *"Gleason's Theorem and Its Applications"*, Kluwer Academic Publ., Dordrecht, and Ister Science Press, Bratislava, 1993.

[Dvu 1] A. DVUREČENSKIJ, *Tensor product of difference posets*, Trans. Amer. Math. Soc. **347** (1995), 1043–1057.

[Dvu 2] A. DVUREČENSKIJ, *Tensor product of difference posets or effect algebras*, Inter. J. Theor. Phys. **34** (1995), 1337–1348.

[Dvu 3] A. DVUREČENSKIJ, *Frame functions and completeness of inner product spaces*, Ann. Inst. Henri Poincaré Phys. Théor. **62** (1995), 429–438.

[Dvu 4] A. DVUREČENSKIJ, *A note on the Nikodým boundedness theorem*, Tatra Mt. Math. Publ. **8** (1996), 231–240.

[Dvu 5] A. DVUREČENSKIJ, *Remarks on Nikodým's boundedness theorem*, Novi Sad J. Math. **26** (1996), 25–51.

[Dvu 6] A. DVUREČENSKIJ, *On a representation of observables on fuzzy measurable spaces*, J. Math. Anal. Appl. **197** (1996), 579–585.

[Dvu 7] A. DVUREČENSKIJ, *The Radon-Nikodým theorem for fuzzy probability spaces,* Fuzzy Sets and Systems **45** (1992), 69–78.

[Dvu 8] A. DVUREČENSKIJ, *Gleason's theorem and Cauchy's functional equation,* Inter. J. Theor. Phys. **35** (1996), 2687–2695.

[Dvu 9] A. DVUREČENSKIJ, *A measure theoretic characterization of inner product spaces,* Tatra Mt. Math. Publ. **10** (1997), 273–280.

[Dvu 10] A. DVUREČENSKIJ, *Commutative BCK-algebras and lattice ordered groups with universal property,* Indian J. Math., to appear.

[Dvu 11] A. DVUREČENSKIJ, *On categorical equivalences of commutative BCK-algebras,* Studia Logica **64** (2000), 21–34.

[Dvu 12] A. DVUREČENSKIJ, *Fuzzy set representation of some quantum structures,* Fuzzy Sets and Systems **101** (1999), 67–78.

[Dvu 13] A. DVUREČENSKIJ, *Ideals of commutative BCK-algebras,* Ricerche Mat., to appear.

[Dvu 14] A. DVUREČENSKIJ, *Commutative BCK-algebras with product,* Demonstratio Math. **33** (2000), 1–19.

[Dvu 15] A. DVUREČENSKIJ, *Measures on commutative BCK-algebras.* Atti Sem. Mat. Fis. Univ. Modena (2000), to appear.

[Dvu 16] A. DVUREČENSKIJ, *Loomis-Sikorski theorem for σ-complete MV-algebras and ℓ-groups,* J. Austral. Math. Soc. Ser. A **68** (2000), 261–277.

[Dvu 17] A. DVUREČENSKIJ, *On Loomis-Sikorski's theorem for MV-algebras and BCK-algebras.* In: Contribut. to General Algebra **12**, Proc. Vienna Conf. June 3–6, 1999, Verlag Johannes Heyn, Klagenfurt, 2000, pp. 165–180.

[Dvu 18] A. DVUREČENSKIJ, *Measures and $\perp$-decomposable measures of effects of a Hilbert space,* Atti Sem. Mat. Fis. Univ. Modena **45** (1997), 259–288.

[Dvu 19] A. DVUREČENSKIJ, *Fuzzy representation of effect algebras and the moment problem.* In: Seventh IFSA World Congress, Prague, June 25–29, 1997, Eds. M. Mareš et al., Academia, Prague, 1997, pp. 146–151.

[Dvu 20] A. DVUREČENSKIJ, *Solér's theorem and characterization of inner product spaces,* Inter. J. Theor. Phys. **37** (1998), 23–29.

[Dvu 21] A. DVUREČENSKIJ, *Measures and states on BCK-algebras,* Atti Sem. Mat. Fis. Univ. Modena **47** (1999), 267–284.

[Dvu 22] A. DVUREČENSKIJ, *On partial addition in pseudo MV-algebras.* In: Proc. Fourth Inter. Symp. on Econ. Inform., May 6–9, 1999, Bucharest, Eds. I. Smeureanu et al., INFOREC Printing House, Bucharest, 1999, pp. 952–960.

[Dvu 23] A. DVUREČENSKIJ, *Bell-type inequalities and orthomodular lattices.* In: Eds. M.L. Dalla Chiara et al. Kluwer Acad. Press, 1999, pp. 209–218.

[Dvu 24] A. DVUREČENSKIJ, *Commutative BCK-algebras and quantum structures,* Inter. J. Theor. Phys. **39** (2000), 653–664.

[Dvu 25] A. DVUREČENSKIJ, *Gleason's theorem in applications.* In: G. Kalmbach: *Quantum Measures and Spaces,* Kluwer Acad. Publ., Dordrecht/Boston/London, Chapter 4, 1998, pp. 39–50.

[Dvu 26] A. DVUREČENSKIJ, *Test spaces and characterizations of quadratic spaces,* Inter. J. Theor. Phys. **35** (1996), 2093–2106.

[Dvu 27] A. DVUREČENSKIJ, *Remarks on representation of fuzzy quantum posets,* Math. Slovaca **44** (1994), 429–440.

[Dvu 28] A. DVUREČENSKIJ, *MV-observables and MV-algebras,* J. Math. Anal. Appl. submitted.

[DAK] A. DVUREČENSKIJ, S.S. AHN, H.S. KIM, *The Ker-Coker sequence in BCI-algebras,* Demonstratio Math. **30** (1997), 844–850.

[DCR] A. DVUREČENSKIJ, F. CHOVANEC, E. RYBÁRIKOVÁ, *D-homomorphisms and atomic σ-complete Boolean D-posets,* Soft Computing **4** (2000), 9–18.

[DCR 1] A. DVUREČENSKIJ, F. CHOVANEC, E. RYBÁRIKOVÁ, *Observables, measures, and atomic Dedekind σ-complete BCK-algebras.* In: Proc. EUROUSE–SC'99 Budapest, Hungary, May 25–28, 1999, Eds B. De Baets et al., pp. 89–94.

[DDP] A. DVUREČENSKIJ, P. de LUCIA, E. PAP, *On a decomposition theorem and its applications,* Math. Japonica **44** (1996), 145–164.

[DvGr] A. DVUREČENSKIJ, M.G. GRAZIANO, *Remarks on representations of minimal clans.* Tatra Mt. Math. Publ. **15** (1998), 31–53.

[DvGr 1] A. DVUREČENSKIJ, M.G. GRAZIANO, *On representations of commutative BCK-algebras,* Demonstratio Math. **32** (1999), 227–246.

[DvGr 2] A. DVUREČENSKIJ, M.G. GRAZIANO, *Commutative BCK-algebras and lattice ordered groups,* Math. Japonica **49** (1999), 159–174.

[DvGr 3] A. DVUREČENSKIJ, M.G. GRAZIANO, *Dedekind complete commutative BCK-algebras,* Order, to appear.

[DvGr 4] A. DVUREČENSKIJ, M.G. GRAZIANO, *Measures on Dedekind complete commutative BCK-algebras,* Atti Sem. Matem. Fis. Univ. Modena, to appear.

[DvKi] A. DVUREČENSKIJ, H.S. KIM, *On connections between BCK-algebras and different posets,* Studia Logica **60** (1998), 421–439.

[DLY] A. DVUREČENSKIJ, P. LAHTI, K. YLINEN, *Positive operator measures determined by their moment sequences,* Rep. Math. Phys., **45** (2000), 139–146.

[DvLa] A. DVUREČENSKIJ, H. LÄNGER, *Bell-type inequalities in horizontal sums of Boolean algebras,* Found. Phys. **24** (1994), 1195–1202.

[DvLa 1] A. DVUREČENSKIJ, H. LÄNGER, *Bell-type inequalities in orthomodular lattices I. Inequalities of order 2,* Inter. J. Theor. Phys. **34** (1995), 995–1024.

[DvLa 2] A. DVUREČENSKIJ, H. LÄNGER, *Bell-type inequalities in orthomodular lattices II. Inequalities of higher order,* Inter. J. Theor. Phys. **34** (1995), 1025–1036.

[DvLe] A. DVUREČENSKIJ, LE BA LONG, *On representation of fuzzy quantum posets,* Acta Math. Univ. Comenianae **58–59** (1990), 239–246.

[DvLe 1] A. DVUREČENSKIJ, LE BA LONG, *Observables in fuzzy quantum posets,* Acta Univ. Comenianae **58–59** (1990), 247–258.

[DvPu] A. DVUREČENSKIJ, S. PULMANNOVÁ, *Test spaces, Dacey spaces, and completeness of inner product spaces*, Letters Math. Phys. **22** (1994), 299–306.

[DvPu 1] A. DVUREČENSKIJ, S. PULMANNOVÁ, *Difference posets, effects, and quantum measurements*, Inter. J. Theor. Phys. **33** (1994), 819–850.

[DvPu 2] A. DVUREČENSKIJ, S. PULMANNOVÁ, *D-test spaces and difference posets*, Rep. Math. Phys. **34** (1994), 151–170.

[DvPu 3] A. DVUREČENSKIJ, S. PULMANNOVÁ, *Tensor product of D-posets and D-test spaces*, Rep. Math. Phys. **34** (1994), 251–275.

[DPSa] A. DVUREČENSKIJ, S. PULMANNOVÁ, S. SALVATI, *Meanders in orthoposets and QMV algebras*, submitted.

[DPS] A. DVUREČENSKIJ, S. PULMANNOVÁ, K. SVOZIL, *Partition logics, orthoalgebras, and automata,* Helv. Phys. Acta **68** (1996), 407–428.

[DvRi] A. DVUREČENSKIJ, B. RIEČAN, *Fuzzy quantum models*, Inter. J. General Systems **20** (1991), 39–54.

[DvRi 1] A. DVUREČENSKIJ, B. RIEČAN, *Decompositions of measures on orthoalgebras and difference posets*, Inter. J. Theor. Phys. **33** (1994), 1387–1402.

[DvRi 2] A. DVUREČENSKIJ, B. RIEČAN, *Weakly divisible MV-algebras and product,* J. Math. Anal. Appl. **234** (1999), 208–222.

[DvSv] A. DVUREČENSKIJ, K. SVOZIL, *Product of partition logics, orthoalgebras and automata,* Inter. J. Theor. Phys. **35** (1996), 2377–2390.

[DvTi] A. DVUREČENSKIJ, A. TIRPÁKOVÁ, *Sum of observables in fuzzy quantum spaces*, Appl. Matem. **37** (1992), 40–50.

[EdRu] C.M. EDWARDS, G.T. RÜTTIMANN, *On the facial structure of the unit balls in a JBW*-triple and its predual*, J. London Math. Soc. **38** (1988), 317–332.

[EHS] E.G. EFFROS, D.E. HANDELMAN, C.L. SHEN, *Dimension groups and their affine representation*, Amer. J. Math. **102** (1980), 385–407.

[Eli] G.A. ELLIOTT, *On the classification of inductive limits of sequences of semisimple finite-dimensional algebras*, J. Algebra **38** (1979), 29–44.

[Emc] G. EMCH, *"Algebraic Methods in Statistical Mechanics and Quantum Field Theory"*, John Wiley, New York, 1972.

[Fal] K. FALCONER, *"Fractal Geometry, Mathematical Foundations and Applications"*, John Wiley and Sons, 1990.

[FeWi] D. V. FELDMAN, A. WILCE, *Abelian extension of quantum logics,* Inter. J. Theor. Phys. **37** (1998), 39–43.

[FiFi] D. FINKELSTEIN, S.R. FINKELSTEIN, *Computational complementarity,* Inter. J. Theor. Phys. **22** (1983), 753–779.

[Fle] I. FLEISCHER, *Every BCK-algebra is a set of residuables in an integral pomonoid,* J. Algebras **119** (1988), 360–365.

[Fle 1] I. FLEISCHER, *Subdirect product of totally ordered BCK-algebras*, J. Algebra **111** (1987), 384–387.

[FRT] J.M. FONT, A.J. RODRÍGUEZ, A. TORRENS, *Wajsberg algebras,* Stochastica **8** (1984), 5–31.

[Fla] J. FLACHSMEYER, *The difference poset of monotone functions,* Tatra Mt. Math. Publ. **6** (1995), 27–33.

[Fou] D. FOULIS, *Coupled physical systems*, Found. Phys. **19** (1989), 905–922.

[Fou 1] D.J. FOULIS, *Mathematical metascience*, J. Natural Geomet. **13** (1998), 1–50.

[FoBe] D.J. FOULIS, M.K. BENNETT, *Tensor products of orthoalgebras,* Order **10** (1993), 271–282.

[FoBe 1] D.J. FOULIS, M.K. BENNETT, *Effect algebras and unsharp quantum logics*, Found. Phys. **24** (1994), 1325–1346.

[FoBe 2] D.J. FOULIS, M.K. BENNETT, *The transition to unigroups*, Inter. J. Theor. Phys. **37** (1998), 45–63.

[FBG] D.J. FOULIS, M.K. BENNETT, R.J. GREECHIE, *Test groups and effect algebras*, Inter. J. Theor. Phys. **35** (1996), 1117–1140.

[FoGr] D.J. FOULIS, R.J. GREECHIE, *Specification of finite effect algebras,* Inter. J. Theor. Phys. **39** (2000), 665–679.

[FGB] D.J. FOULIS, R.J. GREECHIE, M.K. BENNETT, *Sums and products of interval algebras*, Inter. J. Theor. Phys. **33** (1994), 2119–2136.

[FGR] D.J. FOULIS, R.J. GREECHIE, G.T. RÜTTIMANN, *Filters and supports in orthoalgebras*, Inter. J. Theor. Phys. **31** (1992), 789–807.

[FoPt] D. FOULIS, P. PTÁK, *On the tensor product of a Boolean algebra and an orthoalgebra*, Czechoslovak Math. J. **45 (120)** (1995), 117–126.

[FoRa] D.J. FOULIS, C.H. RANDALL, *Operational statistics. I. Basic concepts*, J. Math. Phys. **13** (1972), 1667–1675.

[FoRa 1] D. FOULIS, C. RANDALL, *Empirical logic and tensor products.* In: Interpretations and Foundations of Quantum Theories, A. Neumann ed., Wissenschaftsverlag, Bibliographisches Institut, Mannheim, 1981, pp. 9–20.

[Fri] R. FRIČ, *On observables*, Inter. J. Theor. Phys. **39** (2000), 677–686.

[Fri 1] R. FRIČ, *Łukasiewicz tribes are absolutely sequentially closed Bold algebras*, submitted.

[Fuc] L. FUCHS, *"Partially Ordered Algebraic Systems"*, Pergamon Press, Oxford-London-New York-Paris, 1963.

[Fuc 1] L. FUCHS, *Riesz groups, Annali della Scuola Norm Superiore, Pisa* **19** (1965), 1–33.

[GeHa] M. GELL-MANN, J. HARTLE, *Alternative decohering histories in quantum mechanics.* In: Proc. 25th Inter. Conf. on High Energy Physics, Singapore, August, 2-8, 1990, (K.K. Phua and Y. Yamaguchi, eds.), World Scientific, Singapore, 1990, pp. 1303–1310.

[GeHa 1] M. GELL-MANN, J. HARTLE, *Quantum mechanics in the light of quantum cosmology.* In: Proc. 3rd Inter. Symp. Found. Quantum Mech., Tokyo, 1989, (S.Kobayashi *et al.*, eds.,) Physical Society of Japan, Tokyo, 1990, pp. 321–343.

[GeHa 2] M. GELL-MANN, J. HARTLE, *Quantum mechanics in the light of quantum cosmology.* In: Complexity, Entropy and the Physics of Information, SFI Studies in the Science of Complexity, Vol VIII, W. Zurek, ed., Addison-Wesley, Reading, Massachusetts, 1990, pp. 425–458.

[GeIo] G. GEORGESCU, A. IORGULESCU, *Pseudo-MV algebras,* Multi-Valued Logic, (2000), to appear.

[GeIo 1] G. GEORGESCU, A. IORGULESCU, *Pseudo-BL algebras,* (1999), preprint.

[GeIo 2] G. GEORGESCU, A. IORGULESCU, *Pseudo-MV algebras: A non-commutative extension of MV-algebras.* In: Proc. Fourth Inter. Symp. on Econ. Inform., May 6–9, 1999, Bucharest, Eds. I. Smeureanu et al., INFOREC Printing House, Bucharest, 1999, pp. 961–968.

[GeLe] G. GEORGESCU, I. LEUŞTEAN, *Convergences in perfect MV-algebras,* J. Math. Anal. Appl. **228** (1998), 96–111.

[Gil] R. GILES, *Łukasiewicz logic and fuzzy set theory*, Inter. J. Man-Machine Stud. **8** (1976), 313–327.

[Giu] R. GIUNTINI, *Quasilinear QMV algebras*, Inter. J. Theor. Phys. **34** (1995), 1397–1407.

[Giu 1] R. GIUNTINI, *Quantum MV algebras*, Studia Logica **56** (1996), 393–417.

[Giu 2] R. GIUNTINI, *"Quantum Logic and Hidden Variables"*, **8** B.I.Wissenschaftfsverlag, Mannheim, Wien, Zürich, 1991.

[Giu 3] R. GIUNTINI, *Unsharp orthoalgebars and quantum MV algebras.* In: The Foundations of Quantum Mechanics, (C. Garola, A. Rosssi, eds), Kluwer Acad. Publ., Dordrecht, 1996. pp. 325–337.

[Giu 4] R. GIUNTINI, *Quantum MV-algebras and commutativity,* Inter. J. Theor. Phys. **37** (1998), 65–74.

[Giu 5] R. GIUNTINI, *The logic of weakly linear quantum MV algebras,* to appear.

[Giu 6] R. GIUNTINI, *An independent axiomatization of quantum MV algebras.* In: C. Garola, A. Rossi (eds), Foundations of Quantum Mechanics. World Scientific, Singapore, 2000, to appear.

[GiGr] R. GIUNTINI, H. GREULING, *Toward a formal language for unsharp properties,* Found. Phys. **19** (1989), 931–945.

[GiPu] R. GIUNTINI, S. PULMANNOVÁ, *Ideals and congruences in effect algebras and QMV-algebras*, Comm. Algebra, **28** (2000), 1567–1592.

[Gli] J. GLIMM, *On a certain class of operator algebras*, Trans. Amer. Math. Soc. **95** (1960), 318–340.

[Gle] A.M. GLEASON, *Measures on the closed subspaces of a Hilbert space*, J. Math. Mech. **6** (1957), 885–893.

[Gol] A. GOLFIN, *Representations and products of lattices,* Ph.D. Thesis, Univ. of Massachusetts, Amherst, 1987.

[GoMa] S. GONZÁLEZ, C. MARTÍNEZ, *Order relation in Jordan rings and a structure theorem*, Proc. Amer. Math. Soc. **98** (1986), 379–388.

[Goo] K.R. GOODEARL, *"Notes on Real and Complex C^*-algebras"*, Shiva, Nantwich, 1982.

[Goo 1] K.R. GOODEARL *"Partially Ordered Abelian Groups with Interpolation"*, Math. Surveys and Monographs No. 20, Amer. Math. Soc., Providence, Rhode Island, 1986.

[GoHa] K.R. GOODEARL, D. HANDELMAN, *Rank function and K_0 for regular rings,* J. Pure Appl. Algebra **7** (1976), 195–216.

[GrFo] R.J. GREECHIE, D.J. FOULIS, *Transition to effect algebras,* Inter. J. Theor. Phys. **34** (1995), 1369–1382.

[GFP] R.J. GREECHIE, D.J. FOULIS, S. PULMANNOVÁ, *The center of an effect algebra*, Order **12** (1995), 91–106.

[GrZa] A.A. GRIB, R.R. ZAPATRIN, *Automata simulating quantum logics,* Inter. J. Theor. Phys. **29** (1990), 113–123.

[GrZa 1] A.A. GRIB, R.R. ZAPATRIN, *Macroscopic realizations of quantum logics,* Inter. J. Theor. Phys. **31** (1992), 1669–1687.

[Gru] J. GRUSKA, *"Quantum Computing",* McGraw-Hill, 1999.

[Gud] S. GUDDER, *"Stochastic Methods in Quantum Mechanics"*, Elsevier, North Holland, 1979.

[Gud 1] S. GUDDER, *Event structure in nonstandard quantum mechanics,* J. Math. Phys. **35** (1994), 6332–6343.

[Gud 2] S. GUDDER, *Effect algebras and tensor products of S-sets*, Inter. J. Theor. Phys. **34** (1995), 2395–2407.

[Gud 3] S. GUDDER, *Morphisms on semiorthoposets,* Inter. J. Theor. Phys. **34** (1995), 1417–1429.

[Gud 4] S. GUDDER, *Total extension of effect algebras*, Found. Phys. Letters **8** (1995), 243–252.

[Gud 5] S. GUDDER, *D-algebras,* Found. Phys. **26** (1996), 813–822.

[Gud 6] S. GUDDER, *Examples, problems, and results in effect algebras*, Inter. J. Theor. Phys. **35** (1996), 2365–2376.

[Gud 7] S. GUDDER, *A hierarchy of quantum structures*, J. Natural Geom. **9** (1996), 77–102.

[Gud 8] S. GUDDER, *Lattice properties of quantum effects*, J. Math. Phys. **37** (1996), 2637–2642.

[Gud 9] S. GUDDER, *Semi-orthoposets*, Inter. J. Theor. Phys. **35** (1996), 1141–1173.

[Gud 10] S. GUDDER, *Chain tensor products and interval effect algebras*, Inter. J. Theor. Phys. **36** (1997), 1085–1098.

[Gud 11] S. GUDDER, *Effect test spaces*, Inter. J. Theor. Phys. **36** (1997), 2681–2705.

[Gud 12] S. GUDDER, *A survey of E-test spaces*, Inter. J. Theor. Phys. **37** (1998), 75–83.

[Gud 13] S. GUDDER, *Sharply dominating effect algebras,* Tatra Mt. Math. Publ. **15** (1998), 23–30.

[Gud 14] S. GUDDER, *Fuzzy probability theory,* Demonstratio Math. **31** (1998), 235–254.

[Gud 15] S. GUDDER, *S-dominating effect algebras*, Inter. J. Theor. Phys. **37** (1998), 915–923.

[Gud 16] S. GUDDER, *Morphisms, tensor product and σ-effect algebras*, Rep. Math. Phys. **42** (1998), 321–346.

[Gud 17] S. GUDDER, *A histories approach to quantum mechanics*, J. Math. Phys. **39** (1998), 5772–5788.

[Gud 18] S. GUDDER, *"Quantum Probability",* Academic Press Inc., Boston, San Diego, New York, Berkeley, Tokyo, Toronto, 1988.

[Gud 19] S. GUDDER, *Quantum computers*, Inter. J. Theor. Phys., to appear.

[Gud 20] S. GUDDER, *Basic properties of quantum automata*, Found. Phys., to appear.

[GuGr] S. GUDDER, R. GREECHIE, *Effect algebras counter-examples,* Math. Slovaca **46** (1996), 317–326.

[GuPu] S. GUDDER, S. PULMANNOVÁ, *Quotients of partial abelian monoids*, Algebra Universalis **38** (1997), 395–421.

[GuPu 1] S. GUDDER, S. PULMANNOVÁ, *Representation theorem for convex effect algebras*, Comm. Math. Univ. Carolinea **39** (1998), 645–659.

[HaKa] R. HAAG, D. KASTLER, *An algebraic approach to quantum field theory,* J. Math. Phys. **5** (1964), 848-861.

[Hab] E.D. HABIL, *The Brooks–Jewett theorem for k-triangular functions on difference posets and orthoalgebras*, Math. Slovaca **47** (1997), 417–428.

[Hab 1] E.D. HABIL, *Direct limits of effect algebras*, Inter. J. Theor. Phys. **39** (2000), 711–720.

[Haj] P. HÁJEK, *"Metamathematics of Fuzzy Logic"*, Kluwer Academic Publ, Dordrecht, 1998.

[Haj 1] P. HÁJEK, *Basic fuzzy logic and BL-algebras*, Soft Computing **2** (1998), 124–128.

[Ham] J. HAMHALTER, *Restricting pure states on JB algebras to maximal associative subalgebras*, Tatra Mt. Math. Publ. **15** (1998), 165–171.

[Ham 1] J. HAMHALTER, *Universal state space embeddability of Jordan-Banach algebras*, Proc. Amer. Math. Soc. **127** (1999), 131–137

[Ham 2] J. HAMHALTER, *Restricting and extending states and positive maps on operator algebras*, Inter. J. Theor. Phys. **39** (2000), 721–726.

[HNP] J. HAMHALTER, M. NAVARA, P. PTÁK, *States on orthoalgebras*, Inter. J. Theor. Phys. **34** (1995), 1289–1295.

[HaKo] M. HAVIAR, P. KONÔPKA, *Finitely generated free modular ortholattices III*, Inter. J. Theor. Phys. **39** (2000), 727–735.

[Har] J. HARDING, *Decompositions in quantum logics*, Trans. Amer. Math. Soc. **348** (1996), 1939–1862.

[Har 1] J. HARDING, *Regularity in quantum logics,* Inter. J. Theor. Phys. **37** (1998), 1173–1212.

[Har 2] J. HARDING, *Axioms of an experimental systems,* Inter. J. Theor. Phys. **38** (1999), 1643–1675.

[HKPW] M. HAVIAR, P. KONÔPKA, H.A. PRIESTLEY, C.B. WEGENER, *Finitely generated free modular ortholattices I*, Inter. J. Theor. Phys. **36** (1997), 2639–2660.

[HKPW 1] M. HAVIAR, P. KONÔPKA, H.A. PRIESTLEY, C.B. WEGENER, *Finitely generated free modular ortholattices II*, Inter. J. Theor. Phys. **36** (1997), 2661–2679.

[HOSt] H. HANCHE-OLSEN, E. STØRMER, *"Jordan Operator Algebras"*, Pitman Advanced Publ. Program, Boston-London-Melbourne, 1984.

[HePu] J. HEDLÍKOVÁ, S. PULMANNOVÁ, *Generalized difference posets and orthoalgebras*, Acta Math. Univ. Comenianae **45** (1996), 247–279.

[Hei] W. HEISENBERG, *"The Physical Principles of Quantum Theory"*, Dover, New York, 1930.

[Hen] I.R. HENTZEL, *Alternative rings without nilpotent elements*, Proc. Amer. Math. Soc. **42** (1974), 373–376.

[HRS] G. HOFER-SZABÓ, M. RÉDEI, L.E. SZABÓ, *Common cause completability of classical and quantum probability spaces*, Inter. J. Theor. Phys. **39** (2000), 913–919.

[Holl] S.S. HOLLAND, Jr. *Orthomodularity in infinite dimensions; a theorem of M. Solèr,* Bull. Amer. Math. Soc. **32** (1995), 205–234.

[Hol] A. S. HOLEVO, *"Probabilistic and Statistical Aspects of Quantum Theory"*, North-Holland, Amsterdam, 1982.

[Höl] U. HÖHLE, *Commutative, residuated l-monoids.* In: Non-classical logic and their representation to fuzzy subsets (U. Höhle and E.P. Klement eds.), Vol. **32**, Kluwer Academic Publ., Dordrecht, 1995, pp. 53–106.

[HoUl] J.E. HOPCROFT, J.D. ULLMAN, *"Intruduction to Automata Theory, Language, and Computation"*, Addison–Wesley Publ. Co., Reading MA, 1979.

[Hoo] C.S. HOO, *MV-algebras, ideals and semisimplicity,* Math. Japonica **34** (1989), 563–583.

[Hoo 1] C.S. HOO, *Unitarily extensions of MV and BCK-algebras*, Math. Japonica **37** (1992), 585–590.

[HuPu] R. HUDSON, S. PULMANNOVÁ, *Sum logics and tensor products*, Found. Phys. **23** (1993), 999–1024.

[HuPu 1] R. HUDSON, S. PULMANNOVÁ, *Graded tensor product of quantum logics*, Found. Phys. **24** (1994), 109–116.

[Hus] K. HUSIMI, *Studies on the foundations of quantum mechanics.* 1. Proc. Phys.–Math. Soc. Japan, **19** (1937), 766–789.

[ImIs] Y. IMAI, K. ISÉKI, *On axiom systems of propositional calculi,* Proc. Japan Acad. **42** (1966), 19–22.

[Ise] K. ISÉKI, *A special class of BCK-algebras*, Math. Seminar Notes **5** (1979), 107–119.

[Ish] C.J. ISHAM, *Quantum logic and the histories approach to quantum theory,* J. Math. Phys. **35** (1994), 2157–2185.

[Ish 1] C.J. ISHAM, *Quantum logic and decohering histories.* In: Topics in Quantum Fields Theory, D.H. Tchrakian ed., World Scientific, Singapore, 1995.

[Ish 2] C.J. ISHAM, *Topos theory and consistent histories: The internal logic of the set of all consistent sets*, Inter. J. Theor. Phys. **36** (1997), 785–814.

[IsLi] C.J. ISHAM, N. LINDEN, *Quantum temporal logic and decoherence functionals in the histories approach to generalized quantum theory*, J. Math. Phys. **35** (1994), 5452–5479.

[IsLi 1] C.J. ISHAM, N. LINDEN, *Continuous histories and the history group in generalized quantum theory*, J. Math. Phys. **36** (1995), 5392–5408.

[ILS] C.J. ISHAM, N. LINDEN, S. SCHRECKENBERG, *The classification of decoherence functionals: An analogue of Gleason's theorem*, J. Math. Phys. **35** (1994), 6360–6370.

[JaMa] S. JACKSON, R.D. MAULDIN, *On the σ-class generated by open balls*, to appear.

[Jak] J. JAKUBÍK, *Conditionally orthogonally complete ℓ-groups*, Math. Nachrichten **65** (1975), 153–162.

[Jak 1] J. JAKUBÍK, *Direct product decompositions of MV-algebras*, Czechoslovak Math. J. **44** (1994), 725–739.

[Jak 2] J. JAKUBÍK, *On complete MV-algebras*, Czechoslovak Math. J. **45** (**120**) (1995), 473–480.

[Jak 3] J. JAKUBÍK, *Sequential convergences on MV-algebras*, Czechoslovak Math. J. **45** (1995), 709–726.

[Jak 4] J. JAKUBÍK, *On archimedean MV-algebras*, Czechoslovak Math. J. **48** (1998), 75–582.

[Jak 5] J. JAKUBÍK, *Complete generators and maximal completions of MV-algebras*, Czechoslovak Math. J. **48** (1998), 597–608.

[Jak 6] J. JAKUBÍK, *Completeness properties for MV-algebras*, submitted.

[Jak 7] J. JAKUBÍK, *Subdirect product decompositions of MV-algebras*, Czechoslovak Math. J. **49** (1999), 163–173.

[Jak 8] J. JAKUBÍK, *Radical classes of MV-algebras*, Czechoslovak Math. J. **49** (1999), 191–211.

[Jak 9] J. JAKUBÍK, *On intervals and isometries of MV-algebras*, submitted.

[Jak 10] J. JAKUBÍK, *Cantor–Bernstein theorem for MV-algebras*, Czechoslovak Math. J. **49** (1999), 517–526.

[Jak 11] J. JAKUBÍK, *On the product MV-algebras*, submitted.

[Jak 12] J. JAKUBÍK, *Regular representation of semisimple MV-algebras by continuous real functions,* in preparation.

[Jak 13] J. JAKUBÍK, *Convex chains in MV-algebras of an MV-algebra*, in preparation.

[Jak 14] J. JAKUBÍK, *Distinguished extensions of an MV-algebra*, to appear.

[Jani] V. JANIŠ, *Notes on sums of Boolean algebras and logics,* Demonstratio Math. **23** (1990), 699–708.

[Jan] M.F. JANOWITZ, *A note on generalized orthomodular lattices*, J. Natur. Sci. Math. **8** (1968), 89–94.

[Jan 1] M.F. JANOWITZ, *On the *-order for Rickart *-rings*, Algebra Universalis **16** (1983), 360–369.

[Jau] J.M. JAUCH, *"Foundations of Quantum Mechanics"*, Addison-Wesley, Reading, Massachusetts, 1968.

[Jen] G. JENČA, *A note on ideals in generalized effect algebras*, Tatra Mt. Math. Publ. **16** (1999), 81–85.

[Jen 1] G. JENČA, *Notes on R_1-ideals in partial Abelian monoids*, Algebra Universalis, to appear.

[Jen 2] G. JENČA, *Subcentral ideals in generalized effect algebras*, Inter. J. Theor. Phys. **39** (2000), 745–755.

[JePu] G. JENČA, S. PULMANNOVÁ, *On quotients of partial abelian monoids with the Riesz decomposition property,* in preparation.

[JeRi] G. JENČA, Z. RIEČANOVÁ, *On sharp elements in lattice ordered effect algebras*, Busefal **80** (1999), 24–29.

[Jur] M. JUREČKOVÁ, *The measure extension theorem on MV σ-algebras*, Tatra Mt. Math. Publ. **6** (1995), 55–140.

[Jur 1] M. JUREČKOVÁ, *Ergodic theory on D-posets*, Tatra Mt. Math. Publ. **10** (1997), 267–272.

[JuRi] M. JUREČKOVÁ, B. RIEČAN, *Strong law of large numbers in D-posets*, Inter. J. Theor. Phys. **34** (1995), 1495–1500.

[Kal] G. KALMBACH, *"Orthomodular Lattices"*, Academic Press, London, New York, 1983.

[Kal 1] G. KALMBACH, *" Measures and Hilbert Lattices"*, World Sci. Publ. Co., Singapore, 1986.

[Kal 2] G. KALMBACH, *"Quantum Measures and Spaces."* Kluwer Academic Publ., Dordrecht, 1998,

[KaRi] G. KALMBACH, Z. RIEČANOVÁ, *An axiomatization for Abelian relative inverses*, Demonstratio Math. **27** (1994), 769–780.

[KaNe] T. KATRIŇÁK, T. NEUBRUNN, *On certain generalized probability domains*, Mat. Čas. SAV **23** (1973), 209–215.

[Kat] F. KATRNOŠKA, *Logics of idempotents of rings.* In: Topology, Measures, and Fractals (Warnemünde, 1991), Math. Research **66**, Akademie-Verlag, Berlin, 1992, pp. 131–136.

[KePr] T. KELETI, D. PREISS, *The balls do not generate all Borel sets using complements and countable disjoint unions*, to appear.

[KRF] M. KLÄY, C. RANDALL, D. FOULIS, *Tensor products and probability weights*, Inter. J. Theor. Phys. **26** (1987), 199–219.

[KMP] E.P. KLEMENT, M. MESIAR, E. PAP, *"Triangular Norms"*, Kluwer Academic Publ., Dordrecht, to appear.

[KlNa] E.P. KLEMENT, M. NAVARA, *A characterization of tribes with respect to the Lukasiewicz t-norm*, Czechoslovak Math. J. **47 (122)** (1997), 689–700.

[KlWe] E.P. KLEMENT, S. WEBER, *Generalized measures*, Fuzzy Sets and Systems **40** (1991), 375–394.

[Kol] A.N. KOLMOGOROV, *"Grundbegriffe der Wahrscheinlichkeitsrechnung"*, Berlin, 1933.

[Kon] P. KONÔPKA, *Atomic variety of sets with relative inverses*, Inter. J. Theor. Phys. **34** (1995), 1519–1524.

[Kop] F. KÔPKA, *D-posets of fuzzy sets*, Tatra Mt. Math. Publ. **1** (1992), 83–87.

[Kop 1] F. KÔPKA, *Compatibility in D-posets*, Inter. J. Theor. Phys. **34** (1995), 1525–1531.

[Kop 2] F. KÔPKA, *Compatibility in D-posets of fuzzy sets*, Tatra Mt. Math. Publ. **6** (1995), 95–102.

[Kop 3] F. KÔPKA, *Boolean D-posets as factor spaces*, Inter. J. Theor. Phys. **37** (1998), 93–101.

[KoCh] F. KÔPKA, F. CHOVANEC, *D-posets*, Math. Slovaca **44** (1994), 21–34.

[Kur] K. KURATOWSKI, *"Topology I"*, Mir, Moskva, 1966 (in Russian).

[Lac] F. LACAVA, *Alcune proprietá delle Ł-algebre esistenzialmente chiuse*, Boll. Uni. Mat. Ital. A **16** (1979), 360–366.

[LaMa] P.J. LAHTI, M.J. MĄCZYŃSKI, *On the order structure of the set of effects in quantum mechanics*, J. Math. Phys. **36** (1995), 1673–1680.

[LaPu] P.J. LAHTI, S. PULMANNOVÁ, *Coexistent observables and effects in quantum mechanics*, Rep. Math. Phys. **39** (1997), 339–351.

[LaPu 1] P. LAHTI, S. PULMANNOVÁ, *Coexistent observables and effects in a convexity approach*, J. Math. Phys. **39** (1998), 6384–6371.

[LPY] P.J. LAHTI, S. PULMANNOVÁ, K. YLINEN, *Coexistent observables and effects in a convexity approach*, J. Math. Phys. **39** (1998), 6364–6371.

[Leu] I. LEUŞTEAN, *Product MV-algebras and MV*-algebras.* In: Proc. Fourth Inter. Symp. on Econ. Inform., May 6–9, 1999, Bucharest, Eds I. Smeureanu et al., INFOREC Printing House, Bucharest, 1999, pp. 1002–1005.

[Lin] C.H. LING, *Representation of associative functions*, Publ. Math. Debrecen **12** (1965), 189–212.

[Loc] R. LOCK, *The tensor product of generalized sample spaces which admit a unital set of dispersion–free weights*, Found. Phys. **20** (1990), 477–498.

[LH1] P.L. LOCK, G.M. HARDEGREE, *Connections among quantum logics: Part 1. Quantum propositional logics,* Inter. J. Theor. Phys. **24** (1985), 43–53.

[LH2] P.L. LOCK, G.M. HARDEGREE, *Connections among quantum logics: Part 2. Quantum event logics,* Inter. J. Theor. Phys. **24** (1985), 55–61.

[Lom] J.M. LOMSADZE ET AL., *"Logical and philosophical problems of quantum theory"*. Sever. Oset. gos. univ., Ordžonikidze 1981 (in Russian).

[Loom] L.H. LOOMIS, *"The lattice theoretical background of the dimension theory of operator algebras"*, Memoirs of the A.M.S., 1955.

[Loo] O. LOOS, *Alternative tripelsysteme*, Math. Ann. **198** (1972), 205–238.

[Loo 1] O. LOOS, *"Jordan Pairs"*, Lecture Notes in Math., Vol. 460, Springer-Verlag, Berlin-New York, 1975.

[Loo 2] O. LOOS, *"Bounded Symmetric Domains and Jordan Pairs"*, Math. Lectures, University of California, Irvine, California, 1977.

[Lud] G. LUDWIG, *"Foundations of Quantum Mechanics"*, Springer-Verlag, New York, 1983.

[LuZa] W.A.J. LUXEMBURG, A.C. ZAANEN, *"Riesz Spaces, I"* North-Holland, Amsterdam, London, 1971.

[MaMa] F. MAEDA, S. MAEDA, *"Theory of Symmetric Lattices"*, Springer–Verlag, Berlin, 1970.

[Mae] S. MAEDA, *"Lattice Theory and Quantum Logic"*, Maki-Shoten, Tokyo, 1980 (in Japanese).

[Mack] G.W. MACKEY, *"Mathematical Foundations of Quantum Mechanics"*, Benjamin, New York, 1963.

[MaL] S. MAC LANE, *"Categories for the Working Mathematician"*, Springer-Verlag, New York, Heidelberg, Berlin, 1971.

[Mac] M.J. MĄCZYŃSKI, *On a functional representation of the lattice of projections on a Hilbert space*, Studia Math. **47** (1973), 253-259.

[Mac 1] M.J. MĄCZYŃSKI, *Orthogonality postulate in axiomatic quantum mechanics*, Inter. J. Theor. Phys. **8** (1973), 353–360.

[Mac 2] M.J. MĄCZYŃSKI, *Functional properties of quantum logics*, Inter. J. Theor. Phys. **11** (1974), 149–156.

[MaTr] M.J. MĄCZYŃSKI, T. TRACZYK, *A characterization of orthomodular partially ordered sets admitting a full set of states*, Bull. Acad. Polon. Sci., ser. Math. Astr. Phys. **21** (1973), 3–8.

[Man] P. MANGANI, *Su certe algebre conesse con logiche a piú valori*, Boll. Unione Mat. Italiana **8** (1973), 68–78.

[MaLo] C. MARTÍNEZ LÓPEZ, *A new order relation for JB-algebras*, Ann. Sci. Univ. Blaise Pascal, Clermont-Ferrand II, Sér. Math. **27** (1992), 135–138.

[Mat] T. MATOLCSI, *Tensor product of Hilbert lattices and free orthodistributive product of orthomodular lattices*, Acta Sci. Math. **37** (1975), 263–272.

[May] R. MAYET, *Some characterizations of the underlying division ring of a Hilbert lattice by automorphisms*, Inter. J. Theor. Phys. **37** (1998), 109–114.

[May 1] R. MAYET, *Orthomodular lattice in relativistic non-quantum mechanics,* Inter. J. Theor. Phys. **34** (1995), 1595–1599.

[May 2] R. MAYET, *Orthosymmetric ortholattices,* Proc. Amer. Math. Soc. **114** (1992), 259–306.

[MaNa] R. MAYET, M. NAVARA, *Classes of logics representable as kernels of measures,* Contrib. General Algebra 9, Verlag Hölder-Pichler-Tempsky, Vienna, 1995, pp. 241–248.

[MNR] R. MAYET, M. NAVARA, V. ROGALEWICZ, *Orthomodular lattices with rich state spaces,* Algebra Universalis, to appear.

[MaPt] R. MAYET, P. PTÁK, *Quasivarietes of orthomodular lattices defined by conditions on states,* Algebra Universalis **42** (1999), 155–164.

[MaPt 1] R. MAYET, P. PTÁK, *Orthomodular lattices with state-separated noncompatible pairs,* Czechoslovak Math. J., to appear.

[MaPu] R. MAYET, S. PULMANNOVÁ, *Nearly orthosymmetric ortholattices and Hilbert spaces*, Found. Phys. **24** (1994), 1425–1437.

[MaIp] A. MAYET–IPPOLITO, *Generalized orthomodular posets*, Demonstratio Math. **24** (1991) 263–274.

[MeJu] J. MENG, Y.B. JUN, *"BCK-algebras",* Kyung Moon Sa Co., Seoul, 1994.

[Mey] K. MEYBERG, *The fundamental-formula in Jordan rings*, Archiv Math. **21** (1970), 43–44.

[Mey 1] K. MEYBERG, *Jordan-Tripelsysteme und die Koecher-Konstruktion von Lie Algebren*, Math. Z. **115** (1970), 58–78.

[Mes] R. MESIAR, *Fuzzy difference posets and MV-algebras.* In.: Proc. IPMU'94, Paris, 1994, pp. 208–212.

[Mes 1] R. MESIAR, *h-fuzzy quantum logics*, Inter. J. Theor. Phys. **33** (1994), 1417–1425.

[Mes 2] R. MESIAR, *Differences on* [0,1], Tatra Mt. Math. Publ. **6** (1995), 131–140.

[Mes 3] R. MESIAR, *Do fuzzy quantum structures exist ?* Inter. J. Theor. Phys. **34** (1995), 1609–1614.

[MePa] R. MESIAR, E. PAP, *Extensions of real-valued difference posets*, Inter. J. Theor. Phys. **33** (1994), 2317–2322.

[MeRi] R. MESIAR, B. RIEČAN, *On the joint observables in some quantum structures*, Tatra Mt. Math. Publ. **3** (1993), 183–190.

[Mit] P. MITTELSTAEDT, *"Quantum Logic"*, Reidel Publ. Co., Dordrecht, 1978.

[Moo] D.J. MOORE, *Categories of representations of physical systems,* Helv. Phys. Acta **68** (1995), 658–678.

[Moo 1] D.J. MOORE, *Categorical structures in physics,* Inter. J. Theor. Phys. **37** (1998), 121–126.

[MooE] E.F. MOORE, *Gedanken-experiments on sequetional machines.* In: Automata Studies, eds C.E. Shannon, J.McCarthy, Princeton Univ. Press, Princeton, N.J. 1956.

[MoGu] T. MORELAND, S. GUDDER, *Infima of Hilbert space effects*, Lin. Algebra Appl. **286** (1999), 1–17.

[Mul] V. MÜLLER, *Jauch-Piron states on concrete quantum logics*, Inter. J. Theor. Phys. **32** (1993), 433–442.

[MPT] V. MÜLLER, P. PTÁK, J. TKADLEC, *Combinatorial properties of concrete quantum logics*, Inter. J. Theor. Phys. **31** (1992), 843–854.

[Mun] D. MUNDICI, *Interpretation of AF C^*-algebras in Lukasiewicz sentential calculus,* J. Funct. Anal. **65** (1986), 15–63.

[Mun 1] D. MUNDICI, *MV-algebras are categorically equivalent to bounded commutative BCK-algebras,* Math. Japonica **6** (1986), 889–894.

[Mun 2] D. MUNDICI, *The C*-algebras of three-valued logic.* In: Logic Colloquium'88, Valentini and Zanardo (eds.), Elsevier, North-Holland, 1989, pp. 61–77.

[Mun 3] D. MUNDICI, *Averaging the truth-value in Łukasiewicz logic*, Studia Logica **55** (1995), 113–127.

[Mun 4] D. MUNDICI, *Uncertainty measures in MV algebras and states on AF C*-algebras*, Notas de la Sociedad de Matemàticas de Chile, a special issue in memoriam Rolando Chuaqui, 1996.

[Mun 5] D. MUNDICI, *Tensor products and the Loomis–Sikorski theorem for MV-algebras,* Advan. Appl. Math. **22** (1999), 227–248.

[MuPa] D. MUNDICI, G. PANTI, *Extending addition in Elliott's local semigroup*, J. Funct. Anal. **171** (1993), 461–472.

[MyJi] H.C. MYUNG, L.R. JIMENEZ, *Direct product decomposition of alternative rings*, Proc. Amer. Math. Soc. **47** (1975), 53–60.

[Nan] O. NÁNÁSIOVÁ, *D-set and groups*, Inter. J. Theor. Phys. **34** (1995), 1637–1642.

[Nan 1] O. NÁNÁSIOVÁ, *Decomposition of D-sets*, Inter. J. Theor. Phys. **37** (1998), 131–126.

[Nav] M. NAVARA, *A characterization of triangular norm based tribes*, Tatra Mt. Math. Publ. **3** (1993), 167–172.

[Nav 1] M. NAVARA, *On generating finite orthomodular sublattices*, Tatra Mt. Math. Publ. **10** (1997), 109–117.

[Nav 2] M. NAVARA, *Quantum logics representable as kernels of measures*, Czechoslovak Math. J. **46 (121)** (1997), 587–597.

[Nav 3] M. NAVARA, *Indenpendence of automorphism group, center and state space of quantum logics,* Inter. J. Theor. Phys. **31** (1992), 925–935.

[Nav 4] M. NAVARA, *Description of state spaces of orthomodular lattices*, Math. Bohemica **117** (1992), 305–313.

[Nav 5] M. NAVARA, *An orthomodular lattice admitting no group-valued measure,* Proc. Amer. Math. Soc. **122** (1994), 7–12.

[Nav 6] M. NAVARA, *Uniqueness of bounded observables*, Ann. Inst. H. Poincaré, Theor. Phys. **63** (1995), 155–176.

[Nav 7] M. NAVARA, *Characterization of state spaces of orthomodular structures*, Conference Grado, Sept. 1997.

[NaPt] M. NAVARA, P. PTÁK, *Difference posets and orthoalgebras*, Busefal **69** (1997), 64–69.

[NaPt 1] M. NAVARA, P. PTÁK, *Quantum logics with given center and variable state space*, Inter. J. Theor. Phys. **37** (1998), 139–153.

[NaTk] M. NAVARA, J. TKADLEC, *Automorphisms of concrete logics*, Comment. Math. Univ. Carolinae **32** (1991), 15–25.

[Neu] T. NEUBRUNN, *A note on a quantum probability spaces*, Proc. Amer. Math. Soc. **25** (1970), 672–675.

[vNeu] J. von NEUMANN, *"Mathematische Grundlagen der Quantenmechanik"*, Springer-Verlag, Berlin, 1932.

[Ole] V. OLEJČEK, *q-σ-algebra generated by balls.* Inter. J. Theor. Phys. **34** (1995), 1643–1646.

[Ole 1] V. OLEJČEK, *The σ-class generated by balls contains all Borel sets*, Proc. Amer. Math. Soc. **123** (1995), 3665–3675.

[Ovc] P. OVCHINIKOV, *Exact topological analogs to orthoposets*, Proc. Amer. Math. Soc. **125** (1997), 2839–2861.

[OvSu] P.G. OVCHINNIKOV, F.F. SULTANBEKOV, *Finite concrete logics: Their structures and measures on them*, Inter. J. Theor. Phys. **37** (1998), 139–145.

[Pal] M. PALASIŃSKI, *Some remarks on BCK-algebras*, Math. Seminar Notes Univ. Kobe **8** (1980), 137–144.

[Palk] V. PALKO, *Topological difference posets*, Math. Slovaca **47** (1997), 211–220.

[Pap] E. PAP, *"Null-Additive Set Functions"*, Kluwer Academic Publ., Dordrecht, Ister Science, Bratislava, 1995.

[Pas] J. PASEKA, *On some duality for orthoposets*, Inter. J. Theor. Phys. **37** (1998), 155–161.

[Pav] M. PAVIČIĆ, *Quantum logic for quantum computers*, Inter. J. Theor. Phys. **39** (2000), 813–825.

[Per] A. PERES, " *Quantum Theory: Concepts and Methods*", Kluwer Academic Publ., Dordrecht, 1993.

[Pir] C. PIRON, "*Foundations of Quantum Physics*", Addison-Wesley, Reading, Mass., 1976.

[Pit] I. PITOWSKI, "*Quantum Probability - Quantum Logics*", Lecture Notes in Physics **321**, Springer-Verlag, Berlin, 1989.

[PrTi] D. PREISS, J. TIŠER, *Measures in Banach spaces are determined by their values on balls*, Mathematika **38** (1991), 391–397.

[Pta] P. PTÁK, *Jauch-Piron property (everywhere!) in the logico-algebraic foundation of quantum theories*, Inter. J. Theor. Phys. **32** (1993), 1985–1990.

[Pta 1] P. PTÁK, *Some nearly Boolean orthomodular posets*, Proc. Amer. Math. Soc. **126** (1998), 2039–2046.

[Pta 2] P. PTÁK, *Concrete quantum logics*, Inter. J. Theor. Phys. **39** (2000), 827–837.

[Pta 3] P. PTÁK, *Summing of Boolean algebras and logics*, Demonstratio Math. **19** (1986), 349–357.

[PtPu] P. PTÁK, S. PULMANNOVÁ, "*Kvantové logiky*", Veda, Bratislava, 1989 (in Slovak).

[PtPu 1] P. PTÁK, S. PULMANNOVÁ, "*Orthomodular Structures as Quantum Logics*", VEDA and Kluwer Acad. Publ., Bratislava and Dordrecht, 1991.

[PtPu 2] P. PTÁK, S. PULMANNOVÁ, *A measure-theoretic characterization of Boolean algebras among orthomodular lattices*, Comment. Math. Univ. Carolinae **35** (1994), 205–208.

[PtWe] P. PTÁK, H. WEBER, *Lattice properties of closed subspaces of inner product spaces*, Proc. Amer. Math. Soc., to appear.

[Pul] S. PULMANNOVÁ, *Tensor product of quantum logics*, J. Math. Phys. **26** (1985), 1–5.

[Pul 1] S. PULMANNOVÁ, *A remark to orthomodular partial algebras*, Demonstratio Math. **27** (1994), 687–699.

[Pul 2] S. PULMANNOVÁ, *Difference posets and the histories approach to quantum theories*, Inter. J. Theor. Phys. **34** (1995), 189–210.

[Pul 3] S. PULMANNOVÁ, *Representation of D-posets*, Inter. J. Theor. Phys. **34** (1995), 1689–1696.

[Pul 4] S. PULMANNOVÁ, *Congruences in partial abelian semigroups*, Algebra Universalis **37** (1997), 119–140.

[Pul 5] S. PULMANNOVÁ, *On connection among some orthomodular structures*, Demonstratio Math. **30** (1997), 313–328.

[Pul 6] S. PULMANNOVÁ, *Remarks on orthosymmetric ortholattices and Baer*-rings*, Math. Moravica, Special volume (1997), 301–306.

[Pul 7] S. PULMANNOVÁ, *Quantum logics and instruments*, Inter. J. Theor. Phys. **37** (1998), 163–181.

[Pul 8] S. PULMANNOVÁ, *Quantum logics and convex spaces*, Inter. J. Theor. Phys. **37** (1998), 2303–2332.

[Pul 9] S. PULMANNOVÁ, *States on effect algebras with the Riesz decomposition property and AF C^*-algebras.* In: Proceedings of Eurofuse - SIC'99, B. De Baets, J. Fodor, L.T. Kóczy, eds., 1999, 85–88.

[Pul 10] S. PULMANNOVÁ, *A note on observables on MV-algebras*, Soft Computing (2000), to appear.

[Pul 11] S. PULMANNOVÁ, *Effect algebras with the Riesz decomposition property and AF C^*-algebras*, Found. Phys., to appear.

[Pul 12] S. PULMANNOVÁ, *Bell inequalities and quantum logics.* In: The interpretations of Quantum Theory: where do we stand? Eds. L. Accardi, Istitute della Enciclopedia Italiana fondata da Giovanni Trecani s.p.a., Roma. 1994, 295–302.

[Pul 13] S. PULMANNOVÁ, *Boolean powers and quantum measurements*, Rep. Math. Phys. **32** (1993), 235–250.

[Pul 14] S. PULMANNOVÁ, *A remark on states on orthomodular lattices*, Tatra Mt. Math. Publ. **2** (1993), 209–219.

[Pul 15] S. PULMANNOVÁ, *On conditional probability functions on von Neumann algebras*, Atti Sem. Mat. Fis. Univ. Modena **42** (1994), 477–483.

[Pul 16] S. PULMANNOVÁ, *Logico-algebraic structures I*, Inter. J. Theor. Phys. **32** (1993), 1661–1674.

[Pul 17] S. PULMANNOVÁ, *Corrigendum to "congruences in partial abelian semigroups"*, Algebra Universalis **42** (1999), 59–60.

[Pul 18] S. PULMANNOVÁ, *A quantum logic description of quantum measurements*, Tatra Mt. Math. Publ. **3** (1993), 89–100.

[Pul 19] S. PULMANNOVÁ, *Quantum measurements and quantum logics*, J. Math. Phys. **35** (1994), 1555–1572.

[Pul 20] S. PULMANNOVÁ, *Quantum logics and Hilbert space*, Found. Phys. **24** (1994), 1403–1414.

[Pul 21] S. PULMANNOVÁ, *A quantum logic description of some ideal measurements*. In: Quantum Communications and Measurements, eds V. P. Belavkin, O. Hirota, R. L. Hudson, Plenum Press, New York, London, 1995, pp. 187–192.

[Pul 22] S. PULMANNOVÁ, *Automorphisms of orthomodular posets*, Atti Sem. Mat. Fis. Univ. Modena **44** (1996), 415–425.

[Pul 23] S. PULMANNOVÁ, *Axiomatization of quantum logics*, Inter. J. Theor. Phys. **35** (1996), 2309–2319.

[PuMa] S. PULMANNOVÁ, V. MAJERNÍK, *Bell inequalities on quantum logics*, J. Math. Phys. **33** (1992), 2173–2178.

[PuRi] S. PULMANNOVÁ, Z. RIEČANOVÁ, *A remark to orthomodular lattices with almost orthogonal set of atoms*. In: Proc. Second Winter School on Measure Theory, Liptovský Ján, 1990, pp 175–176.

[PuRi 1] S. PULMANNOVÁ, Z. RIEČANOVÁ, *Logics with separating sets of measures*, Math. Slovaca **41** (1991), 167–177.

[PuRi 2] S. PULMANNOVÁ, Z. RIEČANOVÁ, *Compact topological orthomodular lattices*. In: Contributions to General Algebra, eds. Dorninger, D., Eigenthaler G., Kaiser, H.K., Müller W.B., Hölder-Pichler-Tempski, Wien, B.G. Teubner, Stuttgart, 1991, pp. 277–282.

[PuRi 3] S. PULMANNOVÁ, Z. RIEČANOVÁ, *Modular almost orthogonal quantum logics*, Inter. J. Theor. Phys. **31** (1992), 881–888.

[PuRi 4] S. PULMANNOVÁ, Z. RIEČANOVÁ, *Block-finite atomic orthomodular lattices*, J. Pure Appl. Algebra **89** (1993), 295–304.

[PuRo] S. PULMANNOVÁ, V. ROGALEWICZ, *Orthomodular lattices with almost orthogonal sets of atoms*, Comment. Math. Univ. Carolinae **32** (1991), 423–429.

[PuSv] S. PULMANNOVÁ, K. SVOZIL, *Ideals in ortholattices, Bell inequalities, and simultaneously defined properties*, Inter. J. Theor. Phys. **36** (1997), 1547–1563.

[PuWi] S. PULMANNOVÁ, A. WILCE, *Representation of D-posets*, Inter. J. Theor. Phys. **34** (1995), 1689–1696.

[Pyk] J. PYKACZ, *Fuzzy set ideas in quantum logics*, Inter. J. Theor. Phys. **31** (1992), 1767–1783.

[Pyk 1] J. PYKACZ, *Fuzzy quantum logics and infinite-valued Łukasiewicz logic*, Inter. J. Theor. Phys. **33** (1994), 1403–1416.

[Pyk 2] J. PYKACZ, *Quantum logic as partial infinite-valued Łukasiewicz logic*, Inter. J. Theor. Phys. **34** (1995), 1697–1710.

[Pyk 3] J. PYKACZ, *Triangular norms-based quantum structures of fuzzy sets.* In: Proc. 7th IFSA World Congress, Prague, June 26–29, 1997, Eds. M. Mareš et al., Academia, Prague, 1997, vol. IV, pp. 159–164.

[Pyk 4] J. PYKACZ, *Fuzzy quantum logics as a basis for quantum probability theory*, Inter. J. Theor. Phys. **37** (1998), 281–290.

[Pyk 5] J. PYKACZ, *Quantum logic as a basis for computations*, Inter. J. Theor. Phys. **39** (2000), 839–850.

[Rac] J. RACHŮNEK, *A non-commutative generalization of MV-algebras*, Czechoslovak Math. J., to appear.

[Raf] J.G. RAFTERY, *On prime ideals and subdirect decompositions of BCK-algebras*, Math Japonica **32** (1987), 811–818.

[RaFo] C.H. RANDALL, D.J. FOULIS, *An approach to empirical logic*, Amer. Math. Monthly **77** (1970), 363–374.

[RaFo 1] C.H. RANDALL, D.J. FOULIS, *Operational statistics. II. Manual of operations and their logics*, J. Math. Phys. **14** (1973), 1472–1480.

[Rao] M.M. RAO, *"Foundations of Stochastic Analysis"*, Academic Press, New York, 1981.

[Rav] K. RAVINDRAN, *On a structure theory of effect algebras*, PhD theses, Kansas State Univ., Manhattan, Kansas, 1996.

[Rav 1] K. RAVINDRAN, *States on effect algebras that have the ϕ-symmetric property*, Inter. J. Theor. Phys. **37** (1998), 175–181.

[Red] M. RÉDEI, *"Quantum Logic in Algebraic Approach"*, Kluwer Academic Publ., Dordrecht, 1998.

[ReSi] M. REED, B. SIMON, *"Methods of Modern Mathematical Physics, I"*, Acad. Press, New York, 1975.

[Rie] B. RIEČAN, *On the almost everywhere convergence of observables in some algebraic structures*, Atti Sem. Mat. Fis. Univ. Modena **44** (1996), 95–104.

[Rie 1] B. RIEČAN, *Upper and lower limits of sequences of observables in D-posets of fuzzy sets,* Math. Slovaca **46** (1996), 419–431.

[Rie 2] B. RIEČAN, *On the convergence of observables in D-posets*, Tatra Mt. Math. Publ. **12** (1997), 7–12.

[Rie 3] B. RIEČAN, *On the strong law of large numbers for weak observables in MV-algebras,* Tatra Mt. Math. Publ. **15** (1998), 13–21.

[Rie 4] B. RIEČAN, *Weak observables in MV algebras*, Inter. J. Theor. Phys. **37** (1998), 183–189.

[Rie 5] B. RIEČAN, *On the product MV algebras*, Tatra Mt. Math. Publ. **16** (1999), 143–149.

[Rie 6] B. RIEČAN, *On the probability theory on product MV-algebras.* In: Proc. Fourth Inter. Symp. on Econ. Inform., May 6–9, 1999, Bucharest, Eds I. Smeureanu et al., INFOREC Printing House, Bucharest, 1999, pp. 1032–1036.

[Rie 7] B. RIEČAN, *Probability theory on some ordered structures*, Atti Sem. Mat. Fis. Univ. Modena **47** (1999), 255–265.

[Rie 8] B. RIEČAN, *On the L^p space of observables on product MV-algebras*, Inter. J. Theor. Phys. **39** (2000), 851–858.

[RiNe] B. RIEČAN, T. NEUBRUNN, *"Integral, Measure, and Ordering"*, Kluwer Acad. Publ. Dordrecht, Ister Science, Bratislava, 1997.

[RiZ] Z. RIEČANOVÁ, *Applications of topological methods to the completion of atomic orthomodular lattices*, Demonstratio Math. **24** (1991), 332–339.

[RiZ 1] Z. RIEČANOVÁ, *Topological and order-topological orthomodular lattices*, Bull. Austral. Math. Soc. **47** (1993), 509–518.

[RiZ 2] Z. RIEČANOVÁ, *Pták sum of a Boolean algebra with an effect algebra and its completeness*, Inter. J. Theor. Phys. **34** (1995), 1719–1725.

[RiZ 3] Z. RIEČANOVÁ, *On order continuity of quantum structures and their homomorphisms*, Demonstratio Math. **29** (1996), 433–443.

[RiZ 4] Z. RIEČANOVÁ, *On proper orthoalgebras, difference posets and Abelian relative inverse semigroups*, Tatra Mt. Math. Publ. **10** (1997), 119–128.

[RiZ 5] Z. RIEČANOVÁ, *Strongly compatible atomistic orthomodular lattices and modular ortholattices*, Tatra Mt. Math. Publ. **15** (1998), 143–153.

[RiZ 6] Z. RIEČANOVÁ, *Lattices and quantum logics with separated intervals, atomicity*, Inter. J. Theor. Phys. **37** (1998), 191–197.

[RiZ 7] Z. RIEČANOVÁ, *Generalized effect algebras, D-posets and D-algebras.* In: G. Kalmbach, "Quantum Measures and Spaces". Kluwer Academic Publ., Dordrecht, 1998, pp. 236–241.

[RiZ 8] Z. RIEČANOVÁ, *Compatibility and central elements in effect algebras*, Tatra Mt. Math. Publ. **16** (1999), 151–158.

[RiZ 9] Z. RIEČANOVÁ, *Mac Neille completion of D-posets and effect algebras*, Inter. J. Theor. Phys. **39** (2000), 859–869.

[RiZ 10] Z. RIEČANOVÁ, *Generalization of blocks for D-lattice and lattice ordered effect algebras,* Inter. J. Theor. Phys. **39** (2000), 231–237.

[RiZ 11] Z. RIEČANOVÁ, *Subalgebras, intervals and central elements of generalized effect algebras,* Preprint FEI STU, Bratislava, 1998.

[RiBr] Z. RIEČANOVÁ, D. BRŠEL, *Contraexamples in difference posets and orthoalgebras*, Inter. J. Theor. Phys. **33** (1994), 133–141.

[Rog] V. ROGALEWICZ, *Any orthomodular poset is a pasting of Boolean algebras*, Comment. Math. Univ. Carolinae **29** (1998), 557–558.

[Rom] A. ROMANOWSKA, *Commutative BCK-chains with one generator*, Math. Japonica **30** (1985), 663–670.

[RoTr] A. ROMANOWSKA, T. TRACZYK, *On commutative BCK-algebras*, Math. Japonica **25** (1980), 567–583.

[RoTr 1] A. ROMANOWSKA, T. TRACZYK, *On the structure of commutative BCK-chains*, Math. Japonica **26** (1981), 433–442.

[RoTr 2] A. ROMANOWSKA, T. TRACZYK, *Commutative BCK-algebras. Subdirectly irreducible algebras and varieties*, Math. Japonica **27** (1981), 35–48.

[RoTr 3] A. ROMANOWSKA, T. TRACZYK, *Simple commutative BCK-algebras*, Math. Japonica **29** (1984), 171–181.

[Ros] K. I. ROSENTHAL, *"Quantales and their Applications"*, Longman, New York, 1990.

[Rud] O. RUDOLPH, *Consistent histories and operational quantum theory*, Inter. J. Theor. Phys. **35** (1996), 1581–1636.

[Rud 1] O. RUDOLPH, *On the consistent effect histories approach to quantum mechanics*, J. Math. Phys. **37** (1996), 5368–5379.

[Rud 2] O. RUDOLPH, *The representation theory of decoherence functionals in history quantum theories*, Inter. J. Theor. Phys. **39** (2000), 871–884.

[RuWr] O. RUDOLPH, J.D.M. WRIGHT, *On tracial operator representation of quantum decoherence functionals*, J. Math. Phys. **38** (1997), 5643–5652.

[Rut] G.T. RÜTTIMANN, *The approximate Jordan-Hahn decomposition*, Canad. J. Math. **41** (1981), 1124–1146.

[Rut 1] G.T. RÜTTIMANN, *Decomposition of cone of measures*, Atti Sem. Mat. Fis. Univ. Modena **38** (1990), 109–121.

[Rut 2] G.T. RÜTTIMANN, *Jordan triple systems*, Lecture given at IQSA-Meeting, Prague, Nov. 24-27, 1993.

[Rut 3] G.T. RÜTTIMANN, *Weakly purely additive measures*, Canad. J. Math. **46** (1994), 872–885.

[Rut 4] G.T. RÜTTIMANN, *On a σ-finite elements in orthomodular posets*, Atti Sem. Mat. Fis. Univ. Modena **42** (1994), 531–546.

[Sar] T.A. SARYMSAKOV, et al., *"Uporjadočennyje algebry"*. FAN, Taškent, 1983 (in Russian).

[Scha] R.D. SCHAFER, *"An Introduction to Nonassociative Algebras"*, Pure and Applied Math., Series A, Monographs and Textbooks, Academic Press, New York, London, 1966.

[ScSv] M. SCHALLER, K. SVOZIL, *Partition logics of automata,* Il Nuovo Cimento **109 B** (1994), 167–176.

[SchJ] J. SCHMIDT, *Zur Kennzeichnung der Dedekind-Mac Neilleschen Hulle einer Geordneten Menge*, Archiv d. Math. **7** (1956), 241–249.

[Sch] K.D. SCHMIDT, *Minimal clans: A class of ordered partial semigroups including Boolean rings and lattice-ordered groups.* In: Semigroups – Theory and Applications (Oberwolfach 1986). LNM **1320**, Springer-Verlag, Berlin, Heidleberg, New York, 1988, pp. 300–341.

[Sch 1] K.D. SCHMIDT, *"Jordan Decompositions of Generalized Vector Measures"*, Longman Scientific & Technical, Longman group UK Limited, Essex, England, 1989.

[Shi] N. SHILKRET, *Maxitive measures and integration*, Indag. Math. **8** (1971), 109–116.

[Sho] P.W. SHOR, *Algorithms for quantum computation: Discrete logarithms and factoring*, IEEE (1994), 124–133.

[Sik] R. SIKORSKI, *"Boolean Algebras"*, Springer–Verlag, Berlin, Heidelberg, New York, 1964.

[Sol] M.P. SOLÈR, *Characterization of Hilbert spaces by orthomodular spaces,* Comm. Algebra **23** (1995), 219–243.

[Ste] A. STERN, *"The Quantum Brain"*, North-Holland, Elsevier, Amsterdam, 1994.

[Stu] W. STULPE, *"Classical Representations of Quantum Mechanics"*, Wissenschaft & Technik Verlag, Berlin, 1997.

[Stu 1] W. STULPE, *On the representation of quantum mechanics on a classical sample space*, Inter. J. Theor. Phys. **37** (1998), 349–356.

[Stur] T. STURM, *On commutative BCK-algebras embeddable into directed commutative BCK-algebras*, Math. Japonica **27** (1982), 197–212.

[Svo] K. SVOZIL, *"Randomness & Undecidability"*, World Scientific, Singapore 1993.

[Svo 1] K. SVOZIL, *"Quantum Logic"*, Springer-Verlag, Singapore Pte. Ltd., 1998.

[SvTk] K. SVOZIL, J. TKADLEC, *Greechie diagrams, nonexistence of measures in quantum logics and Kochen–Specker type constructions*, J. Math. Phys. **37** (1996), 5380–5401.

[Syk] S.R. SYKES, *Finite modular effect algebras*, Univ. of Amherst, (1996).

[Tka] J. TKADLEC, *Conditions that force an orthomodular poset to be a Boolean algebra*, Tatra Mt. Math. Publ. **10** (1997), 55–62.

[Tka 1] J. TKADLEC, *Greechie diagrams of small quantum logics with small state spaces*, Inter. J. Theor. Phys. **37** (1998), 203–209.

[Tka 2] J. TKADLEC, *Diagrams of Kochen–Specker type construction*, Inter. J. Theor. Phys. **39** (2000), 921–926.

[Top] D.M. TOPPING, *Jordan algebras of self-adjoint operators*, Memoirs Amer. Math. Soc. **53**, 1965.

[Tra] T. TRACZYK, *On the variety of bounded commutative BCK-algebras*, Math. Japonica **24** (1979/80), 283–292.

[Tri] E. TRILLAS, *Sobre functiones de negacion en la teoria de conjuntos difusos*, Stochastica **3** (1979), 47–60.

[Tur] E. TURUNEN, *Boolean deductive systems of BL-algebras*, Archiv Math. Logic, to appear.

[Tur 1] E. TURUNEN, *BL-algebras of basic fuzzy logic*, Mathware and Soft Computing **6** (1999), 49–61.

[vFr] B.C. Van FRAASSEN, *"Quantum Mechanics: An Empiricist View"*, Oxford University Press, Oxford, New York, Toronto, 1991.

[Var] V.S. VARADARAJAN, *"Geometry of Quantum Theory", Vol. 1,* van Nostrand, Princeton, New Jersey, 1968.

[WebH] H. WEBER, *An abstraction of clans of fuzzy sets,* Ricerche Mat. **46** (1998), 457–472.

[WebH 1] H. WEBER, *There are orthomodular lattices without non-trivial group-valued states: A computer-based construction*, J. Math. Anal. Appl. **183** (1994), 89–93.

[WebS] S. WEBER, *$\perp$-decomposable measures and integrals for Archimedean t-conorms* $\perp$, J. Math. Anal. Appl. **101** (1984), 114–138.

[WeO] N.E. WEGGE-OLSEN, *"K-Theory and C^*-algebras"*, Oxford Univ. Press, Oxford, 1983.

[Wil] A. WILCE, *Tensor product of frame manuals*, Inter. J. Theor. Phys. **29** (1990), 805–814.

[Wil 1] A. WILCE, *Partial Abelian semigroups*, Inter. J. Theor. Phys. **34** (1995), 1807–1812.

[Wil 2] A. WILCE, *Perspectivity and congruence in partial Abelian semigroups*, Math. Slovaca **48** (1998), 117–135.

[Wil 3] A. WILCE, *On generalized Sasaki projections*, Inter. J. Theor. Phys. **39** (2000), 969–974.

[Wri] J.D.M. WRIGHT, *The structure of decoherence functionals for von Neumann histories*, J. Math. Phys. **36** (1995), 5409–5413.

[Wri 1] J.D.M. WRIGHT, *Decoherence functionals for von Neumann quantum histories: bounded and countable additivity*, Comm. Math. Phys. **191** (1998), 493–500.

[Wri 2] J.D.M. WRIGHT, *Quantum decoherence functionals and positivity*, Atti. Sem. Mat. Fis. Univ. Modena **47** (1999), 59–66.

[WriR] R. WRIGHT, *Generalized urn models*, Found. Phys. **20** (1990), 881–903.

[Wyl] O. WYLER, *Clans*, Compos. Math. **17** (1966/67), 172–189.

[YoHe] K. YOSIDA, E. HEWITT, *Finitely additive measures*, Trans. Amer. Math. Soc. **72** (1952), 44–66.

[Yut] H. YUTANI, *The class of commutative BCK-algebras is equationally definable*, Math. Seminar Notes **5** (1977), 207–210.

[Zad] L.A. ZADEH, *Fuzzy sets*, Inform. Control **8** (1965), 338–353.

[Zec] A. ZECCA, *On the coupling of quantum logics*, J. Math. Phys. **19** (1978), 1482–1485.

[ZSSS] K.A. ZHEVLAKOV, A.M. SLINKO, I.P. SHESTAKOV, A.I. SHIRSHOV, *"Rings that are Nearly Associative, Contemporary Algebra"*, Nauka, Moskva, 1978 (in Russian).

Index of Symbols

Index

MIX
Papier aus verantwortungsvollen Quellen
Paper from responsible sources
FSC® C105338

If you have any concerns about our products,
you can contact us on
ProductSafety@springernature.com

In case Publisher is established outside the EU,
the EU authorized representative is:
Springer Nature Customer Service Center GmbH
Europaplatz 3, 69115 Heidelberg, Germany

Printed by Libri Plureos GmbH
in Hamburg, Germany